GOETHE
*The Poet
and the Age*

GOETHE
The Poet and the Age

VOLUME I
The Poetry of Desire
(1749–1790)

NICHOLAS BOYLE

CLARENDON PRESS · OXFORD
1991

Oxford University Press, Walton Street, Oxford OX2 6DP
Oxford New York Toronto
Delhi Bombay Calcutta Madras Karachi
Petaling Jaya Singapore Hong Kong Tokyo
Nairobi Dar es Salaam Cape Town
Melbourne Auckland
and associated companies in
Berlin Ibadan

Oxford is a trade mark of Oxford University Press

Published in the United States
by Oxford University Press, New York

© Nicholas Boyle 1991

All rights reserved. No part of this publication may be reproduced,
stored in a retrieval system, or transmitted, in any form or by any means,
electronic, mechanical, photocopying, recording, or otherwise, without
the prior permission of Oxford University Press

This book is sold subject to the condition that it shall not, by way
of trade or otherwise, be lent, re-sold, hired out or otherwise circulated
without the publisher's prior consent in any form of binding or cover
other than that in which it is published and without a similar condition
including this condition being imposed on the subsequent purchaser

British Library Cataloguing in Publication Data
Boyle, Nicholas
Goethe: The Poet and the Age.
Vol. 1, The poetry of desire (1749–1790)
1. Poetry in German. Goethe, Johann Wolfgang von, 1749–1832
I. Title
831.6
ISBN 0–19–815866–1

Library of Congress Cataloging-in-Publication Data
Goethe: The poet of the Age.
p.   cm.
Includes bibliographical references and index.
Contents: v. 1. The poetry of desire (1749–1790)/Nicholas Boyle.
1. Goethe, Johann Wolfgang von, 1749–1832. 2. Authors,
German—18th century—Biography. 3. Authors, German—19th century—Biography. I. Boyle, Nicholas.
PT2049.G66 1991   831'.6—dc20   90-45050
ISBN 0–19–815866–1

Typeset by Pentacor PLC, High Wycombe, Bucks.
Printed in the United States of America.

For
Michael and Rosaleen
who made it possible

Preface

MORE must be known, or at any rate there must be more to know, about Goethe than about almost any other human being. As the age of paper passes, so he comes to seem its supreme product. Not only did he do and think more than most men—he, and others, left more written traces of what he did and thought. It is true, he also left monuments of a different kind. Nearly 3,000 drawings by him survive, as do the villa he built, the palace he rebuilt, and the park he first laid out. He amassed very substantial private collections of mineralogical specimens, incised gems, and prints and drawings, and if his own working library was not bibliographically outstanding that was because he had at his disposal the resources, and the buying power, of one of the largest princely libraries in Germany, which he spent a lifetime enriching. He ran a duchy for three years, a theatre for twenty-five, and a university and an art school for longer still. A shrewd contemporary thought his greatest achievement to be his devoted personal guidance of his sovereign, eight years his junior, whom he educated into one of the most enlightened of Germany's minor rulers in the early nineteenth century, and who was a model for his neighbour and relative, the Prince Consort. Goethe deserves his place on the Albert Memorial. But of course what matters now is the writing.

A national celebrity at the age of 24, a European celebrity twelve months later, Goethe was thereafter, until he died in his eighty-third year, sufficiently prominent, remarkable, and at times powerful, for those who met him to want to record what he said and those who corresponded with him to take care of what he wrote. After he moved to Weimar the daily chronicle of his doings, now being put together for the first time in seven large volumes by Robert Steiger, is practically continuous, especially once he began to keep a regular diary in 1796. Accounts of conversations with him, excluding Eckermann's famous collection, run to some 4,000 printed pages, over 12,000 letters from him are extant and about 20,000 letters addressed to him. His official papers run to four volumes, and Wilhelm Bode filled another three with contemporary gossip about him, extracted from the correspondence and diaries of third parties. There may be parallels to this flood of documentation for an individual life, though few surely can be sustained over so long a period—Voltaire or Gladstone perhaps? certainly not Napoleon. What makes Goethe's case manifestly unique is the quantity, quality, and nature of the literary and scientific writing which caused this interest in him and which expands indefinitely our potential knowledge of his inner life through its unceasing stream of reflection on the events and projects of his outward career. The most important written memorials to Goethe are the literary works in which he sought to make the particular occasions of his individual existence into general symbols whose significance would be appreciated by readers, initially of his own time and place, but increasingly, in his later years, of other times and places as well.

That individual existence had its eccentricities—among them, an English audience may think, that of being German. None the less, this book goes to press at a time when it seems worth remembering that the centre of Europe, if its diameter is drawn from Lisbon to Moscow, lies somewhere between Frankfurt and Weimar. If Germany is now re-emerging from the marginal position into which it was pressed, from the end of the sixteenth century onwards, by the overseas expansion of the western European powers, and by their implacably anti-Imperial policies, particularly their continued fomentation after 1630 of the Thirty Years' War, a first stage in that re-emergence into European centrality occurred towards the end of the eighteenth century. Perhaps because of a different relation between the state power and the middle classes from that which prevailed further west, there occurred in Germany at that time a religious crisis whose explosive energies, channelled through the then uniquely extensive German system of universities, issued in a series of intellectual and cultural innovations which were to have a most powerful influence on the European, and North American, mind of the nineteenth and twentieth centuries. Biblical criticism, historical and classical scholarship, reinterpretative theology, Idealist (and ultimately materialist) philosophy, sociology, neo-classical art and architecture, aesthetics, and the academic study of modern literature, all were decisively influenced, and in some cases actually originated, by the German cultural revolution which began in the 1790s, and among the last fruits of which in the later nineteenth century, replete with the seeds of the age to come, was the work of Marx, Nietzsche and Freud. Probably not coincidentally, it was also the greatest age of German music. A first reason for reading and studying Goethe is that his literary works are the medium in which a superlatively intelligent and unusually well-placed observer discerned and responded to these numerous shifts in the bedrock of intellectual Europe, some of which led to earthquakes in his own time, others only later.

A second reason is that Goethe was a poet. He was a born versifier and phrase-maker, so that *Faust* to a German audience, like *Hamlet* to an English one, seems a collection of quotations, and no issue of a German quality newspaper is without a handful of Goethe allusions, acknowledged, or unrecognized. He had a natural affinity with the rhythms of the German language and throughout his life produced, unpredictably, but with dreamlike facility, lyric poems of unique form and character, many of which have become internationally known through their later musical settings (which however sometimes obscure the specifically poetic merits of the original). *Faust* is certainly the greatest long poem of recent European literature, and it was Goethe's example, not Marlowe's, which inspired the numerous further treatments of the theme in the 150 years after its first publication. But Goethe was not just *a* poet—for the whole Romantic generation, in Germany, England

and even France, he was *the* poet, and through his influence on that generation he affected all subsequent notions of what poets are and poetry does. In 1797, when the first recognizably Romantic movements began, Goethe was already a figure of authority and achievement, a model of worldly competence and success whose ultimate loyalty was none the less only, and avowedly, to his 'art'. Goethe was the first poet who in virtue solely of his poetry, and not of its sublime or sacred subject-matter, or his contingent personal erudition, was also a secular sage. Indeed it would scarcely be too much to say that with the eighth volume of his *Literary Works*, which was published in 1789 and contained a selection of his shorter poems, Goethe created the very genre of lyric poetry as it is practised today: the book of shorter pieces linked not in the first instance formally or thematically, or by their devotional purpose or their suitability for musical setting, but by their origin in the discrete occasions of the poet's life, what he sees and reads and feels and thinks about, and all given meaning and importance not by any transcendent order but by their reference, explicit or implicit, to the poet's self and his activity of poetic making. Only Petrarch offers a comparable concentration on the self and its vicissitudes, and from Petrarch Goethe is distinguished by the wholly secular context of his thought and feeling. More immediate predecessors there certainly were—Klopstock, Gray, or, particularly, Rousseau—but these furnished only elements for the compound which Goethe was the first to synthesize and which he bequeathed to the nineteenth century, and beyond.

To come to understand Goethe properly is to adjust one's understanding both of modern literature, its course and derivation, and of Europe, and Germany's place within it, and that equally for an English and for a German reader. Goethe's name has been much used, and abused, in a search for identity by German nations which came into existence after his death. One of the principal difficulties in reaching a dispassionate view of Germany's painful struggle towards nationhood, and of Goethe's part in it, has been that the main critical instrument available for analysing ideological delusions, namely Marxist social theory, is itself a product of the particular process which stands in need of analysis. Since 1945 the alternative to the manifest inadequacies of Marxist Goethe-scholarship has been, in the Federal Republic—and more recently, though for different reasons, in the Democratic Republic too—a programmatically non-political approach, which has done marvels in editing, annotation, and the accumulation of source-material, but has not always faced the challenging task of interpretation. In the English-speaking world Goethe's reputation suffered in his lifetime from endlessly repeated charges of immorality and irreligion, then from accusations of obscurantism, nineteenth-century progressivism, antinomianism, and literary incompetence, and latterly from benign neglect. Barker Fairley's admirable literary and psychological *Study of Goethe* is now well over forty years old, and the last substantial and original English biography, by J. G. Robertson, nearly sixty. The larger and somewhat earlier biography by Peter Hume Brown was the first such undertaking since G. H. Lewes's great monograph of 1855, a most remarkable work of scholarship for its time, and one of the first Goethe biographies in any language. There has been much good writing about Goethe in English since the Second World War, but most of it has been for a specialized audience, the obvious exceptions being Erich Heller's stimulating essays in *The Disinherited Mind* and W. H. Bruford's *Culture and Society in Classical Weimar*, to both of which my debt is as great as it is obvious.

The present book has been written in the belief that two different needs can be met by a biographical study of Goethe and his works which starts from first principles and assumes as little prior knowledge of its subject as possible. On the one hand the reader with some knowledge of English and French literary history, but unacquainted with the German language or its literature, or anything but the outlines of the nation's political development, should find here enough information to set Goethe's life in the context of his age, and his poetry in the context of his life. On the other hand, those already familiar with Goethe's works will, I hope, learn something from seeing them presented against their biographical, social-historical, and philosophical background, and discussed, as far as possible, in a rigorously chronological sequence. These have not on the whole been features of the synopses of Goethe's achievement published since 1945, even of the three-volume study, both authoritative and sensitive, of Emil Staiger. To say that, however, is not to be ungrateful, for the present work has been made possible only by the magnificent scholarship of the last half-century. Several annotated editions and selections, Femmel's *Corpus der Goethezeichnungen*, Flach and Dahl's edition of *Goethes amtliche Schriften*, the Leopoldina edition of the scientific works, E. and R. Grumach's new collection of *Begegnungen und Gespräche*, Steiger's 'documentary chronicle', and various other projects, variously advanced, are the foundation of this study—along, it must be said, with the indispensable legacies of earlier generations, the Weimar edition of the works and letters, and the many volumes in which Wilhelm Bode recorded his unequalled knowledge of the literary and social world of later eighteenth-century Weimar. What I can offer is only a synthesis of syntheses, whose value will long be outlasted by that of the compilations on which it is based; yet if such a synthesis is not attempted from time to time, and for a particular time, to what end are the compilations made? The secondary writing about Goethe long ago grew to a point at which no one man could hope to encompass it; the primary sources are now not far behind: it is a moment to pause for thought, and to attempt to situate this extraordinary human phenomenon in that widest possible context in which by its own nature it demands to be inserted. For the specialist there may emerge from the exercise a new view of a commanding literary presence: as a free man responding to the social, spiritual and intellectual demands of modernity, as they formulated themselves around him. For the non-specialist there is the promise of a new acquaintance: limited and even peculiar, no doubt, as we all are, but grand and deep and rich as none of us is, and few of our forebears have been. It is my hope too that the following pages may find some readers in Germany, for they are written in the belief that the Federal Republic has represented not only what is best, and oldest, in the nation's political traditions, but also what is closest to the mind of Goethe, and it is time for the rest of Europe to start to say thank you.

To make Goethe accessible, whether to the general reader, to the student, or to the scholar, was my principal concern in the organization and writing of this book. Only passages in verse, or significantly poetic prose, have been cited in German, and always with a translation. Translations, which are my own, are as literal as possible, and in some cases the word order is deliberately closer to the original than normal English usage would allow, in order to hint at important effects that would otherwise be lost. The book has to some extent been conceived as a companion to the Hamburger Ausgabe, as the selection from Goethe's works and correspondence most likely to be readily available to the student, and wherever possible texts have been cited from this edition; where this is not possible recourse is had to the new edition of *Der junge Goethe*, by Hanna Fischer-Lamberg, for Chapters 1–3, and thereafter to the Weimar edition. The text is cited precisely as found in the edition of reference, since there seemed no way of imposing consistency on the different editorial practices, and no purpose in it either. Place names are given in anglicized forms, when these are available and not obsolete, personal names, except for a few well-known monarchs, are left in their original form. 'Erbprinz' is inaccurately but conveniently rendered as 'Crown Prince'. To aid pronunciation, feminine forms in –a have generally been preferred to those in –e (e.g. Amalia). The page has been kept (almost) free of footnotes and references, but sources for quotations and for important facts or assertions are grouped at the end of the volume and identified by page number and a few key words.

The reader will see from the contents list that within each chapter sections mainly devoted to Goethe's life alternate, on the whole, with sections mainly devoted to his works. Not that the two subjects can be separated—on the contrary, it is a main theme of the whole book that they cannot. But to make orientation easier, and to make the biographical narrative more continuous, it seemed better to confine the literary discussion of major works to discrete sections. The biographical sections still contain some discussion of minor and shorter works, such as lyrical poems, and also of Goethe's scientific concerns and of his drawing—enough, I hope, to confirm the general principle of the inseparability of mind and matter. Works composed over a long period are treated as far as possible stage by stage in the course of their development. With Goethe's œuvre the biographical approach has great analytical power, and to abstract from the gradual process of composition is the surest way of reintroducing those ideological preconceptions and conventional judgements which I am trying to eliminate. The headings to the pages, and the index of

Goethe's works, should help the reader to find the main passages relevant to any particular work.

It is always a pleasure to write about Goethe, and naturally it is a pleasure to express one's gratitude to those who have enabled one to do so. I should like to thank Stephen Brook, who first suggested, more than ten years ago, that I should write this book, and Virginia Llewellyn-Smith and Catherine Clarke, of Oxford University Press, who encouraged me to continue with the project. A scholarship from the Alexander von Humboldt-Stiftung made it possible for me, under the benevolent direction of Professor A. Schöne, to write a first draft of Chapters 1 to 3, and the generosity of that remarkable institution deserves the warmest recognition in these chill times. My work was rendered much easier by the helpfulness of the staff in the Niedersächsische Staats- und Universitätsbibliothek in Göttingen and in the Cambridge University Library, and by the protracted loan of many volumes from the Beit Library of the Faculty of Modern and Medieval Languages in Cambridge. I am grateful to my colleagues, and especially to Dr J. Cameron Wilson of Jesus College, Cambridge, who readily took over my normal duties while I was on study leave, and to Wolfgang and Rosemarie Bleichroth, who were the kindest of hosts and who drew my attention to many of the silhouettes which adorn the book. They, and Paul Connerton, proved indefatigably tolerant of Goethocentric conversation and so have left their mark on much of my exposition. A timely gift of books from Anneliese Winkler, to whom I owe any facility in the German language I may have, greatly eased the later stages of my work. Dr M. R. Minden kindly passed on to me a number of books from the library of Trevor Jones, who I like to think would have enjoyed the use I have made of them. I wish particularly to thank those who have given their time and attention to read and comment, often at greater length than I had any right to expect, on parts of the manuscript as it has developed: Profesor J. P. Stern, Professor and Mrs J. C. O'Neill, Professor T. J. Reed, Dr E. C. Stopp, and my dear colleague the late Professor U. Limentani, who gave me invaluable assistance with Chapter 7. They have saved me from many errors, obscurities, and fatuities which the reader will not see: those which remain on view are all my own work.

I am grateful to the editors of *German Life and Letters* for permitting the republication of some material in Chapter 3 which first appeared in their journal. A general word of thanks is also appropriate here to those who have furnished me with illustrative material and given permission for its use, and who are more specifically acknowledged in the List of Illustrations. That there are any illustrations at all is a result of the generosity of the Morshead–Salter Fund of Magdalene College, Cambridge, of the Tiarks Fund and the Jebb Fund of the University of Cambridge, and especially, and once again, of the Alexander von Humboldt-Stiftung, who have all made substantial grants towards the cost of reproductions. My thanks go not only to the institutions themselves, but also to their individual representatives who have helped and encouraged me.

A project of this kind has necessarily passed beneath the fingers of many typists who deserve grateful acknowledgement: Rosemary Baines, Marion Lettau, Nicholas and Deborah Hopkin, and above all my wife, Rosemary, who not only typed Chapters 4 to 8 but was also their most rigorous critic. To her support throughout, but especially in the difficult hours and days of what has been a more demanding undertaking than I hope is now apparent, I owe both the completion of this first volume and the prospect that the second volume will follow it with reasonable expedition. My parents-in-law spared neither their time nor their energy in sustaining us both and in managing a household which, despite including two children under three, at times seemed to revolve around a man dead one and a half centuries ago. This book is their work also, and I dedicate it to them.

N.B.

January 1990

Contents

List of Illustrations

Silhouettes

Map

The author and publishers are grateful to the following for permission to reproduce the illustrations:

Freies Deutches Hochstift, Frankfurter Goethe-Museum, Frankfurt on Main. Photo Ursula Edelmann: plates 1–4, 6, 8, 9

Kunstsammlungen zu Weimar: plates 21, 24, 32

Nationale Forschungs- und Gedenkstätten der klassischen deutschen Literatur in Weimar: plates 5, 7, 13, 14, 16, 18, 19, 25–28, 29, 33, 34, 37, silhouettes on pages 1, 89, 231, 281, 333, 413, 531

Bildarchiv Foto Marburg: plates 10, 11, 15, 23, 30, 31

Trustees of the British Museum: plate 12

Ullstein Bilderdienst: plate 17

Kunsthistorisches Museum, Vienna: plate 20

Städelsches Kunstinstitut, Frankfurt on Main. Photo Ursula Edelmann: plate 22

Bild-Archiv der Österreichischen Nationalbibliothek, Vienna: plate 36

Goethe-Museum, Düsseldorf. Anton-und-Katharina-Kippenberg-Stiftung: silhouette on page 41(*left*).

The silhouette on page 41 (*right*) is from J. Vogel, *Goethes Leipziger Studentenjahre* (Leipzig, 1899). Plate 35 is from O. Harnack, *Zur Nachgeschichte der italienischen Reise* (Weimar, 1890). The map is reproduced by permission of George Philip and Son Ltd.

△ Mountain

‖ Pass

—·—·— Approximate boundary of
the Holy Roman Empire

Königsberg

Danzig

Hamburg

Berlin

BROCKEN △   Dessau

Düsseldorf   Eisenach   Erfurt   Leipzig
Brussels   Cologne   Weimar
                    Gotha   Jena   Dresden
Coblenz   Lahn   Wetzlar
          Frankfurt
Darmstadt   Main   Carlsbad   Prague

Carlsruhe

Strasbourg   Regensburg

Rhine   Lake   Munich   Vienna
               Constance   0 Starnberger See

Lake of Zurich   Zurich   Innsbruck
Lake of Lucerne   ‖ BRENNER
Lake of Geneva   GOTTHARD ‖
Geneva   △
         MT BLANC   ‖ SPLÜGEN
         Lake Maggiore   Lake
                                Como
Milan   Lake Garda   Vicenza
                            Venice

Florence

Rome

Central Europe in the 1780s

I

Germany in the Eighteenth Century

W HEN Johann Wolfgang Goethe was born in the free city of Frankfurt on the Main on 28 August 1749, the Holy Roman Empress Maria Theresa had just, after nearly a decade of war, confirmed her right to rule Austria, Hungary, Bohemia, and, through her husband Francis I, Frankfurt's nominal sovereign, the Imperial German territories, a right which she was to exercise for the next thirty years; Frederick II, not yet 'the Great', of Prussia, just installed in the enjoyment of illegally annexed Silesia, had nearly forty years still before him as 'first servant of his state'; Louis XV was half-way through his sixty-year reign over a wealthy but mismanaged France that was still the supreme power of Europe; George II, Elector of Hanover, and the last English king to lead his troops into battle, had done so on behalf of Maria Theresa, and against the French, at Dettingen, not far from Frankfurt, in 1743 (the occasion for Handel's Dettingen Te Deum)—three years later his own throne had been successfully defended in the last pitched battle on British soil, Culloden. Within living memory England had had a Catholic king, and Vienna had been besieged by the Turks. In 1749 the North American colonies were loyal to Britain; Quebec and Louisiana were French; Clive and Dupleix were rivals in India; and Australia was effectively unknown. The mail from London to Edinburgh took over a week, Moët and Chandon had begun to export the recently invented champagne, and a pineapple cost as much as a horse. In 1749 Pope was five years dead, Johnson had begun, but not completed, the publication of his Dictionary, Voltaire had yet to move to, and from, Potsdam, Rousseau had yet to deny publicly the moral benefits of civilization, and Mozart had yet to be born.

When the ennobled Privy Councillor von Goethe* died in Weimar on 22 March 1832 there had for a quarter of a century been no Holy Roman Empire; France had just passed through a second revolution and was ruled by a citizen-king; and England, having emancipated Roman Catholics, had begun to reform its Parliament. Napoleon, Beethoven, and Hegel were dead, Walter Scott and Jeremy Bentham—an almost exact contemporary of Goethe's, born in 1748— were to die in the same year. In 1832 Alexis de Tocqueville was touring the America of seventh President Jackson, in search of democracy, and Ralph Waldo Emerson, having resigned from the ministry, was touring Europe in search of religion. The first photographs had been taken by Niepce and Daguerre; the first passenger-carrying steam railway line was operating between Liverpool and Manchester; *The Times* had been publishing for nearly

---

* The 'von', the particle indicating nobility, bestowed on Goethe in 1782, is nowadays usually dropped.

forty years. Otto von Bismarck was a student in Göttingen, where Gauss and Weber were about to erect the first electric telegraph, and Dr Arnold was headmaster of Rugby, already the home of a less useful invention. University College, London, had a professorial chair of German, and German philosophy was being taught in the Antipodes. Tolstoy was 3, Baudelaire was 11, Wagner was 19 and composing his first opera. In the next five years *Sketches by Boz* would be published; Victoria would become Queen of England, so putting an end to the personal union with the Electorate of Hanover; the British Raj would displace the East India Company; Germany too would acquire its first railway line, from Nuremberg to Fürth, and Karl Marx would begin his studies at the University of Berlin. Probably the last person with her own adult memories of Goethe, an innkeeper's wife, would not die until 1906.

The nearness to us of the world in which Goethe died may seem only to emphasize the remoteness of the world into which he was born. If we look at Goethe's end he seems to stand on the verge of modernity, if we turn to his beginning we seem to be reaching the limits of a normal historical consciousness. Between Goethe's beginning and his end, striking directly into the middle of his life, only a few weeks before his fortieth birthday, came the event which initiated the modern form of the State and the modern international political system, and from whose consequences no contemporary could escape: the storming of the Bastille by the Paris mob, the French Revolution as almost immediately it was called. This, largely, is the source of our alienation. If Goethe's works, and the German literature of the age that bears his name, the *Goethezeit,* are difficult of access to English-speaking readers, one reason, far from superficial, is that it is difficult for such readers to credit that an age bisected by so colossal an event can be a unitary period at all.

There are other reasons too. Our English-speaking readers are not helped by the lack of a familiar contemporary literature with which they could make comparisons, for the period of Germany's greatest cultural flowering—from about 1780 to about 1806—coincides with a relatively fallow time in their own literature and, understandably, in that of France. They are not used to attributing literary wealth and an exemplary sensibility to the period from the death of Johnson to the youth of Keats, nor even to that from Gray's *Elegy* to the death of Byron: 1798, the year of the *Lyrical Ballads*, may well stay in the mind as the moment of a new start, a literary revolution to accompany the political, but in Germany it is the date of the high point of a movement of cultural reform which had been in preparation for decades and which was the work of mature men. There are also problems of literary genre. The 'classical' literature of modern Germany is a product of the age of Cowper and Crabbe, and that provides us with an approximate orientation in the matter of tone, but the preferred form of that literature is the drama—and what possible standard of comparison have English readers here? The utterly forgotten dramas of the Della Cruscans? the almost equally deservedly unperformed verse-tragedies of the Romantics, *Otho the Great, The Borderers*, even *Cain*?

Conversely, moreover, the most striking feature of eighteenth-century German classicism for readers approaching it from a knowledge of English literature is its failure to produce a single readable representative of that eighteenth-century form *par excellence,* the novel. If in the course of classicism's violent prologue Goethe produced a best-selling novel, *The Sorrows of Young Werther* (1774), with but one fully realized character, and if in the course of the movement's long Romantic epilogue he produced another novel, *The Elective Affinities* (1809), whose shimmering elusiveness seems now that of Blake and now that of Jane Austen, this surely only confirms the difficulty of grasping him and his age in the literary and historical concepts appropriate to the rest of Western Europe.

It is one of the premises of this book that the very notion of the *Goethezeit* is a serious obstacle to the proper understanding of Goethe himself. There never was such a thing as the 'Age of Goethe'. In the first place, our intuitive recognition is correct that Goethe's exceptionally long writing career spans a great divide not only in European, but specifically in German, history. It is true that in Germany the earthquake was so delayed that paradoxically a period of highest intellectual achievement could coincide in the 1790s with a period of extreme political convulsion in the rest of Europe. But this appearance of unworldly calm should not be allowed to conceal the fact that in 1806, the year in which Prussia died and was reborn, a new age began for all Germany, and not just for its intellectual stratum. It is sheer obscurantism to gloss over this great crisis with some such concept as 'the Age of Goethe'. Germany is not so different that it too did not have its Revolution—true, it was a revolution of a peculiar and unobvious kind, but the political and cultural difference between 1780 and 1820 was in the German-speaking world nearly as great as in France, and certainly greater than in England. What did make Germany unique was that what in France was achieved by an intellectual bourgeoisie was in Germany the work of an intellectual bureaucracy. If the German revolutionaries of the 1790s are men of learning and letters, above all of the universities, such that in Germany at this time literary and philosophical schools succeed one another with the rapidity of French ruling cliques, the next generation shows us the pupils of these men, or these men themselves, in new roles, not simply as the reformers but as the new masters of the state machines of Prussia, Bavaria, Saxony, and, to a lesser extent, of Austria. The Prussian revolutionaries, for example, were men like Gerhard von Scharnhorst (1755–1813), the reorganizer of the army, Carl August von Hardenberg (1750–1822), reformer of the entire political and administrative apparatus, Wilhelm von Humboldt (1767–1835), founder of the (present German) educational system, and Friedrich Schleiermacher (1768–1834), not only a theologian but also an important agent in the unification of the Prussian Protestant churches. Their work, like that of their more demonstrative French counterparts, outlasted their, sometimes short, official careers. We should not, however, assume that because in Germany a great, unbloody, political change took place, in which men were instrumental who looked to France as their model, therefore the change was of the same nature, or even occasioned by the same forces, as in France. In name 'liberals', the great reformers were in fact the founders of that bureaucratic autocracy which, partly as a result of the political expansion of Prussia, was to determine the character of German society—and ultimately not only German —well into the twentieth century, perhaps even to the present day. The German revolution was the transformation of a monarchical into a bureaucratic absolutism, preceded and accompanied by an intellectual explosion, partly determined by French initiatives (including of course the violent intervention of Napoleon), but also, and largely, autonomous. In Germany, and as a result of German actions, the *ancien régime* passed away. Its place was taken by a society as peculiarly German as it was clearly post-revolutionary, the middle-class world of Biedermeier—a sort of Victorianism without (until the middle of the nineteenth century) industrial capitalism.

It is Goethe's singular greatness that he responded intellectually, and in the form of literary production, to every stage of this—for a contemporary hardly graspable (Goethe called it 'daemonic')—metamorphosis. At each new birth-pang of modernity, specifically in Germany, but also generally in Europe, he felt the pain, recollected and recovered himself, and attained and expressed understanding. His beginnings are without revolutionary chiliasm and his end without reactionary nostalgia, yet in over sixty years of writing (the standard edition of his works and letters runs to 138 volumes) he never seriously repeated himself. Always judicious, he none the less always grew. If he thereby brought sense, and perhaps also unity, into an experience more profoundly fractured than is now any longer possible (the twentieth century can offer greater disasters than the eighteenth, but not greater revolutions)—that is his personal achievement. It is in the nature of the achievement that the sense and the unity belong to him, not to the several ages in which he lived. To think otherwise is to diminish the man, and to misrepresent the time—his time, and ours.

In the second place, the term *Goethezeit* carries a seriously misleading implication about the nature of Goethe's unceasingly energetic engagement with the literary and intellectual culture of his German contemporaries. With the possible exception of the first five years of his public literary career, Goethe was never a model which a whole generation chose to imitate. Nor, on the other hand, was he ever the most typical figure of a particular literary movement. 'People are always wanting me to join in and take sides—well, then, I am on my side', he remarked in 1831. Even in 1880 Nietzsche was one of the few to have understood what he meant:

Goethe . . . lived and lives only for a few: for most he is nothing but a conceited fanfare trumpeted from time to time across the German borders. Goethe—not just a good and great man, but an entire *culture*—Goethe is an episode without consequences in the history of the Germans: who for example could demonstrate one bit of Goethe in the

German politics of the last seventy years! (While a bit of Schiller has certainly played its part and perhaps even a little bit of Lessing.)

But the more careful phrase of T. S. Eliot is also the more just: 'Goethe is about as unrepresentative of his Age as a man of genius can be.' There must be a limit to the extent to which greatness can be detached from its time, and slight though Goethe's influence on his contemporaries may have been, their influence on him was, and had to be, constitutive. 'The artist must have an origin, he must know where he comes from', he said in the same conversation in 1831, and 'Do not take me amiss—I too am a joiner.'

In one of those moments of 'joining in', in 1798, he wrote, with an appropriate wishfulness, in the inaugural issue of the (short-lived) periodical *The Propylaea:*

after all, the artist too is a part of the public, he too has been formed amid the same times and events, he too feels the same needs. He strives in the same direction, and so he moves on serenely with the mass, that bears him up, that is enlivened by him.

If a poet's age does not give him his subject-matter, his own resources will prove a shallow well. Perhaps it was belated recognition of a kindred spirit that eventually revealed to Eliot the true relation between Goethe and the follies, heresies, and enthusiasms of his time, on which he had had to draw, for there was nothing else: 'The more I have learnt about Goethe . . . the less I find it possible to identify him with his age. I find him sometimes in complete opposition to his age, so complete perhaps as to have been greatly misunderstood.' 'The Age of Goethe' is not least a misnomer because the enormous powers of absorption, which Goethe, so repeatedly and astonishingly and at every stage of life, brought to the understanding of his world were complemented by a spirit of opposition so unyielding as to be without parallel in the literature that is timelessly great, but which makes of him the authentic, the classic, poet of modernity. 'The Age of Goethe' is simply the series of literary and intellectual temptations which, as it happens, Goethe resisted. That series may have its own logic, but it is not that logic which entitles it to bear the name of Goethe. Goethe's association with it was fortuitous—the stroke of undeserved good fortune that genius is—and Goethe's interest in it was to a large extent polemical.

But even Goethe's uniqueness had its origin: Goethe knew where he came from, and never denied it, even though he finally resolved that he could carry out his life's work only by leaving home for good. 'If someone were to ask me where I could imagine my cradle standing in an easier place, more appropriate to my opinions in civil affairs or more suited to my views of poetry, I could name no sweeter town than Frankfurt.' To understand the relationship, simultaneously of dependence and detachment, between Goethe and that native German literary tradition which assorts so oddly with the contemporary literatures of France and England, we shall eventually have to examine his position as a citizen of Frankfurt. To understand Frankfurt, however, we must first examine the relation between civil and poetic culture throughout the rest of eighteenth-century Germany.

Princes, Pietists, and Professors

'Germany' in the eighteenth century was not even a geographical, at most it was a linguistic expression. No natural boundaries, and certainly no single political structure, enclosed all the speakers of the various dialects of the German tongue, scattered in every degree of concentration from Alsace to the Volga and what is now Romania (where there is still a German-language newspaper), and from the Gulf of Finland to the Swiss Alps and the Adriatic. Not even if we confine our attention to the areas which were to be represented by the German and (non-Balkan) Austrian Empires in 1914 can we find a hint of a national framework, or of a national consciousness. Constitutionally speaking, these areas, which in the late eighteenth century were occupied by some 20 and 27 million human beings respectively (at a time when the population of France was 24 million but that of England and Wales just under 10 million), were composed of three distinct, but partially overlapping complexes, which may be given the short-hand titles of the Empire, Prussia, and Austria. The Holy Roman Empire of the German Nation, whose claim to lineal descent not merely from the Empire of Charlemagne but also from that of Ancient Rome was apparent in the Imperial title itself (Kaiser = Caesar), covered approximately the territory of the modern states of Austria, Belgium, Germany, and half of Czechoslovakia, covered them, it may be said, with confusion. In the north and south of the Empire's unbelievably elaborate geographical and constitutional jigsaw, however, two states were emerging in the eighteenth century which could act as identities on the European stage in something like the same sense as France or England or Sweden or Portugal. In each case a substantial nucleus of Imperial territory was extended and supported by a 'tail' of lands stretching away to the east, and outside the Empire. The Austrian tail, notably Hungary, Slovenia, and Galicia, and the Italian satellites, need not concern us further since the German-speaking subjects of the Habsburgs at this time, perhaps a quarter of the total, were largely to be found within the Imperial boundaries. The Prussian 'tail', however, land of German settlement, originally by the Teutonic Knights, from 1226 until the expulsions of 1945, and reaching around the Baltic to Königsberg (now Kaliningrad) and Memel, was to be of profound significance for the cultural development of all Germany. For by expanding in the seventeenth century into the only semi-Imperial Pomerania, and the wholly non-Imperial Duchy of Prussia, the Electors of Brandenburg shifted the centre of gravity of their nascent state towards a land of vast latifundia (now part of Poland and the Soviet Union) where a scattered population of mainly Slavonic serfs was autocratically ruled by an arrogant and untutored nobility (the 'Junkers'). An element of east European agrarian despotism entered the regime in Berlin, which harmonized ill with the traditions of corporate liberty of the towns and small territories in western, Imperial, Germany which over the next two centuries were to be eaten up by the new giant. Indeed, the kings who from 1701 succeeded the Electors deliberately took their title from their independent eastern, 'Prussian', domains, and not from their historic origin, the Mark of Brandenburg, which remained in theory a fief of the Emperor in Vienna.

The Empire itself, the German heartland, would be better described as an extremely loose confederation of states of very unequal size and importance. In the eighteenth century, the nine Electors (the title refers to the office, largely ceremonial, of electing the Emperor and his heir apparent, the King of the Romans), 94 spiritual and temporal princes, 103 counts, 40 prelates, and 51 free cities (including Frankfurt) were all equally sovereign rulers over their own territories, feudally dependent directly and only on the Emperor. To these must be added some 1,000 Imperial knights claiming equal 'immediacy' in the derivation of their authority but ruling altogether over a total of not more than 200,000 subjects. Not merely, however, was the Empire fragmented, but the fragments of which it was composed were themselves fragmented too. Geographical integrity of the possessions united under a single princely dynasty was the exception rather than the rule. From the seventeenth century Brandenburg-Prussia, for example, had ruled several territories on the Lower Rhine near the Dutch border, and one in Württemberg in the south-west. The Archbishop of Mainz, at the confluence of the Main and the Rhine down-river from Frankfurt, was also the overlord of the Eichsfeld two hundred miles to the north-east, on the edge of the Harz mountains, and an island of Catholicism to the present day. The duchy of Weimar, in Goethe's time, consisted of four separate territorial units, with a total area of 700 square miles. The resultant complexity of frontiers, customs dues, and transport may be left to the imagination.

Apart from the person of the Emperor, the institutions unifying this magnificent chaos were few and shadowy. The Reichstag, the Imperial Diet, at Regensburg (Ratisbon), though divided into three chambers of Electors, other princes, and free cities (the Imperial knights were not represented), was not so much an assembly as a standing conference of ambassadors. There were no debates and business was conducted in writing. The Imperial Supreme Court, the Reichskammergericht, which from 1693 was housed in Wetzlar, north-east of Frankfurt, was very expensive—the fee alone for entering an appeal was 1,500 guilders—and slower by far than Dickens's Chancery: in the middle of the eighteenth century there were over 16,000 cases still outstanding. The Imperial Tribunal, the Reichshofrat, in Vienna, was a rival appeal court whose competence was never adequately defined. Certain towns, finally, had a special relationship to the Imperial office: the Crown Jewels were kept in Aachen (Aix-la-Chapelle) and Nuremberg, and Frankfurt was the town in which the election and coronation of a new Emperor had to take place, in accordance with an elaborate ceremonial laid down in the Golden Bull of 1356. On the other hand, there was no Empire-wide system of taxation (nor even of currency) and certainly no standing Imperial army. Nor of course was there a single state Church. The Peace of Westphalia, which in 1648 had put an end to the terrible devastation of the Thirty Years' War, while confirming the three Churches, Lutheran, Calvinist, and Catholic, in their possessions of the 'normative year' 1624, had laid down the principle that each territory should have its own established religion, that of the local sovereign, with a sort of limited right of persecution of the other two ('cujus regio ejus religio').

It would, however, be a mistake to regard the Holy Roman Empire as a hollow sham simply because it did not have the characteristics of a modern nation-state. It provided a historical, emotional, and juridical framework which, in the eighteenth century alone, survived eight major wars fought on German soil, and even at the political level it was a diplomatic mechanism through which a certain necessary balance of power could be maintained. For 'Germany' was not lacking in ambitious, and thus rival, forces aiming precisely at the establishment of such a modern state or states, nor in forces that were anxious to resist these ambitions. In addition to Prussia and to Austria under Joseph II (Emperor with Maria Theresa from 1765, and sole Emperor from 1780 to 1790) there were the Electorates of Saxony and Bavaria, the Duchy of Württemberg and the Landgraviate of Hesse (Cassel), all substantial powers aiming at their own unification, expansion, and enrichment. The smaller territories, the Imperial knights, and the free cities were the carp or the minnows in this world of pike. Their only protection was the Imperial power, and the historic order that the Empire sanctioned. Frederick the Great was of course very willing to assist the city councillors of Frankfurt when they quarrelled with the Emperor about constitutional matters, but his true opinion of the city's traditional freedoms was indicated rather by his arbitrary arrest, through his Frankfurt agent, of the departing Voltaire, whom he happened to wish relieved of certain compromising manuscripts. The old Emperor and the ramshackle Empire were the natural patrons of those who had nothing to hope from the expansion of the busily centralizing enlightened autocracies with their splendid new geometrical palaces and parks, show-cases of absolute power, dominating unfree cities of residence (Residenzstädte)—Berlin, Munich, Dresden, Stuttgart, and, its street-plan, like its rulers, the most brutal of all, Cassel. Cassel is situated in a broad valley, closed off at one end by steep hills. Halfway up this rampart stands the huge and sombre palace of the Landgrave: down from its main entrance a broad swathe is cut in a straight line through the town and across the entire valley. Above the palace rise, along the same axis, the terraced gardens which culminate in a stone spire, the highest point for miles around, surmounted by a colossal statue of Hercules (who gave his name to various Hessian rulers)—with club.

To understand the peculiarity of Goethe's upbringing and of his natural loyalties, it is important to realize how much, and for how long, the mainstream of German culture was determined by the despotic structure of the major states, centred in cultural as in all other matters on the court, and the person of the prince. From the middle of the seventeenth century the combination of incessantly destructive war and protracted economic depression had sent into irreversible decline the great trading, banking, and manufacturing cities of the fifteenth- and sixteenth-century German bourgeoisie—in 1800 the population of Nuremberg was still only one half of what it had been in 1600, and other cities, Augsburg, Cologne, Ulm, and even Frankfurt, similarly suffered losses ranging from the serious to the catastrophic (the average loss was around one-third of the original population). In the period, still far from peaceful, that followed on the Peace of Westphalia—towards the end of the century the megalomaniac campaigns of Louis XIV reduced western Germany to a state reminiscent of the Thirty Years' War—the responsibility for social and economic reconstruction was taken over not by the entrepreneurs but by the princes, whose undemobilized armies gave both political strength and—since they had to be fed, clothed, and equipped—economic impetus to an increasingly *dirigiste,* absolutist order of things. The towns that now prospered were precisely the residence towns where the ruler set up his local Versailles, where the traditional rights of the citizens were suppressed, and where the original artisan and merchant economy was replaced by the servicing of the court. By 1660 the cultural pattern of Germany's bourgeois period—the combination of the local master in the local form (such as the Master-singer Hans Sachs (1494–1576) of Nuremberg), the workshop artist executing commissions for any patron willing to pay (such as the great wood-carver Tilman Riemenschneider (c.1460–1536)), and the travelling humanist willing to settle in any municipality anxious to retain him (such as Sebastian Brant (1457–1521), author of *The Ship of Fools*)—that pattern had almost completely faded away. Its last representatives include some of the best-known writers of the so-called 'baroque' movement in Germany (the word unfortunately fails to distinguish between bourgeois and princely art), such as the realistic novelist J. J. C. von Grimmelshausen (1620 or 1621–76), who was also mayor of the village of Renchen, and the poet and dramatist Andreas Gryphius (1616–64), town clerk of Glogau (now Glogow). For the next two generations the arts that were to flourish in Germany were to be performative arts, largely, or even wholly, dependent on princely patronage or dedicated to princely glory: architecture, masque, and opera. As for German literature, it would not be too much to say that in the half-century from 1675 to 1725 it simply collapsed. A middle class with literary interests had practically ceased to exist. The Renaissance tradition of cultivating and embellishing the vernacular was kept alive, just, by a few learned 'language societies' ('Sprachgesellschaften'), whose productions, apart from voluminous works of theory, were largely of an escapist pastoral nature, which still has a certain faded charm. Otherwise literary talent grew only to be stifled, as is apparent from the abbreviated career of the satirical and 'galant' novelist C. Hunold ('Menantes') (1680–1721), and from the bitterly early death of the gifted Silesian poet J. C. Günther (1695–1723): often described as a Goethe born before his time, Günther could equally be called a Gryphius born too late. For the middle ground once occupied by the alliance of burghers and humanists had disappeared, and a gulf now yawned between popular and princely literature, between mass-produced and ill-printed trash romances on the one hand, and the prestige productions of the court-printer, the folio anthologies of flattering poems and the thousand-page fictional handbooks of statecraft on the other. The gulf was even more crassly evident in the realm of drama, which was divided between strolling players rising no higher than farce or even puppetry, and court-theatres devoted to tragedies in French and operas in Italian. (It is from this polarization, around 1700, that the distinction between 'art' and 'kitsch' ultimately derives, which still plays an important part in German literary criticism.) Cowed, dutiful, and profoundly insecure, such of the middle class as remained contented itself with devotional literature, with books of hymns, and eventually with morally edifying weeklies, on the pattern of Addison's *Spectator*, but much more earnest. Indeed in the late seventeenth and early eighteenth centuries religious revival could be said to have taken the place of literature as far as the German middle classes were concerned. It was a sign of the times that in the 1690s the citizens of Frankfurt were prepared to accept, as part of the cost of a more stringent religious censorship, the transfer of a large part of the business of their book-fair to their age-old rival, Leipzig.

It was, in fact, from Frankfurt that one of the major impulses to revival came, in the figure of Philipp Jakob Spener (1635–1705), who was head of the established Lutheran Church in Frankfurt from 1666 to 1686. In this position, in similar subsequent positions in Dresden and Berlin, and through an immensely extended correspondence, Spener, the spiritual adviser, together with the more worldly-wise August Hermann Francke (1663–1727) whom he distrusted, became the initiator of that form of radical Lutheranism known as Pietism which was to be of immense importance for German religious life for the next two centuries. The particular feature of Pietism which makes it of interest to us is its natural affinity for state absolutism. A religion which concentrates to the point of anxiety, not to say hypochondria, on those inner motions, whether of dryness or abundance, of despair or of confident love of God, from which the individual may deduce the state of his immortal soul; a religion whose members meet for preference not publicly, but privately in conventicles gathered round a charismatic personality who may well not be an ordained minister; a religion which disregards all earthly (and especially all ecclesiastical) differentiation of rank, and sees its proper role in the visible world in charitable activity as nearly as possible harmonious with the prevailing order (such as Francke's famous orphanage in Halle, one of the functions of which was the recruitment of Prussian military chaplains)—such a religion was tailor-made for a state system in which all, regardless of rank, were to be equally servants of the one purpose; in which antiquated rights and differentiae were to be abolished; and in which ecclesiastical opposition was particularly unwelcome, whether it came from assertive prelates or from vociferous enthusiasts unable to keep their religious lives to themselves. It was not chance that both Spener and Francke gravitated towards Prussia, the most energetic and revolutionary of the new absolutisms, and the most hostile to the Imperial order; and that under Frederick William I (King from 1713 to 1740) Pietism became in practice the Prussian state religion. On Frankfurt, however, Spener left little mark—he was disappointed in the town, his flock retired into privacy, no Pietist was given his post again, and the Lutheran clergy remained rigidly orthodox until the late eighteenth century.

Reconciling the individual with a unified rational order by interpreting individual life in terms of pure inwardness is also a theme in the philosophy of the one towering genius intellectual Germany produced in the period of transition from bourgeois to absolutist culture. The thought of G. W. Leibniz (1646–1716), mathematician, logician, inventor, jurist, diplomat, theologian, philosopher, historian, courtier, and librarian, has at least as many facets as his eventful life. But he himself saw the relationship between the individual and the continuum, as he put it, as one of the two major problems of philosophy (the other being the problem of free will), and he laid particular emphasis on the implications for the German social order of his answer to this most abstract metaphysical question. How can a thing both have an identity of its own and yet also belong to a larger order, also be acted upon by other things, for example, to the extent even of being or becoming a part of them? Leibniz's answer contains three features which make of it a prophetic exposition of the entire eighteenth-century German world-view. In the first place, the ultimate constituents of the world, for Leibniz, are not things, but *forces*—forces, however, with an identity, 'substantial forms', he calls them, or 'monads', and his prime example is the human soul. It is in the nature of such a force, or monad, to be always changing its state, to be always developing: its identity lies not in any one particular state but in the rule of its development that describes all its states, like the single algebraical formula that describes all the positions of a point moving along a curve. In the second place, although all the different states of a monad, such as a human soul, may appear to involve it with the rest of the universe, that is, with other monads, so that it appears to come into contact with them, to act upon them or be acted on by them, none the less, according to Leibniz, all these changes are understandable not as the interaction of several monads one upon the other, but as the autonomous development of each monad singly, fulfilling in isolation the law of its being. All *change is internal* and there are no such things as relations. That there appear to be relations, that monads appear to act upon one another, is a result of a 'pre-established harmony' which ensures that the different rules-of-being of all the different monads fit together as perfectly *as if* they acted upon one another: Darius is not defeated by Alexander, rather, the life-lines of Darius and Alexander are so harmonized that at the same moment Darius loses the battle and Alexander wins it. The motions of the body do not cause the motions of the soul, or vice versa, but the two sets of motions are so harmonized that the relation appears to be one of cause and effect. Thirdly, Leibniz believes that each monad reflects at every moment the state of all other monads, though from its own *point of view* (and Leibniz is the first, after Pascal, to introduce the notion of 'point of view' into philosophy). That is to say, each of us reflects in himself the universal order and in his own development the onward development of all. To be sure, our own point of view is a very one-sided affair, and the result is that much in the universe seems to us to be askew, but if we could see the world from a perfectly central perspective— say, that of God—then all would appear to us to be in perfect order. Moreover, since the world that God sees is not a different world from that which we see, that perfect order is already contained in what we now see, however distorted or foreshortened. Our duty, and our nature as developing forces, is to unfold that potential insight that we already contain, to understand ourselves—and that means to understand the world that we reflect—ever more clearly, and so to approach ever more closely to a divine perspective, which, none the less, since the universe is infinite, will always, however rich and great our understanding, remain infinitely remote.

This impressive and inspiring scheme is built on the narrowest of bases—the conviction of the absolute integrity, the logically guaranteed inviolability, of identity, that is, of the human self. The particular strength of the scheme—but also of course the origin of its most serious ambiguity—is that it fuses that apparently atomistic, even anarchistic, conviction with a belief in a universal order by allowing the world of rational interconnectedness to stand *as if* it were real. Really, there are only windowless monads working out their private destinies, but in appearance, and down to the last detail, everything works as if explicable in terms of cause and effect. Indeed the fusion is still more intimate, for each private destiny is a representation, in miniature, and in accordance with a particular perspective rule, of the destiny of all.

It is not difficult to see in this metaphysical compromise a parallel to the social compromise forced on the middle class of the German absolute states at the end of the seventeenth century: a public, political, and independent economic life (a life of cause-and-effect interaction) is denied to them, but for this they have the threefold compensation: firstly, that their private, personal (perhaps originally religious) life, developing according to none but its own inner law, is thereby secured and unexposed to danger; secondly, that the life denied to them is anyway devalued as a 'mere' appearance; and thirdly, that between the humblest citizen and the total state machine there is said to exist a complete, if not always obvious, identity of interest; the grenadier, or the orphan in Halle, carries the Prussian state within him, and lives its life, as much as Frederick William I, if more obscurely. Conversely, the Leibnizian scheme does justice to the two principles of the enlightened despots themselves; that a state can be ordered in accordance with reason (for though the rational interconnectedness of things is an appearance, it is a 'well-founded' appearance and the pre-established harmony knows no dissonances), and that the state is most perfectly ordered around the interest (properly understood) of its sovereign. In particular the metaphor of perspective, of 'point of view', is immensely powerful, for it derives from and reinforces the visible practice of absolutism. To look out from the royal windows on the first floor of Versailles, or at any of the equivalents in Herrenhausen, Ludwigsburg, or Nymphenburg, and to see the flower-beds, the box-hedges, and the coloured gravels, which at ground level seem at best a playful puzzle, coming together and crystallizing into an ornamental, geometrical, or heraldic pattern; or to sit in the royal box of a 'baroque' theatre, directly opposite the perspective vanishing-point, at the one point in the room, therefore, given the stagecraft of the time, from which the proscenium-framed illusion was perfect, is to understand what Leibniz meant by the divine point of view from which the way of the world seems perfectly rational and all seems for the best. And when in his own popularizing writings Leibniz summarized the political implications of his philosophy as content with one's station, eager prosecution of one's own development, particularly intellectual, and confidence that, however confused things may appear to the subjects, the prince is directing all to the best possible end, he may have been coarsening his scheme but he was not falsifying it.

A phrase which Spener could have written, but which comes in fact from St Teresa of Avila, characterizes for Leibniz the condition of the human individual: 'the soul should often imagine that there is in the world only it and God'. Spiritual solitude of course is not a German monopoly, and it is worth a literary digression to draw a parallel and contrast with contemporary England. *The Life and Strange Surprizing Adventures of Robinson Crusoe...* first appeared in 1719, and by 1740 Germany alone had furnished nine separate translations, the first in 1720, and ten imitations, all frequently reprinted. The immense European success of Defoe's novel plainly reflects a general understanding of the experience of isolation (which after all literally means 'being marooned on an island'), yet the story of Crusoe's reaction to his fate is equally plainly a story by and of an eighteenth-century Englishman. If we compare Defoe's book with his model, Henry Neville's *Isle of Pines* (1668), where the narrator is a cabin-boy happily cast away with a clutch of ladies, we notice at once that Crusoe's isolation is complete, and totally unerotic. In this, *Robinson Crusoe* resembles an earlier imitation of Neville, the *Continuation* of Grimmelshausen's *Simplicius Simplicissimus* (1668), in which the hero becomes a hermit living only for his devotions. Like Grimmelshausen's Simplicissimus, and like Leibniz's monad,

Crusoe has initially only God for company. Unlike both his predecessors, however, but like his contemporary Leibniz in his drive to transcend monadic solitude, Defoe shows us a hero whose burning desire is to escape back to the company of men—or at any rate of Englishmen. The erotic force, frustrated by isolation, is not simply, as with Grimmelshausen, sublimated into religious fervour. It also expresses itself in the work's extraordinarily detailed eye for, and sensuous interest in, the material objects and material conditions of Crusoe's little world; in the active and aggressive spirit in which Crusoe both exploits his island and fortifies it against that human contact which would constitute not escape but violation; and in the moving relationship, both benevolent and condescending, with Friday who, though alien, is no threat, but the object of pious pedagogic zeal. *Robinson Crusoe* is thus an exact delineation of coming British attitudes, in the eighteenth and nineteenth centuries, to the material world, to personal and international relations, and to the Empire beyond the seas. Crusoe's self is as secluded, and as theocentric, as any monad, but the seclusion is overcome, as far as that proves possible, in the partial, loveless, aggressive, sensual, inventive, obsessional, constructive, dutiful and benign way in which the British founded and ruled an Empire, while all the time really wanting to go home. It may be that this solution was open only to a maritime nation. At any rate, while the most significant and most successful German imitation of Defoe offers an almost equally prophetic depiction of future developments in German sensibility, the reader does not find himself breathing the salt air of the open sea. On the contrary, the prevailing impression left by J. G. Schnabel's (1692– after 1750) *Fastness Island* (*Insel Felsenburg* 1731–43) is of the foulness, miasmata and constriction of the European parlours, hovels, galleys and torture-chambers from which the migrants are gathered who people his island utopia. The fastness Felsenburg, like Prussia, opens its arms to the persecuted from every corner of Germany, and beyond, and incorporates them into a rationally ordered Lutheran patriarchy. That this is situated on an island in the South Seas might seem no more than a poetic convenience. But we are indeed dealing with a genuine local variant of the Robinson Crusoe theme—we are in the presence of the specifically German compromise between isolation and engagement so shrewdly anticipated by Leibniz. Crusoe's isolation is a fate which it requires all his energies without exception to overcome, and he gains only partial success—what he cannot overcome he has simply to accept; he does not compromise, he is defeated. By contrast, there is from the beginning something half-hearted about every aspect of *Fastness Island*'s totally successful utopia, if we compare it with its literary forebears. By contrast with the Catholic Grimmelshausen, there is, given the mixed nature of the community from the start, no erotic frustration and so no eremitic otherworldliness. By contrast with Neville, however, there is no erotic anarchy either, for all is kept within the bounds of the proprieties of the Lutheran Church. By contrast with Defoe, there is no harshness of fate, for the isolation of Schnabel's migrants is either deliberate or, if accidental, willingly accepted. Above all, and by contrast with all three, the very isolation of Felsenburg is unreal, not only because, whenever the consequences of seclusion become serious, contact is restored with the rest of the world, either by a convenient shipwreck or by an expedition of the islanders themselves, but also because the very problems that the islanders find most serious, and whose solution occupies most of their time, are problems that presuppose so elaborate a social fabric that the analytical force of the myth of the desert island—what is human life when reduced to its simplest terms?—is hardly felt at all. The problems that preoccupy the Felsenburgers are: how are they to avoid intermarriage within the forbidden degrees of kinship? and how are they to secure apostolic succession for their ministers of communion? Once these questions have been answered satisfactorily, Felsenburg seals itself off from the world. But isolation on these terms can no more be called isolation than can the state of a Leibnizian monad that, though inviolable, knows its own order to be identical with the order of the world at large, or the state of a Pietist soul alone with God but in harmony with the will of the Prussian king. It is difficult to see either Schnabel's or the Leibnizian-Pietist compromise as anything but a classic case of bad faith: the Felsenburgers are loyal neither to their island, for they seek all the benefits of Europe that matter to them, nor to Europe, for they withdraw when they have what they want; and similarly Leibniz's ideal subject is neither loyal to the public realm, for that is unreal to him except as an internal modification of his inviolable self, nor, on the other hand, does he possess the inner life of a distinct individual for he acknowledges no disharmony between himself and the general order. Defoe's Crusoe and Grimmelshausen's Simplicissimus draw their strength as individuals from their conflict with an external order of which none the less they make the best while they can; such strength was not in demand in the enlightened despotisms of eighteenth-century Germany.

The strictly monadistic element in his thought shows Leibniz, himself no Pietist, though a correspondent of Spener's, to be following a path parallel to that of Pietism, in its cult of what it believed to be interiority ('Innerlichkeit'). It was, however, rather the rationalistic element and encyclopaedic superstructure that predominated in the version of Leibnizianism publicized and popularized with immense success by Christian Wolff (1679–1754). Wolff maintained a traditional dualism of mind and body and saw the value of the 'pre-established harmony' principally in the solution it offered to the problems associated with this dualism. Because, therefore Wolff's thought does not fully incorporate the monadistic, that is, the spiritual, basis of the Leibnizian compromise, it does not have the unifying power of that compromise, and so the intellectual life of Germany in the first half of the eighteenth century is dominated by the paradoxical spectacle of a feud between Pietists and Wolffians, both of whom represent different aspects of Leibniz's total vision, different aspects of the bourgeois compromise with absolutism. The feud was at its bitterest in Prussia, not surprisingly, given that the ultimate tendency of both movements was to legitimate the Prussian system. The university of Halle, founded in 1694 under Spener's aegis to provide the new state with servants educated in the new spirit, and the dominant influence on German theological studies until about 1780, was notorious for this continuous and unresolved intellectual civil war, and though Wolff became professor there in 1707, in 1723 the Pietists secured his expulsion by Frederick William I (it was alleged that Wolff taught determinism, and that it was a consequence of this doctrine that deserters from the Prussian army should not be punished). In 1740 he returned in triumph, summoned back by the free-thinking young king, Frederick II, who preferred rational service of the State to religious. (In the same spirit, in 1752, Frederick invited Goethe's great-uncle, the notoriously heterodox Johann Michael von Loen (1694–1776) to move from Frankfurt and take up a post in the Prussian administration.) Five years later Wolff was made a baron. By this time practically all German university chairs of philosophy were occupied by Wolffians.

Quite apart from his achievements as a philosopher and teacher, and as a linguistic innovator—he showed that it was possible to write philosophical treatises in simple, pure German, and himself invented a large part of the modern German mathematical and philosophical vocabulary—Wolff's career is a landmark in the development of German society. He is the first representative of a species which in the twentieth century is still under no threat of extinction: the professor, especially of philosophy, who establishes through his pupils an empire extending over many universities and who acquires in the eyes of the public something of the role of a secular preacher, a preceptor of the nation. After Wolff, the pattern is continuous from Kant to Habermas, for of all German institutions the universities, in the last 300 years, have changed least. The phenomenal rise to fame and fortune of this son of a tanner announces a new, peculiarly German, *modus vivendi* between the middle—that is, the mobile —class and the absolute State, the advent of an age in which the backbone of national culture is to be found not in the municipal councils but in the universities. Wolff's career proclaims that the German bourgeoisie has emerged from its cultural paralysis and has rediscovered itself—as a class of state officials.

For philosophy, theology, and literature the most important consequence of the despotic constitution of the greater part of Germany in the eighteenth century was that the 'Beamtenstand', the class of officials, became not only practically coextensive with the politically significant middle class (the words for 'business' and 'man of affairs', 'Geschäfte' and 'Geschäftsmann' usually refer to official, not to commercial life) but coextensive with the intellectual class as well. Especially was this true of Protestant Germany where schoolteachers and clergymen, as state appointees, formed a crypto-officialdom which received its education in the same institutions as the 'cameralists', the students of law and administration. Within the absolutist system the role of the university as the instrument for the recruitment and regulation of the German Third Estate, was crucial. Germany possessed over fifty universities or similar places of higher education at a time when England boasted only two, but this wealth reflected political fragmentation rather than a disinterested love of learning. Territorial rulers, notably the Electors and Kings of Prussia, made repeated efforts to prevent their subjects from studying anywhere other than at their own state universities. The faculties were not self-governing but ruled by a non-academic bureaucrat. The professors were all state employees and they all took an oath of loyalty to the sovereign. Moreover they themselves had often been, or were destined to become, administrative officials or, in the case of theologians, exercised hierarchical authority. Without prospects in the crushed and stifling towns, an ambitious young man regained a carefully circumscribed self-confidence either in the university or, through the university, in the pulpit, the schoolroom, or the office. It has been well said that 'more than anything else, it was this academic-bureaucratic predominance that distinguished German culture in the eighteenth century'. It was not a culture for Robinson Crusoes. The German eighteenth-century intellectual was confined within a one-dimensional system—wherever he turned he found the State. Neither the university nor the Church offered him the possibility of independence, even if he wanted it, and Germany's economic structure made it unlikely that he would be a man of private means. To make an independent living as a man of letters, too burdensome a task even for the apostle of free thought G. E. Lessing (1729–81), was still a hopeless undertaking for a man of intellectual integrity even at the end of the century, as is demonstrated by the tragic examples of the dramatist Kleist (1777–1811) and the poet Hölderlin (1770–1843). To be sure, the relatively extensive network of official and university posts offered more numerous possibilities of a secure yet intellectual existence than were available outside Germany—but the security, as we shall see, was bought at a price.

The Literary Context, to 1770

It was not chance that the first stirrings of a new literary order in the German-speaking world should have been felt, from 1720 onwards, in Switzerland and Hamburg. Switzerland and Hamburg were peripheral to the Empire, and through the Rhine and the sea, and through religious affiliation, they were attached rather to the Puritan north European mercantile world dominated by Holland and, increasingly, by England, to the free-thinking, all-pirating, booksellers of Amsterdam, and to the philosophers and poets of Augustan London. The deist elements in English culture were probably the more important to the Hamburg city-father and nature-poet B. H. Brockes (1680–1747), even though he was also the author of a Passion text used by Handel, Telemann, and J. S. Bach (in the *St John Passion*), and to the Swiss poet, later polymath, Albrecht von Haller (1708–77), a difficult and divided character on whom many of the tensions in early eighteenth-century thought and feeling left their mark. To the Zurich critics and philogists J. J. Bodmer (1698–1783) and J. J. Breitinger (1701–76) England mattered most for the example it gave, in Milton's epics and the *Spectator,* which they praised, analysed, translated, and imitated, of the possibility of communicating religious and moral truth in a more pleasing and secular form than that of the sermon or theological treatise. In these northern and southern enclaves of a self-governing bourgeoisie the English impetus helped the beginnings of a literary culture based on the secular book, biblical or religious in theme (and so more easily substituted for the Bible) and read at home, possibly aloud and in the family circle. Drama was little known, and less approved, in the Calvinist world, Shakespeare influenced neither theory nor practice at this stage, and the novel, prose narrative on non-biblical subjects, was too direct a competitor with Scripture to be acceptable to such highly clerical societies (in 1762 Zurich boasted 382 clergymen). But although Switzerland, because it was socially, politically, and geographically marginal, was thereby particularly open to non-German influences, it could not for that very reason furnish the decisive impulse to the reconstitution of the German national literature. The German book-trade expanded rapidly in the eighteenth century, the only major setback being caused by the Seven Years' War: it was better organized than its English counterpart, it had a potential market four or five times larger, and its big entrepreneurs in the latter part of the period were industrial capitalists in a recognizable nineteenth-century mould. One might therefore have expected it to enter an easy alliance with the discovery by the Swiss bourgeoisie of the secular book, and indeed it did, but the literature that resulted could not express the aspiration to nationhood that for the reading public was implicit in the language which they shared and which in 1692 displaced Latin as the language in which more than 50 per cent of German books were printed. For Germany was not just Switzerland writ commercially large. A literature that did justice to the realities of life for the middle classes of the German Empire and its Prussian 'tail' would have to take account of the fact that the political and economic fate of the nation lay in other hands than theirs. The decisive impulse towards that literature came in the event from Leipzig, from the commercial capital of one of the most powerful German principalities, the geographical heart of the Empire and the centre of the German book-trade, and from an émigré Prussian (because of his giant build he had had to flee the recruiting officers), who after six years of struggle achieved his goal of a professorship at the Electoral Saxon university, of which he was later several times rector. In this as in other respects the continuator in the literary world of the work of Christian Wolff, J. C. Gottsched (1700–66), professor first of poetry (unpaid) and then of logic and mathematics (salaried), and for over thirty years an honoured, feared, and ultimately ridiculed critical tyrant, is rightly regarded as initiating the modern epoch of German literature, or at least of what we shall have to call its 'official' branch.

The simplicity of Gottsched's recipe—in his *A Critical Art of Poetry Essayed for the Germans* (*Versuch einer Kritischen Dichtkunst für die Deutschen*) (1730) he actually issues instructions for the concoction of poems in the 'recipe' form: 'First select an instructive moral . . . Then invent . . . '—has tended to conceal the complexity of his achievement. There is more than meets the eye to Gottsched's principles that Nature is wholly reasonable, that the Poet is a skilful imitator of Nature, that, among the moderns, the perfection of reasonable, and therefore of natural, poetry is to be found in the classical French drama of the previous and the current centuries, and that therefore this drama alone must furnish the model for future writing in German. Many separate strands in the national culture had to come together if this man, in this place, was to utter these principles, however banal they may now appear. First, Gottsched was a Wolffian. To gain his professional stipend he wrote a two-volume compendium of the Leibniz-Wolffian system; he translated Leibniz's *Theodicy* into German; he wrote an obituary laudation of Wolff, in whose philosophy he claimed he had found spiritual peace; and Wolffian rationalism is the foundation of his theory of poetry. Here we may see already the first outlines of two features of later German literature: its hostile relation to Christian piety (in 1730 no Pietist could have avowed as Gottsched did a passion for poetry, and in 1736 Gottsched's wife translated a French satire on the Jansenists which she reapplied to Pietism); and an association between philosophy and literature that at its height, in the 1790s, was more intimate than anything in Cartesian, or even Existentialist, France. Second, Gottsched, in his concern for a German national literature, and in his conception of the German language as a national vernacular, despite all political fragmentation, was consciously continuing the tradition of the 'language societies', the slender thread that connected the early eighteenth century to its great baroque past. Of one such society, founded in 1697, he was a prominent member from his first arrival in Leipzig, and a considerable part of his life's work (in this also a continuation of Wolff's) was devoted to the purification and regularization of the language, through the writing of a grammar and of commentaries on usage and through the composition of model plays and poems. Thirdly, however, in laying all possible emphasis on the imitation of classical French drama, Gottsched took the decisive step—and an essential step if the tamed and hobbled bourgeoisie was to come back to cultural life—of accepting as normative the taste of the court. Gottsched's *Critical Art of Poetry* is the moment at which the German middle class accepts that its intellectual and cultural institutions are not to be its own autonomous creation but that it will have to make do with, though perhaps ultimately usurp, the institutions of its absolute rulers. It is the moment at which the drama emerges as Germany's alternative to the novel, the alternative, that is, to the literary form in which England's increasingly confident bourgeoisie learns in the eighteenth century to represent and interpret itself to itself and for which the social foundation is simply non-existent in Germany at the time. Through the important section on tragedy in the *Critical Art of Poetry,* through his co-operation with the touring theatre-companies of Caroline Neuber (1697–1760) and J. F. Schönemann (1704–82), bitter rivals for appointment as official players to the Saxon court in Dresden, and through the six volumes of his *German Theatre, Arranged in Accordance with the Rules of the Ancient Greeks and Romans (Deutsche Schaubühne . . .)* (1741–45), a collection of translations, imitations and original pieces, Gottsched did more than advocate the courtly virtues of propriety and regularity for a theatre characterized by crudity and improvisation. He also, and far more importantly, identified the drama, a literary form whose appeal was popular but whose full development was possible only at, and with the subventions of, the courts, as the focal point for several different national ambitions: the moral and intellectual edification of all, the improvement of the German language, and the establishment of a national literature of European status. Gottsched's work announces that the institutional foundation of the coming cultural revival will be, not the rapidly expanding commercial book-market, but the subsidized theatre. Finally, therefore, His Magnificence* Professor Doctor Gottsched exemplifies the official nature of the new culture—in at least two senses of that word. Literature and philosophy will join theology and the other, more obviously court-dependent, arts in being produced by and for officials—the class of teachers, pastors, secular civil servants and university professors (which last will also originate another German speciality: demonstrating the value to the State of an official culture by means of *theories* of literature and the arts). The literature that results will be official in another sense too: closely associated with the aspiration to nationhood, both of the German language-area as a whole, and of certain larger states within that area, it will be characterized, even in revolt, by an aura of seriousness and authority borrowed from the benevolently despotic state, which, from a suitable distance, is paying the piper. Comedy will not flourish. Most of all, however, the literary tradition established by Gottsched deserves the name 'official' because the greater part of what is original and internationally appealing in German literature was conceived outside it.

There is of course also an official history, laid down in the later nineteenth century, of official German literature after Gottsched. This version runs approximately as follows. Gottsched's unnatural reliance on French models, it is held, was soon replaced by an English influence, more congenial to the German national character. From 1740 Gottsched was under attack from Bodmer and Breitinger for his refusal to acknowledge the genius of Milton, and later of Milton's German imitator F. G. Klopstock (1724–1803), author of the religious epic *The Messiah (Der Messias)* (1748–73). The death-blow to the rule of Gottsched and the Corneilles was dealt in 1759 by the merciless polemic of

---

* The correct German title for the rector of a university.

Lessing who set up in their stead the model of Shakespeare and drew the attention of his countrymen to the possibility of a national drama based on the story of Dr Faust. Enthusiasm for the establishment of a German national theatre was now general, and during a short-lived attempt to realize this idea in Hamburg Lessing interpreted afresh Aristotle's theory of tragedy in his *Hamburg Dramaturgy* (*Hamburgische Dramaturgie*) (1767–8). Lessing's own plays, partly written in order to exemplify his theories, are 'bourgeois tragedies' or '*drames*' in prose (it is obscure what relation they are supposed to have to Shakespeare) and are the first monuments of German 'classical' literature, a concept of great importance to the official mind. After an excessively exuberant period of Shakespeare-imitation during the 'Storm and Stress' movement of the 1770s, the literary revival bore fruit at last in the mature dramas of Goethe and Friedrich Schiller (1759–1805), on which the example of Greek antiquity is said to have exercised its moderating influence. The nineteenth century, in the theatre of Kleist, Grillparzer, and Hebbel, sees a continuation of this tradition, of which one of the last representatives, usually a contemporary of our official literary historian, is Gerhart Hauptmann, in his later, non-naturalistic phase. (That the true culmination of the tradition is in the music-dramas of Richard Wagner is usually ignored.) Meanwhile a kind of loyal opposition is provided by the movements of Romanticism and Poetic Realism, of the beginning and the middle of the nineteenth century respectively, whose main achievements lie in the supplementary genres of lyric poetry and the 'novella', or short story.

Save for the appropriation to the 'classical' of Goethe and Kleist, and the Austrian Grillparzer, there is nothing to object to in this standard account, apart from its neglect, obviously, of the great outsiders,* but also of some of the most important factors in the development even of official culture. French influences did not cease with Gottsched, and English influences were not confined to Milton (whom it is difficult to see as of serious importance for anyone but Klopstock) and Shakespeare (who had very little to do with what Lessing or the Storm and Stress, or for that matter Gerstenberg, Wieland, or Herder, practised in his name). As important an adversary of Gottsched as the polemical Lessing was another Leipzig professor, the proverbially gentle C. F. Gellert (1715–69), author of fables, plays, model correspondence, and a short novel. Gellert's defence of the new 'bourgeois drama' owed as much to French as to English fashions; and the reason Lessing's dramatic practice owes so little to Shakespeare is that his true mentor was Diderot. Moreover, two powerful impulses from eighteenth-century England, together with their own German consequences, are largely or completely ignored by the official account: the great vogue for the novels of Richardson and later of Sterne, and the associated movement of Sentimentality ('Empfindsamkeit') which we shall discuss a little

---

* e.g., in the nineteenth century, Heine, Büchner, and the young Hauptmann.

later in this chapter; and more profound still in its effect on the whole structure of German intellectual life, a wave of religious doubt in which the English deists and free-thinkers of the first half of the century played a considerable part. To both these developments there were also French contributions of the first rank: Rousseau's novels fed the fires both of sentimentality and scepticism; and liberation from the chains of religion was the avowed goal of Europe's literary despot, Voltaire.

That in Germany the sceptical movement of biblical criticism, represented by John Toland, Anthony Collins, and Matthew Tindal, led to a religious and cultural upheaval, while in England it was simply forgotten until the late nineteenth century, was largely due to the one-dimensionality of the German social and political system. What in England was a series of skirmishes by freelances and eccentrics could within the German university machine—and especially at the new, progressive and highly influential university of Göttingen —be organized into a campaign, a school of theologians, historians, and philologists sailing, from the 1770s, under the flag of 'higher criticism'. Nor was such subversion necessarily unwelcome to secular rulers of the stamp of Frederick II. It was positively in the interest of an absolute, or would-be absolute, state power to encourage Enlightened questioning of the historical or philosophical legitimacy of any faction—whether Church or Jews or town-councils and guilds—which claimed an inalienable, perhaps even a divine, right to exist, independent of state policy. On the other hand there were limits: the constitution, even identity, of most German states had in 1648 been fused with a particular religious confession, loyalty to which was a signal of reliability and could be a condition of office. Theology remained throughout the eighteenth century the best-endowed subject at all universities. The same agency, therefore, encouraged both the production of theologians and the criticism of the intellectual foundations of theology. Not only, though, did the absolute state itself set up both horns of this dilemma: by restricting the possibilities of a middle-class career outside officialdom, and by increasing its own demand for university-educated officials, it ensured that to this dilemma ever more young students should be exposed. Only the man to whom office was indifferent, who did not have to eat from one of the many hands of the State, could avoid the resultant, so to speak officially prescribed, crisis of conscience: the determination of his own position on the scale between orthodoxy and unbelief in the knowledge that it was a question of the most immediate practical, and indeed material, importance. Those happily spared such a crisis were few in number— but they included Goethe. Goethe was almost alone among literary men of his age in owing nothing—through his family, financially, or socially—to the Church, and practically nothing to the university, and this is one of the factors that detaches him most decisively from the official tradition of German culture. From 1740 onwards that tradition is increasingly one of 'reinterpretation'—of nominal acknowledgement of 'Christianity', or at least of 'religion', but of real engagement with some substitute for the lost certainties, to which this imposing name is now given: thus do we come to the theologies which furnish the 'essence of Christianity', and philosophies which mark out the territory of 'religion within the bounds of reason alone', the theories of education, art, and sport which promise the building of a new humanity, and the 'classical German literature' which is an 'expression in new symbols of this humanistic religion'. Herbert Schöffler counted 120 German men of letters (excluding philosophers) born between 1676 and 1804 who studied theology or had a theological home background: 'from 1740 . . . clergy, their sons and young theologians generally, come into literature in shoals, so that from the middle of the century its total aspect undergoes a great change'. When at the end of the century J. G. Fichte (1762–1814) saw the modern philosopher as the true Protestant who prepares the way for the replacement of the Church by the State, or the Romantic thinkers 'Novalis' (Friedrich von Hardenberg 1772–1801) and Friedrich Schlegel (1772–1829) gave the name of 'religion' to a fusion of poetry and politics, they were not so much prophesying a new order as drawing up the balance of a movement two generations old and more prosaically analysed in 1773 by the Göttingen orientalist J. D. Michaelis (1717–91): the sudden flood of interest among students in the theory and practice of education he attributed to 'the fear of the creeds, to subscribe to which many former students of theology now object'. The emotional extremes which circumscribe German official literary culture of the late eighteenth and early nineteenth centuries are anxiety and hypocrisy, desperation and fanaticism: from all this, by the happy accident of his origins, Goethe was blessedly free. He was, of course, profoundly affected by the historic shift in the foundations of belief which took place for intellectual Europe in the eighteenth century—had he not been, it would not be too much to say, his writing would be of little consequence in the history of literature. But his individual circumstances left him free to face the crisis for himself alone and with complete personal integrity; his mind was not mortgaged to Church or State and so he had to ingratiate himself with neither by reinterpreting, and relegitimating, their power-structures when belief collapsed. (In this no doubt lies the ground of the affinity Nietzsche felt for him.) Instead he concentrated exclusively on the consequences of the crisis for his own responsibilities: the choices he had to make in a life that was subject to much less material constraint than that of many of his contemporaries, and the poetry that was pre-eminently the expression of his freedom. Only once, around 1788, do we find Goethe colluding with the public substitutes for belief in which German culture came to be so fertile. The advice that in September 1824 he gave to R. W. Emerson's elder brother William was based on a lifetime's observation of the devastating effects of religious crisis within a closed social system: it did not take account of the possibility that the freedom Goethe had himself enjoyed, exceptionally, and privately, might in another order of things be more general; but then, as another American visitor remarked a few months later, after explaining his country's electoral system to the Privy Councillor of a Grand Duke, Goethe 'was after all not omnisentient any more than omniscient'. E. W. Emerson, son of the American sage, writes:

The German philosophy and the Biblical criticism shook [William's] belief in the forms and teachings of the religion in which he had been brought up . . . To William, beset by distressing doubt at Göttingen, it occurred that, but eighty miles away at Weimar, lived the wisest man of the age. He forthwith sought him out, was kindly received, and laid his doubts before him . . . The counsel which he received was in effect . . . to persevere in his profession, comply with the usual forms, preach as best he could, and not trouble his family and his hearers with his doubts. Happily the youth, at this parting of the ways where the great mind of the age acted the part of the Tempter, turned his back, and again listened to the inward voice. He left the ancestral path, gave up at the age of twenty-four his plan of life for which he had been with diligence and sacrifice preparing himself, and studied law.

'Amerika, du hast es besser' ('America, you are better off') Goethe wrote in a little poem in 1827: William Emerson was able to support himself through his second apprenticeship with literary activity and to become a successful Wall Street lawyer, while his brother, whose intellectual career so closely resembles that of many a young German theologian of fifty years before, preaching from the rostrum what could not be preached from the pulpit, was able to live in a comfortable independence which was hardly conceivable to an eighteenth-century German man of letters without the support of an official post, and in the absence of a law of copyright.

There are in particular two major features of Germany's developing literature in the period of Goethe's youth—say, until 1770—which are interesting precisely because of the extent to which, for all their importance (usually ignored by the official accounts), they do *not* determine the original direction of Goethe's unique talent. The first is the earliest example of the transfer of religious terminology to a secular application: the growth of German aesthetic theory (the ex-theology of an ex-clerisy), the establishment of the concepts 'literature', 'art' in general, 'artistic genius', and, the religious term in which thirty years of philosophizing are eventually focused, 'artistic creativity'. Breitinger's emphasis in the 1740s on the theoretical primacy of the *effect* of a poem; Gellert's and Klopstock's concentration in the 1750s on the role of the poet himself as an agent working on the feelings of his audience; the reflections on aesthetic problems of the philosophers J. G. Sulzer (1720–79) and Moses Mendelssohn (1729–86), are all synthesized and reorganized in Lessing's *Laocoon: or The Bounds of Painting and Poetry* (1766), which, following a lead given by the French theorist Charles Batteux (1713–80), founds the notion of aesthetic experience as a distinct form of knowledge, of which the several 'arts' furnish subspecies. This entire process can be understood as consisting of two steps: the interpretation of art in terms of psychological effect, and the derivation of a psychology from Leibniz's originally purely logical theory of the relation (to which he gave the name 'perception') between each monad and the whole universe. When from about 1760 onwards the description of the artist as a 'creator' becomes fashionable jargon we are thus witnessing a restricted, but fully explicit, reapplication of a comparison cautiously implicit from the start in the Leibnizian scheme. We read for example in § 83 of Leibniz's *Monadology* (1714):

Souls in general are living mirrors or images of the universe of created things, but . . . minds are also images of the Deity or Author of nature Himself, capable of knowing the system of the universe, and to some extent of imitating it through architectonic ensamples, each mind being like a small divinity in its own sphere.

In England Shaftesbury's notion of the 'second Maker, a just Prometheus under Jove', supported neither by a Leibnizian climate of thought nor by any general movement of secularization, remained without consequence. In Germany it could associate itself with the influence of Leibniz which, after 1765, when his *New Essays on Human Understanding* were first published, was strong once more and no longer mediated through Wolff, and the next two decades show the concept of the 'creative' artist at its zenith.

No doubt it was not Lessing's intention to contribute to this development. *Laocoon* is overtly concerned only to distinguish poetry and painting. But why should two such different activities need distinction in the first place? The very act of distinguishing them implies their underlying affinity. By making use of the, not wholly original, argument that painting represents simultaneous events while poetry represents successive events, Lessing reduced both arts to the same status: they are both differently selective representations of components of the same underlying narrative, that all-embracing sequence of sets of simultaneous events which we call 'experience'. Effectively, therefore, Lessing makes painting into a branch of poetry, for he makes it into something that needs to be read and interpreted by reference to that which it represents, as if, like a book, it were inviting us to imagine something other than the signs it offers to our senses. Goethe would one day take strong exception to this assimilation of the visual arts to literature, but *Laocoon* shows Lessing at his most confident in battling for the books, for the public—and commercially published—medium, through which alone the German middle classes could establish their own culture in independence of princely tutelage—the visual arts being to a much greater extent than literature an accessory and ornament of wealthy courts. Lessing was the freest of free thinkers willing even to call himself a Spinozist—which was tantamount at the time to a profession of atheism—and he had no interest in reinterpretative compromises with religion and the old regime, except as intermediate and tactically necessary steps on the path to complete liberation; but the facts of German social life were against him. Literature, the world of the printed public mind, could not be given the autonomy and predominance for which he hoped, he himself had to compromise and accept a position as a state pensioner, and his usually highly tendentious and polemical theories, including that of the comparability-in-distinction of poetry and painting, were made to serve the interests of such post-religious pieties as that the various forms of 'art' are all equally manifestations of 'creative genius'. Meanwhile that part of his aesthetics which saw art as a mode of knowledge went into abeyance, until it was recovered and modified by Kant.

The full impetus to a religious conception of art however was given by another mind, much the most original, and much the least abstract, of this generation of aestheticians, and the principal target of Lessing's criticism in *Laocoon*. The career of J. J. Winckelmann (1717–68), often regarded as the founder of modern art history, is the living proof of the agonizing emotional and intellectual constriction of the German despotisms, above all of Prussia, his homeland, the 'Sparta of the North'. Born into utter poverty, but given by the educational system the prospect of bettering himself, provided of course that he studied theology, Winckelmann experienced all the deprivations that were reserved for the independent and unconventional sensibility, which in this case had fallen in love with the literature of ancient Greece: unable to accept the career prescribed him he had to undergo the humiliations, first of private tutoring, then of five years wielding a ruler as junior master in a primary school, then of scholarly drudgery for a noble patron. Poverty, cold, ridicule, lack of sleep (for he had to have time to read his Greeks), religiosity, obsequiousness, 'interiority', Lutheran propriety—he turned his back on them all. His removal to Rome in 1756, his conversion to a Catholicism of pure convenience, his notorious homosexuality, his devotion to the material remains, above all now to the sculpture, of a distant age which he painted in seductive idealization—everything in the life of the older Winckelmann was the clearest possible 'No' to the land and circumstances of his upbringing, to the asphyxiating compromises of Felsenburg-Germany. One loyalty he retained, however, almost to the end, his loyalty to the German language. In the prose poetry of his 'ecphrases', his evocations of the spirit of individual works of ancient art, Winckelmann made the most paradoxical, and perhaps the most influential, contribution of all to the new aesthetic theory: the language of Pietist self-scrutiny and religious transport, ingrained in him since his childhood, he put in the service of a sensualist paganism, of a worship of a heathen culture from which he was separated not by the metaphysical abyss that separated the Pietist from his godhead but by the misfortunes of history. This much of Winckelmann official Germany could absorb, as the tide of secularization began to flow, and it was what mattered most about him. So Winckelmann stands in a strangely ambivalent relation to the two German traditions he could be said to have founded—of artistic paganism, and of classical, especially archaeological, scholarship—an ambivalence concentrated in the weirdness of his end: a planned triumphal tour through Austria and Germany was broken off as

Winckelmann was seized with panic loathing of the northern landscape through which he found himself travelling once more; and waiting in Trieste for a ship to take him back to Italy he was murdered by a casual homosexual acquaintance covetous of his gold medals. Winckelmann's life and work and death seem almost constructed to be a symbol of modern German sensibility, and as such they did not escape the attention of Thomas Mann who found here one of the sources for *Death in Venice*. Alone among the aestheticians famous in his youth, Winckelmann, the man who said 'No', fascinated Goethe for over fifty years. Lessing's theories Goethe actively opposed.

The second major strand in German literature between 1740 and 1770 derives even more directly than the first from a sense of social limitation and continues the theme of *Fastness Island*, the theme of isolation. To be sure, this is usually regarded as the period of Enlightenment (Aufklärung) *par excellence*, of trust in the power of reason both to attain truth and to spread truth among the multitude; and in the French-speaking Berlin Academy of Frederick II, in the circle composed of the young Lessing, Moses Mendelssohn, and the bookseller-author F. Nicolai (1733–1811), in professors such as Christian Garve (1742–98), the young Immanuel Kant (1724–1804) or the liberal Göttingen historian A. L. Schlözer (1735–1809), and in the more popular moralistic writers such as Gellert or the harmless 'satirist' G. W. Rabener (1714–71), not to mention a vast array of journals of a political, practical, or entertaining nature, Germany seemed to have a movement comparable in scope and intention with that of the French Encyclopedists, if less polemical in manner. But the spirit of the movement faded away in the immediately pre-Revolutionary period; the hope of a general enlightenment through the printing-press was replaced by a more desperate trust in the fermenting work of secret societies, notably of the Freemasons, introduced into Germany in the 1740s, and to whose number, after 1770, practically the entire German intelligentsia belonged. Nor was this recession of public spirit simply the result of the increasing rigidity of the absolutist constitution, a general European phenomenon of the 1780s. If we look at the more strictly literary products of the period of Enlightenment we can see that from the start they are marked by the Leibnizian, that is, by the Wolffian-Pietist, compromise—through their concentration on the importance of feeling, or 'sentiment', they too maroon their protagonists on Fastness Island. The poets, the seismometers rather than the legislators of the nation, know that the public mind to which through the book-market, the 'Aufklärer', the practitioners of Enlightenment, are addressing themselves, has in Germany neither an institutional nor an economic foundation—it is not an assembly of free men. The men—and women—that the poets show us are not inhabitants of that shared world which has been called 'the sphere of association': they conform rather to the Leibnizian picture of the unitary and autonomous soul. In the mid-eighteenth century, European literature was dominated by a cult of feeling which was both a reaction against a materialist and mechanistic tendency in philosophy and covertly at times a continuation of it ('feeling' is after all a faculty of the body, not of the soul). But if a relation, such as love, is reduced simply to a matter of feeling there is a danger that it will cease to be a relation altogether. Both by their social circumstances and by their philosophical inheritance German writers were predisposed to take to an extreme this innate 'Sentimental' tendency to the isolation of the feeling self.

The German name for the cult of sentiment or sensibility, 'Empfindsamkeit', has, unlike the unprepossessing English term 'sentimentality', a specific and wholly literary origin. It is derived from an adjective, 'empfindsam', invented or revived by J. J. C. Bode (1730–93), at the prompting of Lessing, in order to translate the title of Laurence Sterne's *Sentimental Journey* . . . (1768). Bode's version (1768–9) had if anything a greater success in Germany than the original in England, and it provoked many imitations though, as is often the case in cultural history, it came, chronologically speaking, rather towards the end of the development to which it gave a name. Sterne's tale, which Miss Lydia Languish borrowed, along with Mackenzie's *Man of Feeling* and Baculard D'Arnaud's *Tears of Sensibility,* from a circulating library in Bath (she had to conceal them all rather hurriedly on the sudden arrival of Mrs Malaprop) shows Sentimentality at its most condensed and stylish. A series of loosely connected and inconclusive adventures of the heart ('It had ever . . . been one of the singular blessings of my life, to be almost every hour of it miserably in love with some one') are narrated by 'Yorick' in a mercurial, or would-be mercurial, language trembling between whimsicality and innuendo. The emotions and motives surrounding the slightest of incidents—the hire of a carriage, the purchase of a pair of gloves, giving alms to a beggar—become the object of the humorous microscopy familiar to the reader of *Tristram Shandy.* At more seriously eloquent moments (and there is little to the book except moments) it becomes clearer why the modern word 'sentimental' has the connotations it has:

I sat down close by her and Maria let me wipe them away as they fell, with my handkerchief,—I then steeped it in my own,—and then in hers, and then in mine,—and then I wiped hers again;—and as I did it, I felt such undescribable emotions within me as I am sure could not be accounted for from any combinations of matter and motion.

I am positive I have a soul; nor can all the books with which materialists have pestered the world ever convince me to the contrary . . .

Dear Sensibility! source inexhausted of all that's precious of our joys, or costly in our sorrows! . . . Eternal fountain of our feeling! . . . this is thy *'divinity which stirs within me'* . . . that I feel some generous joys and generous cares beyond myself;—all comes from thee, great—great *Sensorium* of the world!

God, soul, and immortality: despite materialist philosophy and science the existence of them all is proved by, or found in, the existence of feeling, of the capacity for the undescribable emotions that accompany tears—or a blush, or a touch, or a glance—but especially tears.

When Sterne lauds the capacity for feeling as the genuinely divine element in man, he identifies it as a capacity for joys and cares 'beyond myself'. None the less there is in his cult—in any cult—of Sensibility a strong impulse towards egocentrism. The feelings may be feelings of sympathy, common humanity, or love, but the feeling centre from which they emanate is one, indivisible, and solitary. Yorick and Maria may commingle their tears, but the undescribable emotions are located within Yorick, and Maria is credited with no emotions, describable or otherwise. It is true that, in the English literature of sentimentalism, the story for example of Goldsmith's *Vicar of Wakefield* (1766), or of Henry Mackenzie's Harley, the man of feeling (1771), are stories as much about the world with which an exceptional sensibility conflicts as about the exceptional sensibility itself. Yet even in these writers the temptation to an excessive subjectivity is obvious. In his elegy of 1770, *The Deserted Village*, Goldsmith laments the consequences for rural life of the new agricultural policies, yet it is only with difficulty that he concentrates his attention on the visible devastation of the village. The enormity of what has happened is expressed largely in terms of its cost to Goldsmith's own sensibility, just as it is said of the village preacher:

He watch'd and wept, he pray'd and felt, for all;

as if the highest, the culminating, form of engagement were personal feeling. Yet, paradoxically, the cult of the sympathetic feelings emphasizes the distinctness of him who feels from the object of his sympathy.

As we have no immediate experience of what other men feel, we can form no idea of the manner in which they are affected, but by conceiving what we ourselves should feel in the like situation

writes Adam Smith in *The Theory of Moral Sentiments* (1759). Goldsmith is not a suffering inhabitant of Auburn, he is a returning visitor. He writes with passion of the misery of the villagers not because he shares their fate, but because he has a lively faculty for sympathy.

Thomas Gray, too, in one of the best known of all Sentimental poems, his *Elegy Written in a Country Churchyard* (1751), exemplifies this tendency to give a position of privilege, and even primacy, to the feeling heart rather than to its object. The climax of the poem is its transition from the fate of the 'unhonour'd dead', with whom the poet has been 'sympathizing', to the fate of the poet himself. Purportedly, the lot of the poet is here being assimilated to the lot of the rude villagers: like theirs, his transient life, 'to fortune and to fame unknown', will be forgotten. Yet the poem in fact ends with a representation of the poet as somebody who is *remembered,* and remembered in some detail, and commemorated by an epitaph in far from 'uncouth rhymes', which are all the memorial the villagers have. The poet, in other words, is not assimilated, he is differentiated.

In the mid-eighteenth century there arises not just a cult, but a problematic of feeling. Is emotion a medium which relates me to things and people 'beyond myself'? Or is the capacity for feeling simply a component of my own personality —my heart secreting emotions as my lachrymal glands secrete tears? If this were the case, then feeling would not offer an escape from my own identity into communication with other identities, but would simply be the brightly painted walls of my own prison. These questions are more explicit in German than in English Sentimental literature, though they clearly underlie the ambiguities of tone we find in Gray, Sterne, or Mackenzie. In Britain (for, in the age of the Scottish Enlightenment, England is too narrow a term) and in Germany the roots of the Sentimental movement go down into a matrix of developing conceptions and preconceptions which are as much philosophical as they are literary. But in Britain the philosophical tradition to which, for example, Adam Smith belongs, is dominated by Locke, whilst the corresponding German tradition is dominated by Leibniz. English sentimentalities developed against the background of a philosophical concern with the knowledge our sensibility might provide about things and minds other than ourselves—German Sentimentalism against the background of a philosophical concern with the unity of the soul. The emphasis on the individual's capacity for (tearful) sympathy and the resultant drift towards egocentrism are common to both versions of the movement. But German Sentimental literature concentrates less on the overt manifestations of sensibility, its social consequences and social penalties, and more on the faculty of sensibility itself, its internal processes and —ultimately—its self-destruction. The presupposition of a Leibnizian psychology accounts—at least in part—both for the narrower, and clearer, focus of German Sentimentalism and for its closer association in literature with the theme of isolation, and in aesthetics, as we shall see in a later chapter, with a cult of heroic genius.

The heroine (and narrator) of *Life of the Swedish Countess of G____* (1747–8), which its author Gellert intended to have the same appeal as a novel of Richardson's but a more edifying effect, is an example of the Leibnizian tendency of German Sentimentalism. The countess lives a life of which the still centre remains untouched not merely by the disasters which fate sends to try her but by the storms of her own emotions which these disasters provoke and even by her own moral shortcomings. From the position of slightly dry detachment from which she recounts her life, all things, even her own feelings and actions, appear simply as thoughts that occur to her. 'Nothing can happen to us but thoughts', says Leibniz, we are 'absolutely sheltered from all external things': the life-story of the countess is that of a windowless monad. But Gellert was not the only German admirer of Richardson. The complexity of the mind of Lessing, who as a young man ran away from theology to try his luck on the stage, and who then became the first and greatest of modern German publicists, appears in the fact that isolation is none the less the lot of all the major characters in his first full tragedy, *Miss Sara Sampson* (1755). As an attempt, however odd it may sound, to treat within the confines of a five-act drama, which more or less observes the three unities, the themes, characters, and setting, for which Richardson required many-volumed novels, it is a logical continuation of Gottsched's project of compromise with the courtly form of the theatre. Into that form Lessing is attempting to pour the most modern self-expression of a middle class whose economic and political circumstances are increasingly far removed from those of their German counterpart, and it is no surprise if the result is bursting with improbabilities. The play also shows up the difficulty experienced by the Leibnizian Enlightenment in explaining or even acknowledging the existence of evil and error in a world that from one point of view at least is perfectly ordered. The isolation of the characters is the natural, if paradoxical, result of their being endowed with almost infinitely receptive hearts. For each one of them, therefore, all his relations with the other characters are brought within the prison walls of his own self. They are so incapable of any failure to understand one another, any hesitation in forgiving one another or any desire to harm one another that they are effectively incapable of any dramatic interaction with one another. All they can do is to suffer, that is, have 'thoughts'. Such plot as the play contains has to be introduced by a (female) villain whose physical and metaphysical constitution is so utterly different from that of the other major figures as to put her beyond the bounds of humanity. Only on condition, that is, of a breach of the dramatic, as well as the world, order can Lessing motivate an irremediable catastrophe, an object of mourning and remorse. Despite their differing points of view, Lessing's other characters agree so nearly in their vision of what is, and is worthwhile, and it is so difficult for a drama, unlike a realistic novel, to depict the material obstacles to their happy unanimity, that the only conflict possible between them is temporary misunderstanding.

The pattern is no different in Lessing's last play, his refined blank-verse comedy, or 'dramatic poem' as he called it, *Nathan the Wise* (*Nathan der Weise*) (1779), which can be seen as inaugurating the last phase in Gottsched's historic compromise: German literature comes to specialize in the philosophically reflective verse play which, though performable on the stage, is most influential as a book. *Nathan the Wise*, the first German classical drama, properly so called, is often disastrously misunderstood as a plea for religious tolerance. The representatives of the three major religions, Judaism, Christianity, and Islam, are not here shown to tolerate one another's differences, for it is only temporary misunderstanding that prevents them from recognizing that they all think alike: they are shown rather to be agreed in a fourth, secret, religion of agnostic humanism, to promulgate which Lessing wrote the play once he had been forbidden by his employer and ruler, the Duke of Brunswick, to engage further in the theological pamphleteering in which in the end his true, and immensely deceitful, genius lay. Between the two landmarks that open and close his career as a major dramatist Lessing wrote two plays which at once continue the theme of isolation and point to its social and political roots. *Minna von Barnhelm* (1767), set in the chaotic circumstances of the end of the Seven Years' War, is a more realistic comedy than *Nathan the Wise*, if also more stilted, and in it irreconcilability of viewpoints prevails rather than identity, but that of course brings us no nearer to interaction. The source of the irreconcilability, however, lies not in the psychology of the characters but in the incompatibility with all decency and morality of the brutal military and financial policies of Frederick the Great, who is (very) surreptitiously caricatured as a comic Frenchman. In Lessing's model tragedy, *Emilia Galotti* (1772), the avoidance of conflict in the denouement could hardly be more explicit: faced with an absolute princeling intent on ravishing his daughter, the middle-class hero, on whose mind and decision all our attention is concentrated, murders not the tyrant but the daughter. Lessing, wily as ever, does deliberately what seemed to his successors imposed on them by a requirement of art: he makes his conclusion appear tragic, or what his age understood as tragic, in order to conceal that it is revolutionary. The 'bourgeois tragedy', which Lessing did so much to make one of the distinctive genres of the new German literature, is characterized by its concentration on internal conflict within the middle class, or within its individual representatives, rather than on class confrontation and its mechanisms. *Emilia Galotti*, uniquely, raises these issues, but judiciously brushes them away in a conclusion which, with a poker-faced irony subtler than anything of Brecht's, presents a constitutional scandal as a tragedy of individual decision, and murder as a refined form of suicide.

Lessing came to understand the iron necessities that dictated the circumstances of German intellectuals, and in the inevitable compromise saved his self-respect through bitter and devious irony. Klopstock may serve as the model of the new 'official' literary man of Imperial Germany, who lacked that degree both of realism and of self-knowledge, and in whose work the tendency to reduce relationships to internal modifications of the self is taken to the point where it is destructive of content. Born the son of a princely official, Klopstock, after a rigorous classical education, was from the age of 26 the beneficiary of a series of princely pensions, yet refused to dedicate his poetry to the 'courtly praise of . . . half-men who, in full stupid seriousness, think themselves higher beings than us'. This paradox he resolved by claiming for the poet, or 'bard', a privileged status and devoting himself instead to the praise of God. 'The dignity of the poet's subject matter elevated his sense of his own personality' was the dry comment of the older Goethe, and indeed Klopstock was not above a certain cult of his own person, forming coteries of friends who bathed and skated together and wrote, and published, poems to and about each other and gave themselves purportedly old Germanic pseudonyms. Yet although he came to be seen, and saw himself, as the first and most representative figure of a new breed, the independent man of letters, it has been calculated that Klopstock's receipts from his literary works made up only 17 per cent of his life's earnings.

He embodied in his own person that illusion fundamental to 'official' literature: that it was the voice of free men, though it was paid for by the State. His twenty-canto hexameter epic, *The Messiah,* was none the less—perhaps for that very reason—the most important, and was certainly the most controversial, work in mid-eighteenth century German literature. The choices of genre, of form and of theme, were all significant, and all characterized by a certain evasion of reality. To set out to write a Miltonic epic was certainly to follow Bodmer's and Breitinger's analysis of what was appropriate to the Swiss context, but it was to avoid the challenge German conditions posed: to decide, or mediate, between the novel and the drama. To choose to write the epic in the German hexameter, a medium hardly tried and indeed largely of Klopstock's invention, was an act of great bravery and considerable originality, but as an attempt at creation *ex nihilo* it again simply ignored the problems of Germany's literary tradition, which was defective and interrupted, but certainly not non-existent. To choose a sacred subject was the gravest decision of all, and Klopstock reflected deeply and publicly on it. The question what to write about became exceptionally acute for eighteenth-century writers throughout Europe as a result of a new historical awareness. The statesmen and heroes, and even the poets and philosophers, of classical antiquity, who since the early Renaissance had been the timeless models of secular human psychology and morality, now seemed as remote from the modern age as in sacred history, bisected by the saving work of Christ, the Hebrew patriarchs and prophets of the Old Dispensation had always been. 'What is the history of the Greeks and Romans to me, however interesting?' Klopstock asked in his review of Winckelmann's thoughts on the imitation of ancient art. The repository of received themes and figures in which Gottsched still confidently believed had, for the thinker, drained away by mid-century. In a remarkable creative act, the European mind at this time, deprived of its secular past, found a new secular material for literature in the present, in itself, and it invented two new forms for dealing with that material which were to dominate literature for two hundred years: the novel, dealing with contemporary social life, and the subjective lyric poem, dealing with the self. If England and France laid the foundations for the first of these forms, Germany can in Goethe claim possibly the earliest and certainly in his time the fullest exponent of the second. Klopstock, however, like other visionaries who to some extent learned from him, such as Hölderlin and Blake, saw the present age as essentially Christian and the task of literature therefore as the creation of a new sacred poetry appropriate to the new historical vision: 'sacred history and the history of my fatherland', he says, in response to Winckelmann, that is to be his theme. The history of his fatherland, however, always remained in Klopstock's treatment of it a shadowy, semi-mythical affair of Tacitus and Arminius, virtuous German tribes—scarcely distinguished from Scandinavians and Celts—pitted against the corrupt might of Rome, all studiously remote from the political reality in which the poet lived. A similar reluctance to grasp the real characterizes his approach to sacred history in *The Messiah:* the principal actors are anyway angels who spend much time in the sun or other parts of the heavens, human beings are on occasion represented by their guardian angels or simply as 'souls', events on earth get little space in comparison to the feelings they are held to evoke in the poet or the angelic observers, even God the Father is concerned to tell us about the intensity of his feelings, and in so far as the events of Christ's Passion and Resurrection are recounted rather than hymned it is for the sake of their effect on the meditating Abbadona, a fallen angel who is ultimately redeemed. The insistent dualism of the poem, which gives far less attention to the physical and historical world than to the supposedly invisible and supernatural world that shadows it, is most evident and most questionable in the treatment of Christ, whose earthly existence is simply a mask concealing from men and demons his transcendent, omnipotent, and omniscient Divine self, which floats away from his body on the cross and converses with God the Father in the moment of his death. Whatever the intrinsic merits or demerits of such a theology it was in one crucial poetic respect detached from reality: it contained no hint of an engagement with the contemporary intellectual processes which were making increasingly difficult any assertion of the divinity of Christ. In 1773, as the last cantos of *The Messiah* were being published, Theophilus Lindsey proclaimed his conversion to Unitarianism and in the following year he was installed in London's first official Unitarian church. The lasting importance of *The Messiah,* as of Klopstock's odes in classical metres, lay not in their contribution to a new sacred poetry but in their cult of emotion and their stylistic innovations, particularly the encouragement they gave to the growth of unrhymed free verse. But although there is plenty of activity in Klopstock's writing—in its neologisms, accumulated verbs of motion, inter-jections, and adverbs of direction—much of the activity is unattached; it is reflected rather than immediate, metaphorical rather than descriptive, hectic rather than vigorous.

The extreme example—Goethe himself called it 'bizarre'—of emotion isolated from any relationship is a play whose obsession with solitude makes it read like an anticipation of the non-dramas of Samuel Beckett. *Ugolino* (1768), by Klopstock's disciple H. W. von Gerstenberg (1737–1823), is often classified by literary historians as one of the first works of the Storm and Stress. Yet it contains nothing beyond its desperate shrillness that is not implicit, or even explicit, in the literature of the previous twenty years. Its five prose acts, which observe the unities of time, place, and action more conscientiously than most classical tragedies, show us a father and his three sons starving to death in a dungeon (the incident being taken from Dante, *Inferno*, xxxiii). There are no other characters and almost nothing happens. There is an ineffectual attempt at escape, and the corpse of Ugolino's wife is introduced into the prison by his tormentors. Otherwise there is talk: talk expressing affection, despair, fear, anger, and pain—and then just talk, and sometimes silence. The constriction, physical and mental, is implacable and unrelieved: the almost total impossibility of any action leaves the characters to express themselves only in a fantasia of increasingly hysterical feeling. The prison-house of the windowless but sentient monad is here presented with the starkest literalness. And it is in that literalness that *Ugolino* takes on after all the contours of a protest. Its neurasthenia is directly related to the panic which, in the very year of its publication, drove Winckelmann back from German soil and into his lurid death.

The question is pressing; was there then no fundamental opposition to the official literature of secularized religiosity and compromised isolation? The despotic system was powerful, but it was not, and in fragmented Germany could not be, all-embracing. Surely some points of resistance can be found? They can, though they are not numerous, and they are—this is the cruellest device of the 'official' system—by definition eccentric.

Much depended on location. Protected perhaps by his profession as a scientist and by Hanover's and his own personal connections with England, G. C. Lichtenberg (1742–99) proved in his posthumously published common-place-books to have been a life-long mordant observer, from his vantage-point in Göttingen, of the quirks of the German public mind, and Germany's greatest aphorist. The 'anacreontic' poetry of the Hamburg businessman Friedrich von Hagedorn (1708–54) and the ironical, psychological, and erotic prose and verse of the itinerant young C. M. Wieland (1733–1813), whom Lichtenberg greatly admired, show spirits unwilling to sacrifice the elusive, material here-and-now for the sake of metaphysical inviolability. Wieland's most influential novel, *The History of Agathon* (*Geschichte des Agathon*) (1767), has come to be seen as the first 'Bildungsroman', or novel of education, a peculiarly German species which keeps alive the theme of isolation even in that most social of genres, but *Agathon* has more than a little in common with *Candide*. The story of the young hero's conversion from his initial idealistic fantasies, which parallels Wieland's own conversion away from Pietism, is told in a way which shows considerable sympathy with the subversive doctrine of materialism, overtly presented as one of the temptations Agathon has to overcome. Was there anywhere in the German world where this attitude of liberal irony might be congenial? Wieland, during a brief spell as professor of philosophy in Erfurt, wrote his political novel *The Golden Mirror* (*Der goldene Spiegel*) (1772) in the hope of securing himself a summons to Vienna from the Enlightened Joseph II: he succeeded only in provoking the offer of a tutorship at the neighbouring and relatively impoverished court of Duchess Anna Amalia, the regent of Saxe-Weimar, but the results were happier for Weimar, for German literature, and perhaps for Wieland himself, than could have been produced by any move to Vienna, which in its own way was as despotic as Berlin.

A more principled and self-conscious opposition than Lichtenberg's which, though it failed to have any lasting effect on the general direction of thought, none the less modified the attitudes of J. G. Herder (1744–1803), Goethe, and

G. W. F. Hegel (1770–1831), was constituted by the circle of friends and literary acquaintances round the Catholic Princess Gallitzin (1748–1806) in Münster. The unifying characteristic of the circle was its hostility, on various grounds, to the prevailing Enlightened philosophy of religion. The Dutch *philosophe* F. Hemsterhuis (1722–90) and the novelist and philosopher F. H. Jacobi (1743–1819) would not alone have given it weight or distinctness but in J. G. Hamann (1730–88) and Matthias Claudius (1740–1815) it included literary figures of considerable importance. Hamann, who was a customs officer in Königsberg but was long in correspondence with the other members of the circle and who in his last years moved to Münster, was converted to biblical Christianity during a visit to England in 1757 and spent the rest of his life shooting off polemical pyrotechnics against the intellectual trends of his time. In particular he saw clearly, but expressed in his pamphlets immensely obscurely, the link between Biblical criticism, the growth of aesthetic paganism, Klopstockian bardolatry and a general deracination of sensibility. Matthias Claudius, editor of the periodical the *Wandsbeck Messenger* (*Der Wandsbecker Bote*), who found more poetry in bathing his children than in Klopstock's flights into the empyrean, maintained for forty years an incorruptible intellectual independence and wrote meanwhile some of the most exquisite and best known of German lyric poems.

Claudius was one of the first to acknowledge the quality of the most substantial opposition movement of all, the complex literary phenomenon of the 1770s known as 'Storm and Stress'. The movement, which in its original impetus can be understood as an attempt to found a national, politically liberal, and above all realist literature after an English model, soon collapsed, partly because its base was too narrow for it to resist the might of absolutist-bureaucratic Germany, partly because it was itself, above all through the concept of artistic 'genius', profoundly interfused with the traditions of 'official' literature. But in Goethe, a founder of the movement who yet only half-belonged to it, 'Storm and Stress' had a representative who throughout the fifty years after its collapse returned time and again to its original and central aim: to build a national literature out of a spirit of opposition to the prevailing forms of national life and thought. It is natural to see Goethe as predisposed to this task by his birth and upbringing in a city whose interests and loyalties conflicted in many respects with the new order. However, the positive ambition of 'Storm and Stress' to work for the nation, for all speakers of the German tongue, deserves emphasis. It allowed for some sympathy between the movement and the striving toward statehood of the Enlightened autocracies, despite its revolt against their despotic structure, methods, and ideology. 'Storm and Stress' was not just the protest of small worlds against engulfment by the big. Goethe's Frankfurt origins did not make of him merely a conservative particularist, like J. J. Möser (1720–94), who saw in his native Osnabrück the pattern of the primaeval and natural form of human association, ruled by precedent and tradition, and who misinterpreted as ageless the Imperial order, in fact established in 1648, which laid down the duties and liberties of his home town. To be a citizen of eighteenth-century Frankfurt was not simply to be a complacent, or a belligerent, local patriot, nor, furthermore, were Goethe's relations with his native city as straightforward as some of his more affectionate remarks may seem to imply. It was precisely the complexity of the relationship which made Frankfurt so suitable a cradle for a poetic talent that was to thrive in sympathetic opposition to the national consciousness, in detachment *and* dependence. Goethe's work owes its classic status within Germany, and its claim on the attention of the world, to its fusion of *all* the emotional possibilities open to speakers of German in its day: it is a middle path between the official and the marginal, between establishment and opposition, between Klopstock and Lichtenberg, Schiller and Claudius; it draws its substance and rootedness from its recalcitrant loyalty to an older, freer, and more fruitful world than that of advancing bureaucratic despotism, but it draws its scope, its modernity, and its preferred forms from the official national literature which the new order created. To find such a middle way—all the while in the full tension of original literary production—required not only intelligence and immense fortitude, but also a native equilibrium, an intuitive felicity of judgement first founded by the ambivalencies of life in Frankfurt as it was also first exercised upon them.

2

Origins of a Poet

Frankfurt and the Goethes

A S A PROSPEROUS free city with its special relation to the Imperial office laid down in the Golden Bull, eighteenth-century Frankfurt was at once provincial and metropolitan, its character both radically German and unselfconsciously international. Its walls and bastions and fifty-five watchtowers might crowd out its four church-spires and give the approaching visitor an impression of solid civic nonentity, but the dumpy little cathedral in which the Emperors were crowned expressed the idea of the Empire, in all its abstractness, better than the pomp of Vienna. The true symbols of the town, however, were the two great municipal cranes on the wharves, and the ancient bridge, with its fourteen arches, across the Main, which served as more than just a link with the old suburb of Sachsenhausen. Frankfurt was the junction of no fewer than twenty-six major roads, far enough up the Main for land-travellers to avoid the rugged Rhineland hills, yet not so far as to be inconvenient for the Rhine itself, northern Europe's principal waterway at a time when sailing down-river was the fastest of all forms of inland transport. In particular Frankfurt linked the roads from the east, from Franconia (now North Bavaria), Thuringia and Saxony (the south of eastern Germany) and Silesia (southern Poland) with the great north–south route that ran from Italy through Switzerland to the Low Countries, Amsterdam, and the sea, taking up on its way major western tributaries from Lyons, Paris, and Lorraine. Round a community of craftsmen, therefore, no different from the kernel of many another old German town, there had grown up a city of merchants, bankers, and—equally international in their outlook—innkeepers, over a hundred of these last out of a population of 36,000, and some of them very wealthy men. The textile trade, especially in English woollen and cotton stuffs and in French silks (and the related craft of the tailors, 200 of these in Goethe's day), and the old spice, wine, and metal trades were the foundation of many a family fortune, joined later in the eighteenth century by dye-stuffs, colonial imports, and porcelain. Banking naturally flourished, and from the time of the Thirty Years' War, which Frankfurt survived better than most, the city was not only Germany's major military recruiting and provisioning centre but also the money market where finance was found for the wars of kings. The eighteenth century brought a steady, if unspectacular increase in prosperity and in the second half of the century the city numbered 183 families with fortunes of over 300,000 guilders, among them eight millionaires.* The rich men of Frankfurt were as cosmopolitan as their interests: much of the trade with England was in the hands of the descendants of Dutch immigrants or of

---

* For the meaning of these sums see below, pp. 49–50.

Huguenots expelled by Louis XIV; a smaller number of Italians had come with the trade in Mediterranean fruits; and prominent among the bankers were Jewish families, concentrated in Frankfurt after expulsion from other cities such as Cologne and Nuremberg. Twice a year, at the great fairs at Easter and Michaelmas, the city spread out the full span of its European affiliations: 350 extra stalls were set up and the value of the goods on display was estimated at 15–20 million guilders.

The inhabitant of Frankfurt could feel the heterogeneity of his home-town, the conflict between domesticity and internationalism, in the language he spoke and in the religion he professed. Frankfurt German was in the eighteenth century, and to some extent still is, one of Germany's many robust dialects with its own unashamed peculiarities of pronunciation and vocabulary, some of which Goethe himself retained to the last (in writing *Faust,* he rhymed 'Tage' with 'Sprache' in 1829, just as in 1774–5 he had rhymed 'genug' with 'Besuch', and 'Das wäre mir die rechte Höhe', written in 1831, is pure Frankfurt). The dialect had the reputation of combining the circumstantial formality appropriate to an old republic—the correct form of address to the town council included no less than eleven honorific adjectives—with a pictorial, often combative, bluntness (merchants were 'peppersacks' or 'barrelsquires') and a liking for proverbial wisdom and often disrespectfully applied quotations from the Bible (Mephistopheles' sententious reference to wine-making, perhaps even to Frankfurt's famous cider, Eppelwoi,* in *Faust,* ll.6813–4, is drenched also in reminiscences of the Frankfurt *patois*). While there was no pressure on an educated citizen of Frankfurt to distance himself from the local form of his native language, as there would have been in similar circumstances in contemporary France or England, each a nation with a single culturally dominant capital city, he was, however, directly exposed to a wide range of other tongues. In the middle of the eighteenth century Latin was still an international scholarly medium (Dr Johnson used it in Paris in preference to French), and a knowledge of French was for everyone else as essential an accomplishment for business, diplomatic, and social purposes as English is today. Frankfurt's Huguenot population and its geographical position made French perhaps more than normally important: during the Seven Years' War it was for four years under French occupation. Peculiar to Frankfurt, however, were its strong connections with the south—the first modern foreign language to which the Goethe children were introduced was not French but Italian—its openness to the north —the English language came up the Rhine with English business—and the presence of a large community (about 2,500) of Yiddish-speaking Jews—a congratulatory poem in Yiddish was read out at the wedding of Goethe's uncle, Johann Jost Textor, in 1756.

Religious variety, like linguistic, reflected the composition of the population. The Jews lived their own religious life within their ghetto (200 houses and a

---

* 'Apfelwein' in High German.

synagogue), whose two gates were shut every night until 1796. The three Christian denominations of the Empire were all prominent, though in different ways: it was popularly said that in Frankfurt the Catholics had the churches, the Calvinists had the money, and the Lutherans had the power. Certainly the Lutherans were in the great majority, the other two denominations mustering between 2,000 and 3,000 each, and, as the established religious party, had the monopoly of the civic administration. There were two religious houses for Protestant laywomen but the sober and unadventurous orthodoxy of the city's twelve Lutheran pastors and the absence of a parochial structure encouraged the formation of more enthusiastic private conventicles—none however as extreme as the Herrnhut Pietist community at nearby Marienborn. Goethe tells us that the curious young found more of interest in the Catholic than in the Protestant services, and in the customs of the Jews. The Catholics themselves, partly lower-grade artisans and partly wealthy southern immigrants, represent-ed the religion of the Emperor not only with the cathedral (then known simply as the church of St Bartholomew) but with three substantial local foundations, houses of six external monasteries and four different orders (the Dominicans with both men and women), and palaces of all three ecclesiastical princes, the Archbishop-Electors of Trier, Cologne, and Mainz, the last of whom had ecclesiastical jurisdiction over Frankfurt's Catholics. Paradoxically, therefore, there were more active Catholic clergy in Frankfurt than Protestant, and though Catholic parish life made no public impression, the splendour of an Imperial election and coronation (such as that of Joseph II as King of the Romans in 1764 when Goethe was 14) was largely of Catholic making. The Calvinists by contrast, mainly the immigrant Dutch and Huguenot merchant families, had no religious rights within the city walls, and had on Sundays to drive out in a procession of magnificent carriages to the village of Bockenheim, a demon-stration of the worldly benefits of predestination which only confirmed the Lutherans in their refusal to make any concessions to these dangerous competitors. It was among the Calvinist merchants, cut off by the city's constitution from public affairs, that the Masonic movement principally established itself, Frankfurt's first lodge being founded in 1742.

Frankfurt was proud of its freedoms, but they were freedoms for the privileged. The oath taken by every new citizen contained a special and explicit undertaking not to compromise his loyalties by accepting service with 'external potentates, electoral princes or lords': his first duty was to the Emperor, his second to the Council and the citizens' committees. But this proud autonomy was neither easy nor cheap to acquire. In the mid-eighteenth century perhaps 8,000 of Frankfurt's inhabitants were mere 'denizens' (Beisassen), without the right to own land or to unrestricted trade, and subject to a double rate of taxation. Of the full citizens, only the Lutherans had the right to a seat in the Council or to employment as one of the city's 500 officials. The Council itself, however, was by no means a representative assembly. Of its three benches of fourteen members each, the third and lowest bench was reserved for selected

members of certain craft associations (the distribution had not changed since
1315 and some of the crafts had died out), while the second and first benches
consisted of members of the College of Graduates (that is, quite simply, citizens
with university degrees), a very few merchants, and Frankfurt's true ruling
class, the old noble families, such as the von Uffenbachs and von Klettenbergs,
gathered into the two exclusive aristocratic associations of the House of
Frauenstein and the House of Alt-Limpurg. The first bench, the bench of
jurors (Schöffen), the city's supreme court, was chosen out of established
members of the second, and elected from its number the Senior and Junior
Burgomasters. The forty-third member of the Council was its secretary, the
Schultheiss, or chief justice, the supreme official of the city and the Emperor's
representative. From 1747 to 1771 the Schultheiss was Dr Johann Wolfgang
Textor, Goethe's maternal grandfather. The exclusiveness of Frankfurt's
oligarchy was modified after a constitutional battle, lasting from 1705 to 1732,
in which the Emperor took the part of the unrepresented citizenry and which
resulted in the erection of a series of citizens' committees (Bürgerkollegien)
parallel to the municipal administration, whose work they supervised and
audited. By this means the merchants and other less-privileged burghers were
given some share of power, and the city's financial affairs were conducted with
efficiency and probity throughout the eighteenth century.

   If the political pyramid was composed of only a fraction of the city's
inhabitants, all were encompassed by the official social grading into five classes.
The first class included the Schultheiss, councillors of the first and second
bench, doctors of a university, the senior lawyers, and members of noble
families of more than one hundred years' standing in the town. The second
class consisted of councillors of the third bench, and the best citizens and
wholesale merchants, with fortunes of at least 40,000 guilders. The third
covered notaries, artists, and retail traders, the fourth small traders and artisans,
the fifth labourers, coachmen, servants, and everyone else. Sumptuary laws
were intended to reinforce these social distinctions but in the eighteenth
century were largely disregarded.

   Another complexity of Frankfurt life is more difficult to define. It is too
readily forgotten that although the term 'Nature' long played an important part
in his thinking, Goethe was a town child. His first impressions were not of
solitude, of the elemental simplicities of forest and mountain, of the discipline
of the seasons, but rather of a crowded, self-ordering, and ambiguous human
world in which material things bore the stamp of human activity, human
decisions, and the human past. The very shape and size of Frankfurt's 3,000
houses—some of them with frontages of no more than sixteen feet, the average
was around twenty–five—reflected not only the varying wealth of the occupants
and the constrictions of life within a ring of fortifications but also bitter
arguments in and with the city Council about such matters as height (the
regulations permitted no more than three floors but were often ignored), the

amount by which the upper storeys might overhang the street (in the Middle Ages overhangs of up to three feet were permitted which, since most of Frankfurt's side-streets were not more than twelve feet wide, could halve the light available) and the shape of the roofs (from the eighteenth century roofs had to slope towards the street so that rainwater should help to wash out the central, open, drain which carried off the effluents from the kitchens). Many of these regulations, such as that which required a solid fireproof wall between the timber-framed houses, and which Goethe's father contravened for the sake of an extra window in his library, dated from the terrible fire of 27 June 1719 which completely destroyed 400 houses in the north of the city. Fire was the natural force from which the townsman had most to fear, and all citizens, except those of the first social class, were required to take part in fighting it, with hand-pump and leather bucket, should it break out in their quarter. Goethe as a youth drew not wholly favourable attention to himself by, among other things, disregarding the privilege which attached to his class and himself standing voluntarily in the bucket-chain. Otherwise, apart from the flooding of the Main, which left a dreamlike memory of boating from barrel to barrel in the family wine-cellar, the non-human world presented itself to the young poet in the tamed and indirect form of the garden—not the geometrical urban gardens which, with land at a premium within the city walls, only the richest families in the newest and largest houses could afford, but the plots outside the gates of which the Goethe family possessed three. Two were simply stands of woodland and pure investments, the third, near the Friedberg Gate in the north was a true practical garden of about an acre, surrounded by a wall, tended by a gardener, and growing vines and asparagus and other vegetables. The grape-harvest and the family wine-making that followed were the occasion for great open-air festivity here. Between the nearby villages to which the city-dwellers drove out for an afternoon's amusement lay coppices of oak and beech in which the 15-year-old, as he recalled half a century later, could yearn for an indefinite wildness and solitude which this park-like landscape was too decorous and too populated to afford. But it is difficult to imagine him returning home on such a summer evening in 1764, hurrying to arrive before 8.30 when the town-gates closed and late-comers had to pay 4 kreuzer* for admission, without an intimate, if uncomfortable, feeling of belonging.

Visitors agreed that the first thing that struck them about Frankfurt was the smell, and the most material attachments are the strongest: the thin mineral poisons that fill the air of modern Frankfurt as of any motorized cosmopolis can stir no one's emotions. If the young Goethe arrived in time to be admitted through the Friedberg Gate he would pass first through a quarter thick with inns, for this was the main freight route to the north, with acrid stables, great open boxes of oats, and the smokey comfortable odour of malt roasting for the

* There were 60 kreuzer to the guilder. See below, pp. 49–50.

beer-vats. He would probably go by the entrance to the ghetto, where the sale of dried fish and salted meats was concentrated, since these products turned the eighteenth-century Christian stomach, and then down the Zeil, Frankfurt's broadest street, thirty yards wide and, though filthy from its daytime use as a cattle market, the site of the town's most elegant hotels. Past a pungent coffee house, and a cool clean waxen draught from the Lutheran church of St Catherine, he would reach a bouquet of dry exotic scents from the magazines of Belli, the city's largest spice-trader. Everywhere there would be the tang of wood smoke, and perhaps the resin of fuel freshly chopped (as was usual) on the street, for even in high summer the kitchens had to cook and bread had to be baked. But the grey inescapable background to all these domestic smells, more insistent than ever on a warm evening after a hot day, would be the stale fetor of the street drains, of the medieval cloaca into which emptied the privies of the better houses, including Goethe's own, and of the river, the ultimate recipient of practically all Frankfurt's sewage, whether from the cloaca or directly from the bridge, where a special parapet was reserved for the discharge of barrows and carts. Here in the north-west of the city, however, in the sixteenth- and seventeenth-century New Town, one was spared the most penetratingly repulsive stench, from the tanneries on the south bank of the Main, and from the infamous Hole of Pestilence, a black open cess-pool, without efflux, situated up against the wall, between the prison and the orphanage. Here, in benevolent or penal confinement, Frankfurt's true non-citizens pined, paled, and died.

Turning into the Kleiner and then the Grosser Hirschgraben (Little and Great Stagsditch) Goethe was on the very edge of the cramped medieval Old Town, for this street, rather broader than the usual, had been built over, and in, the former city moat. Its houses, such as that of the Goethes—whose illegal library window would already be visible, perhaps too the face of the father watching for the return of his son—were as a result not only larger than the norm for the city centre but also had two storeys of cellars. The immense stock of barrelled wine that these held, some of it still from the classic year of 1706, and worth in all one-third of the value of the house, impregnated the whole building with a sweetness that marked it off, as home, from the stinking world outside. Despite the substantial extension and modernization of 1755 which had flanked the front-door with two reception rooms, the ground floor, hot, fatty, and dominated by the kitchen, would already be dark and lit by home-made candles, reminiscent still of the confined and gloomy house of the poet's early childhood, especially since the same long-case clock still struck the hours. On the first floor however, in the main living-rooms, the fine arrays of fashionably large windows installed by his father came into their own. The air too would be fresher: the great presses that stood on the landings of the elaborate new staircase (father's and mother's initials worked into the iron baluster) contained an immense stock of clean linen, for wash-day was only three times a year, and there was nothing unusual in a man having several hundred shirts. On the second floor, where Goethe was born and where his sister Cornelia (1750–77) had her bedroom, there was a more private atmosphere: here too was the library of some 1,700 volumes. Finally on the third floor, flooded with evening light, Goethe reached on the east side the rooms that were his own until he left Frankfurt finally at the age of 26, and on the west, if he braved the putrid smell of his father's silkworms, fed there in season on mulberry leaves that it was his and his sister's duty to collect, the windows that looked out across the courtyard of the family house and the remains of the old town ditch, then across the flowering back-gardens of the wealthy houses on the western continuation of the Zeil, towards the watch-towers on the city-wall, the distant hills of the Taunus, and above them all the sunset. In 1817 Goethe could still recall these moments of release from his mother-city's stifling embrace:

With the fresh young mind of a child, under an urban-domestic education, one's yearning gaze had scarcely any escape but into the atmosphere. The sunrise was restricted by neighbouring houses, all the freer the westward, just as a walk tended to be prolonged into the night rather than anticipating the dawn. The fading of the day on calm evenings, the colourful withdrawal of the gradually sinking light, the onward march of darkness, frequently occupied the solitary idler.

The truth of the matter is that in Frankfurt the Goethes were near to being outsiders. The poet is often described as the son of a patrician but this is in an important sense far from being the case. The patricians of Frankfurt, with constitutionally guaranteed seats in the city Council, were the ancient noble families of the Houses Alt-Limpurg and Frauenstein, specifically the former. There was no question of public social intercourse between these and a parvenu such as the poet's father. Johann Caspar Goethe (1710–82) was the son of a Thuringian tailor of peasant stock, who after years of wandering that had taken him for a time to Paris and Lyons had settled in Frankfurt only after the revocation of the Edict of Nantes (1685), and had there amassed a fortune first by his tailoring and then by his second marriage, to the widow of an innkeeper, whose business he had then taken over. Caspar Goethe, the only surviving child of this second marriage, was certainly a rich man. In addition to the house, his land, his pictures, books, and silver, and the stock of wine inherited from his mother, which brought in no income, but the value of which he estimated in 1770 as 42,500 guilders, he had investments of another 65,000 guilders which yielded an annual income of around 2,700 guilders. This was half as much again as the salary of his father-in-law, the Schultheiss Textor, Frankfurt's highest-paid official, and six times as much as the allowance paid to his half-brother, the artisan Hermann Jakob Göthe (1697–1761), as a councillor of the Third Bench. A country pastor at this time could expect between 250 and 520 guilders a year at the most, a schoolteacher not more than

190, while Schiller's stipend when he was appointed professor of history at Jena University in 1790 (strictly speaking it was a pension, for the post was unsalaried) was 400 guilders plus admission fees. One guilder (nominally about 7½ new pence) represented three days' wages for a builder's labourer, and with it, we can learn from Caspar Goethe's meticulously kept accounts, one could buy ten pounds of pork (in bulk) or six pounds of butter, or two pounds of sugar, or a pound of coffee, or half a pound of tea, or (in season) a thousand pickling gherkins. Prices rose steeply as soon as manufacture was involved. A copper tea-maker cost Caspar Goethe 8½ guilders, a silver sugar-bowl 20, which was the annual wage (in addition to board, lodging, and certain traditional gifts) of his senior housemaid. Clothes were particularly expensive. Twenty guilders was the approximate cost of a simple black coat for Caspar, while the suit in which his son Wolfgang went to study at university cost over 34, and when his wife put on her new scarlet dress and her fur-lined Polish coat she was wearing ten years' worth of her maid's cash income, or a skilled mason's total income for a year. It was not, however, this considerable wealth that put Caspar Goethe into a higher social class than that of his older half-brother. Caspar Goethe owed his position in the highest class of Frankfurt society, where the distance from the ruling oligarchy was all the more painfully obvious, to his membership of the College of Graduates, to the fact that as a man of leisure and the sole heir of his mother (a powerful woman who survived until 1753 and was just remembered by her grandson) he had been able to study law at the Imperial Supreme Court in Wetzlar and at the universities of Leipzig, Strasbourg, and Giessen, where he had acquired a doctorate in 1738. His intention of leaving behind the narrow tradesman's world into which he had been born is apparent even from his boarding-school days in Coburg, when he changed the spelling of his family name from Göthe to the form in which his son, who in this as in all matters followed him dutifully, made it more widely known. His later acquisition of a coat of arms, in which three lyres represented his devotion to higher pursuits, was an expression of the same ambition, as indeed was the rebuilding and restyling of his house which he undertook only when his mother was dead. Equally a step in his social ascent, though also of the greatest importance for his own personal development and later for that of his son, was his grand tour in 1739 and 1740 via Regensburg and Vienna to Venice (where he saw the Doge wedded to the sea), Rome, Naples, Milan, Paris, and the Low Countries. He kept, in Italian, an extensive and methodical journal of the tour, and subsequently never tired of recounting the splendours of the Italian cities or elucidating his prints of Italian views. On special occasions his young son was even allowed to play with the model of a Venetian gondola which was one of his most prized souvenirs. In 1742, however, after his return to Frankfurt, his ambition led him into the most serious mistake of his life. Ignoring the slow and uncertain path to honour that might have been open to a man of his abilities in the city's administration, he sought office directly under

France's puppet-Emperor Charles VII, the rival of Maria Theresa, residing at the time in Frankfurt: from Charles he obtained, with the assistance of influential friends, the title of Imperial Councillor which was otherwise reserved to the most senior municipal officials. With the death of Charles in 1745 and the election of Maria Theresa's husband to succeed him, as Francis I, Caspar Goethe found himself without prospects in the Imperial administration and burdened with a rank too high to bring him office, or much love, in Frankfurt. Moreover, from 1747 his half-brother's presence on the Third Bench excluded him from the city Council, whose rules against nepotism were extremely rigorous. His social advance came to an end in 1748 with his marriage to Catharina Elisabeth Textor (1731–1808), the eldest daughter of the Schultheiss. It was a fateful and fruitful momentary coincidence, not only of two different generations, in the marriage-partners, but of two different social attitudes in their families, for the Textors too were marking time, though voluntarily. The Schultheiss had refused the ennoblement normally associated with his office after correctly calculating that it would leave his daughters too poor for a noble and too elevated for a bourgeois match. Otherwise, though they had been longer in Frankfurt and were more respected, and so could appear less pushing, the Textors were not a fundamentally different family from the Goethes: they too had immigrated from the east, from Franconia, though as lawyers, not as tradesmen, and in the middle, not at the end, of the seventeenth century. Had Caspar Goethe been as careerist as he might hitherto have appeared, however, he might now have moved away from Frankfurt, like the brother-in-law of the Schultheiss, J. M. von Loen. Instead, after the birth of his eldest son, named in honour of his father-in-law, Caspar Goethe now settled down to a long life of leisure, devoted to managing his fortune, building up his collection of books and contemporary paintings, meeting with fellow members of the College of Graduates, and caring for his family, particularly their education. It was through education that he had become what he was, and he plainly intended them to become something better.

The marginality of Caspar Goethe's position and the bitterness of his disappointment can be measured from his political views, the influence of which on Wolfgang was not confined to the loyalties of childhood. With his usual precision in expressing ambivalences the 62-year-old autobiographer recalls how during the Seven Years' War (between England and Prussia on the one side, and France, Austria, and Russia on the other) he naturally belonged to his father's party: 'And thus I too had Prussian sympathies, or to be more correct, Fritzian* for what was Prussia to us?' The loyalty to Prussia had little to do with an assessment of European or even German power-politics, which was why it readily expressed itself as an admiration for the mere personality of Frederick the Great. But it had much to do with the Goethes' position in

---

* 'Fritz', short for Frederick (the Great).

Frankfurt, which was why—as the poet himself says—it led to bitter family quarrels and an estrangement from the Textors. Caspar Goethe was a high-ranking citizen of an Imperial free city to whose economic and constitutional peculiarities he owed, in worldly terms, everything: he had no more interest than the Textors, or any other members of Frankfurt's upper classes, in the expansion of an aggressively anti-Imperial power intent on the extinction of such peculiarities wherever it met them. But in 1756 Prussia seemed remote from Frankfurt, at least to the imperceptive, and when Frederick had shown how long his arm really was, by arresting Voltaire within the city walls, Caspar's view had been no different from that of any other right-thinking Imperial burgher who had taken a solemn oath against service with foreign potentates: time and again he recounted the affair to his son as an example of the unwisdom of coming too close to the courts of princes. He did not approve of the decision of von Loen to enter Frederick's administration, and the newspapers that he took were violently anti-Prussian. Yet for all he owed the Empire, for all his respect for its institutions and traditions, how could he be an enthusiast for the cause of Maria Theresa and Francis I when the victory of that cause in 1745 had put an end to his own hopes for a career and had left him in a position of permanent slight inferiority to the Textors, than whom he was certainly, in his own estimation, no worse? 'My father . . . could not resign himself to the inevitable', wrote his son of an episode during the French occupation of Frankfurt (which began in January 1759) which focused all these emotions. Caspar Goethe had never resigned himself even to his old disappointment, and when the French civilian governor of Frankfurt, the Comte de Thoranc, was billeted in his house, it must have seemed to him that his enemies—including the Textors—had invaded his last refuge. Thoranc's intrusion was less troublesome and more tactful than many billetings and brought distinct advantages, such as French cuisine (including the novelty of ices). But Caspar Goethe's outburst of frustration and anger against his unwanted guest, after the French had defeated the Prussians at nearby Bergen, on Good Friday, 13 April 1759, is more understandable than his son allows, impressed as he is by Thoranc's immense rectitude and his prudence in the management of his own depressive character. Thoranc's argument that the French were in Frankfurt to defend the interests of Frankfurt's Emperor and Empress could only seem disingenuous to the man who fifteen years before had seen the French fight against that same Maria Theresa and, oh irony, in support of the puppet who had bestowed on him his title. The only force that had consistently been on 'his' side—against the Frankfurt establishment—was Prussian, however desirable he might think it to keep despots at a healthy distance. Caspar Goethe was an able man in a situation too complex for his strong but uneducated emotions. His loyalties went naturally, by interest and by tradition, to the Empire and the constitution of Frankfurt, yet he could not support the reigning Imperial dynasty and in Frankfurt he both was excluded from aristocratic circles—by what he called in his travel diary an 'inhuman prejudice'—and was suspect and mildly pitiable to the officials. He could not even, as a Lutheran of the first social class, identify himself with Frankfurt's internal opposition, the pro-Imperial citizens who had fought the great constitutional conflict with the oligarchy, for these were principally Calvinists, and of the second class. The phantom support for Prussia expressed his alienation from Frankfurt, the opposition to courts and courtiers his basic fidelity. In this atmosphere of multiple detachment his son grew up, and was able to give it more adequate expression than mere ill temper. But to the spirit of Frankfurt, for all their detachment, both remained true. Frankfurt had, and encouraged in its individual citizens, the same self-confidence as Hamburg, but a stauncher conservatism and a greater loyalty to the Empire, at any rate as an institution. Like any free city it lived from its opposition to closed frontiers, outside Germany and within. To be true to the spirit of Frankfurt, therefore, could certainly be to favour the growth of a German nation, but not of a narrow statism, not at the expense of international community. The father's contentious Prussianism and the son's removal to Weimar involved no compromise of these principles. The social marginality which poisoned the relations of the one with his native city the other overcame by simply going away. The reactions of both concealed, but did not annul, their obstinate adherence to the best in the local tradition, despite their rejection by its local representatives. The European revolution that bisected the poet's life consisted precisely in the obsolescence of that tradition and all that made it possible, despite the town's proud proclamation in 1792 that General Custine's advancing troops could bring no liberty, equality, or fraternity that the citizens of Frankfurt did not possess already. Even so, Wolfgang Goethe retained his formal status as a citizen of his father's town until 1817.

'More chatterbox than substance': 1749–1765

Caspar Goethe devoted all his frustrated ambition to the private world of his family and to the furtherance of the two children who reached adulthood, Wolfgang and Cornelia. Like Montaigne's father he had certain definite ideas of his own about education, and unfortunately the leisure and the means to realize them. The resultant combination of the obstinate, the personal, and the mildly eccentric his son later and perhaps unkindly described as 'dilettantism', regretting that he had never regularly and systematically learnt anything. Caspar Goethe had no trust in the public educational institutions of Frankfurt, perhaps rightly since there were in the city only thirty schoolteachers for 3,000 children of school age. With the exception of their very first years his children were taught at home—though frequently in the company of others—by privately employed instructors. Wolfgang's first teacher, from the age of 3, was a remarkable lady, Maria Magdalena Hoff, a Calvinist who ran what would now be called a kindergarten and whose disrespect for the Lutheran authorities had brought her several spells in prison. Johann Schellhaffer, who ran the primary school which Wolfgang and Cornelia attended in 1755 while the house was being renovated, was also involved in violent controversy with the municipal authorities because of the freedom of his religious instruction. Still more extraordinary, however, was Caspar Goethe's choice of the private tutors, eight in all, for the years that followed, which seems to have had the deliberate purpose of stimulating the children by exposing them to as varied a set of influences as internationally-minded Frankfurt could provide. The Latin teacher was a Turk, captured after his father had been killed in battle, and brought up as a Christian: a student of theology, he eventually became deputy headmaster of the city's grammar school (his discipline was poor). Italian was taught from 1760 to 1762 by a former Dominican from Naples who, a victim of doubts, had left Italy and converted to Calvinism; and French by an émigré who had similarly converted, though probably for more worldly reasons. Yet another convert, a formerly Jewish clerk in the Municipal Ordnance, taught the young Wolfgang Yiddish, at his special request, and from this it was a natural step to study Hebrew, and probably his other ancient languages too, with J. G. Albrecht (1694–1770), the headmaster of the grammar school, an individualist described by his contemporaries as a hedgehog, who thought schoolbooks useless and said of Frankfurt's supreme ecclesiastical authority, 'The Consistory gives the orders, but I do what I please.' In addition teachers were found for English (a young man who had just come back from some months in England), drawing, Greek, history, geography, handwriting (Goethe always wrote a firm, clear hand), and music (both children learned the piano, to which Wolfgang later added the cello). Caspar Goethe himself gave instruction in most subjects: in the case of drawing however, in which he felt himself inadequate, but which he thought essential if Wolfgang was to have full advantage of his future grand tour (photography not having been invented), he sat down in the lessons beside his children and showed them how to learn. He himself finished off all his son's uncompleted drawings, trusting in the force of remorseless example, just as later he returned him his letters with all mistakes corrected. Dancing he taught by playing the flute while Wolfgang and Cornelia practised together. The regime of private tutors continued, interrupted only by the usual childhood illnesses, of which the worst was an attack of smallpox in 1758, which permanently coarsened his complexion, until the autumn of 1762, when Wolfgang was 13. His main academic commitment was then to the personal sessions with Albrecht, for by that stage Caspar Goethe was already thinking of the need to prepare him for the university.

Their father's combination of affection and earnest inflexibility was undoubtedly exasperating for the children; to the outsider, who can also see the elements of originality in the scheme, it seems endearing and perhaps in the end appropriate, at any rate for the son. For it was soon apparent that he was an exceptionally gifted child. He made rapid progress in all his languages except Hebrew (pointing defeated him) and showed a remarkably retentive verbal memory, being capable of repeating entire sermons word for word. He also possessed from the start a most unchildlike fluency in the writing of verse: his first preserved poem, a New Year's greeting for 1757, addressed to the Schultheiss and beginning 'Sublime Grandpapa', is clearly not the work of an unassisted 7½ year-old, but by the time he was 12 he was writing German alexandrines little worse than those of Gottsched. A little later he started writing an epistolary novel in six languages, completed a biblical epic in prose on the subject of Joseph, and in the winter of 1763–4 put this together with a collection of his verses under the title of *Miscellaneous Poems* to present to his father, who liked to keep his orderly eye on his son's progress. Unlike the 7-year-old Mozart, however, whom in 1763 Goethe saw playing the piano through a baize cloth and identifying the notes rung on clocks and glasses, Wolfgang Goethe was not a child prodigy—he was quick, effervescent and wilful, in need of ballast. At a school he would have learned with more system than he did with his father, but he would have learned less, and he would certainly have been unhappy. There are odd strains in Goethe's childhood, as in his family, but there are no signs of anything but complete emotional security.

He was probably, however, what would nowadays be called 'difficult'. He enjoyed the eldest son's privilege of the total love of an energetic and happy mother, but it is often forgotten that the Goethe family was not small. Six children were born after Wolfgang, one of them a stillbirth. Cornelia was only fifteen months younger than her brother who from the age of 3 to the age of 9½ was one of at least three children, for three and a half of those years one of four. Only from 1761 onwards were Wolfgang and Cornelia alone. However, only one of the other infants survived more than three years, Hermann Jakob, named after his uncle, who died of dysentery in January 1759 a few weeks over the age of 6. Wolfgang remarks in his autobiography that there was little understanding between the two brothers, and his mother recalled that his main emotion at Hermann Jakob's death seemed to be irritation at the grief of his parents. Then, asked by her whether he had not loved his brother, he produced from under his bed a sheaf of papers that he had himself written out for Hermann Jakob to learn from. The imitation of his father, which was to be so marked a feature of the rest of his emotional life, had already begun, and in the not unusual form of an answer to the problem of fraternal jealousy. Even after Hermann Jakob's death, however, we should not think of the Goethe household as small, though there was no one to threaten the unique position of the young master: there were two maids, a cook and a manservant, there was a lodger, H. P. Moritz, a local diplomat to whom, after the departure of Thoranc, Caspar Goethe had rented the first floor of his house in the hope of being spared any further billeting, and there was also Caspar Goethe's ward, the deranged Dr J. D. B.

Clauer, whose sanity had given way at the end of his university studies and who lived with the family for thirty years. He enjoyed taking dictation from young Wolfgang, who probably enjoyed giving it. Some time between 1768 and 1770 Wolfgang drew what is believed to be a portrait—heavily corrected by a drawing master—of Clauer, sitting pensively in a dressing-gown. If we add the numerous tutors, the comings and goings during Thoranc's stay in the house, Caspar Goethe's fortnightly gatherings of scholars and collectors and his wife's circle of lady friends, it is clear that, despite not attending a school, a Goethe child had an open and sociable upbringing, even if converse was mainly with adults. With his contemporaries indeed, if we are to believe the account in *Poetry and Truth*, Wolfgang was not generally on good terms; unlike the older generation who enjoyed his chirpy precocity, his fellows, even perhaps the future theologian J. J. Griesbach (1745–1812), found him too clever by half, too pompous and too distant, though there was at least some healthy scrapping (Wolfgang preferred to be referee).

A young man who insists that his mother should lay out three separate sets of clothes every morning for him to take his choice at his leisure is unlikely to be popular in any century. With one or two individuals, however, intimate friendships grew up, notably J. A. Horn (1749–1806), and, above all, beloved Cornelia. Brother and sister confided totally in each other, sharing even accounts of their studies and the books that they read. In his discussion of their relationship, in the light of Cornelia's later unhappy marriage to the eleven years older J. G. Schlosser (1739–99), younger son of a Frankfurt legal family, Goethe emphasizes, in *Poetry and Truth*, how little of what needs to be said is 'communicable', and how much has to be 'read between the lines'. He seems, that is, to be suggesting that his sister's marital difficulties derived not from her ill looks and poor complexion, to which he draws repeated attention, nor from her total lack of sensuality, as he calls it, but from there being only one man in the world whom she could love: the one man she could not love, her brother. Certainly no intimacy of her later life would seem to have compared with the childhood excitement of their secret reading together of Klopstock's *Messiah*, forbidden by their father, according to *Poetry and Truth*, on the Gottschedian grounds that as it did not rhyme it was not poetry.

One wonders whether Caspar Goethe was really so hostile to Klopstock. In general he seems to have favoured industry, order, example, and positive direction, rather than restriction and prohibition, and as his son entered adolescence he was largely free to roam the town as he wished. Until the French garrison withdrew, after the Peace of Hubertusburg put an end to the Seven Years' War in early 1763, there was a French theatre in the city, and grandfather Textor provided Wolfgang with a permanent free ticket. Comedies and tragedies were both performed, to the great benefit of the young Goethe's knowledge of the French language, and in the amateur theatricals which his father's friend, the lawyer and future burgomaster J. D. Olenschlager (1711–78) liked to see his children put on at home, he played the part of Nero in

Racine's *Britannicus*. The year 1763 was a time of coming out for Wolfgang, who was confirmed at Easter. He and J. A. Horn were soon the centre of a group of friends with literary interests who foregathered, either at home, or, on Sundays, in the auditorium of the grammar school, to read their works—and to consume refreshments whose cost Caspar Goethe met and carefully noted in his account-book. The younger Goethe, however, also began at this time to make more commercial use of his talents, writing to commission those occasional verses which were, and still are, a more prominent feature of German than of English social life. Through these commissions, and the entertainments for which they paid, his circle of acquaintance widened beyond his family connections and led him at the age of 14 into his first love-affair, described at length in his autobiography, which would have been totally innocent but for its connection with a group guilty of some malpractice within the city administration. Whether there really was a definite attachment to some one person at this time, whether, if so, her name really was Gretchen, whether any reliance is to be put on the circumstantial detail of Goethe's narrative, and in particular whether the discovery of the scandal really did coincide with the coronation of Joseph II on 3 April 1764, are all matters subject to varying degrees of doubt. We may be sure that whatever Goethe and anyone else in Frankfurt was doing from the autumn of 1763 onwards was overshadowed by expectation of the coming ceremonies. The opening diplomatic negotiations in January; the earnest self-importance with which the citizens of Frankfurt prepared themselves for their traditional role as hosts and guardians of the event; the arrival of the splendidly attired Electors or, in the case of the temporal Princes who could not be troubled to come in person, their ambassadors; the election itself, a purely formal preliminary to the entry into the city of the Emperor, Francis I, and his son, in a gilded carriage at the heart of an elaborate procession; the solemnity of the day of the coronation; the sight of the Imperial Crown Jewels brought for the occasion from Aachen and Nuremberg; the city thronged as never before; the festivities and illuminations, the roast ox and the free sausages, and the drinking of the oldest family wine— all amounted to the most memorable public function of Goethe's life, and certainly the one most expressive of the old Imperial Germany to which he belonged, and to that extent his symbolic use of it in *Poetry and Truth* is justified. That it may have coincided with an affair of the heart is suggested by the attachment Cornelia conceived in early 1764 for a visiting Englishman, Harry Lupton by name; it would be only natural for it to have been prompted by a recent example from her brother. There is independent evidence that an official investigation into some such irregularities as Goethe describes did begin in May of that year, and his personal reputation would seem to have been affected by the business, which was an embarrassing one for the grandson of the Schultheiss to be involved in, though he seems to have committed no crime worse than staying out all night drinking coffee. At the end of May, and possibly at the instigation of his father, Goethe sought membership of a secret literary society of refined Frankfurt young men, the 'Arcadian Society of Phylandria', but was turned down on the grounds of his 'addiction to dissolute living and other . . . disagreeable failings'.

The application itself is interesting as the first evidence of the attraction which secret fraternities long held for Goethe, and which was at least partly rooted in a social marginality of his family which may explain both his application to the Arcadians and its rejection. This first unsuccessful attempt at 'joining in' (to use Goethe's term of 1831) is also interesting for the survival of much of the correspondence associated with it. From this it is clear how variable an impression, now as later, Goethe made on those who knew him. That he could talk, that he could write, that he overflowed with surprises in word and deed, that he knew what was expected of him and was as capable of doing it as of ignoring it, that he could if he chose enjoy anything, even hard work, and you would enjoy it with him if you did what he wanted, that was already clear, in a greenish, pubescent way, about this forward 15-year-old. But what to make of it? The earnest Arcadians, Alexis, Amintas, Myrtilus, as their pseudonyms ran, were at first impressed by quotations from Rabener and confident criticisms of their theatricals, but by September they decided they were dealing with a 'hyperbolical' verbosity and a 'treacherous' iridescence, 'a good chatterbox, but no solid substance'.

In August 1764 Goethe put all his recent literary exercises together in a quarto volume of 500 pages and for three years he maintained the practice of marking the month of his birth in this fashion. It was his own way of accepting the methodical discipline of his father, who was now intent on preparing him to study law at Leipzig the following year, and started work with him on some of the basic legal text-books. Wolfgang also began to acquire for himself a certain interest in philosophy, particularly the more imaginative and less comprehensible aspects of Plotinus. But social accomplishments were as important as intellectual: Wolfgang's literary abilities could be taken for granted; the piano lessons begun in 1763, though tedious, continued for another year; it was time for serious instruction in the student's arts of fencing and riding. The young man, however, was not lithe or strong, he was ill during the spring, the riding lessons were not particularly successful, and he positively refused to dance, claiming an emotional aversion to it after his unhappy love affair. Perhaps deciding that his son needed some experience of a more independent existence out in the world, Caspar Goethe took him in June on a trip to the spa at Wiesbaden and on to Mainz and at some point there was also a visit to Worms: some of his earliest landscape drawings probably date from these excursions. As Michaelmas approached, and with it the beginning of the new academic year, the 16-year-old Goethe became increasingly impatient to be out of his hometown and off to a new life 'in Saxony . . . the land where the finest and best verses are made', as Horn, who would himself be studying at Leipzig eventually, described it in a farewell address at the last meeting of their literary club. In retrospect, the four-day coach-journey, from 30 September to 3 October, over deteriorating roads, accompanied by a Frankfurt bookseller on his way to the fair, seemed to Goethe to have had 'something of the character of a flight'—'illusion of being able to hide in the wide world', he noted, 'cutting self off from people'.

For the next three years Wolfgang was away from Frankfurt and his family, and this seems an appropriate moment to reflect on his relations with his parents. Even in his lifetime Caspar Goethe was maligned by his son's friends, particularly by the sarcastic J. H. Merck (1741–91) to whom he may have refused to lend money, and he has often been presented as at worst dictatorial and miserly, at best dyspeptic and unimaginative, his relation with his son fraught and tense, by contrast with the sunshine shed by the mother. These are important errors, which should be dispelled. Caspar Goethe's ponderous and methodical manner, and the comparison with his young and lively wife, nearer in age to her children than to her husband, may have made him seem to Wolfgang and Cornelia as pedantic as he was at times undoubtedly infuriating. But no pedant could have selected the perhaps excessively interesting array of tutors he found for his children or have concentrated as exclusively as he on collecting the work of *living* painters, and the melancholy from which he increasingly suffered was not the malady of an unimaginative man. The gentleness of his nature is revealed by the note confided to his travel diary after a great feast in Venice for the crippled young Prince of Saxony—that all the splendour of the ball could only be a source of pain to one who could not take part—and of course by his guardianship of the lunatic Clauer. It was not an unimaginative act to present his boy with a puppet-theatre and so, by his son's own admission, found his passion for the drama, though it was perhaps the most serious disservice Wolfgang ever did to his father's reputation to attribute this gift instead to his grandmother. The lyres in his coat of arms, quite apart from the emphasis in his syllabus on music and drawing, should be enough to make it obvious that his intent concern for his son's education was not limited to imparting book-learning or laying the foundation for a solid professional career. He doted on his son, he lived for him and in him, and that included his literary activities, from the days when he ordered his first poems and exercise books to his old age when he discussed the first acts of *Egmont* and urged Wolfgang on to finish the play. What else but the flexibility of love explains his adding, as a man in his mid-sixties whose preference was for rhyme, the works of the Frankfurt Storm and Stress writers to his library, or his paying for their festivities in his own house, or, of all things, his going with his wife in 1777 to a public reading from Klopstock's *Messiah*? Had he not in the 1770s thought his son's writing more important than his legal practice he would not have paid for a secretary to relieve him of the drudgery. The reproach of miserliness is particularly unwarranted: throughout Wolfgang's period of higher education, from 1765 to 1773, his father spent on him an average of 1,000 guilders a year, at times nearly a half of his annual income. Though he did not entertain frequently, he always did so lavishly. That, in the words of his son who more than once described all his own works as fragments, 'in the matter of completing things my father possessed a particular obstinacy' is not obviously a vice. And a good word should also be said for his patience, manifest not only in his tuition but also in the absence of any record of arbitrary or irrational disciplining of an adolescent who must at times have been very much more than usually vexatious. The relation of father and son was if anything rather better than is usually the case, and there is no justification at all for presenting Caspar Goethe as the typical German paterfamilias of the veiledly parricidal dramas of Schiller and Hebbel. (His son's poetic work is totally free of this particular obsession of his contemporaries, and of its complications; the relations between parents and children in his plays, poems, and novels raise quite different, and more natural, problems.)

A young friend visiting Frankfurt in 1772 before the poet was known wrote in his diary: 'We went to Goethe's house; only his mother was at home and she received us, me too at a word from her son, which for her is final. Shortly afterwards his father, and again the same; I had a conversation with him.' And in 1775, when he was famous, Goethe was commended from afar to one of the most important women in his life in Weimar, Charlotte von Stein (1742–1827), with the words: 'Ah, if you had seen how the great man is the most proper and loving of sons towards his father and his mother you would have found it, ah, you would have found it difficult not to see him through the medium of love.'

Goethe summarized what he owed to his parents in some famous lines—part of an epigram on his own *un*originality—which demonstrate the fine equilibrium he attained in his affection for them:

> Vom Vater hab' ich die Statur,
> Des Lebens ernstes Führen,
> Von Mütterchen die Frohnatur
> Und Lust zu fabulieren.

My father gave me his build, his earnest conduct of life, my mother dear her happy nature and fondness for storytelling.

We should not, however, let the more spectacular character of the gifts Goethe attributes to his mother's influence mislead us into the view of a later caller on Catharina Elisabeth: 'Now I really understand how Goethe became Goethe.' Her exuberant but maternal temperament, her love for the theatre and theatricals (and actors and authors), her combination of household practicality (which did not extend to financial matters) and open emotionality, her genuine story-telling ability and her excitable, and unbelievably heterographic, style in letter writing, her unaffected lack of respect for persons, her willingness to embrace anyone, whether a dowager duchess or a penniless scribbler, bringing news of her son, her old-fashioned trust in Providence, rather than any specifically Christian hope, and her impatience with theological innovators—these are all characteristics that we find echoed in her son, even where they are muted or reversed by an approximation to his father, and they are characteristics which exercised a powerful attraction on German literary men of her own and later generations. The Golden Age of her life was the period of the Storm and Stress in Frankfurt in the 1770s when her house was, as she said, 'from top to bottom stuffed full of *beaux esprits*', who might stay up half the night, gossiping, arguing, reading, improvising. Wieland, Merck, Lavater, they were all 'dear son' to her, for they were all friends of her one son, and to all of them she was 'Mother Aya', from the name of the mother of the Four Sons of Aymon, in the popular German version of that old French romance. (That Caspar Goethe took over this nickname into his account-book is another sign of his affectionate receptivity.) Perhaps in her imaginative temper and artistic leanings we can see traces of her descent from Lucas Cranach the Elder. Her avalanche of love and admiration doubtless founded her son's magnificent self-confidence, but it cannot have been easy to live with, and once he had entered adulthood it threatened the very autonomy it had established. Goethe's relationship with his mother cannot be presented just as an idyll. After he had left Frankfurt, Goethe, in thirty years, visited her only three times ever again—although a lively correspondence continued between them—and she was never seriously invited to Weimar: that was *his* world, and it shows the strength in what might otherwise seem an over-demonstrative character that Catharina Elisabeth accepted the parent's lot of being left behind.

Motherly letters from Frankfurt urging the importance of wrapping up warm in winter or 'the infuriating fatherly tone' in political discussion (as Goethe is reported to have called it in one of his very few expressions of impatience with the Imperial Councillor)—they equally interfered with the son's ambition of becoming his father. This ambition, however, was unlikely to lead to psychological melodrama since it had from the start been shared and instilled by Caspar Goethe himself:

The course of my father's own life had so far proceeded more or less according to his wishes [throughout *Poetry and Truth* Goethe understates the importance for his father, and for the whole family, of the disappointments of the 1740s]; it was his intention that I should follow the same path, but with greater ease, and further. He valued my natural gifts all the more for lacking them himself; for he had achieved everything only through inexpressible industry, assiduity and repetition. He often assured me, then and later, in earnest and in jest, that given my talents he would have behaved quite differently and would not have managed them with such criminal extravagance.

The path Caspar Goethe envisaged, and from which no entreaties of his son could make him deviate, was: study of the law in Leipzig, a doctorate at one of the other universities the father had attended, practical training at one of the Imperial institutions, such as Wetzlar, and at some point the grand tour of Italy,

which had been the high point of his own life and was in his view the final
making of a cultured man (who presumably was then, like himself, ready to
enter upon marriage). This is also, with hesitations, reversals, and shorter or
longer unplanned interludes, the path that Goethe's education did actually take,
even after his father's death. By interpretative acts of genius he managed to give
a general significance, embodied in works of literature, both to his postponement of obedience and to his ultimate conformity with his father's will. In the
case of the first step, to Leipzig, the conflict of wills was not very serious:
Goethe claims in his autobiography that his own desire was to study history and
classical philology at Göttingen, the power-house of modern scholarship. But
the reason for this desire shows he was not really emancipated from his father's
scheme. He felt, rightly, that life in Frankfurt—particularly his father's life in
Frankfurt—was remote from the strongest current then running in the national
life of Germany. He wanted, so far as that was possible, to serve his nation, and
he felt, again rightly, that most of his contemporaries with the same ideal would
find their way to realizing it through study of the new subjects at the new
university, and through the kind of career to which such study was likely to lead:
'an academic teaching position . . . seemed to me the most desirable goal for a
young man who intended to educate himself and contribute to the education of
others'. But what was this desire other than his father's own hope for a life of
public service, which had been cut off by the death of Charles VII? And if his
ultimate ends did not differ from those of his father, was it necessary to quarrel
about the means? Goethe claims that he reconciled himself to going to Leipzig
by making the firm resolution to change his course once he arrived, but in the
event he did not put up much of a fight, and he kept to his legal studies for three
years—though not very diligently.

A Burnt-out Case?: 1765–1770

Leipzig, where Goethe arrived in the full turmoil of the Michaelmas Fair of
1765, was not so different a city from Frankfurt as to be totally alien, but quite
exciting enough for a young man on his own for the first time and willing to be
impressed by the sobriquet of 'Little Paris'. What made Leipzig different was,
naturally, the university—a new element for him in the otherwise familiar
atmosphere of the fair would have been the complaints of the poorer students in
cheaper, especially ground-floor, lodgings who during this fortnight had to
move out into garrets in favour of the merchants—and more important still,
though not unconnected, the fact that Leipzig was not a free city. The choice of
Dresden, seventy miles away, as the Electoral Saxon residence had left the
burghers of Leipzig a freer hand than those at Munich, for example, but the
traces of a baroque absolutist regimen could not be overlooked, and were not all
by any means disagreeable. The old city of tall houses, painted red, pale green,
and yellow, arrayed round a market-square of a spaciousness which in
Frankfurt was unthinkable, was only one-third of the size of Goethe's home

town, yet the total population was almost the same, some 30,000, for unconstricted by a city-state's proud need for military independence Leipzig had levelled its fortifications and grown outwards into elegant garden-filled suburbs. Along the site of the former walls ran avenues of trees, a promenade for the fashionable, when summer dust allowed, and in the old castle Professor A. F. Oeser (1717–99) ran the Academy of Drawing, Painting and Architecture, preaching 'the gospel of beauty, still more of the tasteful and the pleasing' (once again the phrasing of *Poetry and Truth* is utterly exact) inherited from his most famous pupil, Winckelmann. Oeser was much in demand for his pastoral, allegorical, and mythological interior decoration of public and princely buildings, and private villas. Stern Frankfurt, after the departure of the French theatre, gave only passing shelter to strolling players, such as those who, during Goethe's absence, caused so much offence in 1767 (the 250th anniversary of the Reformation) with their enactment of *The Vicious Life and Terrible End . . . of Doctor John Faust of Wittenberg* that the city council had to apologize for the performance to the Wittenberg theological faculty. But Leipzig, itself as Lutheran as Frankfurt, had the benefit of regular visits from the official troupe of the (Catholic) court in Dresden, which from 18 October 1766 had a permanent theatre-building in which to play less controversial works by such as Lessing (*Miss Sara Sampson*, and later *Minna von Barnhelm*), Lillo (*The London Merchant*), and Otway (*Venice Preserv'd*). The prices of seats reflected the troupe's annual subsidy of 12,000 guilders, and could attract students to ride over from Pietist Halle, twenty-one miles away. The absent court, in fact, was the main influence on the tone of Leipzig, on its fashions and its entertainments; the Leipzig beau was even said to have a characteristic, stilted, gait. Nor was the university, which after all was administered from Dresden, unaffected. Rabener satirically listed the main subjects studied at *galant* Leipzig as hairdressing, French-polishing, the removal of ink-stains from laundry and the cutting-out of paper shapes. Gellert's lifelong crusade for propriety certainly reflected an intellectual courtliness: a more worldly variety was that of the courtier-professor Councillor J. G. Böhme (1717–80) whose labours as a historiographer won him one of the best-known local country-houses (ceilings by Oeser) on an estate half an hour's drive out of town. Caspar Goethe, for whom nothing but the best was good enough, saw to it that Wolfgang was commended on his arrival to the care of Councillor Böhme, who discharged his first duties briskly and bluntly by refusing to hear of the proposed change of subject. The vague prospect that he could always change to the classics later if he wished was enough to save the young Goethe for the law. Reluctant permission was extracted, however, to attend Gellert's lectures on literature, and his essay-writing class. And with or without permission—though he hardly needed it, given his father's views on the subject—the lawyer, in the ample leisure hours left by between three and six lectures and one tutorial, was an assiduous pupil in Professor Oeser's Drawing Academy. Goethe drew throughout his life, his preferred media being pencil, charcoal, chalk, and ink wash—it is perhaps a matter for regret that he used water-colour relatively little. His earliest preserved drawings are reasonably executed if slightly caricatured portrait heads and some crowded theatrical scenes. Landscape makes a first serious appearance in copies of models given him by Oeser and in his few early etchings, in a conventional, nascently picturesque form. His most promising characteristic, to be seen in a few quick sketches, is a strong sense of posture and movement, and it is these which point the way to his future achievements. The old castle, where many of these first attempts must have been executed, was also an important social meeting-place, and Oeser and his family became close personal friends of the rich, talented, and well-connected young amateur.

Goethe's first year in Leipzig was excited and swaggering, timid and vulnerable, like freshman years of most times and places. He wrote home enthusiastically of the new sights—'you cannot believe what a fine thing a Professor is . . . I thirst for no honours but those of a Professorship'—and enquired of his Frankfurt friends at other universities about their experiences while boasting of his: he is getting used to the beer; he has seen Gellert and Gottsched, who at 65 has married a 19-year-old three feet shorter than he is; and as for the food, nothing so ordinary as beef and veal is now set before him, it is goose, snipe, or trout at the table of the Professor of Medicine, C. G. Ludwig (1709–73), where, in accordance with the usual practice of German universities, he has been assigned a place to dine (at his own expense). There were so many intellectual possibilities too: perhaps under the influence of all those medical and scientific students at Professor Ludwig's, Goethe decided to attend the lectures in mathematics and physics of J. H. Winckler (1703–70), where he first saw some of the classic experiments in Newtonian optics. Then there were parties, concerts, the theatre, not to mention actresses. In 1766 he made his first acquaintance with one of Leipzig's great talents, the singer Corona Schröter (1751–1802), whom he came to know personally, and also with a new genre, increasingly popular in the commercial (rather than courtly) theatre, the 'Singspiel' or melodrama, an all-German operatic form, requiring only modest resources, in which songs are interpolated into a prose dialogue. Despite his initial cautious denial of any intention of becoming a 'dandy', he disposed of his entire Frankfurt wardrobe early in 1766 and bought himself in exchange a, rather less extensive, set of more fashionable outfits. Yet at the same time the links with home remained strong: the confident exuberance of those Leipzig letters which have survived, mainly to his sister, seems not to have been reflected in others, particularly those to Horn, which he destroyed when they came into his hands in 1828. It was fundamentally a lonely year, with no established friendships, and though the Böhmes were kind to him, it was evidently a pleasure to him to be in Frankfurt in spirit by writing, as he had done for many years, a poem to be presented to his grandparents on New Year's Day 1766—even if this one had the satirical edge to be expected of a bright young thing—and by composing an epithalamium, well-stocked with mytholo-gical figures, for an uncle's wedding. While still at home he had written four acts in alexandrines of a regular, Gottschedian biblical tragedy, *Belshazzar,* and he continued the project after reaching Leipzig, though he now resolved, rather oddly, to write the fifth act in the blank verse which forward-looking authors unimpressed by Gottsched regarded as uniquely suitable for drama, and with which, in its English form, he was by March 1766 becoming acquainted in William Dodd's anthology *The Beauties of Shakespeare.* But this prolongation of Goethe's childhood loyalties was abruptly checked in May when his epithala-mium was severely criticized by Professor C. A. Clodius (1738–84), to whose class in German style, a parallel course to Gellert's, Goethe had submitted it: 'my courage deserted me', Goethe said, and he wrote no more verse for six months. At about the same time, however, J. G. Schlosser visited from Frankfurt, to attend the Easter fair and took Goethe under his wing for a while (together they paid a courtesy call on Gottsched, who donned his full-bottomed wig to receive them). Thanks to this visit Goethe made two new acquaintances who gave Leipzig a quite different aspect, and adjusting to this change may account for Goethe's six months' silence as much as the strictures of Clodius. Schlosser introduced Goethe into a circle that dined every midday at an inn frequented particularly by Frankfurt citizens and run by C. G. Schönkopf (1716–91). Goethe found the company so agreeable that he gave up his dining arrangements with Professor Ludwig and transferred permanently to the Schönkopf establishment: here he met E. W. Behrisch (1738–1809) and Anna Catharina ('Kathchen') Schönkopf (1746–1810), the daughter of the house.

The figures of Behrisch and 'Käthchen' dominated the next two years of Goethe's Leipzig existence and could be said to divide it between them. For the first time we meet in this period a peculiar duality characteristic of Goethe's whole life and indeed of his posthumous reputation. Those who are distant from him, or who know him only slightly, are repelled by what seems like his unfeeling arrogance. Those who are close to him are irresistibly attracted by the warmth and energy of an independent and unspoiled personality. To some extent Goethe made himself prominent among his 600 contemporaries in the usual undergraduate ways: he dressed extravagantly, conversed annihilatingly, took a leading part in the traditional uproar before lectures, frequented dubious taverns, and paid court to an (unidentified) society woman of easy virtue. Yet Behrisch added to this public persona a peculiar cutting edge: a courtier's contempt for the public before which he performed. For the sardonic and whimsical Behrisch, who became Goethe's closest male friend in Leipzig, was not a student at all but the tutor of Count Lindenau, the son of the Chief Equerry in Dresden, and he marked himself off from the university world by his ironically exaggerated show of exquisite taste in literature, etiquette, and fashion, and in particular, by his horror of printed books and publishing: he was a fine calligrapher and made his cult of handwriting the vehicle for a hostility, of whose social origins he was probably unaware, to the middle-class world of Leipzig, both town and gown. In one respect, of course, the freshman Goethe could hardly avoid drawing attention to himself. With a monthly allowance of 100 guilders he was at least twice as well-off as any student needed to be, even at the most expensive universities. So it is not surprising to hear judgements such as 'he was in Leipzig in our time, and a fop', 'the last thing I expected was that he would make the least ripple in the world of literature', 'his friends puffed him fearfully', 'I was not so fortunate as to belong to the number of his chosen friends, but I . . . saw many of his pranks and was often vexed by the laughter and applause with which my comrades greeted his sometimes eccentric ideas as the *non plus ultra* of wit'. Yet were it not for the influence of Behrisch it would be difficult to say how this unenviable public reputation was earned: following his father's instructions Goethe refused to gamble; embarrassed by his lankiness he would not dance; he writes once in a letter the obvious untruth that he is as drunk as a beast; and the riding and fencing lessons in the last months at Frankfurt bore little fruit: we hear only of one quite unserious duel and that he once fell, or rather leapt, from a bolting horse. The courage shown in this last incident is telling, however, and Goethe's own description to Cornelia of the life he was leading, after he had been in Leipzig for a year, shows there was more to this whipper-snapper than his fellow-students realized:

. . . my life here is as quiet and contented as it could be, I have a friend in the tutor of Count Lindenau, who has been banished from the *beau monde* for the same reasons as I. We console ourselves with each other's company by sitting cut off from the human race, as if in a fortress, in our Hotel Auerbach, which belongs to the Count [and in the basement tap-room of which, incidentally, there were seventeenth-century paintings of Dr Faustus flying out on a wine-barrel], and without being misanthropical philosophers we laugh at the Leipzigers, and woe betide them if with mighty hand we unexpectedly sally forth upon them from our castle.

In Leipzig, and in the person of Behrisch, Goethe first experienced the seductions of a courtly culture that was the personal affair of a few select individuals, elevated above the anonymous mass of the public—and the public, as Leipzig showed with especial clarity, was something created by ugly mechanical devices such as printing-presses, which, so Behrisch claimed, deformed their operators, and by the lowly workings of trade.

The other aspect of his life after Easter 1766 Goethe kept completely secret. Not even Horn, who arrived in Leipzig at that time for the start of the summer semester, was told about it, and he was soon so disgusted with Goethe's public appearance of debauched foppery that he decided to have no more to do with him. At this point, in the autumn, Goethe revealed the truth to him, and he wrote in relief to a mutual acquaintance:

We have not lost a friend in our Goethe, as we wrongly believed . . . He is in love, it is true . . . But not with that lady of whom I suspected him. He is in love with a girl beneath his station but a girl whom . . . you would love yourself if you saw her . . . Imagine a woman, well-formed though not tall, a round and friendly though not exceptionally beautiful face, an open, gentle, attractive expression, much frankness without being coquette, a very shrewd mind without having had the greatest education. He loves her very affectionately, with the perfectly honest intentions of a virtuous man, although he knows that she can never become his wife.

Horn could only conclude that the distressing figure Goethe was cutting in public was a deliberate deception, to conceal from the world his *mésalliance*. Largely he was right, but not entirely. Goethe's love for Käthchen Schönkopf, his first serious love-affair, attracted him to a woman similar both in character and status to the woman who would eventually be his wife. But like that much later relationship, his affair with Käthchen set him at odds with his social peers and with the courtly role and mannerisms to which he had subjected part of his own personality. It was as if he needed to create in himself, in his own life, an equilibrium of the forces at work in the world around him, the court, the middle classes, and the university: 'I shall divide my winter into three equal parts', he wrote to Behrisch in October, 'between you, my little one, and my studies.'

In the autumn and winter of 1766 the private, bourgeois, segment of Goethe's divided existence was happily extended by his introduction to Gottsched's publishers B. C. Breitkopf (1695–1777) and his son J. C. I. Breitkopf (1719–94) (Gottsched himself, who lived with the Breitkopfs, died in December) and to the etcher J. M. Stock (1737–73), who in early 1767 moved into the top floor of a new house the Breitkopfs were building near their old one, as an investment. In this private sphere, in the families of the Böhmes and the Oesers, the Breitkopfs and the Stocks, Goethe was received frequently and with affection, especially by the female members of the households. Frau Böhme struggled to teach him card games, and even though he frequently tempted her husband away from his work into some beer-shop, Frau Stock called the young visitor her Frankfurt Struwwelpeter and gladly combed the down out of his long brown hair (usually tied up in a pony-tail). The daughters he teased for months by always bringing sweets for their father's little dog, and never for them, in case their teeth or their voices should suffer. To the delight of Stock and Goethe, and to the desolation of the ladies, the spoiled animal finally gobbled up the sugar Christ-child in the Christmas crib. From Stock Goethe learned the rudiments of etching, and designed and printed a label for the Schönkopf firm, possibly for use on their wine-bottles. He was of course eating at the Schönkopfs' every day, and on Sundays he would go on after dinner to spend the hours from 4 until 8 p.m. in the Breitkopf house. In the comfortable homes of the very families who had suffered the burden of Prussian exactions during the Seven Years' War—the theme of *Minna von Barnhelm*, which appeared in 1767 and of which he helped put on an amateur production—Goethe came to understand the general local hostility to Frederick the Great, and to abandon for a while his rather theoretical 'Fritzianism'. Even before he had seen for himself the ruins of Dresden, bombarded in 1760, he wrote to Cornelia: 'I do believe that in all Europe there is now no place so godless as the residence of the King of Prussia.'

The task of finding a literary medium that would integrate this securely domestic leisure-time existence with his attitudinizing as a courtly rake, a cut above the student norm, was considerable, and it is no surprise that for some months Goethe's muse was silent. But from about September 1766 he began to write in a new vein: light, erotic poems, in the anacreontic style, whose subject-matter reflected, loosely, his infatuation with Käthchen, and whose cynical manner reflected, rather more exactly, his friendship with Behrisch, his striving after the courtly air, and his sense of distance from his student contemporaries. These songs 'To Annette' (alluding to Käthchen's first name) came slowly— only fifteen between September 1766 and May 1767—and meanwhile Goethe continued doggedly working on his old scheme, *Belshazzar*, which was not finished until the spring. Although even now he was planning more biblical tragedies, the shift in his taste and his growing originality became more distinct in the pastoral playlet *The Lover's Spleen* (*Die Laune des Verliebten*), which was begun in February 1767 and which, more evidently than the 'Annette' poems, drew on the emotions prompted by his relationship with Käthchen. In July 1767, rather than maintaining his practice of putting all his works of the past year together into a single volume, Goethe read out to a little conclave, consisting probably of Behrisch, Horn and perhaps one or two others, everything he had written since arriving in Leipzig, and it was resolved that it should all remain unpublished to the world, except for a dozen of the 'Annette' poems of which Behrisch would produce a single splendid calligraphical copy. In October Goethe drew the logical conclusion from this judgement, and from the earlier criticisms of Clodius, and consigned to the flames all his conveniently bound juvenilia and everything he had been writing in Leipzig except the 'Annette' poems and *The Lover's Spleen*. He regretted bitterly, he told Cornelia, that he could not include in the holocaust his very first published work, *Poetical Thoughts on Jesus Christ's Descent into Hell by J. W. G. Drafted by request* (*Poetische Gedanken von der Höllenfahrt Jesu Christi . . .*), written in Frankfurt in 1765 and printed there without his permission the following year. Though that regret may be an echo of Behrisch's hostility to publication in general, it also hints at what Goethe felt he was being liberated from as his literary past went up in smoke: poems written to commission by others as an ornament to social occasions; poems written in the Gottschedian manner, and heavy with the mythological apparatus Clodius had criticized; and perhaps particularly poems on religious and biblical themes: 'Joseph . . . was condemned because of the many prayers he said during his life . . . It is an edifying book and Joseph has nothing to do but pray. Here we have often laughed at the simplicity of the child that could write so pious a work.'

As early as 1755, Goethe tells us, the terrible earthquake in Lisbon, which cost 40,000 lives, and was marked in Frankfurt by a special day of prayer, had clouded his vision of God. He also claims in his childhood to have set up in his room an altar of mineralogical specimens to a nature deity, although the theology of the 16-year-old's *Poetical Thoughts on Jesus Christ's Descent into Hell* is still grimly orthodox. When he was first on his own in Leipzig, he says, scruples about his own unworthiness caused him to refrain from communion and he gradually distanced himself from the Church; certainly a contemporary noted that there 'he already gave evidence of peculiar attitudes in religion'. What these may have been we can learn from one of the Stock daughters who recalled how Goethe, at work on etching with their father, was often present while a desiccated clerk gave them lessons in reading from the Bible:

Now it once occurred [she continues] that we had to read aloud from the middle of a chapter in the book of Esther which seemed to [Goethe] unsuitable for young girls. Goethe had listened quietly for a little while, then he suddenly leapt up from father's work-table, snatched the Bible from my hand and quite roared at Master Clerk: 'Sir, how can you make these young girls read such whorish tales?' Our master quivered and quaked for Goethe continued his denunciation with ever greater violence until our mother intervened and sought to calm him. The master stumbled out something about everything being the Word of God, whereupon Goethe cited him 'Prove all things; hold fast only that which is pure and good'. Then he opened up the New Testament, leafed through it a little while until he had found what he was looking for. 'Here, Dolly', he said to my sister, 'read us that; that is the Sermon on the Mount; we will all listen to that.' Since Dolly stumbled and could not read for nerves, Goethe took the Bible from her, read the whole chapter out aloud to us and added most edifying comments such as we had never heard from our Master.

These 'peculiar attitudes' are clearly not those of an indifferentist, nor of a pagan, but rather of someone determined to make his own way, in religion as in other things. In one respect, at least, the *auto-da-fé* of October 1767 marked a decisive change: Goethe never afterwards had recourse to biblical subject-matter for a major literary work. Conscious or not, the decision was a brave one, a rejection of the example of Germany's greatest living poet, Klopstock, and a gesture of solidarity with Klopstock's opponent in the controversy of the 1750s over the proper subject-matter of modern German literature, the renegade, heathen Winckelmann. Any pupil of Oeser's was in 1767 more pre-occupied than ever with Winckelmann since only the previous year Lessing's *Laocoon* had presented itself as an attack on Winckelmann's entire understanding of the nature of art. The whole Academy, Goethe included, deliberately shunned Lessing when he visited Leipzig in May 1768, but later in the summer was agog at the prospect of the arrival of the master himself on his triumphal tour. (On the day the news of his murder reached Leipzig Oeser turned away all visitors.) Not that Goethe's burning of his early poems inaugurated a particularly irreligious phase in his life. On the contrary, the next three years were the only time in his adulthood when he could be said to have been nearly a Christian.

Goethe's last academic year in Leipzig began in crisis. Immediately after the destruction of his juvenilia came the farewell dinner for Behrisch who had exchanged his tutorial post in Leipzig for a more distinguished one at the court of Dessau, forty miles to the north, and there ensued a particularly tense time in Goethe's relations with Käthchen. From the letters he now wrote to Behrisch it is for the first time possible to follow in some detail the erratic course of this romance, for Goethe had said little of it in his letters to Cornelia; there is no reason, however, to assume that the story was greatly different in the preceding eighteen months. It is a story made up of much flirting in company, infrequent blissful hours alone together, paroxysms of jealousy on Goethe's side, and stormy reconciliations. J. A. Horn unwittingly identified the source of the friction when he wrote on first learning of the relationship that it was perfectly honourable 'although . . . she can never be his wife'. Even at the passionate age of 18, and quite contrary to his general reputation, Goethe took a staid, parental, view of the inseparability of love and marriage. Since Goethe was not prepared to contemplate marriage to Käthchen, for reasons that were probably not only social and financial (he was after all himself the grandson of an innkeeper), the relationship was a pointless infliction on both. 'That is how I am living,' he told Behrisch in November, 'almost without a girl, almost without a friend, half wretched; one more step and I'll be completely so.' His secret flight from the entanglement in January or February of the new year, to spend a fortnight in Dresden inspecting the paintings in the Electoral collection, in which *Laocoon* had aroused his interest, was understandable, but it set the pattern for numerous future flights, some of which would have more serious consequences. In April 1768 the painful liaison was by mutual agreement transformed into 'good friendship'. Goethe was not reconciled to the loss, and after he left Leipzig later that year, kept up a complaining correspondence with Käthchen, which grew ever more desperate as her wedding drew closer: in 1770 she married a respectable Leipzig lawyer, Dr C. K. Kanne, who later became the city's Deputy Burgomaster. Essentially Goethe was in love with love: Käthchen was for him a possibility, rather than a person. It is as certain as such things can be that at the time he left Leipzig he was still sexually inexperienced.

In these troubled circumstances it was not difficult for E. T. Langer (1743–1820), Behrisch's successor as tutor to the young Count Lindenau, to add for a while to Goethe's 'peculiar' religious attitudes a touch of specifically Christian search, if not of faith. Langer himself was no Pietist—Goethe met him, in the winter of 1767–8, in Oeser's Academy—though he was sympathetic to the less dogmatic cultivators of the inner light. His religious position seems to have been similar to, though less definite than, that of Hamann, and his belief in the unique human status of the Bible and in the need to approach the supernatural by analogy with the natural was a belief which attached him neither to the deist nor to the orthodox parties. It was also a belief to which, once he had freed it of any Christian implications, Goethe remained true, with only occasional interruptions, for the rest of his life. For the present, however, Langer saw the Christian implications as necessarily entailed, and he infected his unofficial pupil—for just as with Behrisch Goethe willingly put himself under the tutelage of the older man—with a spirit of religious quest. In 1768 Goethe was receptive to such a stimulus, suspiciously, hectically so. Almost certainly he had contracted some form of tuberculosis, possibly while still in Frankfurt, and his mental excitement grew as his physical resources were depleted. In the few days in October 1767 when Behrisch's future was being decided Goethe had written three long 'Odes to my Friend' remarkable not so much for their Klopstockian free-verse manner, complete with invocation of the friend by name and imprecations on princes (later deleted at Behrisch's urging), as for their freedom from all Klopstockian piety and their drawing directly on their author's strong feelings for a specific person—significant new steps, when the 'Odes' are contrasted with what is known of Goethe's earlier writing, and even with the poems for 'Annette'. During 1768, however, the progress was more rapid still and a number of short poems were written which combine the song-like measures of 'Annette' with the new freedom and new subject-matter of the 'Odes'. Although they have not quite thrown off the need for self-conscious anacreontic witticism, they contain for the first time a strain of melancholy—possibly influenced by English models—and serious allusions to the imminent possibility of death. At the end of July Goethe suffered a serious haemorrhage. His many friends nursed him for a month with moving devotion, particularly Oeser's daughter Friederike (to whom he presented in gratitude a little collection of his most recent poems), but it was plain that he would have to return home. On 28 August, his nineteenth birthday, he set out back to Frankfurt, leaving Leipzig in broken health without a degree. The first great climacteric of his life—physical, emotional, religious, and literary—had begun.

'You are so merry', a Saxon officer said to me, with whom I supped in Naumburg on 28th Aug., 'so merry, and left Leipzig today . . .' 'You seem unwell', he began after a time. 'Indeed I am', I replied, 'very, I have spewed blood.' 'Spewed blood', he cried, 'aye, then it is all clear to me, then you have already taken one great step out of this world and Leipzig could not but lose its interest for you, since you could not enjoy it any more.' 'You hit the mark', I said, 'the fear of losing my life has stifled all other pain . . .'

The other pain, of course, was the loss of Käthchen Schönkopf, but it is indeed not this grumbling wound, but the fear of death, and the sudden immediacy of ultimate questions, that lend the peculiarly earnest tone to the period that Goethe now spent in Frankfurt. At first things seemed to go relatively well. Goethe was installed in Cornelia's sunny rooms on the second floor and with rest the damaged lung began to heal. During his absence in Leipzig the household had established contact with moderately Pietist circles, particularly through an old family friend, Susanna von Klettenberg (1723–74), a relative by marriage of Frau Goethe. Fräulein von Klettenberg's doctor, J. F. Metz, a man respected among the Pietists for his alchemical knowledge, treated a probably tubercular lymphatic swelling on the patient's neck first with corrosives and then with surgery, a source of more discomfort than pain. Walks and outings were possible, friends from Leipzig visited, and Goethe made the acquaintance of a young Frankfurt painter G. M. Kraus (1733 or 1737–1806) just back from studying in Paris, who Cornelia hoped would paint her a portrait of her English friend, Harry Lupton. Goethe also began to visit the assemblies of the Pietists: a new direction seemed to be establishing itself, though to Langer Goethe compared his relation to the believers with the relation of Abbadona, Klopstock's repentant (but in 1768 not yet reformed) devil, to the ring of angels round Golgotha. Clear in his own mind now that he was destined, if he survived, to be a writer, Goethe spent a week in the middle of November drafting a new play, *Partners in Guilt* (*Die Mitschuldigen*), a comedy set in Leipzig: he had begun to grow aware of the possibility of freeing himself of the burden of immediately past experience by transforming it into literature. But the wound was deeper than appeared: on 6 or 7 December Goethe fell very seriously ill again and only by the end of the month did he seem out of danger. It is probable that this collapse was not a direct recurrence of the Leipzig illness but was the first dramatic manifestation of two conditions that were to accompany him for the rest of his life: a chronic latent tonsillitis and a tendency, pronounced in patients of a neurotic disposition, to the formation of oxalic kidney stones. January brought a relapse and only from the middle of February 1769 did a steady, but painfully slow recovery set in.

' "Where is he now?" ', Goethe writes of himself to Professor Oeser on 14 February 1769. 'Since August he has been in his room, which has been the occasion for going on a nice journey to the great sea-straits that all must pass through.' It had indeed been a time of extremity, a time when the finger of God was to be seen in every turn of events. In her despair Goethe's mother had turned to her Bible and opened it at the consoling verse: 'Thou shalt again plant vineyards upon the mountains of Samaria: the planters shall plant and play the pipe the while' (Jeremiah 31:5 in Luther's translation). Pious friends had been the only visitors at the youth's sick-bed, and they had been stalwart—for this Caspar Goethe, otherwise no lover of Enthusiasm, remained permanently grateful, even permitting the brethren to use his house for an assembly and a love-feast—and a mysterious panacea of Dr Metz seemed to have brought about the decisive turn for the better. It is not surprising that the invalid should see in his being spared the workings of a higher purpose: 'the Saviour has at last caught up with me', he wrote to Langer in January. But the same letter tells us that his soul is very quiet, and complains of weakness of faith. The declared Christian commitment lasted no more than eighteen months, perhaps less. The spiritual significance of this journey to the great sea-straits, however, was lifelong. 'Life always comes first: without life there is no enjoyment', the Saxon officer had told him in Naumburg, and this belief, communicated to Käthchen Schönkopf on 31 January 1769, not the redemption by the Hound of Heaven, of which he wrote to Langer, was the firm faith that was to support him in the years to come. If it was heaven's purpose it was certainly also his own that he should not, like his little brother Hermann Jakob, be swallowed up by nonentity. His two chronic conditions would see to it that his health was never unchallenged, and he never forgot that he had looked death in the face and that to whatever had saved him he owed his life, and all enjoyment. Not, however, until he had begun to seek someone, or something, other than the Saviour to thank could Goethe's return from the shadows begin to have literary consequences.

There was, admittedly, little enough sign of the enjoyment to be had from life in the year of Goethe's convalescence. From the moment of his return in the autumn of 1768 we find him complaining about Frankfurt, which is 'too much the antithesis of Leipzig to hold many attractions for him', it lacks the supreme courtly virtue of 'good taste'. J. A. Horn, his Leipzig studies also at an end, wrote in April 1769 to the Schönkopf family: 'Goethe . . . still looks unhealthy and has become very lethargic. The Imperial atmosphere has really infected him.' Moreover, this was the one year which Goethe indisputably had to live in a strained and unhappy family atmosphere. His grandfather Textor, the Schultheiss, had suffered a severe stroke while the Council was in full session in August 1768, and lingered on in a semi-conscious state until his death in 1771. More tryingly still, the first thing his mother asked of her home-coming son was mediation in the smouldering conflict between the 17-year-old Cornelia and her father who, Cornelia felt, as 17-year-old girls will, hardly allowed her out of the house. But, as time passed and one relapse succeeded another and still recovery seemed no nearer, relations between Wolfgang and Caspar Goethe deteriorated too. Wounding words were spoken both by the embittered father, whose patience was fraying at last, and by the know-all, dandified, and religiose son, coddled by anxious mother, rebellious sister, and a bevy of holy women. Yet if Wolfgang spoke sharply it was at least partly because his own weakness chafed and humiliated him: he more than once compared himself to Aesop's fox that lost its tail, a comparison which, particularly in German, suggests a sense of emasculation. 'However healthy and strong you are,' he wrote in August 1769, 'in accursed Leipzig you burn out as quickly as a poor torch. Now, now, the wretched little fox will gradually recover . . . For us men our strength is like honour for the girls, once a maidenhead is to the devil, it's gone.' And Caspar Goethe's black mood had its own complexity; it expressed not only the pedagogue's disappointment at an inglorious return from a university where he had himself been happy and successful, nor only resentment at the intrusion of religion into his household—the bitterness also sprang from a deep fear, for his own elder brothers had died at the ages of 19 and 23. It was anxiety, not vanity, that made him irascible. 1769 was the Goethe family's most difficult year. It did not, however, shift the emotional foundations: brother and sister grew even closer together ('I could not tolerate a brother-in-law', Wolfgang wrote) but their father did not withdraw his love, or at any rate the pocket-money, whose punctual and regular payment was the nearest he allowed himself to come to expressions of affection.

At first neither convalescence nor the lowering family atmosphere hindered the young Goethe's intellectual activity. He continued to read the newer literature, some of it with approval—Wieland's libertine tale *Idris* (1768) and, with reservations, Gerstenberg's *Ugolino*, for example—but with a feeling of detachment from the most fashionable movements. He followed the controversy initiated by Lessing's *Laocoon*, in which the young Herder had now intervened, and he himself wrote an essay dissenting from Lessing's view, which, however, is lost. The preparations for this essay included a visit to Mannheim at the end of October 1769—a clear indication of returning strength and spirit—in order to study the Elector Palatine's extensive collection of casts of ancient sculpture, including the Laocoon group; and the polemical burden of the argument would have been that Lessing lacked practical experience of work in the visual arts and for that reason had overstated the claims of poetry. His only true teachers, Goethe declared at the beginning of 1770, were Oeser (behind whom of course stood Winckelmann), Wieland, and 'Schäckespear', of whom he had by now read not only Wieland's translations, but also some plays (certainly *As You Like It* and *The Winter's Tale*) in the original. This highly sensual canon clearly marks him off from the ideological crusaders for a national literature, the quasi-Christians and concealed deists of Leibnizian, official culture—Klopstock, Lessing (for whom Shakespeare was a slogan rather than a living poet), even Gellert. The wildness of his republican spirit, Goethe says, has been moderated by his stay in Leipzig, but it will not be denied, and it prevents his joining in the current chorus of praise for would-be bardic and skaldic odes. They are doubtless very German but they are merely verbal gestures, lacking concreteness of imagery, thought, or feeling, their historicizing is learned and unnatural. His own literary work at this time was quite different: preparing for the press a selection of the poems he had written in Leipzig, *New Songs* (*Neue Lieder*), published anonymously in 1769 by Breitkopf, whose interest (he was the inventor of movable type for music) was in printing his son's compositions rather than the words by Goethe; and completing *Partners in Guilt*, which was redrafted in early January of 1769. But after his second relapse and his decisive reorientation towards Christianity the remarkably quick and sure touch with which he wrote the verse comedy seems to have deserted him as if his slowly reviving energies were being dedicated not to literature but to religious activity and reflection. He certainly wrote much, mainly plays, during the year of gradual recovery, but nothing further seems to have been completed and it was all burnt in a 'grand principal auto-da-fé' in early 1770. Goethe later adjudged what was destroyed 'frigid, dry and far too superficial in the respects intended to express the various states of the human heart and mind'. All the terms used in this verdict derive from the language of religious, and specifically Pietist, introspection.

There were at this time in Frankfurt, by their own estimate, some 250 'awakened souls' who foregathered in four loose groups for prayer, singing, country walks, and mutual edification. By profession Lutheran, they varied in the extent to which they attached themselves, in theory and practice, to the radical Pietist wing of Count Zinzendorf and the Herrnhut Brethren. The circle to which the Goethe household was linked was relatively speaking the most respectable, and the most wary of Herrnhut, gathered as it was round a city pastor, D. A. Claus. None the less the ecclesiastical authorities eventually forbade Claus to preach, suspecting him of sectarianism. The young Goethe's involvement with the group was a typical act of moderate opposition, stopping short of overt rupture, and expressing both the marginality of his family's relation to the Frankfurt establishment and his own dissatisfaction with the 'Imperial atmosphere' of his native city. Doubtless too there was something of the appeal of Freemasonry about these small half-secret societies scattered from town to town but held together by a network of personal recommendation. The Goethe family had several points of contact with the circle of Pastor Claus: Dr Metz; the lodger H. P. Moritz, whose brother showed Wolfgang round the Herrnhut community at Marienborn in September 1769; and Susanna von Klettenberg. This last acquaintanceship became particularly close during 1769 and was more important to Goethe than his relation with the rest of the pious fraternity. It was Susanna von Klettenberg who took it upon herself to alert him to the state of his soul, and even though he did not share her Jesus-centred piety or her conviction of her, and his, sinfulness, he looked up to her until her death in 1774 as a counsellor in temporal as much as spiritual matters: 'she was usually able to indicate the right path precisely because she looked down into the labyrinth from above and was not entrapped in it herself'. Fräulein von Klettenberg was an intelligent, cheerful, and even humorous character, but her unworldliness was deeply rooted: in her patrician birth into one of the oldest Frankfurt families, in a serious childhood illness that had permanently undermined her health, and of course in a religious temperament that caused her at little more than 20 to break off her engagement and devote herself to 'Christian friendship', a subject on which she published a book in 1754. From about 1757 she became convinced of her special relationship to the Saviour and drew closer to the Herrnhut Pietists whom she had hitherto avoided. But she preserved her spiritual mobility: she disliked Marienborn when she saw it, and in her last years she even allowed herself to come under the influence of the literary Storm and Stress. A lifetime devoted to the observation of the workings of the human soul, but observation which saw no significant distinction between emotions and acts, and which related everything it found to a consciousness reflecting about itself and its eternal well-being, that was what revealed to

Goethe that his first literary exercises were 'frigid, dry and superficial', and it is here that the inimitable, if ultimately narrow, profundity of his later works has its origin. He could not achieve that generosity of sympathy with the yearnings and doings of his fellow-countrymen of which his nature made him capable, he could not remould their world more nearly to the shape of his, unless he first opened himself to that cult of 'interiority' which in various forms, some Pietist some Enlightened, was by the middle of the eighteenth century the distinctive feature of the 'official' national culture.

Fräulein von Klettenberg could not of course reveal to Goethe the purely literary implications of such an approach to the mainstream of national sensibility. Those only became apparent to him once he had himself detached interiority from a Christian content in the fashion already exemplified to him by Winckelmann. But she could offer him an example of how to draw on a national movement without surrendering to it, of how to live on the edge of centrality. As the history of her religious half-allegiances shows, Susanna von Klettenberg was constitutionally heterodox. It was she who introduced Goethe to a new kind of reading-matter which from the spring of 1769 drew his attention away from the recent literary developments that he had hitherto followed closely: the marginalia of science, philosophy, and history, G. Arnold's uncompromisingly individualist history of heretics of every denomination, united only by their opposition to ecclesiastical establishment, his *Impartial History of Church and Heresy* (*Unparteiische Kirchen- und Ketzerhistorie*) (1699–1715), and works of magic, alchemy, and, at least reputedly, Neoplatonic cosmogony, such as G. Welling's *Opus mago-cabbalisticum et theosophicum* (1735). Goethe was briefly fascinated by this notion of a secret wisdom, a notion which inadequately fuses the ideas of general truth and individual vision. He set up his own little alchemical laboratory and toyed for hours at his desk with compasses, concentric circles, angelic hierarchies, and the fall of Lucifer. It is also possible that at this time he first conceived the idea that he might himself write a play on the magician and heretic already marked out by Lessing as a national hero, Dr Faust. If so, this sympathy for the man who turned his back on God in order to gain the world, like his speculations about the true grandeur and creative role of Lucifer, indicated that Abbadona would not much longer be seeking entry to the circle of the angels. Arnold's *History* was a powerful solvent of all ecclesiologies, even that of a Pietist conventicle, and when, on 1 January 1770, Goethe began his first diary, a commonplace-book entitled *Ephemerides*, it was not in order to follow his progress in knowledge of the Light, but to make a series of prosaic notes on his reading in which it is possible to discern an increasingly quizzical attitude to all shared, and so potentially delusive, beliefs. It is but a step from the impartial history of heresies to Voltaire's dismissal, quoted at length, of religious dogmas as 'd'absurdes chimères', among which other notes suggest Goethe would have included theories about an original sin, and the fate of unbaptized infants. A long citation of the view that God and Nature can no more be thought of separately than can body and soul, or a note that those born (like Goethe) under the sign of Virgo can expect to be adept in writing, suggest a man seeking his destiny in the natural rather than the supernatural order. A close study, in March 1770, of the reasons put forward by the Jewish Wolffian, Moses Mendelssohn (1729–98), for believing in the immortality of the soul, seems better evidence for the doubts of a man with recent experience of the mortality of the body than for the certainties of one reborn to eternal life. The physical crisis of December 1768 and January 1769 had given a special urgency to the question: did Goethe wish to live, and if so, what for? and had made it clear that any answer had to be at least as serious, at least as intimate with the inner life, as the piety of Herrnhut. But the crisis also temporarily obscured the insight which Goethe had earlier expressed to Langer: 'my efforts to become, and fairly well founded hopes of eventually becoming, a good writer are now, to speak honestly, the most important obstacles to my complete conversion'.

Wolfgang Goethe could not remain content with a life spent trifling with alchemy, drawing, etching, uncompleted plays, and the inner movements of the soul, and Caspar Goethe's patience was long since at an end. After a brief hope, immediately extinguished, that he might now be allowed to go to Göttingen, the son, by the autumn of 1769, had acquiesced once more in his father's scheme. He was to start reading for a doctorate in Strasbourg, another of Caspar Goethe's universities, and one linked with Frankfurt by a long tradition, in the spring of 1770. The excitement of the new beginning brought with it that second great burning of his juvenilia. The experiences that lay before him were to cost him his religion and by near-disaster to frustrate once more his father's calculations, but they were to define for him a new kind of life and of poetry, they were to bring to expression his denial of death, and by moving him closer to the mainstream of German literature, they were to change its course

First Writings

At the age of 17 Goethe claimed to have been writing poetry since his tenth year, and all the evidence suggests that he wrote a great deal. Yet from the little that survived the two great incinerations of 1767 and 1770 it is clear that Goethe was no Rimbaud. He only slowly emancipated himself from the literary expectations of his time, or, to be more exact, of his father's time. But the process, though slow, started very early: practically from the first his poetry was a poetry of tension. Moreover it is noticeable that only episodically was it a poetry imitative of the current national avant-garde, of Klopstock, Gerstenberg, or even Lessing. Despite the avowed, and genuine, desire to 'join in', Goethe's deeper need was to find his own way. That meant starting from the taste of his father and from the literary loyalties of an old-fashioned Frankfurt childhood, in which, apart from the exceptional case of the *Messiah,* the only modern German work to figure at all prominently was the natural children's favourite *Fastness Island* (read as a companion to translations of Homer, the Arabian

Nights, and Fénelon's *Télémaque*). Much the most important single book of course was the Bible. The notion of what constituted serious secular literature was formed by French classical drama, particularly Molière and Corneille, perhaps also, given Caspar Goethe's Italian leanings, by translations from Tasso, and by the principles of Gottsched's reform: lyric poetry should be clear, reasonable and should rhyme, dramatic poetry should be written in alexandrines, should observe the three unities, and should have a clear moral theme. Everything that the young Goethe took seriously is demonstrated in his *Poetical Thoughts on Jesus Christ's Descent into Hell,* and in the few remaining fragments of the biblical tragedy *Belshazzar,* which evidently fulfilled all the Gottschedian rules, except for the final act, which, written in blank verse, was a sign that in Leipzig the childhood certainties were no more. During his first year at the university Goethe put up with Gellert's doleful elocution for the privilege of having the vigorous Frankfurt German of his essays corrected in red ink into Saxon, and when Schlosser visited he went to pay his half-serious homage to Gottsched. But that visit was the occasion for introducing Goethe, at the Schönkopfs' establishment, to the new and unexpectedly intractable energy of sexual desire, and, coming at the time of Clodius' humiliating criticism of his poetic style, it left Goethe doubly uncertain: of his personal worth, and of his hopes of fame as a poet. The dual frustration of this search for identity appears with the explicitness that a foreign language allows in one of the young polyglot's poems, preserved in a letter written to his sister in May 1766, shortly after his first declaration of love to Käthchen. For the English reader it has at least a certain curiosity value.

A SONG OVER THE UNCONFIDENCE
TOWARDS MYSELF
To Dr. Schlosser.

Thou knowst how heappily they* Freind
   Walks upon florid Ways;
Thou knowst how heaven's bounteous hand
   Leads him to golden days.

But hah! a cruel ennemy
   Destroies all that Bless;†
In Moments of Melancholy
   Flies all my Happiness.

Then fogs of doubt do fill my mind
   With deep obscurity;
I search my self, and cannot find
   A spark of Worth in me.

* = thy.
† = Bliss.

When tender freinds, to tender kiss,
   Run up with open arms;
I think I merit not that bliss,
   That like a kiss me warmeth.

Hah! when my child, 'I love thee', sayd
   And gave the kiss I sought;
Then I—forgive me tender maid—
   'She is a false one', thought.

She cannot love a peevish boy,
   She with her godlike face.
Oh, could I, freind, that tought destroy,
   It leads* the golden days.

An other tought is misfortune,
   Is death and night to me:
I hum no supportable tune,
   I can no poet be.

When to the Altar of the Nine†
   A triste incense I bring;
I beg 'let poetry be mine
   O Sistres let me sing'.

But when they then my prayer not hear,
   I break my whispring lire;
Then from my eyes runns down a tear,
   Extinguish th' incensed fire.

Then curse I, Freind, the fated sky,
   And from th'altar I fly;
And to my Freinds aloud I cry,
   'Be happier than I'

Verses on which Goethe's own comment is: 'Are they not beautifull sister? Ho yes! Senza Dubbio.'

  The agitation of frustrated desire runs through all the surviving literary products of Goethe's Leipzig days, despite the conventionality of their form. If that conventionality is ignored, and with it the element of posturing in the poet's languid pretence to worldly wisdom and to considerably more than nineteen years, it is easy to misread the anacreontic verses of this period as merely smutty. For not Käthchen Schönkopf herself, who remains wholly invisible, but impersonal physical love is almost, though not quite, the exclusive theme of the poems which Goethe began to write, after his six months' silence, in the autumn of 1766, many of which Behrisch wrote out as the collection *Annette* in August and September 1767, and which were judged sufficiently new in style to escape the flames in October. But nearly always this love, presented directly or

---

* i.e. turns to, or lines with, lead.
† The Muses.

indirectly as the ultimately desirable goal, is either broken off by an access of virtue on one side or the other, or rendered impossible by insuperable distance, or deprived of all charm by marriage, jealousy, or weary experience. At most Amor leads us into the bridal chamber and covers his eyes in the last lines. These *erotica interrupta* resolve, or immobilize, a conflict not simply between desire and inhibition but between the impersonal constraint of a conventional genre (pornography) and a literary imagination which is striving to make of a poem the response to a personal state of the poet (in this case—frustration), striving in other words to rediscover the 'occasional poem' (*Gelegenheitsgedicht*) which Goethe in later life was to call 'the first and most genuine of all poetic *genres*' and which he claimed could only be esteemed by a society, such as Germany in the middle of the eighteenth century was not, that respected the noble and independent vocation of the poet. The other theme that from time to time appears in these earliest writings is the desire to be such a poet, to serve the nation by following in the footsteps if not of Klopstock then of lesser lights such as K. W. Ramler (1725–98) and J. F. W. Zachariä (1726–77), whom Goethe met in Leipzig in 1767, and to whom one of the poems in *Annette* is addressed. The dual 'unconfidence' of the English ode to Schlosser recurs in an altogether more complex, and more serious, form, despite the frivolous tone, in the ballad 'Pygmalion', in whose gallantries are concealed a major symbol which was to be of lifelong importance for Goethe—the sculpture for which a human love is possible—and a major theme which was to preoccupy him for the next twenty-five years—the tension between poetry and the fulfilment of desire, more prosaically, between art and marriage. This Pygmalion of 1767, having spurned women in the flesh and turned to his art, falls in physical love with his own product, and so far his story has nothing novel. But now the ballad takes two unexpected turns. No miracle occurs, the statue does not come to life, a friend instead convinces the artist that he should leave his studio for an attractive slave,

> Ein Mädgen, das lebendig ist,
> Sey besser als von Stein

that a girl who is alive is better than one of stone

and with this we might think the rococo jest is complete (Hagedorn, for example, would have finished here). But the second twist in the poem's conclusion leaves us weirdly suspended between the titillating and the didactic: the poet seems to know that life contains both pleasure and seriousness but not to know where to locate which of these, whether in art, or love, or marriage. For Pygmalion now takes what the narrator sees as a second foolish step: cured of the folly of substituting art for love, he now does not just enjoy his 'girl who is alive' but marries her. The misogynist Pygmalion is not simply punished, as in the classical story, by an unnatural or excessive passion, nor is he merely converted from a hermit to a healthy sensualist, he is punished—but with a wife:

Drum seht oft Mädgen, küsset sie,
Und liebt sie auch wohl gar,
Gewöhnt euch dran, und werdet nie
Ein Thor, wie jener war.

Nun lieben Freunde, merkt euch diß,
Und folget mir genau;
Sonst straft euch Amor ganz gewiß
Und giebt euch eine Frau.

So see maids often, kiss them, yes love them too, get used to it and never become a fool such as he. Now, dear friends, mark this, and do exactly as I say; else Love will certainly punish you and give you a wife.

On its own the amoral *pointe* might seem no more than a fashionable convention. Coming as the third possible consequence of Pygmalion's initial resolute chastity, however, it catches for a moment the emotional ambivalence of a young man who, balanced between vitality and earnestness, cannot see himself as satisfactorily fixed in any of what seem at the moment his possible future identities. In this ambivalence the first outlines are becoming discernible of a new kind of poetry, coexistent with and dependent on the indeterminacy of the poet's personal identity. The most interesting figure in 'Pygmalion' is the poem's narrator. What does the narrator want for himself? Neither the stony unresponsiveness of an art that is pure form, nor the complete sexual and social fulfilment of a state of life—marriage—no longer open to change and ambition. Both of these, to him, are comic. Somewhere in between lies a third possibility: an art that is animated by desire, but a desire that never rests in the possession of its object; a life that promises the satisfactions of beauty but never settles into the finality of a single form. Extraordinarily, the narrator already seems to know that there is a price to be paid for this third possibility, though in the ballad Pygmalion fails to pay it and is punished accordingly with a descent into the prosaic, and that price is a life of infidelity.

Goethe's poetry, in its first adolescent origins is uncompromisingly erotic: directly concerned, that is, with the most powerful source of individual volition and feeling, and struggling to relate that energy to a coherent sense of self (the narrator of 'Pygmalion' very definitely has a viewpoint—what it is is more difficult to say) and to an objective and formal context with which that self coheres (the narrator adopts the, more or less comically, didactic pose of one who wishes to share his viewpoint with his 'friends'). In so radical an undertaking he could expect little help from his literary contemporaries. Goethe's first complete play to escape burning was written as Lessing's *Hamburg Dramaturgy* was appearing, yet it shows no trace of either of the new movements with which Lessing was associated: the theoretical revaluation of Shakespeare and the growth of the new genre of 'bourgeois tragedy'. *The Lover's Spleen*, begun in 1767 and finished in the following year, is a pastoral comedy in nine scenes and some 500 alexandrines, set in a world that Oeser might have painted. The main tension in the play is unique to Goethe and identical with that in his contemporary lyric poetry: the attempt to find within an old form, considerably older than Gottschedian tragedy, an objective correlative for personal, erotically driven, emotion. The personal concerns are already closer to the surface than in the poems: in the shepherd Eridon, whose love for the shepherdess Amina is rendered unhappy by his own jealousy, it is not difficult to recognize the Goethe of the affair with Käthchen Schönkopf, if only because the source of the problem is that Eridon is ashamed to dance himself and is therefore jealous of Amina-Käthchen's dancing partners. One might think the solution would be for Eridon to take dancing-lessons. Far from it: the happy ending is brought about by a muted form of the amoral equilibrium between attachment and independence characteristic of the anacreontic poems: Eridon and Amina can live in harmony once Eridon has been seduced into infidelity too. Or, as the *New Songs* of 1768 put it:

> Du junger Mann, du junge Frau,
> Lebt nicht zu treu, nicht zu genau
> In enger Ehe.
> Die Eifersucht quält manches Haus,
> Und trägt am Ende doch nichts aus,
> Als doppelt Wehe

You, young husband, you, young wife, do not live too faithfully, too pedantically, in close matrimony. Jealousy torments many a household and does not in the end bring anything but double pain.

But the motor feeling in these lines, as in the play, is not a wincing from jealousy—that is the conventional husk—but a fear of the 'closeness', the constriction, of marriage, perhaps of any fixed state. That is why the theme of the play is Eridon's spleen, and not his lack of a dancing-master: the play itself is a first obscure attempt to find a dramatic equivalent for a mood, and that makes it, for all its archaic appearance, a more innovative work, at least in what it promises, than any play of Lessing's.

Although the *New Songs* maintain, by and large, the anacreontic manner of the *Annette* poems, they also betray occasionally that they were written up to two years later, in the shadow of Goethe's parting from Käthchen Schönkopf and of his illness, and at a time when Rousseauist phrases and Shakespearean allusions were beginning to creep into his letters. On those occasions, in 'The Night' ('Die Nacht'), for example, eros can be seen pressing on to transform itself into mood, though it then proves difficult to accommodate the poetic self which has been formulated during that process to the objective and social context, that is, to a single conventional genre—the result is an abrupt and unresolved dissonance:

> Gern verlass' ich diese Hütte,
> Meiner Schönen Aufenthalt,

Und durchstreich mit leisem Tritte
Diesen ausgestorbnen Wald.
Luna bricht die Nacht der Eichen,
Zephirs melden ihren Lauf,
Und die Birken streun mit Neigen
Ihr den süßten Weihrauch auf.
Schauer, der das Herze fühlen,
Der die Seele schmelzen macht,
Wandelt im Gebüsch im Kühlen
Welche schöne, süße Nacht!
Freude! Wollust! Kaum zu fassen!
Und doch wollt'ich, Himmel, dir
Tausend deiner* Nächte lassen,
Gäb' mein Mädchen eine mir.

Gladly I leave this hut, the abode of my fair one, and with gentle tread roam through this dead and deserted forest. Luna breaks up the oak trees' night, zephyrs announce her coming, and the birches bowing strow before her the sweetest incense. An aura that makes the heart feel, the soul melt, walks in the cool among the bushes. What a fair, sweet night! Joy! Pleasure, scarcely to be grasped! And yet, O Heaven, I would let you have a thousand of your nights if my girl would give me one.

The speaker is 'glad' to leave the hut because the physical love, which ought to be consummated in such a rustic setting but which, as always in these early poems, is unfulfilled, instead transfigures the natural landscape, which becomes the occasion of the pleasure and joy he is not allowed to feel on his girl's bosom. But the speaker is uncertain what kind of figure he cuts, standing in an erotic relation with a conventionally unerotic object: how is he to relate the self that participates in this emotionally charged encounter to his pose as an 'I' who tells an audience of his doings? What genre does his poem belong to? How is it to end? The questions are resolved by a brisk retreat into a familiar posture, which undoes the work of transference and sublimation in the first part of the poem and returns us to the simple formula of frustrated eros in which the complex mood had its origin, effectively dismissing it as an illusion. It is not likely that such an an answer will be found satisfactory for long.

The struggle to fuse a revolutionary self-consciousness with an equally powerful need for generally accessible, even generally acknowledged, forms of expression is most apparent in Goethe's letters. Goethe was one of the world's greatest letter-writers, in the vigour and inventiveness of his style and in his perfect adaptation to recipient and subject-matter. In a sense the letter, in its intimacy just short, but always short, of solitude, was the literary form most natural to him. 'My letter has all the makings of a little opus', he writes to

* Originally even more self consciously, 'solcher'—'such'.

Behrisch in the middle of a day-by-day record of a crisis in the Käthchen affair, and in *Werther* such an opus was to come into being. Throughout the first half of his life poems grew out of his letters and his letters grew into poems, and in his old age he developed a theory of literature which effectively turned all his works into letters, letters addressed to posterity. It would be possible to read his earliest surviving poems and plays and not notice the original mind working below the surface. It is impossible to read his letters of the same period and not know one is being caught up by a wholly exceptional personality, effervescent, sturdy, profound. Take for example the opening of his first letter from Frankfurt back to the Schönkopf family in Leipzig, written on 1 October 1768. Lively without being importunate it does all that social decorum requires, charmingly but not ingratiatingly, for it makes no distinction between social and personal delicacy:

Your servant, Herr Schönkopf, how do you do, Madame, good evening Mademoiselle, Peterkin good evening.

N.B. You must imagine me coming in by the little parlour-door. You, Herr Schönkopf, are sitting on the sofa by the warm stove, Madame in her little corner behind the desk, Peter is on the floor under the stove, and if Käthchen is sitting in my place at the window she only has to stand up and make room for the visitor. Now I begin to orate.

I have been missing a long time, haven't I? Five whole weeks and more that I haven't seen you, haven't spoken to you, something that hasn't happened once, not once, in two and a half years and henceforth alas will often happen. How have I been, you would like to know. Well that I can tell you, I think, middling, very middling.

By the way, you will surely have forgiven me for not saying farewell. I was in the neighbourhood, I was actually down at the door, I saw the light burning and went as far as the stairway, but I did not have the heart to go up. The last time, how could I ever have come down again?

So I am now doing what I should have done then, I am thanking you for all the love and friendship that you so constantly showed me and that I shall never forget...

This is already more mature than anything written in Leipzig itself: the letters Goethe wrote immediately after his arrival in 1765 and 1766 were markedly, almost unbearably, more excited, though equally individual. A letter of 30 October 1765 to a Frankfurt friend, with first impressions of university life, is in prose and three different verse forms, including German hexameters to satirize Gottsched's ponderosity. The letters to Cornelia are at times comically didactic, all too clearly, though often with a saving breath of irony, passing on lessons that have only just been learned by the instructor, but when Goethe writes on questions of style in correspondence he speaks already as a master of the craft. But the urge to write in verse and more particularly to write—as many of the Leipzig letters are written—in foreign languages, and even to combine the two by corresponding in French and English verse, is not simply an irrepressible virtuosity: it is an urge to flee into an extreme formalism, into what is or ought to be a totally disciplined linguistic exercise, precisely as a compensation for, or defence against, the uninhibited display of self which, when it is given its head, makes the letters so magnetic. It is not chance that Goethe's most directly personal poetic utterance of the Leipzig period, the poem in which the word 'I' is nearest in meaning to the 'I' of the letters, is 'A Song over The Unconfidence towards myself'. That directness is doubtless what makes it—apart from its linguistic inadequacies—a poor poem. What will constitute the birth of Goethe's unique literary genius (as opposed to the fluent talent that he already obviously possesses) will be the moment when self and form are not simply counterbalanced but are synthesized into a new unity in which self is expressed no longer directly but mediately, and form is no longer an alien discipline, a saving of the appearances, but is bent to the needs of the individual occasion. That moment is already near in 'The Night', in *The Lover's Spleen,* and in the letter to the Schönkopfs, which, it is worth noting, is also nearly a drama.

The return to Frankfurt in 1768 did not bring about any overt changes in Goethe's literary allegiances, such as they were. Leipzig had revealed to him the attractions of the courtly manner: it had proved an important channel for his emotional energies and despite the accretion of other influences had not lost its appeal for him, even though the literary world was now listening more to Klopstock and Lessing, Gerstenberg and Ossian. Goethe emphasized his conscious distance from the contemporary mainstream by acknowledging that the Leipzig comedy he drafted in November 1768 is in 'Louis Quatorze' style and therefore 'nowadays contraband on any Parnassus'. Yet *Partners in Guilt,* of which a second draft was probably complete by spring 1769, is very much more than a patchwork of reminiscences from Molière, Shakespeare, and Wieland, as which critics tend to dismiss it (though that combination would alone be original enough). Technically, it is one of the best plays Goethe ever wrote, though the dénouement is weak. It is an important step towards Goethe's discovery of a literary art that is at once autobiographical and objective. Indeed, in finding for the first time in the characters and circumstances of his own life the subject-matter he had been lacking since the great *auto-da-fé* of 1767 he was arguably making the most decisive innovation of his entire literary career, even if he was not yet able fully to exploit it. And, despite the alexandrines in which it is written, and other conventional elements of classical farce, the play contains moments whose contemporary realism is more vivid than that of any other German work of the age, even of *Minna von Barnhelm.* These qualities are mainly the achievement of the first act, whose addition to the play in early 1769 constitutes the main difference between the first and second drafts.

The plot of the two versions is the same. The noble libertine Alcest has arrived at a town inn to visit again his old flame, the daughter of the innkeeper, Sophie, who is now married to Söller. One night Sophie keeps a tryst with Alcest in his room but before it is too late she wrests herself from his arms.

Meanwhile Alcest's room is visited successively by the innkeeper, who is obsessed with politics and, thinking Alcest an important man, wishes to read his correspondence, and by Söller, who steals Alcest's money but believes himself to have witnessed his own cuckolding. The following day suspicion for the theft falls on the innkeeper and Sophie, but Söller, driven by his own false suspicions about Alcest, reveals the truth. The play ends with the mutual forgiveness of all the partners in guilt.

The first version, that of November 1768, is not divided into acts and begins immediately with the scene of the midnight tryst, into which, therefore, necessary exposition has to be crammed, uncomfortably at times. The play, in this form, is quite simply Goethe's revenge on the Leipzig circumstances from which he had just returned. He envisages the Schönkopf establishment as it will be when the dreaded marriage has taken place and pictures himself as the returning cavalier, Käthchen as the suffering wife and poor Dr Kanne as Söller, who is a worthless brute. The relation between Alcest the unsuccessful libertine and Sophie the virtuous assignee is a repetition of that in the anacreontic Leipzig poems, while the dubiously moral conclusion recalls that of *The Lover's Spleen*. The farcical elements are largely mechanical and there is little attempt at a depiction of the milieu.

Goethe's work on the first act in 1769, which also involved making consequential changes in the existing material, now divided into Acts II and III, is a first example of his ability to transform the nature and quality of a literary work simply by adding to it—an ability nowhere more apparent than in the sixty years long development of his *Faust*. The new first act expounds and interweaves the themes of the plot with a subtlety quite outclassing anything in the first version. The most remarkable achievement, though, is the development, or creation, of a poetic atmosphere, oppressive and cynical even, but alive. Against the background of carnival time, of contemporary public events reported in the newspapers, and of an inn, the tensions between father and son-in-law, and between husband and wife, lose the typicality of farce and take on a horribly plausible specificity. Goethe's interest in the subject has extended from the backward-looking spite of the first version to an attempt to depict, and thus free himself from, two present emotional burdens; the unhappy atmosphere of a disunited family and, by a natural association, his fear of the marriage cage. Looking two years into the future he writes to Käthchen Schönkopf: 'I shall have a house, I shall have money. My heart, what do you desire? A wife!' But if Söller were right:

> Denn zwischen Mann und Frau redt sich so gar viel nicht

For there isn't much talking between man and wife

then the young Goethe household envisaged for 1771 might be no better than the old, now, in 1769:

Dies ist nun alle Lust, und mein geträumtes Glück!

So this is all the delight and happiness I [Sophie] dreamed of!

Goethe is in this second version able to find his concerns reflected in the relation between Söller and Sophie as well as in the relation of Sophie and Alcest: the innkeeper takes on traits of Goethe's father, and Söller becomes a little more affectionate and the marriage with Sophie much more credible. The identification of Sophie with Käthchen remains, and is even reinforced by references to her age and to the dates of her affair with Alcest (the time of the action is apparently projected into the winter of 1770–1) but the farcical element in the play now functions not as a means of personal aggression (in this later version Sophie is relieved of blame for the original breach with Alcest) but as a desperate defence against the morning-after mood induced by irreversible wrong decisions.

The conciliatory conclusion of *Partners in Guilt* is in the second version of course far too lightweight for the realistic dystopia with which it begins. General forgiveness is appropriate as a conclusion only to a play which is about the interrelation of minds. But the strangest feature of the first act of *Partners in Guilt*, given Goethe's later development, is that—unlike *The Lover's Spleen*— it presents a mood without presenting a mind within which that mood exists. Despite the allusion to *Le Misanthrope* in the name of Alcest, the play is devoid of dramatically significant introspection. It is this which prevents us from regarding *Partners in Guilt* as the start of new manner: despite the unprecedented importance of its autobiographical content and its use of contemporary material to create an emotionally charged atmosphere, it still belongs to the matrix of diverse and unintegrated elements in which the mature poet had his origins. Had it been possible for Goethe to continue along this path he might have become a remarkable novelist. (He might also have become a mass-producer of voguish trivia.) That this was not a possible course for him we may deduce from two facts: that *Partners in Guilt* contains no reaction to his meeting with death; and that the writing of this utterly untranscendental play at a time when he was otherwise becoming deeply engaged with religious questions implies an extreme mental polarization which he was unlikely to tolerate for long. There is, it is true, something deathly about the finitude of the world in which Sophie and Söller live, but the force in Goethe which denied the ultimate constriction was in 1769 expressed not in his play, but in his religion.

Next to nothing is known of Goethe's numerous fragmentary dramatic writings from February 1769, after the redrafting of *Partners in Guilt*, until March 1770, which must be assumed to have perished in the 1770 *auto-de-fé*. If they lacked in their treatment of the human heart the warmth and profundity that a Pietist expected, we may assume that they continued in the manner combined out of Molière and Wieland which *Partners in Guilt* had begun, and that they did nothing to resolve the polarization of sensibility which that play implies. As the year wears on there are indications that in any resolution the world rather than the Saviour will have the upper hand: when in February 1770 Wieland is listed as one of the three 'true teachers', it is his ability as a satirist of modern ways of *thinking* that Goethe singles out for praise, as if he has already concluded that Pietism does not have a monopoly on understanding of 'the human heart and mind'. A slight poem that probably dates from this time, 'Rescue' ('Rettung'), seems even to imply that after a voyage to the great sea-straits other forms of life are possible beside the religious: a young man on the brink of suicide meets a girl called Käthchen, tells her of his pain, they kiss—'Und vor der Hand nichts mehr vom Tod' ('And for the present no more of death') is the poem's last line. Of course, the very act of burning so much that was thought frigid and superficial is evidence that great authority still attached for Goethe to the Pietist model of introspection. But for a student of Arnold's *History* there were few limits on the beliefs and attitudes compatible with membership of the true Church. Perhaps for a year Goethe was simply testing the limits of what Pietist Christianity could tolerate. If he did at this time first conceive of writing a play on Faust, it was probably in some such spirit of theological derring-do, for he would almost certainly have imagined a play which concluded with the redemption not only of Faust but of the Devil and the entire contents of Hell. The notion of universal salvation is not only to be found in Welling's *Opus mago-cabbalisticum* but, attributed to Origen, was commonly discussed by the most respectable representatives of the Enlightenment (such as Leibniz, Haller, and Lessing), and the emphasis with which the opposite doctrine is stated in his *Poetical Thoughts on Jesus Christ's Descent into Hell* suggests that even the 16-year-old Goethe already had his doubts. Almost certainly too, however, it would not yet have occurred to Goethe to associate with the story of Faust the sensational tale of Catharina Maria Flindt, which reached the Leipzig press during his student days: sentenced to death for the murder of her illegitimate child, she had been rescued from prison by her lover, but overwhelmed by her conscience had freely returned to execution. The true genesis of the peculiarly Goethean version of the Faust story is the moment when these two themes came together. That moment was still three or four years away and there is no evidence that at this early stage Goethe had committed any of his thoughts about Faust to paper.

3

Prometheus Unbound
1770–1775

The Awakening: 1770–1771

'TO ITALY Langer! To Italy!' Goethe exclaims in a letter to his mentor, written on 29 April 1770, some three weeks after his arrival in Strasbourg, from his lodgings with a furrier beside the old fish-market. 'Only not next year.' The young lawyer knows that he has come for more than a professional qualification to this, culturally the most westerly of the old German universities, and since 1679, with the rest of Alsace, administered by France: in coming here, he has taken a step into the international, French-speaking world, a first step on the educational grand tour that his father has always had in mind for him. Marie Antoinette, daughter of Empress Maria Theresa, will shortly be passing through Strasbourg on her way to her marriage with the Dauphin, the future Louis XVI, and in the pavilion erected and decorated to receive her on an island in the Rhine Goethe has had a first glimpse of Rome: tapestries woven to the patterns of Raphael's cartoons for the Sistine Chapel and his frescoes in the Segnatura. But, as so often, while sharing his father's attitude, Goethe tempers it with ambitious reluctance. 'It is too soon. I do not yet know all I need to', he continues,

there is much that I still lack. Paris shall be my school. Rome my university. For it truly is a university; and once one has seen it one has seen everything. Therefore I shall not enter it in a hurry.

Health, strength, and ambition are returning. But he is unwilling to believe that the total of what life has to offer him, and of what he has to be, is as unexcitingly near, and as readily exhausted, as next year. Recently, and it seemed so providentially, escaped from death, supported now too by an assurance of divine supervision (an assurance which is perhaps more durable than the religious convictions that are supposed to sustain it), Goethe is both questing for a vocation and reluctant to accept the obvious proposals for an identity that circumstances put to him. On the day of his arrival (probably 4 April) he opens a Pietist treasury of biblical quotations and lights upon Isaiah 54: 2–3: 'Enlarge the place of thy tent, and let them stretch forth the curtains of thine habitations: spare not, lengthen thy cords, and strengthen thy stakes. For thou shalt break forth on the right hand and on the left.' Where he is to go, what life he is to lead, who he is to spend it with, are all questions that are quite unanswered and that at 21 he is beginning to find pressing—even deliciously pressing. Someone, something, in this increasingly varied and interesting world is to be revealed as destiny, though not yet. Piously he translates this feeling into philosophy and catechizes a younger acquaintance from Frankfurt, about to embark on his studies, with the words, 'we must not seek to *be* anything but to *become* everything.'

Strasbourg, with a population of some 43,000, of which about half were Protestant, was a city larger than Frankfurt: like Leipzig, it had spilled out beyond the alleys and markets of its medieval centre, on which a French administration was only gradually imposing a more rationalist street-plan, but it remained dominated by the great bulk of the mainly thirteenth-century cathedral. The focus of Goethe's social life throughout his stay was the house in the Knoblochgasse ('Garlic Lane') where he took his midday meal with a dozen others, mainly, as in Leipzig, medical students. He became particularly friendly with one of these, F. L. Weyland (1750–85), and, after his arrival in June, with Franz Lersé (1749–1800), a poor theology student from Alsace, a good swordsman and a dry wit; and to the senior member of the dining group, a legal clerk called J. D. Salzmann (1722–1812), he attached himself, as earlier to Behrisch and Langer. The self-consciousness and cohesion of the student group were furthered by the knowledge that they were a German enclave in what was officially a French university, though the rolling fertile countryside round about was largely German in language and customs and indeed appearance. Goethe was determined, however, not to repeat the mistakes of his time in Leipzig and found it easy to make himself popular with a wide circle of acquaintance, from both linguistic groups, though his motley French, in which the language of Montaigne and Racine rubbed shoulders with the military jargon of the Seven Years' War, regularly drew down on him some coolly condescending correction. One of the best-dressed and best turned-out of students, he found room for such minor self-improvement as taking dancing-lessons at last, learning to play whist, and acquiring a head for heights by forcing himself to spend long periods on the (unfenced) topmost platform of the cathedral. Academically, the university, with about 500 students, seemed inferior to Leipzig: in the lecture rooms 'I heard nothing I did not know already', and Goethe's private tutor prepared him for the examinations by rote learning of a set of questions and answers, with most of which he was familiar from his father's instruction in 1764–5. As a result he made easy progress and, particularly in his second semester, had time on his hands: there was no regular theatre, as in Leipzig, but visiting troupes enabled him to see works of Corneille and Racine, and a German-speaking company performed the popular melodrama of *Dr Faust*, to which Marlowe's tragedy had degenerated after 150 years on the German travelling stage. Goethe's medical friends directed him to the chemistry lectures, and to the dissecting classes of the famous surgeon, Professor J. F. Lobstein (1736–84), who specialized in eye operations, and when there came a three-week break in the summer semester, at the end of June, he set out with Weyland and another law student, who had just qualified, to travel on horseback through the Vosges to Saverne and Saarbrücken, visiting not only the big houses on the way, but also the mines and industrial installations hidden in the hills and forests. In the draft of a letter from Saarbrücken on 27 June Goethe for the first time records a deep impression made by a natural landscape, by something outside the human and urban circle within which his life had hitherto been led:

Yesterday we had ridden the whole day, night was coming on and we had just come up on to the hills of Lorraine, where the Saar flows past, down in a delightful valley. As I looked out on the right hand across the green depths and the river flowed so grey and quiet in the dusk, and on the left the heavy gloom of the beech forest hung down from the hillside over me, as the bright little birds circled silently and mysteriously through the leaves and round the dark rocks—then my heart grew as quiet as the place itself and all the burden of the day was forgotten like a dream . . .

It is a moment which shows Goethe has at last made creative contact with the Sentimentalist stream in German culture: the climax of the vividly pictorial and incantatory sentence is not some further feature of the landscape, but a motion in the heart of the observer and a motion which, in respect of its subject and its object, is wholly devoid of any religious reference. What the energies are which lie stored up within that heart, and which may now be released, is suggested a few lines later, when, returning to the thought of his lost Käthchen, he writes:

When I say love, I mean the oscillatory sensation in which our heart floats, moving always to and fro on the same spot, when some stimulus or other has displaced it from its usual path of indifference. We are like children on a rocking-horse, always in motion, always at work, and rooted to the spot.

Goethe's increasingly powerful, yet essentially generous subjectivity was questing for an adequate object. If so far Strasbourg had all been too much like Leipzig, in the autumn of 1770 destiny, a vocation, did after all reveal itself.

Back in Strasbourg, 'I live rather from one day to the next', Goethe wrote, 'and thank God, and sometimes too his Son, when I may, that I am in circumstances which seem to impose this on me.' There was now a growing distance from the central figure of the Christian religion. Goethe had at first eagerly sought out the Pietist community in Strasbourg, he told Susanna von Klettenberg at the end of August, but found them so narrow and unimaginative, so ecclesiastically minded and hostile to the Herrnhut tendency, and also 'so profoundly boring when they once start, that my vivacity could not endure it'. Salzmann's rationalistic, even Voltairean, religion may also have been responsible for this alienation, and Goethe's last recorded Communion took place on 26 August. Almost seamlessly, however, the sacred enthusiasm passed over into the secular. A letter to Horn written at this time, but now lost, showed, we learn from Eckermann and from Goethe's later correspondence, 'a young man with an inkling of great things lying before him', 'but as yet no sign of a whence or a whither, an out or an in'. The vacancy left by religion in Goethe's emotional life was about to be filled, and overfilled, from other sources. He was just strong enough for the necessary academic work in moderation, he told Fräulein von Klettenberg, tactfully, if a little disingenuously, in view of his recent exertions. On 25 and 27 September he submitted himself to the oral examinations for the licentiate in law, which he passed with distinction, and was thus dispensed from attending further lectures and permitted to prepare the second stage in his course, the dissertation—which would usually be of some 20–40 printed pages—without a supervisor. The winter semester of 1770–71 promised to be an undemanding and interesting time. It also brought new faces to the table in the Knoblochgasse. J. H. Jung (1740–1817), a devout Pietist from Westphalia, later to become a well-known eye surgeon, and, under the pseudonym of Heinrich Stilling, the author of an outstanding autobiography, was on his arrival protected by Goethe, nine years his junior, from some bullying chaff, and tells us how this 'excellent man' with his 'big bright eyes, splendid forehead and fine build' naturally but involuntarily dominated the company. (Like many observers, Jung-Stilling does not refer to the asymmetry of Goethe's face, the pronounced over-development of the left of his skull, which no doubt contributed to the striking impression he made). Later in the academic year came other young men with inklings of great literary things before them: H. L. Wagner (1747–79), from Strasbourg itself, whose literary career, though brief, produced one remarkable play, *The Infanticide* (*Die Kindermörderin*) (1776), and J. M. R. Lenz (1751–92), a theologian, from the other side of the German world, from Livonia (now Latvia), tutor and companion to two young barons who were intending to enter French military service. His boyish face and shy but passionate manner perhaps already in the spring of 1771 gave a hint, not only of the most original and productive talent, apart from Goethe's, in the literary movement that was to come, but also of the madness that would destroy him. But in September and October of 1770 the challenge not merely to 'become' but actually to 'be' something crystallized for Goethe not in his fellow-students and mess-mates, but in J. G. Herder and Friederike Brion.

Herder arrived in Strasbourg on 4 September and Goethe introduced himself some time during the following week. Although he was by five years the older and in 1770 was already something of a literary notoriety, Herder had only recently begun to resolve the same personal issues that life was presenting to Goethe. The immediacy to him therefore of problems that for Goethe were incipient though still hardly conscious, the difference in age and reputation, not to mention Herder's colossal learning and his scathing contempt for any existing achievement including his own, these must all have given him an appearance of imposing maturity in the eyes of a 'sparrow-like' Goethe (Herder's own comparison) who was anyway inclined to rely on the approval and patronage of older men. A year after they had met, Goethe wrote to him that he would continue to wrestle with him as Jacob had wrestled with the angel of the Lord, but that he was prepared to become Herder's satellite, a moon revolving round his earth, if need be. The appearance of greater maturity was partly deceptive, however. The period in Strasbourg was in fact a time of decision for both men. Herder's eight-month stay there—until April 1771—

was made necessary by a protracted, painful and ultimately unsuccessful course of treatment, conducted by Professor Lobstein, for an obstruction of a tear gland, an unhappy conclusion to more than a year of aimless wandering. In June 1769 Herder had left Riga, where he was already a prominent churchman, where his writings, notably *On Recent German Literature* (*Über die neuere deutsche Literatur*) (1766–7), and *Silvae Criticae* (*Kritische Wälder*) (1769), had brought him fame as well as suspicion, and where the most tempting offers of positions both in the Church and in the school system were made to him in the attempt to dissuade him from going away. In the course of his long journey, by ship through the Baltic and North Sea to France, and then through Holland back to Germany, Herder had alternately elevated and lacerated himself with thoughts of what might be, or of what might have been, in the Riga to which he assured himself he had every intention of returning. A revolutionary educational system with a modern, practical curriculum; Livonia, with its medieval privileges liquidated, becoming, under his guidance, a loyal servant of the Russian throne; a history he might write of the whole of human culture; a literary expedition to 'strike sparks for a new spirit of literature' that would revive the whole of Germany—such, among others, were the fantasies that Herder confided to his journal during this troubled time. Egoistic though their formulation might be, their substance was a passionate, even a religious desire to serve—to enlighten and to benefit—his country and the men of his time ('disciples who regarded me as their Christ', he wrote in a letter of the youth of Riga). But what was his country? By the end of his stay in Strasbourg Herder had decided. In May 1771 he accepted an ecclesiastical position not in Riga, or in Russia, but in Bückeburg in the small north German principality of Schaumburg-Lippe. He had in effect decided that his loyalty belonged to no existing political power, not to the Prussia in which he had been brought up and which he never ceased to dislike, not the Russia of which he had such high and extraordinary hopes, nor even any grand scheme of his own. He was going to serve the community for which he had always thought and written, hitherto from its furthest geographical margin, now almost from its geographical centre—the community of speakers of the German language. It was a decision for which Lessing's move to Wolfenbüttel in 1770 may have served as a model and which in turn, we may be sure, had its effect on Goethe's thinking. One other factor should be mentioned that helped to put an end to his uncertainty: in 1770, at the Hessian court in Darmstadt, south of Frankfurt, he had met the sister-in-law of an important minister, Caroline Flachsland, who in 1773 was to become his wife. It was partly for her sake that Herder underwent the operation on his eye that kept him in Strasbourg. There was after all one respect in which Herder genuinely was more mature than Goethe.

Herder was not inactive during these months of discomfort, spent largely in a darkened room. The Royal Academy of Berlin had in 1769 announced a prize for the best essay in reply to the then much-debated question: 'Are men, supposing them abandoned to their natural faculties, capable of inventing language?'—the alternative being that language was acquired by direct divine inspiration (a view preferred by the orthodox). The theme could not have come more appositely for Herder, and his *Treatise on the Origin of Language* (*Abhandlung über den Ursprung der Sprache*), which won the Academy's prize, was completed during his time in Strasbourg. Goethe read each instalment as it was written—as fateful a meeting of minds as any that occurred in their long conversations.

The importance of language, and especially of a man's mother-tongue, to philosophical, educational, and literary theory was Herder's earliest character-istic preoccupation as a thinker and had been his particular contribution to the critical debate of the 1760s. The fragments *On Recent German Literature* share the preoccupation common for the previous forty years to German intellectuals, even when they were as different and as opposed as Gottsched and Lessing: how was a German national literature, and in particular a national drama, to be established that could compare with the literatures, either past or present, of the rest of Europe? But Herder decisively broadens the terms in which this question is asked, and that in two ways. What is at stake, he says, is not a matter of literature only, of models and rules and genres and reviews, but of the national culture in its entirety, for no literature can be separated from the peculiar and national circumstances of its origin. And therefore, secondly, Herder is concerned not only with the formal qualities of literary works, with their originality and relation to a tradition of books, but also with the personal qualities of literary men, with *their* originality and their relation to their time and nation. The factor common to both these extensions of the familiar question is Herder's interest in the national language. The spirit, or genius, of a language, he argues, is inseparable from the spirit, or genius, of a nation (as we can see most clearly in those idioms and turns of phrase which one language shares with no other and which spring directly from the way of life and familiar experience of its speakers). Language is a medium permeating all sections of the national life and imbuing them with one unique character. Therefore Herder sees the need for a literary renaissance in Germany as merely a part of the larger need for a linguistic renaissance. The awareness of the division of labour is so earnest as to be quaint:

Will it be soon that, you philosophers, by your investigations—you philologists, by your collections and criticisms—you geniuses, by your masterpieces—make your language into one that, in Pliny's words, can give 'novelty to old things, the appearance of antiquity to new . . . and to all things nature'?

Such a national language can be achieved only by national renewal; foreign models and foreign languages can, by definition, give no assistance. Drawing on his experiences in Riga, Herder castigates the German school curriculum which places such emphasis on the learning and writing of Latin. And he ridicules the ambitions of various of his contemporaries, even in writing their own language, to be the 'German Horace' or the 'German Homer', or even the 'German Psalmist'. The national literature must renew itself from its own resources—by, for example, a German version of such an exercise as Bishop Percy's recent publication of old English and Scottish ballads and folk-songs under the title *Reliques of Ancient English Poetry* (1765), or James Macpherson's purported translations of the Gaelic poetry of Ossian (1762 onwards), which despite their largely plotless emotionalism, as wild and nebulous as their Hebridean setting amid mist, bare hills, and a starlit sea, seemed, particularly in Germany, to be the authentic remains of a Nordic Bard, the equal of Homer.

What is true of literature as a whole is true of the individual writer, for 'an original writer, in the sublime sense of the ancients, is . . . always a national author'. There is no fact about a writer more important than his national language, the living language of his immediate world: can someone be a Pindar or a Horace in a dead language, or a Shakespeare in any other than his mother tongue?' In passages which Goethe, reading them in July 1772, found, as he told their author, 'inspiring with the warmth of a sacred presence', Herder links the unity that exists in true poetry between thought, feelings, occasion, and expression with the necessity, for a poet, of cultivating his native language: precisely because the language is his own he is at liberty to be original in his treatment of it, he 'must remain true to his roots if he would be the master of expression'.

Goethe met these thoughts eighteen months before he read them in *On Recent German Literature,* for they are developed, and developed in a direction even more plainly appropriate to Goethe's condition, in the *Treatise on the Origin of Language.* The time in Strasbourg was to become for Goethe the time of the origin of his own poetic language, the time, as he tells us in *Poetry and Truth*, of his personal dedication to German as a literary medium. Unlike Goethe, Herder in 1770 had a purpose; he knew what it was that 'the age demanded'—original works of German literature—but his purpose was grander than that demand: it was the service of the nation through service of its language, and service of its language by men of outstanding inventive power. This confident missionary for a half-formed cause could illuminate and inspire the unformed young, in Strasbourg as in Riga. No doubt only Goethe could have led German literature along the path that it took in the 1770s, but at a decisive moment it was Herder who showed Goethe the direction in which that path might lie, and showed him too that it was a path that he personally could profitably follow. In his *Treatise* Herder 'answers'—if that is the correct word—the Academy's question by arguing that humanity and rationality and linguistic capacity are synonymous terms, the human race and human language come into existence together, man simply *is* the speaking animal and is not conceivable, as human, without speech. Man cannot think about the Nature that surrounds him in his primal state except through a system of signs ('Merkmale') furnished initially by the sounds of Nature herself. The first words are interjections, exclamatory imitations of natural sounds. The significance of these exclamations, however, is to identify an action, not a thing—if a man imitates the bleating of a sheep he is imitating the making of that sound, not describing or referring to an object—so the ultimate and original part of speech is the verb:

No Mercury or Apollo need be brought down from the clouds like machines at the opera —*the whole polyphony of divine Nature teaches Man his language and is his Muse! . . . sonant verbs are the first elements of power* . . . The first vocabulary was collected together from the articulate sounds of the whole world . . . The child does not denominate the sheep 'a sheep' but 'a bleating creature' and thus makes the interjection into a verb.

Language is not only the most human and rational thing there is, therefore, it is also both itself intrinsically active and the most immediate bond between man and the overpoweringly active natural world about him. And so it is of the essence of language that it should change and develop. Specifically, as the means of communication between groups of men (initially, within families) it will tend to differentiate itself into separate languages as those groups mark themselves off from one another. But within the groups too language is created anew 'with every original author, yes with every mind that brings a new note into society', for what is language but 'a sum total of the operation of all human souls'? A single language is on the one hand a tradition that embodies the labours of all who have spoken it before, and on the other hand it is renewed and kept in existence by the inventive power of those who learn, and adapt, it now. The first man, the first speaker, 'began to invent, we have all invented in his train' ('haben ihm nacherfunden'). The 'chain of culture' that joins together the nations of the human race is a chain whose links are originalities. Herder's theory of language both justifies the propagation of a national culture, embodied in a national language, by giving it the status of participation in a natural and universal human cause, and at the same time encourages individual linguistic and literary spontaneity as the means by which this great cause is served.

Alsace was a good place in which to discover this dramatic expansion of the ambitions of official German literature. Every day Goethe saw before him, in the Gothic masterpiece of Strasbourg Cathedral, a monument to a religious and architectural tradition, shared by millions and over centuries, which his own age none the less scorned or ignored, a tradition waiting for an individual to call it back to life. Out in the country he could still find, relatively unaffected, the German language and customs that in the city were coming under pressure from the ruling French institutions—a language also awaiting its individual saviour? How far Goethe had been won over by Herder we can judge from the tours he undertook at Herder's suggestion in the first half of 1771 to collect old songs and ballads still being sung in rural Alsace. The upshot of these expeditions (which make Goethe one of the very first field-workers in German folklore) was a set of twelve songs collected from 'the oldest of old dears', whose grandchildren, the tale is a familiar one, knew only fashionable numbers of no regional interest. Goethe sent his trophies to Herder with the words, 'they were intended for you; intended for you alone; so even my best companions have not been allowed to take copies however insistently they have asked'. The same letter promises the despatch of translations of Macpherson's Ossian, on which Goethe was then working. The acknowledgement of discipleship could hardly be more explicit. Three of the ballads from Alsace formed the nucleus of the German component in Herder's collection of folk-song from many lands, which began publication in 1778. Some were also published by Matthias Claudius in the *Wandsbeck Messenger*—an indication that the Herderian expansion of official literature came very close to unofficial opposition.

There was an element of sheer kindness to an invalid in Goethe's almost daily attendance on Herder during the winter of 1770–71—other members of his student circle came, but he was much the most regular and, for all that Herder passed no compliments, much the most welcome—but it was also a most important part of his education: effectively a private course in modern literature and criticism from the best-informed of all conceivable tutors. (Herder however was too formidable a critic for Goethe to discuss his own poetical production with him, and when in April Herder travelled to Bückeburg via Darmstadt and Frankfurt and called on the Goethe household he was surprised to learn how extensive it was.) Goethe was introduced to the work of Hamann and Möser; Shakespeare was read at length and Wieland's prose translation criticized by a scholar more polyglot even than Lessing; Herder encouraged the reading not only of Ossian, but of the *Edda*, in translation, and shamed Goethe into trying out his Greek again on Homer, though the progress he made he owed to Samuel Clarke's parallel Latin version. But it was probably in the indeterminate frontier territory between literature and religion that Herder's influence was most decisive. He always sat loosely to Trinitarian doctrine and during his time at Bückeburg would struggle unsuccessfully to reconcile his universalist cultural sympathies with orthodox claims for the historical uniqueness of Christ. Even had Goethe not already been drifting away from the Pietist fold, it is improbable that his tentative conversion and insubstantial spirituality would have withstood such learned seductions, and from a cleric to boot. Herder's theory of the creative individual linked by language to his national culture was a powerful counter-model to the Lutheran pattern of individual election and salvation within the Church, and his rejection of all supernatural explanations of the origin of language—and so, *a fortiori*, of supernatural explanations of any less mysterious features of human history— accorded well with Goethe's own suspicion that God and Nature, body and soul, were synonyms, rather than distinct and independent forces that occasionally interfered with each other's domains. The Bible itself, which Goethe had consulted in April as the voice of the providential power that was directing his life, had to yield before this onslaught on the miraculous: doubtless Goethe already had his own intuitions, but Herder's development of Robert Lowth's (1710–87), literary approach to the Old Testament, though published only later in the course of the 1770s, must already have had great systematic cogency in its treatment of the Scriptures as the work, not of one divine finger, but of many human hands, a poetic achievement comparable to Homer's. The substitution of literature for religion as the medium of Goethe's self-understanding was achieved in 1770 and 1771 with remarkably little theoretical fuss, since for Goethe neither his profession nor his livelihood was at stake, but rather the question: by what means was he to express his marginal centrality, his loyal opposition to Germany's developing national culture? Herrnhut Pietism, deeply though it could engage the self, had proved imaginatively inadequate: now Herder was offering an alternative of seemingly inexhaustible diversity. In later life Goethe laid particular weight on Herder's propagation of his own favourite modern authors, Swift, Sterne, and Goldsmith, whose *Vicar of Wakefield* he read aloud to his visitors in instalments in December 1770 and January 1771. The 'irony' which Goethe found particularly in Goldsmith and Sterne was essential to the most important lesson he believed they had to teach him: that it is possible to write works of literature which incorporate the author's own feelings and experiences, but which, taken as a whole, as a 'poetic world', express an 'attitude' which transcends such accidents of life and fortune. The theoretical formulation belongs of course to the later Goethe, who admits that in 1771 he could not account for the magnetic power of Goldsmith's book, but the admission may stand as evidence that Goethe at 21 was coming to sense in the very process of literary representation a completely secularized manifestation of the autonomous and invulnerable self on which Leibnizianism and Pietism equally focused their attention and one which might enable him both to share in the passions of his age and to stand reflectively aside from them.

In his autobiography Goethe deliberately, even flagrantly, uses *The Vicar of Wakefield* to cast an idyllic glow over the story of his love for Friederike Elisabeth Brion (1752–1813), the younger daughter of the pastor of the village of Sesenheim, about a day's journey out of Strasbourg. (Goethe writes 'Sesenheim' though 'Sessenheim' is more correct.) The country setting, the antiquated parsonage at Sesenheim, the slightly eccentric (though not actually unworldly) father, the good-hearted mother, taking visiting young men to her bosom, the beautiful but practical daughters wearing the old-fashioned peasant costume, Goethe himself as the sophisticated and perhaps unreliable suitor from outside, all make the parallel an obvious one. Chronologically, however, the association, as *Poetry and Truth* describes it, is impossible, since Goethe was introduced to Sesenheim by his friend Weyland, whose sister was married to Frau Brion's half-brother, in the middle of October 1770, at least two months before Herder began his readings from Goldsmith, and there is no contemporary evidence that the parallel occurred to any of the parties (though Goethe may himself have made it, *in pectore*). Goethe's entire account in *Poetry and Truth* of that first meeting has been shown to be a fabrication, and much of the subsequent detail is as misleading as one would expect from that most misleading of autobiographies, but perhaps less so than some later speculations about the affair. The indubitable facts are that immediately after meeting Friederike, Goethe on 15 October 1770 felt impelled to write a clever and charming letter to his 'dear, dear friend', expressing the assurance that they would meet again; that in November and December they did; that on Goethe's initiative—not, as *Poetry and Truth* implies, Friederike's—a correspondence began; that there may have been further visits to Sesenheim in the first months of 1771; and that from 18 May to 23 June of that year, Goethe stayed with the Brions, possibly preparing some architectural drawings of the parsonage, to help in its restoration and extension. From the few letters that survive, mainly from Goethe to Salzmann, from the later recollections of the Brion family, and from the more reliable parts of *Poetry and Truth*, we have an impression of a self-sufficient, competent, humorous, even rather earthy, 18-year-old girl, and of a highly intelligent law student—with no public reputation, of course, and socially on a par with a clergyman's family, something of a town-cousin, in fact —making himself popular by his liveliness and helpfulness, his impromptu fairy-stories about knights and princesses told to little gatherings in the barn, and his remarkable ability to turn all mealtime conversation into verse; and we glimpse names cut into the bark of trees, a picnic by the Rhine broken up by swarms of mosquitoes (Goethe, like many people resentful of intrusion, seems to have had a particular dislike of biting insects), and a Whit Monday festivity at which Goethe, despite having a cough, and despite an illness which has laid Friederike low, dances from two in the afternoon until midnight (the Leipzig days when he was embarrassed by his own ungainliness are well and truly over). A bizarre anecdote is told which may stand for many concerning the next few years of Goethe's life, indicating as it does the combination of verve, humour, and imperiousness (and generously managed wealth) with which he was to conquer affection and loyalty to the point of devotion, and which no doubt had its effect on Friederike.

From time to time in November, as every year, the lads of the village [Sesenheim] made a bonfire behind the church and enjoyed themselves leaping over it with poles. 'Herr Goethe' too was present on one such occasion and noticed among the spectators six women with old and tattered straw-hats. He told Farmer Wolf to throw the straw-hats on to the fire. He did so at once, only one refusing absolutely to give up her hat. Now when the five straw-hats were blazing merrily, Herr Goethe got out his purse and gave four guilders each [a total therefore of nearly a quarter of his monthly allowance] to the five women, who were standing there fairly bemused; at this the sour looks were transformed into jubilation; now the sixth voluntarily offered her hat; when Goethe paid her no attention at all, full of vexation, she threw her hat into the fire herself, but for this heroic expression of dissatisfaction she earned only the ridicule of the whole village. Farmer Wolf, who told me this story, added as its moral: So fine a gentleman you just have to trust.

The mood of exhilaration in Sesenheim could change rapidly however—'It is raining outside and in, and the horrid evening breezes are rustling in the vine-leaves round my window, and my *animula vagula* is like the little weather-cock over there on the church-tower', he wrote to Salzmann in June 1771, 'about turn, about turn, it goes on all day'. At the start of that long holiday in early summer he was unsettled by a feeling that his own thoughts or behaviour were not worthy of the generosity being shown him—'the *conscia mens*, and alas not *recti*, that follows me around'—perhaps for that very reason he threw 'the whole *me* into the dancing' on Whit Monday, and by the end he was acknowledging that to get what you want is not always to get what you expected: 'Are not all the dreams of your childhood fulfilled? Are these not the enchanted gardens for which you yearned?—They are, they are! I know it, dear friend, and know that one is not a jot happier when one attains what one wanted. The little extra! the little extra! that fate dispenses for us with every happiness' (to Salzmann, 19 June). It is likely that Goethe paid a last visit to Sesenheim in August 1771 and only made it plain that there would be no further meeting by means of a letter from Frankfurt, after his final departure from the university that month. Friederike had perhaps still not recovered from her illness, and after the shock of this parting went into a serious decline; although her health was eventually restored, she never married.

The episode, it has to be said, reflects little credit on Goethe. The long visit to Sesenheim in May and June, and the virtual certainty that then, and perhaps on earlier occasions, he and Friederike were left alone together for long periods, make it very likely that he was by then regarded, in accordance with south German custom, as Friederike's fiancé, and this was evidently a role that, by his behaviour, and perhaps also by his words, though admittedly not by any ceremony, he had brought upon himself. Breaking with Friederike, in these circumstances, seriously compromised her standing—to say nothing of her feelings—and Weyland, who later set up a medical practice in Frankfurt, refused to have anything more to do with Goethe afterwards. It was a betrayal, and in comparison with that fact about the relationship it is insignificant what degree of sexual intimacy the partners had achieved before they separated: complete or not—and the exertions of the Brion family in the nineteenth century to have one of Goethe's letters to Salzmann destroyed unprinted tell their own tale—it is most unlikely that, as has at times been asserted, Friederike bore an illegitimate child. Had that been the case (and there is absolutely no evidence for the assertion), it would be difficult to imagine Goethe returning to Sesenheim and spending a night in the parsonage, as in 1779 he did, now famous, and a Privy Councillor, and being received as he appears to have been.

He describes in a letter the warmth of his welcome, how he gathers that he is still talked about from time to time, how the neighbours are called in and it is agreed that he looks younger, how together they go over in recollection the good times of eight years before and look out various mementos. There seem really no grounds for questioning Goethe's own rather sober assessment in this letter: it had all been a painful and unflattering incident, but one that people more affectionate than he, and less interested in self-esteem and its loss, could find forgivable:

The second daughter in the family had loved me once, better than I deserved, and more than others on whom I expended much passion and devotion, I had to leave her at a point where it almost cost her her life, she hardly referred to any sickness that still lingers on from that time, she behaved so kindly and with so much genuine friendliness from the very first moment when we unexpectedly came face to face and nose to nose on the doorstep that I felt quite at home.

Throughout the first part of 1771 Goethe's distance from Christianity continued to grow. Only 'Providence', not Christ, is mentioned in a letter of condolence he wrote to his grandmother after the death of Schultheiss Textor on 6 February, and Jung-Stilling, whose earnest and simple faith perhaps made it clear to Goethe how little they shared, felt that he was the only 'enthusiast for religion' in the 'circle' made up of himself and Goethe, Lenz, and Lersé, who all 'thought more freely', though were not 'mockers'. While in Sesenheim, Goethe of course attended the services conducted by Pastor Brion, and probably kept most of his thoughts about the sermons to himself. If so, he was not being hypocritical, but consistent with his general intellectual development towards a complete religious individualism. His work on his dissertation, which his father, anxious to see more of his son in print, had been urging him on to write, began seriously in December and took on more of a theological than a legal character, reflecting both what he had been learning from Herder and his own ever freer thinking. It was complete by the spring: a study, as far as can be told, for no copy survives, of the secular origins of ecclesiastical law, of Judaism, and probably of Christianity itself ('no more than sound politics', according to one report), issuing in the far from ingenuously Lutheran principle that religious observance is entirely a matter for the State to regulate, while religious belief is entirely private. This deliberately outrageous—but not necessarily original—piece of Enlightened demythologization was rejected by the university as too heterodox to be published under its name. Goethe had rapidly to arrange to obtain his licentiate instead by the alternative means of public disputation on a set of previously printed Latin theses—among which the main conclusion of his dissertation appears as No. 42.

It is difficult to know how seriously to take the theses as a guide to Goethe's legal and political thinking at the time. Their purpose was after all to give the proponent the opportunity to show off his forensic skill; Goethe was no doubt glowing with self-satisfaction at outwitting the university establishment (and his father) by forcing them to suppress a dissertation which he could now threaten to print privately; one of his opponents at the disputation was his crony Lersé, who is reported to have enjoyed turning the screw; and in *Poetry and Truth* Goethe comments that the disputation, on 6 August, was conducted 'with great good humour, indeed with frivolity'. Certainly, given Goethe's university career hitherto, thesis No. 41 must have been propounded tongue-in-cheek ('The study of law is by far pre-eminent'). Many of the theses relate to technical issues which would have permitted a serious show of the candidate's competence, but there are also several fundamental principles of jurisprudence raised in a form which requires him to defend the position of an extreme 'Fritzian' absolutist: 'The good of the State should be the supreme law'; 'Not tradition, but the interest of each nation is the basis of the law of nations'; 'The Prince is sole legislator, and sole interpreter of the laws.' Yet the prevailing tendency of all these propositions, it will be seen, is to reduce the pretensions of the legal profession (or of the Church), under an appearance of the most loyal service of the State, and with such a half-mischievous intention they can be taken as only a partial representation of Goethe's views. Not a single thesis touches on the feudal law which in Germany determined who or what was 'Prince' or 'State', and to which Goethe's reading of Möser must have given new actuality. The most significant feature of the more fundamental theses is that they show Goethe giving substantial thought to currently controversial matters relating to absolute government, and the same can be said of the social and criminal issues on which he was prepared to speak: on two that were being energetically discussed he defended conservative positions ('Capital punishment should not be abolished'; 'Slavery is a part of natural law'); on two he merely asserted that the matter was open to argument (the judicial use of torture and the capital punishment of infanticides). An air of the prank hangs over the whole examination.

Though Goethe thus immediately acquired the public reputation of a half-baked Voltairean know-all, 'with something missing (or too much) in his upper storey', the university authorities seem to have treated him with unusual benevolence, and there was even talk of an academic future for him if he wanted it. But things had changed in the six years since he had first thirsted for such honours in Leipzig—*inter alia*, he had met Herder. Surrendering the prospect of proceeding to the doctoral degree (and in Germany convention anyway bestowed the title of 'Doctor' on holders of the licentiate), and taking an inconclusive and not wholly honest farewell from Friederike, he left Strasbourg on 9 August. After a second visit to Mannheim to view the casts of ancient sculptures in which Herder will have revived his interest, he was back once more in Frankfurt on the fourteenth, feeling happier, healthier, and more successful than on a similar occasion three years previously, and a fortnight later, on his birthday, he petitioned for leave to plead as an advocate at the

Frankfurt bar. In leaving Alsace, in dropping any plans for further study in Paris, in returning to Frankfurt, his 'fatherland' as he called it in his letter of application to the bar, in taking up immediately the family profession, Goethe was following a lead already given by Herder. He was turning away from the polyglot, international world of Strasbourg and the grand tour, and turning towards his mother-tongue and the tradition into which he had been born and which it was the task of gifted men to reawaken. In the autumn and winter of 1771–2 Goethe remained in close contact with Herder by letter and by December he was writing to Salzmann that he was heartily sick of his legal role, which he was keeping up only for the sake of appearance. In all he handled only twenty-eight cases in nearly four years, much of the work being taken from his shoulders by the clerk J. W. Liebholdt (1740–1806) whom he shared with J. G. Schlosser and who was paid for by his father. It is clear that from early on Goethe's career as a lawyer was far less important to him than his vocation to serve, probably as a writer, the cause that Herder more than anyone had defined for him: the Frankfurt courts certainly thought so, issuing a formal reprimand to him in 1772 for the rhetorical and inflammatory style of his submissions. (In a letter to Langer of 1773 Goethe attributed his adoption of 'learning and the arts', rather than of professional life, to the state of his health, always a good excuse, it would seem, when writing to the religious.)

The exuberance of the last weeks in Strasbourg and the return home seems to have continued for another two months or so. During September Goethe, still in regular correspondence with Salzmann and his 'circle', arranged that on 14 October, in the Protestant calendar the name-day of all Williams, a celebration of Shakespeare's memory should be held simultaneously at his family's house in Frankfurt and among his friends in Strasbourg. In Frankfurt the numerous guests were entertained with speeches on Shakespeare and Ossian, supplied by Goethe, and with drink and music, supplied by his father, at a cost of over six guilders. At about the same time he began to plan a drama of his own on Julius Caesar, and acquired a copy of the autobiography of the sixteenth-century Franconian robber-baron Gottfried (Götz) von Berlichingen, which also filled him with the enthusiastic desire to write a play, though his sister Cornelia mocked him for never putting down on paper the vivid scenes he retailed to her from his imagination. But in the middle of October Goethe's mood changed abruptly. Once back in Frankfurt he had written the perhaps rather carefree letter that had put an end to the relationship with Friederike Brion, and he would appear to have imagined that the affair would have an after-life rather like that of his liaison with Käthchen Schönkopf, in which he could continue to write to his good friend, telling her of his new interests and feelings, sending her fourteen pages of translations from Ossian, and even the occasional melancholy poem about their separation. At some time after the Shakespeare celebrations, however, he received her reply to his original letter of rejection, and with it perhaps the news of the breakdown of her health, and he was shaken as never before: 'now for the first time I felt the loss she had suffered and saw no possibility of making it good, or even of merely alleviating it . . . worst of all, I could not forgive myself . . .'. A continuation of the correspondence was impossible, and in these very weeks there was acted out before Goethe's eyes in Frankfurt an example of what might have happened—what, in terms of the emotions involved, had happened—to the woman who had loved him too well and whom he had abandoned. In the second week in October an infanticide unmarried mother, Susanna Margareta Brandt, a serving-maid at a Frankfurt inn, was being interrogated about her crime in preparation for her trial. She was imprisoned only 200 yards from the Goethes' house and her public execution on 14 January 1772 brought the city to a standstill. In all stages of the proceedings relatives and acquaintances of Goethe were involved—she was periodically examined by Goethe's own Dr Metz, and Schlosser acted for the executioner, who thought himself too old to be certain of decapitating the criminal at one blow and passed the responsibility on to his son.

With his sudden insight into what he had done to Friederike, and the hideous caricature of its moral implications being daily impressed on him, there began for Goethe what his autobiography calls a period of 'sombre remorse'; 'here for the first time I was guilty'. With the admission of a guilt that could not be remedied, or even alleviated, Goethe entered territory hitherto unknown to the moral sensibility of the rationalist Enlightenment, for which sensual desires were but an obscure form of rational desires, disappointment an obscure form of fulfilment, and the only ultimate evil temporary misunderstanding. This was not how love affairs ended for the Swedish Countess or Miss Sara Sampson. On the other hand, Goethe had now consciously detached himself from the Christian Saviour who atoned for the irremediable guilt of helpless men. Goethe was now alone, and in the darkness he had to find his own way. He had done so once before, after a fashion, when he had released himself from oppression by his Leipzig experiences through transforming them into a play, *Partners in Guilt*. And in early November 1771 he began to repeat the cure. This time, however, the play that he wrote was as much more complex as his life and personality had become in the intervening three years. This time the play was, to an extent, autobiographical, as *Partners in Guilt* had been, and rich, much richer, in objective correlatives of mood—but it also contained, if only partially integrated, because split between two principal characters, a symbolic representation of the self, finding its way through that richly objective world. Because this was to be the pattern of Goethe's major works for the next twenty years it was a more important feature of the new play even than that most of its form and themes and material are hardly conceivable without the influence of his meeting with Herder in Strasbourg. Suddenly Goethe started to write down the scenes of which he had so far only spoken to Cornelia. Within six weeks, before the end of the year, Goethe had completed a full-length, a more than full-length, prose drama, *The History of Gottfried von Berlichingen with the Iron Hand, Dramatized* (*Geschichte Gottfriedens von Berlichingen mit der eisernen Hand, dramatisirt*). The first thing he did with the completed manuscript was to send it to Herder for his opinion: 'I did it in order to ask you about it', he wrote in the accompanying letter. But the true occasion for the awakening of Goethe's poetic genius was his meeting with, and parting from, Friederike Brion.

Life and Literature: Works, 1770–1771

The coincidence of Goethe's detachment from Christian belief with his briefly happy encounter with Friederike had momentous consequences. He felt the force of one of the most powerful currents in the national culture at the same time as he underwent a personal moral experience of such intensity that it left plain traces in the themes of nearly all the major literary works he conceived over the next decade—not until he returned to Sesenheim in 1779 was the ghost laid. This coincidence reinforced the exceptionally personal nature of his response to the religious crisis, his freedom from any concern with accommodating the state power through philosophical reinterpretation of church doctrines or institutions, his concentration instead on issues of individual faith and hope, and his association therefore of the religious motive with the force of eros, of personal desire. Once the sense of a personal vocation to literature had absorbed all separate sense of a personal religious vocation—a development which could be said to have been publicly and definitively announced with the insertion of Shakespeare's celebration into the liturgical calendar—the way was open for the peculiar, if not unique, symbiosis of life and art that characterizes Goethe's subsequent career. This pattern was first established by the events, at once personal and literary, of 1771.

Although there are no historical grounds for Goethe's assimilation, in *Poetry and Truth,* of the Sesenheim episode to the model of *The Vicar of Wakefield,* the poetic device points to a real and important feature of the affair: its artificiality. It was indeed based at the time on a literary model, though not one provided by Goldsmith: rather, the old parsonage and its attached farmlands, the Alsatian scenery and its rustic population, with its German language and habits contrasting with the French city, the Christmas preparations during the autumn evenings, and the dancing and merry-making during the summer, had the appeal of being a living image of what Herder meant by a 'people'. Here was the German nation, formed and defined by its language, here was the original and authentic culture of which Goethe collected the poetic monuments in the form of folk-songs, and in Friederike it was embodied in an individual person with whom he could fall in love. Even the few contemporary documents relating to the affair leave one wondering whether, for all her sterling qualities, the romance surrounding Friederike in Sesenheim would have survived transplantation to a lawyer's parlour in Frankfurt—and Goethe explicitly raises the question in *Poetry and Truth,* even though the family visit to Strasbourg which is there his occasion for doing so is probably an invention. Much of the passion in the affair was, according to Goethe's autobiography, generated not in Sesenheim at all but in the thirty or more letters that he wrote from Strasbourg to the Sesenheim that lay before his mind's eye. Yet Friederike was no phantom: although she and her surroundings fitted the pre-existing Herderian pattern she was a person of such charm and character that Goethe could love her, and she could and did love him. With her, perhaps even in a physical, certainly in an emotional, sense, Goethe entered briefly the enchanted gardens where dreams and desires reach a real fulfilment. With Friederike, Goethe attained what he wanted because he convinced himself, and her and her family too, that he intended to marry her, as he had never been able to convince himself, in the days before his illness when life seemed long, that he intended to marry Käthchen Schönkopf. For as long as he held to that intention Goethe was in love, not with a mere possibility, such as Käthchen had been, nor yet a literary fantasy, such as Friederike had perhaps been to start with, but with another, equal, human being. For that moment, Goethe transcended the limits of the monadic soul and stood in a moral and literary world at which most of his contemporaries could hardly guess. But with its fulfilment desire passes, and the essence of Goethe's genius, the original magnetism of his poetry as of his personality is desire—of no man, of no monad, would it be truer to say that he is not a substance, he is a force. Possession, fixity, was something which in 1771 neither his character nor his art was yet mature enough to accommodate. It is possible to account in pragmatic terms for Goethe's leaving Friederike: it must have been evident to his ambition that with a wife and family at 22, and even assuming that in thirty years' time, as his father perhaps planned, he secured his grandfather's office of Schultheiss, he would have no chance in his home town of achieving that broad experience of contemporary society and political affairs which fell to him in the event and for which he must in some sense have hoped. Goethe himself, in his autobiography, overtly offers only an explanation in terms of immediately tangible emotions: the affair was an infatuation, and burned itself out like a firework. But in the light of Goethe's gradually more definite literary ambitions we can see both why the infatuation came about and why it ended: Sesenheim appealed in the first place because it corresponded to a literary model, and Goethe shied away from marriage to Friederike because the fixity of such a commitment was incompatible with the only poetry that he could write, a poetry of continuing desire. This was in effect the interpretation Lenz adopted when in 1772 he also trod the path to the old parsonage, paid his own court to Friederike, and concluded that Goethe had sacrificed her to his genius. But this interpretation of Lenz's was itself dependent on his reading of the poems Goethe had written for Friederike the previous year and is neither psychologically nor morally sophisticated enough. It would be truer still to say that the Sesenheim affair is the first manifestation of Goethe's ability, which he would demonstrate many times over the coming twenty years, to transform, in the very moment of their occurrence and particularly by the means of poetry, incidents that in another life would be commonplace or only half-intended, into objective symbols of central truths about himself. Friederike was, in a commonplace sense, a victim—a victim of Goethe's emotional incapacity for a fixed relationship—but Goethe made her into a symbol, the real fulfilment of his poetic prophecy, in 'Pygmalion' and *The Lover's Spleen,* that the condition on which his sensibility could maintain its equilibrium, and his art could survive, was infidelity. As time went on she, and her literary reincarnations down to Gretchen in *Faust,* became a symbol of all that to which Goethe had to be untrue in order to lead the life and write the works into which he threw himself after deciding not to marry her. Whether or not there was a causal connection between that decision and the prophecies that came before it and the events that came after it, Goethe retrospectively established a figurative connection. For this habit of selecting, developing, and moulding as they happened, those features of his life which could be given significance in the light of earlier trains of thought, he later coined the terms 'symbolical existence' and 'poetic anticipation'. His life was not to be an unstructured and unintended sequence of episodes, but each major event in it, foreseen or not, was to be pondered and given its place in a newly interpreted whole. The principal, but not the sole, repository of those unceasing self-reinterpretations was Goethe's literary production.

What was the origin of this sudden access of meaning-creating power? It is surely not difficult to see it in Goethe's simultaneous and conscious detachment from religious belief. Released from pursuit by the Saviour, and so from any specific obligation to 'imitate' Christ, or to appropriate either the great symbolic acts of His Life, or the symbolic rites which the Church derives from them to articulate all lives, the self-moving monadic soul is free to define its own sacred times and places and actions, to mark the stages of its endless desire, or 'appetition', The mechanism for the construction of meaning remains that of Christianity—a life with symbolic episodes, a literature referring to that life both in prophecy and in retrospective interpretation—but it becomes available for the soul to use only if Christ is displaced from His privileged position: the soul has a meaningful life of its own only in so far as it is *not* a follower of Christ —it is necessarily antagonistic to His rival claims. The rejection, however, of so powerful and established a model creates what might be called a problem of objectivity. The significant events of Christ's life are grounded in secular history and are the fulfilment of the Law and the Prophets, while the preaching and sacraments which the Church has based upon them have an application proved in myriads of lives. What confidence can Goethe have that the meaning he finds in his non-Christian life and literature and reputation is similarly well-founded? What guarantees that his permanently renewed efforts at self-understanding do issue in truth, that he is not just telling endlessly adaptable and multipliable stories about himself, not just painting the walls of his Sentimentalist prison? Even at the time of his affair with Friederike Goethe's confidence that there is such a guarantee is expressed in his belief in binding moments of insight. It is a belief which defines the point beyond which he will not go in accepting the official, emotionally solipsistic, culture of his secularizing age, which has wished on him an ever-changing weathercock of a soul, dangerous to himself and to those around him. For a moment, if only for a moment, Goethe believes desire can meet with fulfilment and so its seemingly endless quest can be guaranteed a goal; for a moment, the monad can emerge from solitude and find its identity confirmed by its meeting with another like itself; for a moment, the busy activity of interpretation can give way to the serene contemplation of truth. Such confirmation that there is fixity, solidity, and truth in life can only be momentary, for the state that is glimpsed is contradictory of the process that leads up to it, but it is sufficient to sustain the lonely and yearning soul, when it resumes the long toil of understanding, with the hope of another moment, or the memory of the last. The moment, and its memory, is not necessarily pleasurable. The emotion which inspires many of Goethe's literary works between 1771 and 1779, and which proves that he does not think that he is alone in the world, or that the world is perfectly harmonized to his wishes, is the remorse engendered by his parting from Friederike. Its first unambiguous expression is found in *Götz von Berlichingen*, but the poems Goethe wrote for Friederike in 1770 and 1771, while they also embody one of those privileged moments when the activities of desire and interpretation are both suspended, already contain hints of its fragility.

In the autumn of 1770 it was some time since Goethe had written poems, but now they came thick and fast, mainly occasional ejaculations included in letters to Friederike, collected by the family and later copied (and interpolated) by Lenz—but in all of them a new and unconcealedly personal note, is sounded, and some of them are works of marked originality. In 'Awaken, Friederike', ('Erwache, Friederike') when it is freed of Lenz's additions (strophes 2, 4, and 5 in the longer version), we can see how the anacreontic vocabulary is transformed when its object is no longer a fantasy or a formula but a person, in this case a person oversleeping:

> Die Nachtigall im Schlafe
> Hast du versäumt,
> So höre nun zur Strafe,
> Was ich gereimt.
> Schwer lag auf meinem Busen
> Des Reimes Joch:
> Die schönste meiner Musen,
> Du, schliefst ja noch.

In your sleep you missed the nightingale so now as a punishment you can listen to what I have rhymed. The yoke of rhyme lay heavy on my bosom, for the fairest of my Muses was still asleep—you.

Because Goethe's humour teases both Friederike and himself, both parties are equally strongly present in the poem. In the slightly earlier fragment of an epistolary novel, *Arianne to Wetty* (*Arianne an Wetty*), we still find the familiar self-indulgent conception of a love that is fulfilled but does not entail personal restriction. But now at last Goethe is meeting a reality, no longer a mere possibility, a person as solid as himself who thus gives him solidity too. And when this new sense of the reality of the poetic object is applied in the new poetic world that Herder has made available, there occurs something like a revolution.

> Es schlug mein Herz. Geschwind, zu Pferde!
> Und fort, wild wie ein Held zur Schlacht.
> Der Abend wiegte schon die Erde,
> Und an den Bergen hing die Nacht.
> Schon stund im Nebelkleid die Eiche
> Wie ein getürmter Riese da,
> Wo Finsternis aus dem Gesträuche
> Mit hundert schwarzen Augen sah.
> Der Mond von einem Wolkenhügel
> Sah schläfrig aus dem Duft hervor . . .

My heart pounded. Quick, to horse! And off, wild like a hero to battle. Evening already cradled the earth and night hung down the mountains. The oak now stood there in a robe of mist like a towering giant where darkness looked out of the bushes with a hundred black eyes. From a hill of cloud the moon looked sleepily out of the vapour [at this point the manuscript breaks off]

In that series of interjections taking the place of verbs, in the Ossianic-Homeric image of the hero going into battle and in such Ossianic words as 'Nebelkleid', these few lines, perhaps from a letter, perhaps describing a ride out from Strasbourg to Sesenheim, seem to do all that the author of the *Treatise on the Origin of Language* could have expected of a literary revival. In the daring personification of the later lines 'the whole polyphony of divine Nature' is the poem's Muse. The sheer activity, the eventfulness, of the rhythm announces the peculiar intimacy with the German language that, once found, will remain in Goethe's versification for the next sixty years. Through the encounter with Friederike, Goethe seems to have learned not only that he *is* someone (if not yet who that someone is) but also that there is in the world, after all, vastly more for him to do than he had feared when he came to Strasbourg a year before. Goethe was not falsifying when in his autobiography he gave Sesenheim the prominence that he did.

But on that encounter followed withdrawal. Goethe seems always to have retained the belief that, as he put it in 1779, he 'had to' leave his dear, dear friend, and at some point, probably in 1771, though whether before or after the parting it is difficult to say, he wrote the rest of 'My heart pounded'—either a troubled retrospect or, more probably, an ominous intimation. He continues:

Die Winde schwangen leise Flügel,
Umsausten schauerlich mein Ohr.
Die Nacht schuf tausend Ungeheuer,
Doch tausendfacher war mein Mut,
Mein Geist war ein verzehrend Feuer,
Mein ganzes Herz zerfloß in Glut.

Ich sah dich, und die milde Freude
Floß aus dem süßen Blick auf mich.
Ganz war mein Herz an deiner Seite,
Und jeder Atemzug für dich.
Ein rosenfarbes Frühlingswetter
Lag auf dem lieblichen Gesicht
Und Zärtlichkeit für mich, ihr Götter,
Ich hofft' es, ich verdient' es nicht.

Der Abschied, wie bedrängt, wie trübe!
Aus deinen Blicken sprach dein Herz.
In deinen Küssen welche Liebe,
O welche Wonne, welcher Schmerz!
Du gingst, ich stund und sah zur Erden
Und sah dir nach mit nassem Blick.
Und doch, welch Glück, geliebt zu werden,
Und lieben, Götter, welch ein Glück!

the winds beat gentle wings, whistled weirdly about my ears. The night created a thousand monsters but more thousandfold yet was my courage. My spirit was a consuming flame, my whole heart melted away into fire. I saw you, and gentle joy flowed from your sweet glance upon me. All my heart was at your side and every breath was for your sake. The rosy light of a spring day rested on that charming face, and tenderness for me, oh ye gods, I had hoped for it, I did not deserve it. The parting—how oppressed, how gloomy! Out of your eyes spoke your heart. In your kisses what love, oh, what ecstasy, what pain! You went, I stood, my eyes on the ground, and followed you with moist eyes, and yet what bliss to be loved, and to love, ye gods, what bliss!

The imaginative drive of the opening lines has been rewarded with a meeting: to the flowing ardour of the poet there responds the flowing tenderness of the beloved, indeed it is more than a response or a reward, it is something whose unprompted generosity shames him into the *'conscia mens* and alas not *recti'*. His most powerful hopes have been fulfilled, despite all the self-created terrors that accompanied them (evoked in the lines he first wrote), but the very fulfilment shows up those hopes in all their unworthy selfishness. Such is the generosity with which he has been received, however, that even in the moment of parting, awkward and unfulfilled as it is, a completely shared and communicated emotion is achieved; the oppression and gloom of the moment belong equally to him and to her, and the next line is *about* communication. We note the uncertainty about the nature of the separation and even about who is responsible—both parties, then, are individual agents, but they act together. The poetry of longing has been replaced by the poetry of meeting. And yet it is impossible to ignore the decline in vigour of the poem's language as it reaches the point where dreams become reality, where the terrors of the night give way to the rosy light of spring, and that vigour is recovered only in the last two lines when separation has been confirmed and accepted. The miracle of acceptance by Friederike has clothed a hitherto naked, anonymous desire in the knowledge of its object. But even when circumstances seem most propitious, even when life's purpose seems clearest, even when Goethe has found himself in the encounter with another truly independent being and the whole world seems to share and voice his feelings, it is only for a moment that the poetry of desire can represent giving and receiving, and then the monad must withdraw, whatever the cost, into a separateness that permits either new desire, or the remorseful recollection of emotion that is past.

The assimilation of this painful moment of self-discovery to the literary model provided by Herder is most complete in a poem of which no accurate version survives. 'Rose upon the Heath' ('Heidenröslein') was written before June 1771 and printed by Herder in 1773, though only, he says, from memory (as if he knew it only from oral transmission): the version now current dates from a revision by Goethe in 1788–9, which has a significantly different conclusion. Goethe could hardly have come closer to the impossible, to writing a true folk-song. With the material he had collected in Alsace still ringing in his ears he picked up a line or two from a German seventeenth-century song-book and produced a tale so reminiscent of the terse, pictorial style of the Scottish ballads he knew from Percy's *Reliques* that Herder may actually have taken it for a folk-song with no known author, and certainly printed it as such. Yet the confused emotions of the last strophe, in which a touch of anacreontic wish-fulfilment only strengthens the sense of a sexual encounter between near-equals, without losing a guilty awareness of violence, and of pain to come, are exactly what one might expect of Goethe at a point a little earlier than when he wrote the conclusion of 'My heart pounded', say, at or shortly before the beginning of his long stay in Sesenheim, in May 1771. A personal confession here insinuates itself into the shared language and collective literary repertoire of a 'people' to the point where it is probably successfully concealed from the very man who had alerted Goethe to the possibility of that intimate union of genius and tradition. Alternatively, one could say that here the poet ruthlessly subordinates his engagement with a woman to an intellectual gesture, a passing literary fashion:

> Es sah' ein Knab' ein Röslein stehn,
> Ein Röslein auf der Heiden.
> Er sah, es war so frisch und schön,
> Und blieb stehn, es anzusehn,

Und stand in süßen Freuden.
Röslein, Röslein, Röslein rot,
Röslein auf der Heiden!

Der Knabe sprach: ich breche dich,
Röslein auf der Heiden.
Das Röslein sprach: ich steche dich,
Daß du ewig denkst an mich,
Daß ichs nicht will leiden.
Röslein, Röslein, Röslein rot,
Röslein auf der Heiden!

Jedoch der wilde Knabe brach
Das Röslein auf der Heiden.
Das Röslein wehrte sich und stach.
Aber er vergaß danach
Beim Genuß das Leiden.
Röslein, Röslein, Röslein rot,
Röslein auf der Heiden!

There was a boy saw a little rose grow, a little rose on the heath. He saw it was so fresh and fair and stood still to look at it, and stood in sweet joy. Little rose, little rose, little rose red, little rose on the heath.

The boy said: I will pluck you, little rose on the heath. The little rose said: I will pierce you, so you always think of me and remember that I will not allow it. Little rose, little rose, little rose red, little rose on the heath.

Nevertheless the rough boy plucked the little rose on the heath. The little rose resisted and pierced him. But afterwards in the pleasure he forgot the pain. Little rose, little rose, little rose red, little rose on the heath.

The most remarkable fruit of the Herderian concept of popular literature, which at the same time contained Goethe's first extensive deployment of the motif of remorse, was the play he wrote in November and December 1771 so that he could ask Herder about it, *The History of Gottfried von Berlichingen with the Iron Hand, Dramatized.* Herder's opinion, expressed in a letter which has not been preserved, seems to have been favourable but caustic—too much Shakespeare, too much 'merely thought'—and he agreed with Goethe's own feeling that it needed a 'radical rebirth' if it was to come to life. (In fact when the play was published in June 1773 it was tidied up, but not greatly altered.) At first the modern English reader may be puzzled by these criticisms: there seems little of thought, and not much more of Shakespeare, unless it be of *Henry VI,* in this confused tale of battles and burnings, sieges, oaths, and poisonings, gypsies, monks, and secret trials in south-west Germany in the early sixteenth century. But Shakespeare, at the time translated only into German prose, meant more than the multiplicity of scenes and plots in which he most obviously differed from French classical, and indeed from French or German bourgeois, drama, and to which Herder's negative criticism no doubt refers. Shakespeare is also for Herder, as we can see from an essay on the subject that he was writing at this time, the model of the writer who makes a national literature by drawing on the nation's traditions about its past. It is specifically by analogy with Shakespeare's 'histories' that Goethe's play originally carries the word 'history' in its title, and as a tableau of late medieval German life *Götz von Berlichingen*— Goethe changed the title in the course of revision—has indeed performed a function for German audiences similar to that of Shakespeare for English playgoers who like the Duke of Marlborough find in him 'the only English history of those times that I ever read'. In seeing *Götz von Berlichingen* as a Shakespearian play at all, Herder was seeing it in the light of that discovery of folk-literature and national culture, 'characteristic' culture, to use Goethe's name for it, which he had himself communicated to Goethe in Strasbourg. In the very setting of the play, in Franconia, well east of the Rhine, and so unadulteratedly German, but in the same latitude as Strasbourg, in middle or 'individualized' Germany, as the sociologist Riehl calls it, we can see the mark of the Alsatian experience. The discovery of the nation, of its present, and its living past, as the subject-matter of literature was the most revolutionary feature of German drama in the 1770s and it was largely Goethe's doing. Klopstock's commitment to 'the history of my fatherland' took on a wholly new meaning, more suggestive and challenging by far than the concentration on the semi-fabulous barbarian period of Arminius. An alternative was created to pastoral, mythological, chivalric, classical, or religious themes, an alternative which might in different circumstances have founded a tradition of realistic novel-writing. (Indeed, in different circumstances it did, through the agency of Walter Scott, one of whose first exercises as a writer was the translation of *Götz*.) At the same time, if at first, in *Götz*, more tentatively, the parallel discovery that the self also can form the subject-matter of modern literature came to the fore, displacing, for a more secularized generation, the 'sacred history' which for Klopstock had been the complement to the national. The revolution, moreover, that bore the name of Shakespeare in its banner was not simply literary. Goethe's interest, in the early 1770s, in the history, language and art-forms of the German sixteenth century, in the Germany of Luther, Faust, and Paracelsus, of Hans Sachs and Dürer, of the street-ballad and the woodcut, is explained not only by his own search for the modern material of literature, or by the influence of Herder's theories, but also by a desire to identify and return to a German bourgeois cultural tradition independent of courtly absolutism and its intellectual buttresses, Pietism and the Leibnizian Enlightenment.

But Shakespeare had yet another meaning for Herder and Goethe at this time—he also meant Poetic Genius, and this fact can ultimately help to explain the criticism that much of *Götz* is spoiled and that it is 'merely thought'. In Goethe's address to the Shakespeare festival on 14 October Shakespeare is presented as first and foremost a great human being whose memory is immortal, unlike that of most of us ephemeral creatures, and even when his works are alluded to, their main merit is to demonstrate the 'promethean' nature of the 'colossus' who created them. These are similar terms, not only to those used on occasion by Herder in his own essay on Shakespeare, written at this time, but also to those that appear in the contemporary sections of Goethe's panegyric of Master Erwin, the architect of Strasbourg Cathedral, which appeared with additions at the end of 1772 under the title *On German Architecture (Von deutscher Baukunst)*. In both these prose hymns of Goethe's the link seems in danger of being severed which, in Herder's theory of language, binds the inventive individual, the original talent, to the tradition that he renews. There is, in the early sections of *On German Architecture*, little about the cathedral itself, near which Goethe had lived for over a year, and little in the 'Shakespeare Day Speech' about the content of Shakespeare's plays, most of which by now he had read. Indeed Goethe describes the plays as being essentially about 'characters', who are themselves would-be colossi, that is, would-be Shakespeares, frustrated by fate and the course of history, when they reach 'that mysterious point (which no philosopher has yet seen or located) at which the particularity of our Self, the supposed freedom of our will, collides with the necessary course of the whole'. While this sheds little light on Shakespeare, it suggests much about *Götz von Berlichingen*, which was preoccupying Goethe at the time and which he was about to begin writing. Clearly this concept of colossal genius has deeper roots than Herder's theory of culture, which in accommodating the concept has merely established a *modus vivendi* with a powerful and alien force. These roots lie close to those of 'the particularity of our Self' and will concern us further. For our present purpose of examining *Götz von Berlichingen* what is important is that the notion of Shakespeare as the colossus, creator of characters aspiring to immortality but bound for destruction, is in conflict with the notion of Shakespeare as the inheritor and revivifier of his own local tradition and local language. Genius, which in Herder's theory balances and fructifies folklore, now threatens to obliterate it.

Götz von Berlichingen is a sprawling work, generously and exhilaratingly so if one compares it with the fussily monothematic plays that predominate in German official literature from Lessing to Hebbel and beyond. Indeed, only in his *Faust* does Goethe himself recapture its combination of an open dramatic form (fifty-nine scene-changes in the original version) with a unifying poetic vision—in that sense, in its openness to life, in what Emil Staiger calls its lack of frame, it is genuinely Shakespearean. In private, though not to Goethe, Herder expressed the view that there was no modern drama to compare with it. None the less the criticism of disunity has some substance. There is an obvious respect in which *Götz* is two plays rather than one. There is the first, the main, story of Götz himself, a historical figure (iron hand and all) whose memoirs were Goethe's source. Götz is an Imperial knight, a territorial lord owing feudal allegiance directly to the Emperor alone, even though the land over which he rules consists simply of his castle of Jaxthausen and little more besides. The period of the play is just before and just after Luther's revolt against the Church in 1517, Luther himself appearing briefly as Brother Martin, a young monk who recognizes in Götz a 'great man', and who, it is perhaps implied, goes on to become one himself. At this time south-western Germany was fragmented into scores of such minuscule states and Goethe presents them as fighting a losing battle for their independence, their 'Territorialhoheit', against the bureaucratic centralized states which grow by swallowing up their smaller neighbours— these are typified in the play by the ecclesiastical state of Bamberg, but their model is obviously the eighteenth-century secular absolutisms such as Prussia. Indeed the political problematic of the play is throughout that of Goethe's time, rather than Götz's: as in the constitutional theory of J. J. Möser, to whose work Herder had introduced Goethe in Strasbourg, the power-hungry autocracies appear as late arrivals and perversions of the natural order. It is they who win the day nevertheless. The scenes chronicling Götz's political defeat make up a tableau rather than a plot, but they are lively and richly varied. We see Götz in the modest domesticity of his castle, we see the respect and loyalty he inspires in his retainers and in his few subjects, to whom he dispenses natural and uncontroversial justice. We see him joined in his feud with the city of Nuremberg by other Imperial knights, his old friends (the most loyal of Götz's companions-in-arms goes by the name of Goethe's Strasbourg acquaintance, Lerse). We see, in a series of particularly vigorous scenes, the siege of his castle, the treacherous breach of his safe-conduct when he surrenders, and the renewal of the fight. The most famous scene in the play, which has made its title into the German equivalent of the *mot de Cambronne*, shows Götz slamming shut the window through which he has listened to the offer of peace-terms, after bellowing at the herald the words: 'Tell your captain: I have as always due respect for the Imperial Majesty. But he, tell him, can lick my arse.' Götz, however, who had been protected by the Emperor, himself no lover of the greater powers nominally subject to him, has eventually to give his word to keep the peace. All might be well, did he not allow himself to be involved in the Peasants' Revolt—the excuse for more picturesque theatrical turbulence— with the intention, he says, of preventing violence, but he cannot control the mob and, bereft now of his honour, he is confined to his castle. Here eventually (the play conceals the fact that for the historical Götz the process took nearly forty years) he pines away, having in the absence of freedom nothing left to live for.

This, however, is only half the play. There is what amounts to a second protagonist, Adelbert von Weislingen. His story has one important political aspect but is otherwise essentially private, and the scenes that retail it are largely closet drama. Weislingen, a friend of Götz since boyhood, is an Imperial knight too, but he has betrayed the fellowship of Götz and his allies by accepting service at the episcopal court of Bamberg, thus exchanging his independence for greater power and a wider field of action. It is he who receives the Imperial commission to bring Götz to order; at one stage he holds Götz's death-warrant. Weislingen's drama is thus largely personal; the pull of friendship towards Götz and loyalty to his own past is countered by his political ambition and above all by his passion for Adelheid, a beauty of the Bamberg court and the most fatal of *femmes fatales*. Weislingen is briefly reconverted to independence by Götz, who appeals to his better nature and their boyhood memories, and he engages himself to Marie, Götz's sister. Not content with successfully tempting Weislingen to break this engagement and marry her instead, Adelheid, who subsequently wearies of him, seduces both another of Götz's fellow knights and, at the same time, Weislingen's squire, whom she eventually sets on to poison his lord. She herself is condemned for her crimes by the secret Vehmic Court and made away with.

Weislingen's indecision and Adelheid's intrigues make up a sensational tale, thin and rather timeless, which might seem to assort ill with the robust and colourful chronicle-play which carries the political theme centred on Götz. Yet it is not ultimately detachable from Götz's story, and it clearly meant much to Goethe. When he asked Salzmann to send a copy of the published version of the play to Sesenheim he commented: 'Poor Friederike will find herself a little consoled when the unfaithful [Weislingen] is poisoned.' Weislingen's betrayal of Marie, clearly intended as a reflection of Goethe's own act of betrayal, repeats at the personal level that political infidelity to all Götz stands for for which there are perhaps also parallels in Goethe's own life. Götz and Weislingen are in a sense different aspects of the same person, representing different possible reactions to the same threat to the way of life in which they both were brought up: Götz resists that threat and Weislingen collaborates with it, just as Goethe himself was coming to sense the attractions both of Herderian opposition to official culture and of Sentimentalist co-operation with it. Both Götz and Weislingen suggest, therefore, different ways in which 'the particularity of our Self' may 'collide with the necessary course of the whole': Götz through identifying himself with the resolute defence of a particular, externally and concretely defined, political interest; Weislingen through abandoning external consistency in order to preserve the continuity of an inner identity. Integrating these two representations of the Self, one from without and one from within, is the great unsolved problem of the play. At a narrative level a unity of a kind, which may have a certain symbolic appropriateness, is achieved through the figure of Adelheid, who is responsible, directly or indirectly, for the downfall of both protagonists. In *Poetry and Truth* Goethe tells us that the Weislingen plot bulked so large because he had himself fallen in love with Adelheid—once again a useful but deceptive hint. Anyone less lovable than Adelheid it is difficult to imagine, she is the quintessence of the designing female. And this is surely her significance, if we are looking for biographical significance in the play. She bears out the picture of womankind drawn in a weird song sung by a group of gypsies at the start of the fifth act—she is a werewolf, and if Goethe, unlike Götz and Weislingen, has preserved his identity, it is because, however reluctantly, he has broken through her toils and escaped from her. One only hopes that Goethe had not clearly seen the caricature he was drawing in this figure when he sent the play to Friederike.

Götz von Berlichingen contains, then, in addition to a wealth of material drawn from Goethe's social and personal experience, not one but two conflicting representations of possible attitudes of the Self towards it. Although this is a quite decisive innovation, if we compare *Götz* with *Partners in Guilt*, there is a disunity in the play at a level more profound than the merely narrative, and the clue to it is given by a disharmony of dramatic language. On the whole Götz's story is 'shown' rather than 'told'—the characters do not tell us what they are doing or feeling, we see them doing it and we deduce their feelings—in so far as they are important—for ourselves. One notes how often some secondary physical action goes on while the characters talk mainly of something else, playing chess, pouring wine, melting lead for shot, for example: they are not on their best behaviour, they are being overheard. This 'documentary' method is most apparent in the way the characters speak. Consider this scene from Act III, brief but complete, set in the castle kitchen at Jaxthausen (itself probably an unprecedented locale in serious drama) once the siege has begun:

GÖTZ: Du hast viel Arbeit, arme Frau.

ELISABETH: Ich wollt, ich hätte sie lang. Wir werden schwerlich lang aushalten können.

GÖTZ: Wir hatten nicht Zeit, uns zu versehen.

ELISABETH: Und die vielen Leute, die ihr zeither gespeist habt. Mit dem Wein sind wir auch schon auf der Neige.

GÖTZ: Wenn wir nur auf einen gewissen Punkt halten, daß sie Kapitulation vorschlagen. Wir tun ihnen brav Abbruch. Sie schießen den ganzen Tag und verwunden unsere Mauern und knicken unsere Scheiben. Lerse ist ein braver Kerl; er schleicht mit seiner Büchse herum; wo sich einer zu nahe wagt, blaff! liegt er.

KNECHT. Kohlen, gnädige Frau!

GÖTZ: Was gibt's?

KNECHT. Die Kugeln sind alle, wir wollen neue gießen.

GÖTZ: Wie steht's Pulver?

KNECHT. So ziemlich. Wir sparen unsere Schüsse wohl aus.

GÖTZ: Poor wife, you've got a lot of work.

ELISABETH: I wish it was going to be for long. I can't see us holding out long.

GÖTZ: We didn't have time to get in provisions.

ELISABETH: And all the people you've been feeding since. Our wine's already low, too.

GÖTZ: If only we can hold out to a point so they offer terms. We're really getting at them. They shoot away the whole day and damage our walls and dent our windows. Lerse's really got guts; he creeps around with his gun; if anyone ventures too near, smack, and he's down.

SERVANT. Coals please, mistress.

GÖTZ: What's up?

SERVANT. We've run out of bullets, we're going to cast some more.
GÖTZ: How's the powder?
SERVANT. So so. We're being careful and rationing our shots.

By contrast with the language of the Storm and Stress dramatists Müller and Klinger, for whom Götz was something of a source-book, this is neither particularly abbreviated and breathless nor particularly violent and coarse. And by contrast with the language of the English Romantic drama that directly or indirectly owed much to German models (such as Wordsworth's *The Borderers* or Keats's *Otho the Great*) it is not at all archaizing. It is a condensed but contemporary colloquial language: some phrases ('Was gibt's?', 'Die . . . sind alle') could quite well be heard in Germany today. This effect, secured by an easy sentence-length, an undemanding (but not non-existent) syntax, and an absence of metaphorical adornment, is a quite remarkable achievement in an age when the dramatic norm was still plays in French or German alexandrines. The gift of rhythm, made to Goethe in Sesenheim, shows itself here in prose. Such abbreviation as there is suggests the phonetic vagueness of careless speech rather than an attempt to force the emotional pace. Such peculiarities of vocabulary as there are ('zeither' for 'seither', 'verwunden', 'auf der Neige' for example) suggest either those shades of dialect that are so important in giving spoken German a welcoming and personal sound, or Herder's beloved 'idioms', or simply a lively, though not reflective, appreciation of what is going on. They do not suggest effort, either poetic or archaeological.

As a contrast we may take an extract from a scene in Act II, in which, though it is comparatively long, only the recently married Adelheid and Weislingen appear:

ADELHEID. Scheltet die Weiber! Der unbesonnene Spieler zerbeißt und zerstampft die Karten, die ihn unschuldiger Weise verlieren machten. Aber laßt mich Euch was von Mannsleuten erzählen. Was seid denn ihr, um von Wankelmut zu sprechen? Ihr, die ihr selten seid, was ihr sein wollt, niemals, was ihr sein solltet. Könige im Festtagsornat, vom Pöbel beneidet. Was gäb eine Schneidersfrau drum, eine Schnur Perlen um ihren Hals zu haben, von dem Saum eures Kleids, den eure Absätze verächtlich zurückstoßen!
WEISLINGEN. Ihr seid bitter.
ADELHEID. Es ist die Antistrophe von Eurem Gesang. Eh ich Euch kannte, Weislingen, ging mir's wie der Schneidersfrau. Der Ruf, hundertzüngig, ohne Metapher gesprochen, hatte Euch so zahnarztmäßig herausgestrichen, daß ich mich überreden ließ zu wünschen: Möchtest du doch diese Quintessenz des männlichen Geschlechts, den Phönix Weislingen zu Gesicht kriegen! Ich ward meines Wunsches gewährt.

ADELHEID. Denounce women then. The unthinking gamer bites and stamps the cards to pieces, that innocently made him lose. But let me tell you something about menfolk. What are you all anyway, to talk of fickleness? You, who are rarely what you want to be, never what you ought to be. Kings in festal robes, envied by the mob. What would a tailor's wife give to have a string of pearls round her neck from the hem of your cloak contemptuously kicked by your heels.

WEISLINGEN. You are being bitter.

ADELHEID. It is the antistrophe to your song. Before I knew you Weislingen, I was like the tailor's wife. Rumour, hundred-tongued, and that is no metaphor, had so cried you up like a mountebank that I let myself be persuaded into wishing: Would that you could get your eyes on this quintessence of the male sex, the phoenix Weislingen. I was granted my wish.

In the later version of the play Weislingen becomes more passionate and the form of address becomes more familiar (changing from 'Ihr' to 'du'):

WEISLINGEN. Könntest du mich lieben, könntest du meiner heißen Leidenschaft einen Tropfen Linderung gewähren! Adelheid: deine Vorwürfe sind höchst ungerecht. Könntest du den hundertsten Teil ahnden von dem, was die Zeit her in mir arbeitet, du würdest mich nicht mit Gefälligkeit, Gleichgültigkeit und Verachtung so umbarmherzig hin und her zerrissen haben—

WEISLINGEN. Could you only love me, could you only grant my burning passion one drop of relief! Adelheid, your reproaches are most unjust. If you could guess at the hundredth part of what has been working in me all this time you would not have torn me up hither and thither so mercilessly with complaisance, indifference and contempt—

The greatly increased length and complexity of the sentences, including rhetorical repetitions and accumulations, the elaboration of rather artificial metaphors, even to the point of impenetrability, and when these run out, the recourse to the threadbare ('my burning passion'), the self-consciousness in the vocabulary about what one is saying, and above all, about what one is feeling (culminating in Weislingen's allusion to its incommunicability)—all this does not suggest the language of a different social stratum in the same world as that which contains Götz's kitchen, it suggests a different world, a different play, altogether. The odd phrases that establish continuity with Götz's world ('let me tell you something about menfolk') are uncomfortably out of place. The disparity of languages is even greater than that between the earlier and later lines of 'My heart pounded'. How on earth is this abstract, introspective, moralizing drama of feeling and identity to be related to Götz's struggle at Jaxthausen against wealthier neighbours, superior forces, and shortening commons?

Götz von Berlichingen is the script of two possible dramas, with two possible heroes, corresponding to two possible roles Goethe could envisage for himself in German life: the saviour of the nation, champion of Germany's best, oldest, and most creative traditions; and the man of feeling, inadequate to his task, a traitor to the cause, to his friends, and to himself. The duality in the plot mirrors the cultural dilemma of a generation: the exceptional originality of Götz's story, without precedent in form or in subject-matter, but with its subterranean links to the burgeoning European art of the novel, has somehow to be related to Weislingen's drama of vacillation and self-doubt, with its links to the domestic *drame* of 'official' German literature, in which so much is 'told', so little 'shown'; indeed Weislingen himself bears some resemblance to Mellefont, the seducer-hero of Lessing's *Miss Sara Sampson*, and Adelheid's speech to that of the Countess Orsina in his *Emilia Galotti*. Goethe's solution to a problem at once personal and national is not wholly convincing, though it correctly foresees that in the coming conflict it is the party represented by Götz which will have to yield ground.

Goethe's solution is to make Götz's story itself somewhat problematic, to introduce into it—though the process bears all the marks of being instinctual rather than deliberate—reflective elements more characteristic of the Weislingen plot. To begin with, Götz's political position is shown as anomalous, anomalous incidentally in precisely the same way as Caspar Goethe's 'Fritzianism'. Götz's enemies are not only the despots and their courts—any Frankfurt citizen could approve of that—but also the self-satisfied, mean-minded merchants of the free city of Nuremberg—which might equally well be Frankfurt. These two hostilities are not in themselves incompatible, however: if we accept that Goethe felt stifled by the Imperial atmosphere of his native city not because of its constitution but because of its stagnation, its displacement from the centre of national life, we can even see foreshadowed in *Götz von Berlichingen* and the movement it briefly inspired the combination of emotions characteristic of nineteenth-century German liberal nationalism. But no such distinct ideal as that of a unified non-autocratic Germany emerges, even by poetic implication, from the struggle of a Götz whose political allegiances are as phantasmal as Caspar Goethe's cult of the person rather than the government of Frederick the Great. Jaxthausen is far too rudimentary an institution to represent any political or social interest. Götz is clearly fighting for his independence but it is not at all clear what he intends to do with his independence when he has it, except continue fighting. The social ideals he professes are those of peace and good-neighbourliness—it is because they do not, in his view, respect these that Nuremberg and Bamberg are his enemies. But in the case of the merchants of Nuremberg, good-neighbourliness seems to consist in allowing occasional baggage-trains to be plundered for sport by roving Imperial knights such as Götz. Out of this considerable political and moral confusion the last act endeavours to help us. In the slaughter and burnings and social turmoil of the Peasants' Revolt we see Götz's free-booting existence carried to an obviously unacceptable extreme. Götz's career is finally broken, his honour is lost, and his loyalty to the Emperor, which has hitherto been his only concrete political conviction and the mainstay of his self-esteem, is subverted by a logical extension of the principles on which he has lived and which have given strength and purpose to the story told in the first four acts. This has, and is presumably intended to have, the effect of making Götz's ultimate tragedy, death as an enfeebled man who has, largely, destroyed himself, surprisingly like that of Weislingen.

The play ends with a denunciation of the age, and the posterity (that is, the eighteenth century), that has no room for a Götz. Such a denunciation would have been a natural conclusion to a chronicle that showed an obviously good, or at least justifiable, political cause—such as that of the survival of the old Empire —being defeated by overwhelming external force. Götz's story would then have ended like the siege of the Alamo, or the burning of Njal. Alternatively, even in the absence of a recognizable political cause, the dying Götz could have gained our sympathy had he remained to the end a resolute outsider whom we none the less recognised as a morally intolerable anachronism.* In this case we would have to endorse, not denounce, the social order that cannot accommodate him. In either case, however, whether we found ourselves giving our approval to Götz's stand against his society, or to society's stand against Götz, we—and that means the play with all the means of theatrical and linguistic excitement that it commands—could expect to be concentrating on Götz's external relations with his time and on the tableau of the age of which Götz is a part, as indeed happens in the first four acts. We should not expect to be concentrating on Götz's 'character' and its internal contradictions. But that is precisely what, in the last scenes of the play, we find ourselves forced to do.

If we can apply Herder's criticism 'merely thought' to the Weislingen intrigue, it ought to apply to the last stages of Götz's life too. The sense of that criticism may be that the interest in genius, the Self, the 'colossal' character, takes over from an interest in the presentation of national and local life and the cultivation and extension of national traditions. Weislingen consciously seeks greatness, Götz—as the admiration of Brother Martin or Lerse is intended to show—unconsciously embodies it. But when the language of the characters goes over from forceful colloquialism into rodomontade it is not the substance of greatness that is predominating (the model of Shakespeare's plays) but the mere aspiration to greatness (the model of Shakespeare the man). With that stylistic shift goes a shift in the action, from an emphasis on external conflict, which we are 'shown', to an emphasis on internal self-contradiction, about which we are 'told'. It is a shift from a play centred on Götz to a play centred on *both* Götz and Weislingen, for in the end Götz's story too is brought within the compass of the Weislingen theme. A kind of unity is thereby achieved. *Götz von Berlichingen* comes nearer than anything Goethe has so far written to a fusion of the personal with the objective, and of Goethe's idiosyncrasy with the dominant official culture, yet the fusion is not perfect. The Götz-intrigue of the first four acts is told with the exultant objectivity of the second version of *Partners in Guilt*, though the material is, in terms of the literary politics of the time, of a truly revolutionary novelty. The Weislingen intrigue overshadows the play's conclusion with an alliance of the themes and forms characteristic of official, 'bourgeois', domestic drama and the theory of genius. Yet the two intrigues

---

* This is the solution adopted by John Arden in his very free, but atmospherically authentic, adaptation of the play, *Ironhand*.

remain largely separate, as do the two 'heroes'. Unable to organize his play around a national, publicly shared, identity which would either endorse Götz's revolt or reject it (it does not matter which), Goethe adopts the exactly contemporary solution of Lessing in *Emilia Galotti* and attempts a reorganization round the hollow centre of his supposed individual 'interiority'. But Götz himself is the embodiment of opposition to all the social and political factors which have allowed that hollow centre to form. His opposition to Weislingen is not terminated by allowing Weislingen to take over his soul. The play's final collapse into self-consciousness is not a synthesis of these opposites but an image of their continuing, unresolved, conflict.

For all its unfinished business, *Götz von Berlichingen* is the work in which the 22-year-old Goethe's talent emerges from mere precocity into unique and lasting achievement. By the end of 1771, though practically no one yet knew it, he had changed the prospects of German, and indeed European, literature. He had also established the pattern by which, with and without the involvement of the public, he would live and write over the next two decades, and in some respects for even longer. Fundamental to both his life and art would remain, as it had been since his earliest Leipzig erotica, an exceptionally powerful and unremitting desire for an adequate object, never completely identified, always more or less absent. But in 1771 the crucial decision had been taken—or had gradually made itself apparent—that although this search was to be undertaken within a culture moulded by Christianity it would be without Christianity's intellectual and moral support. The place of a life filled with Christian meaning, such as the Herrnhut Brethren led, would be taken by a life led as a quest for a perfect, this-worldly fulfilment, punctuated and interpreted by symbols generated out of the material of the life itself. A symbolical existence would gradually develop in which names and dates and places, and weighty decisions and events, would by association and coincidence become replete with a meaning drawn from and given to this one life, possibly by analogy with Christ's symbolic actions and the Church's calendar, but not contributing to, or derivative from, the Christian scheme. Already in 1771 28 August and 14 October had been hallowed in this way. Sesenheim had been transfigured into an image of literary fulfilment and Friederike had become not merely a victim but, in her fate, a symbol of unresting desire. At the same time a new art would emerge which could, in a highly refined sense of the term, be called autobiographical: literary works drawing their meaning, not from some external stock of themes and figures and principles, not from their representation of a social or religious order, but from the interrelation of symbols, found within the works and within the non-literary material generated by the life they interpret, of the self thirsting for its perfectly adequate object. In the first version of *Partners in Guilt* Goethe discovered how literature could be made out of the crises of his own life. In the second version he discovered how depiction of the contemporary world about him could be used to create a mood. In *Götz von Berlichingen* he completed the pattern to which he would adhere for years to come by inserting into this objectively depicted but mood-filled world the dual figure of Götz-Weislingen, an image of the particular self at once both a character and a consciousness, that is that world's creative and reflective centre. *Götz,* and the poems Goethe wrote for Friederike, also however show up the conflicts and dilemmas to which Goethe was exposing himself by abandoning an external religious order of things in favour of a symbolic life and an autobiographical art. Most profoundly, there was the question of objectivity: is so autonomous a self capable of recognizing an object? Will it ever reach the mysterious point where it is checked by an external necessity? Is there a fulfilment to insatiable desire? Is there a truth beyond the symbols of life and art and to which those symbols point? To some extent these questions were also raised by the more concrete and visible conflict of cultural loyalties: where did Goethe stand in the conflict between subversive realism and subservient interiority, between the oppositional and the official strands in the literature of an increasingly absolutist body politic? And this conflict in turn raised the most practical and immediate question of all: where was Goethe to find his audience and how was he to address it? Was he to remain within the restricted courtly circle favoured by Behrisch, seeking to commend himself in manuscript to a few select, though possibly influential, patrons? Or was he to commit himself to the anonymous and bourgeois, public and commercial world of print, and if so, how? Through journals, or books? And if books, what sort of books? With the completion of *Götz von Berlichingen* Goethe had the beginnings of a literary reputation, and with that all these issues, from the most fundamental to the most immediate, became both more pressing, and more dangerous.

Between Sentimentalism and Storm and Stress: 1772–1774

Georg Schlosser was an earnest man. Ten years older than Goethe he already had an extensive legal practice, in the Frankfurt where his grandfather had been burgomaster, but his thoughts were on better things, and not only for himself. During a spell as tutor in Pomerania he had seen at first hand the condition of Germany's poorest and least educated peasantry; when he visited young Goethe in Leipzig he was already composing an *Anti-Pope,* an attack, from a moral, but anti-rationalist, Christian point of view, on the optimism of the *Essay on Man;* and in 1771 he published the first fruits of a lifetime's concern for the primary education of the poor, his *Catechism of Morality for the Rural Populace* (*Katechismus der Sittenlehre für das Landvolk*). He was already well connected in the literary world, was acquainted with Herder, and in December 1771 he introduced Goethe to another friend of his, a well-placed official—Military Paymaster—of the Landgrave in nearby Darmstadt, J. H. Merck, a 'singular man', Goethe observes in his autobiography, 'who had the greatest influence on my life'. From January 1772 the Frankfurt publisher J. C. Deinet, who lived only a few doors away from the Goethes, had asked Merck to take over the editorship of a languishing twice-weekly journal of reviews, the *Frankfurter*

Gelehrte Anzeigen (*Frankfurt Literary Advertiser*) and Merck was building up a new team of reviewers, to include L. J. F. Höpfner (1743–97), a law professor in Giessen, Herder, and Schlosser.

If Goethe attributed to Merck such influence over his own development, it was not simply because by the end of February 1772 Merck had invited him to join his corps of reviewers, who also made up a collective editorial board, this commitment to the *Frankfurter Gelehrte Anzeigen* being Goethe's principal, and probably his only, literary activity throughout the year. Merck was also the last in the sequence of acerbic older men whose intellectual patronage Goethe sought during his adolescence—'a man of leather', he called him. His wit, however, was far more cruel than that of Behrisch or Salzmann, let alone Langer, and his abilities were altogether of a different order. His was a many-sided administrative talent, fertile in progressive and charitable schemes, though with a less certain executive touch; he had both range and depth of scientific knowledge and a successful courtier's network of connections; though mainly an essayist, he also wrote verse and narrative prose and showed acute critical insight into the factors differentiating Germany's literature from that of England and France and directing its development away from the realistic novel; he was personally acquainted with Voltaire, and to his Swiss wife he always wrote in a French rather better than the usual international dialect of the day. In one vital respect, though, Merck as a mentor surpassed even Herder: his literary experience and understanding enabled him, like Herder, to recognize the quality of Goethe's writing, but his appreciation was not impaired by that hypercritical, even envious, element which made it difficult for Herder to give unambiguous encouragement. One reason for Goethe's relative inactivity as a writer during 1772 was Herder's long delay in commenting on the first version of *Götz von Berlichingen,* together with the mixed nature of those comments when they finally arrived in June or July. Merck's abrasive instruction to stop dithering and get on with it gave the decisive impetus, not only to the publication of *Götz,* but to Goethe's adoption of a literary career.

At first, however, Goethe's relations with Merck stood under a rather different sign. The Landgrave of Hesse-Darmstadt had no cultural interests, but his wife, who subscribed to the confidential and very expensive *Correspondence littéraire* by which Baron Melchior Grimm (1723–1807) kept the courts of Europe informed of literary developments in France, patronized a little circle of *beaux esprits,* among them Merck, Herder's fiancée, Caroline Flachsland, her sister and her brother-in-law, and the Crown Prince's tutor F. M. Leuchsenring (1746–1827). Leuchsenring saw himself as the apostle of sentiment, teaching the world, or select souls in it, to weep, and carrying on an extensive correspondence in which his adepts shared with one another accounts of poems and journeys, meetings and partings, and the tremulous motions of their hearts—among them we find Wieland, in Erfurt and Weimar; Sophie von La Roche (1731–1807), Wieland's cousin and once his fiancée, and after 1771 famous as the author of an epistolary novel, *History of Lady Sternheim (Geschichte des Fräuleins von Sternheim)*; J. G. Jacobi (1740–1814) a minor poet and the elder brother of the novelist and philosopher, F. H. Jacobi in Düsseldorf; and Georg Jacobi's particular friend J. W. L. Gleim (1719–1803), a prebendary canon of the joint Lutheran-Catholic foundation in Halberstadt, near the Harz mountains, who regarded himself, after Klopstock, as the patron of young literary talent in Germany—his own verses were diffusely anacreontic, with a touch of Prussian patriotism—and who bore from the early years of a long and leisured existence the epithet 'Old Father' Gleim. In *Poetry and Truth* Goethe offers as a deliberately prosaic explanation for the blossoming of sentimental correspondence in Germany at this time the high quality of the mail service provided across Europe by the house of Thurn and Taxis, and the complete absence from the letters of any political content that might have attracted the attention of the censors. In 1771 Gleim visited Darmstadt and wept, according to Caroline Flachsland, 'a tear of joy' to find himself among the little group humorously—but not too humorously—known to one another as the 'communion of saints'. The group had already indicated its cultural loyalties by printing, in a limited edition of thirty-four copies, one of the first collections of the odes of Klopstock, hitherto scattered in journals or circulating only in manuscript. Klopstock's own revised edition of these poems appeared some months later, in the autumn of 1771, and it was probably at first with the later version of Klopstock's ode to skating on his lips that in the winter of 1771–2 Goethe took up the sport which Klopstock made suddenly fashionable. In February 1772, however, Merck and Leuchsenring both came over to Frankfurt, Goethe met the missionary for Sentimentalism, and at the end of the month he accompanied Schlosser to Darmstadt for a five-day visit which introduced him to Leuchsenring's circle and brought him Merck's invitation to review for the *Frankfurter Gelehrte Anzeigen* and so his entrance to what Deinet the publisher called 'the invisible church'.

At the start of April, despite wet weather, Goethe walked the seventeen miles from Frankfurt to Darmstadt, where he stayed for a week and by his hardiness earned himself the evocative title of 'the Wanderer', a name he had already applied to Shakespeare in his speech for 14 October, and so an indication, since he was now reading scenes from *Gottfried von Berlichingen* to the circle, that he was regarded as pre-eminently Shakespeare's disciple. Caroline Flachsland— who went in the group by the name of 'Psyche'—took to him immediately, and not only because of his affection and loyalty towards her future husband—he recited by heart Herder's translations of English ballads and of Shakespeare's 'Under the greenwood tree', this latter while they all sought shelter on a woodland walk from the pouring rain—but because she felt in him 'a certain similarity in tone or language or somehow' with Herder. In mid-April Goethe and Merck left Darmstadt together on a 'madcap journey' (presumably on foot again) to Homburg, twelve miles north of Frankfurt. Here they were received by the Landgrave of Hesse-Homburg himself, who was only too willing to show them his recently landscaped park, and they called on two other court ladies in the circle, the sickly Henriette von Roussillon, from Darmstadt, known as 'Urania', and Luise von Ziegler (1750–1814) from Homburg, known as 'Lila', who recovered from the restrictions of life at an even smaller court than that of Darmstadt in her private garden at the edge of the woods, where she kept a pet lamb and which she had already decorated with her own tomb. Goethe seemed able to share energetically the enjoyment of any enthusiasm. 'I am beginning', Merck told his wife, 'to fall seriously in love with [him]', and on their way back through Frankfurt, where they were to meet with Sophie von La Roche, whom he then accompanied to Darmstadt, Merck stayed as Caspar Goethe's guest in the family house. From the end of April until 7 or 8 May, 'our heaven-sent friend Goethe' was in Darmstadt again, where 'Urania' and 'Lila' had now joined 'Psyche', and all three eventually received from him Klopstockian free-verse hymns in honour of the feelings, musings, and such happy little events of these days as the poet's climbing an inaccessible rock to carve on it his name. 'If Goethe were of noble birth', Caroline wrote to Herder, 'I could wish he would take [Lila] away from the court where she is quite unjustifiably neglected . . . Goethe is an extremely good person, and they would be worthy of each other.' An obscure allusion in *Poetry and Truth* suggests that though Goethe did not know anything of these dreams, 'Lila' perhaps did. Men, it would seem, could speak the language of Sentimentalism but did not need to take it too seriously.

Herder however was less than pleased at Goethe's sharing emotions, and even a parting kiss, with his future wife, and sent Caroline a savage parody of the lines written 'to Psyche'. Neither he nor Merck, and certainly not Schlosser, could really feel at ease in the febrile atmosphere of the Leuchsenring cult, indeed of all the contributors to the *Frankfurter Gelehrte Anzeigen* only Goethe seems to have been fully able to encompass both worlds, and that was at some cost to his personal coherence. For the new wave at Deinet's journal was the uncertain beginning of a new movement which, as Goethe noted in his autobiography, never offered a theoretical account of itself, but which in practice defined itself by its opposition both to the rationalist Enlightenment and to Sentimentality. It is not perhaps surprising that Goethe should have proved a hostile reviewer of the aesthetic writings of J. G. Sulzer (1720–79), a pupil of the Swiss critics Bodmer and Breitinger, who was, however, teaching at the spiritual and political antipodes of Switzerland, Berlin. Goethe would not tolerate the rationalist premiss, shared by Sulzer with Lessing, and with Batteux before him, that 'the arts' are comparable phenomena, let alone that they can be reduced to a single, teachable, principle. More fundamentally still, however, he also questioned a dogma sacred to Wolffians and Sentimentalists alike: that of the goodness and reliability of Nature (whose original beauties are, according to Sulzer, further adorned by the arts). Nature, he protested, with a rawness born perhaps of his own skirmish with death, is indifferent to human sufferings or sentiments, and artistic genius does not adorn her, but displaces her. Something of the robustness of Robinson Crusoe's attitude to his natural surroundings echoes in these lines. There is a strongly pragmatic and empiricist tone about the *Frankfurter Gelehrte Anzeigen* not previously heard in German controversy: 'God preserve our senses, and protect us from the theory of sensuality!', Goethe exclaims. His first review was probably (all the reviews were anonymous and their attribution is still largely a matter of conjecture) a sarcastic dismissal of the second instalment of J. G. Schummel's imitation of Sterne, *Sentimental Journeys through Germany* (*Empfindsame Reisen durch Deutschland*), (1771) and one of his last was a scurrilous attack on Georg Jacobi and his supposedly effeminate correspondence with Gleim. The contributors clearly wished to say more about the political condition of Germany than was possible in the rarefied and essentially courtly correspondence of the Sentimentalists, and they wished to be more down-to-earth in what they said than Klopstock, whose 'patriotic emotion', Goethe later commented, 'lacked an object on which to exercise itself'. But where was the nation of which the young journalists were to speak? In social terms, it was evident to Goethe, reviewing an English work which purported to characterize the several nations of Europe, that the character of Germany was a middle-class affair which was not to be found by talking to 'fine lords and ladies' but by observing 'the man in his family, the peasant on his farm, the mother amid her children, the craftsman in his workshop, the honest burgher at his jug of wine and the scholar or merchant in his club or coffee-house'. But politically it vexed him that 'the race of poets and philosophers' could not 'grasp that nowadays only the nobility keeps the balance against despotism' or that the essential role of the nobility 'in our constitution' should not be mocked, since all it needed was 'a better and more enlightened education', in order, presumably to perform a function rather like that of its English equivalent, which, through its association in Parliament with the middle classes, restrained the despotic powers of the State. The opposition of Deinet's reviewers to all the conventional wisdoms of the day led, when religious works came under scrutiny, to conflict with the ecclesiastical authorities in Frankfurt; Merck passed on the general editorship to Schlosser halfway through 1772; and the entire new corps resigned at the end of the year, handing over to two professors at the university of Giessen, one of them the notorious rationalist theologian C. F. Bahrdt (1741–92). In their year as journalists, however, they had contributed in their own way, as Lessing in the same year did in his *Emilia Galotti* to that process by which the reading public was, according to *Poetry and Truth*, encouraged to regard itself and its opinion as the only and ultimate tribunal, and to stand in judgement on the acts of princes and their officials; and of that process, it seemed to the older Goethe, the results in the end, and in France, were truly revolutionary. Goethe's later analysis of the motives of his fellow-reviewers, however, as 'combined of poetry, morality and a noble endeavour', and his assessment of their efforts as 'harmless but also fruitless' is a lapidary judgement, whose accuracy and incisiveness it would be difficult to surpass, on the entire movement which came to be known as Storm and Stress and of which the policy of the *Frankfurter Gelehrte Anzeigen* in 1772 was a first manifestation. Certainly nothing less than the attraction of a powerful cultural counter-current can explain Goethe's hostility over the next two and a half years to some of the most prominent representatives of Sentimentalism, to the Jacobi brothers, and to Wieland, whose originality and subtlety had once led Goethe to put him beside Shakespeare as one of his guiding stars.

In mid-May 1772 Goethe, once more postponing thoughts of Paris and Italy, took lodgings in Wetzlar, thirty-two miles north of Frankfurt and close to the Hessian university town of Giessen, in order to carry out that part of his father's educational scheme which saw him widening his legal experience at the Imperial Cameral Court. Wetzlar was a dirty little medieval free city, all steps, half-timbered houses, and middens, some said to have been growing for as long as the backlog of cases at the Supreme Court, which was transferred there in 1693. Goethe seems, despite his hymn to the architect of Strasbourg Minster, to have paid no attention to the fine Gothic cathedral. The population of Wetzlar was a mere 5,000, half that of Darmstadt, and the Imperial Court, with all its attached representatives of the litigant powers within the Empire, accounted for a further 900. Three hundred more officials arrived from 1766 onwards after the reforming co-Emperor Joseph II had ordered a visitation of the Court to investigate its appalling delays and the prevalent rumours of corruption—during the summer of 1772 however the visitation itself was temporarily suspended, and Goethe saw nothing of its operation, though he caught something of the atmosphere of fear and intrigue which it had created and felt more repelled than ever by the law. The ordinary business of the Court was conducted entirely in writing, but all documents, including all submissions of evidence, had to be read out aloud and some cases were of more than two centuries' standing. A few lectures and practical classes were put on for the benefit of the matriculated probationers ('Praktikanten'), of whom there were eighteen, including Goethe during his stay, but the main purpose of a spell in Wetzlar for the wealthy or nobly-born young lawyers who could afford it was to become acquainted, not only with the details of the Imperial constitution, but with one another, for they were the rising generation of Germany's diplomats and state officials. In those first few years after leaving university, when individual careers were not yet finally established, the friendships and habits of student days continued a fading after-life. Goethe found himself a member of a group of some two dozen meeting for dinner at midday at the 'Crown Prince' inn, who had formed themselves into a mock-chivalric order, inventing ceremonies and giving themselves names from the old romances, mainly based on a French play with a medieval theme which had been performed in Wetzlar the previous year. Goethe of course took the name of Götz von Berlichingen.

The translator of the play, F. W. Gotter (1746–97), who in Göttingen in 1769 had helped H. C. Boie (1744–1806) to found a literary annual, the *Göttinger Musenalmanach* (*Göttingen Almanac of the Muses*), and who was to become best known as the author of 'Singspiel' libretti, was still in Wetzlar as the representative of his employer, the Duke of Saxe-Gotha, and was for the time being Master of the 'order'. Gotter succeeded in this role the inventor of the entire game, the mystical, or mystifying, Freemason A. S. von Goué (1743–89), who had been dismissed for idleness from his post in the Duke of Brunswick-Wolfenbüttel's legation to the court. Goué's successor in his official post was also a member of the dining-group, K. W. Jerusalem (1747–72), a thoughtful, moneyed young man, son of a famous theologian and friend of Lessing's. He and Goethe had known each other slightly at Leipzig, and had not got on, and Goethe saw little more of him now.

Goethe, however, was anything but assiduous. There is no record of his engaging in any legal activity whatever while he was in Wetzlar, and he seems to have occupied his mind with Homer, Pindar, the translation of Goldsmith's *Deserted Village*, and continued reviewing for the *Frankfurter Gelehrte Anzeigen*. At the start of June he made the acquaintance of J.C. Kestner (1741–1800), since 1767 one of the members of the Hanoverian delegation, eight years older than Goethe, with whom he shared a birthday, and a more solid character than many with whom he was associating. Kestner had walked out with Gotter from Wetzlar to the village of Garbenheim, and here they found Goethe lying on his back in the grass under a tree presiding over a dispute between a Stoic, an Epicurean (Goué), and an Aristotelian. Kestner's later pen-portrait of Goethe at this time was based on close observation:

He possesses what is called genius and a quite extraordinarily lively imagination. He is forceful in his emotions. His cast of mind is noble . . . He loves children and can long occupy himself with them. He is *bizarre* and there are various things in his deportment, his exterior that could make him disagreeable. But he is none the less well regarded by children, women and many others. He acts as it occurs to him to do, without concerning himself whether others like it, whether it is fashionable, whether convention permits it. He hates all compulsion . . . he has a high opinion of Rousseau but is not a blind worshipper of him. *He is not what you would call orthodox. But not from pride or caprice or to make some sort of show.* And of certain principal matters he speaks only to a few; is unwilling to disturb others happy in their views. He hates scepticism and strives for truth and for certainty in particular principal matters, and believes he has achieved it in respect of the most important, but as far as I have observed has not done so yet. He does not go to church, nor to communion, and rarely prays. For, he says, I am not enough of a liar for that . . . He has respect for the Christian religion but not in the form in which our theologians would present it. He *believes* in a future life, a better state. He strives for truth but values the feeling of it more than its proof . . . He has made the fine arts and sciences his principal activity, or rather all sciences, only not those that are called vocational.

The reference to Rousseau suggests that Goethe's reading-matter at this time also included *La Nouvelle Héloïse,* which came to be something of a leitmotif and literary model for him and Kestner during the events of this summer, in the way in which, at least in retrospect, *The Vicar of Wakefield* was the literary symbol of the Sesenheim episode.

Goethe's grandfather Textor had spent ten years in Wetzlar and had found his wife in a local legal family. Her youngest sister, Goethe's great-aunt, S. M. K. Lange, was still living there with her three daughters from two marriages, and Goethe naturally took a part in the family's social life. On 9 June he was sent in his great-aunt's carriage to a country ball, twelve men and thirteen women, in a hunting-lodge in the hamlet of Volpertshausen and on the way, and at the ball, which lasted until after 4 o'clock the following morning, he had the agreeable company of Charlotte Buff, blue-eyed, 18 years old, and the daughter of the bailiff of the Wetzlar estates of the Teutonic Order of Knights. When he paid a courtesy call on her home the following day he discovered her among ten brothers and sisters (another sister, Helene, was away), to whom she had become a replacement for her mother, who had died the previous year. The domestic side of her character appealed to Goethe as did the huge and rambling buildings of the 'Teutonic House', which were grouped round a farmyard and were home also to the legal officer of the Teutonic Order and his twenty children. Goethe, however, had then to learn that since 1768 she had been informally betrothed to Kestner, and that though he was a welcome guest amid the throng at the Teutonic House his own prospects were strictly limited. It was a fine summer, the longest period in his life so far that Goethe had spent effectively in the country. Every Saturday afternoon after dinner he spent with the Buff family, as in Leipzig he had spent it with the Breitkopfs, and as he had nothing in particular to do—the legal vacation began in the middle of July but probably made little difference to his routine—he was often to be found with them shelling peas, slicing beans, or entertaining the children.

Relations between Kestner and Goethe in the ensuing months were cordial —they went on long walks together and had conversations about the world— but naturally a little strained. Kestner, who kept full office hours, confided to his diary his irritation at young Dr Goethe's continued visits to converse alone with his Lotte, 'meinem Mädchen'. On about 13 August, however, after a 'little quarrel' with Lotte about a kiss she had given, presumably to Goethe, he resolved that he must write to her and offer her her freedom, 'for what is love, what is affection, born out of duty?' The matter was decided without hesitation —Goethe's visits on the next two days, and his little bouquet of wild flowers, were received 'with indifference', and there is an understandable trace of satisfaction in Kestner's note of 16 August that Lotte had told Goethe frankly that they could only be good friends ('daß er nichts als Freundschaft hoffen dürfe'), whereupon Goethe 'grew pale and very crestfallen'. Goethe then spent a few days in Giessen talking with Merck and Höpfner and others of the only twenty professors of that small university, distinguished for the rowdiness of its students, about the future of the *Frankfurter Gelehrte Anzeigen*. Lotte also was discusssed with Merck, who knew all about her from Goethe's letters, and Goethe was most anxious for Merck to meet her when she came with Kestner to join the party. Merck thought Lotte everything Goethe claimed for her but saw the hopelessness of the position and encouraged Goethe only to leave, using perhaps as an incentive the far from welcome news that during his absence Schlosser had proposed to his sister Cornelia and had been accepted: had he not been in Wetzlar, Goethe thought, suddenly filled with jealousy, his friend would not have made such progress. Unable all the same to persuade Goethe to return home immediately Merck arranged to meet up with him and Cornelia in September at the household of Sophie von La Roche near Coblenz.

Perhaps Goethe was persuading himself that he wished only to stay on in order to celebrate his and Kestner's birthday. But beans were sliced in the Teutonic House until midnight on 27 August, and tea was made and passed round with congratulations, and still Goethe did not go. He had warned, however, that his departure when it happened, would be sudden: he was presumably waiting for the 'right', the symbolically significant, moment, and on 10 September it came. The summer was ending, he had at last resolved to go, and that evening as the three friends sat under the stars Lotte began a conversation about death and the after-life. Goethe was overcome by a sense of the finality of the occasion, the rooms he would not return to, the last time Lotte's father would accompany him to the gate, the end of the little refrain from *La Nouvelle Héloïse* that had accompanied their fruit-picking and kitchen-work and picnics throughout the summer: 'and today, and tomorrow, and the day after tomorrow and his whole life long . . . '. Before they all parted that evening Goethe wrote a dedication in Kestner's copy of *The Deserted Village*— an edition privately printed by Merck 'for a friend of the Vicar', as the title-page has it, that is, Goethe. The few lines express all the awkward warmth of their relationship:

> Wenn einst nach überstandnen Lebens Müh und Schmerzen,
> Das Glück dir Ruh und Wonnetage giebt,
> Vergiß nicht den, der—ach! von ganzem Herzen,
> Dich, und mit Dir geliebt.

If one day you surmount the travail and pains of life, and fortune gives you days of peace and bliss, do not forget him who—ah, and with all his heart—loved you, and with you.

Back in his rooms Goethe wrote notes to Kestner and Lotte—'He is gone, Kestner, when you get this note he is gone'—and at 7 a.m. his few things were packed and he rode away down the picturesque valley of the Lahn. Great-aunt Lange was most put out at his failure to take his leave properly and said she would write to his mother to complain of his behaviour.

A solitary wanderer once more, Goethe had perhaps three days to regain control of his feelings. After passing through Ems, where he bathed in the waters, he reached, on 14 September, the cluster of dwellings at Thal, at the foot of the eminence of the Ehrenbreitstein, opposite Coblenz, where the Lahn flows into the Rhine. Here, in a large house with magnificent views from every window, lived G. M. F. von La Roche, a free-thinking and wordly-wise Catholic, Privy Councillor to the Prince Bishop of Trier and reputedly a natural son of Count Stadion, the flamboyant and enlightened minister of the Archbishop of Mainz, together with his intellectual Protestant wife, as strangely self-possessed as the Swedish Countess of G——, and their four good-looking children: Maximiliane, at 16 the elder of the two daughters, was the particular object of Goethe's attention, for Cornelia had not after all come but had remained behind in Frankfurt. Merck, however, was there as agreed, proving a congenial companion to Herr von La Roche, if only because of his ability to add unedifying personal details whenever the conversation turned to some Sentimentalist luminary, and a week of mild sunshine passed agreeably enough in walks in Coblenz and the neighbourhood. Goethe, Merck, and his wife set sail up the Rhine on about 20 September to return past some of Germany's most spectacular scenery, which the two men busily sketched, to Frankfurt and Darmstadt.

Once back at home Goethe gravitated toward the Sentimentalist world at the very time when he was writing of it most cuttingly in the pages of the *Frankfurter Gelehrte Anzeigen*. In June, while he was away, a distant relative, Johanna Fahlmer (1744–1821) had moved to Frankfurt from Düsseldorf: she was no beauty, but she was intelligent and a keen reader of contemporary literature. She also had a half-sister with two children older than she was herself, the brothers Georg and Friedrich Jacobi. Goethe started to give her English lessons so that she could pursue her literary interests, and soon found himself confiding in her. Although he must have remained silent on the subject of her nephews, when he called her 'Aunt' Fahlmer he was using a title derived from her relation not to him, but to the Jacobis. A satire on the Düsseldorf group which he began to plan before he was acquainted with Johanna Fahlmer, and in which she figured, 'The Doom of the Jacobis' ('Das Unglück der Jacobis'), was destroyed once their friendship made it unacceptable. Goethe thus had connections halfway established with the Jacobis and, through Sophie von La Roche, with Wieland, and in the middle of November he left for Darmstadt to spend a month in Leuchsenring's sphere of influence. He was of course looking for friendship and consolation after the devastating conclusion to his singularly fruitless (and expensive) stay in Wetzlar, but if he found it it was more through the distractions of Merck's energetic company than through the sharing of feelings: 'The stay here,' he wrote to Kestner from Darmstadt,

has poured much well-being through my limbs, but in general things are not getting any better. Fiat voluntas. How good your mood is, and how little in the way of self-shooting, . . . I have seen from your letter and please God in saecula saeculorum.

A correspondence with Kestner, and to a lesser extent with Lotte, sprang up as soon as Goethe left Wetzlar, for relations continued to be cordial. There was a warm, and unexpected, reunion with Kestner at the end of September when he came to Frankfurt for the Michaelmas Fair, and from 6 to 9 November Schlosser, who had business there, accompanied Goethe to Wetzlar where he stayed with the Buffs. Although the three friends were together practically all the time and Goethe was received with all the love and warmth he could have hoped for, the visit was not altogether a success, Goethe spending the last night on the couch with thoughts 'of hanging myself, or that deserved hanging'. (From these however he was delivered by a brief visit, on the return journey, to 'Lila' in Homburg.) This recurrence in his letters of allusions to suicide, or at least to his intention of not committing suicide, was not simply spontaneous. On 30 October the quiet K. W. Jerusalem, slightly known, it will be recalled, to Goethe and Kestner, and, it now turned out, not more than slightly known to anybody, had borrowed Kestner's pistols, on the pretext of going on a long journey, and had shot himself. There was talk of a hopeless love for the wife of a friend: Wetzlar was aghast, and Goethe was deeply shocked. The parallel to his own situation made Jerusalem's act a warning of what might have been, or might still be, as the execution of Susanna Brandt had been a warning of what the Friederike episode might have led to. 'Poor young fellow! when I came back from my walk and he met me coming out in the moonlight I said there's someone in love. Lotte must still remember me smiling about it.' Goethe's reaction was violent and he instinctively attributed the suicide to the forces operative in his own life: disappointed love, inner solitude, and the frustration of personal and spiritual yearnings by an established religion in which he could not believe—had not Jerusalem's own father been an apologist for it?

But the devils! Who are these scandalous people who enjoy nothing but the chaff of vanity and have love of idols in their hearts and preach idolatry and stifle unspoiled Nature and exaggerate and ruin men's powers [- they] are to blame for this misfortune, for our misfortune . . . God knows, solitude undermined his heart . . .

'You complain of solitude!' Goethe wrote to Sophie von La Roche as he discussed the terrible news: 'Alas that it is the fate of the noblest souls to sigh in vain for a mirror of themselves'; and he again associated Jerusalem's action with the struggle for a non-religious morality by citing the words of a Wetzlar colleague: '. . . "the anxious striving for truth and moral goodness so undermined his heart that unsuccessful ventures in life and love forced him to his sad decision".' Goethe's visit to Wetzlar was perhaps prompted by his need to find out more about the shadowy figure who had suddenly become another dead brother to him. Kestner prepared a full account in writing of Jerusalem's personality and the circumstances leading to his death which reached Goethe in December in Darmstadt and was shown round in the sorrowing circle there. When, over a year later, Goethe came to write *Werther* he took over many details and phrases from Kestner's report.

If for the present Jerusalem's death simply remained in Goethe's mind as a piece of unfinished emotional business, it must have been for the same reason for which the execution of Susanna Brandt had not yet received any literary expression: not merely did Goethe need some distance from the event in order to determine its significance, but his literary art had not yet advanced to the point at which he could make use of such explosive material. In Darmstadt in November and December, under the immediate influence of Merck, Goethe began a redirection of his energies and within months his creative imagination matured to a productivity it would never surpass. 'Since he lacked all virtues, he said',—so wrote Caroline Flachsland to her fiancé on 5 December—'he intended to go in for talents instead. Now there's a mind that might turn into something.' Goethe was drawing more busily than ever before, and giving instruction to Merck ('there is quite an academy at Merck's'), but his friend's sights were firmly set on literature. After giving up the editorship of the *Frankfurter Gelehrte Anzeigen,* whose struggles with the Church authorities showed it was no easy thing to form public opinion, Merck had devoted more of his attention to the private press he had set up in a village near Darmstadt. After reprinting *The Deserted Village* he undertook a two-volume reprint of Ossian, for the title-page of which Goethe drew and engraved a vignette, and in November 1772, in collaboration with Deinet, he had brought out a pamphlet, the essay *On German Architecture* on which Goethe had been working since Sesenheim days. If Goethe could be persuaded, however, a much bigger original venture would be possible: Merck did not have Herder's reservations about *The History of Gottfried von Berlichingen* and was willing to pay for printing it if Goethe would pay for the paper. First, however, Goethe had to be brought to carry out the revision which in July, in response to Herder's criticism, he had acknowledged to be necessary but at which he had so far made no attempt. There is no record of what Merck said on the subject to him in Darmstadt, but four days after returning to Frankfurt Goethe wrote to Kestner that he was stopping work on the *Gelehrte Anzeigen* and that, once he had written a postscript taking leave of the public, he was turning to 'a really proper bit of work', the rewriting of his play. It was an announcement of the end of a period in which 'I lost myself in aesthetic speculations since I could not succeed in any aesthetic productions', and of the beginning of a conscious attempt to find in art the compensation for desires that in life remained unfulfilled: 'I am revolving new plans and ideas, none of which I would do if I had a girl.'

For the whole of 1773, apart from a fortnight in Darmstadt at the end of April, Goethe was in Frankfurt: not until his old age was he again so static for so long, and never again was he so inventive. It was as if Frankfurt was a pressure vessel, accelerating an extraordinary literary chemistry: memories of a similar year of confinement, 1769, may have played their part. While Goethe remained still, his life changed about him and every change seemed only to accentuate the solitude of a noble soul sighing for its mirror. It was the year in which, more than in any other, he experienced the imprisonment in self to which the main tradition in German 'official' culture condemned the sensitive souls which it produced, and that experience enabled him to perfect the new kind of literature foreshadowed in *Götz* and his Sesenheim poems.

In 1773 three marriages were impending in Goethe's immediate circle, and of these it was Schlosser's marriage to Cornelia that promised to affect his life most profoundly. Its date was also the least definite: Schlosser had first to make his arrangements with the Margrave of Baden for a post in the prince's service which would give him a steady income, and the negotiations proved unexpectedly tiresome. Only in September was he finally appointed to a post from which he could expect to advance to a Privy Councillorship, and with that Caspar Goethe was satisfied. Cornelia's dowry was settled as a capital sum of 10,000 guilders, of which she would initially enjoy only the interest, a generous trousseau, and a cash sum of 1,350 guilders; the betrothal took place on 13 October; and the wedding was fixed for 1 November. As Cornelia packed for her final departure for Carlsruhe on 14 November, 'I look forward to a deadly solitude', Goethe wrote to Johanna Fahlmer, 'you know what my sister meant to me'. The whole year was spent in gradually more certain expectation of that grim moment. At least the wedding of Kestner and Lotte was certainly fixed— Easter Day, 11 April—though towards the end of February came the further bitter news that in June the couple would be moving away from the area altogether, so that Kestner could take up a new post as archivist in Hanover. Goethe's numerous letters to Kestner in the first half of the year—on average about one a week—were mainly filled with friendly gossip about his and their doings, and other members of the Buff family and their neighbours, but were always spiced with reminders to Lotte of his love, and like the letters written to Käthchen Schönkopf during her engagement were occasionally embarrassingly explicit about his jealousy or self-pity. As the wedding approached they became more frantic in tone and more questionable in their content. Kestner knew and could tolerate that Goethe kept Lotte's silhouette on the wall of his room, at first beside the door and then over his bed, but it must have strained his patience to be told that on Good Friday before the wedding the picture would be taken down and placed in a holy sepulchre from which it would not be resurrected until Lotte was in childbed. Goethe's insistence that he should himself have the wedding rings made, and his concern to provide Lotte with material for her nightwear and maternity garb, must have been uncomfortable. A demonic flicker in a letter responding to an attempt by Kestner at remonstration reminds us that with his unpredictable energies and huge talents the young Goethe must have been at times a frightening person to know: 'and I am telling you, if it starts occurring to you to get jealous I reserve the right to put you on the stage with all your most telling features and Jew and Gentile shall laugh at you'. Despite Goethe's assertion that he would not come to the wedding who could know what he might do? It is understandable that Kestner and Lotte took him by surprise and married a week earlier than he expected, on 4 April.

The marriage of Herder and Caroline Flachsland on 2 May was in itself emotionally less trying than those of Cornelia and Lotte and it was the occasion for a reunion with Herder, whom Goethe had not seen for two years. All the same, it meant the loss of Caroline to the Frankfurt area, and when Goethe went over to Darmstadt on foot on 16 April with the remains of Lotte's bridal bouquet in his hat it was in the knowledge that another and heavier loss was impending: Henriette von Roussillon, 'Urania', was on her death-bed, and on the twenty-first he was at her funeral. Merck too was leaving and his wife was returning for a while to her family in Switzerland: the Landgravine was taking her three daughters to Russia, with the intention, already agreed with Catherine the Great, that one of them was to be selected for marriage to the Tsarevich Paul, and Merck was required as purser on the journey. As a result, a plan that Goethe should join Merck on a trip to Switzerland, which might have offered some relief at a trying time, had to be dropped. Even Leuchsenring was leaving Darmstadt for Paris; the personal loss was slight, but the move meant the end of his circle. Goethe sped him on his way with *A Farce of Father Porridge* (*Ein Fastnachtsspiel vom Pater Brey*) in which a Tartuffe, or 'false prophet', tries to take advantage of an innocent girl, modelled on Caroline Flachsland, but is defeated by her returning fiancé. Read as an entertainment on the night before his wedding, it did not appeal to Herder, who did not take to its insinuations about Leuchsenrings's hold over the ladies of the Darmstadt court. Relations between him and Goethe remained cool for months.

'My poor existence is petrifying into a desolate rock,' Goethe wrote to Kestner and Lotte on 21 April and to a Strasbourg acquaintance he wrote asking for letters to relieve his loneliness,

Everyone is leaving this summer. Merck . . . his wife . . . my sister, Mlle Flachsland, you, everyone. And I am alone. As long as I do not take a wife or hang myself, you can say I am really fond of life . . .

'I am alone, alone, and more so every day,' he told Sophie von La Roche, and went on with what is possibly an allusion to his relations with Herder and Caroline, or theirs with each other:

And yet I could put up with souls that have been created for each other so rarely finding each other and usually being separated. But that they should fail to recognize each other in moments of the happiest union, that is a strange enigma.

Frankfurt held little in the way of distractions for such a soul, and little promise of a soul mate. Goethe could skate in the winter again, and on 2 February he joined the College of Graduates that meant so much to his father, but only once is he recorded as attending a meeting. *Poetry and Truth* recounts a strange series of incidents which Goethe sets in the middle of the 1760s but which cannot be earlier than 1769 and probably date from the period 1772 to 1774, the only independent evidence for them coming from the first half of 1773. In the course of a society game among the young people of Frankfurt the couples into which the company had naturally settled were broken up and reassigned by lot, and whenever the group met together they had to remain in the coupling chance had determined, even pretending to be husband and wife. The confusion of game and reality was such, Goethe remarks in his autobiography, that after a while he and his partner—Susanna Münch (b. 1753) whose birthday, 11 January, was the same as Lotte's—would have been quite prepared to marry in earnest, and he comments with the driest understatement, 'it was easy to see that in this insignificant action a variety of passions were in play'. It was a game of and for the Age of Sentiment. Lichtenberg noted in his private journal in 1777: 'If another and later species comes to reconstruct the human being from the evidence of our sentimental writings they will conclude it to have been a heart with testicles.' The implied analysis—that Sentimentalism interprets as 'feeling' what is no more than impotent desire—is far more than a coarse joke. The cult of feeling was not just an intellectual movement: it had clear implications for personal and even physical behaviour, which went beyond a greater readiness to weep or read together or to dress and speak informally, and which were bound to conflict with the prevailing notions of decorum—and so to generate frustration and misery. Goethe's personal experience at this time was, both by a natural momentum and by his deliberate choice, approximating to, and becoming symbolically representative of, that of a whole generation. After Lotte's marriage he hung up a second silhouette over his bed which he now called, in a letter to Kestner, 'as sterile as a sand-dune', though 'I do not know why I am such a fool as to write so much at the very time when you with your Lotte are certainly not thinking of me'. It was the picture of the woman he half-seriously thought might become a second Lotte: Helene Buff, the one sister he had never seen. It was as if he and circumstance were conspiring to compound his individual emotional destiny, reflected in nearly all his writing so far: he could love as few men could, but only at a distance, only in unfulfilment. It was cruel and not quite correct, though the cruelty was that of true insight, for Herder in January to call Goethe 'the cold misogynist'. The truth was hinted at by Goethe, at about the same time, in the title he gave to a virtuoso piece of nonsense verse, imitative of musical form: *Concerto dramatico composto dal Sigr Dottore Flamminio detto Panurgo secondo.* For Rabelais' Panurge, despite his resounding name, and his extensive contribution to the longest set of epithets ever applied to the human testicle, is effectively impotent, his fear of cuckoldry undermining his desire for marriage and leaving him buffeted by an equally long set of contradictory commands: 'Mariez-vous donc', 'Point ne vous mariez'. Goethe was afraid of course, not of cuckoldry, but of the end to poetry that seemed a necessary consequence of the fulfilment of desire. A little dramatic poem of 1773, *The Artist's Earthly Pilgrimage* (*Des Künstlers Erdewallen*), shows inspiration weighed down by the cares of marriage and the need for money; a compensating sequel of 1774, *The Artist's Deification* (*Des Künstlers Vergötterung*), is unconvinced and fragmentary.

It might seem strange, in the circumstances, that Goethe stayed in Frankfurt at all, and Herder mocked his loyalty to his philistine home-town. As the year wore on Caspar Goethe did his best to involve his son in civic affairs and Goethe 'did not resist' these attempts to point him towards a local career. His legal work increased considerably, as he took over cases from Schlosser, and from his old friend Horn who had now also obtained administrative office. Like Gulliver he tolerated these Lilliputian bonds, he said, since he felt in himself the strength to break them when necessary. But when asked whether he did not wish to follow the example of Herder or Schlosser and take service with some prince, he indicated that there was for him, as for Götz, a more important freedom which the old order preserved, and which he would forfeit if he became a Weislingen:

I need too much for my own use the talents and powers that I have, I have always been used to acting simply in accordance with my instinct and that would be no value to a prince. And then, for me to learn political subservience—'the Frankfurters are a cursed lot', President von Moser [the newly appointed Chancellor of the Landgrave in Darmstadt] likes to say, 'you cannot use their pig-heads for anything'.

The *Concerto dramatico*, though it is wholly contentless, is the first manifestation of Goethe's ability to combine the colloquial versification he had already sometimes used in his letters with a tightness of expression worthy of a fully literary composition, and a remarkable rhythmical variety. Goethe's new wave of literary production began, however in a very different form. In the winter of 1772–3, as he gradually recovered from the double blow he had been dealt by Wetzlar, in the persons of Lotte and Jerusalem, and started to write again, he turned first to theology. He wrote two pamphlets, which were anonymously published by Merck, and which first drew their author to the attention of the enthusiastic Zurich clergyman, J. C. Lavater (1741–1801), whose founding works in the pseudo-science of physiognomy were sceptically reviewed by the the *Frankfurter Gelehrte Anzeigen*. The conceptual liberation which the two little treatises express is of a complex kind. The second of the two, *Two Important and Hitherto Undiscussed Biblical Questions . . .* (*Zwo wichtige bisher unerörterte biblische Fragen . . .*), is probably in some respects the older, for it may contain material extracted from Goethe's lost Strasbourg dissertation, printed at last, as he had once threatened. Goethe slyly indicates his source: the writer of the pamphlet, who purports to be a Swabian country pastor, says he has a son who represents the reductive academic philosophical theology of the age and whose views will eventually mature into the broader perspective of his father; the essay, however, is dated 6 February 1773, and we are told that the son has been an M. A. for eighteen months, that is, since 6 August 1771, the date of the disputation which for Goethe had to take the place of a dissertation. The two questions are indeed important, for the answers to them reflect the two main intellectual influences on Goethe as he returned to literary activity. The answer to the first—'what was written on the tablets of the covenant?'—is 'not the Ten Commandments', but certain ritual prescriptions peculiar to the Jews. The notion that the Bible is the vehicle of a universal rational and moral religion (such as the Ten Commandments might be held to summarize) is thereby rejected in favour of Herder's cultural theory which gives value to particular and local traditions. The second question—'what is the meaning of speaking with tongues?'—deals with the individual inspiration which, in Herder's theory, interacts with tradition. Speaking with tongues, Goethe's amateur exegete argues, is speaking with the voice of the Spirit, that is, speaking the inarticulate, onomatopoeic, language of immediate feeling, which contemporary rationalists—'cameralists', they are called, that is, absolutist bureaucrats of the mind—wish to organize out of existence. But the individual, like Arnold's heretics, will continue to speak true to his feelings in his own little hermitage, though Goethe gives him the warning that he cannot do so continuously and indefinitely: 'No mortal can maintain himself at the highest pitch of sentiment.' Yet if the essay on *Two . . . Biblical Questions* seems to suggest that the Christian religion has nowadays been replaced either by Sentimentalism or by Herder's historical and literary approach to the Bible, the earlier essay, *Letter of the Pastor of *** to the New Pastor of ***. From the French (Brief des Pastors zu *** an den neuen Pastor zu ***. Aus dem Französischen)* shows a position in religious matters which, like that which Kestner noted in Goethe in 1772, is not at all that of the typical Enlightenment deist. The Pastor who writes the letter has sympathy with all points of view but one: Calvinists and Catholics, enthusiasts, heathens, and even philosophers, should all be allowed to live and die by their own lights—missionaries, conversions, and ecumenical reunifications are equally undesirable. Only the dogmatic defenders of intolerance, of the hierarchy of the visible Church, are excluded from toleration. The Pastor himself holds only to faith in eternal love, for there is so much that is unknown or cannot sensibly be spoken of, but he thinks he can appreciate the belief in the inseparability of body and soul which underlies the theology of the sacraments, and he sees no *conceptual* difficulties attaching to the doctrine of incarnation, the doctrine that 'divine love so many hundred years ago wandered around as a human being for a short time on a little bit of the earth, under the name of Jesus Christ'—he merely adds that 'since God became Man, so that we poor sensuous creatures should be able to conceive and grasp him, we should avoid nothing more carefully than making him into God again'. The purported writer is thereby distinguished from Rousseau's 'vicaire savoyard' (with which the sub-title invites comparison) and diverted from the main highway of German theological speculation in the last quarter of the century. The problems of religious observance are essentially moral and practical not metaphysical. Such a pastor does not of course represent Goethe's own position—a mistake Lavater made, not realizing that the author of the pamphlet had written 'with Lessing in mind'—but rather the kind of Christian pastor whom Goethe would prefer to deal with and the beliefs which he could himself tolerate in turn. If Goethe is represented within the treatise at all it is when the Pastor asserts that the teaching of Christ has been nowhere more oppressed than in the Christian Church and calls on anyone who is prepared to be a 'successor to' (or 'follower of') Christ to dare to let it be seen in public 'that he is concerned for the salvation of his soul. Before he realizes it, he will be branded an outcast and a Christian congregation will cross itself as he passes.' From a literary point of view the most significant feature of these two pamphlets is Goethe's discovery that his powers of impersonation can encompass a sustained, and uninterrupted, first-person discourse, expressing views on matters of great and immediate importance to him which he none the less does not himself wholly share.

Theological concerns run through practically everything Goethe wrote in 1773, once he had, by the beginning of February, completed his new version of *Götz von Berlichingen.* It is not entirely surprising that one with so unorthodox a Christology should at this time have started to study the Koran and should have found in Mahomet the central figure for a drama, who can raise, in a wholly non-Christian context, such issues as the relation between the sensitive soul, Nature, and God, and the relation between the intellectual revolutionary and his followers. But, with the exception of one remarkable free-verse hymn, Goethe's *Mahomet* did not progress beyond a few fragments in partly biblical, partly Ossianic prose. His inventive capacity was now showing itself rather in the discovery of the poetic potential of the sixteenth-century doggerel verse used by Hans Sachs ('Knittelvers'), and the traditional forms of popular entertainment associated with it. *The Farce of Father Porridge* is in the format of a Shrovetide carnival play: shortly before it, and probably in parallel with the *Concerto dramatico,* Goethe was writing *Lumberville Fair (Jahrmarktsfest zu Plundersweilern)* which both is and incorporates such an entertainment. In a fairground setting a vivid and satirical kaleidoscope of figures, quacks, gypsies, peasants, pedlars from Nuremberg and the Tyrol, an Italian barrel-organist, and one or two more characters from the refined classes, all conversing in a lively rhyming farrago, form an audience for a play within the play: the biblical folk-story of Queen Esther and the villainous Haman. Goethe had written nothing like it before. Alone in old-fashioned Frankfurt he seems to be giving practical proof to the Herderian belief that the individual genius can find his resources in the popular and national culture without the importation of bookish and alien forms. The masque, the farce, the doggerel are not mere antiquarian exercises. Even in Lumberville we are not far from those most recent concerns of the German intelligentsia which are also the background to the tale of Father Porridge. The themes of the inserted Esther play are transferred from pre-Christian Judaism to the present. Haman is a militant rationalist who rejoices in having put an end to belief in Christ but now finds that a substitute religion of Sentimentality has sprung up:

> Was hilfts daß wir Religion
> Gestoßen vom Tyrannenthron
> Wenn die Kerls ihren neuen Götzen
> Oben auf die Trümmer setzen.
> Religion, Empfindsamkeit
> Ist ein D[reck] ist lang wie breit.
> Müssen das all exterminiren
> Nur die Vernunft, die soll uns führen.

What good is it for us to have pushed religion from its tyrant's throne if the fellows put their new idol up on top of the ruins. Religion, Sentimentality are all the same filth, as broad as it's long. We must exterminate all that. Only reason shall be our guide.

Lest we should think, however, that Goethe's play takes sides as readily as its biblical original, Mordecai, now the high priest of Sentimentality, turns out to be as comically loathsome a figure as Father Porridge, and like him no doubt a caricature of Leuchsenring. The true loyalty of *Lumberville Fair* is not to rationalism, nor to Sentimentalism, nor to Christianity, but to its own popular form and the vernacular in which it is written. It is the form in which, by June at the latest, Goethe had begun to commit to writing the most overtly theological of all his works: *Faust*, a retelling of an old puppet-play with a modern meaning, destined to absorb the poetic powers and moral reflection of a lifetime.

In mid–1773 however the most substantial monument to Goethe's Herderian loyalties was still *Götz von Berlichingen*. In February Merck had prevailed on him not to undertake yet another revision but to publish and be damned—and get on to something new. A young Frankfurt man, Philipp Seidel (1755–1820), who had become Goethe's valet in the autumn of 1772, wrote out a fair copy, and in April the printing began in Darmstadt, the first stage in the joint private publication on which the two friends had agreed. Unfortunately Merck now had to make his unexpected departure for Russia with the Landgravine, leaving the distribution of the copies, which arrived in the second week of June, to Goethe. He was probably not the ideal person for the task—150 copies were sent for sale by the Göttingen bookseller Dietrich without any prior arrangement for payment—a pirate edition soon appeared, and far from making a profit Goethe was having to borrow money before long to cover the costs of the paper. From the beginning of July, however, it became clear that in every sense other than the financial *Götz* was a success. The first reactions came of course in private letters. 'Boie! Boie! The knight with the iron hand, what a play! . . . What a totally German subject! What daring treatment! Noble and free like his hero the writer treads the miserable rule-book beneath his feet . . . God! God, how alive, how Shakespearean . . . Le grand Corneille? The Sh—s! Sh—s! all

Frenchmen! This *Götz von Berlichingen* has inspired me to three new strophes of *Lenora!*'—thus, on 8 July, to the editor of the *Göttinger Musenalmanach*, who had lent him his copy of the play, the poet, later magistrate and aesthetician, G. A. Bürger (1747–94), whose lurid ballad *Lenora* (1774), despite its extreme obedience to Herder's principle that primitive language was onomatopoeic, became one of the best-known and most widely translated poems of modern German literature. (Scott's version is called 'William and Helen'.) Bürger showed a childish delight in larding his letter with repetitions of what was already Götz von Berlichingen's most famous phrase—Klopstock, like his disciple Christian Count Stolberg (1748–1821), who reported his view, and his brother Friedrich (1750–1819), 'wholly repudiated the *compliment* to the trumpeter, and other such words', and thought there were too many scene changes, but otherwise found the play 'good and original'. Even the 75-year-old Bodmer, though he thought his own Swiss national dramas were 'more historical and more dramatic', found in *Götz* 'here and there something of the spirit of Shakespeare' and hoped the Germans would prefer this manner to 'the chimeras' of Klopstock and Gerstenberg. Gerstenberg himself wrote to 'the German Shakespeare' to say that 'the approval you find everywhere gives me courage to hope that you are the man to produce in Germany a public of Germans'. Goethe was not so certain that he could create a public, but 'when I published my Götz it was one of my most agreeable hopes that my numerous friends in the wide world would look out for me', and the play did indeed bring him news from all corners of his acquaintance. Officials in Wetzlar wanted copies, the Buffs were delighted, a letter came from Langer, and greetings could be sent through him to Behrisch. Lenz in Strasbourg, whom Goethe had been advising by letter about his adaptations of Plautus and who was preparing his first major drama, *The Tutor* (*Der Hofmeister*) (1774), was prompted to write to the man whose life he had been shadowing (having had his love-affair with Friederike he too had now abandoned her) with a proposal of spiritual marriage. The private reception of his play fully corresponded to the hopes Goethe expressed to Gerstenberg: 'My highest wish has always been to be in relation with the good people of my age', and it was presumably only the reviewers who led him to add, 'but that is so thoroughly poisoned for one that one quickly creeps back into oneself'. Not that the reviews were particularly hostile, though the very first one revealed the name of the author of the anonymously published play; that in the *Frankfurter Gelehrte Anzeigen* was positively, if unsurprisingly, laudatory, and although Goethe was always convinced that the twenty-page review in the *Teutscher Merkur* (the *German Mercury*), Wieland's paper launched at the start of the year, was essentially negative, it spoke affectionately of 'the most beautiful, most interesting monster', and Wieland, in postscript, dissented from his reviewer's more critical remarks and promised to furnish a full account of his own opinion of the work in a future issue. Everyone agreed, however, that *Götz* had plainly not been written for performance and indeed was unperformable.

However hesitantly, Goethe was venturing out before the vast and faceless German reading-public. From now on, for the rest of his long life, he was a public figure, and he was very soon seen as the most prominent representative of a movement. For once, this was not wholly a misapprehension. The years from 1773 to 1775 were more than any other the time of Goethe's 'joining in'. In the wake of *Götz* he sent poems for publication in the journals of sympathetic spirits: Matthias Claudius and Boie, who was preparing the 1774 issue of the *Almanac of the Muses* (published in time for Christmas 1773). The impact of *Götz* was magnified when in August there appeared a collection of essays, edited by Herder under the title *Of German Character and Art* (*Von deutscher Art und Kunst*), which amounted to a manifesto. It opened with two pieces by Herder, one an essay on folk-song, and Ossian in particular, which included the first printing of Goethe's poem 'A Rose upon the Heath', and the other his essay 'Shakespeare', which concluded with an apostrophe of Goethe as the author of *Götz* and so the worthy modern and German parallel to the great but, alas, obsolescent Englishman. Goethe's essay 'On German Architecture' was reprinted, though Herder, who never really appreciated the Gothic style, followed it, true to his own spirit of contrariness, with a disparaging essay on the same subject by an unknown Italian. An article on 'German History' by Möser, extracted from his *History of Osnabrück*, gave a schematic picture of the German past closely corresponding to that implied by *Götz*. The whole collection, a concentrated summary of the interests established for Goethe during his stay in Strasbourg, was clearly intended to announce and introduce a general shift in literary loyalties. The movement had as yet no name, but the continuity with the programme—and the personnel—of the *Frankfurter Gelehrte Anzeigen* was apparent, and nineteenth-century literary historians dubbed it 'Storm and Stress'. With or without a name, however, the relation of the new enthusiasms to Germany's dominant culture of Sentimentalism was necessarily problematic, and has remained so.

Götz brought Goethe not only letters and reviews and requests for contributions, but also pilgrims, and Caspar Goethe proved a willing and generous host. Summer was over, Cornelia's wedding day was fixed, and grapes from the family garden were on the table when G. F. E. Schönborn, (1737–1817), in the Danish diplomatic service and on his way to a post in Algiers, spent a week in Frankfurt from 10 October, bringing Goethe a letter from Boie and compliments on *Götz* from Gerstenberg and Klopstock. Professor Höpfner from Giessen was also there, staying with the Goethes, and he introduced Schönborn. Despite the preparations for his sister's solemn betrothal on the 13 October Goethe spent much time with his visitor, who described their conversations in detail to Gerstenberg:

He is a thin young man of about my height. He looks pale, has a large, slightly bent nose, a longish face and medium dark eyes and dark hair . . . His countenance is earnest and sad, with however a comical, jocular and satirical mood flickering through. He is very eloquent and overflows with *bons mots* that are very witty. Indeed he possesses, as far as I can tell, an exceptionally visual poetic gift that feels itself totally into its objects, so that in his mind everything becomes local and individual. Everything is immediately transformed for him into something dramatic . . . He seems to work with exceptional ease. He is now working on a drama called *Prometheus* of which he read me two acts in which there are quite excellent passages extracted from Nature's depths . . . He draws and paints well. His room is full of fine casts of the best antiquities . . . He wants to go to Italy in order to have a good look at the works of art there. He is a terrible foe of Wieland & Company . . . This life [of Götz, that is, Goethe's play] has been his guide to the finer points of the German character

Höpfner, in his own briefer report on these days, adds: 'Goethe and Merck [who was still in Russia] throw up at [the Jacobis].' The year 1773 was quite unusually productive for Goethe—his *Prometheus* will concern us later—but everything he wrote is marked by the difficulty of reconciling Herderian 'Storm and Stress' with Sentimentalism. In the little farce *Satyros, or the Wood-Devil Deified* (*Satyros, oder der vergötterte Waldteufel*), completed by September but not published until 1817, a kindly though sensually alert Rousseauist hermit, living the simple life, is surprised one day by a wild man of the woods, an emissary of an altogether more violent and immoral Nature than any he has known hitherto. The hermit is robbed, his little shrine desecrated, and he himself sentenced to death for blasphemy against the new deity. For the hairy, naked, satyr with the long fingernails is a master of music and of the music of words: when he sings of yearning and swooning love the girls tremble and do his will; when he rhymes majestically in mysterious neologisms about the origin of all things the people acclaim him as an omniscient god (he eventually sits Satan-like on their altar) and they adopt with deferential enthusiasm the new religion he commands— eating raw chestnuts. Discovered in the act of ravishing the wife of the priest of the old religion, Satyros makes off for a new and more appreciative congregation. Although the self-parody verges on bitterness, particularly in Satyros' quasi-lyrical and quasi-metaphysical songs of seduction, the tone throughout remains, just, rumbustious (which is not true of its derivative, Brecht's *Baal*). The conflict for our sympathies between the gentle but ineffectual hermit and the immoral but fascinating Satyros is precisely the conflict which Goethe was undergoing and which inclined him now to denunciation of Wieland and the Jacobis in the name of Herder, German character and art, and the politically conscious opposition of the *Frankfurter Gelehrte Anzeigen*, and now to intimacy with their friends and relatives when he felt the pains of solitude and sought its addictive consolations. In November Goethe began a 'Singspiel', *Erwin and Elmira* (*Erwin und Elmire*), based on the ballad 'Edwin and Angelina' from *The Vicar of Wakefield*, which may also have contributed to the picture of the hermit in *Satyros* (it is the source of the lines, 'Man wants but little here below/Nor wants that little long'). The prose play with its numerous inserted songs was not complete until January 1775, but its main lines must have been clear from the beginning, and they amount to a dilute version of the issue in *Satyros*. Erwin has disappeared, no one knows where, after being coldly treated by his beloved Elmira. Filled with remorse she is advised to make her confession to a local hermit. This she does, the hermit proves to be Erwin in disguise, and the lovers depart from the isolated hermitage for a happy life together. The counterpoint to this simple action is provided by the social and historical placing of the characters: Erwin and Elmira have enjoyed the benefits of a 'modern' education, and 'my feelings, my ideas', Elmira says 'have always been the joy of my life'; but the two other characters, a family friend, Bernardo, and Elmira's mother, Olimpia, belong to an older generation and—like Goethe's mother perhaps—see things in more realistic and practical terms. Modern education, says Olimpia, simply gives the airs and graces, and inhibitions, of the nobility to middle-class girls who have simpler pleasures and more real responsibilities: 'What are all the most noble desires and sentiments when you live in a world where they cannot be satisfied, where everything seems to work against them! does that not prepare you for the deepest discontent, occasion endless complaining?' Slight though the operetta is, it has a particularly clear eye for the class basis of the opposition to Sentimentalism which was the defining characteristic of Storm and Stress. But between the two forces of realism and sentiment Goethe places two intermediate emotions which belong in both camps at once and so preserve his drama from one-sided, didactic, irony: first, the buffoonery attendant on Erwin's disguise, and second, the moment of nostalgia on which the play ends —for although Erwin and Elmira are learning to enjoy social and unsentimental happiness Erwin cannot help shedding a tear for the isolation of his hermitage that he is leaving behind. Doubtless some details of the action—certainly several of the songs—but above all this mildness and equilibrium of tone will date from the later stage of Goethe's work in 1774–5, but the central image of the hermitage, at once loved and ridiculous, must belong to the original conception of 1773.

The altogether more ferocious tone of the skit *Gods, Heroes, and Wieland* (*Götter, Helden und Wieland*) shows it to be unambiguously co-eval with *Satyros* and *Prometheus* (and probably the first scenes of *Faust*.) It too is a slight affair— Goethe himself says it was written on a Sunday afternoon, probably in September or October, with the help of a bottle of burgundy—but its author has been touched on the raw and its comedy is in deadly earnest. Earlier in the year Goethe had actively sought subscribers for the *Teutscher Merkur*, but his attitude suddenly changed as Wieland began to publish in it a series of letters in defence of his play *Alcestis* (*Alceste*) (1773) which compared his work favourably with Euripides. Formally speaking, *Alcestis* was a significant achievement, for it was not really a play but an opera libretto, the first serious opera libretto written in German, and Wieland's choice of a classical theme and his use of mainly iambic blank verse were both steps destined to have a considerable influence on

Germany's literature, if not its music. Indeed Goethe himself would eventually find in Wieland's work a model for courtly drama. However, it was not the attempt at an opera that aroused his anger in 1773, nor even the spectacle of Wieland passing schoolmasterly strictures on Euripides in a 'paternal tone' that Goethe confessed he could not stand: rather Goethe's lampoon concentrated on what made that tone so unjustified, Wieland's essentially frivolous attitude to death. *Gods, Heroes, and Wieland* is in the Lucianic mould of the *'dialogue des morts'*, a satirical conversation in the underworld, but death is its subject-matter, not just a conventional part of the backdrop. Wieland, safely alive and sleeping in Weimar, is summoned down to Hades and called upon to justify himself to the great figures of antiquity whom he has maligned and travestied in his play, and in the subsequent letters. But the charge against him is not simply that the real Hercules is an amoral 'colossus' while 'my Hercules', as Wieland puts it with prim self-satisfaction, 'appears as a well-formed man of medium height'. It is that in his treatment of the story of the queen Alcestis, who dies in substitution for her beloved husband Admetus, but is recovered from Hades by Hercules, Wieland has presented men and women supposedly great who have no sense of the value of this sensuous and finite life with all the goods and happiness it can offer, but are, in abstract virtue and nobility of spirit, willing, every one of them, to die for one another. But how could any man who knows his life and all that he can do with it not wish that it should last for ever? How can he even bear to contemplate death? 'Have you ever died?' Admetus asks Wieland, and the conventional Lucianic genre takes on a new and sinister reality, 'Or have you ever been really happy?' Wieland replies with all the confidence of a rationalist Stoicism unaware that his grand abstractions such as 'the dignity of humanity' are but the impoverished, secularized remnant of an obsolete religion:

WIELAND. Only cowards fear death.
ADMETUS. A hero's death yes! But common or garden death, everyone fears that, even a hero [. . .]
WIELAND. You speak like people from another world, a language whose words I hear but whose sense I cannot grasp.
ADMETUS. We are speaking Greek. Is that so incomprehensible to you? [. . .]
EURIPIDES. You forget that he belongs to a sect that tries to persuade all hydroptics and consumptives and all the fatally wounded in hip and thigh: once dead their hearts would be fuller, their spirits mightier, their bones more flourishing. He believes that.
ADMETUS. He is only pretending.

Not until Nietzsche's attack on D. F. Strauss will Germany hear again so bitter a repudiation of the Enlightened and Sentimental substitutes for Christianity that play so large a part in its official culture.

Goethe was desperately concerned, and in public, for the salvation of his soul, but he would not pretend. In the autumn of 1773 his increasing prominence before the reading public brought him the attentions of the man who so nearly understood his need that in the end, of all Goethe's myriad acquaintances, close and distant, he was the only one to be cast off in a spirit of enmity that never mellowed into reconciliation or indifference. Johann Caspar Lavater was in some respects too like Goethe, and in some respects in which he was like Goethe too much more successful, to be easy company on the road. He had a pastor's breadth of sympathy, which to some extent paralleled Goethe's own instinctual curiosity and tolerance. He was an extremely prolific writer in many genres (though always in much the same exclamatory style), who energetically pursued his interests in theology, literature, science (if physiognomics and parapsychology may be called science), and art. In addition to his practical involvement with the task of assembling the illustrative material for his massive studies of human physiognomy, he was a close friend of the painter J. H. Füßli (Henry Fuseli, 1741–1825), with whom at a very early age he mounted a successful public campaign against a corrupt official. (Although Goethe never met Fuseli, who transferred to the English art-world and became a mentor of William Blake, he particularly admired his work and copied it from an early date.) But Lavater was unusual among Goethe's friends in having something of his freedom of spirit, his freedom from the prevailing determinants of thought and feeling in Germany—not, however, because he had a wealthy father and came from an Imperial free city, but because, as Deacon of the Orphanage Church in Zurich, he was a successful member of a highly regarded profession in a self-governing republic. The compromises and pretences forced on men in an absolutist constitution meant little to him—he could brush aside both rationalist deism and Sentimentalist secularizations and pose the dilemma with stark clarity: 'either an atheist—or a Christian! I scorn the deist . . . I have no God but Jesus Christ.' He was also wholly at ease with his public, far more so than Goethe: publishing was for him a natural extension of his preaching and his correspondence. Goethe had founded his own sense of identity by dethroning the divine Christ; if someone possessed of Lavater's advantages, similar to his own in kind if not degree, devoted himself exclusively to that divinity, that could only be interpreted as an intolerable threat. Lavater became Christ to Goethe and, eventually, paid the penalty.

Before he even knew Goethe's name Lavater saw in the *Letter of the Pastor of* *** a work of genius. His own generously nebulous theology could embrace all of the Pastor's attitudes—except his abstention from missionary activity. Lavater's zeal for conversions was already notorious. In 1769 he had called on the Jewish Leibnizian philosopher Moses Mendelssohn to refute certain arguments for Christianity or make his submission, and two years later he had presided noisily over the baptism of two Jews, so exposing himself to a scathing satirical attack from Lichtenberg. He was already in correspondence with Deinet and Schlosser, and his relations with Goethe began when he wrote in August 1773 to express his thanks for a presentation copy of *Götz* which had come to him through Deinet. His physiognomical interests were beginning to develop and his belief in the unique status of Jesus Christ, whose face, if physiognomy meant anything at all, must have revealed the most perfect form of humanity, had led him to make a collection of imaginary portraits of Christ drawn by anyone who was himself remarkable. To this collection he wanted Goethe to contribute, but it was a sore disappointment to him when the author of the Pastor's letter wrote back to him that, in so many words, 'I am not a Christian'. For the portrait Lavater was referred to Susanna von Klettenberg, who claimed a closer acquaintance with the sitter. Lavater was not only puzzled —a lesser mortal than Goethe who was not a Christian, could not, he felt, have written the *Letter* without some dishonesty—he had also received a challenge: 'I shall show you Christ,' he wrote at the end of November 'or I shall take up my pen against him . . . you shall become [a Christian]—or I shall become what you are'. That was precisely not what the Pastor had said, for he believed rather in everyone's staying as they were, and different. Goethe wished neither to give himself up, nor to share himself with someone else. He told Lavater in January 1774 to remain content with the Messiah he already had, and not to change loyalties and seek one in him. But the threat to his identity became explicit when Lavater took to signing himself 'from eternity to eternity your brother' and wrote:

If Jesus Christ is not my God—then I have a God no longer—and Goethe and . . . Lavater are dreamers, not brothers, not children of *a single father*—not immortal.— Then friendship is nothing, all is an illusion, no existence etc.

Goethe liked brothers he could lead and teach, but not brothers who were rivals and wanted to make his personality and very existence dependent on their sharing his status as son of his father. 'People say the curse of Cain is upon me,' he had written to Kestner in June 1773, 'but I have not murdered a brother!' He had, however, discarded the Mediator, who Christians said had made them joint sons and heirs of his Father, and he had certainly in his life had murderous thoughts, 'that deserved hanging', about at least two other human beings— Kestner himself, and Hermann Jakob Goethe, who at the age of 6 had succumbed to the fate of Abel. Lavater had unwittingly, or instinctively, chosen the worst possible terms in which to appeal to him. The life on which he had embarked, Goethe told his 'brother' in April 1774, was a life devoted to 'the word of men' as an explicit replacement for 'the word of God'—he could not, that is to say, give the Christian Scriptures an ultimate authority in determining the meaning of his life without destroying himself as a poet, a man of words. Lavater was willing, even on these terms, to continue his struggle: 'we are symbols', he had written, alluding to the terms of a letter from Goethe which has been lost, 'and our words and works are too. Let us symbolize, then, since we must, for as long as we can—and, yes, we need not "fix any goal." ' Perhaps Lavater hoped that in the end the long chain of symbols would catch on a fixed and authoritative truth. He was destined instead to learn the wisdom of the words Schlosser wrote to him as his correspondence with Goethe began:

'Continue to love him. But I tell you in advance: a certain strength of soul is necessary to be his friend.'

It was perhaps partly as a result of Lavater's onslaught that in the early winter of 1773 Goethe showed signs of moderating the explosive contempt for Sentimentalism displayed earlier in the year in *Satyros* and *Gods, Heroes, and Wieland*. A more conciliatory attitude to that alternative to religion was implied in the scheme for *Erwin and Elmira* begun in November. Moreover from September onwards his personal links with the enemy had grown closer: as visitors of Johanna Fahlmer and then good friends of Cornelia, first Charlotte Jacobi (1752–1832), a half-sister of the two brothers, and then Friedrich Jacobi's wife Betty (1743–84), whom he later described as a splendid Rubensian Dutchwoman without a trace of sentimentality, both became part of his circle. He said nothing of the brothers, whose 'friendship I do not want', but when Betty Jacobi returned to Düsseldorf he continued to write to her and send her such pieces as *Lumberville Fair*, which of course did not remain a secret from her husband. Sophie von La Roche also visited Frankfurt in July and August, with her daughter Maximiliane, now 17. A marriage was being arranged with the 38-year-old businessman of Italian origin, P. A. Brentano (1735–97), who already had five children from his previous wife, and it duly took place at Ehrenbreitstein in the new year. On 15 January 1774 the bride arrived in Frankfurt with her husband and her mother, to take up her new role. Merck, whose return from Russia in December had made Goethe 'mad with joy', commented that it was a strange experience for him to seek his friend Mme La Roche 'amid barrels of herrings and cheeses'. With her sharing his burgher world, Goethe found it easier to share her noble spirituality, perhaps unconsciously they recognized in Brentano a common enemy. When she left at the end of January Goethe felt that 'I am far more to her, and she is far more to me, than two years ago, yes, than six months ago'. It was natural for him, as Merck put it, to console 'la petite Madame Brentano' for her transplantation from a country estate to the gloomy counting-houses of Frankfurt. He played with the children, let Maximiliane accompany him on the cello, and allowed himself for the first time to be drawn into the company of the city's Catholic families. Although Brentano was pleased to see him at first, he was not however as accommodating as Kestner of sentimental friendships with his wife. By the end of February there were some ugly scenes and Goethe resolved, or was commanded, not to visit the house again.

With Merck's return the sombre solitude of 1773 was relieved at last. It was a hard winter, the Main froze, and Goethe was out skating even before the ice was firm, getting a soaking as a consequence. Maximiliane Brentano seemed in January to be 'a compensation' fate had sent him for the loss of his sister, she too enjoyed skating, and there were some lively parties on the ice, with benches on duckboards for spectating mothers, and tables laid out with hot pies and hams, chocolate, coffee, and wines. 'No branch of my existence is lonely',

Goethe wrote at the beginning of February, and in the literary world too he could feel the gathering momentum of a collective endeavour. A second edition of *Götz* appeared, published this time in the normal way, by Deinet, and then, extraordinarily, the unperformably irregular play was in April given its first performance, and in Berlin, the heart of francophone absolutism. Lenz's *The Tutor* was expected at Easter: Goethe commended it in his correspondence, having reluctantly given Lenz permission to publish *Gods, Heroes, and Wieland* and so to start a pamphlet war (which came to include a reply about which the best thing was its title, *Men, Beasts, and Goethe*). There were other young contemporaries showing promise in Frankfurt itself: the musician P. C. Kayser (1755–1823), and particularly F. M. Klinger (1752–1831), the son of a widowed laundry woman, who by his abilities had secured himself a place in the grammar school and now wished to study law at a university. Goethe gave him 100 guilders and recommended him to Höpfner in Giessen. Increasingly Goethe felt reconciled to the business of publication: he had a request for manuscripts from the publisher Weygand in Leipzig and there seemed an opportunity to make some money to cover the debts incurred during the printing of *Götz*. With his spirits lifting and a sense at last of distance from the events of 1772, Goethe asked Merck to return to him the letters he had written from Wetzlar, and the day after Sophie von La Roche had left, on 1 February, he set to to turn them into a novel, as he had once envisaged doing with the letters he had written to Behrisch about his sufferings during the Käthchen affair, a novel that fused his own experiences of that idyllic summer with the shock of Jerusalem's terrible end. But as he wrote reality caught up with his fiction and in the Brentano household in February and March he found a third model for the pains of unhappy love between a man and a woman pledged to another. In *The Sorrows of Young Werther* (*Die Leiden des jungen Werthers*), which after two months of intensive work was completed in April, at the printers by June, and published at Michaelmas, and which made Goethe a European celebrity, the woman Werther loves may bear the name of Lotte Buff, but she has the dark eyes of Maximiliane von La Roche.

Detonating the Bomb: Works, 1772–1774

The name 'Storm and Stress', applied to the young writers of the 1770s, is taken from the title proposed in 1776 by Christoph Kaufmann, an itinerant intellectual charlatan, who called himself 'God's Bloodhound', for a drama of his friend F. M. Klinger that was originally, and perhaps more appropriately, entitled *Confusion* (*Wirrwarr*). An older name for the period was 'Geniezeit' or 'age of geniuses'. But the labels tend to overstate the homogeneity of the literary revolt. There were numerous eccentric or marginal individuals, such as Kaufmann, the infinitely worthier Jung-Stilling, Lavater, or Bürger, the author of *Lenora*, who later put the national culture rather more deeply in his debt by translating back into German the stories about Baron Münchhausen (which, though collected of course in Germany, were originally published in English). The last and most passionate of all the representatives of that passionate age, Friedrich Schiller, was destined to secure his own literary development by an annihilating review of Bürger's poems, in which he vicariously put behind him that period of his own life represented by the plays *The Robbers* (*Die Räuber*) (1781), *Fiesco's Conspiracy in Genoa* (*Die Verschwörung des Fiesko zu Genua*) (1783), *Cabal and Love* (*Kabale und Liebe*) (1784), and his *Poetical Anthology for the Year 1782* (*Poetische Anthologie auf das Jahr 1782*). Apart from these more isolated figures, however, not one but two main centres of literary innovation can be distinguished in the mid-seventies, the one more or less 'official', the other genuinely revolutionary. In Göttingen there was a group, essentially of lyrical poets, gathered round H. C. Boie, the editor of the *Musenalmanach*, and J. H. Voss (1751–1826), later to become known as a translator of Homer. Their best poet was L. C. Hölty (1748–76); their number also included Friedrich Leopold Count Stolberg and one dramatist, J. A. Leisewitz (1752–1806). Usually known as the 'Göttingen Grove', or the 'League of the Grove', from the moonlit gathering in a grove of oak-trees at which the founder-members swore eternal fealty and to keep the minutes of their weekly meetings, they owed allegiance primarily to Klopstock. Their emphatic patriotism ('Ich bin ein Deutscher! (Stürzet herab/Der Freude Tränen, daß ich es bin!)', wrote Stolberg—'I am a German! (Cascade, o tears of joy, that I am)') fed on the hope of bringing back to life the Germanic-cum-Celtic past of Klopstock's vision. It had little to do with the attempt to make literary use of the contemporary spoken language that was the distinguishing feature of the second group, consisting essentially of dramatists and centred, though rather less exclusively than was the case with the Grove poets, on Frankfurt, and on Goethe. Goethe himself maintained good relations with the Klopstockians and being a lyric poet as well as a dramatist kept rather above the mêlée, midway as it were between Göttingen and Frankfurt, Hölty and Lenz. If the behaviour of the Grove poets was comically earnest and ritualistic, that of the second group, the 'Storm and Stress' proper, was at times deliberately anti-social. Of Klinger, nicknamed the 'lion's-blood-drinker' (Löwenblutsäufer) it was said: 'If it is good character to despise all other fellow human beings, to do nothing in company but play the great man yet have to borrow one's suits, to straightway threaten daggers and poison if one's wishes are not fulfilled every day, then he too has a really good heart.' Friedrich Müller (1749–1825), generally known as 'Müller the Painter' from his major occupation, or 'Müller the Devil'—he was the author of an interminably uneventful and bombastic *Faust*-fragment—had no qualms about accepting the proceeds of a collection, partly arranged by Goethe, to enable him to study in Rome for the benefit of German painting, and then staying in Rome for most of the rest of his life, leaving to their fate the girl he had seduced (she died shortly afterwards) and her illegitimate child. His appearance at the Frankfurt book-fair in 1778 was described thus: 'At the auction in Frankfurt Müller the Painter behaved almost like Klinger, was extremely rude, like a true genius said of anyone whose physiognomy did not appeal to him [the pious Lavater had something to answer for] "I'd like to have the fellow's head off, he's a rogue".' Yet although both these figures received financial support from Goethe, and Klinger had a special place in Goethe's affections, neither of them had any feeling for what it was that inspired *Götz*, so much of Goethe's lyrical and satirical verse, and the first version of his *Faust*— the poetic potential of the spoken German language of the time. That insight was reserved to what we may call the 'Strasbourg group', those who had shared with Goethe the Alsatian experience of 1770–1, Herder, Lenz, and H. L. Wagner. Klinger's imitative talent enabled him to show traces of this inspiration in his first two plays, *Otto,* the first chivalric drama written on the pattern of *Götz,* and *A Woman in Suffering* (*Das leidende Weib*), which has a contemporary German setting, and is full of contemporary (mainly literary) allusions, but is inconceivable without its model, Lenz's *The Tutor.* It is not the linguistic but the historical—or, better, mythical—aspects of *Götz* that are imitated in *Otto,* as in the fifty or so similar chivalric dramas written by various poetasters over the next three decades; and problem-plays of the period—including all of Klinger's numerous other dramas—prefer a remote, often Italianate, setting to one in contemporary Germany. The dramatic language of Klinger and Müller was vividly characterized by Lichtenberg, when he wrote in his commonplace-book: 'It is as if our tongues were confused: when we want to have a thought they give us a word, when we ask for a word, a dash, and where we expected a dash stands an obscenity.'

In his fantastical satire on current literature and criticism, *Pandaemonium Germanicum* (written 1775), Lenz confesses, in his own person: 'Alas I had resolved to go down the mountain and became a painter of human society . . . '. His first play, *The Tutor or The Benefits of Private Education* (*Der Hofmeister oder Die Vorteile der Privaterziehung*) (1774), despite the ramifications of its sub-plots, and the bizarre main plot, which culminates in the self-castration of the tutor who has seduced his female charge, has many individual scenes that achieve just that ambition, communicating the thrill of authenticity that an English reader is used to looking for in social novels. *The Soldiers* (*Die Soldaten*) (1776), is his masterpiece; as recent criticism has shown, it is tragedy—though Lenz calls the play a comedy—traced through, and only through, the different ways in which the characters speak: as the girl Marie is morally debauched so her speech is progressively corrupted. As she—like everybody else—has only language in which to express herself this is a total and unredeemable corruption. This sense that language is not just a medium that might be arbitrarily different, but is the very 'form of life', is developed in Lenz as in no other writer of his generation. Wagner shares it to a certain extent, as is shown particularly by the first act of *The Infanticide,* though the later part of that play tails away into Sentimental bookishness. It is a measure of his intelligence and mimic gift that the young Schiller, writing more or less in isolation, was able to show this sense too, in a large part of *Cabal and Love*. Herder of course possessed it in full, though as a theoretical acquisition, and increasingly it was to alienate him from the dominant philosophical fashions of the late eighteenth century. Only Goethe possessed it with the sovereign control necessary to make poetry out of it, supremely so in the story of Faust and Gretchen, an achievement which has remained unique. German writers who might be seen as later inheritors of this, Lenzian, aspect of the Storm and Stress, such as Büchner and Hauptmann, have been dedicated to prose—only Brecht has realized some of its possibilities in verse.

It is only from the point of view of the 'Strasbourg group', and its preoccupation with the language of the people, that we can understand the political and cultural implications of the literary programme of the Storm and Stress. On the one hand these implications were nationalist. The common language appears as the ground of a cultural unity more profound than any realized in the existing or imaginable institutions of the German-speaking world. This is why there is a void in *Götz* at the point where we expect to find a positive enthusiasm. On the other hand it would be a mistake to see the ambitions of Storm and Stress as purely cultural ambitions towards a 'republic of learning'—Klopstock's work of this title (*Die deutsche Gelehrtenrepublik*) was announced in 1773 and attracted 3,599 subscribers. National unity alone was not enough, for that threatened to be achieved—and in the course of time was achieved—by enlightened absolutism, which was Möser's *bête noire*, the bureaucratic enemy of local tradition and creative individuality, of the twin pillars of Herder's theory of language. The Storm and Stress writers were looking for something to oppose to autocracy that was as politically and socially concrete as Möser's self-governing 'home towns', but they wanted it to share the aspiration to national unity that, at the time, autocracy alone provided: of this paradoxical ambition their devotion to the language of the people was both a source and the potent, if transient, artistic symbol. It was a symbol, however, of something that in the circumstances of the time was an unreality: it was as if late eighteenth-century Britons had had to build a national culture without London and on the basis of the work of Burns and Crabbe. Lenz acknowledged in *Pandaemonium Germanicum* that contemporary Germany simply did not provide him with the material for his portrait of human society. All too often in the works of the Storm and Stress that dangerous vacuum comes to be filled by a preoccupation not with language but with genius, not, that is, with society, but with the pure, untrammelled, and so ultimately featureless and unpolitical individual. The 'big man' ('großer Kerl'), to quote Müller's preface to his *Faust*, 'who absolutely desires to transcend himself' comes to occupy all of Müller's, and most of Klinger's work. Even the genuine achievements, even *Götz* and *The Tutor*, *The Infanticide* and *Cabal and Love* (though not *The Soldiers*) are flawed or partial: the introspective themes and artificial language of domestic drama eventually break in, the drama of revolt (like that of Götz) gives way to the drama of self-destruction (like that of Weislingen). For absolute individualism belongs only in a theatre of inner conflict, not in a theatre of external conflict between worldly forces, and so is in the end the hand-maid of state absolutism.

The cult of genius was the Trojan horse by means of which Sentimentalist, and ultimately Leibnizian, notions were imported into the Storm and Stress movement and the beginnings of a realist literature about contemporary society were subverted. There was nothing 'unofficial' about the idea of genius. J. G. Sulzer became best known as the author of a *General Theory of the Fine Arts* (*Allgemeine Theorie der schönen Künste*), which appeared between 1771 and 1774, an aesthetic encyclopaedia more remarkable for its conceptual, bibliographical, and indeed alphabetical, completeness than for any great penetration in its content: the radicalism of the *Frankfurter Gelehrte Anzeigen* is particularly apparent in Goethe's repudiation of this later phase of Sulzer's thinking. Yet in the 1750s Sulzer produced a series of original essays, his most interesting work, which demonstrate with great clarity the links between Leibnizian philosophy and the foundations of German Sentimentalism. His *Enquiry into the Origin of Pleasant and Unpleasant Sensations* of 1752 offers a characteristically Leibnizian dynamic portrayal of the soul as first and foremost neither a thing nor a passive organ but an active force. And the one activity in which the soul engages, which Leibniz described as 'perception', Sulzer describes as 'the production of ideas': all sensations, Sulzer claims, can be subsumed under the heading 'ideas', all desires are desires of the soul to have ideas, that is, to exercise more extensively its single faculty; all pleasures proceed only from the soul's desire to feed itself with thoughts, even the delights of love and friendship, or the pleasure taken by the fop in dressing himself well or by the ambitious politician in the success of his plans. Sulzer makes no distinction between receptivity and productivity, between passivity of the soul in sensation and its activity in expression. And in 1757, when he set about analysing the fashionable concept of 'genius', Sulzer did not find—because he could not find—the peculiarity of genius in its possession of special or unusual faculties. Because Sulzer believed in the existence only of one faculty, the peculiarity of genius had to be simply the possession of that faculty to an altogether more intense degree, a greater strength of soul ('Stärke der Seele') than that of ordinary mortals:

All capacities of the soul originate in that basic force which, as the great Leibniz has remarked, constitutes the essence of all substances, and in particular the essence of the soul . . . In it we must also seek the origin of genius.

The identification of receptivity and productivity as simply aspects of the soul's one basic force, the interpretation of that force in the relatively empirical terms of thoughts and feelings (rather than the logical terms favoured by Leibniz), and the definition of genius as an abnormally forceful soul: these are the theoretical foundations of the psychology espoused by German Sentimentalism and passed on by it to the Storm and Stress movement, even though Sulzer's abstract formulations were not always appreciated by less philosophical minds.

We have already seen how the Leibnizian strand in official German culture of the eighteenth century reflects social developments peculiar to Germany and issues in a literature of compromised isolation. If the German Crusoes were less rigorous in their solitude than the English original, German Sentimental literature, from Gellert to Gerstenberg, shows less concern than the English movement with the social mechanisms which might release the castaway from his desert isle, or his not-so-desert Felsenburg. In the poems which Klopstock wrote in the 1750s and 1760s the egoistical twist which we noticed in the English poetry of feeling becomes repetitively and even disablingly apparent. Mere reference to the poet's self (like references by name to friends and fellow-poets) is used, repeatedly, to heighten the emotional tension, yet in the very moment of use this device distracts from the specific nature of the emotion invoked. Even a vividly concentrated poem such as 'The Early Graves' ('Die frühen Gräber') (1764) hesitates vaguely in its last strophe when the poet, in the moonlit night, for the first time speaks directly of those who have passed over before him, and for the first time mentions himself. The reference to his own feelings is supposed to make the moment more intense, but in fact makes it less precise.

> Ihr Edleren, ach es bewächst
> Eure Male schon ernstes Moos!
> O wie war glücklich ich, als ich noch mit euch
> Sahe sich röten den Tag, schimmern die Nacht.

Alas, you nobler ones, sombre moss already overgrows your monuments! O how happy was I when still in your company I saw the day's red dawn, the shimmering night.

The excitement Klopstock derives from mentioning himself is naïve and so not distasteful, though it is sometimes unintentionally comic (as in his frequent addresses to the Almighty). But there is always a hiatus, a discontinuity between the references in his poems to objects, or to people as the objects of his feelings, and the references to himself, or to the heart, the asserted source of the feeling. The second line of the first strophe in his poem on 'The Lake of Zurich' ('*Der Zürchersee*') (1750) is broken-backed in just this way:

> Schön ist, Mutter Natur, deiner Erfindung Pracht
> Auf die Fluren verstreut, schöner ein froh Gesicht,
> Das den großen Gedanken
> Deiner Schöpfung noch einmal denkt.

Fair, O Mother Nature, is the splendour of your inventiveness, strown over the fields, fairer still a happy face thinking again the great thought of your creation.

In Goethe's poetry of the early 1770s there is no such hiatus. Its glory is the immediacy of the contact between the feeling heart and the object of feeling.

> Wie herrlich leuchtet
> Mir die Natur!
> Wie glänzt die Sonne!
> Wie lacht die Flur!

How gloriously Nature glows for me! How the sun sparkles! How the fields laugh!

What is striking about the second line of 'May Celebration' ('Maifest') (1771), a Sesenheim poem first published in 1775, is not, as is often asserted, the emphatic placing of the first person pronoun at its start, but the exact counterbalancing of that pronoun, in the same line, by the reference to 'die Natur'. Where in Klopstock's poem there is antithesis, in Goethe's there is reciprocity. Throughout 'May Celebration' subjective and objective terms interpenetrate, at times to the point of indistinguishability:

> O Erd', o Sonne,
> O Glück, o Lust,
> O Lieb', o Liebe . . .

O Earth, O Sun, O bliss, O pleasure, O love, dear love . . .

The unprecedented fluency of this rhyming litany seems at a single stroke to render obsolete the gawky sentiment of the previous quarter of a century. It is no wonder that the received chronology of modern German literature dates its beginning from 1770. And yet the interpretation of Goethe's first adult works, and of the literary movement of Storm and Stress to which they are usually assigned, is not possible unless we take account of the survival in them of the cult of sentiment. It was through 'Empfindsamkeit', and the theory of genius, that Goethe had his most intimate access to the heart of the national culture. Goethe at this time in his life, as never again, spoke directly from and to the situation of his contemporaries: they recognized in his work the same problematic of feeling that to them was a burden, even if it also seemed to them that for Goethe it was as light as air. In fact for Goethe the burden was not at all as light as 'May Celebration' may make it appear, and had he been able to discard the burden he could not have received from his public the acclaim that he did. The forces in 'May Celebration' are in perfect equilibrium: that is not to say that they are non-existent.

During his long and often solitary journeys on foot between Frankfurt and Darmstadt in the early months of 1772, which earned him among the 'communion of saints' the name of 'the Wanderer', Goethe would chant to himself impromptu verses, 'strange hymns and dithyrambs' in which his reading of Pindar (thought at the time to be a writer of free verse) mingled with the rains and mists of an inclement April to produce what he afterwards called 'half-nonsense'. Among them was 'Wanderer's Storm-Song' ('Wandrers Sturmlied') (not published by Goethe until 1815) which, far from being nonsense, is an urgent meditation, in the rhythm of a foul-weather hike, on the soul's capacity, as Leibniz and Sulzer would have put it, for producing ideas—especially poems.

The thematic image is that of inner and outer heat: the poet's genius is an inner sun, but a sun which strives to put itself in an equal and reciprocal relationship with the outer sun, Phoebus Apollo, who warms the natural world.

> Weh! Weh! Innre Wärme,
> Seelenwärme,
> Mittelpunkt,
> Glüh' entgegen
> Phöb Apollen,
> Kalt wird sonst
> Sein Fürstenblick
> Über dich vorübergleiten,
> Neidgetroffen
> Auf der Zeder Kraft verweilen,
> Die zu grünen
> Sein nicht harrt

Woe! Woe! Inner heat, heat of soul, central point, glow out towards Phoebus Apollo, or else his imperious glance will coldly pass over you and, struck with envy, rest upon the might of the cedar which does not wait for him before greening.

These lines have been found difficult of interpretation, but they are quite consistent, and consistent with the Sentimental theory of genius. According to that theory, genius is an exceptional force of soul, and this exceptional force can be regarded either as intensely receptive or as intensely productive. Goethe, the Wanderer, urges on his inner furnace to 'glow out towards', to meet and to rival, Phoebus Apollo—not only the patron of the Muses but also, as sun-god, the vivifier of all the external world. For the Wanderer can expect the favour of the god only if he demonstrates that he does not need it, only if by his proud independence he can arouse the envy of Apollo, whose light seems to linger on the evergreen that ignores his seasonal moods. Nature, that is, will furnish the poet with inspiration only if he proves himself capable of producing for himself, from his own inner resources all she has to offer. Only on condition that he produces can he receive. This self-imposed demand is of course impossibly extreme. 'Wanderer's Storm-Song' ends with a decline from hectic excitement into a rueful and breathless humour as the poet struggles to the shelter of his mountain hut:

> Glüht deine Seel' Gefahren,
> Pindar,
> Mut.—Glühte—
> Armes Herz—
> Dort auf dem Hügel,

> Himmlische Macht,
> Nur so viel Glut,
> Dort meine Hütte,
> Dort hin zu waten.

Your soul, Pindar, glows out courage against dangers.—Glowed—Poor heart—there on the hill, heavenly power, only so much fire, there, my hut, must wade up there.

'Wanderer's Storm-Song', a dogged chant, is clearly a poem of late winter. 'Ganymede' ('Ganymed'), a meditative incantation, is clearly a poem of early spring (actually, the spring of 1774). Not only is its theme very close to that of 'Wanderer's Storm-Song', however, 'Ganymede' is a working example of the symmetrical process which 'Wanderer's Storm-Song' discusses and describes: Nature (whatever that may be) reaches out to embrace the soul only with the same energy, and no more, with which the soul drives itself into the arms of Nature. As Ganymede's yearning drives him up into the heavens, so the clouds of heaven sink downwards in acknowledgement of this power of love:

> Hinauf, hinauf strebt's,
> Es schweben die Wolken
> Abwärts, die Wolken
> Neigen sich der sehnenden Liebe,
> Mir, mir!

Upwards, upwards the drive, the clouds float downwards, the clouds bow down to the yearning of love, to me, to me!

Goethe has thus decisively modified the classical story which gives his poem its title, for Jupiter's love for Ganymede, which caused him to pluck the boy up into heaven to be his cup-bearer, was certainly not dependent on Ganymede's love for the god. In this very emphasis on interdependence we may detect a note of strain which differs from the serene confidence in reciprocity which we find in the Sesenheim lyrics, in 'May Celebration', and in the last lines of 'My Heart Pounded':

> Und doch, welch Glück, geliebt zu werden,
> Und lieben, Götter, welch ein Glück!

And yet what bliss to be loved, and to love, ye gods, what bliss!

The mood of the poem of 1774 is sometimes described as pantheistic, as if, that is, it were based on the belief that all things are really one, and as if the one thing that all things really are were presented indifferently either as Nature or as God. The fact that Goethe is known in 1773 to have become acquainted with some of the ideas of Spinoza is held to give colour to this interpretation. As late as 1785, however, Goethe confessed that he had never yet read Spinoza so methodically as to have an adequate grasp of his whole system. Moreover it is a serious misrepresentation of Goethe's poem to imply that at any point in it 'Ganymede' loses his identity in that of some universal substance, Nature, or his 'all-loving Father'. 'Ganymede' remains throughout the poem an independent individual whose 'strength of soul' is always equal to that of the surrounding world, never subordinate. Even at the climax of the poem it is as true to say that he absorbs the Other into himself, as that he is absorbed into the Other.

> In eurem Schoße
> Aufwärts,
> Umfangend umfangen!
> Aufwärts
> An deinem Busen,
> Alliebender Vater!

In your [the clouds'] lap, upwards, embracing embraced, upwards, upon your bosom, all-loving Father!

'Embracing embraced'—if a philosopher is to be chosen as patron of these lines it must be, not Spinoza, but Leibniz, the philosopher who believes that every identity is invulnerable but also that every identity has the single task of representing all the rest of the universe from its own point of view. In 1785 Goethe still thought that the Spinozistic vision, as far as he understood it, threatened to eliminate the singularity of individual things.

'Wanderer's Storm-Song' is a poetic statement, 'Ganymede' a poetic enactment, of the dilemma posed for Sentimentality by its assumption that at the centre of each unique individual's world lies a unique faculty of sensibility. This faculty is—supposedly—both completely private and yet the source of all the relations that hold between the individual and the world about him. When that is the prevailing way of picturing oneself, how on earth is a man who intuitively knows his emotions to be dependent on, and continuous with, things as they really are, to write the poetry he intuitively knows to be possible—a poetry of *objective feeling*? Lesser poets would have been satisfied with writing about the dilemma—Goethe, throughout his life, also struggles to solve it. By accepting the vocation revealed to him by Herder, that of being a poet in the German national language, Goethe has accepted also the task of assimilating those elements in the national culture which are alien to his origins and even perhaps difficult to reconcile with the vocation itself. The year 1773 shows him opening himself more and more to the dominant literary and social currents of the age, to secularized religiosity, subjectivity, and Sentimentalism, and so gradually approaching the greatest crisis of all, the task of integrating the conflicting forces, new and old, into one man's productive life. In 1774 and 1775 that crisis is expressed and resolved in three decisive acts which create and fix—in a sense, for ever—what we know as Goethe's literary personality: the writing of *The Sorrows of Young Werther* and of the kernel of his life's work, *Faust* (for the present unpublished), and the complex symbolic act that was the breaking-off of his engagement to be married and his removal from Frankfurt to Weimar.

In 1773, and particularly after the decision to publish *Götz*, Goethe's confidence in his literary powers grew, despite the shadows falling over his personal life. The ease with which he was thinking his way back into the sixteenth century, when cities like his own Frankfurt had had their prime, and was mastering and adapting new poetic forms in order to create an ideal image of an authentic German tradition, was a little breath-taking even to him. However disparate the material, he seemed able to make of it a whole. Just as his new wealth of verse-forms incorporated as much of the rhythms of eighteenth-century operatic and occasional writing ('madrigal verses', they are called) as of the sixteenth-century 'Knittelvers', so *Götz* contained recognizable allusions both to Reformation Germany and to modern conditions, and yet, as one of the play's first readers noticed, 'none the less everything fits together'. As Goethe's personal isolation increased so did the challenge to his all-synthesizing imagination. He created for himself (and the account in his autobiography is supported by contemporary evidence) a mental audience of those whom he was trying to address and win over: not those who were physically and emotionally nearest to him at the time, but dear ones who were distant, transient acquaintances from the past, still remembered with their characteristic gestures, even figures, like Helene Buff, whom he had never met. In his ghostly conversations with these partners only semi-independent of himself, so similar, he felt, to the process of writing letters, he endeavoured to find roles for himself to act out which both had some general moral or historical significance and could be filled by him with a sense of selfhood: roles which fused both a character and a consciousness, the two functions of a personality which he had separately distributed to Götz and to Weislingen. The wood-devil Satyros represents Goethe himself, but as a Herderian, perhaps even as Herder, while the hermit in that farce represents him as a Rousseauist. In the figure of Faust, and (probably) also of Wilhelm Meister, first given written shape during this extraordinary summer, even that residual division of functions is overcome, and the result is in each case a character whose identity is defined not psychologically, but symbolically, even mythically: while seeming concrete and specific, they have the universal uniqueness of the self. 'My ideal figures' ('meine Ideale'), Goethe wrote to Kestner and Lotte in September, 'grow and develop daily in beauty and grandeur, and if my vigour does not abandon me, or my love, there is a lot to come for my loved ones, and the public too will have its share.' If H. M. Wolff is right in identifying Wilhelm Meister as the hero of the novel on which Goethe says in the same letter that he is working, he is if anything a more remarkable creation than Faust: a wholly original symbolic figure who is at once both character enough to be a young German mid-eighteenth-century bourgeois, and consciousness enough to be (as Faust is not) a poet—a body and a name, therefore, for Goethe's self in its most distinctive activity. But progress was slow in the enormously, perhaps impossibly, difficult task of making so fully reflective a self-consciousness into the centre of a socially realistic novel with a contemporary setting, and Goethe instead summoned up in verse a symbolic representation, or 'ideal', of himself as poet. For the brief duration of an ode, a concrete person and situation, the voice that says 'I', and the activity of poetic creation, are all brought together into a single image of searing simplicity. Its name is Prometheus.

Prometheus, the Titan who formed the human race from earth and stole fire from heaven for their benefit, is the very type of the solitary creator: unlike Shaftesbury's 'second maker' he is neither just nor tolerant of his subordination to Jove. The subject-matter of the ode which he dominates is the theological revolt which has made poetry—this poetry—possible, and which, divorced from the issue of poetry, is also the theme of Faust. The solitary 'genius', however, protests at his solitude. As Goethe made the concerns of the German public mind ever more nearly his own, so his personal sensuality and his Frankfurt 'Republican spirit' were roused to conscious resistance. The laughter in *Satyros* at contemporary gullibility is brutal rather than wry, and beneath the comic surface of *Gods, Heroes, and Wieland* is a passionate rejection of Wieland's trivially noble attitude to death. Bitterness and anger without parallel in the rest of his work prevail in his Prometheus' cry of defiance against the heaven from which he has stolen fire, in his explicit and savage repudiation of the God of the Pietists and their bogus consoling Saviour, and his endorsement of the process of secularization in a spirit of ferocity undreamt by Winckelmann (though Lessing, a subtle and militant atheist, recognized himself in the poem immediately). Goethe later permitted the ode to be published only as a companion-piece to 'Ganymede', which seems in comparison feebly program-matic. 'Prometheus' marks the point at which Klopstock's answer to the question: what shall I write about? finally becomes unavailable to Goethe. For a literary man of the Protestant Enlightenment to adopt biblical, and particularly Christian, themes is, as Klopstock's example shows, to drift into a dualism which so separates spirit from body that all that the poet really wishes to represent—the divine nature and activity, the hearts and feelings and even identities of his human characters—is consigned to an unrepresentable world accessible only through similes. But, Goethe's demigod cries, the human heart, the inner furnace of the Wanderer's sensibility, is the only divine fire, this earth and what a man builds here the only abiding city, the will to survive, not any supernatural intervention, the only defence against death; the radiance of heaven is a delusive reflection of men's own activity, a mirage on which only fools and weaklings think themselves dependent:

> Bedecke deinen Himmel, Zeus,
> Mit Wolkendunst!
> Und übe, Knaben gleich,
> Der Disteln köpft,
> An Eichen dich und Bergeshöhn!
> Mußt mir meine Erde

Doch lassen stehn,
Und meine Hütte,
Die du nicht gebaut,
Und meinen Herd,
Um dessen Glut
Du mich beneidest.

．　　．　　．　　．　　．

Da ich ein Kind war,
Nicht wußte, wo aus, wo ein,
Kehrte mein verirrtes Aug'
Zur Sonne, als wenn drüber wär'
Ein Ohr, zu hören meine Klage,
Ein Herz wie meins,
Sich des Bedrängten zu erbarmen.
Wer half mir wider
Der Titanen Übermut?
Wer rettete vom Tode mich,
Von Sklaverei?
Hast du's nicht alles selbst vollendet,
Heilig glühend Herz?

Cover your sky, Zeus, with vaporous cloud and like a boy beheading thistles exercise yourself on oaks and mountain-peaks. You will still have to leave my earth standing and my hut, which you did not build, and my hearth whose glowing heat you envy me. When I was a child, did not know coming from going, my lost gaze turned to the sun as if above it were an ear to hear my complaint, a heart like mine to take pity on one hard pressed. Who helped me against the arrogance of the Titans? Who rescued me from death, from slavery? Was it not all your own achievement, sacred glowing heart?

Neither Feuerbach nor Nietzsche will add anything of emotional or intellectual significance to this outburst of an Antichrist. Indeed Goethe imagined with a deeper seriousness than his posterity the replacement of the old God by the human heart of Sentimentalism and by the human creativity celebrated in the theory of genius. Aware of the exceptional power of the symbol he had stumbled across, he started to develop his 'ideal' from the speaker of a monologue into the protagonist of a drama. 'I am working up my situation into a drama in defiance of God and men', he told Kestner in July. The ode concludes with a tableau that already announces the setting for the two acts which Schönborn heard read in October:

Hier sitz' ich, forme Menschen
Nach meinem Bilde,
Ein Geschlecht, das mir gleich sei,
Zu leiden, weinen,
Genießen und zu freuen sich,
Und dein nicht zu achten,
Wie ich.

Here I sit shaping men in my own image, a race to be like me, to suffer, weep, enjoy and to be happy, and to disregard you, as I do.

Although superficially fragmentary, the drama *Prometheus,* personally one of the most searching of all Goethe's early works, presents more completely than any eighteenth- or nineteenth-century philosopher the limitations of systematically secular literature.

The drama takes up the themes of the Leipzig ballad 'Pygmalion', but treats them with a new profundity, rendered possible by Rousseau's 'scène lyrique' *Pygmalion,* which Goethe had read and admired in the winter of 1772. Surrounded by the statues he has made, Prometheus defies the gods—his first words are 'I will not'—who seem to him no better than himself, and in himself he puts all his trust:

> Vermögt ihr mich zu scheiden
> Von mir selbst?

Are you able to divide me from myself?

he asks Mercury the divine messenger, who is seeking his obedience,

> Vermögt ihr mich auszudehnen,
> Zu erweitern zu einer Welt?

Are you able to extend me, expand me into a world?

That only he can do, and in his statues, he has done it:

> Hier meine Welt, mein All!
> Hier fühl ich mich;
> Hier alle meine Wünsche
> In körperlichen Gestalten

Here my world, my all! Here I feel myself; here all my desires in bodily shapes

He has made his own world out of his own sensations and emotions, he has need of no other, and he resents the gaze of the stars that suggest the existence of something outside 'the circle that my action fills'. At one with his own creation, he feels himself to be as eternal in relation to it as the gods claim to be in relation to him: he denies the possibility of his own mortality. Having no recollection of his beginning, how should he expect an end? Yet the world created by this solipsistic genius lacks one thing, in one respect it is not his equal: it does not live. That supreme satisfaction of the artist, to make something blessedly, actively independent of himself, is denied him. And Jupiter offers to bring his statues to life if he will desist from his rebellion against the divine order: it is, Prometheus admits, the one consideration that gives him pause. It is a pause of the deepest significance in the history of modern sensibility. Is the mortal, transient artist the servant of an independent world-order which stretches before and after him, to whose laws he acknowledges himself subject and parts of which are imitated in his works? If so, a realistic, objective, living art is possible, of the kind we associate with

Homer or Shakespeare (or perhaps, outside Germany, with the nineteenth-century novel). But if not, if the artist remains an autonomous creator, acknowledging no ordering force except that which he finds within him, can his work ever escape from its dependence upon him? Must it not remain, as at the start of Goethe's drama, stone statues scattered through Prometheus' grove, free but motionless—frozen icons of the artist's self?

Goethe's answer to these questions, the most general and abstract form of the dilemma presented by the conflicting claims of national tradition and individual genius, is to be typical and formative for all the subsequent sixty years of his unique and contentious achievement. Goethe's Prometheus says 'No'. He will not, even offered this temptation, surrender the rebel fortress of his self. But the refusal is not what it seems: it does not finally shut off the fountain of life. For Prometheus is not wholly alone, the world of his activity is not wholly devoid of an objective principle. Minerva, 'my goddess', he calls her, something between a Muse and a tutelary deity, she does not abandon him despite his intransigence:

> Und du bist meinem Geist,
> Was er sich selbst ist;
>
> .   .   .   .
>
> Immer als wenn meine Seele spräche zu sich selbst,
> Sie sich eröffnete
> Und mitgeborne Harmonieen
> In ihr erklängen aus sich selbst:
> Das waren deine Worte.
> So war ich selbst nicht selbst, . . .

You are to my spirit what it is to itself; . . . whenever it seemed that my soul was talking to itself, opening itself to itself, and harmonies it was born with were resounding out of it to itself: those were your words. And so I myself was not myself . . .

Ultimately even creativity must have a source. Even Prometheus' autonomous activity must appear to him a gift, though he may not know who or what has bestowed it, and the world that he has made in that activity is not simply a reflection of himself, it has the twilight objectivity of 'myself not myself'. Bowing not to the gods, who are no more than his peers, but to the Fate, the unknown dispenser of all creativity, that is their master and his, Prometheus is guided by Minerva to the source of life. Rousseau's monodrama ends with the vivifying of Pygmalion's statue who in the last lines of the play is allowed three utterances: touching herself, Galatea (it is Rousseau who invents this name for the statue) says, 'I'; touching another statue she says, 'Not I'; touching Pygmalion she sighs, 'Ah! I once more.'

Goethe, however, while accepting this representation of modern creativity, pursues, with a characteristic sceptical honesty, the question: what kind of an art has here been given to the world? For when Prometheus' statues come to life

—and here the parallel with the ambiguous objectivity of so much of Goethe's later work becomes insistent—they are not so much living as resurrected, undead. It is precisely in their deathlessness that they are indistinguishable from their creator, the solipsistic genius who knew neither his beginning nor his end. The statues, now men and women, are certainly no windowless monads: they enter busily into every kind of relation, they build, they establish property, they quarrel and fight, they love. Here, surely, is the material of a living and realistic art. But Prometheus has to tell them also of death and as he does so the one definitive limitation of Goethe's mind and art is briefly but clearly revealed. Pandora, Prometheus's finest, completest and most loved creation, ignorant as yet of death, asks her maker the meaning of the transports of love that she has seen among her fellows. Using one of the oldest of conceits, Prometheus tells her that this is death, the summation of all joy, all pain, all sensation, all weariness, all hopes for further experience, this is the moment that fulfils everything.

PROMETHEUS. Wenn . . . all die Sinne dir vergehn
  Und du dir zu vergehen scheinst
  Und sinkst, und alles um dich her
  Versinkt in Nacht, und du, in inner eigenem Gefühle,
  Umfassest eine Welt:
  Dann stirbt der Mensch.
PANDORA [*ihn umhalsend*]. O, Vater, laß uns sterben!
PROMETHEUS. Noch nicht.

PROMETHEUS. When . . . all your senses fail and you yourself seem to fail and you sink and everything around you sinks away into night and you in your own inner feeling embrace a world: then Man dies.
PANDORA [*clasping him round the neck*]. O, Father, let us die!
PROMETHEUS. Not yet.

Pandora's response indicates that Goethe still knows this seductive misrepresentation to be a conceit, but this is no longer the case in the lines that immediately follow, the last lines of the work and an illegitimate extension of the comparison of love and death:

PANDORA. Und nach dem Tod?

PROMETHEUS. Wenn alles—Begier und Freud und Schmerz—
  Im stürmenden Genuß sich aufgelöst,
  Dann sich erquickt in Wonneschlaf,—
  Dann lebst du auf, aufs jüngste wieder auf,
  Aufs neue zu fürchten, zu hoffen und zu begehren!

PANDORA. And after death?

PROMETHEUS. When everything—desire and pleasure and pain—has dissolved in tempestuous enjoyment and then refreshes itself in a sleep of bliss—then you revive, revive most young, to fear once more, to hope and to desire!

Overwhelming though this sensual drive to enjoy the world may be, such an evasion of finality must tend to deprive any representation of the things and people and experiences of the world of the uniqueness they have through being irrecoverable. Ultimately the fruit of the cult of creative genius must be a poetry of ideals inhabiting a world of eternal life.

Will the race that the genius Prometheus shapes in refusal of submission to Jove truly be his like and image? Will it live? *Can* it live, supported only by the 'creative', monadic soul, the feeling heart? And is any soul (even that of Nietzsche) capable of this exertion? The extraordinarily forceful diction of the 'Prometheus' poem is in part due to Goethe's resentful awareness that he is having to cut a path through a jungle of public, not personal issues, which none the less are fraught with extreme danger for the mind of the individual. (This is why Zeus, like Nietzsche's God, is for Goethe-Prometheus an enemy, not a nonentity.) 'What it costs to dig wells in the desert and knock together a hut', Goethe exclaimed to Kestner while working on the drama. The Wanderer learned, chanting his song into the storm, that the white heat of an autonomous sensibility cannot be indefinitely maintained. 'You *cannot* always *feel*' ('Man *kann* nicht immer *empfinden*'), said Goethe to Lavater. Sentimentality's picture of the human soul, with all the demands that it imposes, is simply wrong. Awakened, if it is fortunate, in time from the delusion, the exhausted heart, the 'armes Herz', will retire to shelter in its hut before it is finally drained and shattered.

It was because it so perfectly understood and represented the pathology and the crisis of contemporary Sentimentality that *Werther* became a European success. Goethe understood that crisis because it was his own. Never again in his writing life was there so exact a coincidence of personal and general concern. But the crisis was imposed, and the coincidence was only momentary, it was neither stable nor straightforward. We shall not have a complete understanding of *Werther* until we have taken account of its—for Goethe— uniquely close relationship to its public. Only then can we grasp the importance, throughout the rest of Goethe's life, of the memory of the experience chronicled and exorcised in his novel.

It mattered to Goethe that his *Götz* was successful with the public and that Gerstenberg thought him capable of contributing to the public's formation. It was in a rueful tone that he wrote to Kestner in August 1773, 'I think it will be some time before I again do something that will find a public'. In April 1773, it is true, a visit to Frankfurt from one of the Buffs' neighbours had caused all the memories of his time in Wetzlar to 'bubble up': 'I'd like to tell the whole story of my life with you, down to the clothes and postures, that vividly', but he did not at that stage have the understanding of himself or his art to see how that story could be made into something of general, public significance. Although *Werther* was, like *Götz*, a composite of elements both intimate and alien, it was not quite the straightforward affair that Goethe suggested in his, understandably rather delighted, summary of April 1774: 'I have lent my feelings to [Jerusalem's] story and that makes up a marvellous whole.' In order to fuse Goethe's feelings—that is, his letters to Merck—with Jerusalem's story—that is, Kestner's report on his last days and death—it was necessary to create the figure of Werther, at once a character and a consciousness, a human symbol who both represented a social and cultural phenomenon of his time, as Götz did, and spoke, with a voice that said 'I', of internal longing and division, as did Weislingen. That meant, firstly, taking the next logical step onwards from *Götz*, the 'Germanic drama', and expressing the national identity in terms not of its past, but of its present—Germany as it was in the 1770s, grasped with the same concrete imagination that had felt its way into the sixteenth century and that now put itself in the service of the very genre that German circumstances seemed to exclude: the realistic novel. Secondly, it meant that Werther's ability to say 'I' was essential to his characterization and to that of the cultural movement which he represented, and therefore that 'his' life could not just be 'told vividly' in the third person, as Goethe had at first thought of telling the story of his time in Wetzlar, or as in 1773 he was probably attempting to tell the story of Wilhelm Meister. It would have to be impersonated, be acted out by Goethe in one of those ghostly conversations with his absent dear ones in which he created images of 'myself not myself', and which he said bore so strong a resemblance to the process of letter-writing, a half-way house between monologue and dialogue. The decision to write *Werther* as an epistolary novel may well have been influenced by the example given by Sophie von La Roche—while Goethe was himself drafting the opening pages of *Werther* he was advising Frau von La Roche on the construction of her second such novel—but the decision was essential if *Werther*, like *Götz*, was to become a 'national' work, given, that is, the prominence of Sentimentalism in the national life.

Thirdly, the symbol that was Werther could only be created if Goethe treated with complete seriousness and as equal partners, and with no trace of embarrassment or contempt, the anonymous reading public to whom Werther's story is addressed and whose condition Werther symbolizes. This crucial step away from the courtly fear of print as a bourgeois medium, which had been strong in Goethe since his Leipzig days, was undoubtedly prompted by the public success of *Götz*. 'I am quite pleased with myself that it is I who am breaking down the paper wall between us', Goethe wrote to Bürger, using a multiply ambiguous metaphor, as he initiated their correspondence by sending a copy of the second edition of his play. The two poets had so far known each other only through their published works—but the division between writer and reader is now seen to have been only as thin as a sheet of paper. Far from being vast, anonymous, and solitary, the world of the reading public turns out to be scarcely different from the intimate world of a personal exchange of letters. Although *Werther* is a book, it is also a collection of letters which appear, since they never receive a reply, to be written to the reader. The purported editor of Werther's papers speaks to his readers with a concern to inform them accurately, and certain of their collective sympathy. Goethe himself approached the publication of the work in a confident and businesslike way, seeking a normal commercial arrangement, and saw no contradiction between that approach and the ravishing thought that the name of Lotte is now being 'spoken with awe by a thousand holy lips', as if what had been published was not a book but a personal circular. The monologue-cum-dialogues of the solitary poet can be addressed either as letters to friends who are known but distant, or as books to those who are unknown but who, in virtue of their reading, are drawn into the extended circle of Sentimental correspondence and so become friends too.

Fourthly, *The Sorrows of Young Werther*—the title could equally be translated *The Passion and Death of Young Werther*—only became a possibility with the completion of Goethe's religious emancipation. Religious vocabulary in this book does not provide a meaning for the story: rather the story gives meaning to any religious terms it may contain; the 'word of God', as Goethe wrote to Lavater, is no more than the 'word of men'. It is Lotte's name that makes sacred the lips that utter it, not vice versa. If Werther is compared to Christ, or if he speaks of a future happiness in heaven, it is not because our attention is being pointed to external realities that transcend these little events. On the contrary, the reader is being told that the fullest meaning of such terms as 'Christ' and 'heaven' lies in the story of Werther's sufferings. And if the reader then asks what the meaning is of *that* story only two answers are possible: either, nothing at all, it is just a story; or, that it is the story of the suffering of a whole age, exemplified and identified pre-eminently in the experiences of the Promethean creator who shaped these 'men in my own image, a race to be like me'. What 'significance' does it have that Werther's birthday is 28 August or that he flees his Lotte on 10 September? Either none—or that his personal liturgical calendar runs in parallel with Goethe's. Werther dies as Christ is born—is the conclusion, then, that if Goethe is to be born, Christ must die? 'All this time,' Goethe told Lotte as his work on the novel progressed, 'and perhaps more than ever you have been with me, *in, cum et sub*'—he was using the terms definitive of the Lutheran doctrine of the consubstantiation of Christ and the Eucharistic species. Goethe's life—his time in Wetzlar, for example—has become sacramentally significant, for his literary works have raised it to a level at which it stands for some general feature of the age; the literary works themselves, however, have become the sacramental vehicles of his life.

In all these four respects *Werther* is a creation of Goethe's intellectual development in 1773 rather than of his Wetzlar experiences of 1772. Forty years further on, reflecting in his autobiography, Goethe took care to situate the genesis of *Werther* amid the sufferings and frustrations of an entire generation:

The effect of this little book was great, indeed immense, and principally because it hit exactly the right moment. For just as little priming is needed to detonate a powerful bomb, so the explosion which ensued among the public was so violent because the young people had already undermined themselves, and the shock was so great because each erupted with his own exaggerated demands, unsatisfied passions and imagined sufferings.

The reference to 'exaggerated demands' may be misleading. Goethe is not thinking of the revolutionary's demands that the world should be different, as he makes clear when he contrasts the melancholy of his own generation with the great suicides of ancient Rome:

We are not here talking of such persons as led an active and significant life, employed their days in the service of some great kingdom or in the cause of freedom . . . We are dealing here with those who lost the taste for life essentially for want of action, in the most peaceful state imaginable, through exaggerated demands upon themselves.

What destroyed these young people, undermined them from within, was a requirement they imposed upon themselves and were unable to fulfil, a requirement to produce—from their own resources—feelings, ideas, perhaps even works of art, a requirement to be original, creative, sensitive, to be, in short, a Promethean genius.

The story of Werther is not, in the first instance, a love-story. It is the story of the self-destruction of a feeling heart, a sentimental soul, and a love-relationship is only one of the elements in that process. Werther, a young man of accomplishment and means, comes to a south German town partly in order to settle a family matter concerning an inheritance and partly in order to escape the aftermath of an unhappy entanglement at home. It is May, and in the countryside and villages round about, in the country folk and their children, in his volume of Homer and the sketch book in which he attempts to record his impressions, he finds a source of intense delight, though his weakness as a draughtsman causes him some regret. Through the medium of his letters to an old friend (whose replies are not given) we learn how he lies in the grass watching insects, how he helps a woman to carry water from the well, and how he is totally won over at a country ball by Lotte, daughter of a local bailiff ('Amtmann'—the office occupied by Lotte Buff's father). Lotte is not only lively and practical, but she has a similar sensibility to Werther's, revealing at the ball her taste for Goldsmith and Klopstock. Her domestic life echoes *The Vicar of Wakefield,* and with increasing ecstasy Werther watches and shares her round of simple duties and virtuous actions: cutting bread for her younger brothers and sisters before they go to bed or visiting an old parson or a dying friend. But, as Werther had known since before he met her, she is 'as good as betrothed' to Albert, who now returns from a business trip, and with him, excellent man that he is, friendly, active, and sensible, a shadow of frustration enters Werther's life. The conversations Werther reports turn more frequently on subjects such as death, infanticide, and suicide; nature causes him no more delight. Finally, on 10 September, he breaks away and leaves the area. As his friend, we gather, has advised him, he takes up a diplomatic post in another part of Germany. But though he meets feeling and intelligent men in public life, the shallowness of most, the tedium of affairs, and above all the social humiliation to which he, not a born aristocrat, is exposed, bring him to resign within the year. A period as house-guest of a count and a visit to the scenes of his childhood are no more successful as distractions from the obsession which draws him back to Lotte. Lotte and Albert are now married, not perhaps as happily as they had hoped (shades of *Partners in Guilt*), and another winter is drawing on. In an atmosphere of increasing gloom and friction with Albert, Werther gives himself up completely to a hopeless passion. At this point in the novel an editor intervenes to provide a narration linking the records of Werther's last days. For, recognizing the pain he is causing to Lotte, whose love for him is transparent, but unfulfillable, Werther discovers in suicide a means of relieving the personal tensions without renouncing his desire. After a last hysterical interview with Lotte, whom he fleetingly embraces, he shoots himself with Albert's pistols, dying, after a twelve-hour agony, at noon on Christmas Eve.

The most prominent formal feature of *The Sorrows of Young Werther* is that, although it is an epistolary novel, the letters in it come, all of them, from one correspondent. We hear, not the polyphony of Smollett, Richardson, or even Rousseau, but the single—and, till the end of the second book, uninterrupted—voice of Werther. Even the editor who holds the narration together during Werther's disintegration does not offer an alternative point of view: his interpretation of character, Lotte's, Albert's, Werther's even, is Werther's own, though his knowledge is more extensive. He shares Werther's interest in the details of his dying and the location of his grave (both Werther's anticipation and the editor's narration of the event recall the close of Gray's *Elegy*). He simply provides external confirmation of the truth of Werther's perceptions; assures us that the unease between Albert and Lotte is real and that Werther's belief that he represents a serious threat to Lotte's virtue is well founded. The catastrophe affects all three principal characters, but it has a single source in Werther's sensibility, and that sensibility is indeed, we gather from the editor's intervention, as it has appeared to us in Werther's letters. We are dealing therefore essentially with a monodrama, and it is not a development in the plot that holds our fascinated attention but a development in the mood. In this novel—perhaps it should be called a sequence of prose poems—feeling is all (to cite the words of Faust). Ten years later, when much had changed, Goethe still marvelled at its 'white-hot expression of pain and joy, irresistibly and internally consuming themselves'.

In his first weeks of happiness Werther rejects with a laugh the suggestion that he should undertake a more active life: is he not as active now as he could be in any employment (letter of 20 July 1771)? Werther is completely committed to the life of sentiment, to life as if the theory of the active autonomous sensibility presented an attainable ideal. From the start, however, if only he would attend to them, there are indications that he is trapped by a ghastly delusion. In the glory of Maytime his soul quests in vain for an object, for an idea that will give the full satisfaction of the knowledge: 'this outward thing I have produced from within me'. But no specific, individual thing, however small, will accommodate itself perfectly to his feeling, and the only alternative for his heart is to subside into impotence:

Wenn das liebe Thal um mich dampft, und die hohe Sonne an der Oberfläche der undurchdringlichen Finsterniß meines Waldes ruht, und nur einzelne Strahlen sich in das innere Heiligthum stehlen, ich dann im hohen Grase am fallenden Bache liege, und näher an der Erde tausend mannigfaltige Gräsgen mir merkwürdig werden. Wenn ich das Wimmeln der kleinen Welt zwischen Halmen, die unzähligen, unergründlichen Gestalten, all der Würmgen, der Mückgen, näher an meinem Herzen fühle . . . Aber ich gehe darüber zu Grunde, ich erliege unter der Gewalt der Herrlichkeit dieser Erscheinungen.

When the mists rise from the dear valley all around me and the sun high above lies on the roof of the impenetrable darkness of my forest and only scattered rays steal through into the inner sanctuary, and I lie in the tall grass beside the falling water and closer to the earth a thousand various grasses become objects of my interest; when I feel, closer to my heart, the teeming of the little world between those blades, the innumerable, unfathomable figures of the little worms and flies . . . But it is destroying me, I swoon at the violence of the splendour of these appearances.

Those contemporaries who found Werther's sensibility distasteful objected in particular to his habit of appropriating everything about him by means of the possessive adjective ('my forest', 'my Homer'). Yet that very act of appropriation indicates a discrepancy between the heart and its object, the feeling does not *quite* reach out and embrace the 'appearances'. And when Werther's mood changes, as it does after the arrival of Albert, the unchanged glory of the natural world proves incapable of sustaining him in his earlier ecstasy. On the contrary, the requirement that all life should be open to, and suffused by, his own feeling becomes a torment when his own feeling is one of misery:

Mußte denn das so seyn, daß das, was des Menschen Glükseligkeit macht, wieder die Quelle seines Elends würde?
Das volle, warme Gefühl meines Herzens an der lebendigen Natur, das mich mit so viel Wonne überströmte, das rings umher die Welt mir zu einem Paradiese schuf, wird mir jezt zu einem unerträglichen Peiniger . . . Es hat sich vor meiner Seele wie ein Vorhang weggezogen, und der Schauplatz des unendlichen Lebens verwandelt sich vor mir in den Abgrund des ewig offnen Grabs . . . Der harmloseste Spaziergang kostet tausend tausend armen Würmgen das Leben . . . Mir untergräbt das Herz die verzehrende Kraft, die im All der Natur verborgen liegt . . .

Was it necessary, then, that what makes man's happiness should also become the source of his misery?
My heart's full, warm feeling for living nature, that used to flood me with so much bliss, that all about me made the world a paradise for me, is now become my unbearable tormentor . . . A curtain has as it were been drawn back from before my soul, and before

my eyes the theatre of endless life is transformed into the maw of the ever open grave . . . The most innocent walk costs the lives of a thousand thousand poor little worms . . . My heart is being undermined by the consuming force that lies hidden in the natural universe . . .

The keyword in *Werther* is 'heart' (Herz). 'What is the heart of man?' is the second sentence of Werther's first letter. His own heart, he tells us, is his only pride.

was ich weis, kann jeder wissen.—Mein Herz hab ich allein

what I know, anyone can know—only I possess my heart

Unstable though it is, he indulges its every whim, as though it were a sick child. It is therefore with something like the shock of tragedy that we learn that this power of feeling is no more, that overtaxed by excessive and conflicting demands Werther's heart is now dead—

Und das Herz ist jetzo tot, aus ihm fließen keine Entzükkungen mehr . . . ich habe verlohren, was meines Lebens einzige Wonne war, die heilige belebende Kraft, mit der ich Welten um mich schuf. Sie ist dahin! . . . o wenn da diese herrliche Natur so starr vor mir steht wie ein lakirt Bildgen, und all die Wonne keinen Tropfen Seligkeit aus meinem Herzen herauf in das Gehirn pumpen kann . . .

And the heart is now dead, no ecstasies flow from it any more . . . I have lost that which was the only bliss of my life, the sacred lifegiving power with which I created worlds all around me; it is gone . . . oh, when this glorious Nature stands frozen before me like an enamelled miniature and none of this bliss can pump a single drop of happiness up from my heart into my brain . . .

The detachment of feeling from its object is now complete. Werther never quite knew the reciprocity of Ganymede's relation with Nature—but now reciprocity has disappeared altogether. And the hypertrophied heart takes its revenge— Werther has lived by feeling and must die by it. Freed from the control of any object, either natural or human, his emotions swirl into hysteria: the clear Homeric vignettes of the early letters give way to wild but indefinite Ossianic landscapes; in the place of the thousand details Werther experienced in the springtime of his heart there is now only moonlight, floodwater, and winter wind. He dies a drained and dismembered travesty of his former self, yet still, hideously, recognizable as the man who sought to grasp, and create anew, the world with his heart.

Even simply as a story in which excessive demands on the capacity for feeling lead with irresistible logic to a loss of all contact with the objects of feeling, *Werther* is still a story of and for our own time. It is true that, with the passage of two hundred years, its immediacy as a documentary novel has paled. It requires some historical empathy to recognize in it the picture of that specific age when the names on the lips of the avant-garde student were Sulzer, Batteux, and Heyne, when friends circulated their portraits in the form of silhouettes, and when Werther's characteristic dress—blue frock-coat, buff waistcoat and trousers, and boots—marked him out as one who followed the English fashions, rather than the French, and lived an outdoor, not a salon, life. (Whether Goethe also knew of the association of 'the buff and the blue' with uncompromisingly middle-class English politics is unclear.) The precision in the dating of Werther's experiences—in 1771 and 1772, when Ossian has 'arrived' but it is not yet essential for a thinking and feeling man to express his view of Shakespeare—is likely nowadays to go unremarked. But in the late eighteenth century the overt theme and its documentation were equally contemporary: their combined effect was overwhelming. Until the last two decades of his life, when *Faust* began to occupy the foreground, Goethe owed his European reputation to *Werther*. Within a year of its publication two French versions and a French dramatization had appeared. The work, first translated into English in 1779 (there were at least seven more English editions in the next twenty years), was by 1800 available in most European languages. On his visit to Italy in 1786–8 Goethe found himself plagued, as he had been plagued in Germany, with questions about the autobiographical background of a novel that he then wished rather to forget. When in 1808 he had a series of interviews with Napoleon the conversation turned mainly on *Werther*, which the Emperor claimed to have read seven times. (One may note that—coincidentally?— *Werther* was also the first, and favourite, reading matter of the monster-hero of Mary Shelley's *Frankenstein* (1818).) In Germany the effect was immediate: by the end of 1775 no less than eleven editions (mostly pirated) had appeared, and only three months after the publication of the first the reviewers had to acknowledge that the novel was too widely known for them to affect its popularity. Two camps quickly formed: those who stammered out their adoration in sentences such as 'Criticize, should I? If I could, I'd have no heart', and who regarded the shower of parodies and 'corrected' versions with happy endings as little short of blasphemy; and those who followed the lead of the redoubtable Pastor Goeze of Hamburg in seeing the blasphemy in *Werther* itself, a book calculated to encourage the mortal sins of adultery and suicide and a sure sign that contemporary Christendom was about to suffer the fate of Sodom and Gomorrah. (It is of course impossible to prove that *Werther* caused any suicides, but cases of suicide associated with a reading of *Werther* are reported until well into the nineteenth century.)

Werther, in short, became a fashion. (The Chinese porcelain manufactories executed commissions for services decorated with scenes from it.) This, however, is no accidental feature of the book's 'reception', an appendix for the delight of literary antiquarians: rather, it corresponds to an essential and almost completely new feature of the book itself, to its ultimate, concealed, theme for which its analysis of Sentimentality is only the excuse. *Werther* became a fashion because it was about a fashion. It is the first novel in European literature after *Don Quixote* to have such a theme, and this is both the real secret of its modernity and the reason for the precision of its location in place, time, and culture. Werther's innermost life is determined by a public mood, he lives out to the last, and inflicts on those around him, the loyalties which—because they are literary, intellectual, in a sense imaginary loyalties, generated within the current media of communication—most of his contemporaries take only half-seriously. His obsessions are not gratuitously idiosyncratic—they belong to his real and socially determined character, not just to a pathologically self-absorbed consciousness. In the, in his, bitter end, Werther's identity is swallowed up by his reading matter, by Ossian, the vehicle of his last declaration of passion, and by *Emilia Galotti*, which lies open on the desk beside his mangled body. It is not simply to the self-destructiveness of Sentimentalism that Goethe is referring when he writes of *Werther*: 'I myself was in this case and know best what anguish I suffered in it and what exertion it cost me to escape from it.' He is referring also to the fact that, like the confrontation with the God of Pietism, so this very impetus to self-destruction is being imposed on him by the German public mind, commerce with which he cannot avoid, or wish to avoid, if he is to express himself at all in the language and literature that are both his property and the nation's.

How very specifically Werther's case is that of Goethe—Goethe at grips with his age—and not simply that of any 'typical' contemporary we can see by comparing the novel with another work of Storm and Stress to become the object of a cult, Schiller's first play *The Robbers* (1781). (The auditorium on the first night, in 1782, was said to be like a madhouse, with sobs, swoonings, and strangers falling into each others' arms.) Partly borrowed from Goethe's *Claudina of Villa Bella*, the story of Carl Moor, who believes himself disinherited by his father, who gets together a band of student friends, and who, until his conscience gives him insight, leads a life of crime which he interprets as a revolt against an unjust society, had a contemporary appeal in the 1780s which is hardly less in the Age of Terrorism. But the contemporary theme is given a historically indefinite setting and so its very contemporariness, the dependence of the central characters' attitudes on the attitudes generally fashionable in their society, the fact that in their minds and feelings they too are people of a certain time and place and set of circumstances, that remains totally unexpressed. True, Schiller presents a milieu and attitudes more central to the national culture than anything in Goethe's book—the world of the university students with its two most widespread substitutes for a dispassionately realist view of contemporary German society: the fulmination against all forms of paternal authority, and the compensating cult of blood-brotherly friendship—but he presents them as if his play were bringing them into existence for the first time, with no understanding, or even awareness, of the world of which they are a part. The self-consciousness of Carl Moor, and of his evil brother Franz, is inexplicable, it has no earthly origin: in this respect they, and practically all Schiller's subsequent heroes and heroines, are unmistakably the monadic souls traditional in German official literature from the Countess of G＿＿ to

Odoardo Galotti and beyond. Werther is the victim of the delusions fostered by that tradition, in its Sentimentalist form, but the literary method which presents him as a young German of 1771 and 1772 has presuppositions quite different from those of Sentimentalism, and of Werther himself. They are none other than the presuppositions of Lenz when he set out to paint human society; the Sentimentalist content of the novel is in perfect but momentary balance with a Storm and Stress aesthetic which determines the manner of its presentation. This is most obvious, in this first edition of *Werther,* in the editor's report to the reader, in the contrast between the dispassionate registration by the editor of the detail of times, actions, and even feelings, and the nebulous emotionalism of the participants. But the whole book is sustained by a premiss of social objectivity. The novel's setting contrasts—indispensably—with the historically remote world of the protagonist's favourite authors, Homer and Ossian. Werther, like Goethe, may not be living his life in Germany's intellectual centre, in the universities which he almost contemptuously disregards, but his confrontation with the determining social realities is direct: through the social and economic station which gives him leisure, through the barrier to his advancement constituted by aristocratic privilege, through the real restrictions imposed by the convention of marriage, through books and conversations about them. Carl Moor's confrontation is indirect and unspecific, one is tempted to say uninterested in what it is opposing: it is a confrontation through the melodramatically deceitful intrigues of his brother, through highly metaphorical verbal denunciation, and through the fantasy of a violent revolt against the world-order in general. Matthias Claudius's suggestion that rather than kill himself Werther should go on a journey to Peking may have been intended as a criticism of the book's morals but is in fact a compliment to its realism. Saint-Preux, the lover-hero of Rousseau's *La Nouvelle Héloïse,* is indeed rescued from suicide by his English friend's arranging for him to accompany Anson on an expedition to the South Seas. The difference from Werther's case, however, is not a difference in the morals of the authors but a difference in the real circumstances of colonial England and provincial Germany, and it is the strength of Goethe's novel that on such differences the story depends.

But the realism of *Werther* is of a peculiar—and peculiarly modern—kind because the real circumstances of Werther's case include, and are predominantly, intellectual, literary, and cultural circumstances. The epistolary form is essential to the novel, for through it Werther's intellectual ideals and cultural loyalties, his conscious moulding of his experience, as he writes, into the patterns given him by Rousseau, Klopstock, or Goldsmith, become as much a part of the subject-matter as the events he recounts. Werther's awareness of the significance of his sitting shelling peas and reading his Homer constitutes the 'mood' which is the book's true subject quite as much as those actions themselves. The real effect of Werther's intellectual attitudes is most clearly seen in the last pages of the book, when his readings from Ossian exacerbate beyond endurance the emotions of an already tense household. It is certainly important that by having recourse to the events of his own life Goethe has found an answer to the question: what can be the subject-matter of a new, contemporary secular literature? But it is equally important that, in the peculiar form of autobiographical literature of which he will become the supreme and perhaps only master, 'the events of his own life' furnish neither a reservoir of impersonal knowledge nor a string of anecdotes accidentally attached to this particular historical personage. The events are not detached from the life so as to become examples of the way of the world, fitted decoratively, as circumstances dictate, into artistic structures obeying their own laws (the relationship which obtains between the notebooks and the novels of many nineteenth-century writers). Nor on the other hand are Goethe's works simply encoded memoirs. It was not in the first instance his recollections on which Goethe drew in writing *Werther*. He did not, in that sense, draw directly on events in which he had been involved: he drew on his formulation of those events two years before in his letters to Merck, and he wrote a novel about the mind that wrote those letters, as well as about the man that met Lotte Buff. Throughout his subsequent literary career the new subject-matter—his own life—will never be available to him simply as material but always in the only half-objectified form of 'myself not myself'. The realism of Goethe's writing after *Werther* will always be of the kind which includes among the realities of the world the consciousness that grasps them. But never again will that consciousness be so closely identified with a current, determinable, datable public reality—with an intellectual fashion. The coincidence of the mind of the poet and the mind of the age could only be momentary, given the power of the poet and the character of the age, if Goethe was not to end as his hero did. The only conceivable third possibility was to live altogether less independently, and be content with a long dotage, writing second-rate works that suited exactly the taste of the time.

The Messiah and his Nation: 1774–1775

After *Götz*, and even more after *Werther*, everyone wanted to see Goethe, and many of those who did made a note of what they saw. The little old beer-brewing town of Einbeck, in the Hessian forest, 'a town where nobody reads anything', was electrified on one occasion by the rumour that Goethe was visiting, though the visitor was really Boie, who had been put up to the deception by a mischievous Klopstock, his travelling companion. There are innumerable testimonies to the impressions Goethe made, though evidently the witnesses themselves were not without influence on what they were observing: the young Princes of Saxe-Meiningen found that 'he has a lot to say that is good, unusual, original and unspoiled and is astonishingly amusing and comical', but it is not wholly surprising that to the stiff and censorious burghers of Zurich it seemed that 'Goethe is a man of few words' who was 'too arrogant and dogmatic' to gain any friends. Goethe appeared as a 'chameleon' or 'Proteus' who responded, in action and reaction, and with violent energy, to his circumstances: 'he is entrancing company; everything he says bears the mark of genius'; 'in the most vigorous conversation he can suddenly take a notion, get up, and go and not reappear. He is wholly his own man, follows no one's customs. When and where everyone appears in the most ceremonial clothes he is to be seen in the most informal attire, and *vice versa*'; 'I wish I had noted down his table-talk.' The silent papers still bear faint traces of that long-dead magnetism, and we can guess at what it was to have known a 'genius'. Challenged by his partner Susanna Münch at one of the weekly meetings at which young Frankfurt played its marriage game, probably on 13 May 1774, Goethe by the twentieth—and with some time to spare—wrote a five-act tragedy, *Clavigo*, on the basis of a story in the *Teutscher Merkur* which had appealed to the gathering. His power of reciting verse, his own and others', from memory was remarkable, though supported at times by an equally impressive capacity for improvisation, as can be seen from a version of *The Artist's Earthly Pilgrimage* which he wrote out from memory in an autograph book nearly a year after its original composition. But it is the passing of the entrancing conversation that one regrets most. One reconstructible feature of it that struck contemporaries, and is borne out by his letters, was its wealth of unusual comparisons: 'as soon as you are in company [Goethe said] you take the key from your heart and put it in your pocket—those who leave it in the lock are fools'; Lavater's *Diary of an Observer of Himself* (*Tagebuch eines Beobachters seiner selbst*) reminded him of 'someone with his handkerchief always ready in his hand to blow his nose and irritated because he can never get to blow'; when the ducal palace in Weimar and much of the ducal art collection was destroyed by fire on 6 May 1774, he was prompted to describe his attack on Wieland's *Alcestis* as, by comparison, merely 'burning down his summerhouse'; 'you want me to write without feeling, to give milk without having given birth', he complained to Lavater; and he summarized 1774 as 'rolling on the moral snowball of one's self for another year'.

Lavater's doctors had recommended that he should take the waters in Ems, and in June 1774 he set off from Zurich on a progress which revealed to him how well-known and highly regarded he had become in both the ecclesiastical and the literary circles of south-west Germany. On the twenty-third he reached Frankfurt at 8.30 in the evening and went straight to the Goethes' house where it had been arranged that he should stay for a few days. After an embrace and a mutual inspection of physiognomies—Lavater being more surprised by Goethe's than Goethe by his—there was time before going to bed for the poet to tell the pastor 'many hundreds of things . . . everything he said was spirit and truth', and in Lavater's diary there follows a stream of topics of theological and literary and physiognomical conversation, interrupted by visits to Fräulein von

Klettenberg and the Calvinist church at Bockenheim for the Sunday morning service, by readings from the manuscript of *Werther*, and by Goethe's designing a garden maze. On 28 June Lavater went on, but Goethe decided to accompany him to Ems, thirty-six hours' journey away. Lavater's notes of their first day's travel suggest how strenuous such companionship could be. Having risen at three, they left Frankfurt at 4.30 a.m. and as the coach made its way through the quiet countryside, sunlit but threatened by rain, Goethe talked at length of Spinoza, more of his life, character, and correspondence than of his philosophy, though he noted that all the more recent deists had merely plundered his works. Stopping for refreshment—wine for Goethe, raspberry juice for Lavater—they wrote a joint letter, and then resumed their journey and with it a conversation about Goethe's plans for a drama on Julius Caesar, and about the terrible scenes at the destruction of the towns of Oppenheim and Worms in 1689 by Louis XIV. By 11 o'clock they were in Wiesbaden where the thermal baths, as yet undeveloped, inspired only 'disconsolate melancholy': the newspaper was read, a passage from *Werther* copied out, and over dinner Goethe spoke of his dramas and observed the other customers. After strawberries Lavater, who had been recognized by the innkeeper and complimented on his writings, had to pay a courtesy call, and then, before the coach left at two, he and Goethe stood at the window of the inn and discussed the Resurrection. Lavater dozed for much of the afternoon, despite a bumpy journey and Goethe's declamation of long passages from a 'Knittelvers' epic he was writing, *The Wandering Jew* (*Der ewige Jude*), a treatment of the legend of Ahasuerus, the cobbler who vilifies Christ on the way to Calvary and is condemned to wander the world until the Second Coming. At half past five they were in Schwalbach, where they were to spend the night. The local waters were tasted and assessed and more letters written: Goethe recited a Scottish ballad, allowed Lavater to read more of *Werther*, and, probably in explanation of Werther's enthusiasm for Homer, embarked on a summary of the whole of the *Iliad*—outside the scope of Lavater's pious education—interspersed with extracts from Clarke's Latin translation of the epic, his stand-by since his Strasbourg days. Lavater's plans for a biblical drama on Abraham and Isaac—a genre which Goethe avowed he could not understand—occupied them until they retired to bed.

In Ems where they arrived the following afternoon—pickled eels for lunch and Goethe reciting long passages from Voltaire on the way—Goethe stayed only one night before returning to Frankfurt, though he had time to hear the company in the hotel dining-room, unaware of his identity, discuss *Gods, Heroes, and Wieland,* much to Lavater's amusement. The skit, published in March, was arousing laughter and disapproval in equal measure—Bürger was glad that he himself had nothing of the kind published, and the earnest poets of the Göttingen Grove feared a loss of their dignity through association with it. Wieland's public response was masterly: an issue of the *Teutscher Merkur* containing a warm if rather patronizing defence of *Götz von Berlichingen* and a humorously tolerant and appreciative notice of the 'heroical-comical-farcical pasquinade', *Gods, Heroes, and Wieland,* both by Wieland himself. Goethe on reading it had the good grace to blush and exclaim that he had been well and truly pilloried ('prostituiert'). In private, however, in his correspondence, Wieland remained angered and embittered by the playlet and no doubt agreed with the words with which F. H. Jacobi concluded his report of Goethe's reactions (passed on to him by Johanna Fahlmer): 'he is still an unbridled, immoderate man'. Yet the lesson of *Werther* had not been lost: though so complete and satisfying a fusion of Sentimentalism and Storm and Stress, literary establishment and literary revolt, could hardly be repeated, there was evidently more to be gained by collaboration than by the confrontation of 1772 and 1773. From the spring of 1774, if not earlier, Goethe can be seen reaching out and seeking reconciliation, in so far as his explosive temperament will allow, with those forces in the national cultural life with which he had recently felt in conflict. With Wieland relations remained difficult. But in May he initiated a direct correspondence with Klopstock without the usual formality of sending letters through a mutual acquaintance, and after the first clear demarcation of boundaries his willingness to correspond with Lavater, and to let himself be wooed, shows a desire to collaborate with a Christianity that was open to the secular and material world and was prepared to take him as its guide and to allow him to 'infuse' into it 'a few drops of independent spirit.' The territory over which he and Lavater could exercise a condominium proved to be physiognomy, and for the next two years Goethe was of major assistance to Lavater in seeing through the press the four profusely illustrated quarto volumes of the *Physiognomical Fragments* (*Physiognomische Fragmente*) (1774–8), to which he also contributed as a writer. The key to an understanding of this surprising co-operation and at times very warm friendship is given by the prayer Lavater told Goethe he often prayed: 'If You exist, then show me You exist.' 'Scripture gives not a single example of faith which does not have *sensuous* experiences as its basis.' The belief that, whatever divinity is, it must be visible in the material world, that the traces of the spirit can be seen in every modification of the flesh, was shared by both: Goethe even wrote a paragraph for Lavater on the manifest distinctness of the human osseous structure from that of beasts, as an indication of the higher destiny of man. It would be some years before he repudiated that view and when he did it was a sign that this strange friendship could not long continue.

The group with which it was most obviously desirable to establish new and more friendly links was that of the Jacobis: Goethe was after all on the best of terms with the wife, aunt, and sister of the two brothers, and with one of their closest acquaintances, Sophie von La Roche. After his return from Ems Goethe did not stay in Frankfurt for long. On 9 July another famous visitor arrived at the house in the Hirschgraben, J. B. Basedow (1723–90), an educational reformer and publicist, who believed in the value of modern and practical, rather than classical and academic, studies, had collected some 30,000 guilders to support his scheme for a new and extensively illustrated primer of elementary education, and had found in the Prince of Dessau a patron for his model school, the 'Philanthropinum'. Basedow's fondness for drink, tobacco, and contentious anti-Trinitarian theologizing made him demanding company, but he too was on his way to Ems and Goethe resolved not to miss the opportunity of sharing a holiday with him and Lavater. Passing over his legal work to Schlosser's brother, he left on 14 July for Ems, 'to take the waters', as he told the court in a request for an adjournment. It is difficult, however, to believe that he had not foreseen something at least of the subsequent events.

On 18 July a boat set off from Ems carrying a party of about a dozen, including Goethe, Basedow, and Lavater, down the Lahn towards the Rhine. Goethe improvised rhymes as they went, and dictated the ballad 'High on the ancient tower stands' ('Hoch auf dem alten Turne steht') as they passed the dramatically situated ruins of Castle Lahneck. After dinner in Coblenz most of the party returned to Ems but Goethe's group continued on down the Rhine to Neuwied, capital of a diminutive principality where Basedow chose to stay for the following week, and where the author of *Götz* had the pleasure of meeting again the Frankfurt friend who had once turned down his application to join the Arcadian Society of Phylandria, and who was now flattered to be lent a manuscript of some of his recent poems. On 20 July, leaving Lavater at Mülheim, Goethe pressed on to Düsseldorf and the Jacobis' country seat at Pempelfort: nominally, his intention was to visit Betty Jacobi, but such an intention could not have excluded the strong likelihood of making contact at least with her husband and his brother. Fate, however, seemed unfavourable: Betty Jacobi was in Aachen, her husband Friedrich in Elberfeld. Nothing daunted, Goethe set off to stay the night at an inn in Elberfeld, where he knew his old Strasbourg friend the Pietist doctor Jung(-Stilling) was now practising. A pretence of indisposition the following morning, 22 July, brought Dr Jung to the inn and to a patient with a puzzlingly strong pulse who suddenly leapt out of bed to embrace him. This reunion, and the meeting at Jung's house with the poet J. J. W. Heinse (1749–1803), who was a close friend of J. G. Jacobi and editor of his new journal *Iris*, was compensation for the disappointing discovery that F. H. Jacobi, hearing Goethe was in Düsseldorf, had already hurriedly left Elberfeld to return home. Meeting Lavater on the way, however, Jacobi realized his error and rode back to Elberfeld, only to find that by now Goethe and Heinse had left the town by another gate. Jung arranged for them to be overtaken and brought back, and so in the midst of a Pietist conventicle, foregathering to hear Lavater, occurred the meeting between Goethe and 'Fritz' Jacobi which had by now lost some of the spontaneity originally intended for it. Within minutes, however, Goethe's *attrattiva* was taking its effect, he was praising Klopstock and Herder and Heinse, dancing round the room, and kissing Lavater, to the astonishment of the sober elect, and when he and Fritz

Jacobi and Heinse rode back to Pempelfort that evening they did so as firm friends: 'and he and I and I and he', as Goethe wrote afterwards to Betty Jacobi, perhaps echoing Montaigne's famous words about his friend La Boétie: 'parce que c'était lui, parce que c'était moi'.

The following day the party was joined by Georg Jacobi, still smarting under the abuse of the *Frankfurter Gelehrte Anzeigen,* but willing to shake the hand that had written *Götz,* and the day after that, Sunday 24 July, was a day that everyone involved remembered for the rest of their lives. Goethe, Heinse, and the Jacobis set off at 5 a.m. in a carriage for Cologne. Their first stop was in Bensberg, where they dined in an inn high up on the hillside with views across fields and woodlands towards the Siebengebirge and the Rhine. The table was laid in an arbour near a garden full of flowers, and here Goethe and Friedrich Jacobi had their first heartfelt conversation about Spinoza. The younger Jacobi had a somewhat similar background to Goethe, although, being six years older, and a married man with children, he was a more settled character. His father was a wealthy businessman and his wife had also brought him money, so his career as a financial official in the administration of the duchy of Jülich-Berg did not need to preoccupy him completely. His philosophical interests—he had spent some years in Geneva and had a close knowledge of the French Enlightenment —and his dissatisfaction with the prestige accorded to deism led him to the study of Spinoza, the only fully consistent rationalist, as it seemed to him. But whereas Jacobi saw Spinoza as the hidden ally, perhaps even progenitor, of the enemy to which he was struggling to articulate his opposition, Goethe—almost certainly on the basis of less knowledge of the system—revealed the possibility of seeing him as a mystic and moralist of unique intellectual independence. As with Lavater, a condominium was created in which two enthusiasms of different origins and with different intentions could, for a while at least, coincide and allow two young men to feel brothers in a single cause. After dinner the party visited Castle Bensberg and then drove on to Cologne, where the art collection of the Jabach family, which had died out in 1761, was also inspected in their eerily uninhabited house. As dusk fell the four travellers retired to their inn, the *Holy Ghost,* and while they watched the moon rise over the Siebengebirge Goethe sat on a table and recited Scottish ballads and some that he had recently composed himself: 'The King of Thule' ('Der König von Thule'), 'Once there was a lusty lad' ('Es war ein Buhle frech genung'), with its terrifying ghostly conclusion, and the one he had dictated to Lavater on the Lahn. After the party had broken up, Goethe came to Fritz Jacobi's room and in a midnight conversation hopes and ambitions were revealed and tears of friendship were shed in the moonlight. Jacobi felt he had found the soul mate that he needed. As yet he had not published anything himself, but Goethe encouraged him to write and make his contribution to the great movement that was beginning and in which, after their unnecessary estrangement, both could play their part. A month later Jacobi wrote to Wieland: 'What Goethe and I were to be, *had* to be, to each other, once we landed up out of the blue beside each other, was decided in a flash. Each thought he was receiving from the other more than he could give him. Poverty and wealth on both sides embraced each other; and so there came love between us.' 'I have seen Auntie [Johanna Fahlmer]', Goethe had told his new friend on the fourteenth, 'and am happy that the dam is gone . . . she can talk to me about her Fritz—Today for the first time.'

On the Monday morning the Jacobis and Heinse went back to Düsseldorf while Goethe, via Neuwied, returned to Ems, on the way rejoining Basedow and Lavater, who was now setting his face towards Switzerland again. Goethe stayed in Ems, among the picnics, casinos, and not-so-invalids that make up spa life, until 2 August and then moved on to Thal-Ehrenbreitstein and the La Roche house for a few days before taking the overnight mail coach to Frankfurt on 12–13 August. As a campaign into Sentimental territory the trip had been outstandingly successful, but before it was over there came a stark and bitter reminder of the true character of the Nature which the Sentimental heart presumed to vivify—the truth which Werther glimpsed in his despair, but of which Wieland in his *Alcestis* had been totally oblivious. On the night of 30 July four boys in Ems, out fishing for crayfish, were drowned, and Goethe remembered the sight of their bodies, unresponsive to all efforts at revival, until the end of his days. 'Only in such moments', he wrote to Sophie von la Roche, 'does man feel how little he is, and how ineffectual with all his fervid arms and sweat and tears.' 'What is the heart of man? Are there not real ills enough? Does he have to add to them by self-created imaginary ones?'

On 27 September another important link was drawn tighter. The *Frankfurt Imperial Post* proudly announced on 1 October: 'Herr Klopstock, the favourite of German and foreign supreme princes, arrived here on Tuesday evening, lodged with his friend, our Dr Goethe, and on Thursday morning continued his journey to Carlsruhe.' Travelling from Hamburg through Göttingen, where he gave his benison to the adoring Grove, Klopstock—'short, plump, neat, of a very proper and diplomatic bearing, with noble manners, verging on the pedantic, but with a more intellectual gaze than in any of his pictures'—was responding to an invitation from the Margrave of Baden (Schlosser's employer) to come and lend distinction to his court. Goethe accompanied him as far as Darmstadt, where he showed a true diplomat's ability to talk with zest to the local intellectuals about secondary issues—skating, riding and fencing—and say nothing of literature. Good relations however had been established, and that was what mattered. There was now an answer to the uneasy questionings from within the Grove: 'how far is Goethe our friend? how closely is he linked with Klopstock?' After Goethe's return to Frankfurt on 10 October there came a visit from Klopstock's permanent under-secretary, Boie. It was Boie's first meeting with Goethe—'whose heart is as great and noble as his mind . . . very pale, intelligence in the face, especially in the bright brown eyes . . .'—and it sealed the bond with Göttingen.

'I made him read me a great deal', Boie noted in his diary,

both complete and unfinished, and in everything there is that original tone, an individual force, and with all its peculiarity and lack of correctness everything is marked by the stamp of genius. His *Doctor Faust* is nearly finished and seems to me the greatest and most individual of all.

The dating of the first stages in the composition of *Faust* is too contentious and speculative a matter to concern us here, but if Boie thought the play nearly finished it can only have been because the story of Faust's love for Gretchen, the only element in those first stages which provides anything like a plot, was by October 1774 largely complete. Goethe must therefore by then—and, in all probability, by the preceding spring, when he wrote, in *Clavigo*, a somewhat similar story to Gretchen's—have taken the decision to join the old tale of the magician who sells his soul to the devil with the new tale of a seduction leading to infanticide and the execution of the mother. That decision, which linked an original theme of theological revolt, with which he may have toyed in 1769, with his feelings of guilt at leaving Friederike Brion and the terrible example given by the torment and execution of Susanna Margareta Brandt, was a stroke of creative genius fully comparable to his linking of sixteenth-, and eighteenth-century themes in *Götz*, or his conjunction in *Werther* of his own experiences in Wetzlar with those of Jerusalem. In each case a new unity emerges from the composition of disparate elements: a single, multi-faceted character (in *Götz*, of course, two characters), whose ambiguity symbolizes a force or potentiality, at once for greatness and for destruction, which belongs to Goethe's age as a whole but is in the literary work concentrated into a self that reflects, but is not identical with, the self of the poet. Götz-Weislingen, Werther, and Faust (and several other characters in lesser works) are not *typical* figures—they do not have the objectivity of case-studies in political, theological, or sentimental pathology—rather they are symbols of temptations offered to Goethe by his world, symbols of what would become real if he were to yield to them. For all their rootedness in Goethe's age and in his experience, they retain the residual unreality of Prometheus' statues, of the unattained object of desire, of the absent conversation-partners in Goethe's lonely garret.

It was precisely this unity at the symbolic level that Kestner and his wife were unable to see in the novel which drew so heavily on events in which they had been involved—not surprisingly, for the ability to retain that vision while being involved in the events was a large part of what made Goethe into the poet that he was. Although publication did not occur until the middle of October, at about the time when Boie was visiting him, Goethe had already in September sent one of his three advance copies to the Kestners—another went to Sophie von La Roche with instructions to pass it on to the Jacobis. Kestner's response to the strange gift was intelligent and remarkably, even admirably, measured— he recognized that there was nothing he could hope to do about publication now and that, in the words of an acquaintance, 'il est dangereux d'avoir un auteur pour ami'—but it was quite uncomprehending. The different elements that he recognized Goethe had woven together into his characters seemed to him simply contradictory and to result in figures that were untrue to life. As a result the real models of those characters were 'pilloried' through their association with falsehoods. In particular Lotte's sharing of Werther's passion was a dishonourable trait which was not and could not be found in Lotte Kestner, or even in the married woman Jerusalem had loved. And did Albert have to be such a blockhead, 'just so that you can swagger up to him and say "look what a fine fellow *I* am"?' Kestner was not perhaps as distressed as he implied: 'I am inclined to forgive him', he wrote to a third party, 'but he is not to know that, so that he takes care in future', and he emphasized that Goethe's behaviour in real life had been considerably more proper, more magnanimous, and more attractive than that of the hero of his novel. But Goethe's defence of the integrity and autonomy of his creation was passionate:

Werther must—must exist! You don't feel *him*, you feel only *me*, and *you*, and the thing [no doubt, Jerusalem's death] which you say is 'just *stuck on*'—but, despite what you— and others—say, is *woven in.*

Werther is not Goethe and Jerusalem and glue. His end is not Goethe's end, and his motives are not Jerusalem's motives (a fact which Lessing, embittered at what he saw as a libelling of his stoical young friend, appreciated as little as Kestner). He has an identity of his own, and intrinsic to it, and to his ability to stand for some general aspiration or danger, is the extent to which he is *not* identical with his model. 'If I am still alive', Goethe wrote to Kestner, 'it is you I have to thank for it—so you're not Albert.'

'One does not need a key in order to read the whole with enjoyment', Deinet remarked, noting, possibly enviously, the number of pirate editions the novel was attracting. And those who did not possess a key to its personal background could recognize the generality and significance of the tale it told. 'It is a truly national book,' said Christian Stolberg in December, 'no one but a German could have written it, and no one else could fully sympathize with it'. (A similar judgement was passed on the unpublished *Faust*: 'a work for *everyone* in Germany'.) *Werther* had succeeded in the task which its author was still struggling to complete for himself personally: to touch the national life at all its growing points, to create a coherent, personal vision of what it was to be German at this time. Christian Garve, the shrewdest and most generous-minded of the Enlightened 'popular philosophers', despatched Lessing's (and Kestner's) arguments with a minimum of pretension:

Even if Jerusalem is not Werther, the latter is still an interesting person. And as a philosopher Jerusalem can scarcely have thought any more profoundly, even if he was more systematic and cold-blooded about it. The voice of the public . . . has decided in favour of the merit of this book, and my own feeling subscribes to this judgement.

Indeed Lessing, even if unwittingly, identified that profound vein in the modern and national culture which Goethe had tapped when he wrote:

Do you imagine a Roman or a Greek youth would have taken his life in *that* way and for *that* reason? They had a quite different protection from the folly of love . . . To bring forth such minutely gigantic, contemptibly admirable 'original' beings was a privilege reserved to Christian education which is so beautifully able to transform a physical need into a spiritual perfection.

Even in the depths of his 'classicism' Goethe never wrote for, or even, really, about, Greeks and Romans. Precisely in its derivation, and detachment, from Christianity *Werther* is a mirror of its place and time. 'You know my love for English', Augusta Stolberg, sister of Friedrich and Christian, wrote to Boie,

That has not changed; but *Werther* has greatly weighed down the German side of the scales. As a masterpiece of genius that novel is . . . so infinitely different from all English ones, even Richardson's, that they cannot be compared at all.

In 1774 the intense need for a focus to the German national literature suffered a great disappointment. Klopstock's *German Republic of Learning,* one of the first German ventures in subscription publishing, had proved to be not at all what its subscribers (at some 10 guilders a time) had expected. The literary world was treated by Klopstock in his book as an entirely self-contained, though middle-class, affair, of which he produced a utopian allegory, couched in an archaic and fragmentary style, in the terms (historically quite appropriate, as Goethe recognized) of sixteenth- and seventeenth-century guild regulations. In its own way—which was almost as opaque as Hamann's—the treatise analysed the dilemmas of the time accurately enough, but it did not make literature of them. *Werther* did. And with *Götz* performed in Berlin, *Clavigo* selling, in Deinet's words, like hot cakes, and the Leipzig autumn fair seeing the publication not only of *Werther* but of *Lumberville Fair* and *Father Porridge* and a couple of other squibs, all passed on to Klinger to help him finance his studies and assembled under the title *Newly-opened Moral-political Puppet-play* (*Neueröffnetes moralisch-politisches Puppenspiel*), it began to seem as if German literature had acquired a new focus altogether. 'Everything I have read of yours', Goethe was told by the patriotic and democratic journalist C. F. D. Schubart (1739–91), a persecuted national liberal before his time, 'delights me, swells my heart with noble pride that we can oppose to foreign nations a man whom they do not have and to judge by their yearning for fossilization will never have.' It is all too often overlooked how ready Goethe was at this time to take part in collective ventures: whether reviewing for the *Frankfurter Gelehrte Anzeigen,* collecting folk-songs for Herder in Alsace, attempting an author's co-operative in publishing *Götz* with Merck or *Of German Character and Art* with a similar enterprise in Hamburg run by J. J. C. Bode, collaborating with Lavater on his *Physiognomical Fragments,* or simply being generous of his money, of his own efforts, and even of his finished works to support individuals such as Klinger or H. L. Wagner (to whom he gave some poems and other pieces to print as an appendix to a book to make it more saleable). The very fact of his writing plays indicates a willingness to take part in the attempt to establish a national theatre and national dramatic tradition. But that willingness to take part in a national movement exposed Goethe to enormous pressure to accept the role formulated by national expectation—to become the German genius, the German Shakespeare, the German Messiah even, casting the money-changers out of the Temple of Fame (as Lenz represents him in *Pandaemonium Germanicum*).

The testimonies to the impression that Goethe, in his early twenties already 'the most terrible and loveable of men', made upon his contemporaries are themselves evidence for the sanity necessary to withstand this spate of adulation.

The more I consider it, the more keenly do I feel the impossibility of writing something comprehensible about Goethe to someone who has not seen or heard this extraordinary piece of God's creation. Goethe is, in Heinse's phrase, a genius from top to toe; a man possessed, I would add, who on almost any occasion is incapable of acting arbitrarily. You have only to spend an hour with him to find it ridiculous in the extreme that anyone should require of him that he should think or act otherwise than he does think and act

—this from Fritz Jacobi to Wieland. To Goethe Jacobi reported Heinse's view that he was 'the greatest man the world had ever produced', though for himself he made an exception of the only-begotten Son of God. Lavater was more moderate, and shrewder: 'Goethe would be a splendid figure acting at a prince's side. That's where he belongs. He could be a king. He has not only wisdom and *bonhomie*, but also force.' But others could not be restrained. 'Every one knows that Goethe has divine force in his nature.' 'Goethe is a god, and an even better human being.'

This Goethe, of whom and of whom alone from the rising of the sun to its setting and from its setting to its rising again I should like to talk and stammer and sing and rhapsodize to you [etc.]. Never before could I so easily have explained and shared the feeling of the disciples of Emmaus in the gospel: 'did not our heart burn within us as he spoke with us?' Why not make him our Lord Jesus and let me be the least of his disciples. He spoke so much with me and so superbly; words of eternal life that as long as I have breath shall be articles of my creed [etc.].

Lichtenberg noted the enthusiasm, but did not share it: 'Goethe has come by the name of a Shakespeare as the woodlouse by the name of a centipede— because no one could be bothered to count the legs.'

The readiness with which contemporaries bestow on Goethe the attributes of divinity is not simply a matter of lack of taste. It is a reflection not only of a general tendency to substitute the life of feeling for the life of religion while retaining, only half-frivolously, the old language, but also of the wholly serious aesthetic beliefs of an age in which the words 'creative' and 'creator' are for the first time generally applied not to God but to men, specifically, to literary and artistic men. In the very essay in *Of German Character and Art* in which he hails Goethe as a second Shakespeare, Herder exclaims of Shakespeare himself:

'Poet—dramatic god! . . . A world of dramatic history, as much and as profound as Nature; yet its creator gives us the eyes and the point of view to see as much and as profoundly.' In writing, and above all in writing plays, Goethe had experienced a miracle by which something for which he was fully responsible acquired a life of its own and independence. That experience was a hint of the adequate satisfaction, the adequate object, sought by his restless, unapplied—and so, for others, utterly fascinating—energy. An adequate object—produced from within the subject: what better analogy for that than creation *ex nihilo*?

> Mein Busen war so voll und bang,
> Von hundert Welten trächtig.

My bosom was so full and apprehensive, pregnant with a hundred worlds.

These lines could stand as an epigraph to all of Goethe's work in the period 1770–75, in its multifarious variety and poignant fragmentariness.

Goethe's attitude to the other preferred religious analogy—the comparison with Christ, the Messiah—was more complicated. For he accepted the comparison, with its implication that he was a creative demiurge, but the figure that attracted him and with which he identified himself was not Lavater's miracle-working Saviour casting his spell over thousands, and certainly not Klopstock's disembodied heavenly prince.

If only the whole teaching *about* Christ [that is, the religion supposedly invented by the Church Fathers, in contrast to the teaching *of* Christ himself—the distinction was given wide currency by Lessing] were not such a damned filthy affair that makes me as a human being, as a limited, needy thing, furious, I could feel fond even of its subject.

The Christ of whom Goethe could feel fond, who did not threaten but confirmed his identity, was the Son of Man who came eating and drinking, who subverted the expectations of Pharisees and Zealots alike, and who, though quite free of malice, could not help out-thinking and out-laughing his good-hearted but slow-witted disciples. Strange how the Germans, whether they were apocalyptic enthusiasts like Lavater, or methodical rationalists like Basedow, seemed not to notice the life that was in their midst and passed them by, while they sought salvation in words:

> Zwischen Lavater und Basedow
> Saß ich bei Tisch des Lebens froh.
> Herr Helfer, der war gar nicht faul,
> Setzt' sich auf einen schwarzen Gaul,
> Nahm einen Pfarrer hinter sich
> Und auf die Offenbarung strich
>
> .   .   .   .   .   .
>
> Ich war indes nicht weit gereist,
> Hatte ein Stück Salmen aufgespeist.
> Vater Basedow unter dieser Zeit,

Packt einen Tanzmeister an seiner Seit'
Und zeigt' ihm, was die Taufe klar
Bei Christ und seinen Jüngern war,
Und daß sich's gar nicht ziemet jetzt,
Daß man den Kindern die Köpfe netzt

. . . . .

Und ich behaglich unterdessen
Hatt' einen Hahnen aufgefressen.

Between Lavater and Basedow I sat at table enjoying life. Mr Vicar was not idle, sat himself on a black nag, took up a pastor behind him and made off for the Book of Revelation. I meanwhile had not gone far, had eaten up a piece of salmon. Old Basedow, the while, grabbed a dancing-master beside him and showed him what baptism plainly was for Christ and his disciples and that it's quite improper to wet the heads of children now. And I in the meantime had leisurely downed a chicken.

Lines to which Goethe a couple of days later added:

Und wie nach Emmaus weiter ging's
Mit Sturm- und Feuerschritten:
Prophete rechts, Prophete links,
Das Weltkind in der Mitten.

And so we went on in fire and tempest as if to Emmaus: to right a prophet, to left a prophet, the playboy in the middle. [The allusion is as much to the Transfiguration as to the Resurrection.]

By concentrating on the legend of Ahasuerus, which plays little part in the poem he actually wrote, Goethe in his autobiography conceals that his epic fragment of 1774, *The Wandering Jew*, which tells in the main of a return of Christ to eighteenth-century Germany, is another and major example of his self-identification with the Son of God. In the context of a parody of Klopstock's Christology a more elegiac tone is here given to Christ's relation with 'his' world: we catch a glimpse of the most profound problem that for Goethe lay concealed behind the acclamation that he was receiving from his nation and even behind the theory of the interdependence of national literature and individual genius with which Herder had fired him in the first place—how, namely, to be a poet in this time and with this people? how, in the circumstances of this age, to satisfy, to express objectively, one's longing for the immensely loveable world?

O mein Geschlecht wie sehn ich mich nach dir

O my race how I yearn for you

says Goethe's Christ, as he approaches the earth once more, and we realize that Goethe's relations with his countrymen, clamouring for their national saviour, cannot be simply relations of amused indifference:

O Welt voll wunderbarer Wirrung

Voll Geist der Ordnung träger Irrung
Du Kettenring von Wonn und Wehe
Du Mutter die mich selbst zum Grab gebahr.
Die ich obgleich ich bey der Schöpfung war
Im ganzen doch nicht sonderlich verstehe . . .

O World full of strange confusion, full of the spirit of order and weary wandering, you circular chain of bliss and woe, you mother that bore me for the grave—world which, although I was present at the creation, I do not on the whole understand particularly well . . .

The irony in the words 'although I was present at the creation' tells us something important about the identification with Christ, and even about the vocabulary of 'creative' artistic activity—that Goethe recognizes these too as part of the world of ideas that his age and nation are seeking to impose upon him, a limited and needy being. His authority for accepting these titles is simply that others are willing to bestow them.

In so far as Goethe' assimilation of himself to the Messiah—his arrogance, as his critics saw and see it—was simply outrageous, its purpose was not to indulge some megalomaniac fantasy but to secure an ironic detachment from all the religious enthusiasms and neuroses of his contemporaries. What is astonishing about Goethe's career is that he should have become a great poet in an age of theological turmoil—a turmoil without parallel in Germany since the Reformation (and the Reformation brought forth no Goethe), and in the rest of eighteenth-century Europe without parallel at all. Throughout the period from 1750 to 1810 (at least) German theology was secularizing itself into philosophy, and practically all of Goethe's literary contemporaries became at some point or in some respect tributary to one or other of the religious philosophies that resulted—but no orthodoxy, old or new, not Kant nor Fichte nor Schelling nor Hegel could seduce Goethe, not even in his youth the deistical Enlightenment of Voltaire or Lessing, or Basedow. His secularized religion remained what it always had been before secularization—a wholly, even obsessively, personal piety: *his* relation with *his* heavenly Father, and no one else's.

'Now there's a man that none of you can grasp', wrote Klinger of Goethe in *A Woman in Suffering*. What is the equivalent of taciturnity for a man who writes, who is under no compulsion save the irresistible one to express himself? What is the equivalent of addressing only a few about the things that matter most, for a man who has published a best-seller by the age of 24, whose every utterance is a public utterance, treasured and passed on by a bevy of busy correspondents, and who in society as much as in books is expected to meet, and wishes to meet, the entire intellectual class of the nation? Goethe cannot simply pass over in silence the principal concerns of his countrymen—to do so would be to deprive himself of something that is more important to a writer even than style or the classics or national tradition, though it is related to all these, namely simply subject-matter, the objective, publicly recognizable, correlative of his feelings.

Goethe distanced himself from Satyros, for example, by allowing him to be unmasked, because he wanted to avoid the temptation, to which Thomas Mann often succumbs when dealing with similar subject-matter, indeed, when dealing with Goethe, the temptation of presenting his relationship with his public as one of contempt and confidence-trickery.

> Wer sein Herz bedürftig fühlt,
> Find't überall einen Propheten.

He who feels his heart in need will find a prophet anywhere,

says the hermit, but Goethe shares the need while Satyros simply exploits it. Goethe's engagement with the theological and philosophical themes offered him by his place and time was, from start to finish of his literary career, through ironical detachment—but irony is engagement too. In an essentially retrospective *ars poetica* of 1776, 'Commentary on an Old Woodcut Representing the Poetical Mission of Hans Sachs' ('Erklärung eines alten Holzschnittes, vorstellend Hans Sachsens poetische Sendung') Goethe draws a picture of the poet-demiurge with none of the Herderian grandiloquence:

> Er fühlt, daß er eine kleine Welt
> In seinem Gehirne brütend hält,
> Daß die fängt an zu wirken und leben,
> Daß er sie gerne möcht' von sich geben.

He feels that he has got a little world brooding in his brain, that is beginning to act and live, that he would like to bring it forth.

And when the Muse of Hans Sachs appears she gives the reason for the comic and earthy tone in which this deep wisdom is communicated. (English poetry, whose vehicle is the language of a profoundly class-conscious culture, lacks the middle tone between verse and doggerel in which this poem, like most of *Faust*, is written.)

> Die spricht: 'Ich hab' dich auserlesen
> Vor vielen in dem Weltwirr-Wesen,
> Daß du sollst haben klare Sinnen,
> Nichts Ungeschicklichs magst beginnen.
> Wenn andre durcheinander rennen,
> Sollst du's mit treuem Blick erkennen;
> Wenn andre bärmlich sich beklagen,
> Sollst schwankweis deine Sach fürtragen . . .'

She says, 'I have chosen you above many others in the world's confusion to have clear senses and not to start up anything silly. When others are running all over the place you are to discern it faithfully; when others complain pitifully you are to put forward your matter in comical form . . . '

Goethe's description of himself in terms of incarnation, and as an unfathomable divine visitor, expresses the belief, abandoned sooner or later by so many of his contemporaries, that it *is* possible to be a whole man, and one's own man, in this land and in this age, provided that one does not allow oneself to be the creature of this land and this age and its follies and simplifications. The literary instrument of this belief is humour.

But how was the balance to be struck in practice? For the year after the publication of *Werther* Goethe's fame in Germany was at its height. (The apogee was perhaps reached in February 1775 when the theological faculty of Leipzig University condemned the book, and its sale in the city where it was first published was forbidden on pain of a fine.) Goethe later complained to have felt rather distant from the clamour and turmoil that *Werther* aroused. The novel had been for him, he says, a kind of general confession ('Generalbeichte') which had liberated him for a new and happier life. But in fact the period which began in October 1774 was exceptionally strenuous, putting Goethe squarely before the question how he was to live with the nation he had succeeded in speaking to so directly. At its most immediate, the question concerned simply his domestic arrangements, as a visitor noted:

The atmosphere in Goethe's home is impressive and you feel as if you were entering the house of a minister of state. His reception room, or rather audience chamber, is never empty, as one goes another always comes. And Goethe is really seriously inconvenienced, for every passing traveller wishes to make his acquaintance. But now he has made it his rule to give audiences only four times a week, in the mornings; for the rest he belongs to his friends, and to his affairs.

'Everything is in a ferment and we are near a revolution', Boie and Goethe agreed at their meeting in October, and the locus of the ferment was Goethe. If there was any coherent sense to be made of all the conflicting forces it would have to be reflected in Goethe's own life, for he had made himself into the mirror of the nation:

Another thing that makes me happy is all the noble people, along it is true with many who are insignificant or intolerable, who come into my area, to me, from every corner of my fatherland, sometimes passing on, sometimes staying. One only knows one exists when one finds oneself in others.

The public, he said, was the echo answering his voice. If he was hostile to Wieland, as the editor of the *Teutscher Merkur,* it was perhaps because the rival locus of public literary consciousness that was Wieland's journal lacked that last element of personal commitment and passion that he alone could provide. But would Goethe 'break up' first, before the coherence was found, as Schlosser thought might happen? 'People fear his own fire will consume him'. The thought was in Goethe's own mind as he came back in the coach from Darmstadt on 10 October and he composed on the way an exuberant, semi-articulate free-verse hymn 'To Postilion Chronos' ('An Schwager Kronos') to dismiss it. The poem likens his life to a coach journey, in rhythms now labouring and clotted for the times of difficulty, now smooth and sovereign as it reaches the heights, now rattling rapidly down to a conclusion: let the end be quick and ecstatic, this young Achilles exclaims, a fiery consummation that compels the applause of a pagan underworld. The reaction was not long in coming, however: ten days later he was writing to Sophie von La Roche

I have been lying curled in on myself and silent and questing around and about in my soul whether there was the power in me to bear all that brazen Fate has dictated in the future for me and mine; whether I could find a rock to build a castle on that I could flee to with my belongings in the direst need.

Was Frankfurt such a rock? The very fact that Goethe was searching suggests that he felt it was not. There was talk of a career for him, if not in the city administration, then as the resident agent of one or more of the larger German states. But Frankfurt was possibly already too restricted an arena for sympathies as broad as his had now become. His conception of the nation he belonged to now ranged from 'the common people' who 'are the best human beings', to the middle classes of whom, he noted, his current literary circle was exclusively composed, to the nobility whose favour he had gained through his defence in *Götz* of the old constitution of the Empire. If Frankfurt could not contain him and he were to leave for somewhere significantly different, yet still wished to remain in the German-speaking world, there were only two practical possibilities. Either he could abandon the Empire altogether, as beyond redemption even by the ferment that had now begun, and move to Switzerland. Or he could follow the example set by Lessing and Herder, and now most recently by Klopstock, and look for a prince interested in making use of his talents. His great-uncle von Loen, and the new Darmstadt Chancellor K. F. von Moser, also with Frankfurt connections, had both written books on the opportunities for honest and able men from undistinguished backgrounds to make substantial reforms within absolutist governments: such a move need not be a betrayal of the national and anti-despotic spirit that had animated the *Frankfurter Gelehrte Anzeigen*. As it happened there was an occasion only two months after Klopstock's visit for investigating this second possibility, though it entailed, as Goethe perhaps intended anyway, bringing to an end his one outstanding literary feud, with Wieland.

The Prince of the Thuringian duchy of Saxe-Weimar, Carl August (1757–1828), would on his eighteenth birthday, 3 September 1775, attain his majority and as Duke take over the government of his little state from his mother, the Dowager Duchess Anna Amalia (1739–1807). In December 1774 he was completing his education with a journey to Paris, on which it was hoped he might find himself a wife, two of the Princesses of Hesse-Darmstadt being still available after the Tsarevich had taken his pick in 1773. On the eleventh he was in Frankfurt, on his way to Mainz, in the company of his principal tutor Count Görtz (1737–1821) (whom Wieland had in 1772 been appointed to assist), of his brother Prince Constantin (1758–93), and Prince Constantin's tutor C. L.

von Knebel (1744–1834), and of his Chief Equerry Josias, Baron von Stein (1735–93). Knebel came from the lesser nobility of Franconia, and had served for a while as an officer in the Prussian army, but he was a man of literary tastes and could not pass through Frankfurt without calling on the author of *Werther*, and *Gods, Heroes, and Wieland*. He was immediately won over: 'one of the most extraordinary phenomena of my life . . . he pulls the manuscripts out of every corner of his room'. Goethe in turn was interested to hear about Weimar's personalities, Wieland and the production of his *Alcestis*, the effect of the palace's destruction, and the changes to be expected from the new regime. It was not difficult to secure an offer from Knebel to present him to the future Duke, it may anyway have been with that intention that Knebel visited him. It was, however, certainly Goethe's doing that when he met the Princes in their hotel much of the conversation turned not on literary matters but on a recently published collection of Möser's political writings, on the role of small states within the constitution of the Empire, and on the history of Saxony, issues calculated to show that he was not merely a flighty littérateur. He made a good impression on the Princes and their entourage, particularly by his praise of Wieland, and was invited to spend a few days more with them in Mainz. Caspar Goethe disapproved of this hobnobbing with potentates, but Fräulein von Klettenberg, now on her sick-bed, gave her blessing to the new connection. On 13 December Goethe set out with Knebel, who had stayed behind to accompany him, and so 'enjoy the best of men', to Mainz, where he stayed until the sixteenth, drawing and skating while his hosts were at court, and then regaling them with his literary enthusiasms and hatreds, and explaining how momentary and impulsive had been his attack on Wieland. He was prevailed on, probably without difficulty, to write to Wieland to repair relations: Wieland was at first unimpressed by the letter, but such a wave of approval reached him in the travellers' correspondence that in January he announced a 'radical' cure of any 'displeasure with this strange and great mortal . . . It will not be for any lack of good will on my part if we cannot become friends.' The wife of Baron von Stein also wrote to her correspondents to enquire about this new star which seemed to be entering the Weimar firmament, but her friend the society doctor, psychiatrist, and popular philosopher, J. G. Zimmermann (1728–95), who was an old acquaintance of Lavater's but himself did not yet know Goethe personally, would tell her only: 'You do not consider how dangerous this loveable and enchanting man could be for you'. Meanwhile the Weimar cortège had passed on to Carlsruhe, for the wife of the Margrave of Baden was the aunt of the Princesses of Hesse-Darmstadt, who were now staying at the court, and here on 19 December Carl August and Princess Luise (1757–1830) announced their intention of becoming betrothed. Goethe's return to Frankfurt, however, was distressing: on the very day of his departure for Mainz Fräulein von Klettenberg had died. He had refused to believe in the seriousness of her illness, perhaps because of the inconvenience of the moment, but her last words

—'Say goodbye to him. I was very fond of him'—showed no resentment of his absence. Thanks to his days in Mainz a new life was beginning for Goethe but in it Fräulein von Klettenberg would never be forgotten, and whatever one may think of his conduct at the time of her death, more than one literary monument would be set up to her memory.

A new air of confidence runs through Goethe's increasingly hectic activity of the winter of 1774–5. There was more skating of course, and he took up chess. The new contacts were vigorously cultivated. Bürger and Boie continued to write. Fritz Jacobi stayed in the Hirschgraben for the greater part of January and again at the end of February and beginning of March, and Goethe agreed to complete *Erwin and Elmira* for publication in the March issue of *Iris*, to which he was now a regular contributor. A new full-length drama, *Stella*, was on the way, and Fritz and Johanna Fahlmer were kept informed of its progress. (Unfortunately, its overt allusions to the triangular relationship of Johanna and Fritz and Betty Jacobi were to lead, later in the year, to a cooling in the friendship at the very time when in his letters, and to some extent in his first attempt at a novel, Fritz achieved his own version of the Goethean fusion of Sentimentalism and Storm and Stress.) A correspondence began with Knebel, to whom Goethe had lent some of his fragments, including *Faust*, and, having already the previous October tried his hand at oil-painting, Goethe in February started to take drawing lessons with his old acquaintance G. M. Kraus, who had just returned from a visit to Weimar and nearby Gotha, and was to take up the post of Director of the Weimar Academy of Drawing in October. Kraus thought well of Goethe's portraits and silhouettes, but Goethe seems to have valued also the opportunity of seeing Kraus's drawings and etchings of Weimar scenes and personalities, and wrote to Wieland to congratulate him on the happy family revealed by Kraus's portrait of him with his wife and children (though he told Kraus he thought the furniture too luxurious for an author's establishment). Goethe was plainly very embarrassed when a new satire on Wieland, and on numerous hostile reviews of *Werther* appeared in March, written in 'Knittelvers' and with the title *Prometheus, Deucalion* [Prometheus' son, i.e. Werther] *and his Reviewers* (*Prometheus, Deukalion und seine Rezensenten*), and generally attributed to him. It was not well received by the Weimar Princes and their companions—Görtz thought it an 'obscenity' and Goethe 'a common fellow' and was sure the two of them would never be found in the same room again— and on discovering that the author was in fact H. L. Wagner Goethe went to the lengths of printing and publishing a notice of disavowal and making sure that a copy went to Knebel. Even Klopstock's decision to return from Carlsruhe to the more republican Hamburg did not reduce Goethe's interest in the possibilities offered by a small but cultured principality. Only a fortnight after Klopstock had again stayed with him, on his return journey, Goethe was exhorting Knebel to 'write a lot to me about yourself. About our dear Duke.' The attention Goethe paid to the two Princes of Saxe-Meiningen, who visited Frankfurt in February, may also reflect their position as well-informed, and possibly influential, neighbours of Weimar, and conceivably an alternative destination. But older friends were not neglected either. Herder had been cool and distant ever since their last meeting in 1773 at the time of his wedding and the recitation of *Father Porridge*, but in January a reconciliation was completed by letter. Jung-Stilling also came to stay in March, though Goethe found it difficult to tolerate the religious agony and ecstasy that accompanied the failures and successes of the cataract operations the pious doctor had come to perform.

Not that Goethe's life was devoid of religious, or quasi-religious, emotion at this time. The belief that Wieland was after all on the wrong side, and that collaboration with him would be impossible, recurred in March when the *Teutscher Merkur* unfavourably reviewed Lenz's most recent work, and in his dejection Goethe returned to the conviction that 'I will come to a bad end'. 'So much happiness and misery I don't know whether I am in the world or not,' he also wrote, '. . . don't abandon me . . . in the time of sorrow that might come when I could flee from you and all my loved ones . . . save me from myself . . .', and then, three weeks later 'the dear old thing they call God, or whatever it's called, is really taking great care of me'. The recipient of these confidences was the higher-born of two new female acquaintances made in January who for a year divided between them the world of Goethe's emotions. Although some of Goethe's most personal and demonstrative confessions were addressed to her, they never met, and indeed her first two letters were sent as from an anonymous admirer of *Werther*. Almost immediately, however, Goethe learned, probably through Boie, that he was in correspondence with Augusta, Countess Stolberg. It was surely the extreme case of communication with an absent object of confidence and affection, a monologue in dialogue form sustained for longer and with far greater intensity than the imaginary relationship with Helene Buff. Goethe's side of the correspondence was eventually written as a diary and couched in the 'Du' form, suited only for brother and sister, or lovers. Augusta, or 'Gustchen' as she affectionately became, herself noticed the similarity in form and content to the letters of Werther, and the relationship shows clearly that Goethe's 'general confession' was very far from resolving for him all the issues raised in the novel.

The complement to Goethe's purely spiritual relationship with a noble woman known to him only as a member of that public which simply echoed his solitary feelings and to which *Werther* had been addressed, was the wholly real, and so for him at this stage in his life impossibly difficult, relationship with Anne Elisabeth Schönemann (1758–1817). 'Lili', as Goethe called her, though she later preferred the abbreviation 'Liese', was the daughter of a Calvinist banking family partly of Huguenot extraction, whose best days were already over, though in 1775 neither she nor Frankfurt knew this, the final bankruptcy not coming until 1784. Her father died in 1763; she and her four brothers were brought up by her mother, who also managed the business. Goethe met her at a concert in the massive family house, whose construction began the decline in the Schönemanns' fortunes, as the carnival season of 1775 began, and within weeks he was playing whist and dressing up for balls in order to see more of her. More to his taste was the semi-rural life of her uncle and cousin in the garden suburb of Offenbach, along the banks of the Main, where he stayed repeatedly between March and May, in the house of the composer Johann André, and where little festivities were held and there was much music-making and recitation of verse (including Bürger's *Lenora*). 'Lili' was no shallow society beauty, however, but a woman, as her trials during the French Revolution were to show, of great courage and strength of character. Her family had no tradition of education or of cultured interests, but she was aware of the deficiency. She was at once serious-minded and physically attractive, and had more intelligence, and more fibre, than any of the other women to whom Goethe had so far given his affection. A measure of her difference is the thoroughness with which she destroyed every single piece of paper relating to her liaison with Goethe, which cannot therefore be followed in detail. What is certain is that their courtship was intoxicated with the promise of happiness, and is that not a great part of happiness itself? 'It was a thoroughly brilliant period', says *Poetry and Truth*, 'a certain exaltation prevailed in the company, one never arrived at a sober moment.' Around 20 April an informal engagement took place: Goethe bought not rings but two golden hearts so that each could wear one on a necklace, and for a little while, he says in his autobiography, he enjoyed the enchanted state of betrothal. But almost immediately there were warning signs. Goethe had little patience with Lili's philistine and materialistic brothers, and his parents were troubled by the religious difference between the families and could not strike up a cordial relation with Frau Schönemann. In his occasional verse for the family circle—much of it of course destroyed—Goethe referred to himself as a bear from the woods, tamed but out of place in the neat rococo zoo run by the imperturbable 16-year-old Lili. One such piece that has survived, written either in the spring or in the autumn of 1775, 'Lili's Park', concludes with the threat from the wild beast that his Storm and Stress is not over and his docility is not to be trusted:

> Nicht ganz umsonst reck' ich so meine Glieder:
> Ich fühl's! ich schwör's Noch hab' ich Kraft.

Not quite in vain do I so stretch my limbs: I feel it! I swear it! I still have strength.

It was not, however, obvious why the engagement should be doomed. Goethe's comments to Eckermann at the end of his life have great poignancy: 'she was . . . the first woman whom I deeply and truly loved. And I can say also that she was the last . . . I was never so near to my own proper happiness as in the period of that love for Lili. The obstacles that kept us apart were not really insuperable—and yet I lost her.' Indeed, the real obstacle was precisely that the difficulties were not insuperable. Lili was completely available and very suitable. She was not absent, distant, playing a game, or married or engaged to someone else: there were no great social or, at the time, financial disparities—in 1778 Lili's dowry was the same as Cornelia Goethe's. There was a strong and mutual attraction and Lili eventually made a better wife than most, if not all, of the other women Goethe admired. The match would have been a good one for any ordinary Frankfurt lawyer, and that no doubt is why from the start Goethe was uneasy. Lili presented him for the first time with the fully real possibility of marriage and for that reason the crisis in which he rejected her was the most serious, and most decisive, of his life. Lili meant marriage, and marriage meant more than coarse and mean-minded brothers, and parents-in-law who did not get on—marriage meant Frankfurt. Not merely because the social and financial pressures to remain in the home town and accept life as an advocate or consul would then have been practically irresistible. Lili herself and her part-Huguenot relatives with their French names, embodied Frankfurt far less ambiguously than the oddly marginal Caspar Goethe, and that is the true significance of the religious difference between the families. As Calvinists and members of the second social class, the Schönemanns were far more committed than the Goethes, by their occupation and by the political struggle of their caste since the turn of the century, to Frankfurt and its constitution. Their connections were cosmopolitan, they entertained frequently and lavishly, but government they were content to leave to the aristocrats: a solid burgher could do his duty by public affairs in the Citizens' Committees and in the Masonic lodge (an invitation to join which Goethe turned down). Goethe's parents, like their son, had less than total sympathy for this spirit of establishment at its most wealthy, uncombative and self-satisfied: their political and cultural perspectives had the width and independence of those who possess wealth and position, rather than achieve them. But even had a Lili and Wolfgang Goethe been able —or been forced, as Lili was later—to leave the burgher world of the free cities, and been able to live like Kestner or Merck or Zimmermann, in a city of residence or at the court of some prince, marriage would still have meant something else for Goethe which, ever since he wrote the ballad 'Pygmalion', he had time and again shown himself unable to accept: the end of desire. He still believed it better to burn than to marry, for that was the condition on which he could write his poetry, and to that belief Lili was sacrificed as Friederike before her. Goethe, however, had not forgotten the pains and self-delusions of Sesenheim and in 1775 he ensured that it was he, not the woman, who suffered the more. Lili later confessed that she had been ready 'to surrender to him her virtue', but that Goethe had refused, and by this refusal he had become 'the creator of her moral existence'. He had, that is, taught her moral responsibility, an earnestness about body and soul and vocation that he had himself been learning for ten difficult years. Had he accepted her advances and then abandoned her she would, she recognized, have lost all social standing and with it the happy and fruitful, though sorely tested, marriage that she in fact enjoyed. Goethe may also of course have shrunk from a bond that would have made marriage almost inevitable, but in sexual matters no motives are straightforward.

The temptation Lili offered Goethe was to conform his life, and so the poetry which, in the absence of a religious motive, fed on his life, to a defined role within the existing social order. Not until he was 57 did he accept such a role: for another thirty-one years he remained unaccountable, unpredictable, uncommitted. But the temptation was also physical, and resisting it, for whatever purpose, exacted a physical cost. Goethe at times felt that he was a simple animal, with obvious natural needs, being asked to occupy a superhuman role for the sake of the rather abstract needs of the German public heart. The exasperated reaction is manifest in the highly obscene fragments of *Hanswurst's Wedding* (*Hanswursts Hochzeit*) written at the time of his engagement, in the spring of 1775 (when, incidentally, his hair was beginning to fall out). Hanswurst, the bumpkin jester of stage comedy and the puppet-plays, who is a good deal more fly than the nobler or more serious characters, has become a literary 'genius', and Kilian Brustfleck, his guardian, the representative of German bourgeois virtue, is crowning the work of educating his ward by settling him in a comfortable, conventional, respectable and, he hopes, fertile marriage.

KILIAN BRUSTFLECK. Die Welt nimmt an euch unendlichen Theil
    Nun seid nicht grob wie die Genies sonst pflegen

    .   .   .   .

    Was sind nicht alles für Leute geladen

    .   .   .   .

    Es ist gar nichts an einem Feste
    Ohne wohlgeputzte Vornehme Gäste.
HANSWURST. Mich däucht das grösst bey einem Fest
    Ist wenn man sichs wohl schmecken lässt
    Und ich hab keinen Appetiet
    Als ich nähm gern Ursel auf n Boden mit
    Und aufm Heu und aufm Stroh
    Jauchtzen wir in dulci iubilo
KILIAN BRUSTFLECK. Ich sag euch was die deutsche Welt
    An grosen Nahmen nur enthält
    Kommt alles heut in euer Haus
    Formirt den schönsten Hochzeit schmaus
HANSWURST. Indess was hab ich mit den Flegeln
    Sie mögen fressen und ich will vögeln

KILIAN BRUSTFLECK. The world is infinitely interested in you. Now do not be crude as geniuses usually are. All the people that have been invited! A feast is nothing without fine well-dressed guests.
HANSWURST. I think the biggest thing about a feast is having something tasty and I've only got an appetite for taking little Ursula up to the loft and on the hay and on the straw we'd rejoice in dulci jubilo.

KILIAN BRUSTFLECK. I'm telling you, all the big names of the German world are coming
  to your house today and making up the finest wedding-feast.
HANSWURST. But what have I got to do with the louts, they can feed and I'll fornicate.

In 1770 Goethe had come to Strasbourg uncertain whether his vocation was to
fleshly satisfaction, to a quiet and comfortable marriage, to poetry, or to a
successful career as a man of the world. The Herderian literary programme had
suggested a role for him that removed the immediate pang of that uncertainty.
But buried within the literary programme was a theory of genius that in the long
run could only exacerbate anxiety about his individual vocation. What benefit
does the Hanswurst of 1775 have from the willingness of all the great names of
Germany to assure him now that he is a genius? True, they all seem to want him
to settle down and get married, but the one thing marriage does not mean to
them is the one thing that might drive him to it. Perhaps the little men are all
conspiring with social convention to thwart his desire and even to stifle the
genius they profess to admire. In the later years of his Storm and Stress period,
from 1773 to 1775, the terms in which Goethe wrote of the creative man
became more extreme, emphasizing his exceptional status, not always to the
benefit of the work concerned. 'Song of Mahomet' ('Mahomets-Gesang')
(1772–3) was a magnificently controlled rhythmic crescendo, an ode in which
the career of the prophet was likened to the course of a great river, from its first
tinklings amid the mountain rocks to the swelling estuary that presents the
waters of all its tributaries to Father Ocean. That the brother-rivers of the one
great stream should have had so prominent role in the poem was appropriate to
a period in Goethe's life when he saw the genius as a participant in a collective
enterprise. In *Clavigo* the focus is already on the individual, and in the *Golden
Words of Solomon, King of Israel and Juda . . . (Salomons Königs von Israel und Juda
güldne Worte . . .)*, written in 1775, the humorous impatience of *Satyros* has
degenerated into a stagey arrogance:

A cedar grew up amid fir-trees, they shared with it rain and sunshine. And it grew, and
grew up over their heads and looked round far into the valley. Then the fir-trees cried: is
this our thanks that you are now puffed up, you who were so small, you whom we have
nourished! And the cedar said, Take your complaint to him who bade me grow.

Vacuous genius entered Goethe's life in person, in May 1775, announced,
appropriately enough, by the absent audience herself, Augusta Stolberg. Her
two brothers, Christian and Friedrich, together with a third nobleman, a
student friend from their Göttingen days, were on their way to Switzerland and
intended to call on Goethe as they passed. Arriving in Frankfurt on 8 May and
lodging at an inn, though taking most of their meals at the Grosser
Hirschgraben, they quickly managed to persuade Goethe to accompany them,
at least as far as Emmendingen, south of Carlsruhe, where Schlosser and
Cornelia were now living. Goethe's agreement threw them into ecstasy: 'the
four of us, by gad, make up such a company as you would seek in vain from

Peru to Hindoostan', wrote Christian, and they all four had suits of 'Werther's uniform' made up for them, so the world would know who was coming—blue frock coats, buff waistcoat, and trousers, as in the book, and in addition grey bowler hats. Goethe gave Lili little notice and no explanation of his departure: it may have been, as he says in *Poetry and Truth*, an attempt to test the strength of the relationship, though if so it was at the least imprudent; it looks, more simply, like a flight from the tensions to which the engagement had committed him; more profoundly, it was a first obscure essay in a genre in which he would exercise himself several times over the next fifteen years: a striking gesture, usually a journey, often commemorated in a literary work and intended to give time and distance to reflect on his life and purposes, to liberate him to follow unresisting the current of events, or 'fate', and to stand as a symbol or landmark of some turning-point in his career. A day or two before he left on 14 May Goethe wrote to Herder in full awareness that he was deliberately putting his life in the hands of forces not subject to his conscious control:

I recently thought I was approaching the port of domestic bliss and a firm footing in the true joy and sorrow of the earth but am now wretchedly cast out again on to the deep . . . On the wire known as inborn fate (fatum congenitum) I am dancing my life away . . . Its will be done!

The journey began in a manner appropriate to geniuses, schooled by Klopstock: naked bathing in a pool near Darmstadt; toasting Friedrich Stolberg's beloved, an Englishwoman by the name of Sophie Hanbury, and smashing the glasses against the wall; sharing his tears when the letter came to say that he and Miss Hanbury could only be good friends. Merck thought the whole affair idiotic and told Goethe so as they passed through Darmstadt: 'You will not stay with them long', Goethe, in *Poetry and Truth*, recalls him saying, 'you strive . . . to give a poetic form to what is real; the others are trying to give reality to what they call the poetical, the products of the imagination, and that results only in stupid rubbish.' A more decorous manner was called for in the minute residence of Carlsruhe, which had a population of only 3,000, where the party stayed from 17 to 23 May. Here Goethe met Carl August of Weimar and his companions once more, when they arrived on 21 May to make ready for the formal betrothal to Princess Luise, and was encouraged by the illustrious couple to visit them when they had returned to Thuringia in the autumn. Five days followed in Strasbourg bringing two happy reunions: with Lenz and with the Minster. 'The old place, now so new again—the past and the future', Goethe wrote to Johanna Fahlmer, as if he were coaxing it all into having meaning and had not yet quite succeeded, 'nothing unexpected on the journey, but . . . everything is better than I thought. Perhaps because I am in love I find everything lovely and good. So much this time from the runaway bear . . . I have seen very very much. A splendid book to get wiser from, the world, if only it was some use.' He was now in two minds whether to continue the journey beyond

Emmendingen. Italy was beginning to beckon again and, though there had been no agreement with his father before he left, he perhaps thought he could get his father's approval were he now to set out on the grand tour to the ultimate university. On the other hand such a step would definitively sever relations with Lili. A possible compromise (but what does it mean to seek a compromise between *these* alternatives?) was to carry on with the Stolbergs as far as Zurich, where he could see, and perhaps draw help from, Lavater. (The account in *Poetry and Truth,* which suggests Caspar Goethe wished his son to go to Italy at this point, so soon after the engagement, is intrinsically implausible and is not borne out by the contemporary evidence.) On 28 May Goethe arrived in Emmendingen, where Schlosser, as the Margrave of Baden's highest-paid official, administered the lives of 20,000 people in twenty-nine towns and villages, but where Cornelia led a desperately lonely existence, deprived of the busy social intercourse she had been used to in Frankfurt. The visit of her brother, closely followed by Lenz and then the Stolbergs, was an unequalled moment of excitement. Did she, as Goethe says in his autobiography, advise him to break with Lili? We may at least be sure she had no desire to see her brother married to anyone, and perhaps in her own way she said so. While staying with her Goethe completed, and sent to Knebel (and so to Carl August) the 'Singspiel' *Claudina of Villa Bella* which he had 'disinterred' in April, which may have originated in the same circumstances as *Erwin and Elmira,* and which tells of the delights and agonies of the vagabond life. When he left the Schlossers on 6 June he headed not home to Lili but southwards to Switzerland, 'for I still feel the main purpose of my journey is unfulfilled and if I come back the bear will be in a worse case than before'.

Passing through the south of the Black Forest to Schaffhausen, where of course the great waterfall was admired, the party pressed on rapidly via Constance and Winterthur to Zurich, which was reached on 9 June. Lavater's welcome was warm; Goethe naturally stayed with him and felt for the first time the peculiar combination of calm, elevation, and unremitting practical activity that prevailed in his household. He must soon have met Lavater's confidante, the homely and comfortable Barbara Schulthess (1745–1818), who shared the pastor's literary enthusiasms but not his proselytizing zeal, and the young engraver J. H. Lips (1758–1817), who was an essential collaborator in the *Physiognomical Fragments.* There were also some familiar Frankfurt faces to greet him: Philipp Kayser, the musician, who had recently taken up a position in Zurich as piano-teacher on Goethe's recommendation, and J. L. Passavant (1751–1827), who something over a year previously had arrived to act as Lavater's amanuensis and who now almost immediately proposed to Goethe a joint tour of the historic cantons round the Lake of Lucerne. It was an attractive idea, but first there were some sights to be seen in the Zurich area itself. All progressive tourists had to visit Jakob Gujer, called 'Kleinjogg' (d. 1785), a model farmer, known as 'the philosopher peasant', and supposedly an example of the applicability of Enlightened principles to the lowest levels of the social and economic structure. The young and Sentimentalist Count of Lindau (1754–76), who had taken refuge with Lavater from an unhappy love-affair, also attracted the attention of the travellers, themselves a little love-lorn, for his decision to adopt a peasant-boy, Peter im Baumgarten, as an experiment in Enlightenment pedagogics. Goethe turned down Lindau's proposal that he should accompany him and Passavant into the mountains, but was interested enough in the experiment to agree to act as Peter's guardian should anything happen to Lindau. To the 77-year-old Bodmer Goethe and the Stolbergs were jointly introduced by Lavater at Bodmer's home, one of Zurich's most magnificently situated houses up above the old city with views out to the mountains. They discussed Julius Caesar, the 'great fellow' about whom Goethe was still planning a drama, and whom Goethe defended against the 'contemptible' conspirators who had stabbed the tyrant from behind. Bodmer was a close and sharp, and unashamedly ungenerous, observer of the literary scene, and his comment on the discussion went straight to the empty heart of the cult of genius: 'It is strange that a German, who tolerates his servitude with an extreme of impassivity, should have such ideals of boldness.'

On 15 June, Corpus Christi, the tours out from Zurich began. A boat containing Lavater, Goethe, Passavant, Kayser, and some others set out across the lake and met up with another boat containing the Stolbergs' party in order to re-enact the picnic of twenty-five years before, which had been the occasion of Klopstock's ode, 'The Lake of Zurich'. Goethe had paid little attention to landscape on the journey so far and he had not kept a diary. Lavater perhaps encouraged him to make a record of his tour and the notebook he now took with him opens with a meeting between the diarist and the natural landscape which has all the force of a discovery. On the boat, and during a walk along the lake-shore, under the spreading branches of the walnut-trees, their green fruit already set, Goethe, in two stages, scribbled a poem which caught his mood as it shifted on that glorious summer morning and which has claims to be regarded as one of his finest works:

> Ich saug' an meiner Nabelschnur
> Nun Nahrung aus der Welt.
> Und herrlich rings ist die Natur,
> Die mich am Busen hält.
> Die Welle wieget unsern Kahn
> Im Rudertakt hinauf,
> Und Berge wolkenangetan
> Entgegnen unserm Lauf.
> Aug mein Aug, was sinkst du nieder?
> Goldne Träume, kommt ihr wieder?
> Weg, du Traum, so gold du bist,
> Hier auch Lieb und Leben ist.

Auf der Welle blinken
Tausend schwebende Sterne,
Liebe Nebel trinken
Rings die türmende Ferne,
Morgenwind umflügelt
Die beschattete Bucht,
Und im See bespiegelt
Sich die reifende Frucht

Now through my navel-string I suck nourishment from the world. And splendid all around is Nature, holding me to her bosom. The wave lifts and rocks our boat in the rhythm of the oars, and mountains, cloud-girt, counter our course. Eye, my eye, why do you fall? Golden dreams can you return? Away, dream, gold though you be; here too is love and life. On the wave there glitter a thousand drifting stars, dear mists all around drink up the towering distance, morning wind surrounds with wings the shadowed bay and in the lake is mirrored the ripening fruit.

The contrast with Klopstock's poem of 1750 does not need to be laboured: this is the young Goethe at his purest and most concentrated, a sensibility at once pictorial, energetic, and intensely reactive. Goethe is above all the poet of the unique form and the unique rhythm: 'On the Lake' (as Goethe later titled the poem) magically combines the opening regular iambic verses with a sudden constellation of trochaic forms (at least five different varieties). Each of these rhythmic modulations is equivalent to a change of mood. The tautness of the poet's feelings is apparent even in the regular opening, through the strange contradictoriness of the images. There was always a paradox implicit in Ganymede's cry of 'Embracing embraced', and that paradox is now explicit in the different relations here said to obtain between the poet and Nature: he is within her, as in the womb, he is on her breast, he is outside her but attached, he is surrounded by her but at a distance. The simple harmony, even identity, of subjective and objective, of the personal and the natural, in 'May Celebration' is certainly gone in 'On the Lake', but this is not to say that Goethe has fallen back into the dualism of Klopstock. It would be truer to say that *identity* has been replaced by *relationship*. Contradiction is perfectly compatible with reciprocity. The first eight lines of 'On the Lake' are about complex and variable two-sided relations: being fed by, being surrounded by, being in time with, coming out to meet. The umbilical cord is not snapped here. On the contrary, the poet's sensibility appears as a living thing in its own right, actively feeding and passively cradled, met and caressed and adventurous and exploring. In twenty lines earth and air and sky and water, light and motion, the animal and the vegetable, are drawn into a single vortex of love and life. Only the four lines in the middle of the poem articulate a threat to the reciprocity, a threat constituted by the very life, the physical independence, bestowed on the active sensibility at the centre of the vortex. The hidden contradictions of the first two quatrains come to conscious expression as a single antithesis between the poet's individual self-consciousness—the man with a particular, if unspoken, past, and with particular personal dreams—and the here and now in which all readers are invited to join. From the magic circle of private self-contemplation the poet ejects himself by an act of will into the landscape of the last eight lines, in which he is present, not as an active participant, but in the silent unobtrusive manner of the traveller who contemplates a view and finds in it a reflection of his own thoughts and feelings. Yet although 'On the Lake' contains autobiographical elements, it is an autobiographical poem only in a rather complex sense. It is essentially a poem about having an autobiography, about being in a landscape when you have a life-story that you tell yourself, an incomplete life-story of your own. As far as the poem is concerned, the final image is not an image of incompletion, of ripening to come, but of completeness and fulfilment now. The image at last fixes on a single, non-contradictory, physical correlative for the poet's particular personal sensibility—the ripening fruit—and also states with perfect clarity the relationship between that sensibility and the landscape in which it finds itself, the relationship of reflection.

By different routes Goethe and the Stolbergs climbed up to Einsiedeln in the afternoon, Goethe meeting a procession of pilgrims on their way to celebrate the feast at the shrine of St Meinrad, the hermit. The following morning the friends separated, the Stolbergs returning to Zurich, Goethe and Passavant making for the road to the St Gotthard pass. Despite his thoughts of a descent into Italy, Goethe seems to have been under some external pressure, it is unclear from which quarter, though probably from his parents, to return soon to Frankfurt, and he was anxious, given that he had to return, to see the frontier from which he did so, to clarify, as it were, the possibility he was leaving unfulfilled. After two days walking and drawing round the lake of Lucerne— 'splendid high sunshine, for sheer pleasure saw nothing . . . in clouds and mist, and round about the splendour of the world . . . back at 8 . . . baked fish and eggs . . . bells clinking, waterfall rushing, water-pipes trickling, hunting-horn . . .' —they were ferried up the lake to Altdorf, noting the sacred places of the Tell legend as they went. It was early in the year to be trying to pass and they had been warned to expect snow: between Amsteg and Wassen the route ran over an old avalanche, hollowed out below by the roaring waters of the Reuss—by the time the travellers returned the snow bridge had collapsed. On 21 June, with the high peaks becoming 'omnipotently terrible', they reached the green valley of Urseren through the tunnel at Göschenen, but after Andermatt the scenery was of Ossianic bleakness: 'bare rock and moss and gale and clouds. The noise of the waterfall, tinkling of the mules. Desolation as in the valley of death—strewn with bones, sea of mist.' The welcome in the Capuchin friary on the Gotthard was, however, as homely as could be desired—bread, cheese, and heavy Italian wine. The following morning Passavant and Goethe turned back to Andermatt, but only after Goethe had drawn the southward road that, for whatever reason, he could not take for the present.

Goethe's long-standing interest in drawing had by 1775 become a passion—a sketch he made of his garret-room in March shows it cluttered with artistic paraphernalia: an easel, plaster casts, and drawings pinned to the wall—and the journey to Switzerland was one of the most productive periods in his early career as a draughtsman. No fewer than five, possibly seven sketches have survived that can all be dated to 17 June, two days after he had written 'On the Lake'. Like everything he drew at this time they stand out for their directed but carefree energy, to which training, or the lack of it, is scarcely relevant, their sense of motion in a landscape, sometimes reduced almost to abstraction, and their carrying to an original extreme the emphasis, characteristic of the Picturesque painting fashionable at this time, on the 'point of view'. A 'picturesque' landscape, according to the Reverend William Gilpin, the authority in these matters, was one that appealed because 'it would look well in a picture'. When such a landscape is itself depicted in a painting, a highly self-conscious art results. In interiors Goethe locates himself behind a stove or the back of a seated figure, producing a domestic and personal effect similar to that of the interior scenes in his plays; in landscapes the eye is in an emphatically high, at times almost airborne position. There is a strong sense of individual locality, whether famous, like the hospice on the Gotthard seen across rising snowfields and bare rock, and the view of the footpath winding down past the snow-flecked northern mountain-faces to unseen Italy, or anonymous, like a group of huts on a steeply cornering downward track. Here, and in random portrait heads, snatched on the way, one feels the sense for a unique but meaningful physiognomy that had made Goethe into a collaborator of Lavater in his attempt to find the laws underlying human individuality.

On 25 June, Goethe was back with Lavater in Zurich, after visiting Küssnacht and Zug, and calling on Count Lindau, who had retreated in Wertherian fashion to a mountain hut. The Stolbergs left on 4 July to begin their own tour of the mountains, and two days later Goethe set out for Basle, where he spent three days before arriving in Strasbourg on the twelfth. Here he stayed for a week, probably with Lenz, whom he saw every day, and renewed his acquaintance with Salzmann. The day after his arrival he went to the cathedral and, leaving Lenz behind on the ground, climbed the tower alone. Here, at a place associated with the very beginning of his public literary career, and with a personal past that already seemed remote, he wished to cut himself off from men as effectively as if he had been on the Gotthard, and reflect on past and future and the present turning-point that linked them. In Switzerland he had found a nation of free men, with a proud tradition of defending their freedom, some of the shrines of which he had visited: true it was 'only human beings on either side of the border', but the Swiss were 'a noble race, not wholly unworthy of their forefathers'. When he wrote to Knebel in August saying he had 'been on a pilgrimage through the dear old Holy Switzerland of the German Nation' the facetious parallel to the title of the Empire indicated the serious thought that here was a genuine political alternative to the constitution under which he had been brought up. To Sophie von La Roche he confided that he had now indeed found the rock on which to build his castle in time of need: 'I am glad that I know a country like Switzerland; now, however things go with me, I have always got a place of refuge there.' But Switzerland is a last resort, to fall back on only in case of retreat: it is no answer for the man who wishes to make something of *his* nation, and who feels in himself the power to do it. The need to return to Frankfurt and Lili, drawn by love, social obligation, parental injunctions, proves to have a deeper significance for his entire vocation, for he is being drawn back to his fatherland as the proper scene for the unfolding of his talent. Here is where his destiny lies, and the moment of this mystical illumination calls out to be given written form. High up on the cathedral spire Goethe started to write another prose hymn to Erwin, the architect buried beneath it, and this time he chose to make the post-Christian sense of what he was doing explicit by using the form of a pilgrim's meditation, such as he had perhaps observed in use during the Corpus Christi procession to Einsiedeln. *Third Pilgrimage to Erwin's Grave* (*Dritte Wallfahrt nach Erwins Grabe*) is divided into 'stations' as the poet ascends from one level to the next of the great structure, meditating on the identity of the creative power in the artist and the creative power in Nature. But it is at the same time an intensely autobiographical piece, seeking to understand the significance of this specific moment, when the eyes of the man who is now for the third time at the spot where he wrote *Of German Architecture* turn 'towards my fatherland, towards my love' ('vaterlandwärts, liebwärts'). Indeed the little piece—one of those given by Goethe to Wagner for printing in 1776—ends as a fragment because Lenz, we are told, tired no doubt of waiting, has climbed the tower too and interrupted the moment of 'devotion'. There could hardly be a clearer example of the reciprocal relation between the significant—that is, potentially literary— gestures that make up a symbolical life and the autobiographical literature that draws meaning from that life, or of the dependence of both on a secularized religion of destiny.

The present moment in Strasbourg was resonant with the future as well as the past. Dr Zimmermann was in Strasbourg, visiting his son who was studying at the university. Goethe had already exchanged letters with him, and he was of interest both for his collection of physiognomical silhouettes and for his correspondence with Weimar. He showed Goethe the silhouette of his particular informant, Frau Charlotte von Stein (1742–1827), wife of the Chief Equerry whom Goethe already knew, and when Goethe wrote beneath it the words, 'She sees the world as it is, and yet through the medium of love', Zimmermann told him so much about her that, the doctor later wrote to Frau von Stein, 'he could not sleep for three nights'. At the end of the month Goethe sent Lavater a detailed character analysis of the silhouette. As he journeyed back to the reality of Lili, it would seem, he still needed to protect himself by the construction of another, safely absent, safely married, ideal love.

Leaving Strasbourg on the nineteenth, and meeting up with Herder and his wife in Darmstadt on the twenty-first, Goethe was in Frankfurt on 22 July. Only ten days later in Lili's room in Offenbach, with Lili dressing next door, to go out riding, and his eyes lingering on her hatboxes, dresses, handkerchiefs, and boots, he was writing to Augusta Stolberg that his months of wandering in the fresh air, 'sucking into all my senses a thousand new objects' had all been in vain; he was back in Offenbach 'as simplified as a child, as constricted as a parrot on a pole' with the 'woman who makes me unhappy, through no fault of her own, with the soul of an angel, whose fine days are clouded by *me*, by *me*'. 'I am stranded in the shit again', he told Merck the following week, regretting the lost opportunities of Switzerland, 'and could box my ears a thousand times for not going to the devil while I was still afloat.' By 10 September it was such common knowledge that Goethe and Lili would have to part, and that he would have to leave Frankfurt again, that it could be clearly alluded to in a chorale he wrote for the wedding of one of the Calvinist pastors in Offenbach, a member of the young people's circle there. For Goethe who, with Lili, helped perform it, it was 'the most cruel, most solemn, most sweet situation of my whole life', and afterwards there were the 'burning tears of love' as he and Lili spoke plainly together at last in the moonlight 'and in the distance the hunting-horns, and the noisy merriment of the wedding guests'. The wedding that in Goethe's life would never be. Yet 'I cannot leave the girl'; 'I am letting myself drift and just holding the rudder so I do not go aground'; 'I tremble at the thought of the moment when she could become indifferent to me, when I could lose hope'. He could not bear to live without hope, but he had learnt in the last months that he also could not bear to live with the certainty of possession. Goethe continued to visit Offenbach, to look out for Lili at the theatre, at a big ball, even though there was nothing to say, even though she positively avoided him. He seemed to be trying to drive himself into illness. It was the crisis of that morbid volatility which Knebel had recognized when they first met—'Goethe lives in a state of continual inner warring and turmoil, for all objects have a most violent effect on him'—and which even in 1776 led Friedrich Stolberg to write, 'if God does not work some miracle upon him he will become one of the most wretched of all. How often have I seen him melting and then raging within the same quarter of an hour.' But in the midst of the turbulence there was a granite confidence that came from prayer to the only mediator he knew: 'as I saw the sun I leapt out of bed with both feet, ran up and down the room, besought my heart so kindly, kindly, and I had relief, and an assurance was given me that I was to be saved, that something was still to become of me'. It was the confidence expressed in 'On the Lake' that even though the eye sank into reflection the fruit would continue ripening, and expressed again, if more theatrically and self-consciously, and so colluding more with the pretence that the highest reality is permanent desire, at the end of the letter to Augusta Stolberg which is our main evidence for the events and emotions of these later days in September:

What a life. Shall I carry on? or end this one off for ever. And yet, dearest, . . . I feel that in the midst of all this nothingness, so many skins are none the less detaching themselves from my heart . . . my view out over the world is growing more serene, my converse with men more secure, firm, extensive, and yet my inmost self remains always and for ever dedicated only to holy love . . .

Goethe at the time was writing his own translation of the *Song of Songs*—'I have become pious again', he told Lavater—as if even in this extremity he wished to assert his belief that the word of God speaks of no holy love unknown to the word of Man.

As soon as Goethe was aware that his Swiss tour had resolved nothing in his relations with Lili he began to think of escape. The decision that the love for which he had returned from Switzerland was an impossibility required a new definition of his mission to the fatherland, for whose sake also he had (he thought) returned. In the first days of August he wrote to Princess Luise a letter which has not been preserved, and to Knebel a note reminding him of his existence and asking for a few words about 'our Duke'. But Weimar was only one possibility: at the same time he asked Lavater to list the points he would like him to investigate 'if I were to go to Italy', and in the middle of the month he actually thought it more likely that 'the invisible lash of the Furies' would drive him in a southerly, rather than a northerly direction. He was most explicit about his situation to Merck:

I am again watching for an opportunity to decamp: only I should like to know whether in that case you would be willing to help me with some money, just for the first stage. Could you in any case demonstrate clearly to my father at the coming meeting that he must send me to Italy next spring, that is, I *must* be gone by the end of the year?

Caspar Goethe had not intended that his son should go to Italy until he had spent some time at the Imperial institutions in Regensburg and Vienna, but he would have been much more willing to finance even a premature Italian journey than any trip to waste time and money and reputation at some princely court. Goethe's request to Merck to consider a loan is a clear indication that he intended if necessary to go against his father's will. He continued to pay attention to the Princes of Saxe-Meiningen, but on 22 September he was again pressed to come to Weimar by Carl August, who was passing through Frankfurt on his way to his wedding, which was to take place in Carlsruhe on 3 October. 'I saw with my own eyes that the Duke had quite fallen in love with Goethe', Zimmermann, who at the time was staying with the family, wrote to Herder, and one can almost see what he meant in the wide-eyed and fascinated portrait-sketch that Goethe drew of the young ruler.

Caspar Goethe, of course, opposed the suggestion. He feared that the Weimar princeling would soon humiliate his brilliant son as Frederick II of Prussia had humiliated Voltaire. No, he thought the time had come, after all, for the educational tour of Italy. His son resisted: he returned the young Duke's affection, his admiration for Wieland was of ten years' standing and the damage done by his pasquinade had been repaired, and Zimmermann himself, 'an achieved character! born Swiss and free, and modified at a German court', may have seemed proof that the republican spirit and court life could fruitfully be married. On 28 September Goethe, who had succumbed to a cold, stayed in bed until 10 o'clock, 'father and mother came to my bedside, there was a more intimate discussion, I drank my tea and so things are better'. Some compromise would seem to have been reached—perhaps it was agreed to wait for a fully formal invitation from Weimar before taking a decision. But by the end of the first week in October that too would seem to have come, perhaps from one of Carl August's gentlemen-in-waiting, J. A. von Kalb (1747–1814) to whom Goethe wrote on the third. Between 7 and 11 October Goethe told Merck, Augusta Stolberg, and Sophie von La Roche quite unambiguously that he was going to Weimar with the Duke—for how long was unclear: to the Stolberg brothers, who were also intending to call on the new Duke and Duchess while on their way to Hamburg, he held out the prospect that they might then all three travel on together. On 12 October Carl August and Luise were in Frankfurt and jointly reaffirmed their expectation of seeing Goethe in Weimar soon. The arrangement was that von Kalb would come from Carlsruhe with a new coach to collect Goethe and transport him to Weimar in the wake of the main party. Goethe said his farewells, but von Kalb did not come. To occupy himself while waiting, Goethe threw himself into a new play he had begun in the summer, possibly under the impress of his Swiss experiences, telling of the struggle for freedom of another mercantile and semi-republican nation, the Netherlands. *Egmont* is also concerned with the destiny of the exceptional individual, and with the place of love in his life, themes to the forefront of Goethe's mind throughout 1775, but it was the political substance of the play that particularly appealed to his father, who urged him on to complete it. After all, as the delay lengthened and still von Kalb did not come, so it began to seem that Caspar Goethe had been right and the ways of courts were not to be trusted—certainly not by any right-thinking burgher, whether Swiss, Dutch, or German. Finally, on 30 October, father and son decided that a departure could be delayed no longer and Goethe and his manservant, Philipp Seidel, set off for Italy. 'I packed for the North, and am travelling to the South; I accepted, and am not coming; I refused, and am coming', he wrote in the first pages of a new travel-journal, struggling to make some meaningful pattern out of the confusion of the last month. 'What now really is the political, moral, epic or dramatic purpose of all this?' he asked. 'The real purpose of the affair, gentlemen . . . is that it has no purpose at all.' But beneath the strained echoes of *Tristram Shandy,* and a biblical quotation, also somewhat Sternean, put in his father's mouth, a note of despair can be heard. There is indeed no purpose, and the meaning-creating mechanism is breaking down, but one certainty is beyond denial: the anguish of the creature at the centre of it all. 'Am I then in the world only in order to writhe in never-ending innocent guilt?'

The Goethes did not know, however, that Weimar too was perturbed by the delay. 'For eight to ten days, from day to day, from hour to hour, we have been waiting longingly for Goethe', Wieland told Lavater on 27 October. 'He has still not come and we now fear he is not coming at all.' Von Kalb had simply been held back by the coach-builders: having sought Goethe in vain in Frankfurt he rushed to overtake him on the southward road, and on 3 November found him in Heidelberg. All plans were reversed; Goethe found himself retracing his steps and, after another unexpected glimpse of Frankfurt, his last for four years, he was on the road to Weimar, which was reached on 7 November. On 10 November, Wieland wrote, overwhelmed by their first meeting: 'The divine man will, I think, stay with us longer than at first he thought himself'.

'Innocent guilt': Works, 1774–1775

The conflict between Sentimentalism and Storm and Stress, between the introspective egocentrism that led to the cult of the 'colossus', and the enthusiasm for the newly-opened vein of colloquial language and native tradition, was decisive for Goethe's literary career between 1771 and 1775. The conflict was first resolved, aesthetically speaking, in Goethe's most un-equivocally tragic work, the novel *Werther*. The dramatic form, however, permitted a representation of conflicting embodiments of the self which did not positively require so destructive a conclusion. 'Oh, if I were not now writing dramas I should be a ruined man', Goethe told Augusta Stolberg in March 1775. In the eighteen months after writing *Werther* Goethe completed (more or less) four substantial dramatic works which attempt to express, though not always in the form of a tragedy, the tragic potential in his experience: *Clavigo, Claudina of Villa Bella, Stella,* and the first version of *Faust*. (Too little is known of the genesis of *Egmont* for us to include it with any confidence in the discussion of this period.)

Even if the view of some critics be allowed that Goethe is incapable of writing tragedy, this must not be taken to mean that Goethe is ignorant of the tragic, or that he regularly seeks some purely ideal consolation. The most effective counter to the temptations of genius—more effective certainly than the tendency to objectivity that Prometheus found in creative activity itself—is a moral strength which more than once in these early dramas brings us close to tragedy and despair. The origin, or just the evidence, of that strength is Goethe's awareness of the possibility of guilt, first impressed upon him by his remorse at abandoning Friederike Brion, and revived by his open recognition of responsibility for his and Lili's unhappiness in 1775. Just as in religious matters the rigid, Enlightened deistical distinction between rational truths and human (that is, usually priestly) regulations meant little to him, so in moral matters he knew from his own experience that life offered more than an interplay of a naturally rational and therefore naturally good human soul with restricting and corrupting social conventions. 'Sombre remorse' was something inexplicable both to an Enlightened eudaemonist such as Lessing, for whom moral questions were in the end questions about the rational organization of society, and to an amoral genius such as Müller 'the Painter', for whom morality was a trammel put by lesser men upon greater (a view not as different from Lessing's as may at first appear). Remorse was possible only to one who could conceive that for all the great process of the secularization of intellectual Germany in the mid-eighteenth century there might be something of human value in the old ways, something that might in that great process be wounded and lost. Of this the domestic tranquillity of a country parsonage might well seem a symbol to one who was not convinced that he naturally belonged in the group of irreverent, and mainly bachelor, geniuses who were claiming him as their own. *Clavigo* and *Stella* may be minor in comparison with *Götz*, *Werther*, and even *Gods, Heroes, and Wieland*, but together with these they make up the greater part of the corpus by which Goethe was known to the German public by 1775, and their special interest lies in their concentration on the theme of the woman loved and abandoned, partly against his will, by a man possessed of an inner energy which he does not himself wholly understand.

Clavigo, written, it will be recalled, in a week in May 1774, is manifestly under the shadow of *Werther* both in its unambiguous conclusion and in its origins in a conflation of two sources: an anecdote by Beaumarchais and an English ballad which, according to Goethe, provided the last scene. But the play already shows an attempt to go beyond *Werther* by including in its scope the problems of specifically literary genius. Clavigo is an ambitious outsider at the eighteenth-century Spanish court, hoping to make his way to influence by means of journalism. In fact he is something of a literary revolutionary, for he is introducing new English models to a decaying imperial culture. In his early days of obscurity he formed an attachment to Marie, daughter of a French family, and herself therefore on the margins of polite society. As Clavigo's star has risen, so he has been embarrassed by this liaison and has left it behind. Marie, however, pines away for disappointment. Clavigo is forced by Marie's brother to resume relations with her, as he half desires to do anyway, but a sinister friend, Carlos, persuades him that the demands of his genius justify his ignoring common loyalties and he abandons Marie once again. Marie dies and Clavigo is killed over her coffin, full of remorse and acknowledging the justice of his end. It is an eminently stageable well-made play in the closet-drama tradition, and Merck's comment on it seems appropriate enough: 'Don't write me such rubbish again; the others can all do that sort of thing'.

But *Clavigo* is not simply a product of its time. For all its woodenness, it never colludes with the temptation with which Carlos corrupts his friend:

Möge deine Seele sich erweitern und die Gewißheit des großen Gefühls über dich kommen, daß außerordentliche Menschen eben auch darin außerordentliche Menschen sind, weil ihre Pflichten von den Pflichten des gemeinen Menschen abgehen.

May your soul enlarge itself and may the certainty of a great feeling come over you, the feeling that exceptional men are exceptional men precisely because their duties diverge from the duties of the common man.

This is not to say that Marie is simply a victim and Clavigo simply a villain. Clavigo's is a genuine dilemma and at least potentially tragic. It is the dilemma from which Goethe was struggling to escape, the dilemma of Storm and Stress itself. Clavigo is a man of ambition, ability, and a kind of patriotism, yet none of these talents can be fully realized *either* by accepting the restricted, private, and unadventurous world of Marie *or* by adopting a pose of exceptional genius, which might seem to justify transgressing the limits of that world, but which any sane man must know to be at least artificial and probably wicked and self-destructive too. The real weakness of *Clavigo* is its dramatic language. It contains no trace of the idiom of the people, and is written as if *Götz von Berlichingen* had never been. Yet this very weakness has its own diagnostic interest. For at the time of writing the play Goethe described it as 'a pendant to Weislingen in *Götz*, or rather Weislingen himself, completely in the round, as a principal character'. Nothing could show more clearly the total separateness of Clavigo's dilemma, and of the entire Weislingen intrigue, from the linguistic revolution that makes the novelty of *Götz*. The revival of national tradition and of popular language can in no way resolve that conflict of Goethe's which he has represented in the agony of Clavigo, indeed, in so far as Clavigo is himself a frustrated patriot, such a revival could only make matters worse for him. *Clavigo* is written throughout in the language of the Weislingen intrigue. The Weislingen drama and the Weislingen language belong to sentimental, 'official', literature, entirely to one side of the realistic art of the 'Strasbourg group'—but they are intimately linked to the cult of genius. Despite Herder's synthesizing *tour de force* in his *Treatise on the Origin of Language*, national tradition and individual genius are not cognate and do not mix.

The consistently natural and lively prose dialogue of *Claudina of Villa Bella* (*Claudine von Villa Bella*) is one of the operetta's many attractions. Thematically, scenically, and atmospherically rich and varied, lyrical, humorous, and with a lucid plot that produces one *coup de théâtre* after another and the most complex stage action in any of Goethe's plays, and with musical settings by Seckendorff, Reichardt, and Schubert to choose from, it deserves to be better known and more often performed than it is. The setting for its six scenes is an imaginary Spain, with touches of southern Germany. Crugantino (whose real name is Alonzo) is the elder brother of Pedro, but has forfeited his birthright by choosing to live the wandering life of a bandit and Don Juan. Pedro has become a successful official, who does not normally despise the court to which he is attached, but who is temporarily disabled from affairs by his love for Claudina, which she reciprocates. She is the only daughter of the elderly Gonzalo, and so naturally his favourite, but Gonzalo is also bringing up two lively nieces whose plausible irritation with their cousin adds to the comedy and intricacy of the action. Sebastian, an old friend of Gonzalo's family, arrives with the commission to arrest Alonzo-Crugantino who has been reported roaming in the vicinity with his band. Crugantino in fact serenades Claudina on his zither, wounds Pedro, without realizing who he is, in a resultant duel and takes him prisoner, and under his pseudonym gains access to Gonzalo's house to continue paying court to Claudina. Sebastian recognizes him but he makes his escape with pistols blazing. Claudina disguises herself as a man in order to visit Pedro in his captivity and is herself then taken hostage by Crugantino when the watch arrives to arrest him. Crugantino is overpowered and all parties are taken to prison. Here identities are revealed at last, but although Crugantino embraces Pedro as his brother and accepts his prior claim to Claudina, he remains truculent to the end and refuses, or appears to refuse, reintegration into his family and the society it represents. Claudina swoons when her father arrives to find her in such degrading circumstances, and she is feared dead, but the general rejoicing at the discovery she is after all alive permits the 'Singspiel' to end in an appropriately harmonious atmosphere.

Light-hearted though this little romance may be, the progenitor, with *Götz*, of the entire species of robber-plays, it has Mozartian depths. Claudina herself, the woman who dresses as a man, and who symbolically dies, only to be resurrected as an ideal with the power of unifying contraries, is an anticipation of other figures in later and far more earnest works of Goethe's. But all the delightfully individualized characters have their part in the depiction of the play's central conflict, that between Crugantino and the 'civil' (or 'bourgeois', 'burgerlich') society he refuses to join. Pedro, and the family of Gonzalo into which he hopes to marry, are not actually stuffy or bourgeois at all. They are minor nobility, dependent on a court, and they put a Sentimentalist's value on poetry, moonlight, and heartfelt emotion. The older generation of Gonzalo and Sebastian is not simply representative of the authoritative tone against which Crugantino rebels as Goethe rebelled against the fatherly manner of Wieland— it also possesses the practical realism of the older generation in *Erwin and Elmira* and, since that makes it alert to the limitations of Pedro's and Claudina's Sentimentality, it has a certain covert sympathy with Crugantino, who treats their effusive love-making with scorn. For the secret of Crugantino's complex and potentially tragic character—and in this, as well as more superficially, he resembles the bewilderingly electric, 'terrible and loveable', Goethe of 1775— is that he is not simply an outsider. He is the only true poet in the play, singing what is by far its most striking piece, the ballad 'Once there was a lusty lad', and rebuking himself for leaving his zither behind when he escapes from Gonzalo's house. And it is his poetry which fascinates and entertains Gonzalo's Sentimentalist family (there is thus a parallel between him and Satyros). He acknowledges the guilt attaching to his anti-social 'behaviour' (a word he abominates) but his refusal to be preached at has a certain justification, for the society that condemns him does not scruple to enjoy the art which his way of life produces. He has in a sense to pay the price of others' pleasure, and the price is high:

Do you know the needs of a young heart like mine? A young madcap? What theatre can you offer for my life? I find your bourgeois society intolerable. If I want to work, I have to be a slave; if I want to have fun, I have to be a slave. Must not someone who is worth anything at all rather go out into the wide world? . . . But this I will grant you, that once started on the wandering life, a man has no purpose and no bounds, for our heart, oh, our heart is endless, as long as its strength lasts.

The last scene of the operetta brings no answer to Crugantino's lament, as similar to Werther's as Crugantino's love for Claudina is to Goethe's love for Lotte Kestner. The dilemmas are not resolved, perhaps because the only possible resolution is again that which Werther found. Instead they are transfigured—one might say, ignored, though they can hardly be forgotten—as in the final tableau and chorus we turn to celebration of a life that, for everyone but Werther, just goes on—unjustifiably, inadequately, miserably, perhaps, but happily too. In this too *Claudina of Villa Bella* is an anticipation of many future works of Goethe's, most immediately of *Stella*.

While they were at Ems in the summer of 1774, Lavater may have told Goethe of a proposed marriage *à trois* in the court at Carlsruhe, on which his pastoral advice had been asked, and the lives of Jacobi and for that matter of Bürger (as later of Wordsworth), show that the problems of a man between two women had a certain topicality. But it is not to these external models, nor to such literary predecessors as *Miss Sara Sampson*, that we should look for the source of *Stella. A Play for Lovers* (*Stella. Ein Schauspiel für Liebende*) but to Goethe's own experience in 1775. For in this year, so unbearably tense and so extraordinarily productive, Goethe too found himself between two women, or rather, between two loves. For Goethe's relation with Augusta Stolberg was not really a relation with a woman: it was a relation with an ideal possibility of endless holy love, which might bear the name of her, or Helene Buff, or any other married or otherwise unattainable or even unknown object of desire. What defined her was not any quality or identity she had in herself, but her opposition to the real and named and present Lili Schönemann—present in the next room while Goethe struggled with a desire for a known object, to which he dared not yield.

Stella has been living the life of a recluse on her estates since the man she loved abandoned her and their child died. She decides to engage a lady companion and Lucia arrives at a nearby posthouse-inn to take up this position, together with her mother Cecilia who sees in Stella's life-story a reflection of her own experience. Blissfully married, Cecilia was deserted by her husband and misfortune has reduced her circumstances to the point where mother and daughter must separate. Also at the inn is Fernando, who after years of wandering now returns to his beloved Stella. But in him Cecilia recognizes her own long-lost husband. Fernando changes his mind—it was perhaps in order to return to Cecilia that he left Stella in the first place—and after the transports of the reunion Stella has to reconcile herself to losing Fernando once again. But Cecilia suggests that there is no real reason why all three (and Lucia) should not live together happily ever after, and on this triangular reconciliation the curtain falls.

The apparently revolutionary morality of *Stella*—which led to a furore involving, for example, prohibition of its production in Hamburg—is in fact, for Goethe, a retrogression: a retreat to the velleities of his earliest works and of the novel-fragment *Arianne to Wetty*. A life of travel and a life of rural seclusion, two marriages, one with offspring and one without, one to an older woman, one to a younger, one rich, one poor, one practical and of the day, one ethereal and of the night, but both deeply and unselfishly in love with him, Fernando combines them all. In much the same way Goethe in early 1775 must have dreamed of combining presence with absence, endless desire with specific fulfilment, love with fatherland, Frankfurt (second class) with an Imperial Countess, an advocate's practice in the Grosser Hirschgraben with a writer's mission to the German nation. There is in the play the material of tragedy, but one crucial element is missing that might bring us to feel that a tragic outcome—rather than a mere failure to achieve gratification—is a serious danger. For *Stella* also reverts to the pre-Strasbourg period in having no sense of the rootedness of its dilemmas, either in the historical data of the national life (as in *Götz* and *Werther*), or in the unasked-for destiny of the literary genius (as in *Clavigo* and *Claudina*), which gives to the fate of Goethe's other heroes a specific and bitter necessity. Only the colourful early scenes in the posthouse (themselves reminiscent of *Partners in Guilt*) suggest a humorous distance from the flight of fancy in which the play concludes, as when the post-mistress remarks of Fernando's first abandonment of Stella:

Aber wie's geht. Man sagte, der Herr hätte kuriose Principia gehabt; wenigstens kam er nicht in die Kirche; und die Leute, die keine Religion haben, haben keinen Gott und halten sich an keine Ordnung. Auf einmal hieß es: Der gnädige Herr ist fort. Er war verreist und kam eben nicht wieder.

Well the way things go. They said the gentleman had curious principles; at any rate he didn't come to church; and people who have no religion have no God and don't follow any rules. Suddenly you heard: The master's gone. He had gone on his travels and just never came back.

There is nothing in the later part of the play and its hymns to love—their style anticipated in the parodies of Satyros and the ambivalent litanies of Prometheus —to compare with or to supplant this dry and finely-phrased assessment. At the end of the play it is as if Goethe were trying to shake himself free, to regain the disponibility of 1770 by pretending that the entanglements and discoveries of the intervening years had never been.

If, however, we turn to the work which, more continuously than any other, preoccupied Goethe during the last two years before he settled in Weimar, we can see not only a clear consciousness of the need for a terminus, for a closing of the books, but also a penetrating, a too penetrating, awareness of the cost of such a terminus. The first version of Goethe's *Faust*—preserved by a happy accident in a manuscript copy discovered in 1887, and generally known as *Urfaust**—is a fragment. But it is not as incomplete as has from time to time been asserted. It is perfectly possible for it to be staged, more or less as it stands, as a play in its own right, and those of Goethe's contemporaries who were allowed to read it, or who heard it recited, described it variously as 'nearly finished', 'finished', and 'half-finished'. Nor was it the end of the story that was felt to be lacking—the most obvious lacuna was in the middle. It is precisely the end of the *Urfaust* which has most to tell us about the character not only of the work itself but of all that Goethe had written in the previous five years.

The *Urfaust* is the transcendent poetic masterpiece of the young Goethe, the quintessence of his literary personality. Zimmermann's judgement of it—'A work for *everyone* in Germany'—referred at once to its achievement of the central ambition of Storm and Stress, and to its tonal variety—as Lavater had remarked to him, 'Usually people have only one soul, but Goethe has a hundred.' The first act of *Partners in Guilt* and the greater part of *Götz von Berlichingen* had shown a powerful, concrete, impersonal, imagination taking the field against its place and time: *Werther* had shown an equally imaginative assimilation of contemporary interiority, the drama of the self. In writing *Werther* Goethe had been able momentarily to identify his own mind with the public mind of Sentimentalism through his recourse to a literary realism unprecedented in Germany. In certain poems—such as 'May Celebration', 'Song of Mahomet', the 'Prometheus' ode, 'On the Lake'—and now at length in the *Urfaust* too Goethe achieves a similar equilibrium, the equilibrium of all his mature work, a poetry as near as may be of individual, autobiographical, objective feeling. Clearly we are not here dealing with what F. R. Leavis called that 'profound impersonality in which experience matters, not because it is mine . . . but because it is what it is, the "mine" mattering only in so far as the individual sentience is the indispensable focus of experience'. Equally, however, Goethe's work now and later does not at its best offer us the unappealing alternative that Leavis constructs, 'in which experience matters . . . because it is to me it belongs, or happens, or because it subserves or issues in purpose or will'. Rather we are dealing with what we might call a personal impersonality—'myself not myself'—in which the quality experience has of being 'mine' has itself become an experience, an experience which is not at all as simple as the image of the mere 'individual . . . indispensable focus' may suggest, but which is, or can in poetry become, subtly and widely differentiated,

* Literally 'the original Faust'.

can also matter 'because it is what it is', and can be treated with an objectivity to which the attainment of purpose or the satisfaction of will is at least as irrelevant as it ever is in a work of art. Such an impersonally personal poetry has its limitations, beyond a doubt. There is perhaps a profundity, and there are certainly experiences, it cannot encompass. The *Urfaust* contains a broad spectrum of ways in which experience can be 'mine', from abstract contemplation, through shared misunderstanding to moral responsibility, but it stops short at renunciation, at the submission to the threat of death which, according to Hegel, is the moment at which the distinctively social relations of power and domination come into existence. To those relations, without which the tragic drama of Greece and England and France, and much that is best in the European novel, is inconceivable, Goethe's work, even his *Faust,* only ever gives marginal attention. But that is not to say that *Faust* is detached from the social and national circumstances of its time: on the contrary, the symbolic figure at its centre, more than any other of Goethe's creations, is the vehicle for his engagement with the temptations of his age, and accompanied him, for that reason, throughout his life. Where *Werther* is the tragedy of official Sentimentalism, the *Urfaust* is the tragedy of the collapse of the Storm and Stress historicist and realist opposition. In the *Urfaust* Goethe fuses the monologue form of *Werther* with the interactive drama of *Götz,* the interest in the national tradition, shown in the story of the sixteenth-century wizard, contemporary of Luther and Paracelsus, with the eighteenth-century Richardsonian realism of Faust's love-story. But if the rhythmic vitality and multiplicity of the play's diction reflect five years of intoxicated hope that the entire national world will let itself be grasped by any and all of a hundred souls, its fragmentariness, the inexplicitness of its tragic conclusion, its insistence that death merely bounds life and does not suffuse and determine it, reflect the defeat of that hope and a flight into solitude and self-preservation. Before we consider the limitations of the *Urfaust,* however, we need to be quite clear about the volcanic originality of its conception.

   In itself it was not wholly surprising that Goethe should have decided to write a play about Dr Faust. The choice of subject-matter was rather less original than in the case of *Götz,* for example. Since the early seventeenth century, when English wandering players had acted, or mimed, before the German public, a debased version of Marlowe's *Dr Faustus,* the story had been popular. But it had only been popular—by the end of the seventeenth century the play was almost exclusively the province of the chap-book and the puppet-theatre (a form of entertainment much more widely distributed in Germany than in England), though it will be remembered that Goethe may have seen a performance of the story by strolling actors when he was in Strasbourg. It was not until the awakening of literary nationalism in the middle of the eighteenth century that it began to seem a possible theme for serious literature. In a famous issue of the periodical *Letters Concerning the Most Recent Literature (Briefe, die neueste Literatur betreffend*), Lessing, in 1759, had simultaneously declared that Shakespeare was a better model for German drama than the French classical theatre, and that if a German Shakespeare were ever to arise he might well find his subject-matter in the story of Dr Faust. Lessing at the same time offered a scene from a Faust drama which he was himself writing but of which now only fragments remain. As one might expect from one of the magi of the German Enlightenment, Lessing shows himself in these fragments less concerned with the religious implications of Faust's pact with the devil and more attracted by the theme of Faust's aspiration to universal knowledge. Equally, however, an enlightened mind cannot regard this aspiration of damnable—'the godhead has not given man the most noble of desires in order to make him eternally wretched' announces an angel at the end —and so a tale from the age of gloomy religious fanaticism is saved for the Age of Reason by the simple device of changing the last scene: Lessing's Faust is the first German Faust who does not go to hell. Similarly, Paul Weidmann, whose play *Johann Faust* was performed, unknown to Goethe, at Prague in 1775, shows himself particularly anxious that his irrational subject-matter should not encourage superstitious beliefs in his audience: angels and devils are for him purely allegorical figures. Faust has committed many crimes and so deserves punishment if he does not repent, but no diabolical pact is mentioned from which he could not retreat if he wished to. Weidmann's Faust implores God for mercy at the end and his prayer is heard, 'lest God's enemies should say: See, he has made beings in order to torment them.' It is difficult to recognize in Weidmann's family drama many traits of the puppet-play, let alone of Marlowe's *Faustus*. The most complete reinterpretation of the Faust story, however, is that of a writer not overtly belonging to the Enlightenment but to the Storm and Stress—Goethe's acquaintance, Friedrich Müller, 'the Painter'. Müller too overlooks the pact and, like Lessing (but unlike Weidmann), seems to have little conception of Faust's sinfulness—whether that sin be theological or moral, against God or against man. Müller's play was never completed, but in those scenes which he published (from 1776 onwards) Faust appears simply as a Storm and Stress Hercules, a 'big fellow', a 'großer Kerl' pestered by the philistine pygmies of his home town. It is difficult to believe that Müller could ever have found grounds for his eternal damnation.

To a greater extent than any of his predecessors or contemporaries, Goethe kept his dramatization of the Faust story close to his immediate source, the puppet-plays. This fact about Goethe's *Urfaust* is often overlooked because much of the action of the *Urfaust* revolves round an episode that is indeed wholly of Goethe's invention: Faust's love for a simple town-girl, Margarethe (Marguerite) or Gretchen, and her subsequent execution for the infanticide of their illegitimate child. However it is a notorious feature of the Faust theme that it is a story without a centre: the drama lies in the opening and closing scenes, and the middle is occupied by a series of disconnected episodes. In the *Urfaust* there are two scenes that are clearly of this kind: one in which Mephistopheles, disguised as Faust, gives some diabolical advice to a naïve young student; another in which Faust uses magical powers to create havoc in Auerbach's tavern in Leipzig. A traditional episode in all versions of the story is the devil's use of a resurrected Helen of Troy to entangle Faust in fleshly lust and it may be possible to regard Gretchen as a reinterpreted Helena. Even if it is not, the introduction of the story of Gretchen improves on Marlowe, by providing his plot with a middle, and does not affect the basically Faustian construction of the play, for the opening—and, arguably, the closing—scenes are outwardly as traditional as could be desired. There is an introductory monologue in which Faust the scholar yearns for more than life has so far given him; there is an invocation of a greater devil (traditionally Lucifer, here called the Earth Spirit); and there is a second invocation of a lesser devil, Mephistopheles, with whom the pact is made. In the *Urfaust* the scene is missing in which this definitive agreement is signed, but there are clear allusions to it later on. Moreover, the last scene of the *Urfaust*, in which Faust attempts unsuccessfully to persuade the condemned and distraught Gretchen to join him and Mephistopheles and escape from her prison, bears signs of being the last scene altogether in Faust's earthly life. The very least that can be said is that after Gretchen has bid Faust an eternal farewell and commended herself to God rather than to him and his companion, Mephistopheles imperiously summons Faust and the two disappear together on the devil's horses. The hero of the *Urfaust*, like Müller's hero, is something of a Storm and Stress Hercules; like Lessing's hero, he has a thirst for universal knowledge; like Lessing's and Weidmann's, he incurs no significant theological guilt; like Weidmann's he does incur serious moral guilt —but unlike all of them he is in all probability carried off to hell. Goethe's *Urfaust* is unique in achieving the secularization of the traditional subject-matter that the eighteenth century required, without sacrificing its traditional tragic outcome. It has been called, with some justice, 'the one supremely great tragic drama of modern German literature'.

It is the great weakness of Sentimental literature, though not a surprising weakness in a movement so closely linked to the Leibnizian Enlightenment, that it has difficulty in accounting for the existence of moral evil in its major characters. Sheer misfortune, the frailty of nature, or the intervention of a secondary character in the role of villain: these are the substitutes usually offered when the author feels, adventitiously, the need to introduce moral categories. But such categories are alien to Sentimentalism—the unitary, autonomous, soul acknowledges no tribunal but itself—and in *Werther*, the most rigorously consequential of all Sentimental works, they have no place. Goethe argued later that there was no need for moral comment within *Werther* —the story alone was lesson enough: 'the end of such fantasies is suicide!' he said. But that is dangerously close to the argument that the one value which all indisputably acknowledge is the importance of personal survival. At least since his parting from Friederike Goethe knew that self-destruction is not the ultimate and only evil. *Werther* does not embody this knowledge, it knows nothing of the 'sombre remorse' which enveloped Goethe after the Sesenheim episode. That sense of the seriousness of personal responsibility was bound to alienate Goethe from the mind of his age, and if *Werther* is the moment of his nearest identity with his public, *Urfaust* is the first major work in which he establishes the initial distance of his own subjectivity from that of his contemporaries which will be characteristic of all his subsequent writing. In the *Urfaust*, both the problematic of feeling and the problematic of genius are thought through remorselessly to a conclusion in guilt and despair that has more ethical substance than any of the suicides or quasi-suicides with which Enlightenment tragedies, such as those of Lessing and Schiller, seek to conceal their moral nihilism.

'*Faust* came into being along with my *Werther*', Goethe remarked in 1829, and although it is clear that some of the material of the *Urfaust* is based on Goethe's experience in Leipzig, and probable that the idea of writing a play on the subject of Dr Faust had come to him by the time he was in Strasbourg, the greater part of the work on the manuscript must be dated around 1774, in the period of *Werther* and *Clavigo*. *Urfaust* shares the themes of both these works— and more besides—but fuses them into a new unity. Altogether it is important not to understate the unity of *Urfaust* and not to fragment it into loosely connected scenes from different periods and reflecting different presuppositions. There is an intimate connection between Faust the frustrated scholar and would-be magician of the opening scenes and Faust the lover and seducer of Gretchen—and the last scene, 'Prison', terminates both of these careers.

Immediately, on Faust's first appearance, as a man of about 30, it is clear that he has much in common with the geniuses of the early 1770s. On the one hand he is as emancipated in moral and theological matters as Satyros, Götz, or Werther in his more supercilious moments.

> Zwar bin ich gescheuter als alle die Laffen,
> Doktors, Professors, Schreiber und Pfaffen,
> Mich plagen keine Skrupel noch Zweifel,
> Fürcht mich weder vor Höll noch Teufel

True I am cleverer than all the knowalls, doctors, professors, scribes and priests, no scruples or doubts torment me, I fear neither hell nor the devil

On the other hand he longs for sympathetic contact with the whole of nature as intensely as Werther burying himself in the Maytime forest. Mere contemplation of the mystical notion of unity, such as Spinoza offers perhaps, or the alchemical and cosmological speculation of a Welling, is not enough for him— he yearns for the directness, the intimacy, of Ganymede's embrace:

> Welch Schauspiel! aber, ach, ein Schauspiel nur!
> Wo faß ich dich, unendliche Natur?
> Euch Brüste, wo? Ihr Quellen alles Lebens . . .

What a spectacle! but, alas, only a spectacle! Where can I grasp you, infinite Nature? Where, you breasts? You fountains of all life . . .

One might think of Goethe's drawing, dated between 1773 and 1775, of an astronomer in his study poring, by the light of an oil-lamp, over a globe, while beside him lies the telescope that could instead take his eyes out into the infinite of the unrepresented natural universe. Faust longs for all the world that is not himself to give him the pleasure that his own life gives him, and without that interfusion of subject and object life itself is insipid to him. He seeks directly to conjure up this identity of life and world, which to Ganymede appeared as the all-loving father. Faust knows it as the Spirit of the Earth (though later Goethe specified that the Spirit's features were to be those of Jupiter), and this apparition combines in one terrible moment of vision all the contradictory demands that life imposed on Werther's inadequate heart over a period of two years:

GEIST. In Lebensfluten, im Tatensturm
   Wall ich auf und ab,
   Webe hin und her!
   Geburt und Grab,
   Ein ewges Meer,
   Ein wechselnd Leben!
   So schaff ich am sausenden Webstuhl der Zeit
   Und würke der Gottheit lebendiges Kleid

SPIRIT. In floods of life, in the storm of deeds I move up and down, weave to and fro! Birth and grave an eternal sea, a shifting life! So I work at the whirring loom of time and make up the living garment of divinity

At first Faust can abide this vision no better than Werther; when none the less he struggles to face it, to look into the sun, so to speak, and to see in it his own spirit, or to look into himself and see there the 'sublime spirit's' likeness, he is dismissed as the victim of his own limitations, a Narcissus in love with his own reflection:

GEIST. Du gleichst dem Geist, den du begreifst,
   Nicht mir!   *Verschwindet.*
SPIRIT. You resemble the spirit you can grasp, not me!   *Disappears.*

But Faust will acknowledge no limitations, whether imposed by God or by man or even, ominously, by language. We immediately recognize the Faust of the opening scene in the Faust whom we meet at the very heart of the Gretchen intrigue, in the scene 'Martha's Garden' (in which the lovers make their first assignation and Faust hands over the sleeping draught which will quieten, and eventually kill, Gretchen's watchful mother). Gretchen is troubled by the feeling that Faust cannot be assimilated by the life of simple piety, domestic order and family duty on which he has intruded. She asks him directly and persistently whether he believes in God, and the credo which Faust is now forced to utter echoes not only his own words in the first scene of the play, but also the ecstasies of Werther and Ganymede:

FAUST. Mißhör mich nicht, du holdes Angesicht!
  Wer darf ihn nennen? ...
  Der Allumfasser,
  Der Allerhalter,
  Faßt und erhält er nicht
  Dich, mich, sich selbst?
  Wölbt sich der Himmel nicht da droben?
  Liegt die Erde nicht hier unten fest?
  Und steigen hüben und drüben
  Ewige Sterne nicht herauf? ...
  Erfüll davon dein Herz, so groß es ist,
  Und wenn du ganz in dem Gefühle selig bist,
  Nenn das dann, wie du willst,
  Nenn's Glück! Herz! Liebe! Gott!
  Ich habe keinen Namen
  Dafür. Gefühl ist alles,
  Name Schall und Rauch
  Umnebelnd Himmelsglut.
GRETCHEN. Das ist alles recht schön und gut;
  Ohngefähr sagt das der Katechismus auch,
  Nur mit ein bißchen andern Worten.

FAUST. Do not misunderstand me, my sweet. Who has the right to name him? ... The all-embracer, the all-sustainer, does he not grasp and sustain, you, me, himself? Does not the sky arch over us? Is the earth not firm down here beneath? And do not eternal stars rise all about us? ... Fill your heart with this, great though it is, and when you are completely blissful in that feeling, then call it what you will, call it fortune, heart, love, God! I have no name for it. Feeling is everything—names are sound and smoke, clouding heaven's fire.
GRETCHEN. That is all quite all right; the catechism more or less says that too, though the words are a bit different.

This is a moment of Shakespearian complexity. In what is perhaps the most dramatic scene he ever wrote, Goethe even finds the poetic means to express a conflict in which one of the parties, and the principal, though not the only, consciousness in the play, assails the foundations of poetry. The sound and smoke of Faust's outburst against names culminates in the naming of the unforgettable image of the sun obscured (an image which for Goethe was to retain for the rest of his life normative value, and not only in the context of his theory of optics). To this paradox corresponds the milder comedy in Gretchen's conciliatory understanding of the catechism. The two speakers agree, from utterly different premises, and giving utterly different significance to their concession, that at the level of 'words' a compromise can be reached. Yet the words which for Faust and Gretchen are trivial or elastic are for the spectator crystal-clear and hard. Through the ironies of these two speeches we see the collision of two worlds: on the one hand, Faust, the revolutionary autonomous genius, on the other, the world of Gretchen, the social and religious network from which Faust has emancipated himself. Faust's claim to liberation even from the bounds of language has another significance, however, including but also surpassing its threat to the integrity of all that Gretchen embodies: that claim is the act of hubris with which the Sentimental theory of monadic genius nullified the Storm and Stress's loyalty to concrete national tradition, it is an explicit assault on the origins and foundations of the movement of literary opposition. Nothing could be further than the cry 'names are sound and smoke' from Herder's intoxication with 'idioms' and 'sonant verbs'. Even here Goethe holds the opposing elements together in poetry: the grammar of the phrase 'Umnebelnd Himmelsglut' ('clouding heaven's fire') is Storm and Stress, but the metaphors are Sentimental. Finally, though, the conflict in this scene is, obviously, more than a disagreement about theological formulae, but also more than a moment of extreme tension between historical forces—it is a dramatic conflict, a conflict of persons. For Gretchen now goes on to tell Faust how she loathes seeing him in the company of Mephistopheles.

Mephistopheles is the keystone in the dramatic structure of the *Urfaust*. In a sense he mediates between the world of Faust and the world of Gretchen. He is assigned to Faust—in the *Urfaust* it is not clear on what precise terms but they are evidently not very different from the traditional—and his first concern is with Faust's career and of course his end. On the other hand Gretchen offers us a more reliable understanding of his devilish nature than does Faust: obviously so, for Faust has emancipated himself from what he believes to be Gretchen's literal-minded superstitions. Faust and Gretchen conflict with each other primarily in their interpretation of Mephistopheles' character. To us, the audience, Faust's blindness to the true intentions of his companion is as painful as it is both credible and consistent with the stance he has maintained since the beginning of the play. Gretchen knows better:

> Es steht ihm an der Stirn geschrieben,
> Daß er nicht mag eine Seele lieben
>
> .   .   .   .
> Auch, wenn er da ist, könnt ich nimmer beten.

It is written all over his face that he could not love a soul . . . And when he is there I could not possibly pray

But Faust condescendingly glosses over the matter:

> Du hast nun die Antipathie!

You just don't get on with him.

But Mephistopheles is no jester. It is impossible to doubt his truly diabolical nature when, at the close of 'Martha's Garden', he tells Faust that he will take his own kind of pleasure in the doings of the coming night. The thought of Gretchen being delivered into the power of one whom she so fears, and who will compass and relish the destruction of all she loves, cannot but arouse our horror. Yet to this horror Faust seems largely impervious. He goes through the play from crime to crime unaware of the pit Mephistopheles is digging for him, the immense burden of guilt he is amassing. Of course he is unaware of it, for the belief in the possibility of ultimate sin, of irredeemable offence against an order not of his own making is something that he has put behind him from the start.

The story of Faust's life thus appears as a question, a question not merely about his capacity for 'joys and cares beyond myself', but about whether and how he can be judged by a tribunal that also is 'beyond myself'. Yorick's commingling of tears with an apparently insentient Maria is coarse by comparison with the tension of love which is first set up between Faust's monologues and Gretchen's solitary songs and which is then given words in the scenes in which the two meet, in 'Martha's Garden' and 'Prison'. This fineness of judgement was implicit in Goethe's conception of the play from the moment when, spurred on perhaps by memories of Sesenheim, he decided to link the Faust of the puppet-story with a woman whose fate is that of Susanna Brandt, but whose loyalty to standards by which she knows herself condemned belongs to Catharina Flindt. That was the moment when a distinctively Goethean Faust came into being, for it was also the moment at which any conventional Enlightenment fancies of a conclusion in universal redemption gave way to the possibility of a wholly tragic outcome to Faust's act of hubris. It is with that possibility that Mephistopheles triumphantly confronts Faust at the last. With Gretchen's mother, child, and brother dead (the last probably killed by Faust), and Gretchen herself wandering in mind and awaiting execution, Goethe dares to put the starkly simple question whether Gretchen and her narrow little world have not shown a truer insight than Faust. Even a pact with the devil, Mephistopheles tells Faust, cannot make a man omnipotent, put him above the moral law. To Faust's demand that he should rescue Gretchen he replies:

Ich kann die Bande des Rächers nicht lösen, seine Riegel nicht öffnen. Rette sie—? Wer war's, der sie ins Verderben stürzte? Ich oder du?

I cannot loose the bonds of the Avenger, nor open what he has bolted shut. Rescue her—? Who was it that plunged her into ruin? I or you?

And to this question Faust has no answer. The stage direction runs simply: 'Faust looks about him wildly'.

Only two factors prevent the last scene of the *Urfaust* from being read as the scene in which the hopelessly guilty Faust is carried off to hell while Gretchen saves herself by her rejection of the assistance of Mephistopheles. One is the absence of a completely unequivocal statement at the end that Faust's earthly career is now over (such as we have at the end of Marlowe's *Dr Faustus* or the

German puppet-plays). The other—not unrelated—is that Goethe did not publish the last scene until he had made it plain that he intended to continue the story of Faust beyond this point (though it remained unclear *how* he intended to do so). It is as if Goethe has fully envisaged the possibility of a tragic conclusion for his play, and has indeed completed its inner structure (there is no substantial scene missing *after* 'Prison') but has not quite had the courage to render the tragedy explicit or public. A similar effect is given by the ballad 'Once there was a lusty lad' which Goethe wrote while he was working on the *Urfaust* and which had a powerful effect on the Jacobis when he recited it to them on that romantic July evening in 1774. (Scott published a free translation under the title 'Frederick and Alice'.) It tells the story of a man who loves and deserts a woman who then dies. Overcome with remorse the man returns and on his frantic journey falls into an underground cavern, where he is greeted by a hundred skeletons, gathered as for a feast. The poem ends

> Er sieht sein Schätzel unten an
> Mit weißen Tüchern angetan,
> Die wend't sich—

He sees his sweetheart down below, covered in white cloths; she turns—

From the point of view of internal structure, the ballad is almost complete (every strophe ends with an unrhymed line), yet not quite enough has been said: the last line is frozen in a moment of horror which is not resolved, and indeed not quite explicit.

There are of course accidental lacunae in the *Urfaust* as there are not in 'Once there was a lusty lad', but common to both works is something that could be called a deliberate fragmentariness. It is deliberate in the sense that it could be suppressed only by a deliberate, self-inflicted dislocation of Goethe's now mature poetic cast of mind. What that cast of mind was we know from the drama *Prometheus*. In no production of the Storm and Stress years could Goethe be more truly said to be 'shaping men in his own image . . . to suffer, weep, enjoy and to be happy' than in his first *Faust*. It is truly hundred-souled, star-spangled with life. In world literature there can be few works of its length that are emotionally so strenuous to read. The reactive temper of 'On the Lake' is here maintained over some fifty pages. Rhythm fluctuates in permanent but controlled excitement, from a foundation of three extensive prose scenes, through the rough-hewn, mock sixteenth-century, pantomime 'Knittelvers', to the rhyming iambic madrigal verse which is perhaps the most natural of all forms for the German speaker and which, though colloquial, is malleable enough to encompass the comic, the sententious, and the lyrical. The gradation out of prose culminates in complete interpolated songs, such as 'A king there was in Thule'. The dramatic technique of the most vivid scenes in *Götz von Berlichingen* is both concentrated and extended: time and again an unfathomable symbolic harmony is created when a powerful emotion, generated usually by the past, the future, or by distance, appears in relief against a seemingly independent but equally clean-drawn background—an action, such as Gretchen combing her hair, or a landscape, or (Goethe's special and probably original dramatic signature) a specific time of day or night. Each scene, however brief, is a world of its own, which the poet approaches, touches, and vivifies, and from which he then withdraws. Each seems to grow its identity from one point —an image, a tension, a verse-form, Gretchen's spinning-wheel, a homecoming in the late afternoon, Mephistopheles in a dressing-gown and Gottsched's full-bottomed wig, black horses and the night sky, flowers for the Madonna. And each seems to grow to its natural limits and then stop. Poetic economy, multiple changes of mood and sheer rhythmic inventiveness—these, not the simple intrigue, keep up the dramatic impetus. There is no scene, there is hardly even a line, of *purely* structural significance. Everything lives, even the final scene of despair—provided Faust dies. If he does not, if his tragedy remains only potential—and that is the state in which the *Urfaust* was left— then his sharing in the life of Gretchen's world remains only potential too, like the life of Prometheus' statues—a not fully realized thought. The bitterness of death comes to Gretchen, it is true, for she refuses to leave the 'sphere of association' in which she is given over to public opinion, to her judges, and to her executioners. But Faust, the 'unhoused . . . unman', in his own words, gives death, even the death of his beloved, no dominion over him: the house of human society, founded on the threat of death, stands over against him as a possible, perhaps as the most poignant possible, experience, but it does not exhaust his world. His world is as boundless as that of Goethe's letters, and the *Urfaust* is as uncompletable as self-expression in a letter. Faust's is the subjectivity that refuses to be (or that German social and political circumstances —untouched by the play—will not permit to be) *merely* 'the indispensable focus of experience'. There is none in whose hands he will acknowledge the power over his life and his death to lie—none, that is, to whom he relinquishes the structuring of his experience. Goethe's Faust, therefore, unlike Marlowe's Faustus, is not laid to rest by a chorus of scholars, and unlike Gretchen he stands before no tribunal except that of the audience. There for the present Goethe leaves him, poised between tragedy and idealization, between the story of one man's failed revolution and an Enlightenment parable of the redemption of all. For the present Goethe can perhaps convince himself that one day he will be able to resolve the issue, that Faust can be finished, its identity fixed, and the twilight in its central character finally dispelled. It will be many years before he realizes that its incompleteness is intrinsic.

Goethe had been able to complete *Werther*, to bring it to an irreversible conclusion, because the book was, in the crucial respect of its ethical awareness, at one with its age and not with its author. This was not the case with *Faust*, nor for that matter with the ballad 'Once there was a lusty lad', which in the summer of 1775 Goethe worked into *Claudina of Villa Bella* in such a way that it appears to be unfinished simply because Crugantino, the singer, is interrupted. In the case of the *Urfaust* this concealment—rather than suppression—of the tragic element takes the form simply of an inability to carry the work forward, to decide either to let what has been written stand, or to alter it. The *Urfaust* could not easily be brought in under the shelter of some other work, as had been done with 'Once there was a lusty lad'. It could, however, be ignored. After the first years in Weimar Goethe seems, for half a decade, to have given up public recitations from his 'half-finished' *Faust*. He had, after all, much else to occupy him.

4

Displacement

GOING from Frankfurt to Weimar six weeks before Christmas, and with the celebrations of the Duke's marriage and coming of age in full swing, must have been like leaving the city for a protracted house-party on a country estate. Twice before, in 1765, when going up to university, and in 1768, when returning from it in broken health, Goethe had passed along the dreadful Thuringian section of the route from Frankfurt to Leipzig, but never before had he turned off the main road at Erfurt for the thirteen-mile drive, along tracks so deeply rutted that carriages regularly made detours through the fields to avoid them, to the Erfurt, or western, Gate of the little town of Weimar, where he now arrived at 5 a.m. on 7 November 1775. Was it even a little town? In 1803 Mme de Staël thought it only a large château. Its 600–700 houses, mostly roofed with thatch or wooden shingles, were no more than an appendage to the ducal residence. A mere 100 yards within its double line of walls lay the largest of its squares, the Pottery Market, with a dozen tall gabled houses, among them the family home of Goethe's courier, von Kalb, and of his father, the Chancellor of the duchy's Exchequer, with whom the literary lion was invited to lodge. Another 100 yards would have taken the travellers through the main Market Square and into the courtly and administrative sector which occupied perhaps a third of the entire area of the town. A large part even of this, however, was in 1775 a blackened ruin, after the destruction in the previous year of the ducal palace, the Wilhelmsburg, and with it of the court theatre and the court chapel of St Martin (whose organist from 1708 to 1717 had been J. S. Bach), and it would be many years before the funds would be available for rebuilding—a project of which Goethe himself was to be the principal director. Meanwhile an air of good-humoured improvisation surrounded court life. The substantial building (now the Franz Liszt College of Music) originally intended for the offices of the duchy's Estates General had been quickly adapted to accommodate the Duke and his bride. The Master of the Court Hunt, A. G. Hauptmann, had speculated on the new ruler's need to keep up the standard of entertainment and, with the aid of a subvention, had in 1775 erected some Assembly Rooms (the 'Redoutenhaus) which for the next five years were to be home to the court balls and the amateur theatricals which were for the present all that could be afforded. The least serious feature of court life, its religious ceremonies, were simply transferred to the only other church within the walls (there was also however the Garrison Church just outside them), the Town Church of Saints Peter and Paul, on the Pottery Market. This Gothic structure of 1500 had been much modified in the early eighteenth century, but was still dominated by its magnificent altar-piece by Lucas Cranach the Elder, who spent the last year of his life in Weimar.

Goethe's remote ancestor, in a gesture worthy of his descendant, here depicts himself, as a representative of human kind, stepping out confidently towards salvation, flanked by John the Baptist and Martin Luther.

Fortunately not all of Weimar's attractions had fallen victim to the disastrous fire of 1774. In particular the library, the third largest in Germany, had survived simply because of its size. In 1766 it had been moved out of the cramped Wilhelmsburg into a building of its own, the Green Palace, overlooking the water meadows of the little river Ilm, which flows along Weimar's eastern side on its way to the Saale and the Elbe: in among them lay the formal garden of the Star ('der Stern'), so called from the radial arrangement of its box-lined paths. The Jointure Palace ('Wittumspalais'), built in 1767, partly designed and decorated by Oeser of Leipzig, was also untouched by the fire, and from 1775 until her death in 1807 it was the town residence of Anna Amalia, the Dowager Duchess, and Regent of the duchy from 1758 until her elder son reached his majority. Before his death from tuberculosis, after only two years of marriage, Anna Amalia's husband had already begun the work of filling in the space between the two city walls in the neighbourhood of what was to become the Jointure Palace and of creating there the 'Esplanade', a tree-shaded promenade, with goldfish pond and Chinese pavilion. The Star Garden was private to the ducal family, but bourgeois, that is, untitled persons, though not their servants, were admitted to the Esplanade, which was closed off at either end by heavy iron gates. For much of the nineteenth and twentieth centuries the Esplanade survived only as the line of the Schillerstrasse, which had taken its place. Now that that street has become a pedestrian precinct with trees and high-class tourist shops it has been returned to something like its original function.

Still, in comparison with Frankfurt, the town of Weimar had little to show. It was a simple place with but a single function: well over a quarter of its 6,000 inhabitants was made up by the court, by the court's families, employees, and pensioners. The remainder, the tailors and shoemakers, the bakers and blacksmiths and apothecaries, were all directly or indirectly meeting the needs of the court or of each other. In 1775 not a single individual was engaged in industry or trade with the world beyond Weimar. Weimar was not even on the route of the mail-coach, which did not stop there but at Buttelstedt, about ten miles to the north on the Leipzig-Erfurt road. The social structure was uncomplicated, with 80 per cent of the population in the journeyman class, earning less than 200 dollars a year,* and the upper class, in W. H. Bruford's words, 'stood out from the general mass of the population far more noticeably than in recent times, with little or no real middle class in between'. Culturally, too, life was simple in this heartland of Lutheranism, with none of the religious,

---

* In Weimar the normal accounting and currency unit was the dollar ('taler'). The rate of exchange was variable but may for convenience' sake be assumed to be 2 guilders to the dollar.

international, or linguistic complexities of Frankfurt: true there were two Jews, but they were both bankers who made their fortunes out of lending money to courtiers who needed to pay their gambling debts. Apart from the barbarous French in which the more old-fashioned court ladies wrote to each other, the only language to be heard was the Saxon variety of German: when Lavater visited, his Swiss German, familiar enough in the streets of cosmopolitan Frankfurt, was incomprehensible to the Weimar locals. Even the smells were simple farmyard smells, for cattle were still kept within the safety of the city walls and led out to pasture during the day.

In many ways the town of Weimar in the middle of the eighteenth century was not unlike the villages that clustered near the great houses of wealthy landed families in England in the same period. What made the difference, and made the difference from Frankfurt too, was the *duchy* of Weimar. For, within the constitution of the Empire, Weimar was the capital of a sovereign state, the duchy of Saxe-Weimar-Eisenach. The great curve of wooded hills that, in the form of the Fichtelgebirge, the Thuringian Forest, and the Harz Mountains, once made up the south-western border of East Germany, surrounded in the eighteenth century a mosaic of nearly thirty duodecimo principalities, a buffer-zone between Electoral Saxony in the east and Hesse-Cassel in the west, Brandenburg-Prussia in the north and Bavaria in the south. Of the Saxon duchies, fragmented by inheritance and recombined by intermarriage and the extinction of ruling houses, Saxe-Weimar was the oldest, though not the largest. Since 1741 it had been united with the extinct duchies of Jena, some twelve miles to the east, and of Eisenach, about fifty miles to the west, while thirty miles to the south, deep in the Thuringian Forest, lay Weimar's other major possession, the bailiwick of Ilmenau. Each of these domains had its own distinct character: Jena had its university, Ilmenau, up in the hills, its copper and silver mines, though these had been abandoned in 1739. Eisenach (the birthplace of J. S. Bach) was of more obvious commercial and historic significance than Weimar: its medieval castle, the Wartburg, dramatically situated on a wooded eminence outside the town, had been the scene of the famous contest of the German troubadours, the 'Minnesänger', in 1207, and had sheltered Luther after the Diet of Worms, while he translated the New Testament. Despite the union of 1741 Eisenach retained almost intact its own administrative system.

The union of these different possessions was personal, of course, not geographical. The road from Weimar to Eisenach, for example, led through two other distinct and sovereign territories. First there was Erfurt, one of the extra-territorial possessions of the Archbishop of Mainz, which was governed by one of Germany's most cultured men, the 'Vicar' or 'Statthalter' Carl Theodor von Dalberg (1744–1817), great-great-uncle of the historian Lord Acton. Erfurt possessed an ancient university—where Luther had once studied—now, in Goethe's day, dwindling into insignificance. Fifteen miles west of Erfurt lay

Gotha, where Gotter was employed, the capital of the duchy of Saxe-Gotha-Altenburg, which in 1775 was ruled by Duke Ernst II (1745–1804) and was larger, wealthier, better administered, and to an extent more cultivated than Saxe-Weimar. The Duke's younger brother, Prince August, a dilettante and valetudinarian, spent much time in Weimar and was to become a particular friend of Goethe's, but the Duke himself was a serious, learned, and rather domestic man who dreamed of living as a private gentleman in a republic such as Switzerland or America. His male succession was to die out in 1825, and in the consequent reorganization of the Saxon principalities Altenburg became independent, the miniature duchy of Saxe-Meiningen (jointly ruled by the two brothers whom Goethe first met in Frankfurt) expanded, while Gotha was combined with the most southerly of the Saxon duchies, Saxe-Coburg. Some of the qualities of Duke Ernst II re-emerged in his great-grandson, Prince Albert of Saxe-Coburg-Gotha, the Prince Consort.

Clearly, then, so fragmented a polity as Saxe-Weimar-Eisenach could not in all respects act with the independence of a nation-state. Not only did it depend on its neighbours for free communication between its constituent parts, but the neighbours also had certain rights, notably over roads and waterways, even within its boundaries. The University of Jena was a joint foundation of all four Saxon duchies, and although Weimar, as the principal source of finance, had the principal say in its affairs, all four had to be consulted about serious matters, including the appointment of professors. None the less, internally, at any rate, the duchy was a nation, a 'fatherland', and Carl August, from the age of 18, was *pater patriae*, the absolute ruler of the 106,000 inhabitants of the Weimar territories. The simplicity of the social, and even the material, structure of the town of Weimar is deceptive: here the threads of power ran up through the offices of the central administration, were gathered together by the Privy Council, and placed in the hands of the Duke, who, in name wholly, and in fact to a considerable extent, was responsible for educating, judging, punishing, patronizing, preaching at, defending—there was an army of around 500 infantry, twenty hussars, and half a dozen artillerymen—and, especially, taxing a body of people more numerous than the population of any Imperial free city, and nearly three times the size of Goethe's Frankfurt.

Economically and socially, it is true, the duchy reflected the weaknesses of its capital: there was next to no industry, only stocking-weaving as a cottage-industry in Apolda, north-east of Weimar, and little significant commerce since the roads in this wet and hilly terrain made transport so difficult. Without a navigable river the considerable resources of timber could not be exploited; there were no operating mines; the economy was perforce almost exclusively agrarian and even had it been possible to generate agricultural surpluses, there was no convenient urban market in which to dispose of them. The absence of a middle class was thus even more marked in the duchy as a whole than in Weimar itself, and it is evident that the perhaps 2,000 courtiers, officials, soldiers, and pensioners who had to be supported from taxation, and the further body of landlords who had to be supported by rents and feudal dues, represented a much heavier burden for the population than did Frankfurt's 500 officials for its 36,000 inhabitants, especially since many in Weimar's upper class enjoyed much better incomes than the Stadtschultheiss in Frankfurt. (In 1790 the von Stein family drew from all sources a total income of 6,229 dollars, and expended 4,419.) Goethe memorably expressed the problem in a letter of 1782 after a long journey which had taken him both as a diplomatic emissary to all the Thuringian courts and as a superintendent land agent and tiro geologist through all the fields, forests and hills of the region, so that he felt he knew them as well as his multiplication tables:

So I have been mounting up through all the classes, seeing the farming man extorting from the earth the barest essentials, which would be a decent living if he were sweating only for himself. But you know how it is, when the aphids are sitting and sucking on the roses and have got themselves nice and fat and green, then the ants come along and suck the refined juice out of their bodies. And so it goes on and we have now got so far that in one day the top is always consuming more than the bottom can furnish or produce in the same time.

It would however be a mistake to think that the roots of these ills lay in the court. The ants in Goethe's metaphor were the entire class of landed nobility, whose income derived from their estates (not necessarily in the duchy) and who might or might not spend much of the year in Weimar. Goethe's criticism expresses a burgher's awareness of a profound imbalance in the agrarian duchy's economic and social structure, it is not especially a criticism of ostentatious consumption by the court, the duchy's political figurehead and the *raison d'être* of the town of Weimar. On the contrary, throughout his life Goethe believed that ostentation and ceremony were an essential accompaniment of power, in order both to declare, not conceal, the true locus of government and to remind rulers of the significance of their actions in the lives of their subjects. He never had any time for the superficial analysis which thinks externals unimportant. Already in 1777, although the Prince in *The Triumph of Sensibility* remarks, 'You know you are not to stand on ceremony with me', the lady-in-waiting replies in an aside: 'Only so that he does not have to stand on ceremony with us'. In 1785 Goethe resisted Carl August's attempts to increase speed and efficiency by simplifying the ponderous style of Chancery documents on the grounds that momentous decisions should be communicated with dignity out of respect for those affected by them. And when looking back on this period and making notes for the continuation of his autobiography, he writes sarcastically of the enlightened despots such as Frederick the Great or Joseph II who regarded themselves simply as civil servants, treated their courts with contempt, and by this betrayal of their office delivered entire nations up to the greater evil of revolution: 'finally the King of France considers himself an anachronism'.

Weimar was a sovereign and autocratic state and the Duke was its monarch: the court expressed and demonstrated that fact and Goethe could with complete consistency give full loyalty to the duchy's political structure while at the same time showing grave unease at the distribution of wealth within it.

The court, in the broadest sense of those who were permitted, or might be expected to dine with or be formally received by the Duke and Duchess consisted of three elements. The court proper, the ducal household, though not of minimal size, was not grossly excessive. In addition to two Marshals and two Equerries, one of them the Chief Equerry Baron von Stein, there were for the Duke himself three gentlemen-in-waiting, three gentlemen and a similar number of ladies-in-waiting for the Duchess, a chamberlain, gentleman and lady-in-waiting for the Dowager Duchess and von Knebel as tutor, later gentleman-in-waiting, for Prince Constantin. It is true there were also six pages, but they were simply sons of the nobility for whose education the Duke was providing, in accordance with one of the oldest of feudal obligations. The second element consisted of the principal officials, notably the three or four Privy Councillors and the heads of the Treasury, the Chancery, and the Consistory (responsible for church matters). The largest and most hetero-geneous group, the nobility—titled gentry might be a more appropriate term—temporarily or permanently resident in Weimar, might have no formal link either with the household or the administration, like the Countess Bernstorff, widow of the Danish Prime Minister and patron of Klopstock, who arrived with her secretary J. J. C. Bode (the translator of Sterne and prominent Freemason) in 1779, when her daughter Sophie married Frau von Stein's brother, Carl von Schardt. Or they might hold some unsalaried position, or be in receipt of an *ex gratia* pension from the Duke, like Baron von Imhoff, who sold his first wife to Warren Hastings on the boat out to India, and gave himself the airs of a nabob, and who, after marrying Luise von Schardt, Frau von Stein's sister, was enabled to settle with his second family in Jena by the offer of free lodgings and fuel, the title of a major, and a secret annuity of 300 dollars.

The court expressed its existence principally by eating. Every day the Duke and Duchess, or the Duchess on her own, since Carl August was often away and in his early years shunned formality, presided around midday over a dinner for their household and their guests in their temporary palace. There were two tables, a high or Princely Table for the ruler and the nobility, and a low, the Marshal's Table, for untitled diners. On such formal occasions all men were expected to wear court uniform of green stuff with epaulettes, or something more splendid in silk or satin with gold and silver embroidery. On two evenings a week the ducal pair entertained to supper and cards, as might any noble lady in town during the season. Though the reigning family were no great church-goers, Sundays were special occasions: after a grander dinner than usual, there would be a reception in the afternoon for all the local nobility and in the evening a card party and a concert given by the court orchestra of about a dozen players. The Dowager Duchess had her own establishment in the Jointure Palace, but on Sundays she would come across to have dinner with her children, driving the short distance in a glass carriage so small that her crinoline protruded from the windows on either side. On Wednesdays she maintained the tradition of her regency by entertaining a group of guests selected for their intellect rather than their rank and in the evening she had her own concert too. Apart from the palaces, the most notable locus of the court was Hauptmann's Assembly Rooms where the masked or fancy-dress balls took place every week or two, and more irregularly, depending on the mood of the amateur groups involved, a temporary stage would be installed and a play or operetta put on. A regular date, however, which Goethe quickly established as the high point in the Weimar theatrical calendar was 30 January, the birthday of Duchess Luise, which was singled out, if possible, for a major new production. The elaborate celebration of birthdays was, and still is, a feature of the highly personalized and private nature of German middle-class life, and Goethe's promotion of the habit—which soon spread to include the birthdays of Anna Amalia and of Goethe himself—within the more public life of an aristocratic court indicates the shift of tone which his presence brought about. Another middle-class innovation over which Goethe presided—with a particularly literary flavour, thanks to Klopstock—and which became a popular fixture for the court during the winter months, was skating, on a specially flooded meadow near the walls. In the winter of 1778 Frau von Stein was reported to be skating for eight hours a day. The ice became an open-air ballroom, with a wind-band and masks, and at night torches, braziers, and fireworks: one of the pages recalls being dressed as a demon and skating through the falling snow with a Roman candle attached to his horns, while escorting the non-skating ladies, well wrapped up in furs, in their ornamental sleighs. In the summer, following the time-honoured European rhythm, the court tended to disperse: the Duke had a country residence just outside Weimar, the rococo Palace Belvedere, with park and orangery; Anna Amalia had the somewhat older Ettersburg Palace on the wooded hills, rich in beech and oak, one and a half hours' drive north-west of Weimar. After 1780, however, she returned closer to Weimar for the summer, taking over Tiefurt, grandly named a manor ('Schloss'), but in fact a large farmhouse, originally converted in 1776 into accommodation for Prince Constantin and with charming grounds running along the upper Ilm. The nobility retired to their estates—the von Steins, for example, to the south, to their moated hall in the hills at Gross Kochberg—or, in later years, reassembled far from Weimar in the holiday atmosphere of the German and Bohemian watering-places.

Why Goethe Stayed

This, then, was the world which Goethe entered in November 1775, purportedly only breaking a journey to Hamburg and the north, in order to meet up again with the Stolberg brothers and join them on a pilgrimage to

Klopstock, monarch of a kingdom of the mind. The Stolbergs arrrived in Weimar on 26 November. On 3 December they travelled on, without Goethe. Why did Goethe stay—then, and on the numerous subsequent occasions, notably in 1777, 1788, and 1806, when a severing of his connection with Weimar seemed possible, indeed likely? What had this petty and impoverished princedom to offer the literary genius and burgess of a city republic whose traditions were considerably older than any of the ruling houses of ducal Saxony? Wieland's intuition that Goethe would stay with them longer than officially announced was only a guess, but in the winter of 1775 there were three people in Weimar who had good reasons for wanting a longer stay, and their positions must be considered individually. They were Anna Amalia, Carl August, and Goethe himself.

There is of course another, if secondary question: why did Weimar want to keep Goethe? For Goethe did not come cheap. In his first months at Weimar his father, in unabated hostility to this courtly connection, refused to support him and he was dependent instead on loans from Merck and Jacobi, and direct and substantial gifts from the Duke. In March 1776 Carl August spent 600 dollars on buying for Goethe, and as much again on putting in order, the house on which the author of *Werther* and *Satyros* had set his sights: a tumbledown cottage on the other side of the Ilm, looking back across the meadows and the Star Garden to the town walls and the ruins of the Wilhelmsburg. And in June of that year the Duke finally overcame the opposition of his Prime Minister, J. F. von Fritsch (1731–1814), who was made to withdraw his offer of resignation and accept the appointment of Goethe as by far the youngest member of the Privy Council of three, and at an annual salary of 1,200 dollars. What was hoped for from this investment?

There was actually a not insignificant financial return. Goethe spent considerably more than his official salary in Weimar every year, much of it in anonymous charitable disbursements, the difference being made up, after his father had abandoned his hostility to the move, by contributions from Frankfurt. Not until 1807 could Goethe lay claim to some modest capital of his own. Usually, however, Goethe's appointment is seen as a continuation of the cultural policy of Anna Amalia, and Goethe was certainly a catch, if a controversial one. Anna Amalia, a niece of Frederick the Great, had spent her childhood at the highly cultivated court of Brunswick, which was eventually to be Lessing's patron, and she retained throughout her life an active amateur interest in literature, the theatre, painting, music—she was capable of composing settings for Goethe's playlets and songs—and even in the Greek and Latin classics. Under her patronage the Weimar court had its own professional theatre company until the fire of 1774. In 1772, after the publication of *The Golden Mirror,* she personally appointed Wieland, then, it will be remembered, a professor of philosophy in the nearby university of Erfurt, to be the tutor to the 15-year-old Carl August, and the following year the

somewhat literary von Knebel was made tutor to Constantin. These appointments were partly intended to counteract the influence of Carl August's original
tutor, Count Görtz, who saw his own advantage in encouraging a rebellious
streak in the future ruler, but they were also decisive steps towards establishing
Weimar as a literary centre. It was after all because Knebel was interested in
contemporary literature that he had thought it worth while to introduce Goethe
to the Weimar princes in Frankfurt in December 1774. And the *Teutscher
Merkur*, the literary and political periodical which Wieland set up, on the
analogy of the *Mercure de France*, in 1773, shortly after his arrival in Weimar,
was so successful in steering a middle course between the learned and the
popular, between radicalism and official convention, even in the time of the
French Revolution, that it survived until 1810 as Germany's nearest approach
to a national cultural institution. It was thanks to Wieland that in 1775 Weimar
was already known to the near on 2,000 subscribers to the *Teutscher Merkur* as
the centre of a civilized discussion that aspired to include not just one small
duchy but the entire German-speaking world.

But just as these consequences of her actions can hardly have been present to
Anna Amalia's mind as she cast around for tutors for her sons in 1772–3, so it is
likely that more immediate concerns than the literary future of Germany
motivated her in her emphatic, and probably decisive, support for Carl August
in his battle with Fritsch over Goethe's appointment in June 1776. The
Dowager Duchess was a strong personality, with some imagination and a taste
for the informal, even the rumbustious, which led her soon into a most cordial
and colloquial mother-to-mother correspondence with Frau Goethe in
Frankfurt, whom she eventually visited. She was capable of grasping intuitively
the consistency of one of her son's first major independent decisions with her
own past policy, even though Goethe was a much less respectable figure than
Wieland had been, and his works had been the cause of much scandal. But
there can be little doubt that a principal consideration in the struggles of 1776
was, as is only natural, power. Anna Amalia had ruled her son's duchy for over
sixteen years and at the age of 35 she was vigorous enough to regret the
prospect of a shadowy retirement and the replacement of her influence by that
of flatterers who would encourage Carl August to kick over the traces, dismiss
his ministers, and ruin his subjects. The remarkable ascendancy over the mind
of the young Duke that Goethe had achieved since their first meeting in
Frankfurt was a new factor in the balance of power at the Weimar court and
highly convenient for Anna Amalia. It presaged the end of the Görtz faction,
with its old-fashioned emphasis on the prerogatives of absolute rule and the
privileges of the nobility: Görtz within the year, true to his earlier threat, sought
a larger theatre for his talents in the Prussian service, though Countess Görtz
and her friend Countess Gianini, the principal lady-in-waiting of the new
Duchess, remained the most malevolent observers of events in Weimar for
some time to come. Moreover, Goethe, as a member of the middle class and of

her alternative court of *beaux esprits*, must have promised to be a biddable, and genuinely like-minded, intermediary with her son. The personal relations with Goethe's mother were warm, but in establishing them Anna Amalia was also consolidating her position. And, lastly, the hold Goethe seemed to have over Carl August must also have been welcome to Anna Amalia as a counterweight to the influence of her new daughter-in-law.

Not that Duchess Luise, a depressive and rather prim young lady, showed any signs of political ambitions on her own account: unhappy in herself she was unable to give happiness to her husband, her pregnancies all too often ended in stillbirth, and until 1783 the marriage was blighted by her failure to bear a male heir. But the Duchess had a style of her own, and it was not that of Anna Amalia: she maintained, as far as she could, a strict observance of court etiquette and of distinctions of rank. Not until he was a Privy Councillor did Goethe dine at the Prince's Table and not until he was ennobled was he permitted to sit and play cards with the Duchess's party during a formal court reception. The more colourless and more conservative elements in the court, including of course the implacably snobbish Countess Gianini, seem to have felt secure in her entourage, and she was particularly close to the von Stein family, especially Frau von Stein, whom Anna Amalia could not abide. At court, style is power. Relations between the two Duchesses were never easy and Anna Amalia, like many a mother-in-law, was glad to see Goethe give her son a companionship of which she could approve and which his wife seemed unable to provide.

What of Carl August himself? He was a man—in 1775, a youth—of intelligence, integrity, and considerable psychological insight struggling with an unfortunate heredity. He was born to rule—his letter to Fritsch expounding Goethe's qualities, defending his own decision and reproving his Prime Minister for tendering his resignation, is a remarkably mature and confident composition, without any touch of arrogance—but that also meant, in a scion of German princes, born to do as he pleased. 'You know his good qualities', wrote the elder Stolberg to Klopstock, 'but he is by nature wild and—which is infinitely worse—hard.' His portraits show a most penetrating gaze, a set, stubborn mouth, and flared, almost lascivious nostrils. By what must have been a great effort of will he subdued both the martinet and the playboy within and made of himself a benevolent despot. From the first he showed, perhaps surprisingly, a streak of sympathy with those in distress or difficulty, but he never really convinced himself that he was there to serve his subjects, rather than vice versa—that may have been the residue of Görtz's tuition. But Wieland opened the young man's mind to the possibility of judging his conduct by a more general standard than his own will, by ideas, that is, current with the thinking public. And how seductively inspiring it must have been to spend the first six months of his reign in the hypnotizing company of a leader and maker of the thinking public's opinion, doing just what he really enjoyed, hunting, riding, camping out in the forests and living on baked potatoes, flirting incognito with the peasant-girls, swimming naked in the duchy's chilly rivers to the terror of passing clergymen, playing practical jokes on his companions (and subjects), clay-pigeon shooting, and cracking his long whip in the Weimar market-place, and all in the name of Rousseau, Werther, Nature, genius, and freedom. Goethe had not introduced himself in Frankfurt in 1774 as a literary *bon viveur* but as a man of serious practical interests. Yet on the wedding-night of F. J. Bertuch (1747–1822), Keeper of the Privy Purse and eventually Weimar's only successful businessman, Goethe joined the Duke in playing such a prank that for years afterwards Frau Bertuch could not meet Goethe without blushing. In such company the rather bleak and, as the months wore on, increasingly insistent prospect of a life of difficult, unrewarding, and small-scale government could well be made to seem something exciting and new. There were many reasons for wanting Goethe to prolong his stay: his fame, his magnetism, his willingness, for a while at least, to be an inventive master of the revels, his promise of practical ability. In the course of the years the relative significance of these factors might shift, but in the middle of 1776, as the young ruler began to accept his lot and settle to the regular rhythm of Privy Council business, at least as important as any of them was the fact that Goethe was Carl August's personal choice. While preserving continuity of administration and, to some extent of policy, the Duke wanted to emancipate himself at once both from his mother and from his Privy Councillors: he must, even if obscurely, have felt the need for a guide and support of his own choosing and his own generation. Prime Minister von Fritsch was twenty-seven years older than he, C. F. Schnauss, the other member of the Council, thirty-eight years older. In inviting Goethe to his court—entirely, as far as we know, on his own initiative —Carl August was welcoming the author of a notorious satire on Wieland, his own tutor and his mother's protégé, and in appointing him to his Privy Council he was giving himself a contemporary and natural ally, and firmly establishing the character of the new regime. As acts of self-assertion by new rulers go, it must count as one of the most sensible and fruitful ever made.

In the event, Goethe outmanœuvred everyone. He did what was expected of him, but he remained his own man and—again and again throughout his life— did also what nobody expected. He played the role of court poet: arranging command performances, writing *pièces d'occasion*, and giving exclusive readings from unpublished works. But he retained the freedom also to write what Weimar did not like, want, or understand. He did what Anna Amalia hoped, gained the confidence of her son and over a period of four years, and even beyond, guided him into maturity as a ruler with a style that accorded with her own. He gave Carl August companionship, inspiration, and at times the security of authoritative plain-speaking that had been lacking in the Duke's fatherless boyhood: he even gave him a certain special status, or at least notoriety, as his patron, and he proved an industrious and versatile official. But he refused any confrontation with the existing ministers and was always willing to take their side against the Duke if necessary. And above all—it was perhaps the key to his political position—he did not neglect Duchess Luise. He paid her assiduous court and made of her close friend Charlotte von Stein his own most intimate confidante and his guide to the etiquette, formality, and self-control on which the Duchess, unlike her husband, laid such weight. He thus became for a while the keystone of the ducal family, making stability out of its internal tensions. But what made this elaborate and arduously achieved political and personal accommodation worthwhile? What reasons had he himself for choosing to stay?

It is of course flattering to be wanted by a duke, especially one with salaries, houses, and gardens at his disposal. But at least from 1781 Goethe was prepared to contemplate leaving Weimar and its material attractions for good, and once he was in Italy in 1786 it was not clear to him, the Duke, or anyone, that he would necessarily return. In 1782 he wrote to his mother that a great part of his good spirits was due to the knowledge that his travail and 'sacrifices' in Weimar were self-imposed and that he was at any time free to post back across Germany to the peace of his family home and, presumably, the security of the family fortune. At various times he gave various reasons for not taking that coach back to Frankfurt, but all have in common that he was under no compulsion, either from his own interest or from external circumstance, to serve Carl August and the Weimar State. Goethe *chose* Weimar, and the multiplicity of the occasions on which he made that choice accounts in part for the difficulty of finding any single reason for it.

Looking back, at the very end of his life, on the stages by which in 1775 Weimar had come to seem to him an interesting and desirable place to visit, or even to be, Goethe, writing the last book of *Poetry and Truth,* inclined to what has since been perhaps the commonest explanation for his move. 'Everything pointed to a briskly active literary and artistic life', he says: Weimar was already, thanks to Anna Amalia, the 'court of the Muses', and Goethe went there, it is argued, to help fulfil that early promise. If so, one would have to retort, he failed. It is true that briefly in 1775 and 1776 it looked as if Weimar might be about to graduate into a court of geniuses, gathering to itself the greater part of Germany's literary avant-garde. 'If it is possible for something sensible to come of Weimar it will be an effect of his presence', Wieland wrote on Goethe's arrival, seeing in it the promise of the fulfilment of his own hopes. F. L. Stolberg on his visit was offered a position as gentleman-in-waiting to the Duke and it was many months before Klopstock, who found Goethe's escapades with Carl August unworthy of the high poetic calling, influenced him to decline. There were schemes for bringing the great engraver Chodowiecki to join the circle round the *Teutscher Merkur.* In January 1776 Goethe achieved what he himself described as a 'coup' in persuading the Duke, and through him the Consistory and the Council, to accept Wieland's suggestion that Herder should be brought from Bückeburg to fill the vacant post of Superintendent-General and Principal Chaplain to the Court in Weimar at Saints Peter and Paul, primate, in fact, of the entire local Church. And during the same year Lenz, Klinger, and the shaggy charlatan Kaufmann, 'God's bloodhound', all spent shorter and longer periods in Weimar until they made themselves intolerable not only to the local society but to Goethe and the Duke as well. But for the ten years following, Weimar, without a permanent theatrical troupe until 1784, showed no further signs of developing into an intellectual and artistic centre nor, more significantly, did Goethe show any signs of wanting to assist such a development until shortly before he resolved to go to Italy. To make Weimar a place of brisk literary and artistic activity was certainly Goethe's aim after his return in 1788 and in that aim he manifestly succeeded. It is understandable that in his 1831 retrospect he should have wished to associate with that aim even his earliest years, when he was doing little for the fine arts or the natural sciences in the duchy and nothing for classical learning or philosophy, when he was publishing nothing, and when even the part which he took in the amateur theatre has been somewhat exaggerated. But the eleven years in Weimar to 1786 were too long for Goethe to have been sustained in them merely by the hope of later successes.

The favoured explanation among Goethe's contemporaries for his presence in Weimar, and one with a clear foundation in fact, was his personal relation with the Duke. The ever-percipient Wieland already thought in March 1776 that 'Goethe will doubtless now stay here for as long as Carl August lives, even if that takes until he is as old as Nestor'. As the reference to Nestor suggests, Carl August was not simply Goethe's friend and patron but, after a rather unorthodox fashion, his pupil. Charlotte von Stein and others saw in the wild behaviour of 1776 the means adopted by 'a friendly guiding genius' to achieve his concealed purpose: 'he must carry on like this for a while in order to win over the Duke and then do good', 'and make of him the model of a great prince'. In this role at least Goethe's success was eventually generally acknowledged. Already in 1777 it was credited to his influence that Carl August was said to have remarked to a visiting English nobleman 'I envy you, my lord . . . every one of your fellow-citizens is enough your equal to defend himself against you if you impose upon him. But I—if I box a man's ears here, there is no one who could or would box mine back.' In the following year Wieland found the Duke so 'strong and healthy . . . so noble, kind, affable and princely' that he was 'convinced Goethe has been a good guide to him', and towards the end of his life he felt that not even Goethe's literary works could compare with 'his incredible services to our Duke in the first years of his reign, the self-denial and extreme self-sacrifice of his devotion to him, the great and noble things, still slumbering in the princely youth, that he aroused and brought to develop'. Certainly, in his first four years in Weimar Goethe thought the Duke the only person there showing any signs of development and confided to his diary his satisfaction at his maturing personality and attitude to affairs. But by the time the two of them returned from their long tour of Switzerland in the winter of 1779–80, this process was largely complete and the tutorial relation between them will not even explain Goethe's remaining in Weimar for the next six years, let alone the next fifty. Merck—'the only man . . . who fully understands what I do and how I do it'—urged Goethe's mother in June 1781 to persuade him away from Weimar since 'he has now done the main job—the Duke is what he ought to be, the other rubbish can be done by someone else—Goethe is too good for that'.

'The other rubbish' was what Goethe, writing to Merck only ten weeks after his first arrival in Weimar, had singled out as principally prevailing on him to stay: 'all the business of the court and politics'. 'My situation is advantageous enough, and the duchies of Weimar and Eisenach will do as a theatre to try out how the worldly role would suit one.' 'Even if it is only for a few years', he wrote three weeks later, 'it is still better than the idle life at home where with the best will I can do nothing. Here I have at least got a couple of duchies in front of me.' Here it might seem that we have at last a reason both comprehensible in itself and avowed at the time by Goethe for his exchanging the bustle and complexity of republican Frankfurt for the restrictions, even deprivations, of a remote and backward principality. Weimar offered him an entrée to the court life that he had hitherto seen only briefly and from outside, and like any other autocracy, untrammelled by constitutions and traditions, it offered to young, ambitious, and gifted men the prospect of far more rapid advancement in the exercise of administrative power than could be hoped for in the cautious city-states, where promotion came essentially only with age, and the scope of government was restricted in order to protect the rights of individuals and corporations. Goethe saw in Weimar's offer the possibility of doing something —perhaps even something useful to his fellow men—with the superabundant energies that in Frankfurt had been threatening to destroy him, and the possibility thereby perhaps of outshining his grandfather Textor, certainly of fulfilling the ambitions that his father had had perforce to renounce when the door closed on his own political career. He had had his eye on Weimar for some time and it was surely not chance that in Frankfurt he had talked to the princes only of Möser and politics and practical things. In 1776 he was certainly excited at the prospect of joining the Weimar administration and this may well be enough to explain his refusal at the time to listen to his hostile father or doubting friends. But the explanation works only for a period, and not even for the whole of that. The new responsibilities brought him neither satisfaction nor peace of mind, he told his mother on the eve of the first anniversary of his arrival, and when in 1785 he effectively withdrew from the government of the duchy and gave up his political career, there was still no question of his taking post-horses back to Frankfurt. Weimar, even without high office, was still where he belonged, and where for another forty-seven years he would remain. The prospect of a political and court career, the lure of the 'world', as he frequently calls it at the time, was Weimar's most specific and manifest attraction for Goethe in 1775 and 1776, but beyond and behind that there was a rightness for him about the place which outlasted those ambitions and to which he found it difficult to give any more definite name than 'destiny', 'fate', 'Schicksal'.

Was it just coincidence that at different times Weimar was able to fulfil different functions in Goethe's life? Or was there something about the place that made it peculiarly able to meet the shifting and special, indeed unique, needs of Germany's greatest national poet? Time and again in his early years there he felt he had to acknowledge a sense or plan behind his presence in Weimar, something which kept him there even in misery, even when it seemed to him a 'hole', and even though he was not compelled by any material necessity to remain. What that force was, he could not say. Sometimes it was simply 'Fate, the object of my worship' which had 'planted me here', and of which he could say only that it was 'wholly concealed from men, they can see and hear nothing of it'. Sometimes it is God that he thanks 'for putting me, with my nature, in so close and loose a situation where all the manifold fibres of my being can and must be thoroughly blanched' and to whom he prays, as his thirtieth birthday approaches, to 'help him on the way': 'not my will is being done but the will of a higher power, whose thoughts are not my thoughts'. When the power that rules one's life is so incomprehensible—'I have less and less idea what I am . . . and what I am meant to do'—the reason for being and staying where one is comes to seem an arbitrary, or even merely prudential maxim: 'the safest, trustiest, most tried advice is: stay where you are. . . . Firm and loyal to a single purpose . . . A man who changes his situation always loses his travel and removal expenses, both morally and economically . . .'. The first recipient of this advice, which Goethe gave emphatically to Kestner, to Bürger, and to Herder, when he was tempted to leave Weimar by the offer of a professorial chair in Göttingen, was of course Goethe himself. It is eminently sensible advice, but it presupposes that you know why you are where you are in the first place. In writing to his mother, who was longing for destiny to bring him back to Frankfurt, Goethe could not beg that question and to her he gives the simplest and least metaphysical reason for his staying in Weimar, the truest that he could articulate: there was nowhere better for him to be. In August 1781, shortly after 'an evil genius' has first whispered to him the thought that he might 'save himself by flight', he writes in response to her report of Merck's anxiety that he should abandon the Weimar 'rubbish': 'My position . . . has . . . very much that is desirable for me, of which the best proof is that no other is conceivable to me into which I should at present wish to transfer.' Two years later he repeats the point: 'I could not think of or imagine a better place, since I am after all acquainted with the world and I am not ignorant how things look on the other side of the mountain.'

Goethe could perfectly well know *that* there was nowhere better than Weimar

—he had only to consider the alternatives: Frankfurt? Berlin? Münster? Zurich? Gotha?—without thereby knowing *why* this should be. Our own historical distance from Goethe's age gives us at least the slender advantage of hindsight in answering a question which to Goethe might have seemed speculative and unpractical. Frankfurt, as he reminded his mother in 1781, had become impossible for him in his last months there. The engagement to Lili had been an attempt to fix himself there and it had felt wrong from the beginning. The literary and cultural movement over which he had presided from the Grosser Hirschgraben could not be sustained. Its inspiration had been Imperial, republican, realist, above all national—it could not but conflict with the 'official' culture of feeling and the inner world through which the German middle class accommodated itself to the facts of life in absolute monarchies, jealous of their autonomy. *Werther*—realist in manner, sentimental in theme— had, uniquely, held in balance an extreme expression of both tendencies. But the national aspirations of the Frankfurt Storm and Stress, manifest in its discovery and exploitation of the language of the people, were doomed to futility, given that political power—and so the political, and ultimately the cultural, future of the nation—lay, not with the Empire and not with the burghers of Germany's city republics, but with the Enlightened autocracies ruled by Germany's princes. In the nineteenth century Germany was to be united, and to be given a national culture, by Prussian absolutism, not by Imperial federalism, and Goethe did not lack contemporaries who were already looking to Berlin to found their country's future. When Goethe pointed out to his mother the disproportion between the restriction and sluggishness of the 'bourgeois world' of Frankfurt and 'the breadth and celerity' of his own nature, and when he praised the 'element of the infinite' in his situation in Weimar, he was simply saying that Weimar gave him more to do—but what that meant was that Weimar's political structure allowed him to touch contemporary life at many more points than Frankfurt, and it did so because it, not the constitution of Frankfurt, was the political structure that was already moulding the whole German nation-to-be.

The decisive feature that Weimar had to offer Goethe was the court: the institutional centre round which, in immediate contact with the state power, the 'official' culture of the German middle class had been growing up since the days of Gottsched. That link, between political power—which meant, in the long run, the future of the country—and intellectual life, Weimar could give him and Frankfurt could not. Frankfurt could offer him only provincialism and marginality, the life of a local, and probably increasingly eccentric, curiosity, a second Möser. He had known this at least since he had urged his father to let him study classics in Göttingen and since the days when the 'good taste' of Electoral Leipzig had seemed so superior to that of Frankfurt. If Goethe was to fulfil the hopes, first fully articulated perhaps in his conversations with Herder in Strasbourg in 1770 and 1771, of contributing to the development of a national German literature, he had to come to terms with the real power in the land, and the literature it was generating. For that reason, to take refuge in Switzerland would have been to evade the German issue altogether, and Zurich, despite the presence of Bodmer and Lavater, was out of the question. In the absence of a court it was no better than another Frankfurt: 'Under the republican pressure and in the atmosphere of well-fumigated weeklies and learned journals any man of sense would go mad on the spot.' Catholic Münster too lacked the crucial immediacy with political power, being a remote fief of the Habsburgs, and Princess Gallitzin's circle lost all chance of competing with Weimar when its own candidate for the bishopric was passed over by Vienna. And though Germany had many courts, not all of them were as closely linked to the university, the main institutional home of the new 'official' culture, as Weimar was to Jena. Neither Lessing's Wolfenbüttel (itself deserted by the Brunswick court) nor Herder's Bückeburg could boast such a connection, although in other ways the acceptance of patronage by these two fiercely independent thinkers may have served as an example to Goethe in making his own move. In respect of the university, Weimar had the advantage even over all the other Saxon principalities, including Gotha, where Goethe frequently felt more welcome than at his own court—although the value of the connection with Jena became apparent to him only with the passage of time.

However, the absolutist culture of the courts and their associated bureaucracies, though it might be destined for a leading role, could not express or satisfy the entire German nation. There were other Germanies besides that of Prussian-style monarchy: there was, for example, Imperial Germany and Catholic Germany, the Germany of the city-states and of the 'home towns' with their own privileges and traditions. Most significant of all there was the new Germany of the mind, that escaped the control of any potentate, even the King of Prussia, and that existed wherever the German language was spoken. This Germany, remote from the courts, was egalitarian in spirit and the exclusive property of the middle class; it was the world of the German book, created and sustained by capitalist publisher-booksellers, who were by many decades the first German industrialists producing for a mass market. This was the Germany that the Frankfurt Goethe had taken by storm, the public mind whose internal contradictions and struggle for identity he had embodied in *Werther* and in *Götz*. None of this Goethe wanted simply to abandon, and in seeking a footing in the princely court of Weimar he was not simply going over to the enemy. Rather, with the unconscious sureness of a sleepwalker, perhaps, he was finding his way to the one place where he could bring the centre of gravity of his own life as close as possible to the political centre of gravity of what was destined to be the nation, while retaining the generously broad and multifarious conception of nationhood of his Strasbourg and Frankfurt years. He was groping towards a creative fusion of those components of the national life whose destructive conflict he had depicted in the antagonism between Götz and Weislingen, and in Werther's unsuccessful attempt at a courtly and administrative career. This compromise between the politically and culturally dominant Germany of the princes and all the other Germanies which absolutism threatened to exclude— a compromise which explains why Goethe's appeal has always been broader than, and subversive of, any purely political German nationalism—took many years to establish. The difficulty of working out how to perform this unprecedented task accounts for the sense of bewilderment and failure that runs through Goethe's first eleven years in Weimar, for all his conviction that he was in the right place. Already in 1781, however, he had some conception of what the task involved. In 1780 Frederick the Great had published a treatise, *De la littérature allemande,* in which, true to the taste of the French court, the model for all absolutisms, he had condemned the Shakespearian trend in modern German theatre, and particularly the 'dégoûtantes platitudes' of the 'détestable' *Götz von Berlichingen.* A chorus of representatives of 'official' German literature hastened to defend their right to produce in German works that did not offend against the best French canons, which was hardly the point. Goethe was passed over in silence, as an embarrassment. Only Möser took up his cause and rebuked the King for not recognizing the play as a true fruit of the nation to whose political and practical interests Frederick had for forty years devoted himself. In a letter to Möser's daughter, a masterpiece of statesmanlike modesty, Goethe thanked her father for the defence of his play; indeed he says 'any German must thank him who is concerned for the good cause and for the progress of the labours that have been begun'. And yet, he delicately implies, that progress cannot be a simple continuation of the movement for which Möser was so prophetic a figure and which has since met with insuperable obstacles: 'It is really admirable of the old patriarch to acknowledge his people even before the world and its great ones, for in the end it was he who attracted us into this land and pointed out to us further more extensive regions than it was to prove permissible to traverse.' It will be necessary, in other words, to abandon the expectation, to which Möser plainly still holds, that literary intellectuals, the representatives of the national public mind already established in the world of the book-market, should share in, or even wholly assume, the political power exercised by princely Germany, and that by these means a German national culture can be achieved. In the literary turbulence of the 1770s it might passingly have seemed that kings could be literary critics, and literary critics could be kings. But the relation between the different Germanies will have to be more complex than that, and autocratic monarchy cannot be expected to show a generous tolerance of variety, literary or otherwise: 'A man of many powers, who directs thousands with an iron sceptre, is bound to find intolerable the production of a free and untutored boy.'

To find the precise form in which his nearness to the source of administrative power in a single state could best be used for the 'good cause' of the German national culture cost Goethe much time and many disappointments. His first years in Weimar were spent learning, through two complementary and bitter lessons, the necessity of a compromise. He had, through disappointment as a politician, to learn what the German bourgeoisie had been learning for a century: that that class was not yet powerful enough to impose its own patterns on national government. And he had, through disappointment—albeit temporary—as an artist, to learn that he could not nourish his poetic talent from court or 'official' culture alone but that a fusion with his own bourgeois origins was essential if his art was to survive. His supreme achievement of later years was to create an art which aspired to the intimate relation with a reading public characteristic of his Frankfurt years, but which drew equally on the forms and values of courts and officials, and which, because it was at once both public— that is, bourgeois—and courtly, was the nearest thing Germany achieved to a national literature in realization of the ambitions of the 1770s. Goethe's art could not, from its first maturity, be detached from his life, and the openness of his later work to all the components that might go into the making of a German nation reflects an openness in his personal and social position in Weimar. To have gone to the Berlin of Frederick the Great, even if he had been welcome there, would have been too complete a surrender to the spirit of Enlightened autocracy: too many possible Germanies would then have been excluded. In more ways than the geographical, Weimar was half-way between Frankfurt and Berlin. Though it looked to its powerful Prussian neighbour for leadership in foreign affairs, Weimar, like all the smaller principalities, was loyal to the Imperial constitution and resisted the Emperor Joseph II only when he took a leaf out of the Prussian book and sought to subordinate the Imperial interests to those of his own increasingly secularized and despotic Austrian state. Weimar was Lutheran like Frankfurt, without being secular or Pietist like Berlin, and though its climate was colder and wetter than Frankfurt's, its landscape was still distinctively that of central, rather than northern or southern Germany. But the special merit of Weimar was not that it was actually the centre, geographically, politically, socially, or even intellectually, of what Germany was or might become. It was rather that it was a little removed from the focus of events in Berlin, Göttingen, or Leipzig but still close enough to observe them conveniently and draw upon them as seemed necessary, an excellent place in fact in which to practise that life on the edge of centrality that had first been exemplified to Goethe by Susanna von Klettenberg. Even Frankfurt was only thirty-one hours' travelling away. In coming to Weimar Goethe had perhaps not come as far from his origins as might at first appear, and though nothing can prove that it was the perfect place for a cultural compromise, we may well agree with Goethe that it is difficult to think of a better.

The Minister

The years from Goethe's first arrival in Weimar until his departure for Italy in September 1786 are a period of which he spoke reluctantly and gave next to no account in any of his autobiographical writings. Its obscurity is the greater since practically all the letters he received in this time he burnt in 1797, and the diary which he irregularly kept is largely confined to the barest notes of engagements. It was a period of mistakes and failures, none greater perhaps, or so it must later have seemed, than the intense and prolonged liaison with Charlotte von Stein, and all painful to look back on. Yet it is possible, on the basis of his own letters, the letters and reminiscences of others, and his literary works, to identify several distinct phases within this period and these we shall need to consider in turn in the next two chapters. Yet the period as a whole has a character of its own, certain themes run through it from beginning to end and its coherence would be misrepresented if we did not consider these separately first. Three factors in particular dominate these eleven years: the steadily increasing burden of Goethe's administrative work, his relation with Frau von Stein, and his difficulties as a writer, which we shall find are associated with the growth of his interest in natural science.

To begin with, the work asked from Goethe in exchange for his little house outside the walls, his salary, and his title of Councillor of Legation, was not overtaxing, though from the day of his appointment to the Privy Council, 11 June 1776, he took it most seriously. He told Frau von Stein that he had never missed a session of the Council 'without extreme necessity', and the records of the well over 600 sessions from June 1776 to February 1785 bear him out. The Council met on average three times a fortnight, always at 10 a.m., and usually, though not necessarily, under the chairmanship of the Duke, and would have on its agenda an average of thirty items ranging from the funeral formalities for a Jena student, killed during the forbidden practice of duelling, to the price to be set for the current year on the heads of sparrows and other pests, and from the annual audits of the different administrative regions of the duchy, or a sinister request from the Prussian military authorities to be allowed to recruit on Weimar territory, to an endless paper quarrel with a dismissed and aggrieved official who was libelling the Privy Council from the security of nearby Erfurt. Each item would be the subject of an oral report from one of the three Councillors who had worked through the relevant papers at home, and the final recommendation to the Duke, with whom all power lay, would be decided by a vote. As no minutes were kept, and much of the Council's work consisted in oral discussion, it is impossible to determine precisely the nature of Goethe's participation, but it is clear that on such a body there was no room for passengers. After the meeting Goethe frequently dined with the Duke and continued discussion of the issues, or of the conduct of the meeting. In his first two and a half years of sitting on it, the Council was thus for Goethe a natural, if time-consuming, extension of his principal role as companion and mentor to Carl August. On the one occasion when we have a minute of Goethe's oral intervention in a discussion—the subject was the Prussian request for recruitment facilities—he confined himself to the psychological task of encouraging the Duke to be consistent in facing the consequences of whichever course of action he decided on.

Goethe's first, and, as it was to prove, longest-lasting, distinct administrative commitment also had its roots in the days and weeks that he and the Duke had spent together roaming through the Thuringian Forest with the other young men of the court, partly on business but mainly on pleasure. Early in 1776, on expert advice, Carl August had decided to reopen the silver-mines at Ilmenau and in February 1777, Goethe, who had inspected the workings with the Duke in the previous summer and had of course some legal experience, agreed to serve on a small committee to clear away the formal obstacles to the reopening, especially the negotiations with the old mine's principal private creditor. From then on Goethe was the official director of an enormously protracted fiasco. The committee turned into a fully-fledged mines commission and had a full-time lawyer as secretary in charge of day-to-day affairs, but it was Goethe who had, for example, to conduct negotiations with Electoral Saxony and Saxe-Gotha, the co-owners of the old mine. A company had also to be set up to raise capital for the new venture, and it was eventually in a position, on Goethe's birthday in 1783, to issue 1,000 shares at 20 dollars each. Enough of these were sold for work to begin in February 1784 on a new shaft, and with that began also a host of new problems of a more practical nature: in February 1786, for example, we find Goethe doing his best to placate the engineer responsible for building the main pumping machinery, who has downed tools after a quarrel with the administration. Further injections of capital proved necessary and Goethe had the task of writing the company's annual reports and mollifying the shareholders. Not until 1792 was the ore-bearing seam reached, and then the ore itself proved of inferior quality. Attempts to open a new seam were interrupted by a disastrous flood in 1796, though one shaft was kept open until the mine finally closed in 1812. Goethe cannot be blamed for the original decision to revive the mine, but the tenacity with which a mistaken policy was pursued for over thirty years bears all the marks of his personal direction.

Had the Ilmenau mine proved profitable, of course, it would greatly have assisted the duchy's financial position: it may well have seemed to Goethe at times the only hope. If there is a theme running through his period as a general administrator it is the need for sound finances. A special committee on which he sat in the early years concluded that the annual income available to support the ducal court and stables was 44,000 dollars, but the minimum conceivable annual expenditure was 54,000 dollars. Mr Micawber could have predicted the result. From January 1779 Goethe began to be in a position to do something directly about the matter, for from this date he took up the chairmanship of two of the standing commissions that reported to the Council: the War Commission and the Highways Commission. With this the basis for his presence in the Privy Council changed somewhat: though he was still the junior member, he was no longer simply the Duke's favourite, deputed from time to time to look after pet schemes, but a fully functioning administrator whose responsibilities outside the Council made his work-load more nearly comparable to that of the greatly more experienced and professional von Fritsch and Schnauss. He even had a certain advantage over them in that he knew the ducal territories and the local officials better than anyone else in Weimar: to his frequent journeys to Ilmenau were now added travels in the Eisenach area as he consulted about matters of land-drainage there with the idiosyncratic but effective expatriate Englishman George Batty, whose practical genius he admired, and as War Commissioner he traversed the whole duchy to undertake the triennial selection of recruits for the Weimar army. His opportunities for helping the state treasury were as yet limited: in the Highways Commission he conceived the scheme of improving the roads between Weimar and Naumburg in the east, and Weimar and Erfurt in the west, so as to divert profitable traffic on the Leipzig-Frankfurt road through Weimar. But the legal and political obstacles were great and the resources few. By the time Goethe gave up the task in 1786 he had succeeded in metalling only the road on which he had arrived in 1775, from Erfurt to Weimar, and the road which was to be of such importance to him later, from Weimar to Jena. In the War Commission Goethe could not make significant economies, given Carl August's now growing passion for playing at soldiers, until he reached the next and highest stage in his official career, for which the three years after 1779 were a preparation.

In 1782 Carl August suddenly dismissed the Chancellor of his Exchequer—President of the Chamber was his official title—J. A. von Kalb, the courier who had brought Goethe to Weimar and who had been appointed to succeed his father at the time when Goethe joined the Privy Council. The reasons were not made public, but were probably of a personal nature (the dismissal was honourable and von Kalb received a handsome pension). Financial crisis however was in the air: von Kalb had just declared himself unable to make the repayment due on a large loan from the city of Berne. Goethe was made acting President of the Chamber with responsibility, not for day-to-day management, but for questions of policy and the introduction of reforms. It was the peak of his political career and he used the moment for Weimar's benefit: he persuaded the Estates—the assembly of taxpayers—to fund the national debt, and persuaded Carl August in exchange—certainly no easier task—to agree to cuts in his army to permit a lowering of taxation. The artillery was abolished and the infantry reduced from over 500 to 142. By these means Goethe became one of the few defence ministers who have voluntarily halved their own budget.

The duchy had been saved from bankruptcy and the Duke had been taught a lesson in household management: 'the economics are on a sound basis, and that is what counts'. In the meantime Goethe was gaining experience as an *ad hoc* foreign minister. Once he had been ennobled, in 1782, he was by far the most prestigious emissary Carl August had to send. In this capacity Goethe came once more to the attention of Frederick the Great, this time as a man of influence whose support his diplomats were instructed to secure for the League

of Princes which Prussia was persuading the smaller German rulers to form—
under Prussian protection of course—to resist the predatory aspirations of the
Emperor, Joseph II. Goethe had himself proposed such a League to the Privy
Council years before, but as a security as much against Prussian as Austrian
intervention, and he was unenthusiastic about both the new proposal and Carl
August's eagerness to proselytize for it throughout the courts of Germany. He
declined the Duke's invitation to accompany him on this mission: he had
probably already decided to bring his period of high office to a close, and he had
soon recognized that he did not have the suavity and flexibility to be a diplomat.
His strength was in the persistent and unrelenting pursuit of a goal, and what
that could achieve in Weimar he had done. He lacked the single-minded
professionalism and total familiarity with conventional forms of the other
Weimar civil servants and their devotion to a task that had no conceivable end
or culmination. By contrast with their conscientious but mechanical reports
Goethe's own official submissions have a fluency and clarity of exposition, even
a touch of spontaneity and dry humour, which suggest, in the last analysis, the
amateur, however brilliant. A fourth, full-time, Privy Councillor was recruited
in 1784, and in February 1785 Goethe withdrew almost entirely from Council
affairs, though he continued his work in the Chamber and the Commissions
until he left for Italy, devoting himself to establishing some order in the land-tax
assessments and in the remuneration of the Jena professors. Perhaps the finest
monument to his period in office is the general opinion, already recorded by
Schiller in 1787, that in not a single case had he ever deliberately done any
harm to anyone.

As early as 1782 it had become clear to Goethe that he had only a limited
future as a civil servant, and that Möser's Promised Land was not to be entered
by any of the routes he had so far adopted. The 'good cause' of establishing a
German national culture was not after all greatly advanced by the education of a
single prince, nor by all his efforts on behalf of the administration of a single
petty state—even if he had been successful in his six years' toil, which at times
he doubted. He wrote to Knebel:

> The Duke finds his existence in baiting and hunting. Affairs jog along satisfactorily:
> he takes a willing and tolerable part in them and now and then is agreeable to something
> good, plants and roots up etc. The Duchess is quiet, lives the life of the court, I see
> neither of them often.
> And so I am beginning to live for myself once more and recognize myself again. The
> illusion that the fair seeds that are ripening in my and my friends' existence need to be
> sown in this soil, and that those heavenly jewels could be set in the earthly crowns of
> these princes, has completely left me . . .

By 1786 he was using blunter language for the folly of attempting the purely
political road to Möser's goal, in the absence of philosopher-kings and king-
philosophers: 'anyone who takes up his time with administration without
himself being the ruler must be either a philistine or a knave or a fool.'

Much has been written on the supposed failure of Goethe's attempts at political and administrative reform in these years, but we should beware of the assumption that to recognize your limitations, or even the limitations of your circumstances, is to fail. Taking the long view, we would have to say that Goethe's policy of doing only what he could do was as successful as it could possibly have been. For what he could do supremely well in his time and place was, as everyone eventually agreed, and despite his own periodic reservations, to educate the young Carl August. The ultimate fruit of that policy was seen in 1809, when the Duke gave his people a liberal constitution far in advance of anything to be found in any other German state. If, in the turmoil in which the Napoleonic Empire came to an end, this revolutionary step failed to have any great impact in the rest of Germany, that is not a failure for which any individual, or any institution, in Saxe-Weimar can be blamed.

Frau von Stein

Even at the height of his administrative engagement, however, official business was only a part of Goethe's life. On 9 January 1782, for example, his diary tells us that he spent the morning in the office of the War Commission; at midday he had dinner with an old flame, Weimar's best actress, Corona Schröter; at 4 o'clock he was in the ducal Drawing Institute, of which his Frankfurt teacher Kraus was now the Director, sharing his latest enthusiasm with a capacity audience and delivering a lecture on the anatomy of the foot; at half-past five he was directing the rehearsal of a ballet-masque he had written for the Duchess's birthday. In the evening, however, after about seven, he relaxed in the company of Frau von Stein, with whom he took supper: this was the compensation for the day's mental exertions. 'Mme Stein is the cork-jacket that keeps me above water', he wrote a month later to Knebel, and the years of Goethe's life from 1776 to 1786 belonged as much to her as to his official duties.

When Goethe met the wife of the Chief Equerry, Charlotte von Stein, née von Schardt, within a few days of his arrival in Weimar, probably on 12 November, her face was already known to him from Zimmermann's silhouette, which Goethe now felt did not do her justice. In his original opinion that her features showed a soul that '*sees the world as it is*, and yet through the *medium of love*' there was rather more of Sentimental physiognomics however, than of Frau von Stein's real character. Seven years older than Goethe, and with a country squire for a husband, who could dance and play the flute, but whose simple tastes ran otherwise to sizing up the form of horses and bulls, enjoying his daily dinner at the court, and engendering seven children in eleven years of marriage (four of whom, all daughters, died in infancy, while three sons survived, one of them sickly) she had a different conception of love, and of what —if anything—mattered in life, from the wilder younger generation. Her father, once Chamberlain in Weimar, and then a busy and demanding but rather ineffectual man, had spent little time at home until forced into early and disappointed retirement by Anna Amalia, and had left the care of his family to his pious but unforthcoming wife, in whose sombre personality there were still perhaps traces of her distant Scottish ancestry. Of the five children only Charlotte and Luise had offspring that survived infancy, and Luise, in 1775, was already unhappy in her marriage to Baron von Imhoff, whose fortune was proving less substantial than his two black pages had suggested. Frau von Stein, however, was not, as is sometimes suggested, an ethereal personality—she had been a lady-in-waiting of Anna Amalia's and she was young and open-minded enough to enjoy the invigorating atmosphere of Carl August's new regime in its early years—but at some level she had been wounded, and had composed herself to suffer in order to survive. Knebel, who knew her well, described her in 1788 as 'of an unimpassioned disposition' and thought that thanks to her own industry and intellectual curiosity she had made good use of the 'excellent company she had been able to keep' (that is to say, probably Goethe and Herder): 'She is wholly without pretension or affectation, honest, naturally open, not too serious and not too frivolous, without sentimentality but intellectually warm-hearted, takes an interest in all enlightened and humane concerns, is well-informed and has fine feeling, and even skill, in artistic matters.' Serious, without being religious, and conventional without being refined—decorum was for uneducated women of any but the most powerful character their only defence against the male-dominated world—she was a natural companion for Duchess Luise. Carl August thought her 'no great genius', and preferred the company of her husband.

In January 1775 Frau von Stein had been warned by Zimmermann: 'A woman of the *beau monde* . . . told me that Goethe was the most handsome, most lively, most natural, most fiery, most stormy, most gentle, most seductive and for the *heart* of a woman the most *dangerous* man she had seen in her life.' Yet even so she can hardly have been prepared for the tempestuous wooing that now began. After the first meetings came the letters and notes—the first of some 1,800 to have survived—and already by the middle of January 1776 she was 'dear Angel'—'dear Lady, please allow yourself to be so dear to me'—and Goethe was 'missing' her at balls and receptions of the Duchess and regretting that he could not 'yet' dine alone with her and addressing her, in writing at least, with the familiar 'Du'. Frau von Stein felt bound to protest to him against this last habit in March,

because no one can understand it correctly as I do and he anyway frequently loses sight of certain circumstances. Whereupon he leaps up furiously from the sofa, says 'I must go!', rushes up and down a few times looking for his stick, cannot find it and runs out of the door without taking his leave, without a goodnight.

The situation is reminiscent of another frank interview with another Charlotte, also otherwise engaged, in September 1772, and Frau von Stein's insistence on the formal 'Sie' certainly provides no more grounds than Lotte Buff's declaration of fidelity to Kestner for accusing her of stuffiness. On the contrary, the real warmth of her interest in Goethe is shown by the continuation of the letter to Zimmermann which reports this incident:

> There is an enormous amount on my mind that I must tell this inhuman creature. It is just impossible; with his behaviour he will not survive in the world. If our gentle Teacher was crucified, this bitter one will be hacked to pieces. Why his endless satirizing of people? They are all creatures of the great Being, are they not, and *He* tolerates them. And then his indecent conduct, his cursing, his vulgar, low expressions. Perhaps it will not have any effect on his moral side when it is a matter of actions; but he is a bad influence on others. The Duke is strangely altered . . . That is all, then, about Goethe, the man who has a head and a heart for thousands, who can see all things in so clear and unprejudiced a way, if only he chooses, who can master whatever he chooses! I can tell Goethe and I will never become friends. And I do not like his way of treating our sex. He is really what they call *coquet*. There is not enough respect in his manner.

The signs of yielding are there, for all the resistance, but also the sharp judgement of character and the genuine concern for Goethe's progress in propriety, to which literature and literary fame are somewhat querulously subordinated: these were to be the emotional constants in her contribution to the relationship until it collapsed in 1789. The only new factor that came with the years was an increasing possessiveness on her part, which may to some extent be understood as a desire to suppress the 'coquettish' element in Goethe, the *attrattiva* of which he was perhaps too aware, and so to secure her own position from reproach. From the start the friendship was public knowledge at the court, and Frau von Stein had no desire to be thought an adulteress.

By the summer of 1776 Goethe was a regular guest in the Stein household and, as a member of the responsible committee, was soon discussing in loving detail the restoration and decoration of some new apartments that the Duke was putting at the disposal of his Chief Equerry. From November 1777 the von Stein family lived in the upper floors of the cavalry stable-block beside the old Wilhelmsburg, just within the outer town wall, and only three minutes' walk across the meadows from Goethe's cottage on the other side of the Ilm. He now had to pass their door whenever he went into town to be about his official business, he frequently dined there while Baron von Stein was keeping the Duchess company at dinner at court, and whenever he had a free evening a servant would come across with a note suggesting that Charlotte and he should spend it together. It would not be too much to say that for ten years not a day went by without Goethe's either seeing Charlotte von Stein, or writing to her, or at least devoting much of his thought to her. The nature of the relationship is difficult to assess, since not only do we have no record of their conversations, but after it broke down Frau von Stein demanded the return of her half of the correspondence and destroyed it. From Goethe's letters, however, which she carefully preserved as evidence of his treachery, it is clear that a main part of it was a simple and complete sharing of their daily doings: talking of Charlotte's children, their health and education, of her other relations, affairs on her estates, the colour of her walls and the design of her stoves, and, on Goethe's side—she repeatedly accused him of talking exclusively about himself, even when he appeared to be talking about other things—of his relations with the Duke and Duchess and other members of the court society, of his emotions and anxieties and plans, of places and people seen in the course of his work (though not his official business itself), of his past in Frankfurt and Strasbourg, of other literary figures of the time, and of his own literary works. She saw all his current writing, even took some of it down to his dictation, and various of Goethe's poems of this period were written with her in mind or grew out of letters to her. Little gifts were sent—the first asparagus from Goethe's garden, a framed landscape Goethe had painted, some game from the hunt, bread from the military bakery (one of the perquisites of office at the War Commission)—and books were read together and discussed. She was brought to share in each of Goethe's enthusiasms as they emerged: now they would be reading a new translation of Aeschylus, now he would be expounding a new theory in geology or anatomy, now they would be studying Spinoza together and seeking Herder's guidance. To begin with, he even gave her English lessons. Goethe himself eventually described the relation between them as a marriage—rather than a romance—sending her the gloves he received on his initiation as a Freemason, which he was supposed to pass on to the most important woman in his life, and receiving from her a gold ring engraved with her initials. In 1783, dissatisfied with the education being given to her youngest son Fritz (1772–1844), of whom he was particularly fond, Goethe took him into his own house and instructed him himself for three years, which Fritz later said were the happiest of his life: Goethe's integration into the Stein family could hardly have gone further.

The question naturally arises, and it is germane to our understanding of what this strange ménage meant to Goethe, whether the relationship was ever carnal. The evidence, in so far as there can be evidence for a negative conclusion, is that it was not. A well-placed observer was convinced that Frau von Stein was 'only Goethe's *confidante*', and it is difficult to believe that, in a court where the contents of a scrap of paper left in a pair of trousers sent to the tailor's could immediately become general knowledge, an adulterous liaison could be carried on for any length of time without its being reported by the lynx-eyed Countesses Görtz and Gianini, or without there being some jovial and tasteless remark from Carl August alluding to it. But not a single scabrous anecdote has survived and there is no evidence that there ever were any. Baron von Stein was a complaisant type, who probably thought it was good for his wife to have someone clever to talk to about things that bored him and went over his head. The court backbiters found his position rather ridiculous, but had he suspected the jackanapes was usurping his conjugal rights he was certainly the man to throw him out of his house. Instead Stein raised no objection to the arrangements for Fritz's education, and was even willing to act as postman, delivering to his wife long letters from Goethe, with whom he seems to have been on perfectly friendly terms.

The pedagogical role which Charlotte assigned to herself, and in which from the start there was a defensive element, also suggests that physical intimacy had no appeal for her, and that Goethe would have to learn this if he wanted her friendship. He had to accept the suppression of the 'Du' form in 1776, and for the next five years he emphasizes in his letters to her that he is struggling to attain an inner state of 'Reinheit', purity and clarity at once. In 1780 to 1781 the letters become more urgent and passionate and in March 1781 he succeeds in re-establishing the 'Du', professing himself from now on satisfied that she does indeed believe in his love, and loves him in return, and resisting an attempt by her to revert to 'Sie' later in the year. From early 1781 we hear less of 'Reinheit' and more of 'Liebe', and the emotional force of the letters is quieter: more assured and settled, perhaps, but also more modest and undemanding. This is also the point at which echoes of her possessiveness, even jealousy, begin to be heard, and Goethe's assertions of love take on a repetitious, even monotonous quality. Whatever the proof of their love that she furnished—two poems, 'Night Thoughts' ('Nachtgedanken') and 'The Goblet' ('Der Becher'), certainly state that it was bodily, though the imaginative artificiality of the latter suggests an element of wish-fulfilment—it would seem to have been coupled with a requirement, to which Goethe submitted, of self-denial. (And had the poems been genuinely autobiographical Goethe would hardly have had them published in the Weimar court-magazine, even though they then masqueraded as translations 'from the Greek'.) 'In the midst of good fortune I live in continuous renunciation', he wrote in 1782, and in 1784: 'I am a poor slave of duty, to which Fate has wedded me'. The gold ring was of course symbolic. The promise Frau von Stein later felt he had broken was the promise to be faithful to her and never to marry, and complementary in his letters to the conviction that his beloved is now everything to him is a resigned emphasis on the restriction, even deprivation, of the rest of his life: how small his circle of friends is, how he lacks the consolations of a home, children, 'and a long etc', how she alone has enabled him to 'renounce' his 'favourite errors'. In 1784 he actually pleads with her not to reawaken that sexual desire for her which is now blessedly asleep. For all the repeatedly asserted euphoria of love his physical condition was not that of a happy man: his figure lean, his face sallow and haggard, his constitution prone to illness. It is the only period when he writes freely of the shortness of life and of the limited time remaining to him.

Both parties had their unacknowledged reasons for desiring this strange impasse, with emotion constantly maintained at the highest pitch, and constantly left without bodily expression. Frau von Stein, who may have been what is technically known as a hysterical personality, in whom unconsciously suggestive behaviour may be combined with an overt sense of propriety, needed, not the endlessly repeated assurances of affection, but the affirmation of her own obliterated identity that they contained. There was a hunger there that no bodily fulfilment could ever assuage, for it was through her bodily marriage to Stein that this identity, perhaps lost in a loveless childhood, continued to be denied her. She needed Goethe's affirmation, and she needed to be sure it would never stop, or be diverted elsewhere. And Goethe too needed, or did not want to resist, the enchantment of the Snow Queen. It was not the first, nor even the second, time that he had become obsessively attracted to a sexually unavailable woman: desire would not let him rest, but its fulfilment, above all in the institution of marriage, threatened him with fixity, with 'being' rather than the infinite charm of 'becoming', with a gift of self that, whatever it gave him in exchange, would leave him without something he had possessed before. He needed the impossible union, for fear of the possible. It cannot be chance that the correspondence with Augusta Stolberg tailed away as that with Frau von Stein began, or that, only months after relations with Frau von Stein entered their final and seemingly satisfactory state of frozen longing, the stability of a 'marriage', even if only of soul mates, the 'evil genius' first brought the thought of leaving Weimar to the surface of Goethe's mind.

But there was more to this half-real half-love-affair than its unconscious emotional dynamics, which it may well share with many other unsatisfactory liaisons of other times and places. We must also ask what Goethe, consciously and, ultimately, artistically, made of it: what unique meaning he found in it to weave into the structures of meanings that are his life and his art.

The first level of significance was offered by Charlotte herself: the relationship was to be an educative one, in which she was going, if Goethe was willing, to teach him to modify his 'behaviour' in order to 'survive in the world'. Her force of approval and disapproval would polish his conduct and purify his language, and already in 1778 he is boasting that at court in Berlin not a word passed his lips that could not be printed, while by 1780 we find him speaking admiringly, even fulsomely, to her of those who have 'Welt', 'mondanité', the capacity of the true aristocrat to live his life in any context in a style of fluent informality. After 1781 he is consciously striving to improve his sociability and make himself agreeable company, particularly on his diplomatic excursions, and reports back dutifully on his progress. Frau von Stein, in short, descended from courtiers, and born and brought up among them, is the embodiment of court values, the guide to what will make him acceptable to her friend, the sad and empty Duchess Luise, perhaps even a substitute for her if there is any truth in the rumour that the first unavailable woman in Weimar to be the object of Goethe's passion was the wife of his new companion, the Duke. Frau von Stein is in person, then, what it was that Goethe came to Weimar to find: the way of life of those near the source of political power, the antithesis of his life in Frankfurt, the haven from Storm and Stress. By that token she is also the antithesis of that public world which existed principally in the medium of the printed book, and with which Goethe had established so remarkable a symbiosis in the years when he had written *Götz* and *Werther*. 'I can assure you', he wrote to her in 1784, as if the knowledge would gratify her, 'that apart from you, the Herders and Knebel I now have no public at all.' 'The world' in the public, bourgeois, sense is the very opposite of the personalized aristocratic 'world' to which Goethe felt drawn in 1780, and it is this Frankfurt 'world' that is shown the door in a poem which, more than almost any other of this period, is associated with Charlotte von Stein. In January 1778, the daughter of a titled Weimar family, Christel von Lassberg, drowned herself in the Ilm, it would seem out of unrequited love. A copy of *Werther* was said to have been found in her pocket. Goethe was deeply shaken by the suicide and, though his plans for a grotto to the young woman's memory were not realized (he himself dug frantically into the night to begin the excavations), he gave her a more lasting memorial in the first version of 'To the Moon' ('An den Mond'), written probably within a few months of the incident. Goethe compares the gentle transformation in the landscape of the Ilm valley through the effect of moonlight to the change in his own life brought about by Frau von Stein. His turbulent heart is associated with the river, which also has its moods: in winter —we think of the dark and flooded winter days in which *Werther* ends—the river is 'in spate with death', sweeping away poor Christel as the *Werther* era had threatened to sweep away Goethe himself. The last lines of the poem concentrate on the counter-example to the unfortunate suicide, the spring that comes after winter and the requited love that can turn its back on 'the world' which brings these disasters:

> Selig, wer sich vor der Welt
> Ohne Hass verschließt,
> Einen Mann am Busen hält
> Und mit dem genießt,
>
> Was den Menschen unbewußt
> Oder wohl veracht'
> Durch das Labyrinth der Brust
> Wandelt in der Nacht.

Happy the one who shuts herself off unhating from the world, clasps a man to her bosom and with him enjoys that which, unknown or actually spurned by humankind, through the labyrinth of the breast, paces in the night.

It will be seen that the Frau von Stein of this poem is also the antithesis of the form Goethe's subjectivity had taken in the years from 1771 to 1775. Not that this makes her into an apostle of objectivity—a common and profound error— quite the contrary. Goethe, in *Prometheus, Urfaust, Werther,* and poems such as 'On the Lake', had founded a way of experience and writing in which the thing experienced was always interfused with the emotions of the experiencing subject, and so was never merely objective but always in some way symbolic. Frau von Stein was made uneasy by this cast of Goethe's mind—which was actually the nearest he could come to objectivity without surrrendering his own unique character—because it seemed to involve him in talking all the time about himself. Her reaction was not to encourage Goethe to take an interest in things for their own sake—it was not for the threat to *their* identity that she was repelled by Goethe's subjectivity, but for the threat to her own—rather, she encouraged him to keep his self-expression to himself, to let the world go its own way, unhated, and to enjoy, not the world, but the secrets of a moonlight identity hidden in the labyrinth of the breast. Not objectivity, then, but dualism: the clear distinction between a private, feeling, self and an indifferent material world, a distinction characteristic of the 'official' culture which members of the middle class were expected to adopt at court. Goethe had always known, and to some extent shared in, the culture of the hidden sanctum of the self—that had been an important source of the universality of his appeal in Germany—but now he was being asked to extend the sanctum to cover all the operations by which he had put his feelings into reaction with the world. In 1778, perhaps only a few weeks after the first version of 'To the Moon', he wrote to Charlotte:

My soul used to be like a city with insignificant walls, that had in its rear a citadel on the hill. I guarded the castle and left the town defenceless in peace or war—now I am beginning to fortify it too, even if only against the lighter soldiery . . . alas the iron hoops encircling my heart are being bound more tightly every day so that in the end nothing at all will be able to seep through . . .

'Apart from the sun, the moon and the eternal stars', he told Lavater the following year, 'I nowadays let no one witness what causes me joy or anxiety.' While his own eyes saw that 'my soul is like an everlasting firework display, never pausing for rest', his visitors found his appearance 'too ministerial and cold'. We should not, however, think of this withdrawal into the interior as something for which Frau von Stein should be blamed, as if it were her invention and responsibility: she could welcome it, for her own reasons, judge it fitting to Goethe's station, and further it by her approval, but essentially it was a precondition of life in the absolutist court milieu, and if Goethe imposed it upon himself it was because he chose to continue to live in Weimar and to continue to 'love' Charlotte von Stein, as he called it.

'I fully understand that to save the *dehors* one is supposed to ruin the *dedans*, but I still cannot bring myself to agree to it.' Goethe's own interpretation of the relationship with Charlotte was a defence that turned her weapons against themselves. He gave up more and more for her sake, living at the last in almost complete isolation; he affirmed her identity by making her the only person that mattered to him: she had 'gradually entered on the inheritance of my mother, sister, lovers'; but she mattered to him only as the *sine qua non* of his own identity. It both was and was not what she wanted to hear: 'how little I can subsist on my own and how necessary your existence still is to me in order that mine should become a whole.' Yet it was, at one level, only justice that she who had deprived him—or in whose person the Weimar court had deprived him— of his class, his public, and his intimacy with the world, should 'be my substitute for everything'. The one thing Goethe's mind could not do was conduct a monologue—his diaries are voluminous and eventually informative, but without any literary interest—and as the friends, the public, and even the natural world with which he had been used to converse fell away and ceased to be something he could address with intimacy, if at all, so Charlotte von Stein entered into their inheritance also. 'I had intended to write a little journal,' he wrote to her at the outset of his expedition to Switzerland in 1779, 'but it would not work because it offered me no immediate purpose; in future I will simply record for you day by day what happens to us.' In this relationship, that blanching of the fibres of his being, which he attributed to his situation in Weimar, is carried through to the point where the structure of his soul is laid bare with almost abstract clarity. Immediately before the great change of early 1781 he closes a letter with the words,

Adieu, sweet conversation of my inmost heart. I hear and see nothing good that I do not immediately share with you. And all my observations about the world and myself are directed not, like those of Marcus Aurelius, at my own but at my second self. Through this dialogue, since with everything I am thinking what you might say about it, everything becomes clearer and more valuable to me . . .

Frau von Stein was herself to become the 'medium of love' through which he was to 'see the world as it is': 'you have been transubstantiated for me into all objects; I see everything well and yet everywhere see you'. And the transfiguring glow of this—still—colossal confidence was such that Charlottte almost believed him: 'everything that has once passed through his imagination becomes extremely interesting. That has happened for me with those nasty bones and the dreary world of rocks.'

It was not true, of course. Goethe's mind was a dialogue, certainly: to exist it needed reaction. That was why he came to Weimar, that was why he pursued in Frau von Stein the antithesis of so much that he had been, and indeed still was. He used her very emptiness as the instrument for him to live to the full the life of the principal alternative possibility that contemporary Germany offered to a man of his gifts and background, the life of an official at an absolutist court. He used her to compress into interiority, into the labyrinth of his breast, the entire greedy tentacular sensibility that had 'created worlds' of the extensiveness of *Götz* and the intensity of 'On the Lake', and in his breast with her help, he had preserved it—a ghostly prolongation, as moonlight is of sunlight, but a prolongation none the less. In her, and in her image, that extraordinary power of feeling survived. Without her, Weimar would have been impossible for him after, perhaps, the end of 1777 (and there was, it will be remembered, nothing better for him to do). But that is not love, though Goethe called it that, interminably, and it is not marriage, though Goethe called it that too. As he wisely remarked to Lavater in 1782, 'what a man notices and feels about himself seems to me the least part of his existence', ' the reader has to carry out a special operation in psychological arithmetic in order to draw the true sum out of such data'. What Goethe really thought—thought without knowing it—about love and marriage at this time he had revealed two years previously when Lavater had suggested he might have a weakness for the Marchioness Maria Branconi, who, unusually for anyone, but especially for a former mistress of the Duke of Brunswick, was clever, beautiful, and good. Goethe replied that his daily duty of erecting, on a base that was given and prescribed to him by circumstance, 'the pyramid of my existence . . . as high as possible into the air' did not permit him the 'momentary inattention' of 'a transient desire'. And as for marriage: 'God preserve us from a serious bond, with which she would wind my soul out of my members.' Johanna Fahlmer wrote that 'Goethe can be kind and good, even great; but in love he is not *pure*, and really not *great enough* for that', a remark which another correspondent glossed as a euphemism: 'She did not want to say: in everything else he loves only himself.' The loss of soul which Goethe feared from marriage was the one thing he did not have to fear from the relationship with Charlotte von Stein; the relationship was constructed to maintain his soul, albeit in circumstances so hostile that they may constitute some excuse for this unwitting exploitation of a wounded woman. 'Goethe always read too much into women', was the rough and ready judgement of the 70-year-old Carl August, who in his own way knew him better than anyone, 'he loved his own ideas in them and never really felt grand passion'.

That Goethe, in these difficult years, was not simply a monster of self-satisfaction, a Tartuffe, as Jacobi saw him—incensed by a cruel and insulting joke of 1779, in which Merck was probably the prime mover—is shown by what he wrote, even more, by what he did not write. That poetic activity which, at least since *Partners in Guilt,* had been the true vehicle for the endless conversation of his soul, could not be brought to share in the self-deceptions of the period. His mind had lost none of its earlier pyrotechnic productivity, he noted while on a long journey with Carl August, riding and walking through the remoter parts of the Thuringian Forest in the early autumn of 1780: 'the greatest gift, for which I thank the gods, is that through the swiftness and variety of my thoughts I can divide up a fine day like this into millions of parts and make out of them a little eternity'. But—the complaint is reiterated—he cannot give permanence to these myriad shards of experience and reflection by writing them down. It is not just that it is physically inconvenient to do so, though it is partly that. 'If only the pen of some good spirit could take down to my dictation all that I say and tell you throughout the day' he writes to Frau von Stein, but when he tries to list these fragments of eternity, formulated in what he thinks of as inner discourse with her, they prove to be only prosaic maxims, from which

the lustre of their living context has faded and for which no new poetic context
has been provided:

> One should do what one can to save individuals from ruin.
> But that is to do little; from misery to prosperity there are countless steps.
> The good one can do in the world is a minimum, etc.

Sometimes, quite unpredictably, throughout these years, during which he was
incapable of forming a sustained poetic structure on the scale of *Götz*, *Urfaust*,
or *Werther*, a shard would fly off whose brilliance remained, a context to itself, a
little eternity. On 6 September 1780 at the start of that same tour in the
Thuringian Forest, on a calm evening with a majestic sunset and thin columns
of smoke rising from the charcoal-burners' fires, Goethe, in total solitude,
wrote to Frau von Stein from a wooden refuge on the highest of the peaks
round Ilmenau, 'if only my thoughts of today were written down complete:
there are some good things among them'. This time he was right: one of his
thoughts of the day was written down, by Goethe himself, a little before 8
o'clock on the wall of the hut, and what he wrote has become in the German-
speaking world—for it is untranslatable—the best-known of all poems, by
anyone:

> Über allen Gipfeln
> Ist Ruh,
> In allen Wipfeln
> Spürest du
> Kaum einen Hauch;
> Die Vögelein* schweigen im Walde.
> Warte nur, balde
> Ruhest du auch

Over all the peaks lies rest, in all the tree-tops you can sense scarcely a breath; the little
birds are silent in the forest. Only wait, in a moment you will rest too.

There is little sensible that literary criticism can say about something so delicate
and so matchless. But for the biographer one feature of the poem is very
revealing: its use of the word 'you'. The word refers either to Goethe or to the
reader or to both—that is in the nature of the dialogue in Goethe's soul—but
the 'you' is not—not specifically, or by allusion—the woman to whom Goethe
was writing only minutes before, and after, he composed the poem. Though for
years he could not bring himself to acknowledge it, the sources of his poetry ran
deeper, and purer, than 'the sweet conversation of my inmost heart' that was his
mental discourse with Charlotte von Stein.

Literary Difficulties of a Statesman

When Goethe in 1829 remarked that his first ten years in Weimar had brought
'no poetical production of significance' he was doing his earlier self less than

---

* Goethe originally wrote simply 'Vögel' ('birds').

justice. In 1780 he noted that his unpublished writings already ran to practically as much again as what had been published. But apart from several pieces written to entertain the court, which even then he acknowledged were 'of a conventional stamp', practically nothing of this, when it did eventually see the light, appeared in the form in which it was first written, and much was fragmentary. The most nearly complete of the less conventional works was the classicizing drama *Iphigenia in Tauris,* modelled on Euripides but written in rhythmical prose: although it was produced in this form four times in Weimar, Goethe remained uncertain whether to recast it in verse, and, if so, how. His play on the sixteenth-century hero of the Netherlands Revolt, *Egmont,* he had brought with him in an advanced state from Frankfurt, but it made only occasional progress over the next ten years. *Torquato Tasso,* at this stage also in prose, was conceived in 1780, but Goethe got no further than the second act and let it lie in 1781. To *Faust* little or nothing was added in the entire course of this early Weimar period. In 1777–78 Goethe finished the first book of a novel, *Wilhelm Meister's Theatrical Mission (Wilhelm Meisters Theatralische Sendung),* which may have been begun as early as 1773, and, though the scheme at first seemed to run into the sand, he took it up again in 1782 and disciplined himself to write five more books in the space of four years. But even so the end seemed a long way off in 1785, and not all who read it approved of its low theme and characters. His *Letters from Switzerland* of 1780 were not published complete— if they are complete—until 1808. Other projects were destined to remain fragments for ever, such as the drama *Elpenor* and the epic poem *The Mysteries (Die Geheimnisse),* and yet others, though planned, seem never to have been begun, such as a biography of the seventeenth-century Duke Bernhard of Weimar, and a didactic scientific work provisionally known as the *The Romance of the Universe.* Indeed, from 1783, and, with the exception of the continuing work on *Wilhelm Meister,* possibly from rather earlier, a period of real sterility appears to have set in, which to some extent justifies the severity of the later judgement. 'My son seems to have quarrelled a bit with the Muses', his mother wrote in 1783, and he himself acknowledged in 1784 that he was 'in an unpoetical condition', while hoping that it would none the less prove possible to 'rouse the sleeping genius'.

Most of what Goethe wrote at this time he read aloud to his friends in Weimar and Gotha, and some pieces—*Iphigenia* and *Wilhelm Meister*— circulated more widely in manuscript. But for a variety of reasons, of which their fragmentariness was only one, he felt unable to print these works as they stood for the benefit of the general public. The most striking feature of Goethe's literary life from 1776 to 1786 is that for ten years he completely turned his back on the reading public and the world of books and journals with which he had been so closely involved, and with such spectacular results, for the four years from 1772. The point is not, as has often been suggested, first of all by Goethe himself, that administrative work proved incompatible with poetry, that 'my writing is being subordinated to life', or, as Wieland prophesied on his appointment as Privy Councillor: 'He will do much good, prevent much evil, and that will have to console us, if anything can, for his being lost to the world as a poet for many years to come'. Goethe's official duties did not prevent him from writing a great deal, some of which—whether or not he was right—he thought as good as anything he had done before, and his least productive time came towards the end of his first Weimar decade, when he might have been thought to be more accustomed to his official role and when its burden was actually being reduced. The point is rather that for those ten years Goethe actively sought a different kind of audience and different intellectual company from that which he had had as a public figure in Frankfurt, and that the effect of this ultimately fruitless search for an audience other than the reading public was so damaging that his artistic conscience could not be finally satisfied with anything that he wrote in the course of it. In the end he had to return to publishing his work as he had done before, and it all had to be rewritten.

As early as the mid–1760s, when he was on the verge of the momentous discovery that he could make poetry out of the material of his own life, Goethe had felt that there was something problematic about the transition from private experience to printed and public book, and he had willingly acquiesced in Behrisch's view that his first frivolous love poems deserved to be enshrined in calligraphy rather than in print. In the early seventies those inhibitions had been overcome, and a literary explosion had been the result, but in Weimar the marriage of private and public broke down. His revulsion from the process of writing itself became so acute that in these years he established his lifelong practice of dictating, albeit after lengthy mental preparation: printing would have been yet another alienating step away from the moment of personal utterance in which his works had their origin. 'It has always been a disagreeable feeling for me', he wrote in 1786 when contemplating a first collected edition of them, 'when things that occupied an individual sensibility in specific circumstances are to be delivered up to the public.' But the arrangements for that first authorized edition already mark the end of the early Weimar experiment in divorcing private and public, and are an acknowledgement that the marriage has to be patched up once more. The trying episode in Goethe's literary development that had intervened needs to be considered in more detail.

Although the edition of his works for which Goethe signed a contract in 1786 with the Leipzig publisher Göschen was the first to appear with his permission and co-operation, it had no fewer than ten unauthorized predecessors—a measure of his early success. In 1779 the Berlin publisher Himburg added a fourth volume to the three volumes of *Goethe's Literary Works* that he had pirated in 1775. Goethe gave vent to his disgust in a savage epigram, the purport of which however was not that Himburg was a thief, but that he was a fraud, pretending that the bits and pieces that had been gathered together were current coin, while Goethe had in fact started a new life, of which the likes of Himburg could not be expected to have a notion:

Für die Himburgs bin ich tot

For the Himburgs of this world I am dead indeed.

Goethe had had enough of the Himburg world, competitive, commercial, and intensely public. He had written, in *Werther*, the tragedy of a man ruled by a fashion of the public mind and he had in writing it created just such another fashion and perhaps contributed to other tragedies too. As Hanswurst he had protested at the cost to himself of the public cult of authorial personality. Now he intended to live a life as far removed as possible from the cultural market-place, a life for himself as an individual and for others as individuals, a life at court and in a country town where the only man of Himburg's stamp, F. J. Bertuch, was as yet only a business assistant for the *Teutscher Merkur*. The association of *Werther* with the world of commercial publishing, and the turning away from both, are explicit in several of the entertainments Goethe wrote for the Weimar court, notably *The Triumph of Sensibility* (*Der Triumph der Empfindsamkeit*) (1778), a substantial comedy in six acts, the adaptation of part of Aristophanes' *Birds* (1780), and the squib *New News from Lumberville* (*Das Neueste aus Plundersweilern*) (1781), all of which are—increasingly bitter— satires on literary movements in which Goethe himself had been involved, and on the entire publishing industry, authors, publishers, and readers alike.

Conversely, a particularly attractive feature of the early Weimar years, Goethe's often secret benevolence to those in distress—in accordance with the principle that 'one should do what one can to save individuals from ruin'—also seems to be at once an assertion of the value of the purely personal in his life (as opposed to the relation with a mass established by writing for the public) and, in several cases at any rate, an expiation for the damage that *Werther* and the associated cult of feeling had caused, most nearly and obviously in the death of Christel von Lassberg. His fellow writers, Klinger, Müller, and Bürger all received financial support from him, or through his efforts. A young philosopher, F. V. L. Plessing (1749–1806), and a disgraced official from elsewhere in Germany, whose real name is unknown but who called himself J. F. Krafft, both wrote in states of depression and nervous debility to the author of *Werther*, and received careful personal attention and advice. Krafft, who was in dire material need, received far more: Goethe found him a home in Ilmenau and supported him out of his own purse with an allowance of one-seventh of his own salary, 200 dollars a year, until he died in 1785, when Goethe also paid for his funeral. The most striking case of Goethe's conscientiously discharging an obligation imposed on him by his involvement in the revolutionary literary movements of the 1770s is that of Peter im Baumgarten, the Swiss peasant-boy adopted by the Count of Lindau, to whom Goethe had promised to act as guardian. In 1776 *taedium vitae* took Lindau into the American War of Independence, and in November he was killed in the assault on Fort Washington. The following year Goethe, although he could not abide tobacco, took the pipe-smoking young lout into his little house in Weimar and looked after him for a year until he could be apprenticed as a ducal game-keeper. At the other end of the social scale, Goethe may have felt that in his tutorial relationship with Carl August he was guiding a soul out of Storm and Stress into maturity and so demonstrating to others that that phase was not a hopeless cul-de-sac.

Goethe's belief, in these years, in the importance of individual acts of small-scale charity is a major theme in an austere hymn of 1783, entitled 'Divinity' ('Das Göttliche'), which begins:

> Edel sei der Mensch,
> Hilfreich und gut!

Let man be noble, helpful and good.

The word for ethical nobility, 'edel', though distinct from the word for membership of the noble class ('adelig'), is not wholly dissociated from it. The poem is explicitly addressed to all men, but its model of humanity is the aristocrat. Goethe, who had been ennobled in 1782, might well feel that the élite and individualist ethos of 'Divinity', as of his whole life since 1776, was socially and intellectually, and not just morally, opposed to all that the squalid Berlin entrepreneur Himburg stood for. But 'Divinity', a daring poetical raid on the articulate, not to say the platitudinous, which still just manages to succeed, is the last verse of quality that Goethe wrote for three years. Its form, for all the pallor of its vocabulary and the discipline of its argument, is a final attenuation of that of the free-verse chants, the would-be Pindaric odes, which began in 1771 with 'Wanderer's Storm-Song'. He might feel contemptuous, even bitter, towards the mass culture of the reading public, which was not just the intellectual but the political enemy of the princely Germany he had joined in coming to Weimar, but he could not write for anyone else.

One factor, however, which cannot be separated either from Goethe's inability between 1776 and 1786 to bring any literary work to a satisfactory completion or from his aversion to the public during that time, is the general state of German literary and intellectual life in this period, which could give his reactive mind neither support nor stimulus. As the 'geniuses' of the 1770s dispersed—Lenz back to Livonia, Müller to Italy, Klinger and Bürger and Voss into official careers—and fell silent—Lenz in madness, Hölty (1776) and Hamann (1788) in death—so there came over German literature, before the storm of 1789, something of the sultry stillness in which many flies breed. Literary culture was growing, poetical almanacs, Sentimental and Gothick novels and dramas were thicker on the ground, actor-managers, such as A. W. Iffland (1759–1814) and F. L. Schröder (1744–1816), were consolidating the position of the German theatre as a serious and financially viable institution; but at the higher level of original production the sense of purpose, of some central energy or conflict, was lost—as much an effect, perhaps, as a cause of Goethe's transfer to Weimar. Frederick the Great's treatise of 1780, and the replies to it, showed that the terms of the theoretical discussion had slipped back twenty years, and, after the death of Lessing in 1781, the manifestly failing Klopstock and the indefatigably unimaginative Nicolai were the fixed stars in the literary firmament, as they had been for decades. Goethe was delighted with Wieland's comic romance *Oberon* (1780, published 1781) but that too was an echo of an earlier epoch. Storm and Stress reached a belated culmination, and also its most stageable form, in Schiller's first dramas, from 1781 to 1784, which, like the autobiographical novel *Anton Reiser* (1785) by C. P. Moritz (1756–93), announced a new generation of talent, for whom Goethe was not so much a contemporary as a pre-existing established presence and influence: for that very reason they were bound at first to seem to him like mere reproductions of episodes in his own life that he had been glad to leave behind.

Only in philosophy were events of the first importance occurring—though there is a sense in which Kant's *Critique of Pure Reason* (*Kritik der reinen Vernunft*) (1781) also derives from the problems of the 1760s, rather than of the 1770s when it was being written—but there were very few who could grasp the implications and importance of the new system until about 1786, when the *Letters on the Philosophy of Kant* (*Briefe über die Kantische Philosophie*) by Wieland's son-in-law C. L. Reinhold (1758–1823) began to appear in the *Teutscher Merkur*, and when Kant himself had written some shorter and more accessible summaries of his arguments. Almost as important for the next two decades of German philosophy were two other developments in these years. In 1784 Herder began to publish his attempt at a philosophical account of human cultural history and of humanity's place in the natural world, his *Ideas for a Philosophy of the History of Humanity* (*Ideen zu einer Philosophie der Geschichte der Menschheit*). And in 1785 F. H. Jacobi detonated a little time-bomb left in his keeping by Lessing before he died: Lessing had remarked to Jacobi that his true religion was Spinozism, which was tantamount to a declaration that he was an atheist, and Jacobi published the remark in the context of an exposition of Spinoza's thought, which he was hoping to refute. But for Goethe the *Ideas* were not a publishing event but a work that he and Herder had discussed together at home in Weimar, and to Jacobi—of whom his criticism was muted by his reluctance to jeopardize their recent reconciliation—Goethe professed himself completely ignorant of the public reactions to his book (which, as it happened, were violent). The thinkers were still only in their teens who in ten years' time would fuse Kant's critical philosophy with Spinoza's pantheism and Herder's historicism into the heady synthesis of German Idealism, of which the most fateful bequest to the nineteenth century was to be Hegel's *Lectures on the Philosophy of History*. It was not until immediately before his Italian journey that Goethe, or anyone else in Germany, could have been expected to sense the coming philosophical revolution, and the same might be said of the musical revolution too. Only in 1785, with the first Weimar performance of his first

German opera, *The Abduction from the Seraglio (Die Entführung aus dem Serail)*, did the genius of Mozart make a direct impact on Goethe's life, and one which —for contingent reasons—was disagreeable.

Even had Goethe wished to play a part on the German literary scene in these years, the company he would have had to keep would have been enough to dissuade him. In January 1786, when his attitude had already begun to change and he had started an operetta, never finished, to be called *The Mismatched Household (Die ungleichen Hausgenossen)*, he found that reading the *Gotha Theatrical Calendar,* which listed all the current theatrical companies in Germany and the plays in their repertoire, 'brought me almost to despair'.

In all the most desolating coldness and honesty a balance is drawn up here from which one can clearly see that nowhere, and especially not in the branch that interests me presently, nowhere is there anything worthwhile nor can there be anything worthwhile. I feel sorry for the poor little operetta I have begun, as one may feel sorry for the child of a negro woman that is to be born in slavery.

But had Weimar in the previous ten years been a more rewarding field to cultivate? Had the court and its society and the friends he could reach from it provided a resonance and response for his literary genius at all comparable to the public reactions that had made his apartments in Frankfurt into something like the audience-room of a minister? It is tempting, for example, to think that the later function of the Weimar Court Theatre, as a deliberately chosen instrument of cultural enlightenment, was prefigured in the amateur theatrical productions of 1776 to 1782, put on before an audience of a few hundred from court and town in Hauptmann's Assembly Rooms, and before an exclusively courtly audience in Ettersburg and Tiefurt during the summer. But, although after October 1776 Goethe had nominal responsibility for them, he was not the driving force, nor even a regular participant, in these entertainments—that role was divided between Bertuch and two young noblemen with both musical and literary gifts, Friedrich von Einsiedel (1750–1828), gentleman-in-waiting to Anna Amalia, and Siegmund von Seckendorff (1744–85), gentleman-in-waiting to the Duke. Apart from occasional appearances as an actor, Goethe was mainly involved as the writer and producer of hastily composed ballets and playlets for special occasions, such as the Duchess's birthday. His more substantial works were really only three in number, and the appeal of *The Triumph of Sensibility* and *The Birds*—both, it will be remembered, attacks on the business of literature—lay mainly in their elaborate stage effects and costumes. The same was even true to some extent of *Iphigenia in Tauris*—the novelty of the expensive Greek costumes particularly struck the first audience —which was the only work whose reception by the Weimar public Goethe seems to have regarded as a serious matter. 'Iph. performed, really good effect of it, especially on the pure', he noted in his diary on 6 April 1779. But *Iphigenia* remained without sequel, and possibly the diary note indicates that Goethe was really concerned only with the play's effect on Frau von Stein. It is perhaps understandable that, apart from the operetta *Erwin and Elmira* (with music by Anna Amalia), and the satirical pantomime *Lumberville Fair,* Goethe did not arrange for the production of a single dramatic work of his own, published or not, from the period 1771 to 1775, although both his early unpublished comedies in alexandrines, *The Lover's Spleen* and *Partners in Guilt,* were performed. *Clavigo, Stella,* and even *Claudina of Villa Bella* would have been well within the technical capabilities of the Weimar amateurs, but Goethe presumably wanted to forget that entire phase in his writing, and found it out of keeping with his court existence. But what makes it impossible to believe that Goethe regarded the Weimar amateur theatre either as a serious artistic and cultural institution or as a vehicle for his own deepest concerns, is that he did not make use of his position to put on any play of Shakespeare's, nor for that matter any translation of the French tragedians, nor even Lessings's last play, *Nathan the Wise* of 1779, before which, as 'the supreme masterpiece of human art', he was reputed in 1780 'to have prostrated himself'. Goethe had a busy theatrical year in 1778 and early 1779, until the production of *Iphigenia,* and he certainly enjoyed the smell of the grease-paint and the glare of the (oil) lights. But after 1779 his energies went into more direct service of the Weimar state as an administrator, and his involvement in theatrical affairs was minimal—nor did the opening of the new theatre in 1780, or the hiring of a permanent troupe of actors from January 1784, reawaken his interest, except in operetta, or stimulate him to complete any of his numerous dramatic fragments.

Nathan the Wise, beneath an appearance of pleading for religious tolerance, depicts a secret, universal religion shared by a world-wide fraternity, a freemasonry, so to speak, of rational men, regardless of their overt cultural and political loyalties, and among Lessing's last published works were some dialogues on the true nature of Freemasonry itself. Throughout Europe the last two decades before the French Revolution saw a marked growth in the activities of secret societies devoted to ideals of equality and fraternity, perhaps in response to the restrictions in absolutist states on any general and public participation in political discussion and decision-making, and in Germany the Masonic movement proved particularly attractive to literary intellectuals. Wieland and Herder were both members of the Weimar lodge, named 'Amalia', of which the Prime Minister, von Fritsch, was Grand Master, and in 1780 Goethe also applied for membership. Did he expect to find in Masonry a secret company of noble individuals, a broader world than that of the Weimar court, but an alternative to the public mind from which he had turned away? If so, he was soon disappointed. He passed rapidly through the usual grades to become a Master Mason in 1782 and joined the order within the order, the Illuminists, in February 1783, but within a few weeks he was writing:

They say you can best get to know a man when he is at play . . . and I too have found that in the little world of the brethren all is as it is in the great one . . . I was already saying this in the forecourt, and now I have reached the ark of the covenant I have nothing to add. To the wise all things are wise, to the fool foolish.

These were difficult times for German Masonry, as yet not fifty years old. In 1782 a General Assembly in Wilhelmsbad had failed to resolve the complicated internal quarrels of the movement, which was, in essence, divided between those, notably the Rosicrucians, to whom the mythology and ritual of the movement, and its claim to occult knowledge, were more important, and those, notably the Illuminati, who were more concerned with its ethical universalism and social egalitarianism. Goethe, as usual, was steering a middle course, seeing merits in both currents. But in 1784 the Bavarian government discovered, or believed it had discovered, that at the heart of the Illuminist movement lay a radical republican conspiracy, decreed the death penalty for recruiting to it, and warned other German rulers accordingly. In the campaign that followed, and that effectively obliterated German Masonry for twenty years, the Weimar lodge, which Carl August himself had joined in 1782, was one of the first to close. Goethe felt that his knowledge of human behaviour had been extended by his involvement with Freemasonry, and his experiences had an important effect on his later assessment of the causes of the French Revolution, but their immediate contribution to his literary achievement was slight—some forty-eight stanzas of a projected Rosicrucian epic, *The Mysteries*. Even these were written, not for the brethren, but for his minimal public of the years 1784–6: Charlotte von Stein, the Herders, and Knebel. Unlike Wieland, or Countess Bernstorff's secretary, Bode, or the actor-producer Schröder, Goethe does not at any time after his admission seem to have thought the lodge a uniquely important medium through which to communicate with or assist his fellow men. On the contrary, his Masonic comedy of 1790, *The Grand Kophta* (*Der Gross-Cophta*), suggests that his original expectations may have been considerably higher, and his disappointment correspondingly more intense, than his cool and tactful assessment of 1783, written of course to a fellow Mason, might seem to imply.

In that play a young knight expresses the bitterest disillusion when what he has taken to be a brotherhood dedicated to missionary altruism proves, or seems to prove, to be a cynical deception. What is to become now, he asks, of the unattached idealism of the disappointed friend of humanity? 'Fortunate he, if it is still possible for him to find a wife or a friend, on whom he can bestow individually what was intended for the whole human race.' These words, almost a prose paraphrase of the last verses of 'To the Moon', may be Goethe's true epitaph on his brief venture into Freemasonry: certainly they tell us something of the nature, as well as the strength, of the emotions with which he turned to Frau von Stein.

The social facts were ineluctable: not many people lived in Weimar, not many people came there, and not much happened there, apart from the administration of the duchy. The Mason Bode even complained, in 1784, that there was a shortage of dinner parties. The theatre provided a transient escape into aesthetic illusion, the lodge into ethical fantasy, but Goethe was realist enough to know that inches behind the backcloth stood a blank wall. There were three men of genius in Weimar from 1776—Goethe, Herder, and Wieland—but there was no body of learned, or even systematically educated, men and women, no thinking and writing and artistically active milieu, such as a great city can provide, to support, stimulate, and give variety to their efforts. Germany, taken as a whole, could provide that milieu, but there were only two points of access to it, and from them both Goethe in his early Weimar years turned away: print and the university. When he resolved to venture into publication once more, and discovered the potential of the University of Jena, he began to emerge from the trance into which those years had cast him. Until then, the charm and the constriction of the world he had chosen to inhabit are perfectly symbolized in the *Tiefurt Journal,* the one periodical to which he regularly contributed throughout this time. It was founded by Anna Amalia, when she moved her summer residence to Tiefurt in 1781, and ran for three years, appearing every three weeks or so. Goethe furnished several poems, including 'Divinity', and among the other contributors were gentlemen and ladies-in-waiting from Weimar, Prince August of Gotha, and Statthalter Dalberg from Erfurt. For the *Tiefurt Journal* was a court game, entertaining the Dowager Duchess, as the masques and comedies for 30 January entertained Duchess Luise, and its disadvantage as a medium of communication with a public was that it was handwritten and had a circulation of eleven copies: Behrisch *redivivus,* in fact, as if the years when Goethe had written for the *Frankfurter Gelehrte Anzeigen* had never been. From the *Tiefurt Journal,* once it ceased appearing in 1784, it was but a single step to Goethe's minimal public of four, or even of one. 'You have isolated me in the world. I have absolutely nothing to say to anyone, I speak in order not to be silent, that is all.'

In the circumstances it is remarkable that Goethe wrote as much as he did, unsupported either by the immediate Weimar environment or by the wider world of German literary culture. Everything that he did write bore the marks of those iron hoops being bound ever more tightly about his heart, and of the brutal division, into a 'dehors' and a 'dedans', of a mind that was properly destined to live in subtle and manifold interaction with its surroundings. As yet the fruits of the decision to come to Weimar and stay there, in order to remain associated with the dominant forces in the structure of the nation, were far from ripe, indeed scarcely visible. The years from 27 to 37, which for many might be among their most productive, were for Goethe a period of beginnings. Apart from the quite unpredictable moments of lyrical inspiration, the only works he actually completed were for immediate court consumption, and we have seen that the more substantial these are—*The Triumph of Sensibility, The Birds,* even, to some extent, the poem 'On the Death of Mieding', 'published' in the *Tiefurt Journal*—the more they avoid, or even mock, the public expression of private thought and feeling. The other works, nearly all more or less fragmentary, in which such expression is attempted, were kept by Goethe to a very restricted circle of friends—even of *Iphigenia* he was reluctant to circulate manuscripts until he was actively seeking help in the work of revision. They fall into two groups, a division which itself reflects the division in Goethe's nature. On the one hand there is a group of dramas—*Brother and Sister (Die Geschwister),* a one-act play written in 1776, *Iphigenia, Elpenor, Tasso,* and, though it is a complicated case, *Egmont.* In these—paradoxically, it might seem, given their dramatic form —the interiority of 'official' culture is to the fore. They are all—eventually, if not always in their original conception—concerned with the inner stability of a single central character, with the internal coherence of that character's moral stance, and to that concern the depiction of a material environment, or even of the inner life of other characters, is (with some reservations in the case of *Brother and Sister* and *Egmont*) largely an irrelevance. On the other hand there are Goethe's narrative writings, the *Letters from Switzerland* and *Wilhelm Meister's Theatrical Mission.* These show that Goethe can still evoke a whole material or social world, as he did in *Götz, Urfaust,* or *Werther,* but they do also show that he is having increasing difficulty in suffusing that world with personal meaning, so that in them the symbolic realism of his best poems is present only by fits and starts. It is as if in the prose Prometheus' statues come only fitfully to life, while in the plays the exclusive subject is Prometheus himself: the delicate balance of 'myself not myself', by which Minerva bestowed life on the products of Prometheus' post-religious art, has been unsettled, now in the one direction, now in the other. Frau von Stein was not Minerva, even though Goethe probably thought she was.

Goethe's works of his early Weimar years are often said to be characterized by 'objectivity', when what is really meant is that the relation between self and world, precariously balanced in the previous five years, has been further destabilized. 'Objectivity', in fact, is not the right word. Firstly, it could be said that it is the cultivation of subjectivity, of an interest in the inner man, rather than of objectivity, an interest in the outer world, that distinguishes the literature of the court-oriented official class in Germany. To call *Iphigenia,* 'objective', by comparison, for example, with the main, or chronicle, plot in *Götz,* would be to stand meanings on their heads. Secondly, there is an important sense of 'objective' in which it means 'as things appear to people other than myself'. When a heroine of Jane Austen's prefers sense to sensibility she is correcting her own notions by means of an agreed opinion in her society —not necessarily perfectly represented by any single authoritative figure, but with very varying degrees of imperfection probably represented at that moment by everyone in the book other than herself—about what really matters in human life as a member of her society has to lead it. That kind of social objectivity is also absent from Goethe's works of the early Weimar period. It was not to be found in Weimar society because, even among the nobility, social relations were too transparently dependent on relations to the structure of political power to have much independent substance. There was simply too much that Goethe could do because he was the favourite of the Duke (which in turn, of course, reflects the fact that there was too much that the Duke could do because he was the absolute ruler), and too much therefore that he had, out of his own resources, to teach himself not to do. There was no agreed and common 'sense' that could act as a counterweight to that intensive observation and stimulation of his own sensibility which formed such a large part of his inner dialogue with Frau von Stein, no friendly but unyielding pressure from a social order in and behind the differing and differently valuable views of other people, which could have guided him from the start, protected him from solitude, and furnished him with the material for his books, as it furnished Jane Austen with hers. Goethe had to find his counterweight elsewhere.

In 1782, at the very busiest stage of his official career, Goethe erected, in the park that was now growing up, partly under his direction, on either side of the Ilm, several stone tablets inscribed with brief poems in classical distichs. One of them, on a marble slab now near a spring emerging from the low wooded scarp that edges the valley, appears in printed editions under the title 'Solitude' ('Einsamkeit'). It calls on the 'healing nymphs' that dwell in the rocks and the trees to answer the silent prayers of those in need, and bestow comfort, guidance and fulfilment,

> Denn euch gaben die Götter, was sie den Menschen versagten:
> Jeglichem, der euch vertraut, hülfreich und tröstlich zu sein.

For to you the gods gave what they denied to men: to be a help and a consolation to everyone who confides in you.

The inscription is a formal, even a ceremonial, acknowledgement that Goethe is not looking for a response to the unspoken wishes of his heart from the society of men, and that the objective power that in their stead will instruct and console him—and indeed furnish him with the material of many books—will be the power of Nature.

Around 1780, partly as a result of the geological studies necessitated by his work for the Mines Commission, Goethe began to acquire an interest in matters of natural science which was to prove lifelong. With the one important exception of optics, his involvement with all the sciences that were most to concern him—geology, anatomy, and botany—has its roots in these early Weimar years. Because these interests so graphically exemplify the many-sidedness of his mind, because at so many points they reflect principles important in understanding his poetry, and because Goethe was a suggestive influence on nineteenth-century biological science, particularly in Germany, it is necessary to be clear at the outset that—again with the contentious exception of his optics—his actual achievements in these areas cannot be called substantial. These distractions—for that is what they were—belong unequivocally under the heading of 'false tendencies' ('falsche Tendenzen') to which Goethe himself assigned them in 1797, along with his amateur drawing and painting, with which he from time to time became obsessed as others do with reading cheap novels. But it is equally necessary to understand why Goethe was so powerfully drawn in these wrong directions, and the period in which the deviations originate gives us a clue. The origins of Goethe's science coincide, and are to some extent identical, with his difficulties in bringing his literary works of this period to a satisfactory completion. In the relationship with Weimar court culture, personified for him in Frau von Stein, Goethe found himself encouraged in that separation of inner and outer worlds which is apparent in his writings. While his absorption in the private world constituted by himself and Frau von Stein continually intensified, leaving him, in social terms, more and more alone, he sought company, and relief from that intensity, in an outer world not of society but of science. 'I am living in a solitude and isolation from the world that are finally making me as dumb as a fish', he wrote to Jacobi, telling him at the same time of his preoccupation with botany and the microscope. Although he lacks Jacobi's home and family and his 'long etc', God, Goethe says, has 'blessed me with natural science, that I may be happy in the contemplation of his works, of which he has given me but few to own myself'. 'The silent, pure, ever recurrent, passionless vegetable kingdom often consoles me for the plight of men.' As time goes on, this study becomes more emphatically impersonal, as if to compensate for the increasingly intense and exclusive relation with Frau von Stein. In the years after 1783 Goethe professes himself a disciple of Spinoza in his devotion to the disinterested study of the natural world in which God is revealed: his 'evening prayer', he says, is Spinoza's proposition that 'he who loves God cannot desire that God should love him in return'.

This choice of Spinoza as the patron of a science of nature deliberately drained of emotional reward contrasts profoundly with the purportedly Spinozistic raptures of ten years before, with the experience of a Nature, 'embracing embraced', that came to meet his own approach, in 'Ganymede' and 'On the Lake', and even with the search for consolation with which his turn to science began around 1780. The Spinozan natural science of the last years before Goethe's Italian journey is the objective pole of an extreme, not to say pathological, dualism, of which the subjective pole is the obsession with Frau von Stein. The interplay of heart and nature, that seemed so fluent and dynamic in the poems that first announced it, such as 'May Celebration', has now ceased almost entirely. 'Divinity' effectively retracts those poems when it tells us in 1783 that

unfühlend
Ist die Natur

Nature is without feeling

and the poem's argument is based, not on any assumption of union with Nature, but on man's moral distinctness from her. This is not simply 'objectivity'—it is a betrayal of Goethe's poetic identity forced upon him by the intellectual, social, and emotional conditions of his life over the previous seven years. Not surprisingly, the Spinozan episode in Goethe's scientific activity coincides with the almost complete exhaustion of his poetic vein.

It is evident that by 1786 Goethe's sensibility was facing a crisis quite as acute as that which in 1775 had driven him to begin the journey to Italy. In 1786 he set out again, and this time he did not turn back to Weimar. Weimar, it might well have seemed, had failed him. Yet Goethe's science was a large and complex part of his life and even at this stage it contained an element which could be said to have saved him for Weimar, and which would eventually enable him to resolve many of his own internal dilemmas, and to come as near as he did to success in 'the good cause' of the German national culture. For science, even in the eighteenth century, was already a collective activity. The story of Goethe's involvement in mineralogy, anatomy, botany, is from the start a story of collaboration with others, of pupillage with chosen mentors, of co-operative projects, and of correspondence with the national and international scientific community. It was science that first broke into the 'circle of my existence in which I have cunningly barricaded myself', it was 'stones and plants' that 'connected me to men', and opened him again to a public life. And it was science that first drew his attention to the possibilities lying dormant in the University of Jena.

5

Court Favourite
1775–1780

THE first phase of Goethe's life in Weimar could be said to last until the middle of 1777, and particularly in its first six months it looks very like a continuation of what had gone before: the Swiss journey of 1775 made permanent, but now with a reigning Duke as paymaster and all his territories as the scene of the fun. As on the Swiss journey, Werther costume was worn at first, even by the Duke, who since he had not had a university education, was anxious to taste the pleasures of student life, and he and the young gentlemen-in-waiting and the 'geniuses' who visited during 1776 made up a band like the robbers in *Claudina of Villa Bella*, all on 'Du' terms with one another, and with Goethe, in Lenz's words, as their 'captain'. 'You wouldn't believe how many good lads and good brains are collected here, we keep together, play up to one another, and keep the court at arm's length.' An element in the court was certainly a little displeased when the Duke, Prince Constantin, Goethe, the Stolbergs, and others insisted on playing blind man's buff after dinner, but Anna Amalia—who simply laughed when Goethe rolled on the floor in her presence—was prepared to enter into the spirit of things. Together with Frau von Stein she interrupted the party, and with two five-foot swords borrowed from the ducal armoury they dubbed the Stolbergs knights. Offence of a different kind was caused when Goethe—who refused to go to church on Christmas Eve but borrowed Homer from the pastor to read instead —arranged a charade representing the deadly sins: outraged religion and decency were not mollified though Goethe himself took the part of Pride, adorned with peacock feathers and strutting around on stilts. There was probably a fair degree of less imaginative hooliganism, breaking of glass, defacing of pictures, drunkenness, bellowing, and pulling off of table-cloths, but Goethe seems usually to have played a moderating role and his 'great art was to trample on the conventions but always to look around carefully to see how far he dared to go'. It is not surprising, all the same, that Weimar rapidly acquired an unsavoury reputation in Germany at large: 'Things are frightful there. The Duke runs around the villages with Goethe like a berserk student; he gets drunk and they fraternally enjoy the same girls together.' Goethe was reported to be undermining the Duke's health (it is true that Carl August was crippled with rheumatism after a winter ride to Erfurt, but Goethe had advised against it). Or he was said to have been killed in a duel by a member of the outraged nobility, or to have broken his neck in a riding accident.

Most of the coarser merry-making took place out of doors and away from Weimar itself, and for much of the winter and summer of 1776 Goethe was outside on sledges or horseback, hunting or shooting with Carl August or riding off at a moment's notice, as he was to do many times again over the next few years, to organize the fire-fighting in one of the desperately vulnerable, wood and straw Thuringian villages. Some striking pencil sketches of comrades smoking, reading, camping out, or sleeping in disarray, date from this time, full of an easy vigour and character which is continuous with the manner of the drawings from Switzerland. The summer life in the forests round Ilmenau while they were inspecting the mines could have been modelled on *As You Like It*, with its camp-fires, incognitos, visits to charcoal-burners, high-flown conversations on poetry, and inscriptions cut into trees or caves. Indeed it is to that other Duke's life in the Forest of Arden that Goethe himself compares it in 'Ilmenau', a long retrospective poem of 1783, which contains a remarkably frank characterization of Carl August in these early months:

> Noch ist, bei tiefer Neigung für das Wahre,
> Ihm Irrtum eine Leidenschaft.
> Der Vorwitz lockt ihn in die Weite,
> Kein Fels ist ihm zu schroff, kein Steg zu schmal
>
> . . . .
>
> Dann treibt die schmerzlich überspannte Regung
> Gewaltsam ihn bald da, bald dort hin aus,
> Und von unmutiger Bewegung
> Ruht er unmutig wieder aus.
> Und düster wild an heitern Tagen,
> Unbändig, ohne froh zu sein,
> Schläft er, an Seel' und Leib verwundet und zerschlagen,
> Auf einem harten Lager ein

   And for all his deep inclination to truth, error is still his passion. Reckless curiosity lures him off, no cliff is too rugged for him, no path too narrow . . . then painful over-excitement drives him violently away, now hither now thither, and from distempered motion he takes distempered rest. And sombrely savage on sunny days, unruly yet unhappy, he falls asleep, bruised and wounded in soul and body, on a hard bed.

Only deep affection could draw so brutally honest a portrait, and at the very time to which it refers Goethe was writing, 'The Duke, with whom I have now for nine months been living in the truest and warmest union of souls, has finally attached me to his affairs as well: out of our love-affair has grown a marriage, may God bless it.' Nor is the didactic tone of the lines anachronistic: in January 1777, when he had six months' work on the Council behind him and was taking up the problems of the Ilmenau mines, he was clear, and more than a little confident, about the tutorial relationship: 'It's a strange thing, the government of this world, getting anywhere at all with the politico-moral clean-up of such a mangy noddle, and keeping it in order.' Goethe, to use a favourite term of his own, is 'recapitulating' the boisterous and libertarian emotions of the last five years of his life, but, at last, at the level of social and political eminence to which they had always aspired: the manly Arcadia of Götz's Jaxthausen is coming to life—with some necessary adjustments to eighteenth-century administrative realities—in Saxe-Weimar-Eisenach.

At the same time the Wertherian line in Goethe's sensibility of the last five years seems also to be developing towards an agreeably picturesque realization, thanks again to the resources of the reigning Duke. On Sunday, 21 April 1776, Goethe took possession of his two-storied cottage—or 'garden-house' as such second dwellings outside the city walls were called when they became fashionable in the 1780s. Within the week he had given guided tours of the ruin and attached wilderness to Wieland, Frau von Stein and her family, and Carl August, but before it was reasonably habitable three months were needed of intensive restoration work, all paid for by the Duke and employing up to twenty-six workmen at a time: the roof and floors had to be repaired and the walls painted, the steeply sloping garden had to be cleared, terraced and provided with top-soil, and the court carpenter, J. M. Mieding had to construct an extensive set of furniture, including a large three-part pine dining-table for entertaining, and two beds, one for Goethe and one for Philipp Seidel. Though the house was neither large—about 33 feet by 25—nor well-built—further work on the thin walls, the windows, and the chimney was needed in the following year—it was from the start a place of rustic peace where Goethe could live out Werther's fantasy of the simple life at a symbolic, but not inconvenient, distance from the town, in which he—unlike Werther—had to spend the busy half of his time and in which he always maintained a rented apartment for emergency use. 'Everything is so still. I hear only my clock tick-tocking and the wind and the weir in the distance', he wrote on the first night he slept there, alone, on 18 May 1776, and two nights later, while fumbling around in the dark for a quarter of an hour looking for flint and tinder, he saw across the meadows the lighted windows of the ducal residence and thought how gladly Carl August would have changed places with him in that moment, had he known. In the first year or so of this life beyond the walls belongs a series of—not on the whole very successful—attempts to express in pencil, charcoal, and chalk the silence and mystery of moonlight over river meadows, trees, and ruins. The cottage and garden were also a good setting for Goethe to show the affection that he, like Werther, had for children. It became a special treat for the Stein children to be allowed to spend the night there and to make pancakes with him and Seidel, and at Easter all the children of the court—parents being strictly excluded—were invited to a party and to hunt the painted eggs hidden round the garden. They were encouraged in physical exercises that had previously been thought unseemly for the high-born—swimming, stilt-walking, even tightrope-walking—and Goethe laid for them the basis of a taste for the old German ways with gifts of the chap-book stories of Melusina, the children of Aymon, and others.

In the garden itself there was time in the autumn of 1776, after all the work on the house, only to plant some lime-trees. In the planting seasons of the following year, however, oaks and beeches were added together with spruce, juniper, and Weymouth pines specially ordered from Frankfurt. Hedges were planted to separate the garden from the meadows, jasmine and honeysuckle to decorate the house, and it would not be long before the climbing roses were blooming under the eaves. In the spring of 1776 there were already some established asparagus crowns to be found among the weeds, and later years would bring a proper vegetable-bed, strawberries, fruit-trees, and a vine against the westerly front wall. That Goethe was confidently expecting a stable period in which he could hope to look on the growth and fruiting of what he had planted is shown by the works undertaken in the early part of 1777. A wooden extension was built on the south side of the house to provide, at ground level, a fuel-store against the wall of the kitchen and the servant's room, and, on the first floor, a verandah, reached by newly constructed doors from the main reception room and the study. The verandah was a warm, sunny spot, even late in the year, and Goethe often slept out on it, so completing, as it were, his detachment from town and office life. His intention of staying in Weimar, however, despite the clear division in his existence that the very move into the cottage implied, was expressed in the 'Altar of Good Fortune', as he called it, the first piece of sculpture to be erected in his garden, on 6 April 1777. It consists simply of a sphere, the emblem of mobility, supported by a cube, the emblem of what is stable. But the symbol is perhaps more ambiguous than Goethe consciously intended: the moment of good fortune that brings the sphere to rest over the centre of the cube is frozen into permanence only by the sculptural form, and perhaps the good fortune that has brought this rolling stone to rest in Weimar is also a matter only of a moment.

There had been good grounds for confidence, of course, over the previous year. At the time he acquired the cottage Goethe felt that in five months he had got to know court life, and he was looking forward to a taste of administration, and, once the difficulties over his appointment had been resolved, his work, as yet not too onerous, for the Council and the Ilmenau mines had proceeded well. In that spring of 1776 a new and firmer footing had also been established for his relation with Frau von Stein: he had abandoned the informalities and insistent courtship of the early months and accepted her tutelage and the need for 'resignation' on his part. The responsibility for theatrical entertainments that he took on in October promised to be an agreeable occupation for someone who had had no sustained practical contact with the theatre before. And in this too it had been exhilarating to feel the Mephistophelean magic of money and power giving reality to the dreams of his former existence: a visit to Leipzig in March and April secured the services for the Weimar amateurs of the city's best and most beautiful female singer and actress, Corona Schröter, who moved to Weimar in November. The return to Leipzig after nearly ten years had caused him to reflect on 'all that had had to pass through my head and my heart' in the meantime and on the very different future before him. 'It is as if this journey was intended to draw up the balance of my past life', he had written to Frau von Stein, 'and immediately afterwards a new one is starting. Do I not now have all of you?'

But there were also contrary indications that the little circle now forming in Weimar would not be without its problems. It could not be simply a continuation and fulfilment of what had gone before, and it could not be detached from the past without pain. For Klopstock the German 'republic of learning' had an ideal existence, realizable only in a past and hazy utopia: poets, he told Goethe, had no business soiling their hands with such actualities of power as 18-year-old monarchs who wanted to drink, hunt, and bait pigs. Goethe's perhaps excessively condescending reply to Klopstock's fatherly admonitions led to a complete and bitter breach in relations, which had admittedly never been cordial, and to the excommunication of Weimar by Klopstock's disciples, including, embarrassingly, Fritz Stolberg, already invited to become one of Carl August's gentlemen-in-waiting. Painful for quite a different reason, the departures of Klinger in October to a short-lived appointment with the theatre in Leipzig, and of Lenz in December, were episodes which showed how delicate Goethe's task in Weimar really was. Neither of these old companions was able to make the adjustment to court and administrative life that Goethe made, and both had to be told personally by Goethe to go, Lenz after an unidentified 'asinine trick', probably involving some extra imposition on the strained relations between Anna Amalia, who was fond of him, and Duchess Luise, whom he over-ostentatiously adored. Only Herder had the intellectual and personal resources to stay, partly because, despite sitting very loosely to theological orthodoxy, he retained a high conception of the dignity of the cloth and so could neither join in nor approve even the more innocent entertainments through which Goethe and Carl August were growing together. He stayed, therefore, sadly, in the role that best suited his personality, that of perpetual critic and overweening malcontent: religion, not art, should rule life in Weimar, and he should rule religion. Isolated both from a secularized court and from a conservative clergy and Church administration, he remained for years isolated from Goethe too, who was the only man who did seem able to make himself into a point of reference for the many disparate elements in his society. Yet even for Goethe there were times when the task of reaching some plausible synthesis of the court world and Storm and Stress seemed impossible, when it seemed better to acknowledge the discontinuity and admit that a new life had begun in which there was no place for the old, not even for its poetry. 'What is man that thou art mindful of him, and the son of man that thou visitest him?' he wrote in his diary on the first anniversary of his arrival in Weimar. While in Leipzig he had visited the prolific local playwright C. F. Weisse (1726–1804), who reported that 'Goethe says he has handed over his literary career to Lenz, and it is he who will bestow on us a clutch of tragedies'.

Works, 1775–1777

None the less, despite a sharp falling-off in its quantity—it is partly displaced by a great increase in his drawing and painting activity—Goethe's literary output in this first phase is marked both by the confidence and, on the whole, by the continuity with the past which are otherwise characteristic of the period. The long poem in 'Knittelvers', 'The Poetical Mission of Hans Sachs', and the free-verse rhapsody 'Sea Voyage' ('Seefahrt') from April and September 1776 respectively, maintain the forms of, for example, *Hanswurst's Wedding* and 'Song of Mahomet', and also the theme of those works: the poetic genius who for all his exceptional gifts, stands in a clear, if sometimes ironic, relation to a public and to colleagues. In 'Sea Voyage', however, the poet is depicted as sailing off alone to the alarm of those on shore, who lament that he did not stay with them, but the high winds and the heavy seas only exhilarate the lone figure at the helm:

> Herrschend blickt er auf die grimme Tiefe
> Und vertrauet, scheiternd oder landend,
> Seinen Göttern.

Lordly he looks on the wrathful deep and puts his trust, foundering or landing, in his gods.

Caspar Goethe, his suspicions of Weimar by now somewhat alleviated, partly no doubt by the appointment of the new Councillor of Legation, to which the Duke had shrewdly asked him to give parental consent, copied this poem out for a friend and added to it an epigraph which shows his continuing close and accurate attention to all his son did: 'When his friends grew anxious that he might not adapt himself to life at court, he sent them the following lines of consolation.' Whether—as it seemed to the poet Bürger, one of the doubting shore-bound friends—the rather flaccid verse of 'Sea Voyage' really is a consolation, or grounds for the confidence it expresses, is another matter. That Goethe, for his part, still saw his literary role now he was in Weimar in terms of the common endeavours of the last five years is shown by his support for Bürger's plan for a translation of the *Iliad* into German blank verse. Together with Wieland and Herder he strongly encouraged Bürger's scheme in opposition to Stolberg's Klopstockian proposal for a hexameter translation, on the grounds that the iambic pentameter was the truly national heroic medium and the natural vehicle for a campaign to revive vigorous but obsolete words. The historical justification for giving this pre-eminence in Germany to the verse-form of Shakespeare and Milton was slight—though greater than any that could be adduced for the wholly alien classical hexameter—but the concern for the revival of the national literature by a return to origins was entirely in the spirit of Storm and Stress. Like so much associated with that movement Bürger's translation faded away and did not get beyond a few specimens, while Stolberg's less radical, more 'official', version was completed and published in 1778.

The most striking evidence, however, for a continuity with the immediate past is provided by a new departure: in the winter of 1776–7 Goethe began dictating the first book of a new novel, *Wilhelm Meister's Theatrical Mission*, of which something had probably already been drafted in his own hand. (While, as we have seen, it is quite likely that the figure of Wilhelm Meister was conceived in 1773, it is impossible with any degree of certainty to separate out a level of text that must have been written in Frankfurt.) This was to be a third-person narrative in prose, such as Goethe had not elsewhere attempted; it was obviously planned on a large scale since it starts in the hero's early childhood; and it was to be a story of Goethe's own time and place, opening in a central German Imperial free city, one evening shortly before Christmas some time in the 1740s. We are introduced to Benedikt Meister, a merchant, who has just completed a term as the city's burgomaster, to his mother, who has been making a puppet-theatre to entertain her grandchildren at Christmas, and, indirectly at first, to his wife, with whom we learn his relations are strained, since she neglects her family of five in favour of a paramour. It is suggested that it is partly in this domestic conflict that lie the roots of the passion of Wilhelm, the eldest son, first for the puppet-theatre, and then, as a youth, for other amateur theatricals. But before we meet Wilhelm we are acquainted with his father's habit of staying out in the evening to play cards until the soup has boiled over at home, and with the gossip of his grandmother who is well off but takes care of the pence, and who uses her fur-coat to conceal the work on her well-lit table when her son comes in. Wilhelm is from the start presented as an inhabitant of a richly furnished world of things and people: when, long after the festivities are over, he gains access one still Sunday morning to the larder, he discovers the puppets lying stored amid soap, candles, lemons, dried apples, prunes and orange peel, and the magic that attaches to them when they are used again, at his urgent request, derives partly from the strange mixture of smells with which they are now impregnated. Underneath this bright awareness of the material world runs a more obscure psychological pulse: Wilhelm's first glimpse behind the scenes into the puppeteer's booth gives him the same unsettling satisfaction that children feel as they begin to glimpse 'the mysteries of the difference of the sexes'. These first chapters are quite remarkable: the sensitivity to things, times of day, turns of speech, and their ability to symbolize passing moods or established attitudes, which we find in *Urfaust* or the poems of 1774 and 1775, are here put in the service of a prose work which has all the precise cultural and historical specificity of *Werther* but which seems intended as a vehicle for a much broader mind, both through the third-person narration and through the promise of a wide canvas. Goethe seems to be taking over Lenz's ambition to be a 'painter of human society' and realizing it in the literary form most suited to it, the supremely uncourtly form of the realistic novel. By beginning—or continuing—this work in Weimar he was keeping alive the conviction that those earlier ideals could be realized without substantial modification in the setting or perhaps even with the support of the courtly world.

There is certainly nothing in *Lila,* the one lengthy piece Goethe wrote at this time purely for the court, to suggest that this audience is for him a rival source of inspiration. *Lila* is a quite undistinguished musical entertainment written for the Duchess's birthday on 30 January 1777, in which a depressive count, with Goethean traits, is rescued from melancholy by a Fairy Sunlight, whose name recalls the symbol Goethe used in his diaries for Frau von Stein. In a written-out but unperformed revision of early 1778, possibly influenced by the death of Christel von Lassberg, the depression was transferred to the Count's wife Lila, her rescuer became a psychologist-cum-magician, Dr Verazio, and the play was turned into a transparent parable of the relation between Carl August, Luise, and Goethe, remarkable only for one invigorating song.

But in the poems that Goethe wrote out of his association with Frau von Stein a counter-current to the inspiration of Storm and Stress is beginning to flow, though it is not yet powerful enough to subsist in its own right. A number of shorter poems, as rhythmically inventive and memorable as ever, grasp at the mood of a moment in the attempt to assimilate this new love to the pattern of its predecessors but come up against some mysterious obstacle. In 'Evening Song of the Huntsman' ('Jägers Abendlied'), from the first weeks after his arrival in Weimar, memories of Lili Schönemann mingle uncertainly with thoughts of the distant Charlotte, so that the restless poet's own identity seems to lose its fixity. The need for rest is a motif in several of these poems; it is synonymous with a need for clarity about himself, for a respite from the endless interchange of relationship:

> Lieber durch Leiden
> Möcht' ich mich schlagen,
> Als so viel Freuden
> Des Lebens ertragen.
> Alle das Neigen
> Von Herzen zu Herzen,
> Ach wie so eigen
> Schaffet das Schmerzen!

I would rather fight my way through sorrows than bear so many joys of life. All the inclining of heart to heart, ah, how strangely it makes for pain! ('Restless Love', 'Rastlose Liebe', May 1776).

Love is bliss without rest, and he is looking for both bliss and rest. The supreme expression of this theme is in a minuscule poem of February 1776 sent in a letter to Frau von Stein and later coupled with 'Over All the Peaks' ('Über allen Gipfeln') as the prayer, for a bliss beyond both pain and pleasure, to which the poem of 1780 is the answer, each being given the title 'Night Song of the

Wanderer' ('Wandrers Nachtlied'). The break in the syntax perfectly mimes a break in the voice and it is quite understandable that this single extended sentence of pure yearning, free of all natural imagery, should first have been published, in 1780, in a Christian devotional journal:

> Der du von dem Himmel bist,
> Alles Leid* und Schmerzen stillest,
> Den, der doppelt elend ist,
> Doppelt mit Erquickung füllest,
> Ach, ich bin des Treibens müde,
> Was soll all der Schmerz und Lust?
> Süßer Friede,
> Komm, ach komm in meine Brust!

You who are from heaven, who quieten all sorrow and pain, who doubly refresh him who is doubly wretched,—ah, I am weary of the bustle, what is the meaning of all the pain and pleasure? sweet peace, come, oh come into my breast.

The repression of eros, the recognition, and determination, that his love for Frau von Stein is not to be a love like the others, not another inclining of heart to heart, give a certain sphinx-like grandeur and lucid mystery to the most purely and intimately psychological poem Goethe ever wrote. It is also the only major poem of his in which Frau von Stein is, eventually, directly and personally addressed. Sent to her in a letter on 14 April 1776, just after his return from Leipzig, it dates from the period of readjustment, when Goethe had only recently begun to accept an element of 'resignation' in this relationship, yet it seems to speak as if the relationship were long since established—it speaks of what is hoped for from the future as if it were already past, and so gives remarkably clear expression to that mingling of real and unreal, that nimbus of potentiality and shared fantasy, in which perfectly real and fixed personalities can, and sometimes have to, live. It has no title and is known by its first line, 'Why did you give us the deeper vision' ('Warum gabst du uns die tiefen Blicke'). The 'you', we learn in line 5 of its thirteen quatrains, is, initially, fate or destiny ('Schicksal'), which has brought together two people who have a preternaturally clear understanding of each other and are free of the self-deceptions by which other lovers normally live. 'I cannot account to myself for the significance, the power that this woman has over me except by reincarnation.—Yes, we were once man and wife! Now we have knowledge of each other—veiled, in a spiritual ether—I have no names for us—the past—the future—the universe.' So Goethe wrote in these very days to Wieland, and the central sections of the poem, which address Charlotte directly, propose this explanation for their mutual insight:

> Sag', was will das Schicksal uns bereiten?
> Sag', wie band es uns so rein genau?

* Goethe originally wrote 'Alle Freud' = 'all joy'.

Ach, du warst in abgelebten Zeiten
Meine Schwester oder meine Frau;

Say, what does destiny have in store for us? Say, how did it bind us together so utterly exactly? Ah, in times lived out earlier you were my sister or my wife;

Protected by the fiction that he is talking of a previous existence, Goethe—or perhaps one should say, the dreaming, semi-conscious mind that appears to be murmuring this poem—can speak of experiencing those things which it will not be seemly or permissible to acknowledge as part of the present relationship; in that other life Charlotte knew and understood him totally, by her love she put an end to the turbulence in his breast, yes, he can even allow himself to say, she gave him peace by satisfying his physical craving:

Tropftest Mäßigung dem heißen Blute,
Richtetest den wilden irren Lauf,
Und in deinen Engelsarmen ruhte
Die zerstörte Brust sich wieder auf

.   .   .   .

Welche Seligkeit glich jenen Wonnestunden,
Da er dankbar dir zu Füßen lag,
Fühlt' sein Herz an deinem Herzen schwellen,

You poured moderation drop by drop into the hot blood, straightened the wild crazy course, and in your angelic arms the shattered bosom was restored by rest . . . What happiness could compare with those hours of bliss when he lay in gratitude at your feet, feeling his heart throbbing against your heart,

At this point a series of grammatical ambiguities breaks down the distinction between the present and the past forms of the verbs, and the poem concludes by returning us to the present state of existence, for which the true and full and bodily relationship in the previous life is only an uncertain half-memory. The hypothesis of reincarnation, which was used to explain the complete rational clarity and calm of the lovers' mature and mutual understanding, has revealed the clarity to be itself a mystery. The clarity has been achieved through the lovers' forgetting its origins, but its origins do not thereby cease to exist, and are the unacknowledged source of a—rationally inexplicable—uncertainty and pain which accompany the clarity and which, because their source is unacknowledged, can only be attributed by Goethe to the unfathomable decision of 'destiny'.

To speak of the unreal, however indirectly, is to give it a certain real power, and 'Why did you give us the deeper vision' could be understood, not as an ingenuous and lucid meditation on the structure of shared consciousness, but as a serpentine exercise in seduction. As such it would have, in real life, to be either yielded to or dismissed. There is an indeterminate relationship in the poem between the clarity about the lovers' situation and history, which the poet's meditating mind achieves (and allows the reader to achieve), and the painful twilight clarity which the lovers themselves are said to have as a result of their history. In real life that indeterminacy cannot be maintained, even though it is what gives the poem its fascination. The one-act play *Brother and Sister* (*Die Geschwister*), written in two days in October 1776 during a long absence from Weimar of Frau von Stein, concludes with a recognition that such coexistence of knowledge and ignorance, chastity and fulfilment, courtly life and Storm and Stress, is unsustainable for any length of time outside the work of art, even though it gives the play itself something of the same charm as the poem.

Wilhelm, a businessman, lives alone with Mariana, who is believed, and also believes herself, to be his sister. In reality, however, she is the daughter of Charlotte, a good and noble widow whose love, many years before, rescued Wilhelm from imminent financial ruin by restoring his morale and giving him a purpose to work for. Before, though, he could re-establish his fortunes and make himself worthy of marriage to her, Charlotte died, leaving Mariana in his care. In the course of the play Mariana comes to recognize that her feelings for Wilhelm are not simply sisterly. Wilhelm also has conceived a passionate love for her but has not wished to take advantage of her position, and when he at last feels able to reveal that they are not in fact related there is no remaining obstacle to the marriage that they both desire. Formally speaking this miniature has some of the best features of Goethe's writing in the previous five years: the poetic realism of times of day, plausible money transactions, contingent parallel events such as cooking or a child crying, and some beautifully natural dialogue. Mariana's speeches in particular are Goethe's best dramatic prose ever, though they are possibly a little too derivative from the Gretchen of *Urfaust,* as she herself resembles Gretchen in her fusion of sanctity, passion, and domesticity. Into this form, so redolent of his Frankfurt existence, Goethe inserts the new theme that Weimar has given him, that of eros suppressed, the resigned, platonic, love of brother and sister, and the play is intended to demonstrate the compatibility of form and theme, Frankfurt and Weimar. Ideal love, the play seems to say, can become real, brother and sister can marry, even the 'holy woman' Charlotte, now an inhabitant of a higher and purer world, can become an object of desire within the world of crying children, coffee in bed, and the knitting of socks. For the plot of the play is a very exact inversion of the development of 'Why did you give us the deeper vision': Wilhelm's life with Mariana is seen by him as the fulfilment of his love for Charlotte whom 'destiny has given to me rejuvenated' in her daugher; 'I can now stay and live united with you [Charlotte] as I could not in that first dream of life'. There is a faintly ghoulish *frisson* to this suggestion that so lively and full-blooded a figure as Mariana is valued ultimately as a reincarnation of her virtuous but dead mother, but the chill is lost in the surge of fulfilled desire with which *Brother and Sister* ends. The play represents as achieved reality what the poem represents as multiply remote, and that is the strength of the poem. The play's last words, however, are a threefold questioning by Mariana whether it is really possible that such 'dreams of life' should turn into life itself. Though within the play the words with which she falls into Wilhelm's arms—'Wilhelm it is not possible'—express an ecstatic certainty, for the audience they are an indication that the author recognizes his hybrid of the real and imaginary for what it is. Goethe knows that his agreeable state of indeterminacy cannot last.

Tragedy and Symbolism: 1777–1780

Even so, Goethe was not prepared for the brutal opening of the second and, as we may call it, tragic phase of this early Weimar period, which lasted from the summer of 1777 until his return from Switzerland in the first weeks of 1780. On 16 June 1777, in the midst of what he calls 'such happy times', he went out into his garden at 8 o'clock in the morning to inspect the progress of the saplings he had recently cleared of pests, and to walk up and down the terraces reading in the sun. At nine o'clock a messenger interrupted him with a letter telling him that on 8 June, four weeks after the birth of her second daughter, his sister Cornelia had died: she was 26. 'Dark lacerated day', he wrote in his diary, and he told his mother: 'With my sister I have had so great a root struck off which bound me to the earth that the branches up above that had their nourishment from it must die off also.' Cornelia had not been happy in her marriage to Schlosser 'without one's being able to blame her, her husband or their circumstances'. She, whose life had until four years previously been inseparable from Goethe's, his fellow-conspirator, correspondent, pupil, reader, and nurse, had died at no age, wretched and unfulfilled. This seemed her fate, dealt out to her by the very gods whose will for him he was struggling to discern in his 'happy times' in Weimar and in the obscure clarities of his relation with Charlotte von Stein, and it was a fate that could still so nearly be his own. When he wrote to Augusta Stolberg of the death he fell spontaneously, for a few phrases, into verse:

> Alles gaben Götter die unendlichen
> Ihren Lieblingen ganz
> Alle Freuden die unendlichen
> Alle Schmerzen die unendlichen ganz.

Gods, the endless ones, gave everything to their darlings in full: all joys, the endless ones, all pains, the endless ones, in full.

We may at least make a guess at one of the upper branches now doomed by the gods to die off, for Cornelia must have been the unspoken—and, in so far as Goethe concealed from himself the reciprocity of their feelings, the unacknow-ledged—model for the beloved sister, at once sensual and chaste, courtly and bourgeois, that ambiguous ideal in which he had been seeking the meaning not only of his relation with Frau von Stein, but of his first two years in Weimar altogether. Cornelia's death marked that ideal as itself deadly and incapacitat-ing, and, after a supreme articulation in *Iphigenia* and a remote echo in *Wilhelm*

Meister, the theme of brother and sister disappears almost completely from Goethe's works. With the theme there disappears also the uneasy and indeterminate synthesis that the theme had been expressing. Its place is taken in 1777 by a tragic awareness of the possibility that a human being may have to face an ineluctably wretched destiny and that neither earth nor heaven may offer any response to the cry of the heart for love. There is no room for the fancies of reincarnation, of a 'first dream of life' giving way to life itself, in the lines of condolence that he writes to his mother on her daughter's death. He commends his ailing father to her care with words that warm the heart, as they rend it, by their simple acknowledgement of mortality: 'we are together like this only once'.

Goethe's reaction to this bereavement was powerful. During the summer an almost manic wildness entered his behaviour at times: in Ilmenau with the Duke in August and September there was even more drinking, frantic revelry with the local peasant-girls, and savage practical joking than in the previous year. On one occasion, on 4 July, the entire ducal family, including Duchess Luise and Dalberg, visiting from Erfurt, went off with him at a few hours' notice to admire, and sketch, panoramic views from the old hunting-lodges at Dornburg beyond Jena. The day passed too quickly, and no arrangements having been made for accommodation, the night had to be spent on straw palliasses in the unfurnished rooms of the little castles. The tired party raised its spirits the following morning by letting off fireworks which re-echoed mightily from the steep slopes overlooking the Saale, on which the lodges are picturesquely situated. They got back to Weimar by midday, while Goethe rode on to visit Frau von Stein's children at the family's country home and talk and draw with their tutor. (Frau von Stein herself was taking the waters at Pyrmont at the time.) From there he had to return to take part in a ducal audience with a committee of the Estates General on 7 July and a Council meeting on the eighth, after which he joined the Duke and Prince Constantin at Tiefurt, staying up half the night with them, and spending the following morning there as well, talking, drinking, drawing silhouettes, and reading from his manuscript of *Wilhelm Meister*. There was intense mental activity too, though a certain pause intervened after the shock of the event itself, as it had done before the writing of *Werther*, or, for that matter, *Götz*. In the twenty months after Cornelia's death Goethe conceived and composed *Proserpina*, *The Triumph of Sensibility*, and *Iphigenia*, and started work on *The Birds*, wrote the grand and gnomic hymn 'Winter Journey in the Harz' and the poems 'To the Moon' and 'The Fisherman', and completed the first book of *Wilhelm Meister*.

What Goethe did not do, now or later, was to take any interest in his two nieces. This is only very partially explained by their soon having a stepmother, for within five months Schlosser engaged himself to another member of the Frankfurt circle, Johanna Fahlmer. Just as he had been unable to correspond with his sister during the time of her unhappiness, so now Goethe could deal directly with sorrows that—on his own admission—went to the root of his being only by protecting himself from them, and by devoting his mental and emotional energies instead to expressing and transmuting them in his art. It would be quite wrong to suggest that Goethe just insulated himself from pain, 'avoided tragedy', as if he simply wanted to forget or ignore. On the contrary, to remember pain, to put up a verbal, if often inscrutable, monument to it, was for him an obsessive need, lest all the suffering should be in vain and be swallowed up in meaninglessness. But he could deal practically with distress and alleviate need only when it was sufficiently distant from him not to require that inner involvement which was the condition for him of literary production. There is no literary monument to 'J. H. Krafft', whom he began to assist in 1778, nor to Peter im Baumgarten, who arrived with his pipe and dog at Goethe's door just two months after the news of his sister's death, on 12 August 1777, and who received the attention that her children did not. Much of the responsibility for this 11 or 12-year-old Goethe was able to pass on by sending him for private tuition with the von Stein boys, but he still consented to spend a year in close proximity to a troublesome and smelly house-mate, which was more than any other member of the Weimar Privy Council would have done. Only when he smeared ink over all but the eyes and nose of a white marble bust of Lavater did Goethe decide in August 1778 to settle him in the more robust environment of the Ilmenau gamekeepers. An ingenious arrangement with 'J. H. Krafft', by which this other casualty agreed to keep an eye on Peter's education, seems to have been more successful than the apprenticeship to which the young lay-abed soon proved quite unsuited. He preferred drinking, card-playing and seducing the vicar's daughter, whom he married, and proclaimed his intention of becoming an artist. Goethe secured him instruction in engraving and he was on the point of starting a small business when in 1793, at the age of 27 or 28, and already the father of six children, he disappeared from Weimar entirely, possibly emigrating to America. At no point in his rake's progress did Goethe ever disown him.

Goethe laughed when Peter told him that the first sights he saw in Weimar on 12 August were a military flogging and a man running the gauntlet. 'The Duke is now possessed by the soldier-devil, as last year by the student-devil. He drills and flogs his army all day', wrote Zimmermann in November. A daring and brutal Prussian cavalry officer had been imported to bring discipline and precision to the Weimar Hussars, and Goethe himself assisted in the laying out of a new parade ground near the ruins of the Wilhelmsburg, but this resurgence in his pupil of a hereditary weakness can have given him no pleasure at all. As summer wore into autumn and winter so his hectic activity gave way to inner silence and a sense of isolation. 'I am very altered', he wrote to Johanna Fahlmer in November, on hearing of her engagement to Schlosser. In September and October he spent six weeks with the Duke in Eisenach, but he soon escaped from the busy little palace of Wilhelmsthal and took over a room high up in the practically deserted Wartburg, where he drew and brooded and worked on *The Triumph of Sensibility* and suffered agonies from toothache, probably associated with his chronic tonsillitis. He deliberately avoided a meeting with the uncrowned king of the French-speaking literary world, Baron Melchior Grimm, from whom, as from 'all these people', he felt separated by a 'chasm', and to a visitor he seemed like 'a thoughtful, serious, frigid Englishman in his dress and in his manner'. As 7 November, the second anniversary of his arrival in Weimar, approached he was taking stock of his situation and trying to make sense of it. On the eighth he wrote to Charlotte von Stein a letter which shows how little he succeeded and how much was now open to doubt:

Yesterday coming away from you I had more, strange, thoughts; among others, whether I really do love you or whether it is merely that being near to you gives me pleasure like the presence of so pure a glass in which one can see oneself so well reflected.

And then it occured to me that Fate in transplanting me here had done exactly what you do to lime-saplings: you cut off their tips and all their fine branches so that they make new growth or else they will die from the top down. True, for the first few years they stand there like sticks. Adieu. I happened on my diary of a year ago and there, for 7 November, stood 'What is man that thou art mindful of him', etc.

Three weeks later he had recourse for the first time in Weimar to the device by which he had before, in 1772 and 1775, sought clarity and relief from despair or doubt. He disappeared. Having received leave of absence from the Duke, on 29 November, before 7 o'clock in the morning, he rode off under an assumed name to an undisclosed destination. Only Frau von Stein received letters from him, and until his purpose was accomplished even these contained no place-names by which he might be traced. During short dark days, sad, misty evenings and well into the night, over roads that were more like rivers of mud, through hail, drenching rain, and bitter winds, and even meeting with a landslip, he struck out to the north-west, into the Harz Mountains. For his aim, avowed at first only to himself, was to climb the Brocken, at 3,740 feet (higher than Snowdon) the highest mountain in north Germany, and long reputed to be the haunt of witches and demons. His reason, as he obscurely hinted to Frau von Stein, and as has been clearly shown since, was that he was looking for a sign. He was questioning the Fate that had transplanted him to Weimar, to life as courtier and official of a turbulent autocrat, that had given him a new and strange love, that already seemed to be diluting his poetic inspiration, and that decreed that one as close to him as his sister should live only a brief life of misery attributable neither to herself nor to circumstance but only to the fateful interaction of both. Was he too perhaps destined for some such terrible and destructive disparity between his own desires and potentialitites and the reality of his world and his time? Had not so many others of his generation, the Werther generation, already come to grief in just that way—Jerusalem, Lenz,

Lindau, Peter's original guardian, and young Plessing, who had written to him and whom he could now visit and advise incognito in his home in Wernigerode on the edge of the Harz? What then of the author of *Werther* himself? The sign Goethe asked for from the unknown powers that rule human lives was that, if in Weimar he was on the right path, they should allow his unlikely and irrational expedition to succeed. In so far as he had as yet any specific administrative responsibilities in Weimar they related to the Ilmenau mines, and he spent several days in a circuit of the northern Harz clambering down deep shafts and visiting smelting works to gain experience—and perhaps to give more substance to the question he was putting to the gods. Was this what he was for? Was he for somewhere else? Or was he for Weimar, but in some special, even unique capacity? On 10 December he was guided up through the snow to Torfhaus, a hamlet with a forester's station a thousand feet below the great domed summit of the Brocken. Now it is a marshalling yard for cross-country skiers, but in 1777 there were no paths and in thirteen years the forester had never climbed the mountain in winter. The night of the ninth had been very cold and the following morning the hills were hidden in mist. Over his breakfast the forester told Goethe that in these conditions the Brocken was unclimbable.

So I sat there with a heavy heart and half a mind on going back . . . I was still, and prayed the gods to change the heart of this man and the weather, and was still. And so he says to me, 'Now you can see the Brocken', I went to the window and it lay before me as clear as my face in the mirror, then my heart opened and I cried: 'And shall I not go up there! have you no servant? nobody?' - And he said, 'I will go with you .'

At quarter past ten they left; the snow was a yard deep but with the hard frost it was strong enough to carry them. 'At a quarter past one on the top', Goethe noted in his diary, 'bright, magnificent moment, the whole world in clouds and mist, and on top everything bright. What is man that thou art mindful of him.' The biblical tone and language that permeate Goethe's account of this day in his diary, and in his letters to Charlotte von Stein, show the religious significance that the ascent had acquired for him and had indeed always been intended to have. The phrase from the eighth Psalm that he had so recently associated with his arrival in Weimar bears witness that it was above all confirmation from a higher power of the rightness of his staying there that he sought, and felt he had received, in the enjoyment of that bright, magnificent moment above the world. And we should remember the continuation of that phrase—'and the son of man that thou visitest him'—when we read the letter drunk with happiness that he wrote to Charlotte on coming down from the mountain, or the poem that day by day and strophe by strophe he had been composing since he left Weimar, now known as 'Winter Journey in the Harz' ('Harzreise im Winter'), for in both of these Goethe's self-identification with 'the son of man' is close below the surface.

The goal of my longing has been reached, it hangs by many threads and many threads hang from it; you know how symbolical my existence is . . . I had said: I have a wish for the full moon!—Now, dearest, I can go out of the door and there lies the Brocken before me above the firs in the noble magnificent moonlight, and I was up there today, and on the Devil's Altar [two of the rock formations on the Brocken are still known as the Devil's Pulpit and Witch's Altar] I offered the dearest thanks to my God.

There is a touch here of that provocative Satanism in which Goethe had occasionally indulged since his sympathetic interest in Lucifer in 1769, and it recurs in some of the last lines of the poem, in which he identifies himself with the mountain summit, on which he had stood, and which he had seen as clearly as his own face in a mirror:

> Du stehst mit unerforschtem Busen*
> Geheimnisvoll-offenbar
> Über der erstaunten Welt
> Und schaust aus Wolken
> Auf ihre Reiche und Herrlichkeit

You stand with unfathomed bosom, mysteriously manifest, above the astounded world and look down from clouds on its kingdoms and glory

Goethe, that is, puts himself in the position of Christ, to whom the Devil showed all the kingdoms of the world and the glory of them, but this Christ plays the Devil, accepts the offer, and gives thanks for it to 'his' God. For he is the Christ with whom Goethe always felt most in harmony: the 'son of man' who came eating and drinking, the poet, playboy, and leader of men, who may now have found in the world of politics and administration a new field for his genius but whose bosom remains unfathomed and who is not comprehended by the world he willingly consents to enter.

'Winter Journey in the Harz' will concern us further when we look at Goethe's literary works of this period, but it is important to understand that the poem, like the expedition to the Brocken itself, is not simply a question about the wisdom of Goethe's embarking on an official career (which had not yet fully begun), but a question about his personal destiny in its entirety, about the possibility of his transcending the restricted choices that circumstances seem to offer him, about whether he is still the beloved son, one of the 'darlings' of the gods, the endless ones. Are the events of his life capable of being filled with 'symbolical' meaning, as in his poetry he has so far been able to make 'symbolical' the things of the world? The unequivocally positive answer he felt he had received had led, he told Knebel when he rejoined the court, hunting near Eisenach, on 15 December, to a 'marvellous opening and melting of the heart' and, after completing the first book of *Wilhelm Meister* on 2 January 1778, he seems in the new year to have thrown himself into court life with renewed vigour. It was a specially active skating season, but 1778 was above all the year of Goethe's greatest involvement in the amateur theatre: it was partly by these means that he succeeded this year in finally winning over Duchess Luise, who

* Originally 'unerforscht die Geweide' = 'with unfathomed entrails', another indication that the event, and the poem, is an oracle.

in April so far disregarded the proprieties of rank as to take supper with him in his garden. *The Triumph of Sensibility,* first performed on her birthday on 30 January, is a considerable work, and 398 dollars were spent putting it on, and the humour, the elaborate transformation scene, and Seckendorff's music, gave it an appeal even to those who found its satire too puzzlingly elusive. Goethe himself took one of the main parts, and he acted in three other productions of plays by Cumberland, Gozzi and Molière, before the other great success of the year, his expanded (and intellectually bowdlerized) version of the plotless pantomime, first written in 1773, *Lumberville Fair,* put on in the autumn at Ettersburg. By this time he knew of and had agreed to Carl August's plan to involve him fully in the duchy's administration by giving him the presidency of two Commissions, responsibilities which from the start of 1779 left him with much less time for court entertainments.

But the assurance Goethe had found on the Brocken was an assurance that he could survive, it was not an assurance that he would be spared danger and perplexity, or that the tragic possibilities in his situation would grow less, or did not exist. That the fears which had led him to question the gods were well founded was shown all too soon and all too cruelly after his return by Christel von Lassberg's act of despair. The two poems most directly connected with this event—'To the Moon' and 'The Fisherman' ('Der Fischer')—both use the River Ilm as an image of a death-dealing force which is none the less capable of exercising a hypnotic attraction over the soul: that hybrid perhaps of eros and self-regard that had drawn Werther to his death by feigning an affinity with the transcendent world of heaven. Goethe warned Frau von Stein not to go down to the scene of the suicide: 'This inviting sorrow has something dangerously attractive, like water itself, and we are lured on by the reflection of the stars of heaven that glistens out of both of them.' A month later Plessing, who had not taken long to identify his mysterious visitor, came to Weimar to continue the consultations that had begun in December in the shadow of the Harz. Above all, the political situation in 1778 gave a renewed urgency to Goethe's interrogation of himself and his gods about his role and his future in Weimar. Storm-clouds were gathering over the Bavarian Succession—Joseph II and Maria Theresa were in effect attempting to annexe Bavaria—war between Prussia and Austria was imminent, and Weimar's little boat threatened to be crushed between these two great men-of-war. In May Carl August took Goethe with him to Berlin for a few days, to gather information about likely future developments. It was an opportunity for Goethe to determine how far he was prepared to go in his move from Frankfurt into the world of the monarchical court and the absolutist power-state, and his conclusion was that he belonged, precisely, in Weimar. He saw the palaces and grand military buildings of Potsdam and Berlin, visited the porcelain factory, met poets and artists, including Chodowiecki, and saw a professional theatrical production. But he also saw, and it both fascinated and repelled him, the mechanism of a great state being prepared for war, 'how the great of the world play with men, and how the gods play with the great': from Prince Henry, the brother of Frederick the Great, who was away in Silesia, down to the horses, carts and artillery on the street, he saw them all in their thousands as dolls on some great barrel-organ, moving up and down and playing their tune only as they were driven, through many hidden cogs, by the one great wheel stamped 'Fridericus Rex'. Nor could a court which had recently recruited Count Görtz, and which questioned him at dinner about the genealogy of the House of Weimar—as a *savant,* he surely ought to know, why else was he there?—expect enthusiastic conversation from him. He appeared proud and taciturn, and knew it. 'No wonder Goethe gave such general offence there, and for his part was so impatient with the rotten brood.' It was from Berlin that he wrote to Frau von Stein of the iron hoops around his heart, and it was with relief that he found himself back in his house and garden: 'I feel more comfortable and like it better in my valley than in the wide world,' he told her, 'the gods must think me a pretty picture since they chose to put so extra precious a frame around it.' He clearly felt that there was an inner connection between his ascent of the Brocken and his visit to Berlin, since in writing both to Merck and to Frau von Stein—'in a quite different situation from when I wrote to you in the winter from the Brocken, and in exactly the same spirit'—he puts the two expeditions in parallel, for all their differences. On the Brocken he had learnt how the gods deal with poets, in Berlin he had learnt how they deal with nations. Looking down from the mountain on to the kingdoms of the world and their glory, he had seen 'all Germany below me', and venturing out into the 'wide world', he had confirmed to himself where his place in Germany had to be.

Goethe's renewed commitment to Weimar in 1778, despite, or even because of, these subterranean threats, was also given a pleasantly physical expression. The house on the Ilm was now in good order and its garden burgeoning and Goethe could turn his attention to its surroundings. On the way to and from Berlin the Weimar party stopped for a while with the Prince of Dessau, a cultured man (Behrisch was still at his court) who, with means not much greater than those available to Weimar, had constructed at his new palace in Wörlitz (itself the first neo-classical building in central Germany) Germany's first landscape gardens in the English style. Goethe was enchanted by the 'dream' that 'the gods' had here allowed the Prince to make into reality: the lakes, canals, groves, and shrubberies, all in the first flush of youth, 'melting into one another in the most gentle variety' on a rainy afternoon, without any single dominating point of interest or vantage, 'it has altogether the character of the Elysian Fields, . . . one wanders around without asking where one started from or is going to'. He recorded the mood in a beautifully atmospheric drawing of the lower storeys of the new buildings, glimpsed beneath the canopy of the young trees, and he was soon sketching a classicizing tempietto—never built—for an as yet non-existent park in Weimar. Apart from the 'Star' and the

Esplanade, Weimar had only the square 'Latin Garden', opposite Frau von Stein's new house, and all three were laid out in the old-fashioned formal style. But there was also the valley of the Ilm and though at the moment it was an unnoticed *terrain vague* on the edge of the town, part meadow, part tobacco plantation, part timber-yard, and though it was in places overgrown with dense thickets of ash and alder, was subdivided between various owners and tenants, and had few paths or points of public access, Goethe and Carl August saw in it the capability for a park in the Wörlitz manner, and sought the advice and help of the Prince of Dessau. Systematic work did not begin until 1785, but from 1778 Goethe personally undertook such little improvements as planting interesting trees or ferns or specimens from elsewhere in the duchy, or planning with the Duke a romantic grotto—'Goethe's new poems on the river', Wieland called them. In this, the year of Luise, it was decided to celebrate the Duchess's name-day on 9 July with an open-air entertainment, and this proved the origin of a landmark which was to be the centre of much of the future development. In the least frequented corner of the valley, in the shade of a group of ash-trees, a wooden hermitage, with thatched roof and moss-lined walls, was built, and here the Duke, Prince Constantin, Goethe, Stein, Seckendorff, and Knebel, dressed as monks, greeted the ladies of the court, and invited them in rhyme to a monastic cold collation and a pot of beer laid out on a rough table. This spartan prospect caused some uneasiness to the ladies, particularly Countess Gianini, but suddenly the rear doors of the hermitage opened to reveal a princely table, full orchestra, and artificial waterfall. The picnic, of course, now went with a swing. The discovery, or creation, of this unsuspected picturesque spot in so unpromising an area was in a sense the inauguration of the Weimar Park. Carl August developed a great affection for the hermitage (the 'Luisenkloster'), making it his counterpart to Goethe's 'garden-house', at times sleeping, and even receiving visitors there. The network of paths which grew up along both banks of the Ilm after 1784, once the Duke had acquired the necessary property rights, had the primary function, apart from providing riverside walks, of linking the hermitage with the different features of the park as they were built, or came to be incorporated into the general scheme.

Behind the appearance of renewed energy, enthusiasm, and commitment to court life, however, the spirits so dramatically exorcized on the Brocken continued to haunt Goethe. The greater readiness to take Weimar on its own, courtly, terms exacted a price. In September Goethe even acknowledged to himself that the oracle had been a subtle act of ventriloquism: 'Actually I am not necessary here, but I imagine I am, and that is a part of my life.' In the Harz and on the Brocken his thoughts had, he told her, been continuously with Frau von Stein: if he had a special destiny, he says in 'Winter Journey in the Harz', it is as a poet of love. Yet the 'love' that he feels for Charlotte is binding him with the iron hoops—in 'Winter Journey in the Harz' he speaks of the 'bonds of . . .

brazen thread'—that are squeezing apart his inner and outer life, his soul and his body. As the relation with Charlotte becomes, in his own words, more 'husbandly', that is, unromantic and everyday, so he becomes tempted by other liaisons that seem to promise, or at least allow the idea of, greater physical and emotional satisfaction. Goethe was clearly fiercely, if briefly, attracted to Amalia Kotzebue—sister of the future dramatist and diplomat August Kotzebue (1761–1819)—who played Mariana to his Wilhelm in the 1776 production of *Brother and Sister*. From 1777 onwards he was meeting increasingly often with Corona Schröter, of whom in July of that year, while she was dozing in the morning sun, he drew a delicately linear charcoal portrait, perhaps his best. But Corona Schröter was a cut above the rest of her profession and not to be conquered: she insisted on being accompanied everywhere by a burly chaperone (whom Goethe also drew, with rather more solidity), and moreover Carl August had set his sights on her too. The situation was obviously impossible and after a 'radical clarification' between the two men in January 1779 the relationship settled into a stable triangular friendship. There were also other, less serious, flutterings of the heart. It is perhaps to Frau von Stein's original demand for a separation of spiritual and physical love that we should attribute this penchant for flirtations with the 'Misels' (=Mslles), as Goethe collectively called them, for that seemed to lend these more earthly love-affairs a spurious, if base, legitimacy. But hard on their heels follows what looks like a sustained attempt to suppress his physical desires altogether: throughout these early years, up to and including the Swiss journey, one can see him campaigning to gain control of his body. In one respect, this never succeeded: throughout his Weimar career, whether on the stage or in daily life, Goethe attracted attention for the unaristocratic stiffness of his movements. (In his early years this gracelessness gave rise to malicious amusement, being imitated, a few paces behind him, by the somewhat shorter Philipp Seidel, who so admired his employer that he had to follow him in everything, whether writing elevating letters, dabbling in natural history, or gangling.) But Goethe's naked bathing in the Ilm and elsewhere, even in the winter and before he began to learn to swim in May 1777, his skating and fencing, walking and riding—he rode from Leipzig to Weimar, a distance of some sixty miles, in eight and a half hours— his sleeping out of doors, in the forest or on his verandah, even, eventually the control of his diet—at the end of 1778 he halved his wine consumption and gave up coffee—all left him, though prone to colds and minor ailments, fit and lean and looking taller than his 5 feet 9¼ inches. But all also bear the mark of an obsessional asceticism: 'May the idea of purity, which extends to the morsel I take into my mouth, grow ever brighter within me', he wrote in his diary. It is hard to believe that so physical a need for purity did not have a sexual origin. However, emotions, once repressed, can lead a strange life without necessarily going away, and the feelings that led him to share his house with Peter and other boys—when Fritz von Stein stayed, he and Goethe slept in the same room and whoever woke first roused the other with a slipper—need not have been simply paternal.

Official cares, however, soon provided a distraction from the frivolities of 1778. In July war broke out at last between Prussia and Austria, and Weimar found itself, in Wieland's word, in a 'nasty' position between the two great powers. The Council forbade any discussion whatever of the war in any of the inns of the duchy, or any expression of support for one side or the other, for fear of giving an excuse for an intervention. More seriously, the closing of the Austrian borders deprived the stocking-weavers of a large part of their traditional market, and famine threatened in the wake of unemployment. In February 1779 came the Prussian request for recruitment facilities— tantamount to a polite threat of occupation—and the cessation of hostilities in April was a personal relief for Goethe, as for any member of the Council. In the midst of this troubled time, in January, Goethe took up his appointments to the Highways and War Commissions and had as one of his first tasks the triennial tour of the duchy to select recruits for Weimar's own army—possibly with a view to their being passed on as a sop to Prussia. Yet the need for detachment and efficiency in the hurly-burly of business gave a justification and useful role for the feeling of being 'frozen off from everyone else' that was increasingly frequent with him: 'I am now living with the people of this world . . . but am hardly aware of them, for my inner life goes undeviatingly on its way'; 'solitude is a fine thing when one is living in peace with oneself and has something definite to do'. The accumulation of practical concerns had in fact a settling effect on Goethe's spirits, and a liberating effect on his art: 'the pressure of affairs is very good for the soul; when it is unburdened it plays all the more freely and enjoys living'. During the recruitment tour through a succession of draughty town halls and unoccupied castles, from the middle of February to the middle of March, and in the intervals between measuring heights and listening to excuses from the fit and newly-married and pleas for admission from the sickly and unemployed (all of which he recorded in one of his few satirical drawings), he dictated to his secretary the first three acts of a new drama in elevated prose, *Iphigenia in Tauris*. Rehearsals began immediately after his return to Weimar, the remaining two acts were quickly completed, and on 6 April the first performance took place on the movable stage in Hauptmann's Assembly Rooms, with Corona Schröter, in fifty-six yards of white linen, muslin, and taffeta, as Iphigenia, and the athletic Goethe, in Grecian costume, as Orestes: 'one took him for an Apollo', said an eye-witness. The success was such that a second performance was mounted on 12 April, and a third took place in Ettersburg in July, with Carl August, standing in for Prince Constantin, taking the part of Pylades.

What does seem to have suffered from the sharp reduction in Goethe's leisure time from late 1778 onwards is his drawing, though that too had been changing its character. After 1776 he drew scarcely a single interior: his attention was turned resolutely from society to nature. In his more panoramic landscapes, often laboriously executed, he had been dropping the Picturesque interest in the 'point of view', which he had so strikingly exploited in his indoor and outdoor Swiss sketches of 1775, and the resulting plainer composition was showing up the deficiencies of his uneducated technique. In 1777 he did his last etching, and between 1778 and 1780 he all but gave up drawing altogether. But moments of genius still recurred, analogous perhaps to the equally unpredictable moments of poetic inspiration: by a bridge in the Weimar park, in the estate at Wörlitz, in the early morning sun beside his garden fence, he caught a chance combination of geometrical, humanly constructed, elements with a softer natural pattern, the whole bathed in a light, winter, summer, or misty spring, that seems the medium of the artist's sight and mood. These sketches, mainly in pencil and ink wash, have rightly been called 'proto-Impressionistic'. For all their strong sense of the open air, however, they show Goethe already fixed on the one theme that, apart from a few portraits and a mass of technical drawing, was henceforth to preoccupy him: a human presence in a natural setting, usually indirectly presented through an architectural or similar object.

The prose *Iphigenia* was not only the high point of Weimar amateur theatre— immediately afterwards came the first moves to erect a new and permanent building—it was also the culmination and recapitulation of all that Goethe had thought and felt over the previous three years. In the new atmosphere of the spring and summer of 1779, with the shadow of war lifted, and enjoying at last the satisfaction of a full-time occupation, Goethe began to turn his mind to his approaching thirtieth birthday. He felt the need once again to take stock and mark a new beginning. Already in March he had acknowledged that hitherto he had treated the gift of poetry 'a little too cavalierly' and had resolved 'to husband my talent better if I am ever to produce anything more'. On 17 August we find him sorting through his papers, burning 'old husks', and writing in his diary:

Quiet retrospect on life, on youth's confusion, busyness, thirst for knowledge, how it wanders everywhere to find something satisfying . . . how there has been so little achievement, or purposeful thinking or writing, how so many days passed in time-wasting sentiment and shadow-passion, how little of it profited me at all and, now that half of life is over, how no progress has been made but rather I am standing here like someone who has just escaped from the water and whom the kindly sun is beginning to dry. As for the time that I have spent in the big and busy world since October 1775, I do not trust myself to survey it.

Merck's visit to Weimar in June and July undoubtedly encouraged this retrospection, and it is in the context of Goethe's desire to shake off the burden of the past that we should see the 'crucifixion' of Jacobi's novel *Woldemar*, which took place during the visit. In a somewhat juvenile prank, which cost Goethe

Jacobi's friendship for some years, the book was publicly nailed to a tree in the neighbourhood of Ettersburg (probably by Merck), while Goethe delivered a speech including an impromptu travesty of Woldemar's end, in which he made the devil carry Jacobi's hero off to hell. This parody of the traditional ending to a *Faust* drama and so of his own most purely tragic piece of writing hitherto can only be explained on the principle 'other times, other cares', which Goethe also wrote in his diary.

But Goethe also felt the need for some more tangible rite of passage, some 'symbolical' act comparable with the Brocken expedition, of which the beneficent and reassuring effects were beginning to fade. (The summer saw a series of frank and highly charged discussions with Carl August and Anna Amalia, possibly concerning among other things Goethe's own position at the court, and perhaps the difficult relations between the Duke and the two Duchesses.) He began to conceive the idea of another, and longer, journey, to Frankfurt this time, to see his parents again and to return home not ignominiously, as in 1768, 1771, and July 1775, but, at last, in triumph. 'For the first time I would be returning to my fatherland completely well and happy and as honourably as possible', he wrote to his mother, suggesting that this was the fulfilment of the prophecy which she had found in Jeremiah during his illness ten years before (and which he had recalled, to give him confidence, the night before attempting the ascent of the Brocken): 'since the vineyards have flourished so well on the mountains of Samaria I should like the flute to be played as well'. And there was other unfinished business awaiting him in that part of the world too: he could visit Schlosser and his new wife, and greet his nieces whom he had never seen; Lili was in Strasbourg, now Lili von Türckheim, since her marriage in 1778 to a prosperous banker; and near Strasbourg lay Sesenheim and some more unfinished business. It also began to occur to him that such a journey could be of value to the 21-year-old Duke: for all his three-year reign 'he is still very inexperienced, especially with strangers, and at first has very little feeling how new people stand with him'. On his journey in the Harz Goethe had himself discovered how fresh and direct are relationships for the incognito traveller: in a small party under an assumed name Carl August could learn to meet people in the middle-class way, man to man, and in Frankfurt he could see from within the middle-class world from which his minister, mentor and court-poet had come. The Dowager Duchess had paved the way the previous year with a visit to the Frankfurt Goethes in June, and provided Frau Goethe could be prevented from making the beds too soft and the cooking too elaborate, and provided she removed the chandeliers, which she doubtless thought rather ducal, Carl August could learn the merits of another way of life without finding it ridiculous or philistine. A journey down the Rhine to Fritz Jacobi near Düsseldorf would complete Carl August's introduction to the milieu in which *Götz, Werther,* and *Faust* had been written and from which *Wilhelm Meister* came, and after returning to Frankfurt the

Duke and his minister could for a while go their separate ways, the one to the court in Darmstadt, the other to 'all my friends and acquaintances'. Carl August was attracted by the idea and made his own contribution to the birthday celebrations by telling Goethe on 28 August that it was his intention to promote him from Councillor of Legation to full Privy Councillor, the highest official rank of all. The court hissed with envy, and then with shock, when it learned of the unprecedented journey, and 'incredible public curiosity' was caused when on 11 September the ruling prince, a manservant, and a groom departed on horseback, accompanied only by Goethe, with his factotum Seidel and a young servant, and by gentleman-in-waiting Wedel, with his huntsman Isleib. They were expected to be away for six weeks.

A week later, half a day's journey from Frankfurt, Goethe announced to his companions a change of plan—'by inspiration from the angel Gabriel', Carl August told his mother, who was not pleased at what she thought must have been a secret deliberately kept from her. They would not after all go to Düsseldorf and Jacobi, but on reaching Frankfurt would go up the Rhine to Switzerland and visit another tutelary spirit of Goethe's pre-Weimar days, Lavater in Zurich. It is uncertain whether Goethe always had the intention of going south but revealed it to no one, whether he had already, as the company passed through Cassel, got wind of the violent offence Jacobi had taken at the treatment of his *Woldemar*, or whether it really was an inspiration of the moment, possibly with the thought in mind that they might not stop even in Switzerland but might, as was anyway soon rumoured, press on to Italy. There is no doubt, however, that the new scheme sharpened Carl August's appetite for the adventure. He was in the best of spirits in Frankfurt and on his best behaviour, making a most favourable impression on Frau Goethe, whom in turn he found 'a splendid woman': 'I have become astonishingly fond of the old mother', he wrote to Anna Amalia, who was also receiving letters from the other party:

The 18th September was the great day [Frau Goethe wrote] . . . His Highness our best and most gracious Prince dismounted—to give us a real surprise—a little way away from our house so came quite noiseless to the door, rang the bell, came into the blue room etc. Now let Your Highness imagine Mother Aya sitting at the round table, the door opens, at that moment Coddlejohn ['Hätschelhans', a pet name for Goethe] falls on her neck, the Duke at a little distance watches the maternal joy for a while, Mother Aya eventually rushes drunkenly up to the best of princes half sobs half laughs doesn't know what to do, handsome Mr Wedel too shares fully in the extraordinary delight.—Finally the scene with father, it is just indescribable—I was afraid he would die on the spot.

After three days of being shown off to Frankfurt, the prodigal son and his patron rode on to the Rhine, passing through Heidelberg and crossing to the west bank of the river at Speyer. Waiting for the ferry Goethe started a long letter to Frau von Stein: 'On this journey I am recapitulating the whole of my previous life, seeing all the old friends.' On the balmy, moonlit evening of the following day, 25 September, he slipped away on his own to visit the parsonage at Sesenheim and entering the door almost collided with Friederike. Of what passed between them practically the only evidence is contained in the letter to Charlotte, the relevant parts of which we cited in Chapter 3, and from which it is clear that only now did Charlotte learn about the affair. But, although Goethe naturally wished to present the atmosphere as untroubled, and all passion as spent, there seems no reason to doubt that Friederike, who later wrote a 'good' letter to Goethe, which has not been preserved, was indeed as friendly, natural, and composed as ever, glad, like the neighbours and the rest of the family, to see again someone with whom they had spent happy times. Goethe left the following morning, glad himself 'that I can now think with contentment of this little corner of the world, and live in peace with the spirits within me of all these people with whom I am now reconciled.' It is not of the first importance whether we dismiss this conclusion as smug and self-centred, or whether we think that simply to have undertaken the visit—which could have proved distinctly uncomfortable—shows a refinement of conscience unusual in a government minister. We should see the incident, not as evidence, one way or the other, for Goethe's moral character, but in the context, into which he had probably for some time planned to insert it, of the whole journey, and so as part of the pattern he was trying to weave in his 'symbolical existence'. And what is truly remarkable about the journey is that it mobilizes the resources of an entire duchy, including many weeks of its ruler's time, to bring order and contentment into the thoughts of a middle-class man about his life and his world. The enormous power of the symbolical pattern that results is shown by the contrast between the 'Sesenheim idyll', sentimentally rounded off in this final visit to Friederike, which has become part of German national mythology, and the purely private amours and guilts of Wordsworth and Coleridge a generation later, which have never similarly entered the public consciousness of the English-speaking world. We see here the beginnings of the collaboration, for which Goethe had yet to find a literary form, between his middle-class origins and his new life near the seat of political power, and can glimpse something of its national cultural potential.

The recapitulation continued. Having left Sesenheim on the morning of the twenty-sixth, Goethe arrived with the rest of the party in Strasbourg at midday and in the afternoon called on Lili von Türckheim, whom he found playing with her seven-week-old first baby and who invited him to stay to dinner. The following day, having crossed again at Strasbourg to the east bank, the party lodged at Emmendingen with Goethe's brother-in-law Schlosser: Cornelia's grave was visited, and her children found to be in the best of health, and with Johanna Fahlmer, now Schlosser, Goethe had a long discussion of the sad breach with Jacobi. After reaching Basle on 1 October, however, the travellers spent a month simply touring the recognized sights of western Switzerland and symbolism was forgotten. The lake of Bienne was visited and the Île Saint-Pierre where Rousseau had taken refuge in 1765, the house carefully preserved for the benefit of the pious, or the simply curious. Then through Berne, they passed on to Thun, taking a boat down the lake on a wet and misty day, to arrive in Lauterbrunnen, where the weather improved enough for them to admire the view of the Jungfrau, and to climb up in the sunshine to the Tschingel glacier. From Lauterbrunnen they walked to Grindelwald and across to the valley of the Aare, enjoying, thanks to the mild weather, magnificent mountain views above the rich autumn colours of the trees. Oddly, however, since leaving Emmendingen Goethe had found himself unable to draw 'a single line': poetry proved more responsive to the landscape. When they were back in Thun on 14 October Goethe could send to Frau von Stein the poem now known as 'Song of the Spirits over the Waters' ('Gesang der Geister über den Wassern'), another meditative hymn in free verse, which reflects his impressions of the great waterfalls on the Staubbach and the Reichenbach. Five more days were spent in Berne, whose wealth and constitutional arrangements were of equal interest to the Weimar statesmen, and then, via Murten and Moudon, they moved on to Lausanne, which they reached on 22 October. Having made the acquaintance here of the remarkable Maria Branconi and allowed themselves to be moved by the locale of *La Nouvelle Héloïse*, the party rode along the north side of the lake of Geneva to pay a call on Merck's parents-in-law near Rolle. It was raining again, but their hosts assured them the weather would clear and they were persuaded to ride up into the Jura, along the Vallée de Joux and climb the Vaulion and the Dôle. From here they did indeed see on a hot sunny day the whole panorama of the Alps from Geneva to Lucerne, with the icy pinnacles of Chamonix to the south and in the foreground the neat agricultural countryside round the lake, and Geneva itself, to which they now descended and where they stayed from 27 October to 3 November. Voltaire's estate at Ferney, a monument to the power of the pen that cannot have been lost on Goethe, had of course to be visited, though Voltaire himself had died the previous year, but there was also much to do in Geneva. Carl August had his portrait painted and Goethe inspected the local art collections and had some rather unsatisfactory conversations with a young theologian and classical scholar, G. C. Tobler (1757–1812), who came with a recommendation from Lavater and copies of Lavater's latest works, a paraphrase of the book of Revelation and an epic poem *Jesus Messias*. The natural historian Charles Bonnet (1720–93), whose views Goethe was later to criticize severely, was visited, as was the scientist and alpinist H. B. de Saussure (1740–99), who in 1787 would become the second man to climb Mont Blanc. Saussure had to advise the young adventurers on the wisdom of their plan to go up into the high mountains, to Chamonix and up the Valais to the St Gotthard, at such a late stage in the year—everyone else they had spoken to thought it madness, but Saussure said that with the weather as it was, and provided they took local advice, they would be perfectly safe. Goethe was delighted, but wrote to Frau von Stein that they would need to be attended by good fortune: 'If things were to be there as they are depicted to us here this would be a journey into Hell', but 'if it is possible to get to the top of the Brocken in December, then these terrible gates too should let us through at the start of November'. The oracle was to be consulted again.

Unfortunately Wedel had proved to have no head for heights, so on 3 November he started to take the horses along the lakeside road to Martigny, where the party was to reassemble, while the Duke, Goethe, and huntsman Isleib set off into the heart of Savoy. By hired carriage and on foot they reached Chamonix on the evening of the fourth. The following day was spent inspecting the Mer de Glace, on to which Goethe and the Duke ventured for a few hundred paces, and on the sixth they were guided up out of the valley of Chamonix, over the Col de Balme and down to Martigny. Their route now, up the Rhône valley between the two great chains of Switzerland's highest mountains, the Bernese Oberland and the Pennine Alps, took them into increasingly desolate and impoverished country. On returning from a detour to see the hot springs at Leukerbad they were warned that further along the main road there would be no fodder for their horses. Once again then Wedel agreed to take the horses separately, on a roundabout route, returning to Lausanne and meeting up in Lucerne with the Duke, Goethe, and Isleib, who would press on on foot and hired mules. But should they follow the route recommended by Saussure, across the Simplon pass, down into Italy, across Lake Maggiore and back to the St Gotthard? Or should they carry on up to the source of the Rhône and endeavour to cross the Furka, so reaching the St Gotthard directly? At this distance there was no reliable information about snow conditions on the Furka, but Goethe, who took the decisions, though Carl August thought of them as shared, was superstitiously reluctant to touch Italian soil. They could spend little time there, and, in respect of the Duke, the purpose of this trip was to bring him close to natural grandeur and to some of the places and people that had been important to Goethe, not to show him the home of classical art, for which he was not yet ripe, if he ever would be, nor to partner him in a joint voyage of discovery to a new world. Goethe's clear aversion to the Italian route must have been due to a feeling that it would not be 'symbolically' right. Italy meant very much to him: he had come to Weimar as the alternative to obeying his father and going to Italy. Only when he had proved himself in Weimar—and he had so far been fully in office there for only nine months—would he feel independent enough to make his grand tour by his own choice rather than by his father's decision, and that moment was not lightly to be anticipated. And perhaps, who knows, Italy meant too much for him to want to go there in his present company: when the time came to go, he would go alone.

So when the little party left Brig on the morning of 11 November they ignored the southward road winding up to the Simplon and set their faces up the valley to the Rhône glacier from which a cold east wind was blowing. It was a welcome relief to find a hospitable house at midday where they were well restored, although the devout landlady insisted on telling them the entire legend of St Alexis. (This utterly adventitious moment was to prove one of the most lasting literary influences Goethe underwent.) But the wind was keeping the snow-clouds at bay, and so it seemed worth continuing, after spending the night in Münster, as far as Oberwald, at the very head of the valley. Here, to their surprise, the travellers' tentative inquiries for guides were met with the assurance that the Furka was passable for most of the winter and, after eyeing the physique of the temerarious strangers, two stoutly-built villagers agreed to take them across. There was a good view of the blue ice-cliffs of the Rhône glacier, but after that the grey day with flurries of snow and only occasional glimpses of a pale sun cut them off from any sight of the peaks. They plodded in the wake of their guides through a white waste, devoid of life of any kind, three and a half hours to the top of the saddle, and three and a half hours down from it, through waist-deep snow in which they did not dare to pause for fear of the cold. 'It was the toughest thing I have ever done', Carl August confided to his diary, and they had done it on the fourth anniversary of Goethe's meeting with Frau von Stein. In Realp they spent the night with the Capuchin monks and on 13 November reached their brethren on the St Gotthard, 'the summit of our journey'. Here Goethe rejoined the route that he had come 'four years before, with quite different cares, attitudes, plans and hopes', when 'not guessing my future fate, and moved by I know not what, I turned my back on Italy and unknowingly went to meet my present destiny'.

So the recapitulation was resumed. Passing through Schwyz and Lucerne, where the company was reunited, they reached Zurich on 18 November and 'the seal and highest peak on our journey', the meeting with Lavater, 'the best, greatest, wisest, deepest of all mortal and immortal men I know'. Here too acquaintance was renewed with Barbara Schulthess and the obligatory visits were of course paid to the town's other literary grandees, notably Bodmer, who was now over 80—his main subject of conversation. But Goethe seems to have been particularly attracted by the combination in Lavater's household of domestic peace and affection with an industrious life of the mind, and of course with the pastoral responsibilities of one of the city's principal clergymen. The fortnight that he spent in this atmosphere of warm but purposeful feeling was a revelation, and a restorative. 'Only here has it become really apparent to me in what a state of moral deadness we normally live together, and what the origin is of the withering and freezing of a heart that in itself is never dry and never cold.' Perhaps the prospect of the return to Weimar was already beginning to cast a shadow, though it was with every intention of maintaining the 'openness' of soul that he had learned here, and opening the souls of others, that he set out with the Duke on the homeward journey on 2 December. Via Winterthur and Constance they made their way to Schaffhausen, where Lavater rode to meet them again, to join them in admiring the Rhine waterfall, and to bid them

farewell from Switzerland. There followed a series of visits, of varying degrees
of tediousness, in generally bad weather, to the courts of south-west Germany.
In Stuttgart Carl August and Goethe, though officially incognito, were the
guests of honour at the annual prize-giving at the Duke of Württemberg's
Academy: among those who received an award was the young Friedrich
Schiller, who had already been working in secret for pehaps three years on the
manuscript of *The Robbers*. In Mannheim they met Heribert von Dalberg, the
brother of the Statthalter of Erfurt, who in two years' time would arrange for
the production of Schiller's play, and, to Goethe's present displeasure, they saw
a production of *Clavigo* put on in their honour, with Iffland in the role of Carlos.
Goethe meanwhile had been keeping alive the memories of Switzerland and the
free mountain air by composing an operetta set in the high pastures of the Alps,
Jery and Bätely, which was completed by the time they all reached Frankfurt on
Christmas Day, when he first set pencil to paper to make a visual record of
some of the scenes he had left behind. The Goethes' house became the base for
various forays to the local courts of Homburg and Darmstadt, where Duchess
Luise's relatives had to be visited, and Frau Goethe had another deliciously
exciting fortnight until on 10 January the travellers left for Weimar, which they
reached on the thirteenth. Despite the apprehension and hostility that had
surrounded the journey at the start—Goethe was thought quite capable of
leading the Duke into breaking his neck, and there was still no male heir—and
despite the total cost of nearly 9,000 dollars, the expedition was now generally
judged a success. Carl August was in good health, had evidently matured
considerably, and had left behind him nothing but good reports. Already in
Schaffhausen Goethe had written to Lavater asking him to secure the co-
operation of the admired J. H. Fuseli in designing a memorial to the journey, to
be erected in Weimar and dedicated as a thank-offering to 'Good Fortune', to
the 'good spirit' that had accompanied them, and to the equally good spirit that
had told them at what point to turn back (that is, presumably, not to go on to
Italy): 'get him to do it, and to do it quickly, and crown this year too for me, and
its good fortune, with this last sign'. But Fuseli was not to be persuaded, and
Goethe's thirtieth symbolical year was left without its crown.

Works, 1777–1780

The literary works of these crowded years can be divided into two groups. The
earlier and larger of these consists of the works that can be directly associated
with what we have called the 'tragic' theme of the period: *Proserpina* and *The
Triumph of Sensibility*, the conclusion to the first book of *Wilhelm Meister*, certain
major poems, and *Iphigenia*. The second group is related to the first, but distinct
from it, and consists of the works associated with the journey to Switzerland,
'Song of the Spirits over the Waters', *Jery and Bätely*, and the *Letters from
Switzerland*.

The essence of the tragic insight with which Goethe was struggling in this period is expressed in a little poem probably, though not certainly, written well outside it, as late as 1783:

> Wer nie sein Brot mit Tränen aß,
> Wer nie die kummervollen Nächte
> Auf seinem Bette weinend saß,
> Der kennt euch nicht, ihr himmlischen Mächte.
>
> Ihr führt ins Leben uns hinein,
> Ihr laßt den Armen schuldig werden,
> Dann überlaßt ihr ihn der Pein,
> Denn alle Schuld rächt sich auf Erden.

He who has never eaten his bread with tears, who has never spent nights of sorrow sitting weeping on his bed, does not know you, o you heavenly powers. You lead us into life, you make the wretch become guilty, then you leave him to his torment, for all guilt is avenged on earth.

As early as 1775, however, it will be remembered, he had written in his diary on leaving Frankfurt, as he thought, for Italy: 'Am I then in the world only in order to writhe in never-ending innocent guilt?'—thinking no doubt of Friederike, of Lotte Buff, of Lili, of *Werther,* and all its suffering readers. But the years in which Goethe was most consistently haunted by the terrible fear evoked by his little poem—surely far nearer to the 'terror' which Aristotle thought one of the instruments of tragedy than all the pages of Lessing, or indeed of Nietzsche—were the years when it had become clear that the life he had lived from 1771 to 1775 could not simply be continued in the new setting, and with all the material advantages of the Weimar court, but that a sacrifice was required. The fear that these lines, and all the works of this period in their different ways, express is the fear that the sacrifice might not be effective. Could Goethe perhaps rescue himself from the self-destruction and self-delusion into which Sentimentality and Storm and Stress had been drawing him only by entrusting himself to another delusion—Weimar—no more substantial than the first ('Actually I am not necessary here, but I imagine I am . . . ') and so in the end equally destructive, destructive indeed in the same way: through excessive cultivation of self? There are two distinct elements to this fear, which on their own are pathetic rather than tragic: first, the possibility that the gods are indifferent to the cries of weeping men, and then, the possibility that for the soul the only escape from the self-inflicted misery of guilt may be into the misery of self-inflicted punishment. Taken together, however, they provide the material of such truly tragic paradoxes as that the gods hold aloof from men because of men's distance from the gods, or that, as for Oedipus, our awareness of guilt is the crime for which we are punished. The self-destruction of a sensibility is not in itself tragic—it may even be comic—but it becomes tragic when the soul appeals to the gods for a relief that is beyond its own powers to bring, and yet finds no relief, for the gods have punished it by leaving it to its own devices.

The brief monodrama *Proserpina* is Goethe's nearest approach to the stark, even bitter, simplicity of classical tragedy. First printed in the *Teutscher Merkur* in February 1778, it is most likely to have been written in 1777, immediately after Cornelia had died, but it may be as much as a year older. We find Proserpina in the underworld bewailing her abduction, the torments she sees about her in the kingdom of death, and the prospect of her union with the hateful Pluto (for which Goethe may have had a model in the marriages not only of his sister but also of Frau von Stein and Duchess Luise). But the play is not simply a lament: at its centre lies a decisive action, a turning-point. Proserpina appeals—like Ganymede before her, but also like Werther—to the loving Father, Jupiter in his heavens, who, she is convinced, will raise her from death. Hope floods her heart and refreshes it, and as a sign of hope there stands before her a pomegranate tree, life in the wilderness, planted surely to revive her spirits. She eats the fruit, which does indeed prove refreshing, and which takes her mind back to the happiness of her youth in the upper world. And then a terrible certainty seizes her, which finds confirming expression in a chorus chanted by the invisible Fates: by tasting the fruit she has violated the condition, unknown to her, on which she would have been allowed to return to life. She is now more firmly than ever bound to the underworld, bound in eternity to be queen of a realm of shadows. Not merely, then, has Jupiter deceived her, and proved not to be the all-loving Father she imagined, but her very hope that he was going to save her, embodied for her in the refreshing fruit of the pomegranate, has been the instrument of her deception and final downfall, the crime for which she is now to be punished. Her first state, before she eats the fruit, familiar to us from the darkest moments of Goethe's Frankfurt existence and their expression by Werther, is an emptiness of the heart in a grey world, surrounded by shadows that, like Prometheus' statues, are empty of all independent life:

> Leer und immer leer!
> Ach, so ist's mit dir auch, mein Herz!

Empty and always empty! Alas, my heart, that is your state too!

The hideous possibility, enacted by her story, is that this state is the true one, and that hope for anything else is a hubris of the heart, a self-deceptive belief in the power of our feelings to transform the world, which only makes our desolation worse—perhaps even makes sinful (by the eating of the forbidden apple) a state which before was merely wretched, but which now is rightfully punished. Proserpina is banished from the sunlight not once but twice; her appeal is not heard, rather, her making it is the cause of a yet deeper despair; and this fear of a double condemnation—which he saw exemplified in the fates of Cornelia or Plessing or Lenz, or even Luise—Goethe kept at bay only by the device of leaving the confirming chorus of Fates invisible, and so conceivable as only another twist of Proserpina's self-tormenting fantasy.

At the time when he wrote it Goethe seems to have been appalled at what he was saying in his monodrama, and almost immediately he sought to neutralize it —rather as he did with the ballad 'Once there was a lusty lad'—by inserting it into a new and alien context. In the fourth act of *The Triumph of Sensibility,* the play *Proserpina* is performed by the Queen Mandandana, and so serves to show that she is infected by the disease of Sentimentality, which has made monodrama into a vogue. Yet, though a comedy, and a good one, *The Triumph of Sensibility* has a theme not as different from that of Proserpina as may appear at first sight. Mandandana, wife of the 'humoristical King' Andrason, is the object of the passionate attentions of Prince Oronoro, a worshipper of nature and of the heart, and she is so swept away by these ethereal delights that she is quite alienated from her husband, who seems to be the Pluto to her Proserpina. Prince Oronoro, however, is of so delicate a constitution that he cannot tolerate the dews and vapours of a real landscape, which so easily cause a chill, let alone the ants, midges, and spiders that disturb one's contemplation when lying in the grass or in a grotto on a warm moonlit night. So when he comes to stay at Andrason's palace he brings with him his own 'travelling Nature': artificial bushes, rocks, springs, and moonshine, all packed up in boxes, from which they emerge at the touch of a lever, to transform an interior scene into an ideal moonlit glade, without any of the inconveniences of the real thing. In an arbour, in the centre of the bogus glade, is Oronoro's true love, a stuffed dummy with the features of Mandandana, which he carries about with him in order to be able to sink into rapture whenever he wishes. While Oronoro is away, consulting an oracle in the mountains to which Andrason has also earlier made a pilgrimage, Andrason is able to dissect the dummy and discover the secret of its magnetic power over its owner: it proves to be filled with works of Sentimental literature including, as the 'sludge at the bottom', *La Nouvelle Héloïse* and *Werther.* Rather than destroy Oronoro's illusions, Andrason patches up the dummy, tells his rival, when he returns, that he will after all allow him to possess the object of his heart's desire, and confronts him with a choice between the real Mandandana and her sentimental image. Oronoro of course is delighted to choose his bag of books and sawdust, while Mandandana's eyes are opened and she is reconciled to her husband.

The recurrence of the motif of Pygmalion—the artist who, as in Goethe's early ballad, prefers the image he makes to the life it represents—should warn us that his 'dramatical whim', as Goethe subtitles it, goes deeper than the song and dance and farcicality suggest. In the two performances of 1778 Goethe played the part of Andrason, the King whose idea of nature is a stout staff and a good hike, but there is plainly just as much of him in Prince Oronoro (which, together with the equation of Mandandana with Duchess Luise, may have led to some of the 'stupid interpretations' of the play of which Goethe complained in his diary). Oronoro's Wertherian inability to escape from the creations of his own mind and desires could in a different context be tragic: the play in fact ends with his ignoring the oracle which warns him that, unless he gives up his fantasies, he will suffer 'the fate of Tantalus, both in this and in the next world'. Goethe at this time, as is clear from an expression of sympathy with him in *Proserpina*, confused Tantalus with Ixion, and thought that Tantalus was punished with his famous torments of unsatisfied desire for having attempted, of course without success, to ravish Juno, Queen of the gods. Oronoro's fate therefore is not to attain his Queen, but to be punished as if he had done so; he is consigned to endless frustration both of body and of soul; and in his own way he comes, therefore, under the double condemnation of Proserpina. For in his moment of 'hope', as he calls it, in the final scene, when he thinks that Andrason is arranging for the fulfilment of the divine oracle which promises an end to his cravings, when he thinks he is renouncing fantasy and embracing reality, his heart has betrayed him and he has chosen instead the goblin fruit that will keep him in servitude for ever.

Goethe of course 'is' the Götz-like Andrason as much as he 'is' the Werther-like Oronoro, and the play at times mocks itself and its Weimar audience as much as it mocks Sentimentality in the abstract, located somewhere and nowhere in a Germany outside Weimar. This division of Goethe's self, this uncertainty as to his own position, in relation both to his past life and to his present circumstances, is the subject-matter of 'Winter Journey in the Harz', just as it was the motive that drove him to consult his own oracle on the summit of the Brocken in the middle of the writing of *The Triumph of Sensibility*. In the poem Goethe tries to determine whether his own destiny is to be one of 'Glück' or of 'Unglück', of good or evil fortune, of the happiness of the man who quickly and easily runs his appointed race to a publicly acknowledged goal, or the misery of the man, or woman, closed in on himself by ever-chafing bonds from which the only release is an equally bitter death. He directly opposes to each other now the two possibilities for his life that in his first eighteen months in Weimar he had thought might coexist in incestuous indistinction. There is the path of worldly good fortune, trodden with ease by the crowd of hangers-on in the baggage train of a prince: the description of court life is mildly condescending, but it is not dismissive. For what is the alternative? What is there better to do, away from the 'improved roads' of a modern absolutist state? The evocation of the Harz landscape and the interrogation of the poet's destiny, which have so far in the poem largely run in parallel, now converge in an image of overwhelming immediacy:

> Aber abseits, wer ist's?
> Ins Gebüsch verliert sich sein Pfad,
> Hinter ihm schlagen
> Die Sträuche zusammen,

> Das Gras steht wieder auf,
> Die Öde verschlingt ihn.

But off the highway—who is it? His track is lost in the undergrowth, the bushes close to behind him. The grass straightens up, the wilderness swallows him.

A passionate lament follows for all those whose hearts have been turned in on themselves by misfortune, such as Cornelia, perhaps, but such especially as that whole generation which had refused to take the path being followed by their society and those powerful within it, and had looked instead for their future to the frail new network of the public mind, the new national community of feeling, mediated and supported, as Prince Oronoro knew, by books. That generation, finding themselves in a book, in *Werther*, had trusted in the power of their own hearts, in the power of love, and, when this proved impotent to change the world, were destroyed by the dissolution of their unapplied emotions. Goethe may be thinking of Plessing here—as he was later anxious to suggest—but he was equally certainly thinking of himself, cut off in September as by a 'chasm' from the court life at Wilhelmsthal and from the international courtly culture of Baron Grimm, and increasingly presenting a frigid and fortified exterior to the world. There was here for him still the danger of emotional self-destruction, which had not been exorcized by the writing of *Werther*, nor of *Proserpina*, nor of *The Triumph of Sensibility:*

> Ach, wer heilet die Schmerzen
> Des, dem Balsam zu Gift ward?
> Der sich Menschenhaß
> Aus der Fülle der Liebe trank.
> Erst verachtet, nun ein Verächter,
> Zehrt er heimlich auf
> Seinen eignen Wert
> In ungenügender Selbstsucht.

Ah, who will heal the pains of one for whom balm turned to poison, one who drank misanthropy from the brimming cup of love? First despised, now a despiser, he secretly consumes his inner worth in unsatisfying self-addiction.

In this dilemma, between a life of the heart that cannot be sustained without external support, and a social life that cannot accommodate the heart and is too insubstantial to support it (being merely a following in a ruler's train), Goethe turns to the gods. As in the expedition to the Brocken, so now in the poem about the expedition he seeks a response to his anguish. Like Proserpina he calls on the 'Father of Love', in the hope that to so great a need as that within him there must be a reply from the external order of things. Subjectivity cannot be left to destroy itself, or to die of inanition; the objective world must somehow, somewhere, come to meet it and bear it up. *Proserpina* speaks the unspeakable reply: that there is no Father, or he does not hear and leaves fate to its inexorable workings. The need of the heart for happiness, love, and fulfilment is, the monodrama says, a pure illusion, and the trust that the need will be met is a double illusion, a double self-indulgence, like that of Oronoro who not only made his dummy, but turned to it in preference to the real princess. The ascent of the Brocken was for Goethe personally an objective and tangible confirmation of his *good* fortune, a response from a power outside him, a power of circumstances over which he had no control, and a response which by its very existence confirmed that he was not Proserpina. There was indeed, he could conclude, a benevolent, natural order, however unfathomable, which ensures that there is ultimately an object for the yearnings of the subject, and that something so balm-like as the love that needs fulfilment is not a poison that detaches us from truth and gives us up to the destructive embrace of our own self-regard. In an established bourgeois culture such as that of Jane Austen, secure in its own public self-image, Goethe could have received this assurance from the external and independent power of his own society, his fellow human beings. As it was, he had to create an objectivity of his own, making an oracle out of a task so physically demanding that it could seem not to be speaking with his own voice, making the Brocken into 'myself not myself'. Something of this paradox—an objective emptiness, even banality, into which the observer is strenuously reading a sublime but featureless significance—can be sensed in the charcoal drawing of the Brocken by moonlight which Goethe completed in Torfhaus after his descent, his culminating, and practically his last, attempt at rendering a moonlit German landscape. In the poem the success of the augury in Goethe's life is more effectively imitated by the sudden revelation that there is a third path for the speaker to follow, neither that of the courtier nor that of the solitary, and the threatening dilemma of the poem's central sections is thus resolved. After asking for blessings from the Father of Love on both these alternative ways of life, Goethe turns to himself, and this 'solitary one' is now spoken of for the first time as a 'poet'. But this is not the poet-genius, whom we met for example in 'Song of Mahomet', the leader of a collective fraternal movement who speaks, like Hans Sachs, with the voice of the nation—though there may be a vestigial allusion to that ideal in the cryptic final lines of the hymn. This is a more conventional figure, a poet crowned with bay, and looking forward to being crowned with roses, a poet therefore in a court uniform, and with a conventional courtly subject: he is the poet of love, love, he assured Frau von Stein, of her. As such he proves to have a third route through the wilderness of the Harz, neither on metalled roads nor petering out in the undergrowth, but, borne up in all weathers by the Love that he sings, he follows nature's path—fords, moors, and boggy tracks—up to the natural landmark where he receives his objective, but non-social, confirmation of the bond between gods and men (and, incidentally, has revealed to him the springs in the desert which he had asked should be made manifest to the despairing wanderer). These lines he cut into the window of the forester's station in Torfhaus looking out on to the Brocken, a votive inscription binding together the culmination of his poem, of his expedition, and of the mountain range before him:

> Und Altar des lieblichsten Danks
> Wird ihm des gefürchteten Gipfels
> Schneebehangner Scheitel,
> Den mit Geisterreihen
> Kränzten ahnende Völker.

And an altar of sweetest thanks is now for him the feared peak's snow-hung crown, which half-knowing peoples wreathed with ranks of spirits.

For as long as Goethe remains the poet of this 'love', he can hope that there remains a third path for him in Weimar between assimilation and isolation, and that the response to his appeal to the heavenly powers will not be the tragic rebuff suffered by Proserpina, but the maintenance in his life of a 'symbolical' relation with the external world, both human, such as it is at a provincial court, and natural.

Convinced by the oracle of the Brocken that the path marked out for him did indeed run through Weimar, Goethe seems on his return to have been taking deliberate steps to cut himself off from his previous life. Not only was *Proserpina* deprived of its tragic effect by being inserted into *The Triumph of Sensibility*, but he hurried to dispatch a task which had been begun at a time when Weimar still seemed an agreeable continuation of what had gone before, the first book of *Wilhelm Meister*. On this too the events of 1777 left their mark. As the story moves into Wilhelm's adolescence the narrative texture becomes noticeably thinner: the perspective we concentrate on is principally Wilhelm's, and that principally as it relates to theatrical matters, to the amateur dramatic group set up by him and his young contemporaries, to his first hypnotized visits to the performances of a professional company, to his acquaintance with the members of the troupe, and to his falling hopelessly in love with one of the actresses. There are still many humorous and realistic touches, such as the embarrassing failure of the young actors at their first production, having carefully prepared scenery, costumes, and properties, to think of providing themselves with a script, or the extensive description of the dirt and disarray in an actress's dressing-room. But the breadth of scope of the early pages has gone, the townspeople, Wilhelm's family and their doings, even the physical presence of the house, are largely lost from view, and a generalizing and summarizing narrative manner, telling rather than showing, has taken over. The promise of a realistic panorama has faded, the mind and experience of the central character have become dominant, and these are soon caught up in a tragic illusion. The name of the actress whom Wilhelm loves is Mariana, Goethe's pet name for Amalia Kotzebue, and the affair between this Wilhelm and this Mariana is in some ways a reversal and recantation of that in *Brother and Sister*. For the ingenuous Wilhelm Meister, his relation with the young but experienced

Mariana, which develops gradually into physical intimacy and pregnancy, is the medium for the embodiment of an ideal. He sees himself as becoming, through their planned elopement and marriage, a great actor, and the founder of that national theatre for which all Germany is longing, indeed he sees the theatre itself as fulfiling a religious role, parallel to that of the Church (which has no part in the story at all): 'I won't utter it, but I will certainly hope it: on us great beauty shall descend, and the appearance, desired by all, of the more than human in human form.'

This ambition, purportedly expressed in the 1750s but so closely associated from Goethe's point of view in 1778 with the movement of Storm and Stress, is shattered, along with all Wilhelm's illusions about Mariana and his plans for their future, by his discovering, immediately before their intended departure, when his emotions are at fever pitch, that Mariana is the kept mistress of a local merchant. With this revelation to him (though not to us) the first book of the novel ends. Wilhelm Meister's theatrical mission seems to be over in the moment of its first articulation. Goethe probably drafted the first pages of the second book at much the same time; they let us know of the physical breakdown Wilhelm suffered as a result of his discovery, and sketch out what is effectively a completely new beginning for his story. Wilhelm is now for the first time presented as a poet who has written much, and thought much about his craft, and this, rather than a passion for acting and the physical charm of the stage, is to be the basis for his future involvement with the theatrical world. This was perhaps all that Goethe needed in order to be able to pick up the threads again later, if he chose to. We do not know how far he progressed with his recast novel, but when he dropped it he let it lie for a good four years. He was not in the mood for maintaining continuities with his past, nor missions to the German public: the attempt to found a national culture, such as that of which Wilhelm Meister dreamed, had cost him quite as much as the physical collapse it cost his young hero, and if he was to maintain himself in Weimar (and, it must be said once again, there was nowhere better to go) he had to find other means of serving 'the good cause'. About this time he wrote 'To the Moon', with its exhortation to turn from the 'world' without to the 'breast' within, and after the publication of *Proserpina* in February the last faint signs of an interest in the *Teutscher Merkur* as a national organ disappear. Wieland, who, in his (early) retirement, still lived his life for the German public mind, and whose only contact with the court was with Anna Amalia, wrote in June, with his usual sweet temper, of the new state of his relations with the minister and court poet inspired by a strange love: 'Two years ago [Goethe] and I were still living *with* one another; that is no longer, and *can* no longer be, since he has business, *liaisons*, pleasures and pains in which he cannot let me share, and in which I *ex parte* my position, could not share and would not wish to.'

We have noticed that the conviction of guidance by a benevolent fate which Goethe derived from his journey to the Harz was always fragile and threatened.

Tragedy, such as that of Christel von Lassberg, was an ever-present possibility (and this partly accounts for a certain brutality in his treatment of those of his works in which that possibility was too clearly expressed). Some time between 1778 and 1781 he wrote a quietly-spoken free-verse hymn, 'Bounds of Humanity' ('Grenzen der Menschheit') which, within a humble, even devout, acceptance of the difference between the transcendent, contemplative gods and fate- and storm-tossed human beings, encases a suggestion that there is something almost predatory in the gods' superiority to human life:

> Ein kleiner Ring
> Begrenzt unser Leben,
> Und viele Geschlechter
> Reihen sie dauernd
> An ihres Daseins
> Unendliche Kette.

A little ring rounds our life, and they string many generations on to the endless chain of their own existence.

Even if the image of the chain here suggests that 'chain of culture' which according to Herder links one generation to another and is the only means by which divinity is incarnated in human history, there is in the poem a hint of suppressed resentment here—at our being made only as the fodder for greater, even if possibly benevolent, purposes unknown to us. The resentment is quite explicit in Goethe's letter from Berlin in May 1778, when he had had the opportunity of seeing the same predatoriness in the human world: 'I worship the gods and yet feel courage enough in me to vow them never-ending hatred if they wish to behave towards us as does their image, mankind.' Here we have in embryo what could be called the theology of the play which, drafted some nine months later in the space of about six weeks, was intended as a summary, and has become a symbol, of these early years in Weimar: *Iphigenia in Tauris*. *Iphigenia* is a play about purification—of the heart from passion, of the past from guilt, of relations between men from mistrust and deceit, of relations between men and gods from fear and hatred. The play, indeed, gives a central position to the gods—that is, to the objective background, and support, to human feeling. Like 'To the Moon', it is a work in which the presence of Frau von Stein is felt as a power shutting the doors of the mind on a violent past, haunted by death, sin, and irrational desire.

It is probably, however, the formal innovations which most immediately strike the reader. For *Iphigenia* is also the first play in which Goethe attempts a purification of literary form and language consonant with the new social and emotional certainties that flow from his pondered decision to stay at the Weimar court. It might at first sight seem as if this, one of the few fully serious dramas to be mounted by the Weimar amateurs, is simply an application of Gottsched's recipe for the German middle class's cultural compromise with princely absolutism: Bodmer, who mercilessly identified and listed the play's many artificialities, finally remarked, 'What pride! to want to write only for a part of the nation . . .'. Wieland, conversely, if for the same reason, welcomed it as a contribution to the purification of the national taste. It is a five-act drama, closely modelled on an ancient source, observing the unities of time, place and action, and with a symmetrically aranged cast of only five high-born characters. The language eschews impropriety, though it is marked by strained or strange constructions intended to create an atmosphere of dignity and to echo Greek works or their translations; the dialogue is often sententious, crystallizing into lapidary single-sentence exchanges reminiscent of ancient stichomythia; monologues are frequent and lengthy, even when characters are not alone. But Gottsched's alexandrines are an impossibility in 1779, and no German substitute has yet been found—this *Iphigenia* is written not in verse but in a hybrid medium, a rhythmical prose, purged of the colloquial without attaining the formal, a sign of a fundamental uncertainty of intention, and more than any other factor responsible for Goethe's reluctance to publish the play, or even circulate it in manuscript, until he had rewritten it in 1786. It is as if Wieland's *Alcestis*, the opera so fiercely attacked in *Gods, Heroes, and Wieland*, had been recast in the prose of Lessings's *Emilia Galotti*, and indeed the influence of the blank-verse passages in *Alcestis* can be heard at many points in *Iphigenia*, when its predominantly iambic rhythms concatenate into five-foot units.

Alcestis, a local product, was a natural choice as a model for a courtly drama to be performed in Weimar, and Goethe in 1779 did exactly what he had found so reprehensible in Wieland in 1773. He took a play of Euripides and adapted its plot and its characters to make it a vehicle for modern moral reflection. Iphigenia, the eldest child of Agamemnon, is the great-great-granddaughter of Tantalus, whom, it will be recalled, Goethe thought to have been guilty of an assault on the Queen of the gods after he had been privileged to eat at Jupiter's table on Olympus, and she can trace through every stage of her ancestry the hideous crimes against humanity that have followed on that first offence against the divine order. The world indeed thinks of her as one of the latest victims in her family's bloody history, for she is believed to have been offered by her own father as a human sacrifice at Aulis, to secure favourable winds for the Greek ships sailing against Troy. In fact, however, she was rescued from the altar by the goddess Diana, chaste sister of Apollo, and transported to remote Tauris (nowadays the Crimea) on the Black Sea, where, her identity unknown and cut off from all her family, she has acted as Diana's priestess ever since. Pure and queenly like her patroness, she has acquired great moral authority over the people of Tauris, and their king Thoas, and persuaded them to abandon their custom of sacrificing to the goddess any foreigner unfortunate enough to reach their shores. Despite her beneficent influence on the Taurians, however, she longs to return to Greece and her family. But, when the play begins, Thoas, feigning the excuse that his only son and heir has recently been killed in an otherwise successful military campaign, and that he is under the pressure of public expectation to provide for the future government of his country, comes to her temple and renews his frequent previous requests to her to marry him, allowing her to glimpse the iron fist of the ruler, as of the barbarian, by ordering her, when she refuses, to resume the old practice of human sacrifice. Two roving foreigners have recently been arrested: let her begin with these. The two foreigners are Orestes, Iphigenia's brother, and his childhood friend Pylades, who have come at the behest of Apollo's oracle in Delphi seeking healing for Orestes, maddened by the pursuit of the Furies, the underworld goddesses of vengeance. For Orestes has killed his own mother, Clytemnestra, in punishment for her murdering his father, Agamemnon, on his return from Troy, the conquest of which, it seemed to Clytemnestra, had been achieved only at the price of Iphigenia's death. Apollo's oracle has told them that if they fetch a 'sister' from Tauris back to Greece Orestes' torment will end, and they have come with a shipful of companions, hidden in a nearby cove, to find a means of abstracting the image of Apollo's sister, Diana, from Iphigenia's temple. Out of caution, the Greek captives only gradually reveal their identity, and the history of Agamemnon's family, to the strange holy woman charged with their ritual sacrifice. While Iphigenia sees in the arrival of her brother in her place of exile the culmination of a divine scheme to rescue her, and thanks the gods accordingly, Orestes sees in the impending slaughter of brother by sister a worthy conclusion to his family's sequence of crimes and falls into a demented vision of himself after death meeting his ancestors in the underworld. From this he is roused by Iphigenia, who prays for him to be cured, and he is: 'I think I hear the fleeing chorus of the Furies close the gates of Tartarus with far-dying thunder', he says, and he is spared any further visitations from the spirits of revenge and remorse. How, though, is the Greek party to escape, and with Diana's image? Thoas is demanding that the sacrifice be carried out. Pylades has a scheme for delaying the ceremony and getting the image down to the shore, but it requires that Iphigenia should prevaricate. She sees that there is no other way to obey the oracle and return to Greece, but has scruples about robbing and deceiving the king who has been good to her for so long. How can she bring back blessings on her family home in Mycenae if her journey starts in criminal ingratitude? She almost doubts the goodness of the gods, and repeats a terrible song she heard in her childhood, calling on mankind to fear these higher beings at their golden tables on mountain tops, from which they look down in satisfaction on their bound and suffering enemies. In the last act, Thoas, his suspicions aroused, comes to the temple once more to insist on his will, and Iphigenia resolves to appeal to the good she knows to be in him by revealing everything. Reluctantly he yields to this courageous openness and to her plea for mercy, and when Orestes suddenly sees that, in speaking of a 'sister', his oracle was speaking of Iphigenia, and that only a misinterpretation led him to think it necessary to carry off the Taurians' temple-image of Diana, the way is clear for Thoas, his political excuses now forgotten, to suppress the passions that drew him to Iphigenia, to cease attempting to force her into obedience, and to allow brother and sister to depart together, with his blessing.

Goethe may have been considering a play on this subject since 1776, or even 1775, but if we compare the relationship of brother and sister, on which *Iphigenia* depends, with that relationship as it appears in 'Why did you give us the deeper vision' and the play *Brother and Sister,* it is evident that a great change has occurred: the suggestive confusion of family and erotic relationships which makes for the fascination of the works of 1776 has in 1779 been firmly eliminated. Iphigenia stands between a man, Thoas, who offers her sexual love, and a man, Orestes, to whom she offers sisterly affection, and she leaves the one to cleave to the other. The suppression of any erotic element in the feelings between her and Orestes is all the more decisive for a certain identification between Iphigenia and her mother Clytemnestra, who was so deeply wounded by her daughter's supposed death: Iphigenia as priestess is in a position in which she could return on Orestes the blow which he dealt to Clytemnestra, but instead she brings him a forgiveness for that crime, and a cure for his remorse, so complete that it must in some sense derive from his victim. Their healing embrace may look at first sight like a re-enactment of the embrace of his sister-wife which was so dangerously erotic that the dreaming poet of 1776 located it in an earlier existence. In fact, it confirms for Iphigenia the possibility of escape from the physical love with which Thoas threatens her, provided that the chaste Diana, also a sister who loves her brother, and to whom in this climactic moment of the play Iphigenia addresses her prayer, is willing 'to grant blessed deliverance to me through him [Orestes], and to him through me'.

The central action of the play, then, is a prayer to the gods: Diana is asked to confirm the nature of her purpose in keeping Iphigenia in seclusion and exile by giving a tangible sign, the restoration of Orestes to sanity, which is also a necessary step to attaining that purpose. The cure is granted—but what the sign means is not yet clear, for all oracles have to be interpreted. The bringing together of brother and sister was interpreted by Orestes as merely a cruel jest of arbitrary divine powers who had decreed the ruin of the whole house of Tantalus, and this thought roused him to a peak of fury and delusion. Against that wild and fearsome theology, Iphigenia had asserted the full force of her own conviction of the gods' good purposes and their love of mankind. Orestes has for the present been rescued from his insane belief that the gods are cunning bloodthirsty spirits of vengeance, deserving from human beings only manly hatred, rather than intimates to be trusted, prayed to, and thanked. But as the play continues, so it begins to seem that he might yet be proved right after all. 'Deaf necessity' seems to dictate to Iphigenia a compromise with the purity she has so rigorously maintained; to rescue herself and her brother she must, it appears, show contemptible and deceitful ingratitude to her host, and so her hope of ending in her own person the recurrent cycle of crime and vengeance in the house of Tantalus would be vitiated. Herself a criminal, how could she expect to be spared punishment? Yet the alternative seems to be only the gruesomely ironical end that Orestes in his madness envisaged: as a fratricide priestess Iphigenia would be showing herself a true Tantalid. If the gods lay such traps for mortals, from which no escape is possible without guilt, must they not be as they are represented in that ancient 'Song of the Fates' that Iphigenia recalls? Is it not true that they 'lead us into life, make the wretch become guilty, then leave him to his torment, for all guilt is avenged on earth'? Iphigenia, at the end of Act IV, sees ahead of her the possibility of the tragic fate that engulfed Proserpina: her trust in the gods, represented by her life of determined moral purity in steadfast hope for the day of her release, may prove the instrument for delivering her up to terrible crime. Maybe, as Goethe pondered at the outset of 'Winter Journey in the Harz', the difference between good and evil fortune, happiness and unhappiness, is made only by inexorable fate, and the human will has no part in it but is at tragic best the means by which what has been decreed for man is realized. As the poet turned in that anguish to the 'Father of Love' and sought his blessing, and as Goethe himself sought an augury from the Brocken, so in the last scenes of the play Iphigenia, by casting herself on the mercy of Thoas, challenges the gods to declare themselves finally: do they or do they not support and confirm the intuitions of the heart that trusts in them? Is the benevolent order which the pure and good spirit discerns in the world an illusion? Is Iphigenia's goddess a stuffed dummy?

The answer comes not from the gods, but from a man. *Iphigenia in Tauris* is unique among the works of this tragic period in that the favourable response to the prayer of the heart is revealed not through the symbolic conformation of nature, but through a human moral act. Thoas' permission to Iphigenia and Orestes to depart is not a symbol or promise of their rescue, it is their rescue: it is an event in the objective social world, which men share with one another, brought about by the action of heart on heart, by Thoas' coming to join Iphigenia in her inner respect for divine 'truth'. 'If you are the truthful ones that men praise you as being', Iphigenia calls to the gods before she makes her declaration, 'then show it by your assistance and through me glorify the truth'. And Thoas in his reply, angered and astonished at the same time, recognizes that the challenge is directed, not just to the gods, but to him: 'You know that you are speaking to a barbarian, and trust him to hear the voice of truth!' Such a reply shows, of course, that the battle is won already, that Thoas knows and respects that higher voice—which, he says, has 'so often calmed me' in the past —and can, and eventually will, stay the powers of lust and death with which he has threatened the future pure and passionless lives of Iphigenia and Orestes. The conclusion of the play expresses a confidence, echoed elsewhere in Goethe's work only in one or two poems, including 'Divinity', that, in the specific, aristocratic, social world, the 'part of the nation' in which he finds himself, there lies the fully adequate objective response to his inner needs, a confidence that, through a personal purity that sacrifices eros to agape, lust to benevolent moral action on behalf of others, Weimar can become for him a haven from frustration, guilt, and fear, and the possibility of tragedy can be put behind him.

Yet the inadequacies of the rhythmical prose in which this first version of *Iphigenia* is couched reveal that at the deepest level Goethe is not convinced. He has not found a poetic and dramatic form that can express his commitment to an ascetic life as a ducal official because fundamentally that commitment is not there, neither to an official existence nor to asceticism. He is entering on the long period of a self-imposed self-deception, for which another name is his spiritual 'marriage' to Frau von Stein, and to which he will cling with such iron determination that the only signs of his true state will be his haggard physical appearance and the gradual desiccation of his poetic powers. The prose *Iphigenia* expressed the rationale of a way of life Goethe was consciously adopting as an alternative to the perplexities of life as a publicly visible literary intellectual. For that reason it is not in the end a Gottschedian work at all, and even Wieland's welcome for it, as heralding a shift in the national taste, was misplaced: it is a purely, and privately, courtly work. Publication would have been directly contradictory of its purpose: like the inscriptions he was to compose for the Weimar park it was written for and from a particular place and belonged in a private not a public domain. But it necessarily shared therefore in the inconsequentialities of Goethe's position in Weimar and not only, if most tangibly, in the uncertainty of its dramatic language. There are other weaknesses in this first draft, too, which must have made Goethe shrink from publication, not only on grounds of principle, but because he knew that the play, like his life in Weimar, was not fully thought through. The link between the two great moments of trial—the healing of Orestes and the appeal to Thoas —is unclear, almost as if we were dealing with two plays, one about the brother and one about the sister, one about guilt and one about chastity, one about the reliability of the gods and one about the reliability of the heart. And then Iphigenia's motive for wishing to go back to Greece is uncertain: it sometimes seems no more than a personal nostalgia, which is hardly a sufficient basis for the moral and theological dilemmas generated by the difficulty of return. Most significantly of all, there is next to no representation, either in Act III or in the last act, of the *process* by which divine goodness becomes embodied in human life, no representation at all of the obstacles, whether personal or political, which might be expected to resist that embodiment, the simple difficulty of being good. The conversion of Thoas like the cure of Orestes, takes place behind the language of the play, not through it. Thoas does not speak like a man struggling with passion, he reports on his barbarian passions with a distance and calmness which indicate that he possesses a moral sense and a rational self-control as strong and developed as Iphigenia's. Indeed, when we see him alone, before his final confrontation with her, the source of his wrath with her is shown to be his belief that she has deceived him. Before Iphigenia's great declaration, therefore, we know that Thoas shares her fundamental respect for truth and that, in throwing herself on his mercy, she is not in fact taking as great a risk as she would be if Thoas were a small-minded, self-important, or even lascivious, tyrant, determined to have his way. No such dramatic conflict as that between Gretchen and Faust, or Götz and his adversaries, can arise here, for we are back in the Sentimental world of the Leibnizian Enlightenment, the world of *Miss Sara Sampson* and *Nathan the Wise*, where all good men think alike, where those who are not good are not men, and where the only possible conflict is a result of misunderstanding. Like the message that came to Goethe on the summit of the Brocken, the voice of the gods that, through Thoas, responds to Iphigenia is only the echo of her appeal, and the oracle that she consults utters only by virtue of a higher ventriloquism.

From a biographical point of view there is something retrogressive about the works Goethe wrote during his Swiss tour with Carl August in 1779–80. The ethical, or apparently ethical, conclusions of *Iphigenia*, though they may have settled the mind of the statesman, did not stimulate the poet. Instead, he turned back to the manner of 'Winter Journey in the Harz'—and after all the Swiss tour was in a sense the Brocken expedition writ large—though the manner had been diluted by the passage of time and the increasing separation of the *'dedans'* and the *'dehors'*. *Jery and Bätely* is an attractive little *proverbe*, in a Swiss peasant setting, but no more, and was precisely characterized by Goethe many years later: 'I can still sniff the mountain air in it when the figures come out at me from between cardboard rocks and canvas.' 'Song of the Spirits over the Waters' is a hymnic meditation on the human soul, seen as moving like water in an endless cycle between heaven and earth, tumbling broken over rocks or tossed like spray in the wind. This is clearly not the soul of Iphigenia, and Frau von Stein disliked the poem: 'not quite your and my religion', she wrote to Knebel. She thought of her soul in less material terms. Yet, even though the waterfalls which inspired the poem are strongly present in its eerily monochrome imagery and cascading rhythms, the bond between the poet's mind and the landscape is not as strong as in 'Winter Journey in the Harz': in place of the vigorous metaphors and untroubled juxtapositions of the earlier poem we find at the beginning and end of the 'Song of the Spirits' a tentative, almost unconfident simile;

> Seele des Menschen,
> Wie gleichst du dem Wasser!
> Schicksal des Menschen,
> Wie gleichst du dem Wind!

Soul of man, how you resemble water! Fate of man, how you resemble the wind!

The *Letters from Switzerland*, which Goethe put together in the first part of 1780 from letters written to Frau von Stein during the tour, and from accounts of his travels dictated for her to Seidel whenever the party had paused for breath, are at once a prose masterpiece in the 'symbolical' manner of 'Winter Journey in the Harz' and an attempt at elaborating a new attitude to landscape which, if it were successful, would destroy the basis of the 'symbolical' existence altogether. On hearing it read to the Dowager Duchess, Wieland immediately recognized the merits of this now rather neglected achievement, which he compared to Xenophon's *Anabasis:*

The ladies listening enthused about the *Nature* in this piece, but I enjoyed even more the cunning *Art* in the composition, of which they saw nothing. It is a true *Poëma* . . . The peculiarity that here too, as in almost all his works, marks him off from Homer and Shakespeare, is that everywhere the 'I', the 'ille ego', shimmers through, though without any ostentation and with endless finesse.

Goethe was impressed by Wieland's 'incredible eye for everything one tries to do, actually does, and what works and doesn't work in a piece of writing', but felt he was intending much more than a poem and 'if I succeed I would like to catch several birds with this snare'.

The *Letters* are a highly selective account of the travels of the previous October and November. As the writer at one point remarks—Goethe does not identify himself as the author until the last few pages—he makes 'little mention of people'. The tour of Swiss society and the Swiss intelligentsia has no place in this story, neither have the characters of those making up the party or their interaction. Attention is concentrated on the encounter with the landscape, and not even on all of that. After the entry into the highlands above Basle the entire trip to Berne and Grindelwald is omitted, of the stay by the Lake of Geneva only the visit to the Vallée de Joux is included, and the main part of the narrative then deals continuously with the long trek from Geneva to Chamonix and Martigny and up the Valais to the Furka and the St Gotthard, where the book ends. The first two episodes provide a prelude and introductory exposition of the themes in the main section. The journey out from Basle is conceived as an entry into a new world, both of landscape and of feeling, and up in the Jura Goethe rehearses for the first time the motifs of high-altitude walking and dependence on the weather, which will shortly be so important, and gives us a distant panorama of the peaks among which the journey will take us. The approach to the high places and their scenery is carefully graduated: after the first glimpse of it from the heights of the Jura, Mont Blanc is not described again until, riding into the dark valley of Chamonix after nightfall, the travellers become aware of a strange glow among the stars, brighter than the Milky Way, larger than the Pleïades: eventually they realize it is the peak of the great mountain, glimmering in the starlight. The traverse of the Col de Balme is an intermediate climax before the long slow haul up to Oberwald and the Furka, when snow and cold first enter the narration, and the culmination of the entire story is the passage from the Furka to the St Gotthard, the 'crown' of the Alps, we are told, even if not their highest peak. Here at last, as he revisits familiar terrain, Goethe begins to allude to his past life and to Weimar; he inserts a lengthy, and possibly invented, harangue from one of the Capuchin fathers on the unique teaching authority of the Catholic Church; and his concluding paragraph, which bears a strong resemblance to the last lines of 'Winter Journey in the Harz', describes the geographical situation of the St Gotthard so as to make it a nodal point between Germany and Italy, the eastern and the western Alps, near the sources of the Rhine and the Rhône. As on the Brocken in December 1777, therefore—the comparison is taken over from the letters to Frau von Stein—so, from the high place of Europe in November 1779, the son of man who has a quizzical eye for the doings of priests and scribes, looks down on the kingdoms of the world and their glory and turns towards his 'present destiny'. It is unclear whether Goethe ever intended to continue, or fill out, this his first venture in sustained autobiographical prose, but the work is complete as it stands. By selecting a single narrative strand from the events of the previous winter Goethe has created a powerful image of a journey of the soul.

The first letter of all suggests that the experience of sublime natural objects fills and expands the soul, so that by a process of 'inner growth' it becomes familiar with their calm grandeur and comes to share in it. The inner growth that this journey has brought about could be called a purging of the imagination by the experience of the object. In that first letter, as the party mounts the ravine cut by the river Birs, Goethe seems to be making a direct contrast between this new, objective, sensibility, which is characterized by the pre-eminent virtue of Frau von Stein and Iphigenia, by 'purity', and the Wertherian sensibility that feeds ultimately only on itself:

My eye and my soul could grasp the objects, and since I was pure and nowhere distorted the sensation, they could have the effect they should. If you compare such a feeling with that when we laboriously struggle with trivialities and make every effort to pad and patch them with as much as possible of our own, and to feed and pleasure our spirit with its own creation—then at last you see what a wretched makeshift that is.

Such mountain scenes as the setting sun, or the vista that incorporates both near and distant detail, show by their combination of grandeur and perfect clarity how illusory was the object of Werther's indefinite and insatiable longing: 'one then gladly gives up any pretensions to infinity, since in our perception and our thought we cannot even come to terms with the finite'. This pure experience of the finite object clearly foreshadows the turn to natural science which was to be so marked a feature of Goethe's life from 1780 onwards, and a certain coexistence or conflict of Picturesque or Sentimental with scientific travelling is noticeable as the journey progresses. The party arrange their tour in accordance with what is 'the fashion', seek out the recommended perspectives, and note scenes or episodes that would look well in pictures. At the same time—one of the recurrent motifs in the story—Goethe is observing the natural phenomena, particularly the clouds, with a view to understanding their form and, quite specifically and practically, predicting the weather. One of the great advantages of travelling in the mountains, he says, is that, by contrast with the lowlands where the high and distant clouds are something independent and alien, 'here one is enveloped by them as they are engendered and one can feel the eternal inner power of Nature moving suggestively through every nerve'. He can, as it were, enter the heart of a natural object and understand it from within. The purification of feeling by the confrontation with the object as it is in itself reaches its height in the passage of the Furka: in 'the most desolate region of the world', surrounded by mist and snowfields, and totally uninhabited peaks, it becomes imperative to concentrate the mind on what is, and resolutely shed any interference from the imagination: 'I am convinced that on this path anyone at all mastered by his imagination would necessarily perish of fear and anxiety.' Goethe's purpose in following this evocation of physical and mental extremity with the long report of the Capuchin's apologetic discourse is no doubt to suggest that he who has been through and made his own the great places of non-human nature has all the scripture and authority he needs. Yet one cannot leave the *Letters* without a feeling that the authority of the natural object is as yet a shaky affair: on the one hand, as when penetrating the cloud, Goethe can seem involved in an imaginative identification with the object not far removed from Werther's attempt to embrace it; on the other hand, he can seem so detached, as from the wonders of the Mer de Glace, that he simply recommends his reader to the standard guidebook, which will do the job better than he. He is aware that he is an uneasy amalgam of two people who would fare better if separated: 'one to see it, and one to describe it'. It is a quite understandable irony that, although the *Letters* were clearly originally intended as an emancipation from Werther-ism, when Goethe did eventually publish them in his works he always printed them in the same volume as *Werther* and represented them as a product of Werther's state of mind before he met Lotte. To an extent, the later Goethe was right: in 1779 he had left rather less of his past behind him than he thought at the time.

What was new—and it characterizes the whole of the early Weimar period, from 1777 to 1786—was the *will* to suppress the past, to turn away from tragedy, to believe that there had been a change. The change refused to happen: whatever Goethe's social progress, a new emotional and poetic life refused to begin. But he clung grimly on to his intention: to the court career and court culture, to ethical and scientific activity as the alternative to self-absorption, to the purification of soul and body and behaviour. The most powerful feelings in the *Letters from Switzerland* are released neither by great moments of sensuous receptivity, nor by the physical and moral achievements of the journey, but by one of the few personal encounters recorded in the book —and also, independently, recorded in Carl August's diary—the meeting with the hospitable and anonymous peasant woman, probably in the village of Fiesch, who told them her favourite story from the lives of the saints, the legend of St Alexis. Goethe says he was so moved by the tale, and the heartfelt involvement of the teller, that he had great difficulty in suppressing his tears. Yet, however Sentimental the scene, what he heard was the very opposite of an account of the shared palpitations of a Yorick and a Maria. Two elements in the story of the saint who early vowed himself to perpetual chastity were to engrave themselves on Goethe's mind, and be with him, in one or another form, as emblems of his own fate, until he died. First, there was the wedding, arranged for Alexis by his devout Roman parents and to which, as an obedient son, he submitted—but from which, before it could be consummated, he escaped by ship to lead a life of holy poverty in Asia Minor. Second, there was his return to Rome years later, as a beggar, to live out his last years unrecognized in a cellar of his parents' house. It was no doubt this second period, when Alexis was living a saintly life of prayer and of voluntary poverty and chastity alongside his bereft wife and in the very bosom of his sorrowing family, that particularly touched Goethe in 1779. Had he not too resolved to live in chastity beside the woman for whom he was destined, to hide his heart from all those who surrounded him, and to devote himself to the pursuit of purity and to benevolent and worthwhile activity for others? This was for some years yet to be the pattern of his own life, and obstinate men are sometimes moved to self-pity by the thought of their self-inflicted sufferings. But the tears, and particularly his telling Frau von Stein about the tears, suggest that Goethe's resolution was not in the end as firm as that of the Christian saint.

6

The Baron
1780–1786

WEIMAR was favourably impressed by the change in the travellers on their return from Switzerland. The Duke cut his hair short and Goethe now appeared at court in the conventional fine clothes, made up for him in Stuttgart, and including an embroidered waistcoat. 'Multum mutatus ab illo', in Wieland's eyes, he was beginning to take on something of the appearance of a man of quality, thought Countess Gianini, 'et tant mieux'. Goethe thought so too. Later in the year, reflecting on his first arrival in Weimar, on the crisis of adaptation that was, he hoped, behind him, and on the change now coming over him, he compared himself 'to a well-intentioned bird that has plunged into the water and, when it was on the point of drowning, the gods have gradually started to transform its wings into fins. The fishes busying themselves about it do not understand why it does not immediately feel at ease in their element.' The metaphor was not without a certain literal truth: when Goethe now bathed in the Ilm, he did so in a full-length bathing-costume. The five years after the favourable response of the Alpine oracle were Goethe's most sustained attempt to make sense of his presence in Weimar by accepting court and official life on its own terms and as a wholesome influence on his emotional and social personality. The work of an administrator seemed worth doing for its own sake, and for the sake of the slight improvements that might come with its being done well, and perhaps as a natural sequel to the task of educating the Prince (which he hoped had been completed and crowned by the pilgrimage to Lavater in Zurich). Poetry was to be a private matter, with no pretensions to influence either the world of public affairs or the public demeanour of the poet.

Even though the first meeting of the War Commission after his return appeared 'very prosaic', Goethe devoted himself conscientiously to its affairs, despite the obstructiveness of his immediate subordinate, the fat, idle, and probably corrupt, Major von Volgstedt, whom he was able to unseat only in December 1780, and at the cost of taking on his work himself. By August 1781 he was sufficiently pleased with what he had done to feel confident of his ability to tackle a larger department, or even several more. The conscious effort to take himself in hand, discipline his life, and make himself an efficient social being, is apparent from the start of 1780 in his careful filing of letters, and requests to correspondents to do the same, in his systematic copying of other artists, particularly the rustic scenes of Everdingen, both to improve his technique and to make up for a two-year dearth of inspiration, which not even the Swiss journey had remedied, in his resolve to observe and record the rhythms of his mood and his ability to work, in his attempts to give up drinking wine altogether. Within a month of being back in Weimar he had applied to join the Masonic lodge, having realized during his recent journey, he said in his letter of application, the need for a social medium in which to 'make the better acquaintance of those I had come to esteem'. The Privy Councillor and minister felt himself a member of a nationwide class with which he desired a more organized and material contact than had been provided by the diffuse and uncertain medium of printed literature. In June 1781 he was writing to Möser's daughter on the need to seek entry to the Promised Land by paths other than those attempted by the literary and cultural movement of 1771–5.

It was literature, and Goethe's literary past, that paid the price of this great reorientation. The principal objects of satire in Goethe's adaptation of the opening of Aristophanes' *Birds*, which was completed in the summer of 1780 and performed in Ettersburg, are two unappreciated authors who hope to find a new country where food, wine, and the daughters of the town will be thrust upon them gratis, and 'for some work of genius or other, five, six or eight hundred *louis d'or* will be sent directly to the house by an unknown public without its having to be asked'. But the birds themselves, representing the German reading public, another boundless and insubstantial kingdom of the air, are made no less ridiculous, and the play as a whole seems the work of an author who is shaking from his feet the dust of a public literary career. *The Birds* is the last work of any length that Goethe completed for five years, and it was anyway intended only as a first part. He continued for a while, as he had done for the last two years, intermittently adding to *Egmont*, and he conceived two new full-length dramas in the rhythmical prose with which he had experimented in *Iphigenia*. But *Torquato Tasso*, begun in the autumn of 1780, deals, inescapably, and as the heart of its theme, which is the life of the great Italian poet, imprisoned for years by his patron on grounds of insanity, with the tension between court life and the poetic vocation. And work on *Tasso* came to a halt, for seven years, in 1781. *Elpenor*, begun in that year, was permanently to remain a fragment. It is perhaps not a coincidence that Goethe marked the fifth anniversary of his arrival in Weimar, 7 November 1780, by drawing a lively but sarcastic little sketch, known as 'Cold Shower', showing an amorous young man having his ardour cooled by a toppling bucket of water.

The more vigorous new beginning for Goethe's intellect at this time was in the realm of natural science, which was at first closely connected with his official activities. 'Since I have had to do with mining affairs, I have taken up mineralogy heart and soul.' J. C. W. Voigt (1752–1821), the younger brother of an administrative colleague of Goethe's, and now his secretary at the Mines Commission, had been educated at Carl August's expense at the famous Saxon Mining Academy in Freiberg, in the Erzgebirge, near the Bohemian border, and Goethe in 1780 started to exploit this investment by despatching Voigt on a mineralogical survey of the entire duchy and the neighbouring areas, the results of which were published over the next three years. Voigt also acted as Goethe's scout, selecting features of interest in the local terrain for his superior to visit. In the summer of 1780 Goethe established his own collection of minerals, which

Voigt, furnished by his Freiberg training with a rigorous and up-to-date nomenclature, put in order for him, and he set about extending it by asking for specimens from correspondents all over Germany: from F. W. von Trebra (1740–1819), the expert who had pronounced the Ilmenau mines workable, and who now had a post in the Harz in the employ of the Elector of Hanover, King George III; from Merck in Darmstadt, whose scientific interests were better grounded and of longer standing than Goethe's; from amateurs in Berne and Zurich, and professors in Freiberg; even his old friend Sophie von la Roche was asked to press her wide circle of acquaintances into service. When the dramatist Leisewitz talked with Goethe in August 1780, in a room in the cottage crammed with plaster-casts of classical statues and cases full of geological specimens, the conversation took a long time to become literary but turned quite naturally to the age of the world and the absurdity of the orthodox figure for it of 6,000 years, by comparison with the estimate of 100,000 years published in 1778 by Buffon. The main purpose of all the activity was as yet, however, simply the collection of data, and Goethe's own education: in an interim report to the Duke of Gotha on Voigt's work, he emphasized that they were refraining from any theoretical speculation about the origins of the Thuringian hills. None the less, such speculation was eventually to come.

Isolated though he might feel in Weimar, Goethe's scientific turn brought him a Germany-wide common pursuit of the kind on which he thrived. It gave a new, technical, impetus to his drawing, in which the landscape inspiration had been languishing for years. And it brought him an important new acquaintance: the young professor of medicine in Jena, J. C. Loder (1753–1832). Goethe had already taken an interest in the Duke's acquisition for the university of a natural history collection and a large private library, but Loder, who was also a Freemason and entered the apprentice grade at the same session as Goethe in 1781, and the master's grade at the same session as Goethe and Carl August in 1782, was his first personal link with the Jena professors. (Griesbach, the theologian, was known to him from Frankfurt days, but they had little in common.) About a year after his mineralogical studies had got under way, in the autumn of 1781, Goethe embarked on a new science by spending a week attending Loder's public dissection of two cadavers and discussing anatomical matters with him afterwards. The newly acquired knowledge was immediately put to use. The Weimar Academy of Drawing had led an erratic existence since its foundation in 1774, but in 1781 it moved into more spacious accommodation and here, from November to the following January, Goethe passed on his new enthusiasm through giving exercises in anatomical drawing based on the work he was doing with Loder. Anatomy would soon preoccupy him even more than geology, but the most important fruit of the acquaintance was the discovery of Jena as a place for the peaceful but concentrated learning of new things, and a point of contact with a national, even an international, scientific public.

The new style took longest to establish itself in Goethe's personal life. The late summer of 1780, after the visit of Leisewitz, remained in his memory as a time of emotional confusion and painful adjustment. Much of it was spent in the wilds round Ilmenau and Eisenach with the Duke, who, once away from Weimar, showed a tendency to revert to his earlier immature behaviour, which, Goethe gradually came to fear, 'lies in his deepest nature . . . the frog is made for the water even though it can spend some time on land', and he resolved not to undertake another journey like that to Switzerland with him again. This loosening of the personal ties with Carl August, at the very time when Goethe's official links with the Weimar administration were becoming closer, was not at first compensated by any shift in the relations with Frau von Stein. The letters he wrote to her from the Ilmenau forests, at the time when he composed the second 'Night Song of the Wanderer', telling her of his endless mental converse with her, did not immediately produce a similar response, and the two had a serious quarrel, for which Goethe apologized abjectly, when they met on the Stein estates at the end of the tour. It is possible that Goethe's emotional equilibrium had been upset by the visit to Weimar, just before the tour began, of the attractive and available Marchioness Branconi—though he had emphatically denied to Lavater any interest either in marrriage or in momentary aberrations, from which he claimed to be preserved by the powerful 'talisman' of his love for his married friend. But it was not until March of the following year, 1781, after an absence of a week during which he wrote long and increasingly urgent letters to her every day, that the great change in his relations with Charlotte occurred. An ecstatic, though softly spoken, intimacy enters his daily notes to her, in the course of the next month the 'Sie' form of address gives way to the 'Du', and she is now 'new' and the knowledge of her love has brought about a 'conversion' in his innermost self, he says; it has given him a new openness and peace of mind. A purification is said to be taking place and he believes he is recovering his old spirit of benevolence to all men, recognizing his past readiness to disregard the feelings of others, and determining to remedy it. He daily acknowledges more deeply, he says, that we are earthbound creatures and subject to earthly limitations, and from this insight into our material and bodily condition he does not hesitate, in writing to Lavater, to draw the theological conclusion that 'God and Satan, Hell and Heaven' are but 'concepts which man has of his own nature'. Not that that in any way demeans those concepts, for what is greater than human love: 'what could not the *love of the universe* do, if it *can* love as *we* love?'

We do not know the form of the mutual reassurance that Goethe and Charlotte von Stein gave to each other in that spring of 1781, but a hint of the mental intoxication that resulted for Goethe can be gleaned from an essay called *Nature*, written by the young Swiss scholar G. C. Tobler, who followed up his meeting with Goethe in Geneva by visiting Weimar between May and September of that year. The visit was a success. Tobler quickly had an entrée to Anna Amalia's circle, he no longer saw himself as a theological emissary of

Lavater's, his command of Greek and his verse translations of Attic drama made him particularly welcome to Goethe and Knebel, and his essay was eventually 'published', in the winter of 1782–3, in the *Tiefurt Journal*. It is based, as Goethe himself confirmed, on the many conversations on natural science and natural theology that he had with Goethe during his stay, and has often been wrongly attributed to Goethe himself. Goethe also agreed, many years later, that it captures well the atmosphere of a particular, if transient, stage in his thinking. The natural world is seen, in the Leibnizian terms so fundamental to Goethe's vision that he could never discard them, as a mass of individualities, as many as possible, so as to maximize the pleasure to be derived from so many different viewpoints and from the interrelation between them. The Nature that is said to be responsible for bringing these individualities into existence, and for delimiting them from one another by difference and ultimately by death, is a strange, capricious, playful being and the older Goethe rightly drew attention to the difference between this 'whimsical, self-contradictory' force and his later conceptions. This 'Nature' is evidently a pantheistic godhead still closely related to the Fate which Goethe felt had ruled his first five years in Weimar and in whose greater identity he had said, in the poem 'Bounds of Humanity', we are submerged. There is as yet no suggestion of the ethical humanism of the hymn 'Divinity' written in 1783. Instead there is an impassioned invocation of the power of love, the power which brings individualities together, which therefore can exist only because of their differences, and which is said to be Nature's 'crown':

She makes chasms between all beings and everything desires to absorb everything else. She has isolated everything in order to draw everything together. With one or two draughts from the goblet of love she makes recompense for a life full of toil.

Whatever the draughts of love may have been of which Goethe speaks in the poem 'The Goblet', written in the summer of 1781, they were a recompense for the many toilsome years, past and still to come, of that strangely frozen relationship with Charlotte, in which two individualities were both drawn together and held apart by their own isolation.

The court found the affair incredible, with Goethe 'withering away', and 'his divine Lotte . . . growing uglier as you watch'. And sometimes, in that most private region of his soul where his poetry had its well-springs, Goethe too acknowledged the superficiality of the prosaic cheerinesss, industry, and benevolence with which he was cementing over an essentially unfulfillable love, and the hidden anguish escaped in a lyrical moment of intense imaginative creation. Very probably before August 1781, perhaps some time in the preceding winter, Goethe wrote the most terrifyingly erotic poem of his life, which tells a very different story from the facile wish-fulfilments of 'The Goblet'. Based partly on a recent local incident, partly on a Danish folk-song translated by Herder, and partly on his own memories of a night-ride to Tiefurt in April 1779, with Fritz von Stein in the saddle in front of him, the ballad 'Elf King' ('Erlkönig') shows us the inverted image of that whimsical, playful Nature and that open, peaceful, and purified love, and suggests the 'bitter earnest' that the later Goethe, commenting on Tobler's essay, could still detect behind these games. The father in that poem rides, with his sick son in his arms, through a nature instinct with death—night, wind, streaming mist, dry leaves, grey, bare willows (the German title suggests the etymologically false meaning 'king of the alders'). And out of this dead, dark world rise spirits of lust, a perverted lust which whispers golden promises but is in reality the voice of a cold mist-grey king and his daughters so old that they too are grey. It is the true voice of desire, speaking to the boy with a directness scarcely parallelled elsewhere in Goethe's verse, it will brook no refusal, and its object is unequivocally unnatural:

> 'Ich liebe dich, mich reizt deine schöne Gestalt;
> Und bist du nicht willig, so brauch' ich Gewalt'—
> Mein Vater, mein Vater, jetzt faßt er mich an!
> Erlkönig hat mir ein Leids getan!

'I love you, your fair form excites me; and if you are not willing I shall use force.'—My father, my father, he is getting hold of me! Elf King has hurt me!

The simple childishness of the last phrase makes the death with which the poem ends unbearably poignant.

No one can hope to unravel the complex emotions that made it possible for Goethe to write this poem at this time. Not even Frau von Stein seems to have recognized that in it feelings she would not allow to be applied to herself were being transferred to her son, for would she then have allowed him to live three years with its author? What 'Elf King' shows beyond any doubt, however, is that the 'Nature' and 'love' in which Goethe was consciously taking refuge in 1781 were so far removed from the Nature and love which he had invoked ten years before—for example in 'May Celebration'—that the poetic vein of that earlier time could be tapped only if these newer concepts were inverted and degraded. And perhaps that was a recognition of a true perversity in the substitutes he was now finding for what had earlier inspired him.

Apart from a few days spent on mining business in Ilmenau during the summer, Goethe was fairly continuously in Weimar for the whole of 1781. In November, just after the sixth anniversary of his arrival, the day on which he had started his anatomical lectures, changes began that signalled his final acceptance of the duchy as his predestined place and that established the shape and nature of his life there for fifty years to come. On 14 November he signed an agreement with the garrison doctor in Weimar, P. J. F. Helmershausen, to rent from him a floor that would shortly become vacant in a large house he owned on the south side of Weimar, at the front overlooking the square known as the Frauenplan and with a back garden adjoining the outer town-wall. Three days later, at an interview with Anna Amalia in which all aspects of his present position were discussed, including his mother's desire that he should do as he had just done and get himself a permanent residence in town, the Dowager Duchess told him of Carl August's intention to have him raised into the Imperial nobility. It was an evidently necessary step if Goethe was to continue negotiating with the neighbouring princely houses, as he had for example been doing in the summer over the Ilmenau mines, but by that token it bound him yet more closely for the future to his sovereign patron. It made all the more apparent the symbolic significance of the move into town, a significance which must have weighed with Goethe as much as his mother's anxiety about the effects on his health of walking to and fro between the town and his cottage in the severe Weimar winters, or the sheer practical consideration that he was running short of space. For, even though a special attraction of the new house was that it had an escape-route, a private postern-gate in the town wall, through which, after a minute or two's walk alongside the Latin Garden, he could be at the house of the von Steins, and after a minute or two more be across the river and back in his cottage, the planned move was a clear statement that the Wertherian idyll, uncertainly poised on Weimar's outer margin, was over.

The year 1782, specifically the first fortnight in June, was, at least outwardly, the climax of Goethe's first Weimar period. The year began sadly with the death on 27 January of the loyal court and stage-carpenter J. M. Mieding, while Goethe, short of a new play for Duchess Luise's birthday, was in the midst of hectic preparations for a scenically elaborate ballet, *The Spirit of Youth* (*Der Geist der Jugend*), with little spoken text, but requiring much rehearsal time, for it involved all those Weimar children who took dancing lessons. All the same Goethe was sufficiently moved to give time to writing a long memorial poem in the, for him, unusual form of the heroic couplet, 'On the Death of Mieding', which took up a whole issue of the *Tiefurt Journal*. The poem was something of a concluding retrospect on the history of the Weimar amateur theatre, for there were to be very few more new productions now until the arrival of a professional troupe in 1784. Goethe, for one thing, was far too busy himself, though only a few could know that in the first months of the year he was preparing himself for what was to come. From March to May he was fairly continuously travelling through Thuringia. First, as War Commissioner, he had to repeat the exercise of 1779 and select recruits for the Weimar forces. (He also attempted to repeat the literary productivity of that period, when he had written the first three acts of *Iphigenia*, by taking with him this time the still unfinished *Egmont*, but he returned with little progress made.) Then he was on a diplomatic tour to the Thuringian courts, nominally to settle a matter concerning the University of Jena, but in fact with the purpose of introducing himself as Carl August's future principal emissary. He also seems during these travels to have taken a particular interest in local economic conditions, which suggests he may have had some knowledge of the impending dramatic events. On 2 June he moved into his grand new apartments on the Frauenplan—that of course had been long intended and was no secret—on the following day he received his diploma of nobility from Vienna—that was a happy coincidence, and though it was supposed to be made public only on his birthday, it was soon widely known—but then on 7 June the President of the Chamber, von Kalb, was quite unexpectedly dismissed and within four days Goethe, 'a man who understands rather less of these matters than I do of Syriac', was appointed as his provisional replacement—and that was a scandal to set every tongue in Weimar wagging, among them, regrettably, Herder's:

So he is now Permanent Privy Councillor, President of the Chamber, President of the War Office, Inspector of Works down to roadbuilding, Director of Mines, also *Directeur des plaisirs*, Court Poet, composer of pretty festivities, court operas, ballets, cabaret masques, inscriptions, works of art etc., Director of the Drawing Academy in which during the winter he delivered lectures on osteology; everywhere himself the principal actor, dancer, in short the factotum of all Weimar and, d.v., soon the Majordomo of all the Saxon houses, round which he progresses for the benefit of worshippers. He has been made a baron . . . has moved from his garden into the town and set up the household of a nobleman, arranges reading circles which will soon turn into receptions etc. etc.

Goethe, at not quite 33, was now the most powerful man in Weimar, after the Duke, and if even Wieland felt obliged to note 'I am sure I can have nothing to fear from him', many must have trembled. What would his father have thought at seeing his son succeed where Voltaire had failed? Alas, Caspar Goethe, who had still been able to register delight at the visit of Carl August and his Privy Councillor in the autumn of 1779, had slipped further into dementia, and on 25 May he had died unaware even of the coming ennoblement. It is perhaps an indication of Goethe's sense that an epoch in his life had ended that after 13 June he stopped keeping a diary.

Yet even though he was so near the top of the tree, and perhaps for that very reason, Goethe clearly did not regard his position as other than temporary. The peculiar arrangements for his 'interim' and partial presidency of the Chamber, which themselves astonished the court, can only have resulted from his own refusal to take on the full burden of his predecessor. He saw the task before him not as a career, but as limited, and perhaps as the least that he could decently do in recognition of the distinctions heaped upon him. 'Dating from midsummer I now have to sacrifice two full years before the threads will be drawn together so that I can honourably stay or resign.' He kept very strictly to this timetable and almost exactly two years after he wrote these words, in July 1784, J. C. Schmidt (1728–1807), a cousin of Klopstock's, the tenant of the property adjoining Goethe's cottage and a long-serving Chancery official who in 1788 would finally occupy the still vacant post of President of the Chamber, joined the Privy Council to prepare the way for Goethe's withdrawal from public affairs. In the very midst of the crisis of 1782, on 4 June, Goethe wrote to Frau von Stein: 'How much happier I should be if, cut off from the battle of the political elements, I could, at your side, my dearest, turn my mind to the sciences and the arts for which I was born', and more than once in its aftermath he proclaimed that 'I was truly created to be a private person and I cannot understand how destiny should have chosen to stitch me into a civil service and a princely family'. These years were the point of closest, most immediate, conjunction between the private life of the Frankfurt citizen and the absolute power that determined the political shape and future of Germany. But in such a conjunction Goethe was no more than just another collaborator, just another Weislingen, if he failed to keep open the vein of poetry. If, for the benefit not simply of himself but of the nation for which he had written, his life was to be more than an example of loyal service to a prince, he had to continue to make literature out of it. This proved increasingly difficult. The will was certainly there: 'I am settling down in this world without compromising a hair's breadth of the being that inwardly sustains me and makes me happy', he wrote of the new house on the Frauenplan, and immediately after his appointment as, effectively, Finance Minister he set to and completed the second book of *Wilhelm Meister*, which had long been left on one side, by the end of June. But *Wilhelm Meister*, which he now worked on until 1785 at the deliberate and disciplined rate of one book a year, was essentially a protest. Although Goethe felt he could draw on his experiences as a courtier and administrator for the benefit of his 'politico-moral-dramatical pigeon-hole' these furnish only occasional characters and episodes for his theatrical novel, which in its form is the opposite of courtly, and in its content is concerned not with matters of state but with the life of 'a private person' 'born to be a writer'. From 1782 onwards Goethe's only poems of direct lyrical utterance—that is, those poems in which, characteristically, a moment of deep feeling in a symbolic setting is caught in a strong, often original, rhythm—are scattered through the pages of *Wilhelm Meister*. It is as if the novel provides a bourgeois envelope which protects them from the hostile atmosphere of the court. His other, longer, poems become both less frequent and more arthritic as they take on more regular and more conventional forms. In 1782 he even began a rewriting of his literary past, with a plan for a revision of *Werther* so as to show in a better light the character of Albert, whom he had after all in the meanwhile come to resemble more himself. Werther's and Albert's Lotte now seemed to him, he wrote to Charlotte von Stein, a 'ghostly anticipation' of the woman to whom he was spiritually wedded by gift of a gold ring, just as he was, again in his own word, 'wedded' to the duties of his office.

Goethe's last contribution to the Weimar amateur theatre, *The Fisherwoman*, (*Die Fischerin*), enlivened the summer season of 1782 in the park at Tiefurt. In 1781 the 23-year-old Prince Constantin had set off on a grand tour; Knebel, his former tutor, depressed by inactivity, had temporarily withdrawn from Weimar and settled rather unhappily in his native Franconia; and Anna Amalia had seized her opportunity and taken over the now empty house at Tiefurt as her summer seat in place of Ettersburg. She immediately set out to improve the grounds on the model of the Prince of Dessau's park in Wörlitz, and, as in Weimar itself, the improvements initially consisted in the erection of monuments and little inscriptions—provided by Goethe—and the construction of grottoes and pavilions, from which it was possible to enjoy the view of the Ilm, which there bends in a great loop round the house, and of the tree-covered slopes above it. Goethe's playlet was designed to exploit this setting. It is about a fisherwoman who hides herself in order to be thought drowned and so, by giving them a fright, repay her father and her suitor for staying out too late in the evening. The climax of the open-air performance, held in the gathering summer dusk, was the episode of the search for the missing woman, when torches appeared among the trees and all along the banks of the Ilm. It was a memorable spectacle, but the play had largely been written in 1781 (it provided a context for Corona Schröter to sing the ballad 'Elf King' to a tune of her own invention), and Goethe furnished no successors. Most notably he failed to produce anything more than a congratulatory poem of a few lines for the most important court occasion of his first ten years in Weimar: the celebration in March 1783 of the birth, on Candlemas Day, of Carl Friedrich, son and heir to the Duke. There was however a grand 'Venetian Carnival' to honour the Duchess, with a procession of 139 participants and 100 horses, led by Carl August in cloth of gold and cloth of silver, and including Privy Councillor von Goethe 'in old German costume' of white satin with a scarlet cloak, feathers in his cap and riding a white horse caparisoned in yellow, richly embroidered with silver thread. That was the substitute for the unwritten last three acts of *Elpenor*, which Goethe had intended to have ready for the day. The francophone cosmopolitan court literature represented by Baron Grimm, whom Goethe had refused to meet in 1777, but with whom he conversed at length in 1781, could not inspire him itself, but it could asphyxiate any other inspiration.

Apart from their location, the great advantage of Goethe's new lodgings, on the first floor of a three-storey house with a row of fourteen windows at the front, was the space they gave him to organize his life. In the long suite of spacious rooms it was easy to mark off the public reception area, where he now gave weekly tea-parties for all comers, noble and bourgeois alike, from the area for private study: 'I have completely separated my political and social from my moral and poetical life (externally, that is, of course).' Perhaps the separation was more than merely external, perhaps the very belief that there was a distinction between his external and his internal self was what made it difficult for his private muse to play a political tune. By 1784 he would be telling Frau von Stein 'I see no one and if I do see anyone it is only an image of me that appears in society.' In 1782, however, he was still grateful for the 'most agreeable order' into which he was putting things and felt that even in his social and business life there was a reciprocity of relationship for which Frau von Stein was responsible: 'I am wholly yours and have a new life and a new way of conduct towards people since I know that you believe that'; 'I too am friendly, attentive, talkative, forthcoming towards everyone . . . the calm, the equanimity with which I receive and give rests on the foundation of your love'. It is surely to her influence that we must ascribe Goethe's desire, in these years of high office, to put in order not just his letters, art and natural history collections, and memories of his past, but also his old friendships, troubled by untended wounds. In October 1782 he went to some trouble to re-establish an indirect contact with Jacobi that left his injured friend completely free not to respond if he did not wish to: the letter that he then wrote is a marvel of tact, affection, and informal, so all the more persuasive, regret, and it is no surprise that Jacobi was won over. When he finally visited Weimar for twelve days in 1784, Goethe 'went pale with joy' at the meeting. Earlier in that year Goethe had also 'gone pale as whitewash for joy', and felt 'rejuvenated by nine years', at the first visit since 1775 of the Stolberg brothers, on their way to the spa at Carlsbad, and when they left he asked them to give his warmest regards to Klopstock, in the evident hope of undoing those years of estrangement.

But the happiest and most significant reconciliation of all was that with Herder in the summer of 1783: 'one of the supreme joys of my life'. The year, it is true, did not begin well for them. There was a sharp exchange after Herder's sermon at the christening of the Crown Prince, in which Herder had expressed the hope that when Carl Friedrich grew to manhood he would not squander money on the arts so long as there was still poverty in the duchy. Goethe felt that at least the arts were better than dogs, horses, or diamonds. He had good reason to be touchy: at this very moment Prince Constantin had returned from his travels, and his indiscretions were coming home after him, first a French, and then an English mistress, and both of them pregnant. Goethe, charged with quietly arranging the departure of the ladies and the future settlement of them and their offspring, naturally preferred a more realistic approach than Herder's to the moral improvement of princes. Herder, however, felt that he could no longer have any confidence in Carl August and that he was totally isolated in Weimar. But no one could woo like Goethe. He wrote Herder a long and helpful letter when the text of the sermon was being prepared for publication; he invited the Herder children to a party at the start of June to look for painted eggs in his garden; and in August he asked the parents to come and celebrate his birthday with him and Frau von Stein, while suggesting closer co-operation between Herder and himself in respect of the Weimar schools which Herder was endeavouring to reform. It was a turning-point for both men. Goethe re-established contact with an intellectual equal and so set up that minuscule 'public' of the last years before his Italian journey, consisting of Frau von Stein and the Herders, and also Knebel, who came back to reside in Jena in 1784, but with whom Goethe was anyway in permanent correspondence. Herder was reconciled to Weimar at a crucial moment in his own development, for he was just embarking on his most important single work, his *Ideas for a Philosophy of the*

History of Humanity. As he wrote the first three parts of the *Ideas*, each divided into five books, and published in 1784, 1785, and 1787, they were read, either aloud or in manuscript, in Goethe's circle. The impact of the early books, available from December 1783 onwards, by which time Goethe, Herder and his wife were meeting roughly once a week, was considerable: 'the world and natural history is now a proper rage with us', Goethe told Knebel. It was a rage that coincided closely with the changes in his own interests, for these first books are concerned with the primeval state of the world and man's relation to the animal, vegetable, and mineral kingdoms. The shared intellectual enthusiasm was also a shared consolation for that anguished sense of solitude which, for all the brave words when he moved into town, and all the expressed confidence that Frau von Stein was turning him into a socially efficient being, he knew as well as Herder. 'Goethe . . . suffers too in his soul, but more magnanimously than I', wrote Herder, and Wieland was not deceived either: 'Goethe . . . is in the full sense *l'honnête-homme à la cour,* but all too visibly suffers in soul and body from the oppressive burden he has taken on for our good.' That 'gnawing worm within', as Wieland also called it, has plainly left its mark on the poems Goethe inserted in the fourth book of *Wilhelm Meister,* completed in November 1783, which include the quintessentially tragic 'He who has never eaten his bread with tears' and are mainly put in the mouth of the Harpist, a bard-like figure who stands outside all normal society.

In 1783, however, geology remained in the forefront of Goethe's publicly acknowledged interests. With Fritz von Stein, who had moved permanently into the house on the Frauenplan at the end of May, he spent a month from early September to early October travelling in the Harz, Fritz trotting along on a pony and only occasionally showing boredom when Goethe and his professional companion, von Trebra, stopped to collect specimens. 'Onwards, onwards,' Goethe cried, scrambling from the shoulders of Trebra up a dangerous cliff to knock out a particularly interesting piece of fused granite and hornfels, 'we have still great honours to claim before we break our necks'. The Brocken was climbed again, and Goethe and the forester reminisced about their ascent four years before. Science proved its value as an instrument of social contact too, for on leaving the Harz Goethe passed through Göttingen, where Professor Lichtenberg put on a special lecture for him and other visiting members of the nobility in which, with many a gas-filled bubble and bladder—and consequent bangs and flashes—he demonstrated the principles behind the recent invention of the Montgolfier brothers, the balloon in which, later in the year, Pilâtre de Rozier would become the world's first aeronaut. A visit to Cassel and the biologist S. T. Sömmerring (1755–1830), also preoccupied with balloons, was less agreeable, since it seemed to Goethe that at the Cassel court the pursuit of art and scholarship was little more than a cloak for 'monstrous' misrule. On returning to Weimar Goethe made his own experiments with gas and fire balloons, but without great success. In early 1784 he began dictating the introduction to what was manifestly intended to be a substantial scientific work, perhaps his once-planned *Romance of the Universe,* and possibly a scientific, rather than historical, parallel to Herder's *Ideas.* The fragment—for that is all it ever became—is now known by the title *On Granite* (*Über den Granit*) and is interesting mainly for its explicit announcement that Goethe is turning from literature, and the study of human feelings, 'the most mobile part of creation', to science, specifically geology, 'the observation of the oldest, firmest, deepest, most immovable son of Nature': 'yes, may I, who have suffered and still suffer much from the variations of human sentiments and from their rapid motions in myself and others, be not begrudged the sublime calm given by that silent and solitary propinquity of great and softly speaking Nature'. What is intended is no rhapsody, however, but an informative treatise, and when Goethe undertook journeys to the Thuringian Forest and, once more, to the Harz, in the summer of 1784, he did so in a spirit of systematic inquiry, taking the younger Voigt with him in Thuringia and, in the Harz, Weimar's resident artist, G. M. Kraus.

Kraus's task was to draw rocks and rock-formations 'not in a picturesque fashion, but as they interest the mineralogist', and it was above all questions of geological shape and form that concerned Goethe at this time. The problem which, in *On Granite,* he says he is setting out to solve is whether any regular pattern can be discerned in the surface of the earth, whose mineralogical confusion is such that the observer, seated perhaps on the Brocken and surveying the Harz below him, is tempted to exclaim: 'Nothing here is in its original, ancient position, all here is ruin, disorder and destruction.' This is very similar to Herder's formulation of the problem he faced in trying to find order in the chaos of human history (though Herder, having less experience of empirical, observational science than Goethe, tended to think that the order of things in the world of nature was already fairly well understood). For Goethe, as for Herder, there is a concealed theological significance to the problem: unless there is order—for the clergyman, in God's workings in history, for the poet-geologist, in the manifestation of God or 'the gods' in nature—there can be no 'sublime calm' in the certainty that our own individual existence, in what it does and what it thinks, can reach harmony with, or at least propinquity to, the fundamental forces that shape the universe. That, for Goethe, would be tantamount to a relapse into the tragic perspective of Proserpina. But, like Herder, Goethe thinks in 1784 that the problem can be solved and that the way to solve it is to find a 'clue' ('Leitfaden'), an Ariadne's thread, a single guiding principle that will lead us through the labyrinth of confusion. The metaphor of Ariadne's thread has been shown to be of great significance to late eighteenth-century philosophers of history, notably Herder and Kant, who were trying to relate the welter of rather unpromising empirical detail offered by the facts of history to an overall rational, and ultimately theological, framework. The same metaphor determined Goethe's own formulation of the problems he was setting himself in natural science. He was looking for a single principle of order in the phenomena *as they appear to us now* (for it is the present world which we inhabit and in which we have to make ourselves at home); he was not trying to construct a complex picture of a past world and past events of which the present animal, vegetable, or mineral kingdoms are simply the inert end-products.

Goethe, on his geological expeditions of 1783 and 1784, thought no differently from his contemporaries about the mechanisms of rock-formation: most rocks were believed to have precipitated as crystals out of a primal liquid, though whether that liquid was initially fiery or whether it always was, as it certainly was later, watery, Goethe says in *On Granite,* we do not know. The original deposition was of granite, which formed the core of mountain ranges and, as the primal waters retreated, so they left, crystallized behind them, layer upon layer of newer and different rocks. Geological opinion was to change very greatly in Goethe's lifetime, but he himself never abandoned the conceptual framework within which his studies began. What interested him was not so much that framework itself as the principle of order, in the world as it presented itself to his senses, which that framework had enabled him to discover. 'The simple thread', which 'leads me nicely through all these underground labyrinths and gives me a clear view in the midst of confusion', was 'a simple principle' that 'completely explains the formation of the larger mineral bodies'—by which he meant that it enabled him to see a pattern in the present shape of rocks and mountains; it was not the cause of that shape, he emphasized, but simply 'a harmony in the effects' of a cause which other researchers would have to identify. This harmony was a pattern which could, for example, be drawn to the attention of an artist such as Kraus, and once Kraus had grasped it he would paint more accurate and more revealing pictures than before. Despite Goethe's proviso, however, his pattern necessarily in practice implied some account of how the 'harmony in the effects' had come into existence and forced him, now and later, into definite views about what had or had not happened in the earth's past. His 'Ariadne's thread' was in essence the premiss that not merely are minerals internally crystalline in structure, but the rocks, the massive slabs and strata of mineral material that are the elements of landscape, are themselves crystals, deposited (presumably) as huge regular bodies at the moment when the mineral substance was precipitated out of the primal liquid. (The divisions of the rock mass which Goethe attributed to the process of crystallization are now explained as shrinkage cracks or even entire fault lines.) In the case of granite, these enormous overlapping parallelepipeds, heaped up into the cores of mountains, formed steep cliff-faces, and even overhangs, and to these surfaces the later rocks attached themselves as the primal sea declined. According to this account, strata which lie out of the horizontal—and which are nowadays assumed to have been folded by movements of the earth—were from the start deposited in their present positions against the angled facets of granite 'crystals'. Beneath the surface confusion of the landscape, therefore, lies, for the informed eye, the regular structure of its granite foundation. In Goethe's own landscape drawings of this period this underlying grid becomes almost disturbingly insistent.

It was not only in geology that Goethe looked for Ariadne's clues. At the start of 1784 he was preoccupied with another specific natural phenomenon from which he promised himself systematic illumination of an entire field of scientific study. At the end of February, scarcely returned from the ceremonial reopening of the mines in Ilmenau, he had been despatched to Jena where there had been a disastrous flood. In no other incident of his official career does he appear more evidently and simply the agent of enlightened despotism: bringing relief to loyal subjects by the careful distribution of compensation from the treasury and at the same time peremptorily ordering the demolition of an ancient monument, without consulting the Jena town council, since its materials were needed for improvements to the flood-defences. But in the intervals of this urgent work Goethe was spending as much time as possible with Professor Loder, extending his studies of anatomy to include the skeletons of animals other than man. He paid particular attention to the bones of the upper jaw. In man, left and right maxilla are immediately juxtaposed, without a central intermaxillary bone, from which in many animals the incisor teeth depend. This absence of an intermaxillary bone had been held to be a distinguishing mark of human beings, who were thought not to be related to animal species as those species were to one another, even Herder accepting this interpretation in the fourth book of his *Ideas*. In the palates of the human skulls Goethe examined, however, he thought he could clearly see the sutures which in other animals demarcate the intermaxillary bone. The insight was instantaneous: Goethe immediately concluded that man did indeed possess the missing bone. Though little study had preceded this conclusion, it was followed by an immensely industrious combing of existing anatomical works and the examination of skulls of as many, and as exotic, animals as possible (while attending the meeting in June of the Estates in Eisenach, to negotiate his crucial financial reforms, Goethe spent his spare time closeted with the skull of an elephant: he allowed his landlady to think that the enormous chest contained porcelain, 'so I am not thought to be crazy'), and inspection of human embryos showed them to possess a separate intermaxillary. By the end of October he had finished a little treatise expounding his view which, exquisitely illustrated by a local draughtsman, was circulated in manuscript to some of the leading Continental anatomists, including the Dutchman P. Camper, perhaps the greatest authority of all. His reaction to their unanimous rejection of his conclusion summarizes the whole story of his fifty years' involvement with science and scientists: 'I can quite believe that a professional *savant* should repudiate his five senses. They are rarely concerned with the living concept of the thing, but with what people have said about it.'

The trouble was precisely that Goethe lacked a methodical theoretical framework by which to interpret his observation—what he dismissively called 'what people have said about it'. It is simply not true that man 'has' an intermaxillary bone in the same sense in which he 'has' a femur. Professor G. A. Wells has pointed out that we do not say that the human being possesses gills and a tail, although at an early stage the human embryo possesses both these organs. An evolutionist of the later nineteenth century could happily assert that man showed vestiges of the intermaxillary, but Goethe had no methodical evolutionary theory to appeal to which could give a precise meaning to the term 'vestiges'. As in his geology, so in his anatomy, he was concerned, as professional scientists were not concerned, with finding a principle of order in the world as it presented itself to his five senses. And what in both cases, after brief study and in a flash of insight, he had discerned was an Ariadne's thread to lead him through the labyrinth: the principle of an underlying unity, visible to the informed eye, behind the multiplicity of natural forms. What mattered to Goethe about the intermaxillary was not in *what* sense man could be said to have it, but the implication for contemporary biology, anthropology, and theology, if man could in *any* sense be said to have it. The implication was that, as Herder had said in the second book of his *Ideas*, all living forms, not excluding the human, should be seen as variations of a single principal type, and for this reason Herder was the first person to whom Goethe confided his discovery in March 1784: 'it is like the coping stone on the human being . . . I have thought of it too in connection with your notion of the whole, and how beautiful that makes it'. Man could now be thought of, by means of a 'living concept', as part of a continuous sequence of varying animal forms, a sequence perceptible to the five senses in their common skeletal structure, from which the discovery of the human intermaxillary had removed the last anomaly. In his geological expedition of 1784 we find Goethe applying the same principle of unity in continuously varying multiplicity to mineral formations; the very variations which plague the systematic mineralogist, because they blur the dividing lines between different types, are what he says he is seeking out and has by good fortune found.

This principle of continuity—which has direct Leibnizian antecedents—was already present to Goethe's mind in 1783, before he made the discoveries in anatomy and geology that seemed to bear it out. The poem 'Divinity', which appeared in the *Tiefurt Journal* in November of that year, as Herder was beginning his readings from his *Ideas*, raises an issue untouched by any of its predecessors in the free hymnic form: the relation of man not to the gods (as in 'Bounds of Humanity') or fate (as in 'Song of the Spirits over the Waters'), but to animals. The poem begins by stating explicitly that no physical difference marks man off from the other natural beings of which we have scientific knowledge. The difference between man and animals lies purely in the ethical sphere: man should be noble, helpful, and good,

> Denn das allein
> Unterscheidet ihn

> Von allen Wesen,
> Die wir kennen.

For that alone distinguishes him from all beings that we know.

The gods in this poem, though nominally its subject, are reduced to the status of an ideal reflection of man's moral activity. They are not beings of which we have (scientific) knowledge, we 'sense' or 'intuit' their existence, and the grounds for that intuition are provided by human moral goodness:

> Heil den unbekannten
> Höhern Wesen,
> Die wir ahnen!
> Ihnen gleiche der Mensch!
> Sein Beispiel lehr' uns
> Jene glauben.

Hail to the higher, unknown beings, that we sense! Let man resemble them. Let his example teach us to believe in them.

So it is man's capacity to 'intuit' the gods and by his own goodness to be an image of divinity that alone marks him off from all natural beings, subject as they all are to 'eternal, brazen, majestic laws':

> Nur allein der Mensch
> Vermag das Unmögliche:
> Er unterscheidet,
> Wählet und richtet;

Only man can do the impossible: he draws distinctions, chooses and judges.

The dualism implied by 'Divinity'—man is on the one hand an element in the material continuum of natural beings, subject to precisely the same mechanical laws as they, and on the other hand he is a free moral agent capable of doing what in natural terms is impossible—is an attitude peculiarly characteristic of the German 'official' Enlightenment, and we have seen its origins in Leibniz's notion of the pre-established harmony between material and mental events. That Goethe should now be expressing it in such unadorned and commanding terms shows how far he has moved away, not only from his recent tragic fear of the power of the gods and of fate, but also from the earlier Storm and Stress awareness of the material, social, and historical determinants of human behaviour, and towards the official culture of absolutist bureaucracy.

Among the characteristics of that culture, it will be recalled, are an acute sense of individual isolation and an attitude of, albeit limited, hostility to Christianity. The isolation of a self cut off from a society into which it sends only its mask, or 'figure', is a recurrent refrain in Goethe's letters of the early Weimar period, which only intensifies after his move to the apparently more social world of the house on the Frauenplan. By 1783 even Carl August was troubled by Goethe's 'taciturnity', and at the last, in 1786, Frau von Stein herself could write: 'Goethe lives in his thoughts, but he does not communicate them . . . a happy man utters'. In the years after his return from Switzerland Goethe's attitude to Christianity also underwent a change, evident particularly in his changing relations with Lavater. For the religion of the man who in 1780 was 'the best . . . of all mortals and immortals', who was chosen by Goethe as the prophet to 'anoint' the Duke and complete his initiation, and who revealed to Goethe the 'moral deadness' of his life in Weimar, was by 1782 a source of exasperation which, he said, would have roused him to parody had he not deferred to friendship (and, doubtless, to memories of the Jacobi affair), and by 1783 it was simply a 'quackery' to which Lavater, as 'an active physician', was prepared to stoop in case of need. When Lavater visited Goethe in Weimar in 1786, Goethe passed a judgement on his visit more terrible perhaps in the light it casts on the host than on the guest: 'Not one heartfelt or intimate word was exchanged between us and I am free for ever of hatred and love . . . I have drawn a great line under *his* existence too and now know the balance of what remains of him for me.'

Goethe had of course only briefly, in 1769 and 1770, shared that evangelical Christian enthusiasm with which Lavater mingled a curiosity about any secular evidence for a world of spirits. Later, in Wetzlar and Frankfurt, Goethe marked himself off from the Christian religion by the technique he had learned from Susanna von Klettenberg and Arnold's history of heretics: he too was a heretic, and had to be, if he was to be a poet, but thereby he was truer to the principle of religion than the orthodox, and he felt an affinity with the Son of Man whom the orthodox had crucified. In the first few years in Weimar, when the strange, incomprehensible powers that he called 'fate' and 'the gods' were most active in his life, and he felt they could be directly consulted, he expressed a pitying distance from Lavater's thirst for an invisible Christ: 'you are in a worse case than we heathens—at least our gods reveal themselves to us in our need'. Though the visit to Zurich in 1779 revived his admiration for Lavater's ability to combine awareness of the sublime with a commitment to domestic virtue and the public good, it also revived his earlier sense of opposition to too spiritual and theoretical a Christianity, and he contrasted the religion of the synoptic Gospels with that of the book of Revelation, which Lavater was paraphrasing: 'I am a very earthly man, the parables of the unjust steward, the prodigal son, the sower, the pearl, the penny etc. are for me more divine—if there is to be anything divine at all—than the the seven bishops, lampstands, horns, seals, stars and plagues.' Once Goethe was back in Weimar, and especially once he had come closer to Frau von Stein in 1781, a more cuttingly critical note is to be heard in his analyses of Lavater's religious psychology: Lavater's Christ is said to be an image which does the impossible and concentrates in a single individual all the possible sources for Lavater of moral and aesthetic pleasure: into him 'you can transfer your all and, mirroring yourself in him, in him can worship yourself'. Goethe is now conscious of belonging to a group, identified by no more than a 'we', but older than Christianity, opposed to it, and destined to outlast it, and he reacts angrily to Lavater's attempts to appropriate all that is best in everyone else to a Christianity which he scornfully calls a 'bird of paradise', all the colours of the rainbow. Such eclecticism is permitted only to 'us' who—using a metaphor made current the previous year by Lessing's treatise *The Education of the Human Race*—'put ourselves in the school of every wisdom revealed to man, and through men, and who, as sons of God, worship him in ourselves and all his children'. This sense of humanist solidarity is no doubt derived partly from the Masonic movement and partly from a close coincidence of views with Frau von Stein, but it also reflects that community of attitudes which in 1781 Tobler found prevailing in the whole of Anna Amalia's circle. In 1783 a pious visitor to Weimar commented: 'Wieland has philosophized religion out of the Duke's heart, and Goethe has laughed the rest away.' In the touch of resentment and acerbity in the correspondence with Lavater we catch an echo of the age-old cultural rivalry, acute in late eighteenth-century Germany, between Church and State, between a republican's Christianity and a court secularism which recognizes the concealed but worldly ambitions of its opponent: 'You continue . . . to extend your kingdom in this world, by persuading everyone that it is not of this world.' In July 1782, at the peak of his political career, Goethe declared himself roundly: 'I am not anti-Christian, nor un-Christian, but decidedly non-Christian', 'I for my part could not be persuaded by an audible voice from heaven that a woman has given birth without a man or that a dead man has risen again; on the contrary I regard these as blasphemies against the great God and His revelation in Nature.' It was not likely that Goethe's reconciliation with Herder, a Mason and a deist, who by his own avowal prayed little, would compromise this profession of faith in the impersonal and non-social objective order. Even Goethe thought that Herder's sermon at the baptism of the Crown Prince contained too little that was specifically Christian, and a glimpse of 'official' culture at its most cruelly absurd is afforded by the spectacle of Goethe, in March 1784, consoling Duchess Luise for the death of her 3-year-old daughter by reading to her from the proofs of the first part of Herder's *Ideas*. Jacobi, already embroiled in controversy over his report that Lessing, before his death, had avowed himself a Spinozist, may have been the first, during his visit in September 1784, to revive Goethe's old interest in the Jewish philosopher, and Goethe began to study Spinoza again in November, but it was Herder who presented him with the Latin edition of Spinoza's works on Christmas Day. Herder pronounced himself at one with Goethe in his understanding of Spinoza, which he opposed to Jacobi's, and he told Jacobi that he could no more believe in a personal transcendent God than Lessing could. When Goethe in December called Spinoza 'our saint', he was speaking for a 'we' that defined its religion in conscious opposition to Christianity and had outgrown even Goethe's ealier vestigial affection for the person of the Son of Man.

Goethe's need for a new cultural fund on which to draw, different from the German, bourgeois, and essentially Christian heritage that had inspired his first great works, became more apparent as he himself became more closely identified with the court culture that those early works had largely rejected. Gradually his attention was turning to the classical civilizations of the Mediterranean. From the early 1770s the *goût grec,* all the rage in France, had been spreading to the other courts of Europe. In the autumn of 1782 Goethe ordered for the Weimar library the most recent catalogue of the works of Palladio, the supreme adapter of ancient principles to modern conditions. After his success, with *Iphigenia,* in the courtly art of adapting ancient plays for the stage of a modern prince, Goethe found Tobler in 1781 a particularly welcome guest and prevailed on him, during his stay, to translate the whole of Aeschylus for him. In the following years Goethe read and imitated the epigrams of the *Greek Anthology,* practising the elegiac distich to the point where he felt willing to have his efforts inscribed in stone. But the clearest proof of his alienation from the cultural loyalities of his youth is surely the correspondence with Frau von Stein which in August 1784 he conducted from Brunswick (and even continued when back in Weimar)—in French. The Brunswick festivities, which were the cover for the negotiations setting up the League of Princes, must have been a decisive revelation to Goethe of the emptiness of his court career. He was now playing the role of a senior diplomat, but not merely was he tortured by the necessity of spending hours of the day eating and playing cards, he was also not very good at it, and knew it: 'I am not skilful enough to conceal from society this lack of interest, though I do all I can'. Countess Görtz was certainly not deceived: 'for having a fine suit of clothes he is no more confident in good society' and 'seems very out of place'. He hoped to escape in time to spend his birthday on the Brocken, but this time the augury was denied him. Even when he was in the mountains he had to acknowledge that 'they do not seem as picturesque and poetical as they did', and although he added, 'but there is another kind of painting and poetry that accompanies my climbing now', by the time the expedition was over he had to admit that his sensibilities were being deadened by the single-mindedness of his new mineralogical passion: 'the subject is almost beginning to bore me'. Inwardly totally detached from his official existence, his source of poetry all but dried, with only uncertain enthusiasm for his new interests, to which he was anyway unsuited, and lacking even the belief in the value of his native language that had so long inspired him, Goethe, after nine years in Weimar, was intellectually and emotionally on the verge of bankruptcy. In that dry season in Brunswick in 1784 there can have been only one glint of hope: Schmidt was already in place to take over his responsibilities on the Council in the new year. And beyond that another piece of unfinished business from his past was beginning to beckon, insistently at last, now that his two years at the head of Weimar politics had run their reasonably successful course: maybe the recuperation, indeed rebirth, that he now desperately needed was to be found in the home of those ancient Mediterranean civilizations which seemed to offer an alternative not just to the religion but to the art, landscape, and even climate that had surrounded his life so far. Had he not, in 1782 or, more probably, 1783, in his last truly lyrical outburst for several years to come, written for his *Wilhelm Meister* a poem of such imperious dreamlike yearning for his ideal, Palladian, Italy that its command to pass the 'terrible gates' of the Alpine passes, and not turn back as he had done in 1775 and 1779, could not much longer be resisted?

> Kennst du das Land, wo die Zitronen blühn,
> Im dunkeln* Laub die Goldorangen glühn,
> Ein sanfter Wind vom blauen Himmel weht,
> Die Myrte still und hoch[†] der Lorbeer steht,
> Kennst du es wohl?
>   Dahin! Dahin
> Möcht' ich mit dir, o mein Geliebter,[‡] ziehn!
>
> Kennst du das Haus? auf Säulen ruht sein Dach,
> Es glänzt der Saal, es schimmert das Gemach,
> Und Marmorbilder stehn und sehn mich an:
> Was hat man dir, du armes Kind, getan?
> Kennst du es wohl?
>   Dahin! Dahin
> Möcht' ich mit dir, o mein Beschützer,[‡] ziehn!
>
> Kennst du den Berg und seinen Wolkensteg?
> Das Maultier sucht im Nebel seinen Weg,
> In Höhlen wohnt der Drachen alte Brut,
> Es stürzt der Fels und über ihn die Flut:
> Kennst du ihn wohl?
>   Dahin! Dahin
> Geht unser Weg; o Vater,[‡] laß uns ziehn!

Do you know the land where the lemon-trees bloom, the golden oranges glow amid the dark leaves, a gentle wind blows from the blue sky, the myrtle stands motionless and the bay-tree tall—do you know it, then? To that, to that, O my beloved, I should like to set out with you. Do you know the house? its roof rests on pillars, the hall gleams, the chamber shimmers, and marble images stand and watch me. What has been done to you, you poor child? Do you know it, then?—To that, to that, O my protector, I should like to set out with you. Do you know the mountain and its path strung through the clouds? The mule seeks its way in the mist, in caverns dwells an ancient brood of dragons, the rock sweeps down and over it the torrent: do you know it, then? To that! to that our way goes; O father, let us set out.

---

* Originally (1783) 'grünen' = green.
[†] Originally 'froh' = joyfully.
[‡] Originally 'Gebieter' = commander.

Works, 1780–1784

What indeed had been done to the poor child of Goethe's genius? By the end of 1784 he might well have felt like a tattered vagrant who had wandered by mistake into the marble halls of the literary pantheon. In the five years after the *Letters from Switzerland,* none of his major literary projects, with the partial exception of *The Birds,* came to completion. Work on *Tasso* stopped in 1781, on *Egmont,* tantalizingly close to a conclusion, in 1782, on *Elpenor* in 1783. After 1782 *Wilhelm Meister* had advanced with almost mechanical regularity, but where it was going was unclear. One or two longer poems, a scientific paper, and one or two rather exiguous new ideas were all the evidence that he was not lapsing into impotent silence. In what Goethe wrote in these years of his dramas and of his novel, and in the poems, we find a depiction of his situation, and of the conflicting factors within it, which is increasingly direct and to which the symbolic art of his poetic imagination is ever more incidental, and as symbolism fades its place tends to be taken by allegory.

A particular difficulty in discussing *Egmont,* with which Goethe struggled for seven years, on and off, after moving to Weimar, and the first two acts of *Tasso* is that no manuscript evidence has survived of the first state of these plays. This is not completely disabling, however, since, even with works whose genesis was very slow, Goethe seems usually to have proceeded by addition and correction to what was already written, rather than by deletion and wholesale redrafting. The later stages may therefore be a good guide to the material content of the earlier, though not to its significance. In the case of *Egmont* we know that a not inconsiderable part of it was written and discussed with Goethe's father in the autumn of 1775, and that, although it was not composed methodically and sequentially, but as individual scenes took Goethe's fancy, none the less in 1781 the overall structure was clear enough for Goethe to be sure that 'were it not for the impossible fourth act, that I hate and absolutely must rewrite' the play could be finished in three weeks. When in 1787 he did bring himself to make that final exertion he discovered that whole scenes could be left untouched. It therefore seems likely that the general outlines of the action were from an early stage what we find in the play's printed form. What we know for certain has changed, in whatever parts of the text date from 1775, is the dramatic language, which in 1782 Goethe told Frau von Stein was 'too unbuttoned and student-like' and 'contrary to the dignity of the subject'. *Egmont* evidently was an erratic block from an earlier epoch, too big to be ignored, but too alien to be accommodated: 'it is a strange play. If I had to write it again I would write it differently, and perhaps not at all.' In that respect *Egmont* was like *Urfaust,* but unlike *Urfaust* it was in prose and so less protected from meddling when its author was 'in an unpoetical condition'. But while *Faust* dealt with things Goethe wanted to think about less and less, Egmont dealt with things he could not avoid thinking about all the time, even though in Weimar he now had rather different thoughts from those he had had in Frankfurt. This was why he could neither finish the play nor, until 1782, abandon it. The fourth act was 'impossible' because it was the most political act in Goethe's most political play, and its politics, especially for an aspiring minister, were an embarrassment.

The play's first act unfolds a historical panorama of the kind Goethe invented in *Götz von Berlichingen*, though—at any rate in the 1787 version, which is all we have—more controlled, and focused on the single figure of Egmont, whose personal appearance is held over until the next act. In three contrasted scenes, set, like the rest of the play, in various parts of Brussels, we see an entire nation at a crucial point in its history: the Low Countries in 1566–8 (the events of two years are compressed into a few days) on the verge of revolt against Philip II of Spain, unsettled by preachers of the new, Calvinist, religion and the yet more radical iconoclasts, but resentful of the dictatorial methods adoped by the Spanish authorities against them. In the opening scene, a crossbow contest, crowded with quickly sketched individual characters, we first hear of Count Egmont, a national hero for his military successes against the French, and a man of tolerance in religious matters, while in the next, set in the palace of Margaret of Parma, Philip's regent in the Netherlands, the conversation, after dwelling on the difficulties of benevolent rule, again comes round to Egmont and the danger to good order represented by his popularity and carefree living. After the public world and the world of high politics we move to a domestic middle-class interior and a scene with all the characteristics of Goethe's Frankfurt period: a (largely) natural dialogue between Clara (later Clärchen) and her mother, with interventions from her unrequited lover Brackenburg, some homely background activities—knitting and woolwinding—and a bright little song with some unacknowledged psychological depth to it. In its course we learn that Egmont loves Clärchen and is a frequent nocturnal visitor, while Brackenburg is in despair and has acquired some poison with which to kiil himself. Unlike all Goethe's otherwise similar domestic scenes, however—in *Clavigo, Stella, Urfaust, Brother and Sister*—this is directly connected to a public and political world, for during the conversation Clärchen looks out of her window in alarm and sees the soldiery of the regent going by to quell an unidentified disturbance.

Of this unrest we see more in the first scene of the second act, another crowd scene, but no longer genial and harmonious, as in the first act, and rent instead by jealousies, fuelled by a half-educated agitator, Vansen, and by fears of the Spanish threat to the ancient privileges of the provinces and corporations. At the height of the tumult Egmont appears in person for the first time and calms the throng by his innate authority and with his assurance that an orderly and industrious existence is the best defence of privileges. For the second scene we follow Egmont to his house and, in his conversation with his secretary, see something of the private cost of his public role: the anxious administrative detail with which he is burdened and which he impatiently shrugs off. In the latter part of the scene he also shrugs off the much more serious political cares that are insistently put to him by his old friend, the far-sighted and devious William of Orange: the Duke of Alba is marching on the Low Countries at the head of an army and Orange fears not only repression of the populace but a pre-emptive strike against the nobility. Egmont refuses to listen to Orange's urgent plea that he should leave Brussels and refuse to meet Alba when he comes—he argues that such provocative non-cooperation would precipitate the war they are trying to avoid.

The third act is very short, as if to concentrate attention on the significant position of its two scenes at the centre of the play. In the first, in which Egmont is not mentioned at all, and the mechanisms of intrigue at the highest level are laid bare with total dispassion, Margaret accepts that the arrival of Alba makes her resignation inevitable. Then Egmont appears in Clärchen's house in the full glory of a Knight of the Golden Fleece, a transfiguration emblematic of the magnificence of their love, in which a private and a public personality embrace. After this ecstatic moment the fourth act begins with a striking change of tonality: we are back on the streets, but they are streets patrolled by Alba's soldiers, where conversations between the citizens about the latest restrictions are hurried, whispered, and demoralized. Then we move to Alba's palace where the last preparations for his coup against the local nobility are being made: in response to an invitation which it would be treason to refuse, Orange and Egmont will put themselves into his hands and be arrested and executed on a trumped-up charge. But Orange, shrewd enough to know that the time for polite fencing is past, does not come: only the carefree Egmont rides guilelessly into the trap. Alba contents himself with but one of his intended victims and after a long dialogue in which Egmont's espousal of the cause of 'the people', their fears for their freedom, and their 'old constitution', meets the immovable determination of Alba in carrying out the will of the King, the trap is sprung, as Orange had predicted, the arrest is made and Egmont, disarmed, is led off to prison.

The fifth act alternates scenes centred on Clärchen and scenes centred on Egmont. First of all a distracted Clärchen is shown, in the gathering dusk, attempting, and failing, to interest the cowed citizenry in Egmont's fate. Then in a monologue we see Egmont consoling himself in prison with thoughts of rescue. But this very night his scaffold is being built, as Brackenburg can tell Clärchen after evading the curfew and slipping through the streets to her house. She, who has never been able to love Brackenburg as he would wish, now asks him to act as her brother and commends her mother to his care while, amid lines and emotions that have their—to be sure, incomparably more forceful—parallels in the last scene of *Urfaust*, she takes the poison he had intended for himself. For the last time we return to Egmont, to whom the death-warrant is read and who is left with Ferdinand, Alba's illegitimate son, to make his final dispositions. In their conversation, Egmont discovers that

Ferdinand is one of his warmest admirers and after a brief moment of hope, soon extinguished, that escape may still be possible, he reconciles himself to his death by passing on to Ferdinand the charge to live as he has done. Falling asleep at last, Egmont has a vision of an allegory of Freedom, bearing the features of Clärchen, who gives him a crown of victory as a token that his martyrdom will eventually bring liberty to his country. Awakening refreshed he goes triumphantly to his execution in the knowledge that his death is unleashing a popular revolt which will sweep the Spanish tyranny away.

It is difficult to think of precedents for this extraordinary work, although elements in it derive from Shakespeare's Roman and history plays and were passed on, with their Goethean tincture, to Büchner when he wrote *Danton's Death*. But the core of the play is outstandingly original: the representation, untrammelled by theological scruples about the status of monarchy, of the mechanics of domination in a modern state (for that, however unhistorically, is how the sixteenth-century Low Countries are treated) and of the interaction of the public, the private, and the political worlds in a conflict between the traditional rights and freedoms of a decentralized society and the remorseless ambition of a unifying and rationalizing central power with the instruments of government and military force at its disposal. The intrusion by the central power, not merely into streets and public places, but into the private homes and lives of the citizens, and its reduction of a social organism to a steerless mass of isolated individuals is impressively depicted. The implicit contemporary issues in *Götz von Berlichingen*—the conflict between the burgher society of the towns and the absolutist aspirations of the Enlightened principalities—are here explicit and further clarified, for while Götz is in conflict with the Nuremberg merchants, whose political interests are really no different from his, Egmont can identify himself directly with the national cause of his fellow citizens. It is understandable that Caspar Goethe should have taken so enthusiastic an interest in the play in the autumn of 1775: not only did it represent his profound —rather than his superficial—political allegiances, but it was also hardly the work to commend his son, whom he was trying to persuade to go not to Weimar but to Italy, to the favour of an enlightened despot.

It seems likely—for one can do no more than speculate—that the original scheme for *Egmont*, which throughout his first seven years in Weimar Goethe was struggling to modify, consisted of three elements, or circles of interest. First there was the political kernel, the scenes with the populace and perhaps some of the discussion about the nature of government—the element that gives the play its originality as the first drama of revolutionary nationalism and makes it unique among the literary products of the Storm and Stress for the directness and force with which it addresses the political condition of contemporary Germany. Then there was the figure of Egmont himself, a man of action and achievement like other Storm and Stress figures, a leader with a mysterious power of personal attraction, such as by 1775 Goethe knew himself to possess, and a lover. But in this original stratum of the play Egmont's love for Clärchen was probably more closely linked with his political role than it is in the final version: Clärchen was not so much an allegory of Freedom as the personal embodiment of Egmont's love-affair with the people of his nation. The play at this stage would have been an attempt to state the role of a man of Egmont's, and Goethe's, gifts in a national cause, and Clärchen would have been what Lili would have had to be if Goethe and she were to have stayed together: the intermediary betweeen Egmont's, or Goethe's, individual self and a social struggle of such dimensions that even to conceive of it was a considerable imaginative achievement. In 1775 it was already manifest that that struggle in Germany was lost (and it should be remembered that the revolt of Belgium, in which the play is situated, did not succeed either, unlike that of the Northern Netherlands). A third element which was therefore probably in place by the time Goethe came to Weimar was Egmont's death. However, the play as it now stands has two conclusions, and it is unlikely that the second of these, Egmont's allegorical vision of Clärchen in the role of Freedom, was written even by 1782, let alone by 1775. The original conclusion was probably the conversation with Ferdinand in which Egmont passes on the standard—perhaps in 1775 more explicitly than he does now—to the natural son of his implacable opponent, who is thereby deprived of his posterity, while the apparently childless Egmont acquires a spiritual son and dies looking into a future on which he will have had a decisive influence. (Goethe had already tried out this consolation for mortality in 1774, in the *The Artist's Deification.*) By nature, too, as well as by moral inclination, Ferdinand is destined for the camp of Egmont and Clärchen, for the grimly calculating Alba has to admit that he had been able to beget him only in union with his mother's carefree love: whatever Alba has tried to build will eventually be overthrown by the powers of life and growth which he is trying to repress but which he cannot do without.

Goethe's diary tells us that this last trait was added to Ferdinand's characterization on 5 December 1778. Otherwise we know almost nothing of the detail of Goethe's intermittently reported work on the play in his Weimar years. But we can guess that, because of their closeness to his experiences and preoccupations during this time, certain parts of the play are more likely than others to have been affected by this work, whether of creation or of revision. Margaret of Parma is a very different character from Duchess Luise, but the suggestion of a tenderness between her and Egmont is perhaps another example of what we may call the 'Tantalus' theme of the forbidden love of a subject for his monarch, and the shrewd understanding of intrigue in high places which we find in her scenes may also reflect Goethe's growing political experience. The scene between Egmont and his secretary certainly echoes both the frustrations of the poet—and the outdoor man—chained to administrative office, and those of the War Commissioner struggling with the extravagances of a wayward Duke. But it is also one of the two scenes—and there are really only two—that drive a wedge between Egmont's public role and responsibilities and his private life, and so destroy the original unity in the play of his personality and his political task. The other moment at which this division is made is at the end of the second scene in Act III, when, at the high point of his declaration of love, Egmont distinguishes between the 'irritable, stiff, cold Egmont' who appears in public and is understood by no one who deals with him, and the 'calm, open, happy' Egmont who is known and loved by Clärchen—rather as Goethe, reporting to Knebel on the new order of his life on the Frauenplan in November 1782, distinguished between 'the Privy Councillor' and 'my other self, without which a Privy Councillor may quite happily exist'. Like the division of Goethe's own self, the division of Egmont's—which is unlikely to have been part of the original scheme—was forced upon the poet by his position at a princely court. Completing *Egmont* meant making it acceptable not only stylistically—though to that requirement are due many of the present artificialities and inconsequentialities in the language of the play—but also politically, and acceptable not only to Weimar, but to Goethe himself and his changed understanding of German realities. As is shown by his letter to Möser's daughter in 1781, that now excluded a directly political satisfaction of the needs expressed by literary intellectuals in the first half of the 1770s. The means by which Goethe sought to adapt his drama to his new understanding of himself and his relation to the German body politic were not only the division of Egmont's personality but also a shift in Clärchen's function and a considerable extension of her importance. No longer was she to be a personification of that life of the people which Goethe had lovingly depicted in many more and less central scenes in *Götz, Urfaust, Werther, Brother and Sister,* and book 1 of *Wilhelm Meister*: she was to become a principle in her own right, the expression in her person, rather than in her social position, of an alternative way of life for Egmont, parallel and even opposed to his political existence, which as a result now moved away from the centre of the play's focus. It is at about this point that we must assume Goethe's restructuring of Egmont ran into the sands in 1782: he could not yet envisage a new imaginative unity for his depoliticized play, and he was by now so highly placed in an absolute government that for the present he could not but wish he had never written it at all.

Accommodation to Weimar was probably also the main subject of the 'good scheme—*Tasso*' which occurred to Goethe on 30 March 1780 while he was still determinedly readjusting his life after the return from Switzerland. In October he began writing—in the hybrid prose of *Iphigenia*—the first act of the new play, and set himself as a deadline 12 November, the first anniversary of his crossing of the Furka, and the fifth of his meeting with Frau von Stein. By the same date in 1781 he completed the second act, but for the present nothing more followed. In general outline, as with *Egmont,* or to an even greater extent, the content of these acts is likely to have corresponded to that of the final version. In particular it is likely that Act I already centred on the crowning with laurel of the poet who has just completed his _Jerusalem Deliver'd_ and presented his work to his patron Alfonso, Duke of Ferrara, for only a week before noting his new 'scheme' Goethe had himself sent a laurel crown to Wieland to congratulate him on his _Oberon_ which he had been reading in manuscript. Act II would have consisted of a conversation between Tasso and the Duke's sister, the Princess Leonora d'Este, followed by a more tense meeting between Tasso and a courtier diplomat, known at this stage as Battista, though later to be called Antonio. Their conversation would have turned into a quarrel which the Duke had to allay and which probably culminated in Tasso's drawing his sword, in contravention of the court proprieties, for these are prominent incidents in the accounts of the poet's life on which Goethe drew and figure accordingly in the play's final version.

This material had an obvious appropriateness in a period when Goethe was trying to put behind him the theological perplexities he had resolved in _Iphigenia_ and turn his mind instead to the task of living within his chosen society. The (unhistorical) crowning of Tasso at the very start of the action announced a problem which was evidently pressing for Goethe at the time and may have been intended to be more central to the play than it eventually became: how does the poet live _after_ achieving greatness, after completing a world-famous work such as _Jerusalem Deliver'd_ or _Werther_? Perhaps—so the answer to be given by the original plot may have run—by becoming a successful courtier. The conversation between Tasso and the Princess was no doubt intended as an introduction to the ethos of court life, given by a woman who as an educator of the soul resembled Frau von Stein and in her rank and frail constitution resembled Duchess Luise. Even at this stage in the play's development, she may also have resembled both these women in being an object of the poet's chaste and distant love. _Tasso_ was eventually to become Goethe's most explicit treatment of what we have called the 'Tantalus' theme, but it is quite uncertain —and perhaps rather unlikely—that in 1781 Goethe was fully conscious of this potential in his story. Finally, the relation between the imaginative, and perhaps loving, poet Tasso and the practical social being, Battista-Antonio, directly parallels the division in his own personality that Goethe was acknowledging at the time and probably introducing into the character of Egmont. Here are two men who were surely, as we read in the final version, intended by Nature to be one, and although dramatic tension, if nothing else, requires that at their first meeting they should be hostile to each other, it may have been the original intention of the play to show their reconciliation. That would have been a declaration by Goethe that the division of his self was not necessarily permanent and that Weimar could become a place where, without his abandoning a leading role in the duchy's affairs, his own, currently hidden, inner intentions could be acknowledged and fulfilled.

It is of course possible that in 1781 Goethe already intended for _Tasso_ the tragic, or apparently tragic, conclusion that it now has. But the possibility of a tragic conclusion to an attempt at social, and personal, integration was something that from 1781 to 1786 Goethe was doggedly reluctant to admit. On the other hand, a conciliatory end to the play, with integration achieved, would, as those unrewarding years passed, necessarily have looked increasingly fanciful and unrealistic. So Goethe did not merely pause in his work on *Tasso* in 1781—he actually gave up any intention of completing it, and not until 1787 did he bring it back into the category of works, like *Egmont* and *Faust,* on which he wished to work again. From 1782 the interim President of the Chamber was himself too much of a Battista-Antonio to have much sympathy for Tasso, whose voice is only to be heard in the brief lyrical moments of poems such as 'Do you know the land.' Tasso, though a court poet, was too unequivocally a solitary writer for a reading public, his art too much a written and printed affair, for Goethe to wish to pursue his problems at a time when he was himself—however perversely—trying to forget the period in his life when he had been such an artist, and to expiate, or obliterate, its consequences. So in 1781–2 Goethe shifted his poetic attention to two other projects: the theme of the relation between a poet and his society was transferred to *Wilhelm Meister,* and, as an exercise in the new dramatic form of *Iphigenia, Tasso* was displaced by *Elpenor.* Carl August, too, may have had a hand in diverting Goethe's attention, for at some point in these years he advised Goethe against pursuing the subject of Tasso: he may have thought it unhealthy for Goethe to reflect too self-consciously about the poet's role, or he may have foreseen that a play which dealt with a poet's difficulties at court, whether or not its ending was happy, could hardly bring Weimar anything but embarrassment.

There was something a little unsuitable even about *Elpenor,* given its intended role as a festival play, begun in 1781 with a view to Duchess Luise's approaching confinement—the child was stillborn—and continued for a while (in clear preference to *Tasso*) in 1783 at the time of Carl Friedrich's birth. For the most prominent motif in the action is the death, or apparent death, of a royal child. The principal character is a little older than Fritz von Stein, probably modelled on him, and perhaps intended to be acted by him. In the second half of 1781, his confidence restored by the newly intimate and essentially chaste relationship with Fritz's mother, Goethe was able to see in his young friend an embodiment of hope, a new dawn, as if the terrors and desires of the night, expressed once by Orestes and Thoas, and most recently in the ballad 'Elf King', had never been. The innocent hero of this play, in which any threateningly sexual theme is more effectively repressed even than in *Tasso,* receives a name built on the Greek word *elpis,* meaning 'hope'. The plot, though set in ancient Greece and woven out of elements from various sources, including a Chinese drama, is Goethe's own invention. Antiope, a widowed queen, tells of the death of her husband in a past war and of her own traumatic experience shortly afterwards, when she was ambushed by unidentified brigands and her only son stolen from her. Shortly afterwards a son, Elpenor, was born to her dead husband's brother Lykus, ruler of the second of the two kingdoms into which the family's inherited territory was divided. On visiting Lykus years later Antiope was filled with love for this nephew and received Lykus' permission to bring him up in her own palace. The day arrives when Lykus comes to fetch home Elpenor, whose childhood is over and his adolescence begun. Before she parts from him, Antiope makes Elpenor swear vengeance on whoever may have been responsible for the cruel attack in which she lost her son and tells him of marks by which he can recognize his cousin should he still be alive. Polymetis, an old servant of Lykus, reveals in monologues that his master is guilty of a long-concealed monstrous crime, which it is his intention to denounce, and at this point the fragment breaks off. It is not, however, difficult to imagine the continuation. Elpenor would learn that Lykus had commanded the attack on Antiope in order to remove a rival heir and ensure that the kingdoms would eventually be reunited under the rule of his house, and so he would be faced with the apparent necessity, in accordance with his oath, of killing his father. He or Antiope might or might not be moved to forgiveness, but in any event before it was too late it would become clear that Elpenor was Antiope's abducted son—Lykus' own son having either died in infancy or never existed—and that therefore Elpenor had strictly speaking nothing to avenge.

Although *Elpenor* would have been a regular drama in the classical style— apart from the strongly iambic prose in which it is written—with few characters and the unities closely observed, its two acts have a charming and rather personal atmosphere, largely because of the unusual age of its young hero. His nicely depicted boyishness mingles well with the wisdom of the older characters and their long experience of the art of government, and the predominant mood is therefore didactic rather than tragic. The gods before whom Elpenor makes his vow are not the snake-haired Furies but the family gods of Antiope's, and so his own, household, and the dark imagery used by Antiope and Polymetis when their thoughts turn to death and revenge is more than counterbalanced by images of morning, purification, water, growth, and hope. This is plainly not the self-deceptive hope that betrays Proserpina but a simpler and more natural force. But for being simpler it has perhaps lost the power to sustain a passionate drama, the tension that is so intrinsic to Goethe's poetry. The themes of *Elpenor* are those of *Iphigenia*, but without the theological ambiguity that in *Iphigenia* gave them their cutting edge, their tragic potential. At the centre of *Iphigenia* lay a real crime—Orestes' matricide—and misunderstandings about the nature and intentions of the gods had to be cleared away in order to reveal that only human beings can forgive such human evil, as only they can do human good. Once the misunderstandings had been cleared away in *Elpenor* it would have been revealed that there had not really been a crime at all and that forgiveness was hardly necessary. Goethe could not complete *Elpenor*, not merely because it would have been a repetition, and dilution, of *Iphigenia*, but because it would have represented the final step into Enlightened 'official' drama, where all discord is harmony not understood, all partial evil, universal good. Goethe longed to enter that sunlit monarchy of reason but he was too aware in his own life of the rival powers of sense, emotion, and social circumstance to indulge the longer with any persistency.

From 1782 another project displaced *Tasso* as the work on which progress had to be shown by the significant date of 12 November, and it was not *Elpenor*. It was, like *Tasso,* a work that tried to do justice to the whole life of a poet struggling with himself and his age, and over the next three years the regular dictation of *Wilhelm Meister's Theatrical Mission* was Goethe's only major literary activity. As each book was completed a transcript was sent to Barbara Schulthess in Zurich, and it is to these copies that we owe our knowledge of the first state of Goethe's longest novel. Wilhelm Meister, given Shakespeare's Christian name and a significant surname, was from the start intended by Goethe for some kind of literary mastery. Wilhelm, first identified as a poet and intellectual at the start of the second book, continues to be credited with outstanding literary gifts in everything Goethe wrote after the four-year pause in the composition of the novel from 1778 to 1782. His five-act biblical tragedy *Belshazzar,* the detailed description of which is our only source for Goethe's own juvenile play on the subject, is performed with great success; he proves an able and rapid worker in improvising a court entertainment; and he translates into German, from an Italian original which is not communicated to us, the passionate words of the song 'Do you know the land where the lemon-trees bloom?' But Wilhelm is not a poet as Tasso was, nor as Goethe had been, though he was now doing his best to forget it. Wilhelm suffers no solitary anguish, nor does he share his sufferings with a reading public or posterity. His mission is indeed the literary transformation of Germany, but not by the means by which alone it could be achieved, and by which Goethe and Goethe's generation had begun to achieve it—the printed book. His chosen means is the theatre, the medium of Gottsched's historic compromise with the culture of the courts, but deprived of the specifically bourgeois element in the compromise: the nation-wide audience, supplied by publishers such as Himburg, for whom plays are—like novels or poems—books to be read, not, in the first instance, scripts for relatively private performance. After the narrowing of focus in the course of the first book, so that we find ourselves concentrating not on a portrait of mid-eighteenth-century German society but on the single thread of the main character's development, Wilhelm's story becomes a, necessarily episodic, series of encounters with all the different forms of theatre available to his age, with the exception of the one that really counted—the literary drama that, in so far as it was a book like other books, linked intellectuals from all over the German-speaking world in the study of feeling and in social, moral, and historical reflection. The novel retains some vigour and poetic interest, some admittedly rather simple humour, and a realistic manner, but like all Goethe's works since 1778, including the prose *Iphigenia*, it gives the impression of having been lobotomized, of being blind and numb in the very quarter where *Werther* had been so alert and intimate with its audience.

In the second book it still seems as if Wilhelm is living in a definite time and place: although the book consists mainly of conversations about literary matters, they are such as belong in the context of the 1750s—the nature of tragedy, comedy and pastoral, the status of Corneille and the unities, and of alexandrines—and they originate in discussions of printed books. Wilhelm reads from his plays and notebooks, argues with his brother-in-law Werner, second-in-command in his father's business, about the relative merits of poetry and trade, and sounds like the young Lessing as he does so. We are still aware of the material medium of life, of the curtains round the convalescent Wilhelm's bed, of the coffee, tea, and tobacco that alter his mood, of the sandstone flags in the central courtyard of the house and of the balcony between the gables that Werner hopes to turn into a roof-garden. The political contrast between the free city in which the Meisters live and the neighbouring principality is clearly marked by a meeting at the frontier between the uniformed and disciplined militia of the prince and the disorderly town watch. Their arrival, to transfer the custody of a couple who have eloped and been apprehended in the city, interrupts a literary picnic which Wilhelm, his sister Amelia, and her husband Werner have been enjoying nearby. The man of the couple proves to be an actor, and Wilhelm's intervention on behalf of a fellow artist, as he sees him, first with the legal authorities and then with the woman's parents, provides the only narrative episode in the book and one which catches the atmosphere of Goethe's early administrative experiences. (Goethe's drawing of either this scene or its original is unfortunately not precisely datable.) No little irony attaches to the episode: the man has pretentiously changed his name from Pfefferkuchen (Pepperbun) to Honigkuchen (Honeybun), and Grecized it to Melina, and wishes not to continue in what seems to Wilhelm a blessed profession but to settle down with some convenient office job, which however he is not allowed to do. Wilhelm's theatrical passion remains impervious to the warnings these ironies carry, however, and Melina is destined to play a greater part in his life than his sister Amelia, who does not reappear and is one of the loose ends left by the novel's long period of composition. At the end of the book it is agreed that Wilhelm shall go off on a journey which, though it is intended to be educative, will also have a practical function, in that he will be able to collect outstanding debts from his father's remoter creditors. The journey is the main instrument of the transition which Goethe engineered, when he resumed work on the novel in 1782, into a world, and form of experience, more loosely related to contemporary German realities than the first book had been, and specifically more detached from the German public mind.

The opening of the third book transparently announces that a new and progressive stage in Wilhelm's story has been reached when it introduces the traveller to three simple forms of entertainment that are labelled 'the beginnings of theatre'. While carrying out his task of debt-collecting with a success that surprises him, and so collecting a fund of some 1,500 dollars in cash that will be the indispensable basis for what is to come, Wilhelm is entertained first by a sketch put on by some miners at an inn, then by the amateurs of a wallpaper factory, who use drama to while away their winter evenings, and then by a professional troupe of acrobats. The next step in the development of the theatre Wilhelm experiences only after a transformation which cuts him off from his earlier existence. His coachman loses his way at night and he finds himself in a town where he is unknown and which is not on his itinerary. When he gives his name not as Meister (Master) but Geselle (Prentice), he completes the transition to a new life, acknowledging, like the novice Freemason whom his new name suggests, that there is a whole new wisdom for him to learn. However, one of the first things he meets in the town is something familiar: a playbill announcing a performance by a travelling company of actors among whom he finds the name of Melina. Wilhelm's desire to renew this acquaintance is the first in a series of contingencies which conspire to detain him in the town, against his better judgement, but in accordance with his true wishes, throughout the rest of the book. The attraction of seeing first one performance and then the troupe's whole repertoire, a meeting with the actors and with the powerful personality of Mme de Retti, their director, possibly modelled on Gottsched's collaborator Caroline Neuber, and the lure of having his own *Belshazzar* produced, are progressively more effective incentives for him to stay. The realistic manner remains. The cost of Wilhelm's generous entertainment for the actors, a principal reason for his popularity, is carefully noted. The material preparations for the production of *Belshazzar*, more expensive of course than intended; the difficulties of the male lead, an uncultured and incompetent soak, who owes his position solely to the patronage of Mme de Retti, in enunciating Wilhelm's verse; and sarcastic remarks by the narrator, for example, about the narrowly German literary education of Melina's wife, who writes occasional verses and prologues for the company—all contribute to a simple ironic contrast between the visions and expectations of Wilhelm, 'the old hopeful', and the disordered and unedifying reality with which he is ever more closely engaged. But the scope of this realism is now very narrow: it extends only to the theatrical world, and of the town where Wilhelm is staying we see little. An upright army officer with literary interests, modelled on Knebel, who gives Wilhelm advice and support, particularly when the ecclesiastical authorities temporarily prohibit the production of his play on the grounds of its biblical content, the rising public interest in the new play and its author as the first night approaches, and the landlord of the inn where the troupe is staying, are the only indications of a social context for Wilhelm's obsession. Even the climax of the book, his setting aside all social propriety and consenting to take the leading role in the first performance of *Belshazzar*, since Mme de Retti's favourite is once again incapacitated, lacks definition, for there is no one, or no set of interests, to speak against Wilhelm's step. The theatrical mission is clearly depicted, but the nation for whose benefit it was undertaken is being lost from view.

Were this all there is to *Wilhelm Meister*—a picaresque tale unusual for its time and place of origin, in its eventfulness and colourful detail, and in its economical and unsententious style (it is plainly a story told by a busy man)— then one would have to admit that it is a minor piece of work and evidence for Lessing's view that if Goethe 'ever came to his senses, he would not be very much more than an ordinary person'. By its halfway point the novel still has not engaged its hero in any lasting personal relationship. But the third book does bring a sign of latent genius, the first in a series of figures whose association with Wilhelm—for relationship is too strong a word—will create a unique and intangible symbolic constellation at the heart of the novel, which has little to do, either with its theatrical setting or with its sardonic and increasingly narrow-minded realism. As he goes to meet the Melinas once again, and so, unwittingly, to resume his love-affair with the theatre, Wilhelm passes for the first time a child, whether a boy or a girl he cannot be sure, with a dark complexion and long black hair plaited and coiled round the head, and dressed in waistcoat and loose trousers. She—or he—is the child of one of the acrobats Wilhelm has recently seen who has been bought, for the price of her clothes, by Mme de Retti. She is of Italian origin, speaks little German, and is known as Mignon. Wilhelm is fascinated by this puzzling figure with her agile movements and strange gestures, her intense secrecy about herself but her industrious and soon devoted service to Wilhelm, her passionate desire to buy her freedom and return to Italy, her daily attendance at early Mass. Her sex remains indeterminate; though she is only occasionally referred to as 'he'—once after giving a resounding slap to a vulgar fellow who has tried to kiss her—more often she is 'it', as German usage allows of children. There has already been much verse inserted in the novel, though dramatic verse cited from Wilhelm's plays: Mignon is the first to recite a lyric poem, also supposedly of Wilhelm's composition, and she does so with a ferocity of feeling which tells us that lines which must originally have been intended for Frau von Stein have been appropriated for her:

> Heiß mich nicht reden, heiß mich schweigen,
> Denn mein Geheimnis ist mir Pflicht;
> Ich möchte dir mein ganzes Innre zeigen,
> Allein das Schicksal will es nicht.

Do not tell me to speak, tell me to keep silent, for my secrecy is my duty; I should like to show you all my inner being, but fate will not permit it.

It is Mignon who sings, with 'irresistible yearning', the song 'Do you know the land?' at the opening of book 4 of the novel, and this book, written in 1783, contains in all four striking poems. Three of them are sung towards the end of the book by a new marginal or symbolic figure, the Harpist, an old man with a white beard and blue eyes, dressed in a long dark robe, like a friar or a Jew, the very figure of the bard as we meet him in Gray's ode, or Martin's painting. The Harpist, however, who enters an inn one day to entertain the dispirited company by singing, in a ballad called 'The Bard' ('Der Sänger'), of the pure gratuitousness of poetry, is not fired by wrath: his gift, we are told, is like that of the leader of a Moravian congregation—he can grasp and respond to the atmosphere of the moment, weave it into song, and transfigure it by relating it to higher things. He and Mignon and their songs could be said to have this function in the novel: they suggest—rather than actually create—a further dimension of significance to events that are, at any rate in the opinion of the narrator, increasingly trivial and even squalid. It is with them that Wilhelm comes to feel a deeper affinity than with the strolling players: they are, for Goethe in his own life as much as for his hero in his, the image of a power of heartfelt poetry that may yet draw meaning and magic out of a desperately prosaic existence. For Wilhelm's relations with the theatre company in this book are one long and disagreeable awakening after the intoxicating excitement of the production of *Belshazzar*. Not only has he the expenses of the production to pay, and the maintenance of the actors at the inn, but he has loaned much of his remaining cash to Mme de Retti to keep the company afloat, and Melina warns him to secure his play's takings before they are squandered or used to satisfy other creditors. But disasters now follow thick and fast. Mme de Retti's favourite insists on taking over for the second performance of Wilhelm's play, and is hounded off the stage by an infuriated audience, who pursue him through the streets and start to break up the improvised theatre. Order has to be restored by the local garrison, Mme de Retti and her companion take to their heels, and Wilhelm discovers he is now unacceptable in the town's polite society, indeed his officer friend is wounded in a duel defending Wilhelm's honour. Wilhelm sees his life in ruins once more and relives the despondency he felt after Mariana's betrayal. Only Mignon offers him a strange consolation: during the ominous silence before the riot she closes the shutters of his room, calls in a fiddler, lays out a number of eggs on the ground, and then dances blindfold an extraordinarily elaborate, energetic, and repetitive dance between the eggs without damaging a single one. Wilhelm becomes aware of his powerful paternal feelings for the waif, and his responsiblity for her since Mme de Retti has disappeared is a further burden to him, for he now desires only to escape. The news of the outbreak of war—probably the Seven Years' War— puts a complete end to the prospects of the itinerant company, of which Melina is now the director, though Wilhelm is effectively its owner. Melina resolves to accompany Wilhelm on the remainder of his business journey to the city of H. (no doubt Hamburg), where he hopes to find a position for himself and his wife in the well-regarded local theatre, but when the rumour of this project gets abroad everyone wishes to join in and Wilhelm finds himself setting out with the whole troupe. It is at this point that, on their way, they are joined by the Harpist. His song 'He who has never eaten his bread with tears' seems to Wilhelm to express all his own despair at his entanglement in a destiny which, while giving him moments of happiness such as he has never known before, has weighed him down with anxiety, responsibility, and shame. But now the party crosses the path of a Count who fancies himself as a literary connoisseur, and who has shortly to play host to a Prince who will be a general in the coming war. The Count is happy to engage the troupe for a while to help entertain his illustrious visitor, and Wilhelm knows he will be unable to resist the temptation to join them as their house-poet. 'It is rare that man is able, and fate permits him, after a series of sufferings, a sequence of relationships, to make a completely clean break with himself and with others', Goethe comments, in a passage written almost certainly in November 1783, as the worm gnawed at him within. But to his hero he allows an ecstatically tearful embrace with Mignon, while the Harpist plays outside his door, as a concluding tableau to the book, which tells us that Wilhelm's true loyalty and inspiration lie with them.

One more figure has been added to Wilhelm's immediate entourage in the fourth book: the young actress Philina, a strongly marked individual character, symbolic, if of anything, then of lower rather than higher things. She lives only for the day—and the night—and its pleasures, has the entire male sex at her disposal, and ruthlessly, and laughingly, exploits her attractions and, in relations with the upper classes, her considerable powers of dissembling, in order to secure her own interests. With her train of admirers—including a groom of the Count's and a runaway apprentice, Friedrich, who have a midnight brawl in the corridors of an inn—she brings an element of the *louche* into Wilhelm's surroundings, which emphasizes how far he has come from his respectable origins and helps justify his despair at his plight.

But Philina is introduced with striking casualness—'a merry young actress whom we have either not mentioned, or only in passing'—and an important conversation between Wilhelm and his officer friend on the limitations of the theatre-going public is similarly reported at a convenient moment, only after the officer has left, and with a self-conscious apology: 'Many of our readers, who at the end of the previous chapter were contented that we had at last changed the scene, will perhaps be indignant that we are once again returning to mention various things that occurred at the departure.' Such uneasy moments are probably indicative more of informality, or carelessness, in construction than of Goethe's recent reading of Diderot's *Jacques le Fataliste,* or of any deep reflection on the nature of narrative fiction, and they are more frequent in this book than in its predecessors. What they show is an increasing awareness on Goethe's part that there is no logic to his story: Wilhelm may proceed from one species of theatre to another, and may collect one bizarre companion after another, like the hero of a fairy-tale such as the *Musicians of Bremen*, but it is not clear where he is going, or why, or, since the wealth and security which are the goal of most fairy-tales were Wilhelm's starting-point, what they will all do when they get there. For his story is not being told any longer as a portrait of some aspect of contemporary Germany, of its society, or its theatre, or its collective mind, it is being told as a parable, or metaphor, of Goethe's story, and the only logic behind the sequence of episodes is that which is inherent to them as reflections of issues in his own life. That is why he wishes to be free to obtrude his own personality in the narrration, where he feels it ought to be as much at home as the 'I' in a lyrical poem, and why he attempts to accommodate in the narration original lyrical poems. If the *ille ego*, as Wieland called it, is not in fact at home in the novelistic manner, it is because that manner is fundamentally alien to it, as much a 'false tendency' for Goethe as any scientific hobby-horse. The realistic novel was after all the invention of a self-confident and economically automonous bourgeois culture, particularly in England, and the failure of the Storm and Stress generation, or the impossiblity of the task it set itself, is apparent in nothing so much as its inability to transplant this new form into a German setting. Goethe's concentration on *Wilhelm Meister* in the years immediately before the Italian journey can be seen as a last attempt to fuse the autobiographical art that he had discovered in 1769, when revising *Partners in Guilt*, with the ideals of the national literary revival that had snatched him up and swept him along so shortly afterwards. As work on the novel progresses so the gaps yawn more widely between its constituent elements.

Book 5 of the *Theatrical Mission* opens with the most direct of all Goethe's addresses to the reader, a whole chapter devoted to praise of the 'mighty of the earth' and of the improved prospects for Wilhelm's 'development' now that, after the Count's engagement of the company, he can be assumed to be moving into their sphere of influence. Equally this is the book in which the parabolic relation of Wilhelm's story to Goethe's own life is most apparent. For in passing from the experience of strolling players to the experience of a court theatre, Wilhelm is not merely moving to a more sophisticated branch of the dramatic art, he is also entering the world which Goethe entered when he moved to Weimar and began his own daily converse with the 'mighty of the earth'. And not only are Wilhelm's expectations once again disappointed—so are those which Goethe himself aroused in his first chapter exordium. Book 5, written in 1784 as Goethe's nine years of service on the Weimar Privy Council were coming to an end, is his most extensive and devastating parody of the fate of the national cultural and artistic ideal at a provincial German court.

Wilhelm's dreams are caught between two uncooperative realities: the Count and his secretary, to whom theatre is an entertainment and a hobby, and the company, to whom it is their daily bread, well salted with pretension. From the moment they are engaged, the actors show more interest in their future comfort and status than in preparing their repertoire. As they struggle through the night and pouring rain towards the Count's residence they wonder which of the many illuminated windows belong to their bedrooms, only to discover, on their

arrival, that no provision has been made to receive them at all, and they have to
shelter as best they can in the unheated and unfurnished old wing, closed up
since the building of the new palace. Only Philina, who has paid obsequious
court to the Countess, and becomes as it were a temporary lady-in-waiting, is
quickly and comfortably installed in the new wing. When they too are
established, however, and living briefly off the fat of the land, the rest of the
troupe only become factious and arrogant and a source of dissension within the
Count's household. (In this they may reflect the behaviour of the rather
unsatisfactory troupe of Giuseppe Bellomo which had begun to provide
Weimar's entertainment in January of 1784.) The Count, on the other hand,
has his own notions about art, and Wilhelm, after working into the night to
produce a tasteful alternative to his ludicrous proposal for an entertainment to
welcome the Prince, is taken aback to learn that the Count's wishes are absolute
and everything has to be as he has said. Fortunately it turns out that what most
matters to the Count is that there should be an allegorical conclusion in which
the bust of his guest is adorned with flowers and his name appears behind in
glowing characters, and once this has been provided, and with the assistance of
the ladies of the court, who distract him at awkward moments during the
rehearsal, he is prevented from noticing the difference. Not that the ladies—the
Countess, a Baroness friend of hers, and, in the background, Philina—are
concerned for Wilhelm's art either. But they are much taken with the handsome
young man—in Wilhelm's relations with the Countess the 'Tantalus' theme is
again briefly sounded—they urge him to appear on the stage before them, send
him anonymous presents, and invite him to their rooms to read to them—where
admittedly he has to take second place to a pedlar of trinkets. In an
establishment where the poet is the centre of romantic intrigue and the players
appear alongside performing dogs and horses it is no wonder that Mignon
refuses to execute her special dance in public as Wilhelm wishes: she has a
unique affinity for Wilhelm's heart, and she seems to suffer from cardiac
palpitations—and in a hostile environment, as Goethe knew, the heart will not
perform to order.

   The only person who recognizes Wilhelm's ability and promise is the private
secretary of the visiting Prince, who goes by the strange name of Jarno. Widely
travelled, fully acquainted with German literature and wholly unimpressed by
it, he is a man of affairs and the world who has no illusions, a sardonic manner,
and a bitter, even cruel, sense of humour. Some, at least, of his characterisitics
are those of Herder. When Wilhelm praises the art of Racine, as that of a man
who fully understood and closely observed the court world on whose margins
Wilhelm now finds himself, Jarno asks him innocently if he knows any of
Shakespeare's plays. When Wilhelm says no, Jarno lends him some volumes
and Wilhelm withdraws for a while completely from his surroundings and loses
himself in the Shakespearian world. The initial effect of this overwhelming
encounter is, in Jarno's terms, wholesome: Wilhelm is spurred on 'to make

faster progress in the real world' so that he may eventually offer to the German public 'a few cupfuls from the great sea of true Nature' as the British writer did. Jarno commends this resolve and holds out to Wilhelm the prospect of a good position from which, in the coming wartime period of rapidly changing fortunes, he may expect much useful experience. At this moment Wilhelm seems to be being offered the possiblity of the official career, the role in the *theatrum mundi*, which, for reasons similar to those advanced by Jarno, Goethe undertook in 1776. But Wilhelm rejects the offer, for it has attached to it what seems to him an unacceptable corollary: 'It has often caused me disgust and vexation', Jarno concludes, 'to see how, for the sake of any sort of a life, you have had to attach your heart to an itinerant ballad-monger and a foolish, hermaphrodite creature.'

Wilhelm cannot accept this dismissal from his life of the powers of the heart and poetry, not to mention the rest of his 'strange family', now swollen by the young apprentice Friedrich and an adolescent actor who is given the sobriquet Laertes, and so he opens himself to a much less healthy form of Shakespearian influence than his original enthusiasm for 'the real world'. Their temporary contract with the Count having run its course the troupe reluctantly continues its journey to the city of H. Wilhelm travels with them, feeling now like the disguised Prince Hal enjoying low company at the Boar's Head, and dressing himself for the part in cloak, sash and boots, with an attempt at a ruff round his neck, and a hat with a bright hatband and a feather in it. Resentful at expulsion from their paradise, the actors respond enthusiastically to a denunciation from Wilhelm of the heartlessness of the wealthy and noble classes, which all have had the opportunity to observe at the Count's residence and he particularly in Jarno. By contrast, he says, true friendship can flourish only among those who have little or no material advantage and who have only themselves to give, totally, to those they love. His proposal that during their journey the good companions should constitute themselves into an egalitarian republic of friends is immediately accepted, a senate is established, in which women are fully represented, and Wilhelm is elected first director in place of Melina, who has wisely stepped down at the moment when the going starts to look difficult. One of the first tasks of the senate is to discuss the route the party should take in these troubled times, as marauders have been reported in the vicinity of the little town where they are temporarily staying. Wilhelm persuades them not to make a detour but to take the more dangerous direct route, as originally planned, and having armed themselves, and feeling brave and adventurous, they set out into the country. Shakespearian dress is the order of the day, they camp out in a forest clearing, boiling potatoes like gypsies and singing the praises of the carefree life, and Wilhelm and Laertes practise a duel for the forthcoming production of *Hamlet*. With some chronological implausibility, Goethe tells us that this scene, presumably set in 1756, was the original for all the scenes of 'stout vagabonds, noble robbers, great-hearted gypsies and all the other idealised rabble' that have in latter years become so wearisomely familiar on the German stage—from his own *Götz*, no doubt, to Schiller's *The Robbers*, first performed in 1782. Reality overtakes fantasy, however, when shots are fired and the marauders emerge from the trees to ambush the company and plunder their belongings. Wilhelm and Laertes resist the onslaught vigorously but, wounded by a bullet and a sword, Wilhelm loses consciousness and is unable to protect his friends, and all the goods and money they have recently saved, from a band whose practice of liberty and equality is more serious and more effective than theirs.

By November 1784, therefore, Goethe has brought his hero to the point where, having rejected, in the name of poetry and the heart, the choice that Goethe made in 1775–6, he has discovered that there is in Germany no political alternative, other than anarchy, to the frustrations and absurdities of princely rule. Whether there is an artistic alternative remains to be seen, although it is already clear that simply to reject the courtly culture, without such an alternative, is to fall into an empty mannerism. But the moment of this revelation, perhaps the most important single moment in the novel, when the party sets out for its brief taste of an open-air, Shakespearian existence, also lays bare the peculiar nature, and peculiar limitations, of the novel's realism. For realistic it is, in at least two senses. We know pretty well how the figures are dressed and armed, what they are eating, what their financial situation is, and how and where they are travelling (and it should be remembered that in Germany there is nothing merely conventional about a road that leads through a forest). And secondly, the unpleasant surprise of the attack brings some fanciful self-stylization down to earth, and contrasts the 'idealized' robbers and vagabonds with the real thing. But in a third and crucial respect, a respect in which *Werther*, by contrast, is utterly consistent, there is a failure of realism which seriously limits the effectiveness of the rest of the novel's claim to be showing things as they are. For, in telling us that this episode is the origin of the later literary vogue for robbers and wanderers of every sort, Goethe is raising the question of the relation of his story to the German public mind, but he is palpably misrepresenting that relationship as he does so. Wilhelm and the actors are at this point a critique of a literary fashion as, in his own way, was Werther. But Werther was presented as the victim of a fashion, not as its origin, and his story therefore depicted—realistically—the mechanism by which that public fashion reached the individual mind: the books, the names, the conventional feelings and gestures and themes and gambits of conversation. Goethe not only deliberately flouts the requirements of a realistic perspective by making his fictional characters into the origin of a historical movement: he also makes it impossible for himself to depict the form that the movement publicly took, for it all post-dates the events of the story he is currently telling. Most importantly, perhaps, individual minds can only ever be realistically represented as victims of mass movements, not as their source: it is in more than one sense an unrealistic estimate of the nature and power of a mass movement to attribute its origin to the inventive activity of a single mind. The critique of the intellectual and emotional movement supposedly initiated by Wilhelm has none of the imaginative sharing in the movement itself which made for the power and integrity—and enormous success—of *Werther*. Indeed Wilhelm and his friends are not and cannot be aware of the public and national dimension of their behaviour, nor does Goethe suggest any mechanism of publicity by which their wholly incidental extravagances might come to assume such general significance. But because Goethe has raised the issue in the first place we have an uneasy sense of a great gap unfilled, a sense that the national literary consciousness is a reality worth representing, and indeed that no representation of mid-eighteenth-century German life would be complete, or even accurate, without it, but that Goethe is choosing to leave it on one side. And the only reason that could justify such neglect must be that Goethe is writing for an audience—'only a part of the public', Bodmer would have said—that was, or liked to think itself, unaffected by the vogue, that looked down on it, and so by definition was not part of the national consciousness. The crucial respect in which *Wilhelm Meister* is not a realistic novel is, then, not so much a feature of its content as the fact that it does not expect to be found realistic by many more than half a dozen people, notably the Herders, Frau von Stein, and Knebel; and there has to be a sense in which realism is something that a whole public, and a mass public, ought to be capable of recognizing.

In each of the years 1782, 1783, and 1784 Goethe wrote a long poem that addressed itself directly to the question which was rising to the surface of Book 5 of *Wilhelm Meister*: what on earth was he doing in a place like Weimar? Writing 'On the Death of Mieding' in the first months of 1782 may even have been the spur to resume work on his novel: the sceptical attitude to the theatre in its later books is already suggested by the sober tone, at once humorous and kind, in which the poem assesses the life of the court carpenter, avoiding the hidden pitfalls of hyperbole, facetiousness or condescension. There is in the panegyric much half-mocking, half-affectionate allusion to the specificities of Weimar life, to the tradesmen who supplied costumes or painted the scenery, to the coloured paper and tinsel, the hot glue and last-minute hammering, with which productions were knocked together in Weimar, Ettersburg, and Tiefurt, to some of the great successes—the shadow-plays, *Lumberville Fair*, *Iphigenia*—and to Corona Schröter, who had played Iphigenia and who at the climax of the poem is made to approach Mieding's coffin and lay on it a wreath, itself a product of the artificial flower factory which in 1782 Bertuch was starting as Weimar's first industrial venture. But through it all we glimpse Mieding himself, a sickly, pale, coughing man, still clambering about among the flies as the performance begins, a self-effacing factotum who could never be prevailed on to do things in time but who in his own way was so devoted to the art of illusion that he neglected more profitable occupations, and did not even make enough to buy himself a decent funeral. This gently discordant note is explained once we realize that Goethe's sympathy with Mieding is based, as always, on an ability to feel his struggle and sufferings as his own. Goethe appears in the poem, not as poet but as a 'statesman', with whose own heavy labours, undertaken 'for pleasure in the work rather than in profit', and also destined to end in vanity, Mieding's are compared. Goethe too is expending body and effort in what many would see as a fruitless or insubstantial, or even a morally dubious, task, and he recognizes in Mieding a kindred soul. Is Weimar worthy of the Son of Man in its midst? he asks, adapting once again that familiar allusion:

> O Weimar! dir fiel ein besonder Los:
> Wie Bethlehem in Juda, klein und groß!
> Bald wegen Geist und Witz beruft dich weit
> Europens Mund, bald wegen Albernheit.

Weimar, yours was a special lot: both little and great, like Bethlehem in Judaea. Your name is in all Europe's mouth, now for wit and intellect, now for inanity.

A more emphatic, even troubled, note is to be heard in the remarkable poem 'Ilmenau' which Goethe wrote for Carl August's birthday on 3 September 1783. By the somewhat artificial means of a vision that looms up through the mists during a walk in the forests round Ilmenau, Goethe confronts his life of six or seven years before: Knebel, Seckendorff, the Duke, and Goethe himself appear as at one of their woodland camps in the style of *As You Like It*, a first recurrence of the image that in 1784 will have so important a role in book 5 of *Wilhelm Meister*. Goethe makes his earlier self acknowledge with startling frankness not only, as we have seen, the turbulent confusion of the young Duke, but also his own near-tragic perplexity, 'Unschuldig und gestraft, und schuldig und beglückt' ('Innocent and punished, guilty and fortunate'), unable to explain how he can be publicly successful as a poet and yet meet with personal resistance from court society. In 1783, having shut himself off from the German reading public, Goethe was if anything less able to understand that paradox than he had been in 1776. In the poem he deals with it as he had done in his life: by abrupt, even arbitrary, repression. He makes the figure of his younger self 'unable to tell you where I have come from, who sent me here', when it is precisely his social origins that are the source of the problem. And from the vision itself he is released 'at a word', when the mist lifts and the past fades. The journey to Switzerland is taken to mark the beginning of the sunny present, with the duchy's administration reordered, the weaving industry revived after the war, and work resumed on the Ilmenau mines. How the change has come about is as mysterious as the process by which consequences follow on intentions, or a butterfly emerges from a caterpillar's cocoon, but— look, we have come through! All, however, is not well in this sudden summer, something is missing: only once, in those parts of the poem which refer to the present of 1783, is there a mention of the poetry that was the source of the younger Goethe's happiness and popularity, and on that one occasion poetry is said to be 'old rhymes' which these particularly favourable circumstances have 'coaxed' out of him once more. For all the calm confidence of its beginning and end, 'Ilmenau' shows strain and agitation in its verse-forms: it endeavours, in its first and last lines, but without success, to settle into stanzas or couplets, this latter experiment having been consistently maintained in the less private context of 'On the Death of Mieding'. One feels the madrigal verse, Goethe's easiest and most natural form, trying to break through, and in the vision it does, reaching a peak of rhythmical disturbance in its account of the youthful Duke, before it is suppressed once more by the more regular metres of the close. There is an issue raised by the poem which Goethe cannot bring himself to address but only to gloss over, if necessary by main force.

That issue is so nearly explicit in the fifth book of *Wilhelm Meister* that it is not perhaps surprising that in August 1784, by which time much of that book was probably written, Goethe gave it direct expression at last, though at the cost of adopting a form more constricting than any he had ever used before. He was on his way to the mind-numbing diplomatic conference in Brunswick, from which he would report to Frau von Stein in French, so formal constraint was the order of the day: even so, his intention of writing an entire epic in the stanzaic form of 'Dedication' ('Zueignung') is, to say the least, peculiar, suggesting as it does a remarkable imperceptience about the nature of his own poetry. And the nature of his own poetry is, precisely, the subject-matter of this poem, originally drafted as an introduction to the Masonic romance *The Mysteries*, which in 1784 Goethe was beginning to construct, but later used to open the first volume of every collected edition of his works once he had decided to abandon his project of a prose 'Dedication to the German Public'. By 1784, with work on his dramatic fragments completely closed down for a year or more, with few new ideas, and those uninspired, and with *Wilhelm Meister* offering only some fairly disconsolate reflections on his condition—and no immediate promise of either a structure or a conclusion—the question was insistent: was he still a poet at all?

As far as its content is concerned, 'Dedication' answers that question with a decisive 'Yes.' It describes how Goethe sets out in the early morning to climb a mountain and, though he is soon swathed in mist, strives on towards the diffuse radiance of the sun—the phenomenon itself resembles an experience of Goethe's only three weeks earlier during his geological expedition to the Thuringian forest; as a poetic device it is familiar from 'Ilmenau'. As the mists part, however, the source of the light is revealed to be, not the sun, but an allegorical female figure, later identified as Truth, whom Goethe claims to have been his consolation since childhood. She, who knows all his struggles and weaknesses, rebukes him gently for his recent reluctance to proclaim her to the world, with which she tells him now to live in peace, and hands him 'the veil of poetry' 'woven from the morning haze and the brilliance of the sun'. With this he can in future ease and adorn the lives, whether happy or burdensome, of himself, his contemporaries, and posterity.

   The truth of which 'Dedication' speaks is recognizable to any reader of Goethe's letters as the insight into himself and the world which he particularly attributed at this time to his converse with Frau von Stein, and the injunction to 'live at peace with the world' is one of which echoes can be heard throughout their correspondence. But the language of the poem is more general than its occasion, and what it says about his poetry continued to seem of central importance to Goethe long after the convictions of these years were but a memory. That, indeed, is just what the poem says about his poetry. To clarify this point we must look, not merely at the assertion that poetry is a divine gift 'from the hand of truth', but also at the imagery of mist with which poetry is associated. The mists from which the figure of Truth draws the veil of poetry are the same mists through which Goethe in the opening stanzas struggled up towards the sun, and, in the brief survey of his life which the central stanzas contain, these mists correspond to a period of 'error'. Goethe is therefore suggesting that out of the material of his life, even, or perhaps especially, out of its mistakes—that in it which is not truth, and obscures truth—he can weave something to please and benefit his fellow men, and that is an insight which can in time apply as much to what he said and thought during the ascendancy of Frau von Stein as to any earlier 'errors'.

   It is particularly notable in this poem, and indicative of a profound change that is beginning in Goethe, that it identifies his recent, and continuing, solitude as incompatible with the true nature of his genius. The light towards which he has always striven is something he is intended to share with others; it is the essence of his poetry that it should be communicated to others:

> Für andre wächst in mir das edle Gut
>
> .   .   .   .
>
> Warum sucht' ich den Weg so sehnsuchtsvoll,
> Wenn ich ihn nicht den Brüdern zeigen soll?

The noble fruit grows within me for others' sake . . . Why did I seek the way with such yearning if I am not to show it to my brothers?

For Goethe—who, it will be remembered, overcame his own childhood feelings of fraternal envy by turning them into pedagogical zeal, and was by now an only child—the invocation of his 'brothers' has a resonance much wider than mere Freemasonry. In 'Winter Journey in the Harz' and 'Song of Mahomet' his 'brothers' were all those intellectual companions-in-arms with whom he had fought for 'the good cause' and from whom he now felt cut off, by their own misfortune or development, like Lenz or Klinger, or by his, like Lavater, or by his own fault, like those with whom he was seeking reconciliation—Herder, Jacobi, the Stolbergs, Klopstock—not to mention the unknown army of those who had shared the sufferings of young Werther, and of whom only a fraction had crossed his personal path. Goethe is acknowledging, in these moving lines, that 'his' poetry is not his own, but a gift to a whole generation, that it is, or should be, a medium in which his contemporaries can meet and be bound together in such a way that the interaction of so many spirits, all in their own way pursuing the truth, will still be there to give delight to the distant future:

> Und dann auch soll, wenn Enkel um uns trauern,
> Zu ihrer Lust noch unsre Liebe dauern.

And afterwards too, when grandchildren are mourning us, our love shall endure for their pleasure.

The content of 'Dedication', then is an inspiring, and exceptionally direct, expression of a sense of solidarity with his contemporaries and their endeavours from a poet whose autobiographical art is too often interpreted in narrowly personal, even egotistical, terms. Whether the form of the poem entirely justifies the confident tone of this manifesto is another matter. The allegorical mechanisms—the figure of Truth, the veil of poetry—are dull and, worse, they creak, and they blend ill with the elements of natural description: we are at the furthest possible remove from the symbolic landscape of 'Song of Mahomet'. Moreover, the constraints of the regular metre lead to padded lines, and of the stanzaic form to lines that are completely redundant. If we add that the diction is by turns flat, slipshod and strained, and that the ornamentation consists largely of dead or dying metaphors, it will not be found surprising that this poem contains some of the worst lines that Goethe ever wrote, such as 'Ich kann und will das Pfund nicht mehr vergraben!' ('I can and will not bury my talent any longer'), or 'Mich zu ihr nahn und ihre Nähe schauen' ('come near to her and see her nearness'). The point need not be laboured. Goethe has awoken to the indispensable connection between his poetry and the contemporary audience, but it looks as if he may have awoken too late and, in the long years of being dead for the Himburgs of this world, his poetry may have died too.

'I thought I was dead': 1785–1786

After 11 February 1785, and apart from a spell in the following September when Fritsch and Schnauss were both away and he stepped in at an important time to assist the inexperienced Schmidt, Goethe attended only two of the 127 meetings of the Privy Council before he left for Carlsbad and Italy in July 1786. He had, of his own volition, put a clear and decisive end to his career as Weimar's leading political figure. This is the true turning-point in Goethe's relations with the duchy, more so than the Italian journey itself, and during the eighteen months it now took him to resolve on that journey, and prepare for it, the nature both of the estrangement and of the new life he was trying to define for himself begin to become apparent.

Despite his absence from the central organ of government—to which, however, he continued to give occasional reports and expressions of opinion—Goethe's essentially administrative responsibilities remained. But his heart was no longer in it: 'I am patching the beggar's cloak that is about to fall from my shoulders.' Although there was not much to do in respect of highways or even of financial reform, after the successful completion of the last two years' negotiations, the Ilmenau mine had become more demanding. The new shaft had been begun the previous February, and it had now become necessary to build a water-powered lifting engine to extract the rubble. For Goethe this entailed further negotiations with the neighbouring states, as well as practical problems with the installation and its personnel, and not until the autumn of 1786 was the engine fully operational. At the same time Goethe had been charged with reorganizing the taxation of the Ilmenau district after the depredations of a dishonest official whom he had been instrumental in unmasking. However, in these as in mining affairs he now had the assistance of a thoroughly reliable fellow Commissioner, C. G. Voigt (1743–1819), the elder brother of the geologist, so that the burden was by no means intolerable. In the diplomatic area Goethe remained initially indispensable: even though he did not share the Duke's enthusiasm for it, he had from the start been privy to the negotiations to establish a League of Princes against Austrian expansionism, which the brief War of the Bavarian Succession had not deterred, and in August and September 1785, when Frederick the Great's emissary came to Weimar to secure Carl August's adherence to the League, there was no one better qualified than Goethe to conduct the discussions with him, indeed Goethe had to devote his entire birthday to the visit. But Goethe disapproved of the 'itch' for war that lay behind Carl August's ventures into high politics, all the more since his own antennae were beginning to alert him to a much greater danger in a quite different quarter: at the very time when the League was being set up, in the autumn of 1785, Goethe was reading Count Necker's justifications of his financial policies in France, and was receiving an 'indescribable impression' from the report of the Diamond Necklace Affair. The news of this Parisian court scandal, in which even the Queen, Marie Antoinette, was involved, and which seemed to Goethe to threaten the authority of the entire French monarchy, appeared to his friends to have made him briefly insane. He was later, in 1791, when his intuitions had been borne out by the turn of events, to make the affair the subject of a whole play, *The Grand Kophta*, while the antics of the League of Princes, which petered out in 1788, provided him with no more than some atmospheric touches in *Wilhelm Meister* and *Egmont*.

Goethe's perspectives were already different from those of his lord and one-time pupil, and his withdrawal from the Council was but a beginning acknowledgement of the difference: Goethe's 'maturing sense of what is human in life positively deprives him of all pleasure in his political situation'. 'Unfortunately, in the respect in which our friend has given up hope', Frau von

Stein wrote, 'there is nothing that can be changed, because there is nothing that can be hoped for, and both the key and the time signatures that rule our system are morally incorrect.' The Goethe who believed that noblesse obliges to court ceremonial was irritated when, again in the autumn of 1785, Carl August abolished the ancient practice of the daily household dinner, with its upper and lower tables, and withdrew to a private dining-room; and the acting Finance Minister could hardly approve when at the same time an expert was imported to train the ducal hunt in coursing with hounds. 'The Duke . . . has sent off the courtiers and brought on the dogs', Goethe commented to Frau von Stein, who, like him, was probably vexed that Baron von Stein now had to take his dinner at home. The direction Carl August was taking—towards an absentee absolutism —was clearly indicated at the start of 1786 when he went to Berlin to be presented to his great-uncle Frederick the Great and to the heir presumptive Frederick William, a journey on which Goethe, mindful no doubt of their visit in 1778, did not accompany him. Frederick died in August 1786 and within a fortnight Carl August had set out for Berlin again to discuss his future, and that of the League, with the new King. The following year he entered Prussian military service where he could promise himself a wider field of action—and more soldiers to drill—than at home in his Council discussing an exchange of insults between one of his excisemen and some travelling Jews (Goethe took the part of the Jews). Only days after Carl August left for Berlin, Goethe set out for Italy—a parting of ways that was no metaphor. Yet the drifting apart of these two friends, for such they remained, should not be interpreted purely as an opposition of political principles: whatever changes and improvements Goethe had hoped for in Weimar had always been those that could be accomplished through the education of the prince; he had never operated in accordance with some concealed, theoretical, agenda of his own. And the truth now was that Carl August had outgrown the need for a mentor, and purely personal affection could no longer disguise the real differences in the starting-points and the goals of his life and Goethe's.

If he was not to devote his life to the service of another man, what was Goethe to do with it? Strange to say, in 1785 and 1786 he thought the best thing was to write an operetta. But his obsession with the oddly mechanical little farce *Jest, Craft, and Vengeance* (*Scherz, List und Rache*), and its setting to music by his old Frankfurt friend and fellow-Mason, Philipp Kayser, now living in Zurich, was indicative of a profound reorientation. For suddenly—perhaps partly influenced by the extensive, and by no means purely Italian, repertoire of the Bellomo troupe—Goethe was again writing for a public as wide as the German-speaking world. 'When I wrote the piece I had my eye not only on the narrow Weimar horizon'—what a change that word 'narrow' implies from the days when Weimar was the *theatrum mundi*—'but on that of the whole of Germany, which to be sure is narrow enough.' In some respects this experiment in *opera buffa*, which tells how the two marionette-like figures Scapin and

Scapine deceive and bully a miserly doctor into disgorging the inheritance which is rightfully theirs, is an understandable product of these desiccated years. Goethe was later to regret the 'lack of feeling' in his text, but at the time he wrote contemptuously to Kayser of 'the emotion . . . that the representation of affection immediately arouses and for which the common public yearns so much . . . it's easy to tinkle on those strings'. To this has the dissociation of *dedans* and *dehors* reduced in his mind the art of *Urfaust* and *Brother and Sister*. He saw *Jest, Craft, and Vengeance*, with its cast of impersonal Italianate comedy figures, as a purely technical exercise in a new art form—the fully-fledged opera (or 'lyrical drama') in German. Wieland's *Alcestis* was its only predecessor in the serious mode and in an important sense its inspiration and model: both were attempts at a theatrical mission to the German public which ignored that the public could be reached only by print. At the same time as writing long letters to encourage the dilatory and (rightly) self-doubting Kayser to finish his setting, Goethe was putting out feelers to Munich and Vienna, where in 1782 Joseph II had briefly and unsuccessfully attempted to establish a theatre for German-language opera, in the hope of finding, there or elsewhere, an audience able to appreciate and support his own similar venture. Goethe may have made mistakes in his choice of theme, manner, proportions (the text is far too long for a cast of only three), and composer, but there can be no doubt that he was once more engaging in a national cultural enterprise, even if this time his contribution was to be wholly overshadowed by another's. In 1785 Mozart's 'Singspiel', *The Abduction from the Seraglio*, one of the fruits of Joseph II's experiment in 1782, reached Weimar, and after Goethe had overcome his repugnance for the text and become attuned to the music, he was able to grasp its appeal and, as he remarked with acidulous envy, to 'comprehend the difference between my judgement and the impression made on the public'. In later years he was to attribute to the success of the *Seraglio* the total neglect of his and Kayser's collaborative efforts. At the time, and undaunted by the competition, he started work on a second libretto, *The Mismatched Household*, which was intended to contain those elements of 'feeling' which the contemporary public plainly demanded. In early 1786, however, in despair at the desolate state of the German theatrical repertoire—or perhaps because he was not finding it as easy as he had boasted to 'tinkle on those strings'—he abandoned the project. But in the mere attempt to write a German opera a crucial preliminary step has been taken: Goethe no longer—as, for example, in the poem 'Ilmenau'—sees Weimar as the circle within which his ambitions have to be fulfilled, if they are to be fulfilled at all, but sees again the possibility of a national role for himself, to take the place of the local high office from which he has stepped down. As yet, it is true, he has not found his proper medium—not the stage, not even the musical stage, but the book—and he will not find it until he starts work on the collected edition of his writings.

In 1785, however, Goethe was still excusing himself from any part in the national battle of the books. Jacobi's publication of *On the Teaching of Spinoza, in*

Letters to Mr. Moses Mendelssohn (*Über die Lehre des Spinoza in Briefen an Herrn Moses Mendelssohn*) opened a fierce controversy, of whose course however Goethe professed to Jacobi complete ignorance since, he said, apart from Herder, there was no one in Weimar with any interest in it. From the start he made the Spinozan reading of his little cenacle, whose other members were Herder and Frau von Stein, into a personal devotion to a holy man, which he endeavoured to protect from exposure to the public realm where Jacobi was operating. Even when Jacobi, without his permission, published two of his poems—the odes 'Prometheus' and 'Divinity'—as part of his polemic, Goethe contented himself with a firm paragraph in a letter to Jacobi and made no public statement about the indiscretion. Yet Jacobi's campaign was of great importance, being directed at the very centre of the religious culture of 'official' Germany, and in one sense not being principally concerned with Spinoza at all —though Goethe constantly pretended that it was.

If Leibniz's philosophy is a compromise between the claims of individual identity and those of rational order, the philosophy of Baruch Spinoza (1632– 77) is wholly uncompromising—even down to the mathematical form of its presentation as theorems, proofs, and corollaries. It is a philosophy of pure order. Where Leibniz posits an infinity of substances, or monads, jostling for existence and building a maximal and optimal order among themselves, Spinoza posits only one substance—perfect, infinite, and identical with all existence, both material and mental. Matter and mind are but aspects, Spinoza calls them 'attributes', of this one substance which may with equal propriety be called either Nature or God. God and Nature must be identical, for if they were not—if, for example, God were Nature's creator—there would be something that God was not, and there would thus be a limit on God's infinity and perfection. Individual existences, whether bodies or souls, are simply passing conformations, Spinoza calls them finite 'modes', of this one substance, and so to be a Spinozist is to believe that God is in everything, and everything is God, but nothing has a freedom of action or even an identity (let alone an immortality) of its own. This was long regarded as an atheist belief by Christians, and also by Spinoza's own Jewish congregation in Amsterdam, which excommunicated him. Throughout the eighteenth century copies of Spinoza's writings were exceedingly rare, and it might seem perverse of Jacobi to give inaccessible ideas greater currency by expounding them, even if in order to rebut them.

But Jacobi was after bigger game than the solitary and traditionally vilified Spinoza. He wanted to show that the Leibnizian compromise, and indeed any philosophical position that attempted a compromise between Christianity and rationalism, was inherently unstable and must necessarily eventually collapse into Spinozism, the only purely rational philosophy. Religion, Jacobi thought, had to be based not on the intellect but on faith, and since the entire intellectual system of Enlightenment, even where it claimed to be a friend of religion, had no place for faith, it was all, even if unwittingly, crypto-Spinozism. The significance of the example of Lessing, the central figure in the latest phase of the German Enlightenment, was that he was not unwitting. His provocative avowal of Spinozism to Jacobi personally, and his recognizing the doctrine—which can only have meant his recognizing the denial of a personal creator-God—in Goethe's 'Prometheus', showed an awareness and affirmation of the anti-Christian logic at the heart of contemporary philosophical and theological reflection in Germany. Jacobi was in fact, though he did not realize it, attacking the peculiar compromise by which the system of German absolutism both encouraged a critique of religion, in order to reduce the pretensions of the Church, and set limits to that critique and maintained a large body of clergy, in order to sanction its own legitimacy and retain its control over the minds of its subjects.

From these wider implications of Jacobi's charges Goethe sedulously turned his attention. Being himself now near the heart of the 'official' culture, and having welcomed it into his life to the point of allowing it to suppress his native poetic genius almost completely, he had no interest in seeing it criticized. In his correspondence with Jacobi it is always Spinoza himself that he is defending, not the contemporary German crypto-Spinozism which Jacobi was attacking. As the dispute grew more contentious Goethe rather priggishly washed his hands of it, giving as his excuse (in which one hears the voice of Frau von Stein) his 'disinclination for all literary quarrels . . . What are we all, that we should elevate ourselves much?' But Goethe's Spinoza is, philosophically speaking, and when stripped of the veneration paid to the moral purity of his life, a markedly Leibnizian figure. In the crucial matter of the relation between individuals and the divine, between finite modes and infinite substance, Goethe still gives practical, and to a great extent theoretical, priority to individual existences, conceived as centres of energy just like Leibniz's monads. In a fragmentary essay from this period, published long after his death, and given by editors the title 'Spinozan Study' ('Studie nach Spinoza'), Goethe starts from the concepts, important to Spinoza in defining God-or-Nature, of existence, perfection, and infinity, but within a few sentences he is asserting such centrally Leibnizian themes as the autonomy of individual existences and their independence of any causal connection one with another, any such connection being merely 'apparent'. Throughout the study the point of view of an individual limited existence—of a soul, or monad, in effect—is adopted as the point of reference, and much of the text is given over to a discussion of an issue familiar from the Sentimentalist theory of genius: the relation between a soul and its perceptions, which is seen as the source of our notions of what is sublime or beautiful or merely comfortable (this last plainly a shot at Jacobi). Goethe's professed discipleship of Spinoza, whose work even in 1785 he confessed to reading only desultorily and without full understanding, and which he later dismissed as 'abstruse generalities', amounts to three things. First, Spinoza, as Goethe interprets him, provides the Leibnizian compromise between individuality and order, but without the encumbrance of Leibniz's Christian superstructure. Whatever its value in promoting among the populace the social cohesion of which Leibniz's metaphysics is the image, Christianity is unnecessary for the intellectual élite for whom, as Goethe was later to say, 'science' and 'art' can perform the function of 'religion'. Second, Goethe was deeply impressed by Spinoza's account, in the later parts of the *Ethics,* of the nature of the moral life. Although he did not in the long term hold with full imaginative consistency to any of the Spinozan rejections of a personal God, of freedom, or of immortality, he showed fewest qualms about freedom. Spinoza's belief that the only human freedom was an 'emendation of the intellect' by which, as mere finite modes of divinity, we lost interest in our finite selves and concentrated instead on what was truly divine about us, was a model to which Goethe repeatedly returned. It later took on however, a very different form from the rather unctuous humility of what we may call the Frau von Stein period, and at times it disappeared from view completely. Even this peculiarly Spinozan attitude, however, if it is understood simply as a renunciation of causal influence on the world around us, fits well into the Leibnizian compromise. Third, Spinoza's emphasis on the unity and divinity of Nature— which admittedly is also in principle derivable from Leibniz—provided an important authority for Goethe's approach to scientific study. This involved concentrating—again, in a more Leibnizian way than he perhaps realized—on individual (and self-moving) objects as marked by the red thread of the single divine plan. Here 'in single things', 'in herbis et lapidibus', he could find the divine, and could experience in what he could see ('When you say one can only *believe* in God, . . . I say to you that I think *seeing* is important') the bedrock of objectivity that he could not—and to some extent would not—find in the social world about him. Goethe is usually at his most Spinozist when he is feeling most alienated from the German public.

By the middle of 1785 plants, then, had joined stones (and 'those nasty bones') as the material of Goethe's natural theology, in which Spinoza was 'not an atheist, but the greatest of theists, yes, of Christians'. In fact botany was very soon to displace mineralogy as Goethe's main scientific interest, not only because of its closer relation to anatomy, as a science of living form, but also because he had come to realize that he could make little further progress in mineralogy without a knowledge of chemistry, which was quite beyond his present scope. Some experimental dissection of seeds, including a coconut, was done under a microscope in Jena in the earlier part of the year, but the summer brought some extensive fieldwork. From the beginning of July to the middle of August Goethe, for only the second time in his life, went to take the waters at a spa, Carlsbad (Karlovy Vary), some 120 miles away in western Bohemia. For a fortnight beforehand he and Knebel toured the Fichtelgebirge in the company of a young Weimar herbalist, F. G. Dietrich (1768–1850), who was at this time quite uneducated apart from his phenomenal knowledge of plants and his

ability to name them in both German and Latin, according to the new system of
Linnaeus. The purpose of the expedition was mineralogical rather than
botanical, but Goethe saw some unusual flora, notably native orchids and large
colonies of sundew. During the subsequent time at Carlsbad Dietrich brought
Goethe a new batch of local plants every day, to build up his acquaintance with
a natural kingdom that, for all his interest in gardens and woodcraft, had
hitherto rather repelled him, because of the great effort of memory required to
grasp it all, and the drynesss and artificiality of the Linnaean system. A year
later Goethe was telling Frau von Stein that 'I am agog with the vegetable
kingdom again . . . the enormous realm is simplifying itself out in my soul, so
that I will soon be able to see through the most difficult problems straight away'.
Once again, a short period of study was enough to locate for him an Ariadne's
clue which could be his guide through the labyrinth of the natural world, a set
of what he called 'those first great notions on which I repose . . . in order to be
able to judge correctly and easily of great and new objects in nature and
civilisation'. Now, however, he has a terminology, which he thinks of as
Spinozan, to explain and justify his approach: 'it is no dream, no phantasy; it is a
perception of the essential form with which alone as it were Nature is always
playing and in her play producing the variety of life. [By his use of the term
'essential form' Goethe is here deliberately alluding to a sentence of Spinoza's,
which he took as providing the justification for his devotion to scientific study,
but which he completely misunderstood (it actually refers to reflection about
the nature of thought and extension) and which anyway does not, in its
authentic version, contain the words 'essential form'!] If I had time in my short
span of life I would trust myself to extend it to all the realms of Nature—to all
her realm.' To define what it was that he was, as yet inchoately, seeing and that
united all nature into a single realm, would take some years, but already by the
summer of 1786 the programme was fixed for practically all his future scientific
work, except that in optics.

Goethe needed his trip to Carlsbad in 1785. He had had a bad bout, probably
of tonsillitis, during the winter, and the disorder recurred just as he set off for
the Fichtelgebirge at the end of June and held up his journey for nearly a week
(which he devoted to a close study of *Hamlet*). The Duke noticed the fragility of
his health—and no doubt a certain disaffection from Weimar—and gave him
40 louis d'or (200 dollars) towards the cost of the trip, at the same time as
raising his annual salary by a similar amount. The grant must have been
welcome, for Goethe's expenses in the six weeks ran in total to 530 dollars,
nearly 40 per cent of his salary. Carlsbad was becoming fashionable—though
the Duke and Duchess this year went to the spa at Pyrmont—and perhaps the
previous year's visit by the Stolbergs contributed to the decision of 'a whole
caravan' to set out from Weimar: not only Goethe and Knebel, but Frau von
Stein, the Herders, the Voigts, the Fritsches, Countess Bernstorff and her
secretary Bode, and others. Furthermore, Carlsbad offered much to the
geologist: dramatically situated in a narrow rift, it gave easy access to the local

strata, but it also had the marvel of its pulsing hot spring, spouting up several feet into the air from the river bed, multi-coloured with strange deposits, between which and the steep wooded cliffs ran the two lines of tall houses that constituted the resort. In the mornings the visitors gathered to drink their mug of the steaming sulphurous water, and in the throng of the central European nobility Goethe stood out 'only by his beautiful eyes'. Whatever the medical benefits of the holiday, and these must certainly have included the sheer rest ('fainéantise' was Goethe's word) after six years of ceaseless activity, there can be no doubt that it was for Goethe a social tonic. 'The necessity of always being among people has also done me good. That is the best way to rub off many of the rusty spots that a too obstinate solitude brings upon us.' In addition to the Bohemian and Polish aristocracy there were nobility and courtiers, some with developed literary and artistic interests, from all over Germany, and Goethe suddenly found himself in a large company which was not only stimulatingly new, but which also regarded him as its particular literary adornment. The visitors constituted a shifting society with no fixed structure or hierarchy, in which fame was as important an index of position as birth or title. Goethe had a public once more and an appreciative one, far larger than he had in Weimar, and not at all jaded. There was even a substitute for Fritz von Stein—who was spending the summer in Frankfurt with Goethe's mother—in the person of the 13-year-old son of Count Brühl, from Dresden, whom he introduced to mineralogy, so beginning a lifelong friendship. The familiar dynamics of the holiday-camp are recognizable in the growth of informal little groups for expeditions into the neighbourhood or for the unexpectedly emotional celebration of birthdays—to which Goethe contributed an occasional poem or two—in the tearful farewells, and in the firm promises to meet again next year. The interlude had not only offered a release from the social and personal fixities of Weimar, it had also introduced him to a visible and tangible section of the German-speaking public, which he was now newly hoping to address. It is no wonder that as Goethe was making his plans for 1786 the one certainty he was prepared to recognize was that he would be going again to Carlsbad—'I am indebted to that spring for a completely different existence'—nor that he arranged to take with him for distribution there 1,000 copies of the publisher's announcement of the collected edition of his works. In Carlsbad he had rediscovered not just the atmosphere of his brief visit to the spa at Ems in 1774, but also something of the youthfully confident relation with a nation-wide following that he had felt in those years. When Herder wrote to Jacobi in September 1785, his letter must have sounded like an echo of the hopes and enthusiasms of an earlier age: 'Goethe and I invite all good men and true to Carlsbad, and if heaven grants good fortune, we shall march on it with a mighty host.'

The 'different existence', however, was even beginning to define itself at home in Weimar, or, to be precise, in Jena. In Carlsbad the matter of the new and advancing sciences was a 'great gift for social intercourse' providing on every walk, whether 'along the calm valley or to wild rugged cliffs', occasions for shared 'observation, reflection, judgement and opinion'. How much more could Goethe hope this to be true of the university for which he was feeling an increasing responsibility. In 1783 (probably), during their joint expedition to the Harz, Goethe had told the geologist von Trebra of his vision of a society of observers of Nature who, though scattered round the world, would share their observations for the sake of increasing disinterested knowledge. By 1785 it was becoming clear to him that he had an institutional instrument for realizing that vision on his doorstep. It was one of the few continuing irritants in his relation with the Herders that as interim Finance Minister he took every opportunity to make savings for the benefit, not of Herder's schools, but of 'his own favourite scheme, Jena', to which everything else was 'sacrificed'. This development, however, was a recent one. In 1782 the retired Göttingen professor C. W. Büttner (1716–1801) had followed his great library, sold to Carl August in exchange for an annuity, and had been settled in the old palace in Jena where, as himself a 'living encyclopaedic dictionary', he willingly discussed with Goethe anything from the classification of plants to the geography of the Philippine Islands. But it was not until July 1784 that Jena became a true second centre in Goethe's life when Knebel, who had by now himself acquired a keen interest in natural history and mineralogy, returned from Franconia and also moved into rooms in the Jena palace. While in the two years between his first anatomical lessons with Loder and the floods of early 1784 Goethe did not visit Jena at all, in the two years after Knebel's arrival he was there at least twenty times. Encouraging the Duke to attract Major von Imhoff to settle there, Goethe remarked in 1785 that 'if only a beginning is made, a nice little circle should collect in Jena'. No doubt that circle was to be composed both of courtly amateurs, like himself and Knebel, and clubbable professionals like Büttner and Loder, who already often met together, when the conversation must perforce have been largely scientific. To build up such a circle, however, it was necessary not only to attract outside talent into the society of Weimar and Jena —such as Imhoff, the enormously wealthy Niebeckers, who arrived in 1785, or the Brühls and one or two others who were so interested by the Weimar 'caravan' at Carlsbad that they inquired about the possibility of settling in the duchy—it was also necessary to improve the standard of the university. Since 1737 competition from the new and progressive foundation of Göttingen, on the other side of the Harz, had halved the number of Jena's students, and the poorly paid professors felt themselves in a backwater.

Developing the provision for natural sciences in Jena—in modest emulation of the Göttingen example—quickly became a priority for Goethe. In 1785 he arranged for J. F. A. Göttling, assistant to the Weimar court apothecary Dr W. H. S. Buchholz, to study medicine in Göttingen at the duchy's expense, in order that he should later become Jena's first professor of chemistry, which he duly did in 1789. At the same time Goethe sought to provide Göttling with the nucleus of a laboratory by purchasing the chemical collection and equipment of the eccentric August von Einsiedel, brother of Anna Amalia's gentleman-in-waiting, who had left for Africa in order to prospect for precious metals for the French government (and, it later turned out, in order to elope with Emilie von Werthern, the wife of Carl August's Second Equerry, who had succeeded in covering her escape by staging her own death and funeral). In 1786 he similarly arranged support for the studies, this time in Jena, of the impoverished A. J. G. C. Batsch (1761–1802), whom he intended to design and direct a new botanical garden for the university, of which the head gardener was to be none other than young Dietrich. One of Goethe's last acts before withdrawing from the regular deliberations of the Privy Council in February 1785 was to report in detail on the level of the professors' salaries—then on average about a quarter of his own. His increasingly close official involvement with the university was further demonstrated at the start of 1786 when, in one of his now rare commissions from the Council, he conducted what was effectively a visitation on behalf of the Duke in order to improve the discipline of the student body, much troubled by drinking and duelling 'corporations'. Goethe recommended strengthening the supreme disciplinary body by appointing to it the four professors in whom he had most trust—including Loder—and this measure seems to have met with some success. Seven years later the atmosphere of the university was reported to have changed completely, duelling was quite out of fashion, and student numbers had recovered to their highest level since 1750.

At the same time as Goethe's personal links with it were becoming closer, Jena itself was entering into a new relationship with the German public. In 1785, after careful preparation, Weimar's entrepreneur, F. J. Bertuch, launched the *Allgemeine Literatur-Zeitung* (*General Literary Journal*), a daily paper consisting exclusively of book-reviews, furnished anonymously and largely, but not exclusively, by Jena professors, who were well paid for their contributions. The standard was high and the coverage, both of German and of foreign books, extensive, and within two years there were 2,000 subscribers. Its intrinsic merit, its publicity value, and its effect on the morale of the professorate, all helped to make the *ALZ* one of the most important factors in raising Jena for two decades to the position of intellectual centre of Germany. In 1789 Goethe would call the *ALZ* 'not the work of a single man but, through the participation of so many scholars, the voice and, so to speak, the Areopagus of the public'. In one extremely important respect, however, the journal did bear a personal stamp: Bertuch had chosen as his principal editor the professor of rhetoric, C. G. Schütz (1747–1832), assisted by the Jena doctor Gottlieb Hufeland, both of whom, and Schütz in particular, were convinced of the significance of Kant's as yet largely unappreciated new critical philosophy, and the *ALZ* was from the start a vigorous defender of the Kantian cause. In its first year it published Kant's own censorious reviews of the first two parts of Herder's *Ideas,* and reviews by Schütz of the first edition of the *Critique of Pure*

Reason and the *Prolegomena to Any Future Metaphysics* (1783), together with a short notice of the *Foundation for the Metaphysics of Morals* (1785), of which a full review appeared in 1786. These reviews stimulated C. L. Reinhold to write his *Letters on the Kantian Philosophy,* and the *Letters* in turn led to Reinhold's appointment, first to the staff of the *ALZ,* and then in 1787 to the new chair of philosophy in Jena. This was the moment at which post-Kantian philosophy could be said to have begun, and with every phase of that movement, of such decisive importance for the national culture, Jena was to be intimately connected.

As yet, however, in 1785, Jena's great future was only in germ, and a Goethe who was unwilling to be drawn into Jacobi's Spinoza controversy cannot immediately have seen the merits of a new journal which began with a devastating attack on the *Ideas,* the major work of his only intellectual companion, and one to which he had indirectly contributed much himself (though Goethe seems to have shared Kant's view that Herder's treatment of history did not give enough weight to political institutions.) In too many ways life after his withdrawal from the Council carried on much as before. Though his health improved after the visit to Carlsbad, Weimar after the abolition of the court table was if anything more stiflingly solitary than ever. 'You have to take the place of all I lack', he told Frau von Stein, after the *Theatrical Calendar,* read in the draughty halls of the Gotha palace, had revealed to him the desolating sterility of the German theatrical world to which he had thought to address his operettas; and when there was talk of Herder's moving to Hamburg he saw plainly how much he stood to lose: 'without you and him I should be alone'. To Jacobi he wrote at this time of the isolation that was making him 'as dumb as a fish' and of his compensating intention to devote his 'whole life to the contemplation of things', that is to Spinozan natural science. Yet even with Frau von Stein there were signs that the endless converse of his soul was threatened: to her he now seemed 'ever silent'. The repression of his inner self, the total separation of his emotional life from a formal social exterior—'something fearfully stiff in his whole deportment and hardly speaking, as if his greatness made him embarrassed'—had reached a point where Goethe was in danger of being shut off even from the woman on whose insistence, and with whose aid, he had first begun to divide the inner and the outer worlds. Worst of all was the state of his writing, dwindling and withering still, polarized into meaningless objectivity and objectless longing. Beyond the deliberately mechanical and emotionless cavortings of the figures in *Jest, Craft, and Vengeance* could lie only silence. There was no future there for the Goethe who had once achieved and might still achieve again, a poetry of objective feeling. On the other hand, 1785 brought a half-strangled cry, which was put into Mignon's mouth, but which Goethe told Frau von Stein 'is now mine too':

> Nur wer die Sehnsucht kennt,
> Weiß, was ich leide!

Allein und abgetrennt
Von aller Freude,
Seh ich ans Firmament
Nach jener Seite.

Ach, der mich liebt und kennt,
Ist in der Weite . . .

Only he who knows yearning knows what I suffer! Alone and cut off from all joy I look to the firmament in yonder direction. Ah, he who loves and knows me is far away . . .

The extremity in which Goethe now found himself was not due simply to the distractions of administration. Goethe thought he had reduced those to manageable dimensions, and still no rebirth came. Fortunately he could not read the apologetic epitaph with which Merck passed on to an expert his essay on the intermaxillary bone: 'remarkable . . . for a man of the world who . . . was once a very famous poet'. But the cruel words about him published in 1781 by a young Swiss writer, surveying the contemporary literary scene, may well still have rung in his ears: 'Alas, what he has given, he has given. He is now as unfruitful for the public as the desert sand.' He was now turning again to the public—in the schema for his *Literary Works* that he drew up in June 1786 he planned to open the first of eight volumes with his 'Dedication to the German Public'—but what had he to offer them? Half of what he was proposing to print had been published already, and as for the rest, the schema told its own tale: 'Egmont, unfinished, Elpenor, two acts. Tasso, two acts. Faust, a fragment . . .'. Publishing these bits and pieces was an act of desperation: for three years he had not added a word to any of them. 'When I resolved to have my fragments printed', he wrote to Carl August in December 1786, 'I thought I was dead.'

Goethe's final decision to cut the knot and flee to Italy was probably taken fairly late, at about the same time as his decision to publish his fragments uncompleted, and in a similarly desperate mood. Forty years later he told Eckermann 'that in the first ten years of my life as minister and courtier in Weimar I did next to nothing; I was driven to Italy by despair . . .'. But the roots of the decision ran back deep into his past, well before the moments in 1779 and 1775 when he had turned aside from the supreme educational journey, back to 1770 when, ariving in Strasbourg, he had written to Langer that the true university for him was Rome and that he who had seen Rome had seen everything. It is in the educational programme, and so in the cultural and political loyalties, of Caspar Goethe that lie the origins of his son's fascination with Italy, or rather, specifically, Rome. We miss the true importance of Goethe's flight in 1786 if we think of it simply as an escape to Italy. Rome was not just an important stage in the itinerary of the grand tour, which by the declining years of the eighteenth century was no longer the indispensably fashionable undertaking that it had been in Caspar Goethe's youth: Rome was also the capital of the Empire of the Caesars, from whom the liege-lord of all Frankfurt citizens, the Holy Roman Emperor, claimed lineal constitutional descent. It was, especially perhaps in the centuries after the Reformation, the image of Germany's age-long involvement, both political and religious, with the powers south of the Alps. It had a sizeable German colony, consisting mainly of artists, and it had provided patronage and a milieu for Winckelmann, the model of a German devoted to art, scholarship, and the cultivation of his native language. It was, in a ghostly sense, Germany's non-existent national metropolis. To go to Rome was to go to the centre of the mental world in which the young Goethe grew up, it had always been intended by father and son that visiting it should crown and complete his education, and in the last analysis that is why he went there. Goethe did not go because he was overworked (in 1786 that was no longer true); nor because he needed to get away from Weimar and see more of the world (in Carlsbad in 1785 he turned down a Polish count's proposal that he should accompany him to London); nor because he had acquired a taste for the antique and wanted to plunge himself in what remained of classical civilization (there is no evidence of any radical or programmatic change in Goethe's aesthetic loyalties before 1786—only a gradual, and highly problematic, shift towards courtly values, which of itself and independently of any classical antiquarianism, involved a hostility to Christianity and an exclusion of the contemporary and historical realism characteristic of bourgeois Storm and Stress; moreover, Goethe during his stay in Italy refused invitations to travel on to Greece and classical sites even further afield). Goethe went to Rome because he had always known that one day he had to do so. The question had always been, when?

    After he turned away from Italy in Brig and on the St Gotthard in 1779 Goethe devoted himself to adapting to his chosen life in Weimar. But already in the winter of 1780 we find the first tremors of a desire to be elsewhere when he asks to be sent some views of Rome painted by Franz Kobell (1749–1822), then living there, and in 1782, when he seems firmly cemented into place in Weimar, he is studying Palladio and exerting himself to persuade the Duke of Gotha to support the young German painter Wilhelm Tischbein (1751–1829) during his apprenticeship in Italy. Perhaps a touch of envy for this protégé, who could spend all his time doing what Goethe could do only as an amateur, in rare hours of leisure, and in a place where he had so often himself wanted to be, was the occasion, in that or the following year, for Mignon's song of yearning for the land of lemon-blossom, orange-groves, and Palladian villas. 'For some years now', he would write from Venice in 1786, 'I have not been able to look at a Latin author or touch anything that aroused an image of Italy without suffering the most terrible pain.' In 1784, Tischbein, now in Rome, required further support. Kayser too was in Italy, investigating early church music and that June Goethe wrote openly now to him of his envy 'that you should enter upon and travel through the land that I as a sinful prophet only see lying before me in the haze of distance'. Perhaps the work on *Jest, Craft, and Vengeance*, conceived from the start as 'in the Italian manner', was intended to coax a little southern sunshine from Germany's 'brazen skies'. Only, however, in November 1785 is there a clear indication that he is planning a complete interruption to his official work in Weimar, when Goethe remarks that things are going so well with the new shaft at Ilmenau that 'if I hold out and stay at it for a while longer, then it can manage on its own for a time'. It is evident where his thoughts are turning when, in the same letter, he says that the sight of a picture by Tischbein 'produced on the other side of the Alps . . . will help fortify me for the Thuringian winter'. In January 1786 his despair at the state of German theatre, and at the prospects for a German opera, is such that he wishes that twenty years previously he had devoted himself to mastering the Italian language, rather than his 'barbarous' mother tongue. Had he done so, he writes on 5 May to Kayser, he would even now invite him to take part in a joint expedition south of the Alps and they could seek their fortune in the operatic world. The careful wording of his warning, however, that 'from June onwards I shall not be at home' suggests that this joint project is but a flight of fancy and he has already made his own decision and will not be returning to Weimar from Carlsbad. A fortnight later he attended an Italian lesson in Jena, perhaps in order to test what he could still remember of the first foreign language he had learned: the first step clearly preparatory for the great journey.

Not a word of the inner ferment that must have led to this decision was breathed to anyone, not even Frau von Stein. Goethe's glimpse of a quite different existence in Carlsbad in 1785 seems to have unsettled all his certainties about Weimar, even that one. Not long after his return home he received as it were the second instalment of the visit from the Münster circle that had begun the previous year when Jacobi and Claudius had come: this time it was the turn of the philosopher Hemsterhuis, the statesman F. F. W. von Fürstenberg (1729–1810), and the patron of them all, the Princess Gallitzin. After an initial uneasiness, perhaps caused by the Princess's radically unconventional views about dress, education, and the position of women, Goethe felt 'things turned out very well', 'we were quite natural and open towards each other'—as befitted two people who shared the same birthday, though the Princess was a year older—and his subsequent offer of a correspondence in which he might open up his whole mind to her has led to the not unreasonable speculation that there must at the time have been 'a certain cooling off' in his relations with Frau von Stein. Perhaps he was already casting around for a new spiritual home. Certainly his notes to Frau von Stein, though as insistent as ever in their expressions of affection, were less frequent, and on the whole rather shorter, between the two visits to Carlsbad than in the previous ten years. She learnt nothing of the crucial decision, even after it had been taken: she went to Carlsbad on 2 July 1786 and arranged accommodation for Goethe, who was to follow on as soon as he was free, and still his letters contained no hint of what he was planning. Even when she left Carlsbad in the middle of August and Goethe accompanied her on the return journey as far as the Bohemian border and in parting they talked of their love and Frau von Stein said it was 'once again' bringing her joy—evidence perhaps that any coolness in the last year was not entirely Goethe's fault—even then he told her only that he would not be returning for another six weeks. Back in Weimar, she walked across, on Goethe's birthday, to the cottage by the Ilm to which she had had the keys ever since he moved into town, and left a present in his desk, expecting him to find it on his return from 'a period to be spent obscure and incognito in forests and mountains', as she now believed. But he did not come, suddenly the letters stopped, and even Fritz von Stein, left alone with Goethe's servants in the big house on the Frauenplan, did not know why: for six months he waited and then he found it too lonely and went back to his parents.

The secrecy of it all was obsessive. Of Goethe's friends only Jacobi received a hint of what was up, and that was because he was at a safe distance, on a journey to England, and because it was 12 July when Goethe wrote to him, expecting to leave for Carlsbad at any moment, and detained only by his decision to wait for the outcome of Duchess Luise's latest confinement. Even Carl August knew at this stage only, at most, that Goethe was not going to return immediately from Carlsbad: Goethe asked him in writing for indefinite leave only after their final meeting in the Bohemian spa on 27 August, when the Duke was himself on the point of spending a long period away from Weimar, and even then Goethe did not reveal his destination. All this can surely be only partly explained by what Goethe once called 'the element of the *absolute* in my character . . . I could be silent and patient as a dog for months on end and still hold firmly to my purpose; when I then emerged to carry it out, I had *unconditionally* to reach my goal, using all my strength, whatever might fall to right or left of me'. Clearly Goethe must have feared gossip and objections; even more, perhaps, he feared company. He did not wish to take anyone with him, not Fritz, or Knebel, or Herder, or the Duke. He wanted to be alone for a while, free of the burden of his literary reputation, free of the attentions of the public he was gathering his courage to address once more, and free above all of his noble rank and official status, and that was only possible if his route, and the very fact of his journey, remained unknown. He wanted, perhaps obscurely and unconsciously, to rid himself of all the entanglements that had grown up in the last ten years, so that he could taste life without them and resume them, if he decided to do so, by a deliberate choice. To be alone, he had to keep his counsel. But also he may have feared delays, and for a special reason.

This was not the first time that Goethe had laid plans for a long and unexpected incognito journey and revealed nothing of them in advance, not even to Frau von Stein and the Duke. It had happened before, in 1777, when he had suddenly left Weimar to climb the Brocken, and in 1779, when he had announced, as by the inspiration of the angel Gabriel, that the Duke's party, already on the road, was going not to Düsseldorf but to Switzerland. And, as on those previous occasions, he was perhaps in 1786 seeking a symbolic pattern in his existence at a moment of crisis. The purpose for which he had turned his face away from Italy and towards Weimar on the St Gotthard in 1779 was now fulfilled, in so far as it could be, and the fulfilment had brought him to despair. The time had come to try the will of the gods again. Goethe wrote that it was his original intention to set out from Carlsbad on 28 August, his birthday. It seems very likely that it was also his original intention, perhaps even the form in which the inspiration for the journey first came to him, to reach Rome by the, to him, so significant date of 12 November. He had since 1780 been keeping the anniversary of his passage of the Furka and his first meeting with Frau von Stein by using it to mark his progress in a particularly important literary project. Since 1782 that project had been *Wilhelm Meister,* but by November 1785 he had brought the life-story of the hero of that semi-autobiographical novel up to, and beyond, the stage that he himself had reached. In the first half of 1786 he complained repeatedly of his slow progress with the next book of the novel, and after May *Wilhelm Meister* seems to have been abandoned: by then the plan was probably already in existence to crown 'the wishes and hopes of thirty years' by reaching Rome, no doubt by the landmark date. But would he be permitted this fulfilment of all that his life hitherto had been planned for? Goethe's letters in July are thick with a foreboding complicated by memories of old anxieties, which re-reading his earlier works had revived. He lingered in Weimar in cold and rainy weather working at his usual rate, or sitting by his fireside eating stewed cherries with Fritz von Stein, and waiting for the ducal birth. But the birth, expected in the first days of the month, seemed endlessly delayed. Even though he felt the precious days of holiday that he could be spending with Frau von Stein trickling away, even though he was 'almost as overripe as the princely fruit and equally waiting for my deliverance', Goethe kept by his resolve to 'take as an oracle' Luise's confinement, and not leave for Carlsbad until it was over. 'These days are pregnant with events', he wrote: it seemed he was being forced by a higher power to wait in Weimar until Lavater passed through, whom he had very much hoped to avoid. Eventually, on 18 July, the new Princess Caroline and Lavater arrived together. 'The gods know better than we do what is good for us, therefore they have forced me to see him.' Their meeting confirmed for Goethe how little they now had in common. Now he could draw a line under that Christian existence, and all that it had meant for him, and look forward within the week to joining with Frau von Stein, who had the true hegemony of his heart, in Carlsbad—'if it is the will of the heavenly ones, who for some time now have been ruling me with forcible love'. The omens were favourable at last and he could set out on his third and greatest symbolical journey. 'Favete linguis'—such religious actions are usually begun in a holy silence.

Taking the waters at Carlsbad was of course excellent cover. All the preparations for a long absence—Goethe was originally thinking of eight months or so—could appear quite natural. Thanks to his far-sighted decisions of 1782, and the following two years of vigorous reform, his official work was at a stage where he would hardly be missed: the Chamber looked after routine financial matters for itself; the War Commission had little left to administer, and what there was could, as with the Highways Commission, be left to colleagues; Ilmenau was safe in the hands of the elder Voigt; and the Council could manage without him, as it had done for the last eighteen months anyway. His private papers, correspondence, and copies of his writings were packed up in boxes and deposited, probably together with his will, in the ducal archive. Philipp Seidel, now a clerk employed by the Chamber but still effectively his private secretary, was the only person to know where he was going, or the pseudonym, 'Johann Philipp Möller, of Leipzig, merchant', under which he would be travelling. Seidel had instructions to open all letters and pass them on to the appropriate parties, all personal correspondence going to Frau von Stein, and to return the books Goethe had borrowed from the Göttingen University Library. Money matters were left in the hands of J. J. H. Paulsen, a banker in Jena.

The financing of the secret plan was made not a little easier by a development of the last two months which must have seemed to Goethe further evidence of heaven's 'forcible love'. In May, at about the time when he was resolving on his journey, Bertuch came to him with a warning that Himburg was proposing to pirate his works once more, and secured from him an agreement that there should now be a properly authorized collected edition and that Bertuch should look around for a publisher for it (other than Himburg). Unknown to Goethe, Bertuch had recently gone into partnership with a new publisher in Leipzig, G. J. Göschen (1752–1828), and Göschen opened negotiations in June. If Bertuch was less than honest in the preliminaries, Goethe now proved he could drive a hard bargain himself. Göschen was taken aback at his demand for a flat fee of 2,000 dollars for the eight little volumes of the edition, and at the requirement that the fee should be paid in advance, in proportion as the manuscript was ready for delivery. As Göschen was planning to print 6,000 copies at up to 8 dollars a time, and was hoping for 1,000 pre-publication subscribers, he did not really have much to complain about, provided, of course, he could sell his stock, and provided Goethe kept to the agreed timetable: the first four volumes were to be published by Easter 1787, and the second four by Michaelmas. There was little anxiety about the first half of the edition, consisting largely of material already published—with the important exception of *Iphigenia*—though this was all to be reworked, and each volume was carefully arranged to contain at least something that was new. But among the last four volumes were to be a collection of 'miscellaneous poems', which would take time to assemble, and two volumes of fragments. Goethe promised his best efforts to complete the fragments, but gave no undertaking to do so.

From the agreement with Göschen, the details of which took another two months to complete, Goethe derived two considerable advantages. First, well

1. J. H. W. Tischbein: Goethe at the window of his lodgings in Rome (1787)

2. J. P. Melchior: Johann Caspar Goethe (1779)

3. J. P. Melchior: Catharina Elisabeth Goethe (1779)

4. The Goethes' house in Frankfurt after rebuilding,
print after F. W. Delkeskamp, 1823

5. Goethe: Self-portrait(?) in his attic room in
Frankfurt (1768–70)

6. J. L. E. Morgenstern: Cornelia Goethe
(*c.* 1770)

7. Goethe: Parsonage and farmyard,
Sesenheim (1770–1)

8. J. H. Schröder: Charlotte Kestner, née Buff (1782)

9. E. M. von Türckheim: Anna Elisabeth ('Lili') von Türckheim, née Schönemann and B. F. von Türckheim (*c.* 1798)

10. A. Graff: Herder (1785)

11. H. Strecker: Merck (1772)

12. S. Collings and T. Rowlandson: More of Werter. The Seperation (1786)

13. Goethe: Swiss mountain huts (June 1775)

15. G. O. May: Goethe (1779)

14. J. Juel: Duke Carl August of Saxe-Weimar-Eisenach (1779)

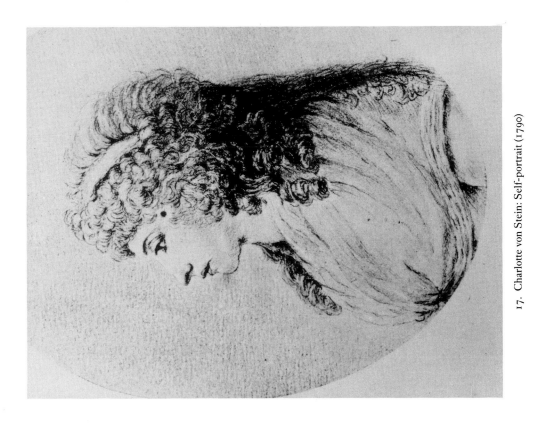

17. Charlotte von Stein: Self-portrait (1790)

16. G. M. Kraus: Duchess Luise of Saxe-Weimar-Eisenach (c.1781)

18. Goethe: The Brocken by moonlight (December 1777)

19. Goethe: Mist rising from the valleys near Ilmenau (22–3 July 1776)

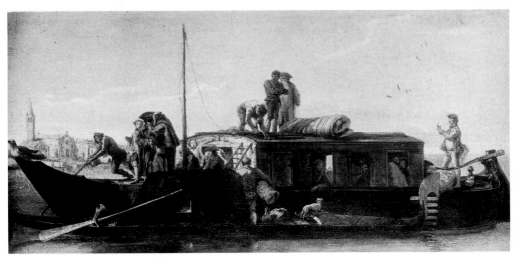

20. Giandomenico Tiepolo: Venetian mailboat (1760–3)

21. J. P. Hackert: St Peter's, Rome, from the Ponte molle or Ponte milvio (1769)

22. J. H. W. Tischbein: Goethe in the Campagna (1787)

24. J. H. W. Tischbein: Self-portrait in his Roman studio (1785)

23. J. H. W. Tischbein: Duchess Anna Amalia among the ruins of Pompeii (1789)

25. Goethe: Balcony with vase (1787)

26. Goethe: Italian seacoast with full moon (June 1787)

27. C. H. Kniep: Temple of Segesta (1787?)

28. Goethe: Schloss Wörlitz (1778)

29. Goethe: Imaginary landscape with bay and castello (autumn/winter 1787)

30. F. Bury: Self-portrait (1782)

31. C. H. Kniep: Self-portrait (1785)

32. J. H. Lips: Carl Philipp Moritz (1786)

33. J. H. W. Tischbein: 'Emma Hart' (later Lady Hamilton) as a Sibyl (1788)

34. Goethe: Trees (1788)

35. J. H. Meyer, copy after Carracci: Ulysses and Circe

37. Goethe: Christiana Vulpius, with shawl (1788–9)

36. C. Küchler, after J. C. Reinhart: Schiller (1787)

before he made any request to Carl August for paid leave, he could know that over the coming year for every two volumes that he made ready he would have 500 dollars at his disposal. And then he had given himself a task to perform while he was away, and a deadline to meet. He was forcing himself to face the facts: that his art could flourish, and achieve what he hoped from it, only in the medium of print. He was turning to the German public now, not just with an operetta and a plan to reform the stage, but with the greater part of his entire literary output and potential, in book form. It was of course a paradox, but a paradox which had always run through his relations with his public, that in the moment in which he brought himself to face the German nation once again he should flee the very soil of Germany and all contact with its inhabitants. In the next two years he would be testing the public, to see how it would respond to the challenge of his life's work, but he would also be testing himself, to see if by putting himself under extreme constraint, though in what he hoped would be the most favourable possible circumstances, he could bring to flow again the springs of inspiration that for the last two years and more had left him as sterile as the desert sand. The relief of a decision, the clarity of the task ahead, the prospect of release from the official treadmill, of relief from the Thuringian winter, and of travel in the land of figs and melons, the happy conjunction of Göschen's offer with his own plans, and the good omen given by the live birth of Princess Caroline, all combined to lift his spirits as on 24 July he drove off to Jena, the first stage of the journey to Carlsbad and beyond. The note of farewell that he wrote from here to Carl August allowed of more than one interpretation: 'I am going in order to correct all kinds of defects and complete all kinds of omissions; may the health-giving spirit of the world stand by me and assist!'

Works, 1785–1786

Not only *Jest, Craft, and Vengeance* but all Goethe's literary production, such as it is, of the last phase before his Italian journey, is marked by a growth and shift in his attitude to the German public: the epic fragment, *The Mysteries*, the sixth, and what can be reconstructed of the seventh, book of *Wilhelm Meister*, and the editorial work on the first volumes of Göschen's collected edition, in which after many years Goethe identifies his public as a reading public once more.

After writing the poem 'Dedication' in August 1784 Goethe continued immediately to compose the opening of the *The Mysteries*, to which 'Dedication' was to be prefixed, in the same eight-line stanzaic form. About fifty stanzas were written by the time he broke off the project in April 1785. Already in December 1786 he must have given up any serious hope of taking it further, since by then he had decided to divert 'Dedication' to another purpose, as the preliminary poem to Göschen's edition of his *Literary Works*, in which forty-four of the stanzas were published in 1789 as a fragment. These introduce us to a monk, Brother Mark (the name became, and may already have been, a sobriquet of Goethe's), travelling on an urgent mission through the mountains in Holy Week. Seeking a place to sleep he comes across a monastery dominated by the Rosicrucian sign, the cross garlanded with roses, and is welcomed by the community, a monastic order of aged knights. They have a rule not to accept any young person whose heart 'bade him renounce the world too soon', and are all men of experience for whom the monastery is a haven of rest after the storms of life. At the moment they are in some anxiety since their founder and master, Humanus by name, has announced that he will shortly be leaving them. We are told a little of his miraculous childhood and virtuous youth and then, after his meal, Brother Mark is shown a large chamber which seems to perform some of the functions of a chapel: it contains thirteen seats along its wall, arranged round the sign of the rose-bearing cross, each with its own emblematic coat of arms and other symbolic accoutrements, such as weapons, banners, or chains. That night Brother Mark is surprised to see some handsome young men hurrying with torches through the cloisters, but the story goes no further. In 1816, however, in response to enquiries, Goethe gave a general account of what his intentions had been. Each of the twelve monastic knights would have represented a different religion and different national culture. ('Mysteries' was in the eighteenth century a term that referred not only to the secret wisdom of Freemasons, but also to the specific doctrines and practices that, in the eyes of deists, marked off historically existing, or 'positive', religions from the one true rational religion, and from each other.) Each knight would in turn have narrated that part of the life-story of Humanus not merely known to him but in a sense embodied within him. The reader would thus have been introduced to the whole range of human religious experience, and the culmination of the poem would have been the passing of Humanus at Eastertide and his replacement as master of the order by Brother Mark, whose narrative viewpoint the reader would largely have shared.

The poetic interest of the fragment is slight: there is the same padded diction, extending at times over the greater part of a stanza, and the same flatness of imagery, as in 'Dedication', together with the same uncertainty of narrative posture that occasionally mars the later books of *Wilhelm Meister's Theatrical Mission*. As in *Jest, Craft, and Vengeance*, the artificiality of the construction prevents Goethe from establishing a dominant mood and relating it to an objective correlative: the prevailing atmosphere is allegorical and emblematic. *The Mysteries*, however, is a significant reflection of Goethe's thinking in 1784–5, though we must be careful not to give too much importance to his expressions of intent uttered thirty years later. Although there are certain links in the plot with the story of Parsifal and the legend of Christian Rosenkreuz, supposed founder of the Rosicrucian order, and obvious parallels with the language and practice also of orthodox Freemasonry, the interest of the fragment lies in its relation of this material to the thought of Spinoza and, especially, Herder, and then to its intended audience. The principle guiding the whole of Humanus' life is, we are told, a saying 'difficult of comprehension' which accords with the view Goethe seems to have held from 1783 onwards that the supreme religious issue is not the relation between men and gods (gods being but enlarged shadows of men), but the relation between man and animals. Spinoza's belief that in a completely determined natural order the only freedom open to men is to reduce the hold over them of their self-image and displace it by an active participation in the work of the infinite divine substance is expressed in terms reminiscent of the opening of 'Divinity':

> Von der Gewalt, die alle Wesen bindet,
> Befreit der Mensch sich, der sich überwindet.

From the power that binds all beings, the man [or possibly Man] frees himself who overcomes himself.

It seems possible that each of the contributions to be made by the twelve knights —the story at once of his own life and of a part of the life of Humanus—would, for all its individual characteristics, have in some way exemplified this one principle. Each member of the human fraternity of culture and religion would have been shown, in the spirit of Herder's *Ideas*, to offer a uniquely perfect embodiment of this ideal Spinozan humanity. The purpose of the poem, however, would not have been simply to retell the *Ideas* in allegorical verse, but to involve the reader directly. The replacement of Humanus, ideal humanity, by Brother Mark, with whose experience and viewpoint the reader is already encouraged to identify in the existing fragment, would have been a symbolic indication that, by absorbing the cultural history of the human race through reading the poem, the reader had himself become the vehicle and focus of all human endeavour. Here perhaps we have one of the reasons why the poem had to remain fragmentary: Goethe could write only for an audience that was in some respect limited and definable, and the audience for *The Mysteries*, like its subject-matter, was in principle the whole human species.

However, much of the intellectual repertory first sketched out in *The Mysteries* was taken over into Goethe's later work, particularly the various stages of *Wilhelm Meister*, and it is possible that Goethe decided on this redeployment of his material in the winter of 1785. Having finished the sixth book of the *Theatrical Mission* by the appointed date, he drew up a plan on 8 December for 'all the six following books', and this suggests a fairly radical new beginning. If what Goethe then planned bore any relationship at all to what he subsequently wrote in *Wilhelm Meister's Years of Apprenticeship* (*Wilhelm Meisters Lehrjahre*) and *Wilhelm Meister's Years of Wandering* (*Wilhelm Meisters Wanderjahre*) (the schema has not survived) it is likely to have included some reference to the motifs which these novels have in common with *The Mysteries:* a quasi-Masonic brotherhood whose members pool their life-stories as contributions to a common wisdom; and the moral principle of self-overcoming. At this time Goethe, his own mind already turning in that direction, may also have envisaged Wilhelm's going to Italy, as he eventually does in the *Years of Wandering*, and as Mignon's song had seemed to demand in 1783. What is further implied, however, by this large-scale plan for the next six books is that Goethe regarded the six books he had already written as constituting a unit, and a unit with which he was now far from satisfied, as we may infer from his omitting it entirely from the negotiations with Göschen. In 1783, when he was in the middle of the fourth book, he had told Knebel: 'I myself cannot take pleasure in it. This work was not written in states of calm, nor have I since had another moment to survey it as a whole.' The problem of plot, of finding a cumulative logic to Wilhelm's story that would make it a whole, continues right to the end of the sixth book. In an attempt at a solution, the autobiographical strand in the novel is, at least temporarily, interrupted, so that Wilhelm's artistic mission to the German nation can develop far enough to include the option from which Goethe himself turned away by coming to Weimar in 1775: participation in a public cultural institution independent of courtly patronage.

The presence of the Bellomo troupe in Weimar from 1784 onwards gave Goethe more than simply an insight into the unsavoury personal side of professional court theatre, even if that was what he had drawn on most when writing the fifth book of *Wilhelm Meister*. It also meant a significant widening of his own cultural experience, for, with a small potential audience and three performances a week, the Bellomo repertoire had to be large. Operas by, among others, Salieri, Cimarosa, Paesiello, Gluck, Benda, and Mozart stimulated Goethe to write *Jest, Craft, and Vengeance*. But plays in German were also performed, including all the best-known representations of 'idealized rabble' that had followed on *Götz*, down to Schiller's *The Robbers*. In particular, 1785 saw the first productions in Weimar of several plays by Shakespeare: *Hamlet, King Lear, Macbeth,* and (considerably adapted) *A Midsummer Night's Dream*. Goethe must have been vividly reminded by his evenings at 'the play', often shared with Frau von Stein, of the ambitions of ten and more years ago for a German national drama aware of its place in a European culture and drawing especially on Shakespeare: almost the full range of potential models processed before him as he had never previously had the opportunity of seeing them. Had that early and generous ideal perhaps not been as misguided as he had sometimes thought since? And if it had indeed been impracticable, what should, or could, take its place—in his life and, for that matter, in the life of the hero of his novel? In the sixth book Wilhelm's progression through the different levels of dramatic art brings him to theatre in its most competent, organized, and professional form, and to a meeting with those who know the German-speaking public for what it is.

At first the book is devoted to the aftermath of the ambush, in the course of which an important new strand is woven into the action. While waiting, in vain, for the rest of the party to bring them help, Mignon, Philina, and the wounded Wilhelm are surprised in the clearing by a group of noble travellers—the robbers' intended victims, it turns out—who include a handsome and compassionate woman on a white horse, and who give them aid. A surgeon is brought to tend Wilhelm's wound and he and his companions are sent down to the nearby village in the charge of a servant since the travellers themselves have to press on. Before they do so the horsewoman lays a warm coat over Wilhelm and her face appears to him transfigured before he faints again with pain. Down at the inn in the village there is an unfriendly welcome from the rest of the troupe for the leader they hold responsible for their misfortune, and also for Philina, the object of enraged and contemptuous jealousy, since she has secured from the brigands the trunk that contains all her belongings by coming to an arrangement in the bushes with their chief. Melina, who has lost most, is in despair, and in an adjoining room his wife, whose pregnancy has accompanied us through the previous books, gives birth to a dead child. Goethe's inability to give proper weight to the stillbirth is the one defect in this remarkable Hogarthian scene, in which the reproaches of the actors meet with a moralizing self-defence and unreciprocated protestations of solidarity from Wilhelm, while Philina sits on her trunk cracking nuts.

The actors press on to the city of H. to try their fortune there, leaving Wilhelm to recuperate in the company of Mignon and the Harpist, and also Philina, though she too eventually departs, much to Wilhelm's relief, since her impure presence troubles the image left in his mind by the fair horsewoman. Wilhelm is obsessed by the memory of this radiant figure, which Goethe associates with the one poem this book contains, 'Only he who knows yearning', and when all his attempts to trace her fail he has only the coat to assure him he is not deluding himself. His longing for her remains and is the only factor in his life that tempts him away from the theatrical world, to which he now feels bound by a debt of honour, and also by memories of Mariana, which have recently revived. To divert himself from these thoughts, in the tedious days of his convalescence, Wilhelm, like Goethe in the summer of 1785, makes a detailed study of *Hamlet*, with the hero of which he has previously identified, though he now finds his earlier interpretation, based simply on the melancholy of Hamlet's monologues, inadequate to this complex character.

Wilhelm, and through him Goethe, considers all the romantic turns his fate might now take, as he lies waiting for recovery, but none of them materializes, and eventually he does the obvious thing, the only alternative to returning home a failure once more, and continues his journey to H. in order to assist the actors to whom he has sworn loyalty. Here he quickly meets up with his old friend Serlo, the director of the commerical repertory theatre (a figure partly based on F. L. Schröder), who is delighted to see him, despite his recommendation of Melina and his fellows, whom Serlo thinks unemployable. Serlo presents Wilhelm to his sister, Aurelia, a professional actress. The state of the theatre is discussed, and soon Wilhelm is expounding his interpretation of *Hamlet* and his defence of its qualities as a 'canonical' work of art, which, broken up into instalments, provide one staple of the several subsequent conversations between the three of them. Serlo is anxious to put on the play, and immediately, and with callous fraternal meaningfulness, casts his sister as Ophelia. Her account of the experiences that lie behind this identification is the other main element in the conversations which together make up most of the second part of the book. The opportunity of showing us something of the 'lively and commercial' world of the city of H. is passed over completely: Goethe merely assures us of the high quality of the productions Wilhelm sees at Serlo's theatre and lets him meet one or two musicians whose virtuosity and self-confidence impress him. Apart from some glimpses of the poverty-stricken members of Wilhelm's old troupe, and of Philina, who alone has managed to get herself engaged by Serlo, the narrative, nominally situated in one of Germany's busiest and wealthiest free cities, which until well after the middle of the century was also one of its most active intellectual centres, consists only of boudoir conversations, which are effectively monologues by one or other party.

Aurelia, however, with her hatred and contempt for the male sex that border on derangement, her hopeless love for the man who has deserted her, her complex feelings for Wilhelm who is to her at once a hated man, an admired poet, and an inexperienced youth whom she desires to educate, and with her strangely impatient inattention to a little boy of whom she has charge, and whom Philina confidently pronounces her love-child, is a remarkably living and individual character—much more so than the demonic females in Lessing's dramas from whom she partially derives. In the absence of any direct depiction of the city of H. and its society, it is through Aurelia's eyes that Wilhelm is shown a portrait of the German nation to whom his theatrical mission is addressed: Aurelia's reflections on her experiences provide the element of generalization which Goethe needs if this book is to furnish a conclusion, rather than simply another episode illuminating another partial aspect of German culture. Aurelia's first observations of German manhood were not promising. An orphan, brought up by a loose-living aunt, she saw 'how dull, insistent, brazen, clumsy was every man she enticed to her, how sated, arrogant and tasteless everyone who had found the gratification of his desires'. She none the less went on the stage 'with the highest conception of my nation . . . What were not the Germans! What could they not become! . . . my public and I understood each other perfectly, were in perfect harmony; and attendant on my public I always saw the nation, all who are noble and all who are good!' But she soon learned that an actress interests the men in her audience as her aunt had interested her visitors, and whatever their rank or occupation—which she lists in comic detail—she saw through them all: 'I began to despise them all from my heart; it was as if the whole nation deliberately wanted to pillory itself by proxies in my eyes. They seemed to me as a whole so awkward, so badly mannered, so ill-informed, so empty of charm, so tasteless . . .' But all this way changed by love. A local landowner enters her life, Lothair by name, intelligent, agreeable, open, and kind in a practical way without being importunate, and, once he is in her audience and she is playing for him, 'as by a miracle my relation to the public, to the entire nation was altered'. She recognizes the fine kernel in the modest German shell and is no longer troubled by an absence of culture or manners—faint praise, no doubt, but seriously meant. The whole episode reflects Goethe's own return in 1785 to a limited enthusiasm for the German national literary cause, but demonstrates a quite new analysis by him of the artist's relation with the public, one which was to make it possible for him, a few months later, to contemplate 'delivering up to the public', in his *Literary Works,* 'things that occupied an individual sensibililty in specific circumstances'. According to this new view, the audience is to be addressed for the sake of the loved one hidden within it; though it is a 'mass' (Aurelia's term) it is affected by art not collectively, but individual by individual; and, even if inappropriate to a theatrical audience, this analysis precisely captures the nature of the reading public. Though the nation as a whole is admitted to be characterless, the reason for that defect is now held to lie simply in its nature as a mass, as a collectivity distinct from the individuals of which it is composed, and the characterlessness is the artist's opportunity, for he, or she, can speak to it, 'lead' it, Aurelia says, and elicit character from it. Whether or not he fully shares the view, Goethe can now envisage the German poet's relation to his people as educative, as inspired by a personal emotion of love, and yet as not dismayed by the public nature of the forum in which the poet acts. Such a relation is, however, best maintained, and maintained on a truly nation-wide scale, by books, not by theatres— Aurelia's generalizations about the 'nation', on the basis of what can only be experience of the city of H., gradually come to seem rather hyperbolical—and Aurelia's reconversion to the German public heralds, though of course she does not realize it, the end both of Goethe's and of Wilhelm's theatrical aberration. Aurelia's account of her abandonment by Lothair leads, not to any further reassessment of the German public, but to a sudden demented attack by her on Wilhelm with an extremely sharp knife which we already know to be her treasured possession. She slashes his hand and the cut severs the lifeline in his palm, an indication possibly that, through his involvement with Mariana, which he has confided to Aurelia, love and the stage have decisively, perhaps irremediably wounded his existence. (Certainly Goethe's own life from 1775 to 1785 could be seen as brought to the brink of madness by the double sterility of his love for Frau von Stein and his relation with, or detachment from, the German reading public.) The incident thus connects with other devices in the sixth book by which Goethe attempts to bring about reflection on Wilhelm's career as a whole, and a mood of summary and conclusion, notably the discussion of *Hamlet.*

For Wilhelm's new interpretation of Shakespeare's play involves no less of an identification of its story with his own than his original indulgent sharing in the Prince's melancholy. Using as the key to Hamlet's behaviour all the evidence the play provides of his character before the death of his father and the command to exact vengeance, Wilhelm concludes that Shakespeare wished to show in him 'a great deed imposed upon a soul that was not adequate to it'. It is not chance that this phrase recalls Goethe's description of the excessive demands that Sentimentalism imposed upon the soul and that led to Werther's self-destruction: the founding of a national dramatic culture, demanded of Goethe and his Storm and Stress contemporaries and their representative, Wilhelm Meister, was also a great deed beyond the powers of the limited and sensitive souls on whom it was imposed. Is Wilhelm, then, to end as Werther did, and as at the end of book 1 it seemed he might, with his life-line severed in the middle? Is he, like Hamlet, to be a pot shattered by the oak-tree planted in it? The example of Shakespeare's play suggests that, if he is not, it will not be because of any merits or strength of character of his own, nor because his task is intrinsically any easier of performance than Hamlet's, but through the workings of an inscrutable fate. For Wilhelm's analysis of the last two acts of *Hamlet* concludes that neither earthly nor supernatural powers (Wilhelm's word for the ghost is 'subterrranean', the preferred epithet for the Greek Furies) succeed in bringing about the 'great deed' of revenge: that remains entirely in the hands of all-powerful destiny. 'The hour of judgement comes. The evil fall with the good. One generation is mown down and cast aside and the next steps in.' These words sound like an echo of the grim verdict on the hopes of the sensitive soul passed in *Proserpina* or Iphigenia's song of the Fates, and they reduce Serlo and Aurelia to a stunned silence. But, as in *Iphigenia,* the cloud passes over, and Wilhelm does not have to learn to fear the gods—not, however, thanks to a creative reinterpretation of the question about the nature of the gods as a question about the nature of man, such as concludes that play, but thanks to a revelation that, in his case at any rate, fate is kind and ensures by its own non-catastrophic means the fulfilment of Wilhelm's desire and the performance of the great deed that was too much for his inadequate soul.

Wilhelm's father has died during his son's long absence, but the fate which thus manifests itself in Wilhelm's life brings to him not, as to Hamlet, obligation and failure, but liberation and achievement, as Serlo points out. He is not needed at home now, where the family business is secure in Werner's hands; his whereabouts are unknown; and he is free to realize the ambition with which he began—for Serlo is most anxious that Wilhelm should join his company. He has learnt from Philina of Wilhelm's earlier acting triumph in *Belshazzar,* and Aurelia has identified Wilhelm to his face as a 'poet and artist, for you are both, even if you do not wish to proclaim yourself that'. So valuable an acquisition does Serlo think Wilhelm will be that he is willing, in order to secure him , to take on as well Melina and his troupe, to whom Wilhelm can thus fulfil his promise of recompense for all they lost in the ambush. As important to Wilhelm is the consideration that, if he accepts Serlo's offer, he will not have to detach himself from 'his' Mignon and 'his' Harpist. And so, after the repeated reflections in this book about the way 'his story' might develop, Wilhelm finds himself 'not at the cross roads, but at the goal': it turns out that, as in his interpretation of *Hamlet*, 'the hero has no plan, but the play has one'. Everything which he had promised himself when he was planning to elope with Mariana, and which had seemed lost when he discovered her infidelity, has been brought to pass, not by his own deliberate efforts but by the 'gently guiding' hand of chance events. He now has the opportunity to decide from the highest of motives, from the love of art—free from compulsion and free from the desire merely to escape the restrictions of life at home—to appear before the German 'nation', as Aurelia calls it, as its educator and leader, on the boards of Germany's most respected and independent stage. Destiny has said yes to 'the hopes and wishes he has long nourished and maintained in his heart'. He can begin to realize that ideal of a national bourgeois culture embodied in public institutions which in 1775 Goethe, if only half-consciously, abandoned as a figment, and which as late as 1781 Möser still thought posssible of attainment. What hinders Wilhelm from recognizing fulfilment when it is there, and plucking the fruit that puts itself into his hand? Nothing, or only 'a something that has no name'. So of course he yields to these weighty arguments, especially to the thought that he need not lose Mignon and the Harpist, and to the combined pleadings of Serlo, Aurelia, and Philina. But in the moment when he accepts Serlo's offer the image of the fair horsewoman comes into his mind, distracts him briefly with its radiance, and, as if banished by his acceptance of a lesser and illusory alternative good, disappears. On this uncertain note Wilhelm Meister's theatrical mission comes to an end.

The inadequacy of this conclusion must have been apparent to Goethe even as he wrote it. The retrospective discovery of logical purpose in the inconsequential episodes of Wilhelm's earlier life is little more than a *deus ex machina*, and a half-hearted one at that. For not only has Goethe in 1785 long since ceased to feel threatened, as Iphigenia and Proserpina were, by the uncertain character of fate, without which the revelation of fate's benevolence loses its power to convince, and to provide the sense of an ending. But also his failure to give a realistic, and novelistic, account of the social milieu of the city of H., in which Wilhelm's ambition is to be fulfilled, is a failure to underpin the assertion that fate is benevolent, and in an absolutely essential respect. Unless Wilhelm's end is as real as his starting-point, unless the productions by which he is to educate the German nation in H. can be made as credible and visible to us as the puppets lying among the prunes and dried apples in the larder in M., the hopes which are now being said to be realized are not the same as the hopes with which Wilhelm began—for those were precisely hopes for the manifestation of the 'more than human' in the concretely envisaged world of that central German Imperial free city. But there was of course no way in which the Goethe of 1785 could produce a literary representation of the fulfilment of his ideals of

1775. The assertion none the less that Wilhelm's life has come full circle is belied by the difference in the literary mode of his beginning and his ending. But Goethe has—as yet—no means to account for that difference or build it into the conscious structure of his novel. The one element in the narrative that questions this half-hearted happy ending, the recurrent image of the fair horsewoman, is too undeveloped to outweigh Wilhelm's reflections on fate in his own life and in *Hamlet*, or to suggest a perspective in which the entire story of Wilhelm's theatrical hopes and desires, from inception to apparent fulfilment, could appear as but a part of some larger theme—and nothing less than that is needed if that story is not to be self-contained but is to have a sequel, and if Wilhelm's belief that he is 'at the goal' is to prove to be a delusion. It is not therefore surprising that Goethe made little headway with the next book before it petered out in May 1786. We may guess that the preparations for a production of *Hamlet* already figured prominently in what Goethe wrote and that the child for which Aurelia cares would have been the object of some interest, though whether it would yet have been revealed that the child is Wilhelm's son by Mariana, and that its nursemaid, whom we have already briefly but unrecognizably glimpsed, is Mariana's former confidante and adviser, is quite uncertain. There are numerous other potential growing points in book 6 which might or might not have been taken up immediately in book 7, but it is most unlikely that Goethe made any real progress in solving the major problem with which the end of the *Theatrical Mission* left him: how to engineer a transition for Wilhelm to a new form of life independent of the theatre, which has hitherto been his ruling, indeed his defining, passion. Such a transition required a full revision and recasting of all that had been written so far, and it required also some certainty about what new life could inspire Wilhelm in the future as the theatre had done in the past; and that certainty Goethe as yet possessed neither for his hero nor for himself.

In June 1786 a much more urgent literary task claimed all Goethe's attention. Göschen was planning to bring out the first four volumes of the *Literary Works* in less than a year and Goethe hoped to have the manuscript of these volumes ready before his birthday and his as yet wholly secret Italian expedition. As he did not any longer have the originals of most of the works involved he had first to collect manuscripts from friends or printed—often pirate—editions of works that had already been published in order to furnish the copy for Göschen. The little Weimar circle lent a willing hand: Frau von Stein helped to copy up the lyrical poems—even though they were to appear only in the last volume, one of Goethe's first tasks was to work out a series of headings under which they could be grouped—and Wieland and Herder agreed to advise on corrections and alterations: 'since I cannot offer much, I have always wished to offer that little well', Goethe wrote to Göschen. Written comments on *Götz* from the two advisers were compared in July and no changes of literary significance were made, though here, as throughout the edition, Goethe was aware that he was contributing to fixing a norm for the German literary language and took as his standard in orthographical matters Germany's first modern dictionary, that published by J. D. Adelung between 1774 and 1786. *Partners in Guilt* was to be submitted in the form in which it had been played in Weimar, accommodated, that is, to courtly manners, with the direct addresses to the audience removed, and what was farcical and obscene in it moderated in the interests of sentiment. Together with *Götz* it made up volume ii of the new edition. Volume iv was also quickly dealt with, indeed it was probably the first to be got ready, in June, when *Stella* was passed as reprintable with little alteration, some scenes in *The Triumph of Sensibility* were reset and rewritten to make the whole 'more producible', and there remained only a little editorial work on *The Birds:* some of the juicier delights which the two vagrant authors promise themselves from their ideal city-state had to be excised. But the major problems for Goethe lay in volume i—devoted, apart from the 'Dedication', to *Werther*—and in volume iii, which was to contain *Iphigenia* (along with *Clavigo* and *Brother and Sister,* which gave rise to no difficulties), and both of these problems accompanied him to Carlsbad.

Goethe had begun to revise *Werther* in 1782 and 1783, when he felt settled at last and at a distance from the turmoil of ten years before, and when he was no longer so productive of new works as to have no time for the old. Two considerations had then been uppermost in his mind: the need to meet the criticism that the novel condoned or even commended suicide; and his obligation to Kestner and Lotte, who had felt themselves put on public display in the book, to remove those elements which, attributed to them, they found particularly offensive. There was another attempt at revision in 1785. Now, at the beginning of July 1786, he received from Herder the opinion that there were weaknesses 'in the composition'—whether he was referring to the original or to the effect of any alterations that had already been made is uncertain. It is, however, clear that the substantial work on the novel that Goethe reported from Carlsbad to Frau von Stein after she had returned to Weimar in the middle of August related to the concluding section entitled 'The Editor to the Reader'. By 22 August the revised conclusion was ready to be discussed with Herder, who was also spending this rainy summer in Carlsbad, and after pondering the two versions for some days Herder decided in favour of the new.

In the event Goethe removed nothing from his original text, except to make room for revisions, but he added a certain amount to the main sequence of Werther's letters and in the final section completely rewrote all but the last few pages and the translations from Ossian. There was also a very large number of small changes in the forms, and in the choice, of individual words, with the intention of muting the turbulent informality, and even dialectal quality, of Werther's style. Many of the additions are short—a sentence or two, or a new letter of a few lines—but their purpose is nearly always evident: to emphasize what is pathological in Werther's love and what, under a thin veil, is simply sensual, and to increase Werther's own clear-sightedness about his self-destructive self-indulgence. He tells Lotte that the sand with which she blotted a note got between his teeth when he raised the paper to his lips, or writes to Wilhelm 'how I worship myself now that she loves me', or even 'I am astonished how wittingly I have entered into it all, step by step'. Werther's tendency to read his own emotions into his natural surroundings is reinforced: 'As Nature declines towards autumn, autumn comes within me and around me.' One or two little touches increase Werther's distance from the court or dull the edge of his criticisms of it. But the most substantial addition is an entirely new sub-plot, running to half a dozen pages scattered across the whole length of the narrative, telling a parallel story to Werther's own, and ending not in suicide, but in murder, arrest, and the beginings of legal process. Werther makes the acquaintance of a young farmhand who has fallen sick with love for his widowed employer: she rejects him, though perhaps not firmly enough, he tries to force himself upon her and is dismissed. When she shows some tenderness for his successor his jealousy knows no bounds and he kills his rival. Werther follows each stage of this parallel life with the greatest sympathy, and even attempts to secure the farmhand's release after he has been apprehended—a scene which Goethe may himself have illustrated in a drawing based on his own experience as an official. Werther listens unconvinced to the magistrate's explanation that such excesses strike at the very foundations of the State—this, one of the major charges brought against Goethe's novel, is now being made by the novel against its own hero.

The tendency of all these changes is to show Werther as his own worst enemy, and in a sense this is profoundly at odds with the original tendency of the book. The Werther of 1774 certainly dies of a shattered self, but that injury cannot be isolated from the intellectual and social world in which, and to a certain extent by which, it is inflicted: from the books that Werther and Lotte read, from the fashions to which Werther succumbs, from the opportunities of escape that are open, and, even more important, that are not open, to him. Nor can it be entirely separated from the people with whom he lives: the concluding section in the 1774 version is utterly realistic in this respect. Werther loves Lotte, and she knows, and Albert knows, that she loves him. The 1774 narrator of the last stages in Werther's life is uninhibitedly familiar with the intimate emotional detail of Lotte and Albert's marriage and leaves us in no doubt both that Werther is a serious threat to it, and that Albert wishes to see the back of him and knows full well why Werther wishes to borrow his pistols. The narrator whom Goethe creates in 1786 is much more tentative about these matters, and more reticent about the sources on which he is drawing. He tells us something about Lotte's feelings, but not everything, he couches what he has to say in hesitant, interrogative, or alternative forms, though his drift is always evident, and always evidently contrary to the original version. Lotte did not love Werther; when she heard him knock on her door for his last visit her heart did indeed leap, as we were told in 1774, but the narrator of 1786 feels that 'we may almost say it was for the first time'; this 'editor' knows 'she would have liked to have him told she was not at home' (whereas in 1774 'it was too late to have him told'), yet he does not know whether her inner turmoil after the visit was due to the memory of Werther's fiery embraces or to 'displeasure at his audacity' (whereas the version of 1774 knows only the first possibility). The new version goes to great pains to eliminate the suggestion of ill-feeling between Lotte and Albert or, as far as Albert is concerned, between him and Werther. Both husband and wife are concerned for Werther as for a friend who is ill, they are completely secure in their marriage—Albert leaves the room when Werther arrives, not (as in 1774) because the two cannot abide each other, but because Albert feels that his presence makes Werther uncomfortable—and their one fault, in relation to each other and to Werther, is to have kept silent about the problem for too long. Without this excessive discretion Werther 'might perhaps still have been rescued'—and so by implication there is no other respect in which his death can be laid to their charge.

Quietly, but with remorseless insistence, this concluding section, in its 1786 version, unbalances the relation between Werther and his surroundings so that his belief in Lotte's love for him becomes a pathological delusion for which there is no objective excuse or occasion at all. As a result it becomes very difficult to relate the Lotte and Albert whom the narrator presents to us to the figures who have appeared in Werther's letters—and that is why, by comparison with the original version, the later version has to shroud what it says about the feelings and motives of the married couple in an appearance of uncertainty: without that the discrepancy between Werther's first-person vision and the narrator's third-person reality would be so great as to be inexplicable. So, strange to say, Lotte and Albert are more phantasmal figures in the conclusion of 1786 than they were in the conclusion of 1774: the later narrative actually offers the reader less relief from the hypnotic spell of Werther's selfhood than the earlier. But the strength of the earlier version was precisely that Werther's selfhood was not presented as all-embracing but as embedded in a particular social and intellectual milieu. That element of realism in the earlier *Werther* is now put in question in the later, as the hitherto apparently objective factors in his story—the books, the people, the political institutions, the natural world, the seasons—all risk becoming mere symbols or symptoms of his inner state.

This shift in the text of *Werther* is wholly in accord with the development we have seen in Goethe's own mind over the twelve years after the novel was written. Far from being a development into objectivity, Goethe's intensifying love-affair with courtly and 'official' culture leads to a concentration on the power of the self to make its own world, for better or for worse, from which the only relief is to be found, not in the realistic contemplation or depiction of a surrounding society, but in the wholly non-literary business of natural science.

The indulgences of Prince Oronoro, 'To the Moon', *Iphigenia,* the division of the character of Egmont, 'Divinity', the realization, in a social vacuum, of Wilhelm Meister's originally social ideal, these are the stations along the route at the end of which lies the rewriting of *Werther*—and also *Iphigenia* once more.

The second task Goethe set himself for his working holiday in Carlsbad proved too much. When on 22 August he turned from *Werther* to *Iphigenia* he thought his play would occupy him for only a day or two. He had just included *Iphigenia* in a series of evening readings from his unpublished works (*The Birds* and *Faust* were also read). It was well received by the fashionable audience, and Carl August, who would shortly, on 27 August, be leaving Carlsbad to test the new political waters in Berlin, was 'strangely affected' by it —perhaps he was remembering the distant days, seven years before, when he himself had played the part of Pylades. The rhythmical prose had by now been cut up into irregular verse and this at first seemed to Goethe all the transformation that was needed. But alterations continued to present themselves, and above all Goethe remained unconvinced by his medium. The pressure of business detained him beyond his intended departure date of 28 August, which was perhaps just as well, since an elaborate ceremony in celebration of his birthday had been prepared by the other visitors to the spa, with four priestesses in white, an altar of fame crowned by a painting of the characters in *The Birds,* and poetic addresses in the name of his unfinished works. Reading Sophocles in the very last days of August, possibly even on 1 September, Goethe now suddenly perceived the grandeur of the long iambic line of the Greek tragedians, by contrast with the brief and stumbling lines of his own play, which required such skill to speak, and he realized the necessity of rewriting it entirely in a regular form. He may initially have thought of imitating the six-footed Greek trimeter directly, and asked Herder for some urgent assistance with the metrics, but Herder showed him the merits of the shorter and infinitely more flexible pentameter, the verse-form of Shakespeare, of Lessing's *Nathan the Wise,* and of an interesting recent experimental fragment by Schiller, *Don Carlos.* By 1 September it was decided that Göschen would have to wait until Michaelmas for the copy for volumes iii and iv, and Goethe would have to take *Iphigenia* with him on his journey—which Herder, like Frau von Stein, thought would be a mineralogical tour of a few weeks in the mountains.

However, a flirtatious Austrian countess of Italian background, Aloysia von Lanthieri, of whom Goethe had perhaps asked some incautious questions about Italy, seemed to have some inkling of what was afoot. Haste was needed, and besides, if he did not leave soon, the weather might deteriorate still further. On 2 September Goethe had a great deal of letter-writing to do. Two letters to Seidel listing points to be attended to, and messages to be delivered, including the arrangements for the delivery to Göschen of the first two volumes of the *Literary Works* and for the payment of the first instalment of the honorarium. A short note to Fritz von Stein giving him permission to have a fire in Goethe's open fireplace whenever he wanted. A long letter to Carl August, speaking clearly at last, as he had not done at their meeting on 27 August, of his impending absence (though still saying nothing of his destination), detailing the necessary arrangements in Weimar, and asking for leave. A letter to Göschen, who also needed to know that his author was about to disappear, accompanying Goethe's signed copy of the contract. A letter to Herder expressing thanks, apologizing for his secretive departure, and offering some advice on Herder's proposed transfer to Hamburg. And of course there was as always the letter to Charlotte von Stein. By 11 o'clock that night he was finished. At 3 o'clock the following morning, Sunday, 3 September, the birthday of Carl August, Goethe, with two light bags, mainly filled with books and papers, was sitting in the mail coach driving up the hillside and westwards out of Carlsbad. The day dawned quietly in mist, and the morning was cloudy, but by noon when he reached the great Gothic market-place of Eger (Cheb) on the Bohemian border it was hot and sunny. Perhaps he was still revolving in his mind the words he had written to Herder a few hours before:

The ten years in Weimar will not be lost if you stay, though they will be if you move, for in the new place you will have to start from the beginning again . . . I know that in our Weimar there is much that is disagreeable for you, as for everyone, but still you have a certain footing and familiar position etc. In the end it is a matter of holding out and outlasting the others. How many circumstances may not be possible in which the aspect of our existence may change for the better!

7

To Italy at Last
1786–1788

The Road to Rome

AFTER Eger the road turned south. The coach passed into Bavaria, through the rolling domains of the Cistercian monastery of Waldsassen, and by the late afternoon had reached the watershed between the rivers running north to join the Elbe and the tributaries of the Danube. On the excellent Bavarian roads, metalled with granite chips, they could now travel 'with incredible speed', making, at some seven or eight miles an hour, twice as good progress as in Bohemia; driving on through the night, Goethe was in Regensburg by 10 o'clock the following morning. His mood was euphoric, his mind feverishly active: as he bowled along he noted the changing geology and agriculture of the terrain, watched the sky and theorized about the mechanisms of the weather, recorded arrival and departure times at the post-houses, sketched the scene when there was a pause, and kept up a continuous internal discussion with Frau von Stein. In Regensburg he stopped for a day, bought himself a proper travelling trunk, visited a local natural history collection and the elaborately decorated churches, and saw something of the annual dramatic production by the pupils of the Jesuit college: 'How glad I am to be entering completely into the Catholic world and getting to know its full extent.' But although he was snatching all he could on the way, his main concern was to keep moving. Besides, his alias might be discovered at any time: only a cool denial saved him from being recognized in a bookshop, and after leaving Regensburg at midday on the fifth another overnight journey brought him to Munich at 6 o'clock in the morning of 6 September. Again there was time to spend only one cold, overcast day, mainly looking at paintings, of which he had seen few since his last visit to Cassel in 1783, and at classical antiquities, with which he still felt very unfamiliar, before setting off once more at 5 o'clock the next morning. The tentative scheme of a detour to Salzburg was abandoned in favour of the direct route to Innsbruck and the Brenner pass: 'How much am I leaving by the wayside, so as to carry out the single idea that has grown almost too old in my soul.' So precious was that idea that superstition would not let him name his destination even to himself.

But the seventh was a beautiful day—the first good day of the summer, said the postilion—and it seemed to Goethe that 'my guardian spirit is saying Amen to my Credo'. It was another good omen when, after passing the Starnberger See in the heat of noon, and beginning the climb beyond Benediktbeuern into the Tyrolean foothills, Goethe gave a lift to the little daughter of a wandering harpist ('so I am gradually meeting my characters') and she predicted, from the tension of the harp-strings, that the good weather would continue. And so it did. After spending the night in Mittenwald, above the Walchensee, Goethe passed early the following morning through the frontier rampart into the

Imperial territory of the Tyrol: there was a bitterly cold wind, but the sky was a cloudless blue, the unchanging backdrop to an ever-varying landscape of grey limestone crags, dark green firs, and high peaks white with the first snows. He would have liked to stay longer in Innsbruck, which he reached at 11 a.m., 'but the inner self left me no peace', and at two he was on the road again, up through the ever narrower valley with whitewashed, slender-steepled churches and hamlets perched in meadows among the overhanging rocks, until there were only cliffs and the road and rushing water, and dusk had come and the moon and stars came out and gleamed on the snowfields above. At 7.30 he was in a clean and comfortable inn on the top of the Brenner. 'From here the waters flow down to the German world and to the Latin world, and these latter I hope to follow tomorrow. How strange that I have already stood twice at such a point, and paused, and did not pass over. Nor will I believe it until I am down. What other people find ordinary and easy is made hard work for me.' In one sense, of course, it was Goethe who made things hard work for himself: it was his choice to make a 'symbolical' pattern of his existence. Yet in a profounder sense he was right: what had held him back on those two previous occasions was the knowledge, which was not just a superstition, that his life could only have a meaning if it was put in the service of Germany, and if Germany made that service difficult for him that was not simply his fault. He had turned back before because love and literature and a political task had seemed more important than his own education. If he now went on and fulfilled his father's educational scheme it was because those other vocations had all gone sour—this time there was nothing to turn back for, and beneath the excitement of liberation and travel and the promise of new pleasures the symbolic meaning of the descent into Italy, as Goethe much later acknowledged, was the desperate search for meaning itself.

Goethe spent a day on the Brenner, poised between the two worlds, putting into order his notes on the journey so far, for the benefit of Frau von Stein. He did not want to send her letters, which would perforce have revealed the direction he was taking, but he could not do without her as the addressee of his inner conversation. His solution was to keep a journal, though one written to and for Frau von Stein personally, rather in the manner of the reports he had written from Switzerland seven years before, but not to send it to her until the unnameable destination had been reached; just as he had kept silent about his intention of climbing the Brocken until he had carried it out. The journal, initially divided into a personal, chronological, section, and a systematic section of little discourses on meteorology, climate, botany, mineralogy and anthropology, has almost exactly the intimately pedagogical tone of the young Goethe's letters to his sister Cornelia: he is a student again, in the university of the wider world, and as before he cannot learn without an audience. The audience, however, must pay for its intimacy with exclusive loyalty. If he will not say to himself where he is going, the audience must not expect to know either.

The unposted journal in letter form is a *reductio ad absurdum* of Goethe's reluctance over the last ten years to communicate with a public despite his inner driving necessity to articulate his experience. It is also a measure of his internal resistance to the resumption of a public role implied by his contract with Göschen. As he set off down from the Brenner at 7 o'clock on the calm fine evening of 9 September he was beginning his last and most sustained venture into privacy.

It was soon dark and in the bright moonlight the coach rattled down beside the cascades of the Adige at a breakneck speed. At 2.30 a.m. they passed through a silent and slumbering Bressanone. Dawn brought the first sight of vineyards, and the sun was already hot when at 9 o'clock Goethe reached the bustling market in Bolzano. He dutifully noticed the signs of the silk- and leather-trade, and of banking business, all recorded for him in the earnest statistical guidebooks, but what really caught his eye were the shallow four-foot wide baskets piled up with peaches and pears, 'for I am now only concerned with the sensuous impressions that no book or picture can give me, so that I can regain interest in the world and test my powers of observation and also see . . . whether the wrinkles drawn and imprinted upon my soul can be ironed away again'. Since his arrival in Regensburg fresh fruit had for Goethe been the measure of his approach to the south: there he had to make do with pears, poor after a cold summer, though he ate them on the spot in the street, but he was borne up by hopes of grapes and figs. Figs he had found in Munich, though dear and not very good, but now he was in the land of plenty: as, on that sunny afternoon, he hurried on down the dusty roads of the ever more fertile Trentino, the broadening valley seemed crammed with vines, and maize growing between them, with mulberries and quinces and nuts and other fruit-trees; and grapes, sprayed with chalk to discourage greedy travellers, hung out over the walls where lizards basked and started as the coach went by. In the warm evening, with the cicadas already beginning their shrill chorus, he finally halted for the night in the ancient city of Trent. Unperturbed by the memories of the great Council, he could imagine that 'I had been born and brought up here and was now coming home from a voyage to Greenland, from a whaling expedition'. In his boots and overcoat he felt like a bear that had wandered out of the north into a land where the men went around bare-chested and the shops had neither doors nor windows but opened directly on to the street. The overcoat was soon packed away in the trunk and he resolved to buy himself some lighter clothes when he reached Verona. It was a liberation to have no servants to mark him out as a man of rank and cut him off from the people around him, and to have to do all his daily tasks for himself—changing money, keeping his accounts, even writing his journal, which he would otherwise have dictated. The sense of physical well-being was immense.

After moving on to Rovereto, however, where he finally left behind the German-speaking area and discovered that his Italian was still more than adequate, Goethe decided not to take the direct route to Verona, but to make a detour to Lake Garda. Crossing the hills from Rovereto he came down through his first olive-groves to Torbole. The inn was of the simplest kind, with which he was to become all too familiar: no locks on the doors, oiled paper in the windows, and when he asked for a privy he was directed to the open courtyard and invited to avail himself of its facilities. Nothing daunted, he ate figs all day and enjoyed the local fish. In his room he moved his table to the doorway and looking down the length of the lake he resumed work on *Iphigenia:* 'it went well'. Lake Garda, mentioned in Virgil's *Georgics,* was the first object described in classical literature to come to life before his eyes. The fulfilment of his education had indeed begun; with the Alps behind him he had seen 'the finest and greatest natural phenomena of the land; now my path lies towards art, antiquity, and the proximity of the sea'. He might well have felt the next morning that Mignon's yearning was satisfied at last, when the boat from Torbole took him past the terraced plantations of lemons, still free of their winter protection of straw. He was held up by adverse winds, however, and had to spend an extra night in Malcesine; here he left the territory of the Empire and entered that of Venice, Europe's oldest republic, now in a golden decline and unaware how few were the years still left to it before its dissolution by Napoleon. The castle at Malcesine, guarding the frontier, had long been a ruin, but as Goethe sketched the picturesque sight he drew suspicion on himself as perhaps an Imperial spy, and he needed all the eloquence he could muster in a foreign tongue to persuade the crowd otherwise, unused as they were to tourists and unwilling to admit that their castle was of artistic rather than military significance—he felt he was enacting a scene from his own *Birds.* In the early hours of the following day, 14 September, he sailed easily down to Bardolino, opposite Sirmione, and then drove up over the flinty hills to find Verona lying in blistering heat in the midst of 'one great garden, a league long and a league wide'.

Verona was the first purely Italian city of any size that Goethe had seen, and he stayed there for five days. He completed his physical adjustment to the warmer climate by buying himself a suit of middle-class clothes in which he could hope to be inconspicuous, though he later deliberately spoiled the effect, in the interests of apparent classlessness, by adding some coarse linen stockings which could not possibly have been worn by a respectable native. Here also there first became apparent what were to be the main features of his tour— apart from the long bone-shaking days (from now on he generally avoided night-time travelling) in the old-fashioned Italian carriages. First of all, there was the search for Roman antiquities: after completing his diary entry for Frau von Stein Goethe's first task in Verona was to visit the great amphitheatre in the centre of the town, the best-preserved after the Colosseum in Rome, and the first ancient structure that he had knowingly seen (he does not mention the Roman tower in Trent). The simplicity of the design seems to have unnerved him a little, but he immediately perceived its pure functionality and its adaptation to a social and political purpose: ensuring that everyone present could see not only the arena but everyone else as well. The notion that in its theatres the populace of the ancient republics presented itself to itself, affirming thereby its own collective character, is found, if in a confused form, in Winckelmann, and was to be a guiding principle in the coming Hellenistic period of German culture, but in 1786 in Verona Goethe saw it, or thought he saw it, in the oval form of the auditorium and in the eye's need for a mass of human faces on the terraces, to give it a sense of proportion.

The second main object of his attention throughout his time in Italy, the public behaviour of the contemporary population, was valuable to him partly as an imaginable substitute for the vanished race that had filled those amphitheatres, and partly for its contrast with life at home in Germany: it was certainly different from what he, north of the Alps, knew as the modern world, and it was perhaps different in the same way that the ancient world would have been different. Goethe was impressed by the extent to which the Italian climate favoured a public life in streets, squares, porticoes, and open-fronted shops (he had no entrée to private and domestic life, so his impressions had some of the one-sidedness of any tourist's view). There was, it seemed, always an audience for a group of noblemen playing the (tennis-like) game of balloon or for musicians or other entertainers, always a throng promenading on the Piazza Brà in the cool of the evening, no end to the shouting and singing and arguing and fighting till late in the night, and anyone who gave himself airs by building an arcade on to the front of his house had to expect this intrusion on the public domain to be treated as a public convenience. Goethe's amused and patronizing respect for the Italian populace reflects not only the belief that out of some such openly shared life as this grew the great legal, political and religious institutions of ancient Greece and Rome, but also the slightly forlorn hope that somewhere within his German public might lie the kernel of a similarly confident collective self-awareness: he often refers to the Italian public by means of allusions to the chorus in *The Birds*, his satire on his own, German, readers.

Thirdly we find Goethe using his time in Italy to increase his familiarity with more modern art, especially that of the Renaissance. Verona was the first city with a museum and galleries where he could begin these studies, but already there a pattern of reaction was being established which was to become characteristic, an Ariadne's clue to which he would cling as the new impressions accumulated and threatened to overwhelm him. On the one hand Goethe was anxious to understand and appreciate the techniques of modern artists, especially painters, and nowhere in the eighteenth century was there a greater wealth of masterpieces in the art of colour than in the churches and palaces of Italy. On the other hand he felt obliged to compare modern art unfavourably with that of the ancient world because of the predominantly Christian themes of the works that he saw. The very subject-matter of Christian religious art seemed to him unsuited for that form of visual representation which antiquity had brought to perfection. In Verona he could admire the younger Brusasorci's *Rain of Manna,* or Paolo Farinato's *Miracle of the Five Loaves,* but the themes themselves—and it will be remembered that by the age of 18 Goethe had defined himself as a writer by his rejection of religious themes —could only be called pitiable: 'What was there in them to paint? Hungry people falling upon a few little grains, innumerable others having bread presented to them. The artists racked themselves to give any sort of significance to such banalities.' At the museum he was able to overcome some of the feeling of remoteness from ancient art that had troubled him in Munich by contrasting some pagan funerary bas-reliefs with what would be their Christian equivalents (especially in the Gothic churches of northern Europe):

Here a man beside his wife looks out of a niche as out of a window . . . here a couple give each other their hands. Here a father, lying on his deathbed, seems calmly to be taking leave of his family . . . I found the presence of these stones deeply moving and could not refrain from tears. Here is no man in armour on his knees waiting for a joyous resurrection, here the artist, with more or less skill, has each time just put down the simple presence of human beings and so continued their existence and made it permanent. They do not put their hands together, do not look up to heaven, but are what they were.

To this incipient conflict between ancient and modern the next stage of Goethe's journey was to offer a temporary solution.

On 18 September Goethe wrote to Seidel in Weimar, saying that all was going to plan ('this journey is really like a ripe apple falling from the tree, I would not have wished for it half a year earlier'), and enclosing letters for Frau von Stein, the Duke, and the Herders that carefully gave no hint of his whereabouts, though the letter to Frau von Stein contained a promise of the journal he was keeping. Early the next morning he set out on the first leg of a slow progression towards Venice, the wonder of the world of which his father had spoken so much and from which, he told Seidel, he hoped to dispatch the completed versification of *Iphigenia.* The vintage was beginning and the road was busy; there were frequent ox-carts carrying great vats full of grapes, with the drovers standing in the vats as in some Bacchic triumph, but none the less by the middle of the day Goethe had reached Vicenza, where he decided to stay, in the end for a week. Vicenza appealed to him: he had a lot of work to do on *Iphigenia* and it was quieter than he expected Venice to be. It was beautifully situated and still had the 'lively public' that he had found in Verona, but there was a more cultivated, less provincial air about it:

one enjoys with the Vicentines the privileges of a great city; they do not stare at you, whatever you may do, but are otherwise talkative, agreeable etc. . . . They have a liberal kind of humanity that comes from a continually public life . . . But how I feel what wretched lonely people we are forced to be in the little sovereign [German] states because, especially in my position, one can speak with scarcely anyone who does not want or desire something. I have never felt so strongly the value of sociability.

There was an opera for Goethe to go to on his first evening and three days later there was a debate organized by the local academy and attended by a good 500 people. Even when he went into a bookshop he was surprised at the friendly readiness of the other customers to discuss his purchases with him and the bookseller. Vicenza, he eventually decided, rather than Verona, must have been the home of his Mignon, 'and for that reason too I must stay here a few days longer'. But what struck him above all about Vicenza, and with the force of a revelation, was the architecture of Andrea Palladio (1508–80).

It should be remembered that in the mid–1780s there were next to no neo-classical buildings in Germany, and moreover there was no great stock of Renaissance architecture to represent an earlier stage of Europe's engagement with its classical heritage. The Palladianism of early eighteenth-century England had, it is true, touched Goethe's life through the new buildings at Wörlitz, but these were unique; the atmosphere at Potsdam he had found unsympathetic; and with the French neo-classicism that was beginning to filter into Coblenz and Carlsruhe he had little contact at all. Goethe had of course his compendious three-volume German guide-book with him—and gave Frau von Stein page references to it in his journal—but it did not prepare him for the sensuous impact of so many masterpieces in so small a compass: 'if one does not see these works physically present, one can have no notion of them'. In his first hours in Vicenza he made a sortie to inspect 'the buildings of Palladio', including the Teatro Olimpico ('inexpressibly beautiful') and recognized that 'Palladio was a truly great man both inwardly and in his outward actions . . . There is really something divine in his designs, altogether the power of the great poet, who out of truth and falsehood builds a third element that binds us with its spell'. The following day he paid the first of two visits to the hilltop Villa Rotonda, half an hour's walk to the south of the town, and admired the commandingly decorative presence in the landscape of its four symmetrical temple-like façades, but had some doubts about the practicality of its internal design. Goethe accepted the contemporary interpretation of Palladio as a learned architect reviving the forms of antiquity, and the associated estimate of his works: the early 'Basilica', as it is called, the façade to Vicenza's town hall, on the Piazza dei Signori, with its two storeys of arch-and-lintel bays, had pride of place, along with the Teatro Olimpico, Palladio's last work, built for the academy which Goethe had heard debating and of which Palladio had been a founder-member. He took the Teatro Olimpico at its face value as architectural archaeology, a reconstruction of an imperial Roman theatre, rather than as a monument to late sixteenth-century humanism. The four giant columns on the Loggia del Capitaniato facing the Basilica across the Piazza dei Signori he found 'infinitely beautiful'. But, without any assistance as yet from scholarly commentary or from Palladio's writings, and guided only by his own visual intelligence, Goethe seems instinctively to have grasped the true character of Palladio's art. He does not refer to such unclassical features of the Loggia del Capitaniato as the windows between the columns that interrupt the architrave, but something like these, or the numerous adjustments of proportion necessary to fit the façade of the Basilica to the building behind it, must have been in his mind when he wrote in his journal: 'The greatest difficulty is always how to use the columnar orders in civil architecture. To combine columns and walls without impropriety is almost impossible, of which more when we meet again. But how he has managed to interweave them! how he impresses by the sheer presence of his works and makes us forget that they are monstrosities!'

Goethe recognized in Palladio an original, rather than an archaeological, genius: a man who could combine the inspiration of antiquity and of formal principle ('truth') with the sometimes very specific demands of modern life and a particular commission ('falsehood') to produce something quite novel: a work rooted in reality but suggestive of an ideal fulfilment, a work that has 'truth' in the fullest sense for it is true to what is, as well as to what ought to be.

One earns little thanks from men for wanting to elevate their inner neediness, to give them a great ideal conception of themselves, to make them feel the majesty of a great and true existence (and in a sensuous mode the works of Palladio do that in a high degree); but if one lies to the birds, tells them fairy-stories, helps them along from day to day, then one is their man, and that is why there have come to be so many churches.

For a little while, until he reached Rome—but that was long enough for the new version of *Iphigenia* to crystallize into its uniquely Palladian form—Palladio gave Goethe a vision of how to enjoy the uninhibited, public, and sensually satisfying life of modern Italy without either sharing in, or being alienated by, the Catholic religion which seems so intrinsic a part of it, how to be creatively true to pagan values without withdrawing into morose antiquarianism: 'Towards evening I went to the Rotonda again . . . then to Madonna del Monte and strolled down through the porticoes to the beloved square again, bought myself a pound of grapes for 3 soldi, ate them under the colonnades of Palladio and slipped home when it began to get dark and cool.'

On the morning of 26 September Goethe stowed himself, trunk, and bags into a single-seater chaise and in four hours was driven over from Vicenza to Padua through luxuriant countryside—'a green sea', he called it, after climbing the tower of the university observatory, with white villas and churches peering out of the trees, and on the distant horizon he could, through the observatory telescope, see St Mark's and the other pinnacles of Venice. He stayed in Padua only two days, visiting church and university buildings and, perhaps more important, buying himself a copy of Palladio's *Four Books of Architecture*. However, in the Ovetari chapel of the Eremitani church (strangely, he seems to have made no attempt to visit the neighbouring Scrovegni chapel with its

Giottos), he was privileged to see Mantegna's frescoes of the lives of St James and St Christopher (now largely destroyed), and he was astonished: 'The sharp, sure presence in these pictures is inexpressible. This presence, complete, true (not just apparent, mendaciously effect-seeking, speaking to the imagination), blunt, pure, light, exhaustive, conscientious, gentle, circumscribed, was the starting point for his successors . . .'. The presence of a 'great and true existence', rather than of something that appealed to the imagination to make up an essential that it lacked, was what had impressed Goethe in Palladio's buildings and what had made him contrast them with the mendacious church art in which the things actually depicted were nonsensical—the coronation of Mary in heaven—or banal—people having bread presented to them—and had to be transformed into significance by an effort of the imagination. That he could none the less find this true 'presence' in religious painting before Raphael should have told him there was something irrational in his hostility to Christian art—after all, his understanding of Palladio had room for the 'falsehood' as well as the 'truth', and it ran deeper than his growing revulsion from Christianity. He had recognized that Palladio was a 'poet', and Goethe's own poetry had never been, and could never be, an art devoid of appeal 'to the imagination' or sensuously rooted in nothing but the material here and now. But his belief in the possibility of such a purely objective art continued to grow. Another manifestation, or symptom, of this belief in objectivity had for some time been his pleasure in the passionless order of nature, and half of his second day in Padua was spent in the botanical garden of the university. His enthusiasm for documenting the sequence of forms in which the leaves of the Palmyra palm emerge is more understandable—given that he could have been spending the time looking at Giottos—when we remember that this was the first botanical garden in Italy Goethe had been able to visit, the bishop's garden in Vicenza being given over, 'as is proper', to the cultivation of cabbages and garlic.

Giandomenico Tiepolo the Younger (1727–1804), whose more realistic paintings of contemporary life Goethe preferred to the grand manner of his father (though when he had seen the work of both in the Villa Valmarana in Vicenza he had taken them for different styles of the same man), has left a charming picture, now in Vienna, of 'il Burchiello', the mail-boat that plied on the Brenta between Padua and Venice. Baggage and provisions are going aboard; at the stern stands the captain, smoking a clay pipe, his legs astride the rudder; in the bow his clerk in a tricorn is taking the fares from two friars, a gentleman, and a pensive middle-class traveller. In between is the cabin, in which the ladies are seated, and on its roof the sail is being made ready. Early in the morning, in some such company, Goethe set off in 'il Burchiello' to drift down-river from Padua, 'and so on my page in the book of Fate it was written that on the evening of 28 September, by our clock at five, coming out of the Brenta into the lagoons, I should see Venice for the first time . . .'. To cover the last stretch Goethe took a gondola, which corresponded in every detail to the model his father had brought back to Frankfurt nearly fifty years before. The portentous opening sentence of the Venice section in Goethe's journal is perhaps addressed to the shade of Caspar Goethe, placated at last.

Immediately on landing Goethe went to look at St Mark's Square, not far from which, on the Calle dei Fuseri, he took lodgings in a hotel now called after Queen Victoria, but then simply after the Queen of England, from which he looked out directly on to one of the little canals running off the Rio Memmo. He intended to stay until *Iphigenia* was finished, the new deadline of Michaelmas which he had set himself being practically upon him, and for the next fortnight the play occupied the first hours of the mornings. After that he usually went out, returning to eat and to make some short notes in the diary around noon, and a full report was written up in the evenings—when, as the housemaid remarked, most normal people went for a walk. Not that Goethe shunned the crowds. Apart from remarks on the main public buildings and their location, his comments on the physical appearance of the city are limited to describing expeditions through its narrow alleys and to the ghetto, and the filth that covered the ground when the drains (in whose improvement he took some technical interest) backed up after a rainstorm. He gave his attention rather to the human face of Venice, as a specimen *par excellence* of the Italian 'people' whose ways he had begun to observe in Verona: the bargaining in the fish-market, betting on a game of balloon, a precisely gesticulating story-teller surrounded by an earnestly attentive audience, a Capuchin preacher competing with the din of the stallholders outside the church, the dramatic manner of the advocates in the law-courts, the grand procession of gilded barges, with the senators in red gowns and the Doge in a golden cope, to the High Mass on the anniversary of the Battle of Lepanto—a whole population enjoying its collective existence in public. The culmination, and stylization, of this public life was in the theatre, which Goethe assiduously attended: *opera seria* and improvised comedy in masks, tragedies by Gozzi and Crébillon, a satire on English tourists, and, shortly before he left, the high point, a local Venetian comedy by Goldoni which gave Goethe the opportunity of joining in the pleasure of 'the people' at recognizing the display of their own characteristics. The continuity of literary and popular culture in Italy was demonstrated to him by gondoliers singing— for a fee—verses of Tasso and Ariosto to each other across the waters in the moonlight, in the manner of the fishermen farther down the coast: artificial though the effect was, it moved Goethe to tears. 'A solitary man singing into the far distance, in order that another in like mood may hear and answer him'—it was an emblem of his relation with Frau von Stein when he went back to tell her about it in his journal, but also perhaps of his relation through his poetry with any soul in like case to his, in a German culture that was neither homogeneous, nor joyful, nor publicly shared. Far from reviving his ambitions for German theatre, the Venetian experience so emphasized the impossibility of achieving them in the circumstances prevailing at home that 'all that' seemed 'empty and null' to him—his *Iphigenia*, he now told Frau von Stein, was not being written for stage performance.

As it had no classical past, Venice did not, however, arouse in Goethe to any painful degree the sense of conflict between ancient and modern he had begun to feel in Verona. He was able to appreciate the modern Venetian arts of church music (in San Lazzaro dei Mendicanti) and painting (particularly Tintoretto and Veronese) with only passing comments on the tastelessness of the subjects depicted or of the churches in which the art was housed. Besides, Venice was another Palladian city, and he was at times working as intensively on his study of Palladio, and then of Palladio's master, Vitruvius, as on *Iphigenia*. Reading the *Four Books*, 'the scales fall from my eyes, the mist parts, and I perceive the objects. Even as a book it is a great work. And what a man that was!' In the churches of San Giorgio Maggiore and Il Redentore he could follow Palladio's solutions to the problem of adapting a Greek temple façade to a Christian church interior and find, as most would agree, the Redentore the more successful. His greatest admiration however was reserved for Palladio's last venture in the archaeological manner, the unfinished cloister court in Santa Maria della Carità, intended as a reconstruction on the grand scale of a Roman house as described by Vitruvius. Through Palladio, Goethe felt, the ancient world was becoming accessible to him. When he visited the collection of plaster casts from the antique in the Palazzo Farsetti he recognized 'how backward I am in knowledge of these matters, but things will move, at least I know the way. Palladio has opened that to me, and the way to all art and life.' What the architect was teaching him—a lesson that was indeed eventually to affect his practice of his own literary art—was the true relation between tradition and the individual talent: not imitation or revival, but sensuously alert adaptation of old-established principles to modern conditions—the art of making 'monstrosities' that none the less communicate an idea, or impression, of the classical. 'Palladio was permeated so by the life of the ancients, and felt the littleness and narrowness of the age in which he found himself, like a great man who will not surrender, but as far as possible will transform everything else in accordance with his own noble ideas.' In the notion that the modern Christian age, for all its littleness and narrowness, could be moulded to accord with an inherited pagan ideal, Goethe was establishing a *modus vivendi* with his contemporary world which enabled him, for a while, for as long as they remained at a certain distance, to reach out to and appropriate the treasures of the ancient past with which as yet he felt unfamiliar.

The one completely new experience that Venice had to offer Goethe was the sea. He saw it for the first time in his life from the tower of St Mark's on 30 September, looking across the lagoon and beyond the Lido to a few scattered sails on the open water, while to the north and west the Alps stood guard on the horizon. A visit on 5 October to the Arsenal, where he saw a warship under construction, brought him in touch with the material basis of maritime civilization, of which of course he had no previous knowledge, and prompted him to general reflections on the course of history. The modern age seemed to him dominated by the demands of trade and the development of technology: the flowering of the arts, to which he was devoting himself on this journey, seemed but a brief interlude, whether in the Renaissance or in classical times, and he felt that when he was back in Weimar he ought to turn to useful sciences and skills, to chemistry and mechanics. This insight into the nature of the modern world—coming, rather paradoxically, in the dockyards of what he recognized as a declining power—shed a brief and strange light on his education so far: it appeared suddenly rather insular and backward. 'Alas, alas, my beloved! all a little late. Oh that I did not have some shrewd Englishman as my father, so that I have had, and still have, to acquire and conquer all this alone, completely alone.' Not for many years, however, would these thoughts recur to trouble Goethe, but then their influence on his writing would be decisive, and modernity and the sea would always stand for him in a close symbolic relationship.

For the present the sea was simply a new and fascinating natural phenomenon. The day after visiting the Arsenal Goethe crossed with the manservant he had hired for a few days—essentially as a guide—to the Lido. For the first time he heard and saw the waves breaking on the shore, botanized among the sand hills and walked along the beach, wet from the outgoing tide, collecting shells. On 9 October he spent the whole day on the Lido, travelling down to Pellestrina to inspect the great sea-defences, but watching also with intense pleasure the movements of the crabs and limpets over the face of the breakwaters: 'What a delightfully splendid thing is something *living*. How appropriate to its condition, how true, how *existent!* And how much my little bit of study helps me and how pleased I am to keep it up!' In the middle of his experience of the urban and totally artificial world of Venice, a triumph of human skill and cunning, kept in existence by ingenious devices such as the carefully designed sea-wall, Goethe is exhilarated by the higher ingenuity of life. He is perhaps contrasting the authentic existence of the real crabs before him with the hollowness of what he judged to be bad art, such as St Mark's Cathedral, whose façade he liked to think of as modelled on a giant crab. 'Truth', 'existence', 'appropriateness' were the qualities he had found in the amphitheatre in Verona, in the buildings of Palladio, in the frescoes of Mantegna. He had begun to construct a conceptual framework which could link together the art of antiquity, the best of modern art, and the natural world that was the object of his scientific studies, and could oppose them all to the mendacious art and artifice of the Christian world. In this framework too belonged a suspicion of republican constitutions, such as that of Venice, which allowed too many extraneous factors to interfere with the completion of an artist's grand design and forced so many compromises, for example, on Palladio. He would later discern in an ancient pagan temple the same quality of 'completeness' that he noticed in the manifestations of the monarchical government of Tuscany, ruled by a branch of the Habsburg family. The latent opposition to Christianity of the absolutist courtly culture to which Goethe had for years been growing closer was being reinforced by all the most prominent elements in his experience of Italy. There was a danger that this powerful alliance would overwhelm his initial appreciation of the Palladian *modus vivendi*, the art of accommodating the classical ideal to modern circumstances, and would divert him instead into nostalgia for a lost pagan past. This danger grew more immediate as his tour progressed and his direct acquaintance with classical monuments increased.

By the time of his day-trip to Pellestrina it must have been clear to Goethe that *Iphigenia* was not going to be completed in Venice, but he was now confident that the play could only benefit from its exposure to 'the southern climate'. He decided to move on. On 12 October he stayed indoors to read the newspapers, prepare his departure, and write another batch of letters home, still with no hint of where he was. 'The first stage of my voyage is over' he wrote, and packed up the journal, with 25 pounds of best Alexandrian coffee beans, in a parcel for Frau von Stein. He seems to have been leaving a month for his second stage, which would have brought him to Rome around the middle of November, and to have been expecting to spend some time in Florence, from which he promised Seidel he would write again and where he perhaps still hoped to finish *Iphigenia*. To Carl August he mentioned the significance, 'for a superstitious man like me', of his dating his 'hegira' from the Duke's birthday: in the private thoughts where alone he named his destination, Rome was still perhaps linked to another privately significant date, 12 November.

However the very act of despatching the diary, of letting out of his hands an acknowledgement of what he was doing, broke the spell for Goethe, and an unreasoning impatience began to take hold of him, as if he feared that he might be prevented from reaching the goal of his journey's second stage. From Venice he travelled by the mailboat down the coast and up the mouth of the Po to Ferrara, spending two nights on deck wrapped only in his overcoat. Yet rather than restore himself in Ferrara he looked out as many pictures as he could from the list in his guidebook, visited Ariosto's grave and the room in which Tasso was reputed to have been imprisoned, and on the following day, 17 October, hurried on across the plain of the Po to Cento. Here there were many Guercinos to see, and again only a day in which to see them, but 'if I were to obey my own impatience I would look at nothing on the way and just hasten ahead'. His immediate goal was Bologna, with its Raphaels, and here too, in one of Europe's oldest university cities, he might have expected to stay longer than three days, but 'the world rushes away beneath my feet, and an inexpressible passion drives me on . . . . I cannot enjoy anything until that first need is satisfied.' He gave himself some respite from the craving to be in Rome 'and slake the yearning of thirty years' by changing his plans and aiming now to reach

Rome in time for the celebrations to be expected on All Saints' Day,
1 November, even though this meant passing Florence by until his return
journey. This hectic impatience, undermining both his temper and his
concentration, may excuse, though it cannot fully explain, an extraordinary
outburst in his journal, which, after expressions of admiration for Guido Reni,
and in particular for Raphael's *St Cecilia,* as well as for the quattrocento painters
Francia and Perugino, gives vent once more to a loathing of the Christian
subject-matter of their art:

What can one say but that with these absurd subjects one eventually goes mad oneself? It
is as when the children of God married the daughters of men and they brought forth
monsters. While you are drawn to the heavenly mind of Guido, a brush that should only
have painted the most perfect things presented to our senses, you would like to avert
your gaze from objects that are so disgusting and stupid, and that all the insults of the
world could not adequately abuse . . . One is always at the dissecting table, the gallows,
the knacker's yard, heroes always *suffering* never acting. Never a matter of present
interest, always something awaited in fantasy. Always sinners or ecstatics, criminals or
fools. While the painter tries to salvage himself by dragging on a naked fellow or a
beautiful woman among the onlookers. And treats his religious heroes as dolls and
covers them with really beautifully folded garments. Nothing there that could give even a
notion of humanity.

There are echoes in this expostulation of the beliefs both of Winckelmann and
of Lessing about the subjects proper for depiction in the visual arts, but these
cannot account for the violence of its tone. Goethe's horror of death certainly
has a part in that. So also perhaps does the strange circumstance of his last
meeting with Lavater in July, which had seemed willed by the gods in order that
his journey might begin under the sign of a firm rejection of Christianity.
Maybe too he was starting to be unnerved by the cultural wealth and
cohesiveness of a 'Catholic world' that was less assimilable to the Protestant
Enlightenment, and also less interested in it, than he had imagined in
Regensburg. But for the full explanation for the specific animus of these words
we must probably look to the hint of sensuality in the suggestion that 'humanity'
is best represented by the nude, and especially to the charge that in Christian
art 'presence' is displaced by expectation 'in fantasy'. Goethe set out for Rome
seeking fulfilment—fulfilment first of all of the educational plan of his father,
dominant in his mind, by his own admission, for thirty years. He was looking for
a place and time in which he could enjoy what all his adult life he had been
postponing, expecting, desiring, and for the explanation and compensation for
all the unfulfilment that had preceded. There may well have been a secret
intention that the journey should bring some more substantial sexual
satisfaction than he had been given hitherto. There was certainly a hunger for a
warmer climate and a more fertile land. There was also the belief that the places
hallowed by ancient civilization and the physical remains of ancient art were the
real objects to which the essentially verbal and literary culture of northern

Europe referred, that only if you came south of the Alps could all those Latin and Greek works become something more than abstract erudition. But fundamental to all these partial interpretations of what he was doing was the assumption that fulfilment, in all these senses, was possible, and had actually been achieved by the ancients, and that the monument to that achievement lay in the art and artefacts of ancient civilization. Their buildings and sculptures, and the modern paintings imbued with their spirit, had 'truth' and 'presence' because any desires that they aroused or expressed they simultaneously and of themselves fulfilled. The subject-matter of so much Christian art, by contrast, was longing for something absent, states of faith, expectation, deprivation, detachment, suffering or, if of glory or joy or possession or satisfaction, then in a world understood as not present but future, not sensuous but transcendent or symbolic. And as Goethe hastened towards Rome, to the place where, if anywhere in this world, he hoped to see and perhaps himself experience the satisfaction which alone could make sense of his thirty years of struggle, of waiting, and of confusion, he found himself met at every turn by images of unfulfilment, and worse still, expected to tarry to admire them. But the source of his irritation ran deeper still. For the literary art that he had practised in those years of waiting and by which he had gained ascendancy over his fellows was an art not of possession but of desire, of a sensuous presence always suffused with recollection, reflection, or anticipation: that unfulfilled desire for the always absent object was the origin of his personal, as of his literary magnetism. We have only to think of Faust, Gretchen, Werther, of Mignon, of Götz's alienation from his age, of the insistent presence of the reactive self in 'On the Lake' or 'My heart pounded'—and even the second of the 'Wanderer's Night-Songs' offers not a moment of rest, but the promise of rest 'soon'. The image of a fulfilled existence that lured Goethe away from Carlsbad was profoundly at odds with the nature of his genius—why else was there no second 'On the Lake' scribbled into his journal in such unprecedented moments of satisfied desire as the boat-trip past the lemon-groves of Lake Garda, or the emergence into the Venetian lagoon? The art which repelled him, which aroused in him such obviously irrational fury, was an unwelcome and insistent reminder of where his true spiritual home lay—not, after all, in Mignon's Vicenza but in a land where only those who knew yearning could know what he suffered.

Goethe recovered control of his mood by hiring a horse on 20 October and spending the whole day on a mineralogical expedition to the foothills of the Apennines. He was glad the next day to be leaving the plains, with nothing taller in them than poplars, and moving up into the mountains again, and his pleasure in geological speculation, and in observing the primitive local agriculture, was not dampened by the wretched village inns of Loiano and Montecarelli, where he spent the next two nights, or by the need often enough to plod behind the carriage on foot. The carrier was going all the way through to Rome, and

Goethe for the first time had some travelling companions: an Englishman 'and
his so-called sister', who complained all the time, and a Count Cesarei, in
service in the Papal Guard and en route for Perugia, with whom he got on
famously, though often he would rather have been quietly pondering his literary
plans. He had had the idea of a drama, *Ulysses among the Phaeacians*, on the story
of Nausicaa, though he cannot have considered how unflattering it was to Frau
von Stein that this subject should occur to him now; for when the wanderer
Ulysses meets Nausicaa he is fleeing from long years of enchantment in the
house of the nymph Calypso—was Goethe too fleeing from an enchantress?
'Why do you think so much?' asked the Count, 'a man should never think. By
thinking you only get old.' And again, 'A man should not attach himself to a
single thing for then he goes mad. One should have a thousand things in one's
head, a whole confusion of them.'

On 23 October Goethe had three hours in Florence before pressing on
towards Arezzo: he could see already that the journal of his 'second stage' was
going to be little more than a brief appendage to the first. So poor was the
accommodation up in the mountains, and so much was his mind now on the
goal rather than on the journey, that he did not write anything for another three
days and his itinerary is not quite certain. Passing by Lake Trasimene he
reached Perugia on the twenty-fifth and endeavoured to catch up on his diary.
But it was cold and the freedom from care of Count Cesarei's compatriots,
which seemed to extend to making no preparations for winter, was wearing him
down, as were 'the different currencies, the prices, the carriages, the wretched
inns': he now wanted only to get to Rome, 'even if on the wheel of Ixion', and
his notes were increasingly disconnected. The circumstances were not
favourable for a visit to the centre of devotion to St Francis. When, the
following afternoon, he left the carriage trundling on to Foligno while he
climbed up to Assisi, it was not in order to visit 'the sacred gallows-hill'—he
omitted the convent and the tomb of St Francis (and so once again passed
Giotto by), since 'like Cardinal Bembo [who refused to read the Bible because
of its poor style] I did not want to corrupt my imagination'. His purpose was to
find Santa Maria sopra Minerva, a Roman temple preserved through its
transformation into a Christian church. Right on the market of the old town he
found the object of his quiet pagan devotion, 'the beautiful holy thing . . . the
first I have seen from the ancient period. So modest a temple, as was fitting for a
little town, and yet so *complete* . . . That was the way of the ancient artists . . . that
like Nature they could make do anywhere and yet could produce something
true, something living'. Exhilarated by his discovery of a classical work so
plainly embodying the principles he had discerned in the buildings of Palladio
and in the natural artistry that had formed the crabs on the Venetian
breakwater, Goethe spent a fine evening walking the ten miles over to Foligno
and talking inwardly to Frau von Stein of what he had seen. In the inn, one
great hall with a central hearth 'in Homerical style', with the guests at long

tables 'as in paintings of the marriage feast at Cana', Goethe sat amid a din so deafening he could hardly think and shared an inkpot in order to write down something of that internal conversation. It was Thursday, and on Sunday he could expect to sleep in Rome, 'after thirty years of desire and hope. A foolish and comical thing is man.' He was signing off with an appropriately, if feebly, secular paraphrase of the psalm that had meant so much to him on the Brocken: 'What is man, that thou art mindful of him?'

Goethe was now sleeping in his clothes in order to waste no time in the mornings and dozing for the first hour or so in the carriage. On the Friday they pressed on past Spoleto where he walked on to the great Roman aqueduct: after the amphitheatre in Verona and the temple in Assisi it was 'the third work of the ancients I have seen, and once again so beautifully natural, purposeful and true'. They reached Terni: 'two more nights! and if the angel of the Lord does not strike us on the way, we are there'. On the Saturday the carriage crossed the Tiber by the Roman bridge at Otricoli; Goethe immediately noticed the transition from the limestone hills to the volcanic terrain. A good firm lava road took them to Civita Castellana that night, and in the early evening of Sunday 29 October Goethe passed through the Porta del Popolo in the great Aurelian Wall into 'the capital of the ancient world'. He took a room for the night in the 'Locanda dell' Orso' near the present Ponte Umberto, where Dante, Rabelais, and Montaigne were all reputed to have stayed, and immediately sent a message in his own name to the unsuspecting Tischbein asking him to call. His first thoughts were of gratitude 'to heaven, for bringing me here', his second of Frau von Stein to whom he scribbled a note in his journal. All he could think to say was, 'I am here'. With these words the inner conversation of ten years stopped, and was never resumed. His last great symbolical journey was over: now he had to find out what it meant.

'Quite as much toil as enjoyment': October 1786–February 1787

Tischbein came round that same evening to the inn where a figure in a green coat seated by the fire stood up and approached him with the words, 'I am Goethe'. A warm understanding immediately arose between the painter from Hesse and his famous benefactor, only two years older than he, who had hitherto known each other only through correspondence and their shared acquaintance with Lavater. Goethe explained his intention of maintaining his incognito for the four weeks that he expected to stay in Rome and of leading a completely private existence so that he could concentrate on art and architecture without being fêted as a literary celebrity or condemned, as an Imperial Baron and Privy Councillor, to an expensive and time-wasting round of visits and receptions. He asked Tischbein's help in finding some suitably modest accommodation, and Tischbein offered him a spare room in the house where he himself had his studio. So the following day Goethe moved into Nos.

18–20 on the Corso, the Casa Moscatelli, a stately corner-house, a few yards to the south of the Piazza del Popolo and with a fine open view on to the Pincio and the gardens of the Villa Borghese. The mile-long Corso, lined with tall houses, the perfectly straight though relatively narrow continuation into the city of the Via Flaminia from the north, was one of the few streets in Rome that was regularly swept and whose cobbles were properly maintained. So busy was it in the evenings that carriages filing along between its raised pavements, which narrowed the roadway still further, had to observe a rule of the road and keep to the left. Its northern end was a largely middle-class area—the Palazzo Rondanini (now the Palazzo Sanseverino) opposite the Casa Moscatelli was something of an exception—and, perhaps because of its proximity to the Porta del Popolo, was particularly favoured by the artistic colony from northern Europe. The room Goethe now came to occupy was small and bare, with plain wood-block floors, wooden shutters on the big windows, a bed, a table for the oil-lamp and some hard chairs, and a shelf or two. There was no stove or fireplace, but there was all he needed—and more, for the bed was double. Here he was at last released from the discomforts of Italian inns: the house was kept by a 72-year-old coachman of Cardinal Carafa's, his wife cooked the meals, and there was a manservant, a talkative maid, and a cat. And here he could cut himself off from the higher society of Rome and spend his precious days with men of similar interests to his own. Two other young German artists, both also from Hesse, shared the house with Tischbein: Johann Georg Schütz (1755–1815), a landscape artist and native of Frankfurt, and Friedrich Bury (1763–1823), later a successful portrait-painter, but at 23 still little more than a student.

'The law and the prophets are now fulfilled, and for the rest of my life I shall be left in peace by the Roman ghosts.' After a first hectic day of 'initiation', seeing the main squares of Rome, St Peter's, and the principal antiquities, notably no doubt the Pantheon and the Colosseum, Goethe's dominant feeling on the evening of 30 October was relief. The images that had haunted his mind for so long had come to life: Pygmalion's statue, he wrote, had come to him saying 'It is I'. However, he went on, the living woman was a very different matter from the stone the artist had formed in accordance with his desires. From the start his reaction to Rome was not straightforward, almost as if, in the passage from image to reality, and desire to fulfilment, he had lost something and did not yet know how to deal with what instead he had found. Not for the first time he perhaps thought, faced with living flesh, that he preferred stone statues. In the 'living' city, with its over 160,000 inhabitants the largest he had yet seen, the centre of some significant diplomacy, of international tourism and an intercontinental church (it contained at the time nearly 280 monasteries and convents and nearly 7,000 priests and religious), he disclaimed any interest, 'in order not to corrupt my imagination'—the reminiscence of Bembo clearly showing the animus behind his attitude. The Republic of Venice had been as decadent a political institution as the Papal States, but it had still aroused in Goethe considerable anthropological curiosity. But apart from some enquiries on Seidel's behalf into the appalling state of the Papacy's finances, sustained only by a fraudulent currency, he gave no attention to this unique administrative entity, and he took no pleasure, as he had done so far on his journey, in the behaviour and amusements of the common people. His circle was limited, not merely to artists, but to German artists, and he made no serious attempt to get to know either local or international talent. For most of the time he was in Italy, he afterwards admitted, he spoke German, rather than Italian. But, if living Rome could not attract him, what of the stones to which he turned instead, scattered, as the English reader knows from Shelley's correspondence, through an overgrown and seemingly under-inhabited wilderness the size of contemporary London? In eighteenth-century Rome the ruins of antiquity were more prominent and commanding than they are today—to that extent Piranesi is not misleading—but they were not for that any easier to comprehend. 'It is a sad and laborious task seeking to extract the old Rome from the new . . . one finds traces of a splendour and of a degree of destruction that both exceed our power to conceive them.' Already by the middle of December he felt the need to leave Rome and 'wash my soul clean of the image of so many mournful ruins'. Instead of the presence and permanence of the ancient world he found himself confronted with its transience and with the need to reconstruct it by an effort of the mind: 'where one would like to enjoy, one finds food for thought'. Throughout his journey Goethe had been reflecting on the reasons why a traveller's enjoyment never measures up to his expectations, and now he found himself, though at 'the goal of my wishes', still 'not so much enjoying things as busy with them'.

Goethe's first four months in Rome were a period in which he was almost uninterruptedly busy. The first hours of the day, until 9 o'clock, were spent on the revision of *Iphigenia,* which was finally completed by the middle of December. Then he was free to go out visiting some of Rome's over 300 churches and over 80 palazzi and villas, systematically following up the directions of his voluminous guidebook and also of the new Italian edition of Winckelmann's *History of the Art of Antiquity.* He could also count of course on personal guides. Tischbein accompanied Goethe on several occasions, particularly in the early weeks, but he could not do so indefinitely as he had to be earning his living, the Gotha pension not being adequate on its own. Goethe, however, was fortunate enough to have the services of Rome's second most sought-after connoisseur, the young art-historian Aloys Hirt (1759–1837), who so impressed him that he strongly recommended him to Wieland as a Rome correspondent for the *Teutscher Merkur.* Hirt did his work so well that by January Goethe was himself acting as guide to German visitors. Goethe soon also made the acquaintance of Winckelmann's friend and pupil, and his successor as Rome's most established cicerone, J. F. Reiffenstein (1719–93), the Rome agent of a number of German principalities and of the Russian court. Having commissions to distribute, and often being required to report on the progress of those supported by pensions, he was something of an unofficial headmaster, and private banker, to an unruly and comparatively youthful community of about eighty German artists. From 14 to 16 November Reiffenstein invited the illustrious visitor, who still insisted on being addressed as Herr Möller, though his identity had become common knowledge within a few days of his arrival, to join a house-party at his villa in the hills of Frascati, with views towards Rome and the sea. During the day the participants sketched assiduously; in the evening they reassembled and discussed the results, and no doubt listened politely to the exposition by their pedantic and elderly host of the merits of his revival of 'encaustics', the classical art of painting with wax.

In Rome itself the meeting-place of the German colony was the Caffè Greco, also known as the Caffè Tedesco, by the Piazza di Spagna, where painters' models could be hired on the great stairway leading up to the church of Santa Trinità dei Monti. From here, particularly on Sundays, expeditions would set out for a collective visit to some chosen monument or gallery (the entry charge of about half a dollar was the same whatever the size of the party). In one such chance grouping, just four days after he arrived, on All Souls' Day, the feast-day of all artists, when the Papal collections in the Quirinal Palace were opened free to the public, Goethe met the man who was eventually to have more influence than any other on his developing taste in the arts: for the present, however, J. H. Meyer (1759–1832), was simply 'an honest Swiss fellow', a pupil of Winckelmann's friend, the inspired Henry Fuseli, who disapproved of his master's extravagances and who drew attention to himself by his solid knowledge of the history of painting. Of the older generation of established Roman artists Goethe soon met the 'earnest, blunt and brisk' Alexander Trippel (1744–93), the teacher of Tischbein and of Meyer, who ran a private art school, with a speciality in sculpture, in the Trinità dei Monti area. Nearby, in the Via Sistina, lived, amid a substantial private collection of older paintings, Rome's finest artist of the time, Angelica Kauffmann (1741–1807), with her husband Antonio Zucchi, in whose house, formerly occupied by Raphael Mengs, Goethe was a frequent and welcome visitor: the comfort, and the intelligent conversation, must have been an agreeable interlude in a life otherwise led in cafés, public buildings, or his own spartan room, and among people of little or no literary culture. 'Angelica', as Goethe soon regularly called her, had lived in England for fifteen years, was a friend of Sir Joshua Reynolds, and a founder member of the Royal Academy. She was also at the time collaborating with Goethe's old acquaintance, the Swiss engraver J. H. Lips, who was spending the years from 1786 to 1789 in Rome. One figure whom Goethe, not surprisingly, seems to have avoided was his former colleague in Storm and Stress, Müller 'the Painter', who had proved a bad investment for all who had contributed to his support in his early years in Rome. All the members of the German community had jocular Italian sobriquets, though Müller's was hardly flattering—he was 'the German horse'. Tischbein was 'the twisted-nose phlegmatical'; Meyer, who was rather short of stature, was 'thundering Jove'; the learned Hirt was the *'letterato'*. Goethe himself was 'the Baron opposite the Rondanini'. Reiffenstein's position was honestly acknowledged as 'God the Father Almighty', while the most successful German painter in Italy, Philipp Hackert (1737–1807), who had now removed to Naples but who had begun the sketching-parties that Reiffenstein continued, was 'God the Son redeemer by free lunches'. Angelica Kauffmann was 'the Madonna' and Zucchi, ambiguously, 'St Joseph'.

Apart from the stay in Frascati, and an outing one day with a hired carriage and horses to Fiumicino, to see the sea—he bought a netful of fish, and found an electric eel among them—Goethe did not leave Rome until February 1787: 'I can almost say, I have not wasted a moment' he then wrote. The weather was favourable to a concentrated programme of sightseeing: throughout November there were light daily showers but it was warm—the sirocco was blowing and Goethe did not feel as affected by it as the natives. December and January brought the best winter in living memory: day after day of bright clear skies, and though the air was cold the sun was warm, particularly at midday. Within little more than a month Goethe had seen all the most important ancient monuments —'aqueducts, baths, theatre, amphitheatre, circus, temples . . . the palaces of the Emperors and the tombs of the great'—and begun to take a second and closer look. He was particularly impressed by the more complete survivals, by the pyramid of Cestius and the circular tomb of Caecilia Metella in the extraordinary landscape of the Via Appia, by the façade of the Pantheon, and by the Colosseum, especially in moonlight. He was not above some private archaeology of his own: on 10 November he was walking with Tischbein in the Farnese Gardens, part park, part wilderness, which then covered the Palatine Hill; in an underground chamber they glimpsed statues buried beneath thick stems of ivy and collected pieces of stone and bronze which they concealed in a summer-house, with the intention of removing them later. Though these were forgotten, Goethe was later pleased to acquire a decorated potsherd from the excavations in front of Santa Trinità dei Monti where Pope Pius VI was having the obelisk erected that still stands there. In the museums and the private collections in palaces and villas, most of which were open to the public, he could see relatively undamaged—or industriously restored—pieces of sculpture, which gave him rather more than the diffuse sense of grandeur which was too often all that he could derive from the architectural remains. One of his very first visits was to the Pope's newly extended gallery, the Museo Pio-Clementino, with its famous statue-court: the Belvederean Apollo, which had impassioned Winckelmann, struck him—far more than the Laocoon group—as 'the greatest work of genius, so that one has to say it seems impossible'. Later, in the Palazzo Giustiniani, he found another complete work that claimed 'all my admiration': a Minerva which had escaped general recognition because the taking of plaster casts from it was prohibited. A colossal bust in the Ludovisi collection, to which Winckelmann had drawn special attention, believing it to be a Juno (it is now thought to be the Empress Antonia Augusta), was however available in cast form and on 5 January Goethe, in the role of Tantalus-Ixion at last, carried it off to his room in the Corso where it stood on a shelf supported by his less-used folio volumes beside a Jupiter and an equally colossal ancient foot. By the new year he had also acquired an interest in ancient carved gems and coins—which had previously left him cold—not only perhaps because these were more portable and could still be obtained at a reasonable price but because on them the images had often been preserved intact.

Renaissance Rome was not ignored, however. Goethe even put the early baroque frescoes of the Carracci in the Palazzo Farnese on the same level as the Farnese Hercules, one of the most admired of all ancient sculptures. Having concluded in Bologna that Raphael was a genius equivalent to Palladio, Goethe hurried in the very first days of his visit to see the Stanze and Loggie, and a little later saw the Galatea in the Villa Farnesina. But here too he found that time and decay had taken their toll and 'the pleasure of the first impression is imperfect', 'it is as if one had to extract Homer from the study of a damaged and partially obliterated manuscript'. Such study did eventually restore 'wholeness' to Goethe's enjoyment he said, but there is no mistaking the pang of disappointment in his immediate reactions. But Raphael soon came under threat from another quarter. On 22 November, St Cecilia's Day, Goethe and Tischbein emerged through the tangle of medieval streets and houses in the Borgo into the astonishing, lake-like expanse of St Peter's Square. Here they walked for a while eating grapes in the warm sun, from which they eventually sought protection in the shadow of the great obelisk, and then in the Sistine Chapel where Goethe was overwhelmed by the ceiling frescoes and the *Last Judgement*. A week later he was able to get a closer view of the ceiling from the high gallery and wrote, 'I am so infatuated with Michelangelo that after him I do not even have any taste for Nature, since I cannot see it with such great eyes as he'. A visit immediately afterwards to the Loggie was a mistake: Raphael's 'arabesques' seemed only 'ingenious diversions' by comparison. Disputing the relative merits of Raphael and Michelangelo was a staple of artists' conversation at the time, as it had been since the days of Vasari, and in July 1787 Goethe was still prepared to take Michelangelo's part in such a discussion. On the whole, however, he seems to have avoided the debate or to have sought some conciliatory conclusion, such as an agreement that all parties could praise Leonardo, perhaps because on the one hand Raphael was, 'like Nature, always right, and most profoundly so where we understand it least', while Michelangelo's was a talent too like Shakespeare's, indisputable, but so powerful an influence that Goethe felt if anything the need to be protected from it. It is perhaps for some such reason that although he repeatedly praises the mask of

Medusa in the neighbouring Rondanini Palace (a Roman copy of a Greek original of the fifth century BC), even admiring the 'fearful stare of death' in it, he never mentions Michelangelo's last unfinished *Pietà* in the same collection.

The association between modern art and Christianity remained problematic for Goethe. This did not prevent him from attending all the major church ceremonies: not only a magnificent concert in a lavishly decorated Santa Cecilia in Trastevere on the day he first saw the Sistine Chapel, but also the Pope's masses on All Souls' Day, and the blessing of horses in the Piazza di Santa Maria Maggiore on 17 January. On the night of Christmas Eve he toured several churches, including St Peter's (in darkness except for a side-chapel, where prime was being sung) and at Epiphany he attended a mass according to the Greek rite: 'the ceremonies . . . seem to me . . . more theatrical, more pedantic, more reflective and yet more popular than the Latin'. But a polyglot celebration in the Palazzo di Propaganda Fide (near the Caffè Greco) on the same day impressed him not with the catholicity of the Roman Church, but with the unique euphony of the Greek tongue, 'like a star in the night' amidst the barbarian gabble. The blessing of candles in the Sistine Chapel on 2 February he could tolerate only a few moments: 'I am quite spoilt for this hocus-pocus.' The 'masquerades' of the priests seemed to him no better than the inferior local theatres, and the Pope simply the best of the Roman actors. Christianity in Rome seemed as much a fraud, and as much an inexplicable cult of that which was not 'whole' or 'present', as it had done on his journey down through north Italy, but now it was associated with that deep disappointment, to which he hardly dared to give explicit utterance, at the ravaged state of the beauties he had hoped to enjoy:

I read Vitruvius [he told Knebel] so that there may breathe on me the spirit of the age when this was all just rising out of the ground; I have Palladio, who in his time still saw many things more completely . . . and so Rome, the ancient phoenix, rises like a spirit from its grave, but it is exertion instead of enjoyment, and sorrow instead of joy.

. . . everything is only ruins, and yet no one who has not seen these ruins can have any conception of greatness. So the museums and galleries are only Golgothas, charnel-houses, chambers of skulls and torsoes—but *what* skulls etc.! All the churches only offer the ideas of torment and mutilation. The new palaces too are all only ravished and looted fragments of the world—I do not like to make my language any fuller. Enough—one can seek anything here, only not unity, not harmony . . . And my letter too is just such a patchwork, as are all the letters that I write from here. When I return my mouth shall have something more complete to give.

Goethe urgently needed someone of similar culture and comparable intellect, with whom he could endeavour to make sense of this welter of new and unexpectedly unmanageable impressions. He started writing to Weimar as soon as he could—the first letter, on 4 November, went to the Duke, accompanied by an open letter for all his Weimar friends, and the first personal letter to Frau von Stein was despatched on 11 November—the post went only once a week, on Saturdays, and letters took some sixteen days to arrive, and a little longer in the other direction, so he could not expect any response for at least a month. Fortunately, on 17 November, just back from Frascati, he took part, along with Bury and Schütz, in a group outing to the Villa Doria Pamphili and its extensive gardens on the Janiculum, and at the Caffè Greco beforehand and during the walk he made the acquaintance of a literary man seven years younger than himself with whom he felt an immediate affinity. Carl Philipp Moritz was already known to him as the author of *Travels of a German in England in 1782* and would seem now to have shown or given him the first three parts of his 'psychological novel' *Anton Reiser,* which had appeared in 1785–6. For Moritz the meeting with Goethe brought 'the fairest dreams of my youth to fulfilment'. In the autobiographical *Anton Reiser,* reading *Werther* is the decisive event in the narrator's early life, and thereafter he is obsessed by the desire to meet its author—in order to do so he attempts to become an actor in Weimar and is even willing to be Goethe's domestic servant. Born in poverty in Hamelin and educated in Hanover, Moritz, originally apprenticed to a hatter, had become a schoolmaster in Berlin. An unstable and depressive character, with a remarkable gift for introspection and a touch of homosexuality, he had there fallen in love with an unavailable married woman, had thrown up his position, sought an advance from one of his many unsatisfied publishers, and set off for Italy in June 1786. He had reached Verona on the day Goethe left it, and arrived in Rome two days before him, completely unaware that he was travelling on the same path as the man he called 'God', 'and not wholly in jest'. They now met frequently on walks through the city and Moritz, who had just published a significantly original treatise on German prosody, was able to give Goethe some encouragement, possibly even some substantial help, during the last three weeks of the rewriting of *Iphigenia.*

On 29 November Moritz joined the party going to spend the day by the sea at Fiumicino. In the evening came a sirocco drizzle and as they were all returning past the Pantheon the horse Moritz was riding slipped and fell on the ancient cobbles, and Moritz's arm was broken against a projecting wall. He was carried back on a chair to his rooms in the Via del Babuino and here, after his arm had been set by a surgeon, Goethe arranged for a roster of the German community to nurse him and watch by his bedside during the nights, and to visit him for a set time each day. For forty days, until 6 January, when the bandage was removed, the system was maintained, Goethe himself taking his turn, visiting the immobile patient more than once a day, and acting 'as confessor and confidant, as minister of finance and private secretary'. It is an impressively practical example of Goethe's concern for his 'brothers', for those like Plessing, Krafft, or Lindau who belonged to the Werther generation and had not been as fortunate as he. His identification with Moritz was particularly strong: 'when I was with him he told me bits of his life and I was amazed at the similarity to mine. He is like a younger brother of mine, of the same ilk, only neglected and

damaged by destiny at the points where I have been favoured and preferred.'
'Moritz is being held up to me like a mirror.'

It does not diminish the worth of Goethe's very considerable efforts that he
was at the time in urgent need of human contact and emotional warmth:
'Tischbein and Moritz . . . do not know what they mean to me.' Every week he
wrote to Frau von Stein and by the beginning of December there was still no
reply: 'How I long to get a word from home again, since tomorrow I shall have
been abroad for three months without hearing a syllable from my nearest ones'.
Anxiously he repeated that she must by now have received the first part of his
journal, unaware that, through an inexplicable oversight of Seidel's, the entire
parcel was lying unopened in his Weimar house. Eventually, on 9 December,
the first mail came: a letter from the Herders, and one from Seidel. In Seidel's
was enclosed a note of a few words from Frau von Stein. She would appear to
have asked whether their correspondence was at an end, and if so whether she
might have her letters back. 'So that was all you had to say to a friend, to a loved
one, who for so long has been yearning for a good word from you. Who has not
lived for a day, not for an hour, since he left you without thinking of you.'
Goethe's 'heart was so torn' by this curt notice of offence that he was unable to
continue his round of sightseeing. It is perhaps in part to these days that a
contemporary report refers saying that 'in Rome he sometimes did not leave his
room for a whole week, and sometimes for another week did not return to it . . .
his letters from Rome breathe exactly the tone to be heard in his *Werther*'.
Matters were not helped by his bewildering surroundings, or by his
preoccupation, now that *Iphigenia* was effectively finished, with other works that
had to be made ready for publication and that revived old memories. When by
20 December there was still no letter from her 'it is seeming ever more likely to
me that your silence is deliberate; I will bear that too and think: well, I set the
example, I taught her how to be silent'. Yet the silence was not as long as it
seemed: Goethe's first letter to Frau von Stein reached her on 27 November. If
she waited a week before replying she was doing no worse than Goethe who had
not written to her until he had been in Rome for nearly a fortnight. By the time
he received that reply on 23 December he was in an agony of unease which the
'painful' truths she now told him turned into a desperate apprehension of loss.
She had been hurt by the secrecy of his departure, and by his subsequent
reticence about his journey, and probably she had been humiliated too at having
to admit ignorance of his doings—at any rate it had all made her unwell, she
said, and unable to write, and she probably added that there was anyway little
point since they were as good as separated now. Goethe passionately besought
her to forgive, though even in this extremity he wasted little time in self-
recrimination: 'I was myself in a life-and-death struggle and no tongue can
express what was going on inside me. This fall has brought me to my senses.'
But she had begun to write every week long before she received these pleas. He
had to endure another two 'bitter-sweet' letters before, on 17 January, he learnt

that on Christmas Day, her birthday, Seidel (who had perhaps been saving the parcel for this occasion) had finally handed over to her the journal with which he had from the start tried to placate his own unquiet conscience. 'Since the death of my sister nothing has made me so miserable as the pain I have caused you by my departure and my silence . . . Why did I not send you the journal stage by stage?' It was obvious why not: he had not wanted her to know where he was. He had been running away from her as much as from Weimar, but having made his escape he was now trying to maintain the fabric of the life that he had built over the last ten years, perhaps because he already knew that there was no real alternative. The obstinacy which kept him true to so many long-term projects—both well and ill conceived—was causing him to reinstate what had begun to crumble the previous summer. From a distance, perhaps, Weimar seemed more tolerable, and unless he was to make a yet more radical break than his scandalous departure on 3 September, he could see that he was going to have to return to it. The pleading with Frau von Stein is a clear indication that such a break was not on Goethe's Roman agenda.

By the middle of January 1787 Moritz was out and about again, the manuscript of *Iphigenia* had been despatched to Herder (on the thirteenth) and Goethe's correspondence with Frau von Stein was restored to stability. At the same time a new note appears in his understanding of his relation to the antiquities of Rome. From the moment he arrived in Rome, he had expected this experience to make of him a 'new man', indeed some such expectation had always been implicit in the yearning for Italy. His letters home are punctuated by repeated assertions that he is being 'born again' and that his whole subsequent life will show the benefits of what he is now doing. The nature of the imagined rebirth, however, shifts somewhat as time passes. In his first month Goethe was still looking for what he had promised himself on his journey and was disappointed not to find: completely fulfilled sensuous enjoyment of the artistic achievements of the classical and Renaissance epochs, an answer to the longings of thirty years. The ravages of time, the unwholesome influence of Christianity, the discomfort of his own circumstances, had always seemed to conspire to prevent his enjoying the fruit of pure pleasure as he wished to, as one might enjoy a ripe fig plucked on the shores of Lake Garda. He had identified Rome as the place, if there was anywhere, where that enjoyment might be possible and partly for that reason hastened towards it so frenetically. When he arrived and began to look around the Golgotha of the classical past he at first tentatively ventured the hypothesis that a sense of the 'solidity' of the things of this world was the unique benefit that coming to these ruinous monuments had conferred on him. The paradox was too acute, and we next find Goethe praising the remains for the 'greatness' that in their dilapidation they still embody. Finally in January he settles for the non-committal belief that 'someone who has truly seen Italy, especially Rome, can never be totally unhappy at heart'. For a rebirth was by now indeed upon him but of a quite different kind from that which he had expected: 'I did not think I would have to *un*learn so much'; 'I am like an architect who wanted to build a tower and laid a poor foundation; he becomes aware of it in time and is happy to knock down what he has already thrown up, the better to be sure of his base'; 'I have been cured of a monstrous passion and sickness'.

Goethe admitted at the time that he could not 'say in what specifically the new light consists', and he never tells us what the disease was from which he had been cured, but we may take it that it was something like the hope of fulfilment in which he had been travelling since September, the yearning that had gnawed at his heart since his youth. But the yearning had been cured not through being satisfied, but through his recognizing, and refusing any longer to tolerate, the unreality of its object. He had come to Rome and not found what he was looking for, and that, he thought, was his great discovery. He had set out on a 'symbolical' journey, and at its end had discovered the limits on the world's willingness to be filled with his own personal meaning. He had come looking for culmination, enjoyment, and a revelatory immediacy of experience, and he had found, or thought he had found, the need for study, informed understanding, and hard work. Weimar was not so easily escaped. When he first arrived in Rome it was with a conceptual framework that linked the products of nature, of classical art, and of the best of modern art as all equally 'true', 'existent', 'present', and 'whole'. He began by regarding Rome as a natural object that would yield up its secrets if he looked at it intently and dispassionately enough: 'You know my old way of treating Nature, that is how I am treating Rome and already it is rising to meet me; I just carry on looking . . .'. But the manifestly fragmentary character of most of what he was looking at, the need for mental reconstruction, and the presence around him of artists and scholars who recognized and took for granted the theoretical principles and historical allusions embodied in works of art, had by the middle of December forced Goethe to the conclusion 'that nature is easier and simpler to observe and assess than art'. The structures of art might seek to imitate the structures of nature, but the two were fundamentally distinct.

The least product of Nature contains within itself the circle of its own perfection and I only need to have eyes to see and I can discover its proportions, be sure that a whole, true existence is enclosed within a little circle. A work of art, by contrast, has its perfection outside itself, the 'best' of it lying in the artist's ideal, that he rarely or never attains . . . There is much of tradition in works of art, works of nature are always like a word direct from the mouth of God.

Goethe has discovered in fact that works of art are not simply forms that either do or do not satisfy the desires of the beholder but are historical phenomena, the product of human intention, and that historical knowledge is needed, not merely to understand and appreciate them, but even in order to see them properly. It was to acquiring that knowledge that he began consciously to devote himself in the first two months of 1787.

Whether such a discovery is best described as a 'rebirth' one may doubt. It looks more like a continuation of that process of suppressing the power of his desire which had been gathering momentum ever since the death of Cornelia. Paradoxically, it was in Rome that Goethe succumbed most completely to the 'official' culture of the German courts, a culture which emphasized the otherness and unworldliness of art, its detachment from the business of life. For the young Goethe who had written the 'Prometheus' ode, it was the perfection of works of art to share in the life, in the desires and loves, the sufferings and joys, of human beings. Art and Nature were not distinct, he had still held while he was driving down through northern Italy. From that Storm and Stress aesthetic, however, Goethe was now in Rome detaching himself, in favour of an aesthetic of disinterested beauty. The chaste figure of the Minerva Giustiniani, whose hand had been kissed white by countless visitors, embodied his new understanding of art—a Pygmalion's statue whose peculiar merit it was *not* to come alive. The caretaker's wife could explain the admiration of English female tourists for the sculpture only on the assumption that it represented a goddess in their infidel religion—that of Goethe on the assumption that it must remind him of some distant beloved: 'the good woman knew only worship and love, but of the pure admiration of a splendid work . . . she could not begin to form a notion.' She could conceive of art only as the creation of objects of desire. Throughout his adulthood Goethe had been sustained by a belief in the symbolic significance of his life which was based on desire, especially desire in its noblest form of hope, the hope for a German literary revolution, the hope for practical achievement in Weimar, the hope that the weary years of service and abnegation were the means or prelude to some future fulfilment such as the perfect and immediately accessible beauty he was looking for as he came over the Brenner. By the time he had been in Rome for a month that hope had faded: partly because he had reached what he wanted, partly because what he had reached had turned out to be something different from what he expected, most profoundly perhaps because the hope had faded already under the long impact of a courtly and administrative existence, and the entire dramatic scenario of the secret flight from Carlsbad to Rome was a desperate reinterpretation of its death-throes, the candle flaring before it went out. Goethe certainly learned something new in Rome, something useful for the rest of his life: he had, in the end, nearly two years of sustained and partially systematic study of a subject, the history and practice of art, under the tutorship, sometimes formal, of professionals and scholars. It was the nearest he came to a true university education and it was the foundation of his mature knowledge and culture. But it was no rebirth for his poetry. On the contrary, nothing demonstrates so clearly the continuity of Goethe's Italian journey with the sterile years immediately preceding it as its failure to stimulate him to any lyrical poems of note, or indeed to any substantial new literary work at all. The ten years after Goethe's Roman 'rebirth' brought him practically no *new* literary flowering: there was one remarkable cycle of poems, and there was, in its own way, the equally remarkable completion of some long-standing projects, but otherwise there was only some of his most obviously mediocre writing, and a resounding silence. These were years of a great illusion, of the belief that the alternative to the poetry of desire was a poetry of possession. Only at their end did Goethe realize that the true alternative, the only alternative that could inspire *him*, was a poetry of renunciation.

Meanwhile 'it is not only the artistic it is also the moral sense that is undergoing a great renovation': not only in literature, also in life Goethe was being 'cured', as he saw it, of the willingness to be directed by unsatisfied desire. The time for possession had come. A humorous sketch by Tischbein shows him pushing aside in vexation one of the pillows on his double bed with the words 'The damned second pillow'. He evidently considered ways of filling the vacancy since he wrote to Carl August in February 1787 that the complaisancy of the painters' models would have been very convenient 'did not the French influence [he means syphilis] make even this Paradise unsafe'. He made a note for himself to read *Fanny Hill*. But the most significant indication of a change came when, in the turmoil of packing for Naples, he brought himself to write to his friend, absent in Weimar, as she had always been physically absent from his life:

It is terrible, the memories that often harrow me. Ah, dear Lotte, you do not know what violence I have done to myself and still do and that in the end the thought that I do not possess you, however I take it and push it and pull it, wears me down and eats me up. Whatever forms I give my love to you, still always, always—forgive me for saying to you again what has been checked and silent for so long. If only I could tell you my feelings, my thoughts of those days, those loneliest hours.

The implication of these words was the very opposite of what Goethe or Frau von Stein can have imagined: they were an epitaph on a friendship. Perhaps the extraordinary abstract effort of maintaining and re-establishing the relationship by—initially unposted—correspondence had helped to reveal to Goethe the false premiss on which it was based: that a marriage was possible without the consummation of desire. It was now only their physical separation that kept them together. Once Goethe returned to Weimar his inability any longer to accept a desire permanently maintained and permanently frustrated was bound to put an end to the unnatural liaison.

'Here as everywhere one cannot be involved with the fair sex without wasting time', Goethe told Carl August, and the theatres, which opened in Rome for a short season from Epiphany to Ash Wednesday, were equally a distraction which could not tempt him away from the 'many solid objects of contemplation' around him (and, apart from the ballets, they seemed anyway not to be very good). In January and February he was fully engaged in 'exercising my eye and my mind' on 'the styles of the different peoples of antiquity, and the different periods within these styles, for which Winckelmann's *History of Art* is a reliable guide'. He even regarded himself as collecting material that would be useful in a new edition of Winckelmann's work. 'Then I exert myself in the study of the different divinities and heroes [of antiquity]'. At the same time a new confidence that he knew where he was going and that he had the support of Frau von Stein and the Duke drew him 'out from behind my barricades' and made him more sociable, as he knew they would both wish him to be. He effectively abandoned his pseudonym, which had served its purpose, during his first weeks, of establishing him as a serious student rather than a tourist; and having on his arrival rejected an offer from the Arcadian Society to crown him with laurel on the Capitol, on 4 January he accepted membership of that slightly frivolous body, consenting to hear a sonnet composed in laudation of his works and to receive the pastoral name of 'Megallio' 'on account of their greatness'. Rather more to his taste were the evening gatherings with Moritz, the German artists, and German-speaking visitors, over which he now presided, and at which discussion and reminiscence carried on into the early hours, sustained by bread, cheese, salami, and locally brewed German beer. The rewriting of *Iphigenia*, which Tischbein and Angelica Kauffmann, both penetrating judges, had pronounced a success (though Tischbein, whose primary schooling had been interrupted, always wrote of it as 'Efigenia') spurred him on to think that *Egmont*, *Tasso*, and *Faust* could all be finished too, perhaps even *Wilhelm Meister*, 'with my entry into my fortieth year', and that possibly Göschen should be thinking of an edition of ten volumes rather than eight. He was even beginning to yield to the temptation to resume drawing, which he had resisted at first, for although it would have greatly helped his studies it was too time-consuming. On 7 February he spent the whole day drawing and trying to lose his 'small-minded German manner', by which he meant exchanging the atmospheric landscapes and sometimes delightfully detailed (if not always original) picturesque views of the last ten years for a firmer and more architectural line, which to the modern eye seems rather characterless. In Tischbein and others he had at last instructors with whom he could make what he thought of as rapid progress; he began to take up water-colours and by the end of February he had ten coloured views to send to his mother, who was to forward them to Frau von Stein. This energetically dilettantish existence seemed both puzzling and suspicious in the Chancery in Vienna, which through its spies was keeping a close eye on Goethe's doings and even intercepting his mother's letters: the presence in the Papal city, and under a pseudonym at that, of the closest confidant of the Duke of Weimar, known to be a moving spirit in the anti-Austrian League of Princes, could surely not have so trivial a purpose. Busy though Goethe was, there is, however, no evidence to support Vienna's suspicion that he had a hand in the appointment as coadjutor, and so successor, to the Archbishop of Mainz of Carl Theodor von Dalberg, his old friend and patron from Erfurt, for whose elevation Carl August was actively intriguing.

It is the Goethe of this period, December 1786 to February 1787, who is represented in the best-known of all his portraits, the over life-size oil *Goethe in the Campagna* which Tischbein began in early December. Despite certain anatomical infelicities it is a fine symbolic portrait, and the head, to which Tischbein gave special care, was held to be an excellent likeness. Goethe in a flowing white cape and wearing a broad-brimmed painter's hat (Aloys Hirt was portrayed in a similar one) is reclining on a series of granite blocks, which in the first state of the painting are recognizable from hieroglyphs as fragments of an Egyptian obelisk. Beside him, crowned with ivy, representing immortality, is an ancient bas-relief showing the recognition scene from *Iphigenia* in what was held to be the high classical style. The remains of a composite capital beside it allude to Goethe's architectural studies and in the background we see, against the volcanic landscape of the Alban Hills, some of Goethe's preferred monuments from the Via Appia, notably an aqueduct and the tomb of Caecilia Metella, which he had, we may be sure, enthusiastically discussed with Tischbein in November. Goethe looks intently to the right, 'reflecting on the fate of the works of men', as Tischbein himself expressed it, though the clouds, plants, and rocks also suggest his scientific interests. The painting is dominated by the head and the right hand, the writing hand, which between them have summoned up in poetry an immortal image, that of Iphigenia and Orestes, out of the ruinous fragments of the past—an achievement which, the dimensions of the painting suggest, has something of the superhuman about it. In fact, as we shall see, *Iphigenia* does not belong in this setting—it was finished too early to be touched by the Roman 'rebirth'—but in January 1787 Goethe began to read it freely to his friends. Though he never saw the portrait in its final state, with the *Iphigenia* relief in place, he would surely have been flattered by the suggestion that he had already achieved in the products of his mind a solidity, permanence, and greatness which made them comparable to works of ancient art, for that was undoubtedly what at the time he wanted to believe he was capable of doing.

A renewed note of uncertainty hangs, however, over Goethe's last days in Rome in February 1787, whatever Tischbein, and for that matter Goethe, may subsequently have thought. To his senior colleague, Prime Minister von Fritsch, Goethe writes on Shrove Tuesday of the three stages through which his reactions to the city's antiquities have passed. The first two are known to us already: 'The first period of a stay here is inevitably spent in amazement and admiration until gradually one becomes more familiar with the objects and as it were aware of oneself. Only then does one learn [that is, as a second stage] to distinguish, judge and appreciate.' But the third stage, which occasions Goethe's final assessment of the weeks of intensive study now coming to an end, suggests how uncertain he still is of his 'rebirth', or at any rate of the part that Rome has had in it: 'But in the end the quantity to be studied is too great, concentration is too much dispersed . . . And so, with the best will in the world and after a stay that has afforded quite as much toil as enjoyment, one finds that one would like to begin again, just when one is compelled to finish.' The compulsion of course was purely internal: the truth was that since the beginning of February, he had been feeling that 'it is time for me to leave Rome and make a break in these too serious contemplations, or at least change their objects. I am looking forward to Naples and, if you can all do without me for even longer, to Sicily.'

Goethe's original intention of starting on his return after about a month in Rome was probably first modified when, around 17 November, he learned that Vesuvius had recently begun to erupt, a natural phenomenon of the highest interest to him. As it then became more difficult to discern in Rome the meaning of the symbolical journey, so an extension of it grew more appealing. A new plan was formed of going on to Naples in the new year and returning to Rome for the Easter ceremonies, before spending two months in north Italy, particularly Florence and Milan, and another two months in Switzerland and Frankfurt, so arriving back in Weimar in August. The departure for Naples was delayed, however—by the crisis in the correspondence with Frau von Stein, by the growing need to deepen his historical understanding of Rome, and by a generous and understanding letter from the Duke giving Goethe completely indefinite leave. If he now stayed in Rome until Ash Wednesday he could both observe the Roman Carnival and take time to clarify his plans. Goethe therefore produced a series of possible itineraries giving himself return dates between autumn 1787 and spring 1788, depending on whether or not he went in the autumn to Sicily, where there were reputedly magnificent temples to be seen and where a visit to Etna would complement his studies of Vesuvius. The whole company of friends in Weimar—the ducal family, Prince August of Gotha, the Herders, Knebel, Frau von Stein—who were still receiving regular reports on his progress, in the form of circular letters, were asked to consult together and advise him. But before Goethe could learn the result of their deliberations a second letter from the Duke, who was away from Weimar, gave him a more specific, if courteously expressed, deadline: he would 'not be required back before Christmas' 1787. This ruled out Sicily in the autumn, but he might still be able to squeeze it in in April and May before the weather became too hot for travelling. Frau von Stein, however, declined to advise him one way or the other: as Ash Wednesday, 21 February, approached, Goethe was certain only that he was leaving behind 'the dismembered world' of Rome and exchanging it for the 'happier world' of Naples and the country round about it, 'which is said to be unspeakably beautiful'.

Goethe had lived, with only one brief interruption, for nearly four months in a city which, however remarkable, was 'in many respects rather provincial' and he was 'glad' to be leaving it. It was not the best mood in which to enjoy the Carnival, which took place in the last week before Lent, and mainly in the Corso, immediately under Goethe's windows. The Carnival, 'not really a feast given for the people, but a feast the people give themselves', was 'the life of Rome', and if one stayed away, as Goethe had done, from aristocratic and ecclesiastical circles, the life of Rome was a coarse and plebeian affair. (In the first four weeks Goethe was there no fewer than four murders came to his notice, and one of the rare public events he attended was the slaughter of 1,000 pigs on the Piazza della Minerva.) The Carnival, the only time when there was any light in the streets at night, was an essentially anarchic occasion. There were dances, and the theatres were still open, but the only publicly organized event was the riderless horse-race every evening down the entire length of the Corso from the Piazza del Popolo to the Piazza Venezia. This brief excitement, and the preparations for it, drew thousands to the street, which ceased for the nonce to be a thoroughfare and was lined with chairs and benches and decorated by the occupants of the houses with hangings and lanterns. Here from midday onwards, after a bell had sounded on the Capitoline, Carnival was king, gradually more and more masked or disguised figures, impromptu actors, and entertainers would be seen, and any non-criminal disorder, ribaldry, or flirtation was permitted. The crush was appalling, especially immediately after the race, and the noise more so. Goethe, struggling in his room with the periods of ancient art and hurrying to finish his series of views of Rome for his mother and Frau von Stein, found it 'tedious fun, especially since the people have no inner jollity and lack the money to give vent to whatever little mirth they have'. One or two visits to balls left him bored after half an hour, and after three days he wrote to Herder that he had had more than enough of the Carnival. From Vesuvius—he told Fritsch—'I am hoping for a more noteworthy natural spectacle than the carnival has so far given us an urban one', and in Naples he hoped 'to recover the desire to finish looking at Rome'. It was one consolation that the weather was 'unbelievably and inexpresssibly beautiful', the almond-trees were in full bloom amid the dark evergreen oaks, and crocuses and pheasants' eyes were out everywhere. After packing up in the penitential silence of Ash Wednesday, Goethe left for Naples on Thursday 22 February, with Tischbein. The only literary work he took with him was the fragmentary manuscript of *Tasso*. Instead he had bought a large quantity of the finest quality paper: there was going, he hoped, to be a lot of drawing on this trip.

A Glimpse of Fulfilment: Iphigenia *and 'Forest. Cavern'*

Once Goethe reached Rome, the ultimate symbolic goal for any quest that a European man could undertake, his own 'symbolical existence' could not last much longer. The goal itself, ruinous and unsatisfactory, turned out to symbolize not fulfilment but only the need for a new start and a new way of life. But by the time he had begun to understand this, in the New Year of 1787, Goethe had already completed the poetic expression of the spirit of the quest that had taken him from Carlsbad to Rome—the second version of *Iphigenia in Tauris*. Goethe's most perfectly Palladian work, and the last to be quite untouched by the distinction between Art and Nature that his stay in Rome brought him, it is animated throughout by the belief that a fulfilment of the heart's desire is possible, and that the real world, however limited and recalcitrant, can become the vehicle for the highest ideals of the human spirit. Never again did Goethe write out of so sustained a belief in the imminent realization of his hopes, and *Iphigenia* remained unique.

The first version of the play centred on the question which drove Goethe to two earlier symbolical journeys: were the gods benevolent? Did fate confirm the intuitions of the heart, and would it show, by its response to Goethe's prayer for an oracular sign, or to Iphigenia's prayer for the healing of Orestes, or for her own release, that it was not, as the Harpist feared, or as Iphigenia herself feared in her moment of doubt, a predatory and incalculable power that made men guilty and then punished them for being so? That question was certainly alive in Goethe's mind once more during the anxious months of 1786 when he laid the secret plans for his journey, waited like a mystic for the sign that would release him to go to Carlsbad, and hurried with increasingly uncontrollable and irrational impatience to reach his ultimate goal by a significant date. But more than seven years had passed since it had last been an urgent question, and there is something perfunctory, almost mechanical, about Goethe's references to the divine oversight of his flight to Rome. In his long toil for the ducal administration he had followed assiduously the injunction with which the prose *Iphigenia* concludes: to seek among men the answers to perplexities about the gods. And, as the poem 'Divinity' shows, in which that injunction is authoritatively expressed, his conception of the gods had changed as a result. No longer was their arbitrary will an object of fear; they themselves were but shadows cast on the sky by human qualities. As a result of the marked shift in the role the revised play assigns to the gods, the tragic potential of the original version fades into the background.

In the prose *Iphigenia* it remains uncertain whether or not there is an additional cast of divine actors—Apollo, Diana, the Furies—who (like Jacobi's transcendent personal God) might at any point appear on the stage and intervene in the action, but happen not to do so. In the new version of 1786 there is a systematic identification of the gods with the human moral attitudes that belief in them expresses. When in the first act Iphigenia objects to Thoas' proposal to reintroduce human sacrifice

> Der mißversteht die Himmlischen, der sie
> Blutgierig wähnt,

The man who imagines the heavenly ones thirst for blood misunderstands them,

she is paraphrasing in verse a sentence in the prose version, which simply defends the character of the gods as if they were benefactors who happen to be absent. But her next words have no prose precedent and correct, in the spirit of 'Divinity', any impression that the gods might have a moral character independent of that of their worshippers:

er dichtet ihnen nur
Die eignen grausamen Begierden an.

he is fancifully attributing to them his own cruel desires.

The principle can be indefinitely extended: a cruel man has cruel gods, a clever man clever gods, and a pure, holy, and healing woman worships a pure, holy, and healing goddess. As Goethe revised the detailed wording of his play so he associated more events, not fewer, with the divine dimension, but this divinity is always an image of a human moral reality. The calculating Pylades, who sees life in terms of opportunities to be seized rather than a moral law to be obeyed, said in prose merely that 'our return hangs by a slender thread', but in verse he says that

unsre Rückkehr hängt an zarten Fäden,
Die, scheint es, eine günstge Parze spinnt.

our return hangs by slender threads which, it appears, a favourable Fate is spinning.

Even Iphigenia, in her moment of gravest doubt at the end of Act IV, no longer simply fears that the Olympian gods may leave her to a tragic destiny: she prays now also that the gods may save their face by saving the mirror in which alone they become visible—her benevolent and believing soul:

Rettet mich
Und rettet euer Bild in meiner Seele.

Save me, and save your image in my soul.

This principle of correspondence between the human and the divine serves to remedy one of the weaknesses of the prose version and to pull together the two halves of the play—that about Orestes and that about Iphigenia. For the symmetry between the earthly brother and sister and the heavenly—Apollo and Diana—no longer seems just a happy coincidence. Rather, the inability of Orestes to understand that, in this respect also, what is in heaven corresponds precisely to what is on earth, is the factor which prevents him for so long from understanding correctly the words of the oracle. Once he realizes that the 'sister' of the oracle is actually Iphigenia, the only true Diana to be found on earth, he has not only resolved the last remaining perplexity in the action but has restored in himself the image of the young Apollo setting out to slay monsters and 'hunt down the joy of life and great deeds'. That image had been obscured only by his failure to rise to the same level of theological insight as his sister and recognize that Apollo and Diana are the ideal reflection of themselves.

Goethe's most substantial addition to the thematic texture of his play is central to its theology and also has a unifying effect. Building on a few scattered references already present in his prose draft, he develops the notion of a divine

curse ('Fluch') on the house of Tantalus into a motif that recurs throughout the
play and establishes an intimate link between its two climaxes—the cure of
Orestes and the release of Iphigenia. In Act I he revises Iphigenia's exposition
of her family's crimes so that Thoas explicitly asks whether her ancestors were
themselves morally culpable or were merely suffering the consequences of
Tantalus's original fault. Iphigenia's reply is ambiguous: she admits that the
gods bound Tantalus's offspring with a 'brazen bond', but attributes the
atrocities her fathers committed to their personal moral defects. The image of
'bonds' returns in the two references to the continuing curse on the Tantalids
which Goethe adds to the 'axial' scene, as he called it, of Orestes' cure:
Iphigenia's prayer that he should be released from the curse, and the words of
Orestes that tell us the prayer has been heard: 'Es löset sich der Fluch, mir
sagt's das Herz' ('The curse releases me, my heart tells me so'). And in the last
scene of all, when we learn the terms of the oracle that has guided Orestes,
Goethe adds to them an explicit reference again to 'release' from 'the curse'.
The result of this new emphasis on an apparent determination by the gods that
generation after generation of Tantalus's descendants shall be made guilty and
left to their torment—for all guilt is avenged on earth—is not only that the
'release', first of brother and then of sister, appear as stages in a process of
redemption from that fate, but that a new and powerful motive has been found
for Iphigenia's longing to return to Greece. She is driven now, not by a rather
selfish nostalgia, neglectful of her acquired responsibilities in Tauris, but by a
passionate belief in her original and divinely appointed mission to return to the
halls of Mycenae and there, as by special divine intervention the only member
of her family to have remained free from sin, to purify and reconsecrate the
ancestral dwelling, crown her brother king, and obliterate the memory and
effects of the bloody history with which the gods had apparently cursed her line.
The question whether Thoas will let her go (and not just the question whether
he will force her into fratricide) thus becomes quite directly a question about
the gods' willingness to extinguish the curse, about their intentions in bringing
Iphigenia to Tauris in the first place, and about the reliability of Iphigenia's
insight into their nature.

The increased prominence of the motif of the curse in no way contradicts the
identification of the divine world with the world of human moral attitudes. Just
as a good and pure woman believes in gods with good and pure purposes, so a
man tormenting himself with guilt and remorse believes in tormenting and
vengeful gods, and conversely a man who believes he has been released from
the brazen chains of a divine curse is a man who has made himself morally free.
At the centre of the idea of a curse is the idea that a man may not be wholly
responsible for his own moral destiny but may, in spite of his own best will, be
forced by an external power into committing crimes. For Iphigenia, who holds
that the gods 'speak to us only through our heart', such a belief is indeed a
curse, the only real curse there is: one that a man inflicts on himself and by that

same token can at any time lift from himself also. The only thing that keeps us from moral perfection is, for Iphigenia, the belief that there is something that keeps us from moral perfection—whether in the form of Orestes' fear of avenging Furies, or of Pylades' prudent respect for the practicable. Orestes adopts not just Iphigenia's vocabulary but her entire way of thought when he says that his heart tells him the curse has been lifted—and it is lifted in his very act of allowing his heart to tell him so, for the curse is to believe that anything other than our heart directs our lives.

In the new version of his play that Goethe constructed in Verona, Vicenza, and Venice, and during his first weeks in Rome, there reigns therefore as intense a preoccupation with the self-sufficient heart as in the revision of *Werther* that he completed immediately before he set out for Italy. If the first version asked, 'Are the gods benevolent?', the second asks, 'Can the heart redeem itself?' *Werther* had been rebuilt around the image of a personality wholly responsible for its own destruction, and the novel's conclusion had been rather implausibly detached from the social and historical context that had originally been so much of its strength. The new *Iphigenia*, though envisaging an equally extreme subjectivity, has, or wishes to have, an opposite tendency. A sensibility can be as self-enclosed as Werther's had now become and still, it is thought, rescue itself from destruction and find itself supported by the world. The possibility of catastrophe seems to remain: Orestes might prove as incapable of conversion as Werther, and Iphigenia might find her hope as deceitful as that of Proserpina. However, the link betweeen the gods and their image in the soul is now so close that, if Iphigenia is deceived, the gods must be, not predatory and unfathomable, as they were for Proserpina, but simply non-existent. If Iphigenia is wrong, there is no order in the world, rather than an order that is hostile or tragic, and the moment in which she throws herself on Thoas' mercy is not so much a moment in which she tests the character of the gods, humbly submitting to the outcome, as a moment in which by the sheer force of her own faith she compels a general acknowledgement of a humane moral order, so displacing the gods and becoming herself the principal agent in the play. But the order she and the other characters come to acknowledge is only apparently independent of her own will and the risk that her appeal will be disappointed is only apparent too. By the time she makes her appeal, in the new version of the play as in the older, the hegemony of the heart is already established in Thoas as in Orestes, and we know it. In the new version, as in the older, the reason for Thoas' wrath is not that he fears to lose Iphigenia, but that he believes he has caught her out in deception and hyprocrisy, and the ideal which once transformed his life now seems to have feet of clay. He is not himself, however, any the less a disciple of truth because he believes Iphigenia to have fallen away—and although that makes his final conversion more plausible, once he realizes she has not betrayed the good cause, it also makes it less miraculous, less indeed a conversion at all. In the new version of the play, with its greater insistence that the gods are secondary, it is if anything clearer than in the old version that at the last no external power comes to succour Iphigenia, any more than Werther. Where Werther's desires destroy him, however, hers are made to appear to recreate the human world in their own image. But this apparent interaction with the world is itself illusory—Iphigenia is in fact as solitary and self-enclosed at the end of the play as Werther at the end of the novel. The last scenes mime a conversion—*as if* Iphigenia were affecting Thoas—which they render psychologically credible only by showing that Thoas is already of Iphigenia's persuasion. The reasons for Thoas' belonging to Iphigenia's school in the first place, however, are as private to him, and as unrevealed to us, as the reasons for his cure from belief in the curse are to Orestes: 'my heart tells me so'. Heart after heart catches fire from the spark in Iphigenia's breast, but the process of communication itself remains unrepresented and mysterious. Out of the material of subjectivity the play thus creates a simulacrum of objectivity, out of pure longing an image of fulfilment.

As Goethe crossed the Brenner into the land where the lemon-trees bloom, and where he began to turn *Iphigenia* into verse, he left behind him ten years of deprivation that could now expect fulfilment, and ten years of a symbolical existence that could now expect their crowning meaning. And for those first miraculous weeks, as the play grew and he ate figs every day, it did indeed seem as if the world was coming to meet his desire, the goal of his life had become tangible, to the point of its lying only a few days' journey away by bumpy carriage, in Rome. The ripe apple was falling from the tree, he had written to Seidel, and the words echo through the lines he gave to Iphigenia when Orestes revealed his identity and she at last glimpsed the possibility of deliverance from her own long years of waiting:

> So steigst du denn Erfüllung, schönste Tochter
> Des größten Vaters endlich zu mir nieder!
>
> . . . .
>
> Kaum reicht mein Blick dir an die Hände die
> Mit Frucht und Segenskränzen angefüllt
> Die Schätze des Olympus niederbringen.
>
> . . . .
>
> so kennt
> Man euch ihr Götter an gesparten, lang
> Und weise zubereiteten Geschenken,
> Denn ihr allein wißt was uns frommen kann,
>
> . . . .
>
> Gelassen hört
> Ihr unser Flehn das um Beschleunigung
> Euch kindisch bittet, aber eure Hand
> Bricht unreif nie die goldnen Himmelsfrüchte,
> Und wehe dem, der ungeduldig sie
> Ertrotzend, saure Speise sich zum Tod
> Genießt.

and so Fulfilment, fairest daughter of the greatest Father, you do at last descend to me! My eyes can scarcely reach your hands that, filled with fruit and wreaths of blessing, bring down the treasures of Olympus . . . and thus one recognizes you, O Gods, by your gifts, stored up and long and wisely prepared, for you alone know what is good for us . . . You calmly listen to our prayer that childishly beseeches you to hasten, but your hand never breaks the golden fruits of heaven when they are unripe, and woe to him who, snatching them by force impatiently, enjoys sour food that brings him death.

Yet even in its new version *Iphigenia* ends with a farewell, and the sails billowing on the ship that is to take the heroine back to the land of her desire. The fulfilment towards which Goethe hoped he was journeying as he wrote the play, though it is welcomed by Iphigenia, and though its possibility is made manifest in the conclusion, is itself no part of the action. The poet of desire for the absent object, who could not write four lines to celebrate being on Lake Garda in the September sunshine, wrote five acts of melodious verse to express a cruelly threatened but increasingly hopeful longing that at the last does not quite spill over into enjoyment. In the second *Iphigenia,* after ten years, the poetry of the young Goethe is reborn. It is the play of 'the heart' as *Werther* was the novel of 'the heart', and we hear in it once again—for the last time and in a refined, almost ethereal form, but *in extenso* as never before—the music of 'On the Lake'. The natural imagery that had almost faded away from Goethe's writing of the last five years is restored and revitalized, and the fluctuating rhythms that had once nervously charted every change of mood return again, supported by the new blank-verse framework.

Stimulating his five senses, exercising his powers of observation, was quite consciously one of Goethe's purposes in coming to Italy, and the very first lines of the play, in Iphigenia's opening monologue, testify to a new firmness and immediacy of experience. In the prose version Iphigenia's phrases are featureless, unstructured, even repetitive, and given weight only by an emphatic epithet or an artificially antique turn:

denn mein Verlangen steht hinüber nach dem schönen Lande der Griechen, und immer möcht' ich über's Meer hinüber, das Schicksal meiner Vielgeliebten theilen

for my yearning lies across to the fair land of the Greeks and I should ever like to cross the sea, sharing the fate of my much-loved ones

The contrasting vigour of the verse is astonishing: the repetitions and redundancies go; a dramatic, even pictorial image emerges; and the clauses follow a clear logical structure, rising at one point to epigrammatic strength. And the lines have been written by a man who has manifestly seen, and heard, the sea:

> Denn ach! mich trennt das Meer von den Geliebten
> Und an dem Ufer steh ich lange Tage,
> Das Land der Griechen mit der Seele suchend,
> Und gegen meine Seufzer bringt die Welle
> Nur dumpfe Töne brausend mir herüber.

For alas! the sea separates me from my loved ones and on the shore I stand for long days seeking the land of the Greeks with my soul, and to meet my sighs the waves bring me only a muffled roar.

More striking still is the vital relationship between this symbolic landscape and the person of the speaker: every one of the lines contains a reference to her, identifying the sea as that which separates her from the object of her desire. The interfusion of subject and object so characteristic of Goethe's writing before 1776 is here revived. The sea becomes something of a leitmotif, one of the most prominent images in the new version of the play, and whenever it recurs it has associations with all that Iphigenia has to overcome—doubt, distance, the burden of past crime—in order to reach her goal. Throughout the new version of the play the imagery of concrete things becomes not only more vivid and alert, but also more closely related to the viewpoint of the person involved. The despairing Orestes, convinced of his imminent destruction, says in prose; 'Den gelben matten Schein des Todtenflusses seh' ich nur durch Rauch und Qualm' ('I see only the yellow, dull glow of the river of death through smoke and fume'). But in verse his vision acquires a hideously personal and dynamic attraction (of such immediacy that its colours can be left merely suggested):

> Durch Rauch und Qualm seh' ich den matten Schein
> Des Totenflusses mir zur Hölle leuchten.

Through smoke and fume I see the dull glow of the river of death lighting me to Hell.

In choosing for his play the Shakespearian blank-verse form, the iambic pentameter in which Bürger had once planned to translate Homer, rather than any more obviously antique form, such as the trimeter, Goethe was building on the enthusiasms of ten years before. But as remarkable as the ease with which Goethe mastered the task of writing in a regular metre at much greater length than he had ever done before is the fluency of the rhythmical variation that he also introduced, and for which his only model was his own earlier practice. There are, it is true, set-piece free-verse monologues, such as the 'Song of the Fates' which Iphigenia chants at the end of Act IV, which seem patterned either on operatic arias or on the choric elements in Greek drama. But it is in the subtler variations, within the staple verse itself, that we can hear Goethe's old art being recovered. Some of these are quite conscious—'I have let some half-verses stand where they may perhaps have a good effect, and been careful to incorporate some alterations in the metre', Goethe wrote to Herder in January 1787—as, for example, the turbulent rhythms in which Orestes tells how the Furies first began to pursue him after the murder of his mother. The pentameter flow is first interrupted by one of the half-lines that Goethe let stand, then an uncertain attempt is made to reassert the regular rhythm, before it collapses completely in the wholly non-iambic lines which reproduce Clytemnestra's cry for revenge:

Wie gärend stieg aus der Erschlagnen Blut
Der Mutter Geist
Und ruft der Nacht uralten Töchtern zu:
'Laßt nicht den Muttermörder entfliehn!'

Up from the blood of the murdered woman seemed to bubble the mother's spirit calling to the ancient daughters of the Night: 'Let not the matricide escape!'

The change of rhythm here—and even more markedly in the next scene, Orestes' demented vision of the underworld—registers a shift in the level of consciousness in just the same way as the metrical shifts in 'On the Lake'. The art is at its most refined, however, when such changes occur without disrupting the framework of the blank-verse line. In Act IV, for example, Iphigenia becomes aware of the moral doubtfulness of Pylades' scheme for the deception of Thoas and her confidence is shaken. Her utterance is slowed down into hesitancy by several powerful and unusually placed caesurae, the penultimate line seems too short, the last too long, and the speech peters out into a series of almost weightless monosyllables:

> Doppelt wird mir der Betrug
> Verhaßt. O bleibe ruhig meine Seele!
> Beginnst du nun zu schwanken und zu zweifeln,
> Den festen Boden deiner Einsamkeit
> Mußt du verlassen! Wieder eingeschifft
> Ergreifen dich die Wellen schaukelnd, trüb
> Und bang verkennest du die Welt, und dich.

The deception grows doubly hateful to me. O stay calm, my soul! Are you now beginning to waver and to doubt? You are having to leave behind the firm ground of your solitude. Embarked once more the swaying waves take hold of you, troubled and fearful you no longer recognise the world, or yourself.

But while rhythmically the speech tails away into uncertainty the imagery of the sea gathers firmly and clearly towards a climax and conclusion: hinted at in the mention of 'wavering', implied by contrast with the 'firm ground', it is first explicit in 'embarked', and fully developed only in the penultimate line. An alien power is growing in strength and threatening to overwhelm all Iphigenia's self-possession. The counterpointing of the fragmented rhythm and the onward sweep of the imagery is a dramatic and psychological effect of a sophistication last achieved by Goethe in some of Werther's original letters and in the Gretchen scenes of *Urfaust*.

Such an effect is of course possible only because of the context of regularity which the blank-verse form provides. In this rewritten *Iphigenia* the poetic spontaneity of Goethe's Frankfurt period is revived and happily combined with the conventions of formality and restraint he had come to value in his first ten years in Weimar. The uneasy experiments of the first prose version, which reflected Goethe's attempt, scorned by Bodmer, to create something peculiar to the little world of Weimar and its few 'pure spirits', have given way to a new synthesis. Time and again, if the versions are compared, the verse speeches will prove not only to be the more vigorous poetically but to be the clearer, the more formal, and the more deeply pondered, in their sequence of thought and imagery. Goethe now knows what he is about, and who he is writing for. Not, as he said in Venice, any theatre—not even the now defunct Weimar amateur theatre—and so, not any court. He is now writing a book—addressing once again the, essentially middle-class, reading public to whom he owed the reputation that had taken him to Weimar in the first place. The verse-form he has chosen has enabled him to tap again the poetic resources of that period, whose national aspirations the form, with its English and Shakespearian connections, tends to reflect. At the same time the book embodies also the values of the courtly culture to which Goethe had more recently devoted himself: although being intended for publication and reading, as if it were a novel or an epic, it purports to be a courtly entertainment, a play, and a play in the French classical manner, with operatic features and accommodations to the fashionable Greek style. In terms of its content, *Iphigenia* concludes with a mime in which monadic subjectivity is represented as capable of having, or as on the point of having, some causal effect on the world about it—the fulfilment from which the German middle classes had been excluded for over a hundred years—provided that the authority of the King is respected, that Orestes sheathes his sword, and that the powerless (here represented by a woman, fully aware of her subordinate position) continue to seek only an interior victory. In its form, too, the play shows how the ambitions of the 1770s for a national literature disseminated through the press can achieve fulfilment, if only momentarily or marginally, by accommodating themselves externally to courtly genres and conventions. For Goethe himself, therefore, the versified *Iphigenia in Tauris* is a resolution of the conflict between his Frankfurt and his Weimar loyalties, a courtly achievement he has built out of Storm and Stress materials, an apparently classical, 'whole', and 'present' work of art, of the kind which he believes to await him in Rome, but which he has none the less constructed from the modern themes of interiority and longing for an absent good—just as Palladio conjured up a classical ideal out of the villas and town-houses, monasteries and churches that his modern patrons required him to build. It was in fact in terms reminiscent of Goethe's own reactions to Palladio that his contemporaries commented on the play when it was finally published in June 1787. It was 'inspired by none other than the muse that animated Euripides', indeed Hemsterhuis thought it better than Euripides, so perfectly 'Greek' was it in spirit, and Schiller used Goethe's preferred vocabulary in reviewing it in 1789: 'one cannot read this play without feeling the breath of a certain spirit of antiquity, which is far too true, far too alive, for any mere imitation, even the most accomplished'.

In the same metre as *Iphigenia,* and probably at about the same time, Goethe wrote another hymn to fulfilment: a magnificent monologue by Faust which in the published versions of the play opens the scene 'Forest. Cavern'. Certainly Goethe later in his life attributed some of his work on *Faust* to January 1787, when *Iphigenia* had just been finished, and when his mind, as he had told Carl August in December, was turning to completing his fragments, *Faust* among them. It was a formidable step to take up again the project so closely associated with his last years in Frankfurt and which Weimar had seemed to cast into an enchanted sleep. But perhaps the preoccupation since the summer with his other earlier works, particularly *Werther*, and his success, in *Iphigenia*, in summoning up again the powers that had once inspired him, and in bringing them to serve in a new synthesis, had given him confidence. The speech opens with a most emphatic statement that an epoch in life has reached a culmination and brought a new beginning:

> Erhabner Geist, du gabst mir, gabst mir alles,
> Warum ich bat.

Sublime spirit, you gave me, gave me everything I asked for.

Addressed, no doubt, to the Earth Spirit, who had once 'turned [his] face to me in fire', it is as definitive an assertion as possible that the yearning which had first driven Faust to call upon the Spirit is a thing of the past. He has, he says, recalling at times the very words of his earlier speeches, been given the intimacy with Nature that he sought, he 'feels' and 'enjoys' her, looks into her breast as into the bosom of a friend and recognizes her various manifestations as his brothers. At the same time he has been given knowledge of 'the deep mysterious wonders' of his own self, into which, as into a cavern, he can when necessary withdraw from the communion with Nature. If in this we can detect Goethe's Spinozan natural science, with its Leibnizian orientation, of the years immediately before the Italian journey, the last lines introduce an element peculiar to the early stages of the journey itself. Out of the moonlit natural scenery—and generated perhaps in accordance with the same principles of 'truth', 'presence', and 'wholeness' that operate in Nature—emerge the 'silver figures of antiquity', and in this converse with the classical past Faust experiences a 'mellowing of the rigorous pleasure of contemplation', a phrase echoed in Goethe's letters of December 1786 and February 1787.

Much less ambiguously than the final version of *Iphigenia*, this speech praises the goodness of the gods and is remarkable for Faust's acknowledgement of his dependence on higher beings: any 'power' of his own is a gift from elsewhere. Yet at the same time Faust has not reached a self-forgetful union, either with the Spirit or with the world around him: the dualism of Nature (the forest) and Self (the cave) is pronounced. Nor is there anything restful about the monologue: it is full of verbs of motion, and the relation of subject and object is changing all the time. There is no way of telling when Goethe wrote the much more pallid concluding section, in which Faust complains that the companionship of Mephistopheles makes even such a moment of bliss a transient pleasure, but the seeds of loss are already present in the first, richly suggestive, twenty lines. Goethe was long to remain uncertain to which episode in Faust's career the speech was to be a response but probably from the start it was, like the verse *Iphigenia* with which it has so much in common, intended to offer only a glimpse of fulfilment.

Uneasy Paradise

After a stop at Velletri to look at the Borgia collection of antiquities Goethe found himself, at the dawning of his second day out from Rome, trundling down the raised causeway of the ancient Via Appia across the bleak expanse of the Pontine Marshes. To the right, on the seaward side, Pius VI had had some success with his scheme for draining this vast area, but on the left, towards the hills, the land was lower and little progress had been made. The Thuringian minister for roads and waterways could look at this scene with eyes trained by George Batty and forty years later it would still be clear in his mind as he came to write the last scenes of *Faust: Part Two*. For Tischbein, the painter, however the long stretches of leafless willows and poplars on a showery winter's day offered few promising motifs, beyond the straw hovel that passed for a post-station, where their carriage changed horses. The malarial marshes were no place to linger and by the afternoon the travellers had passed their southern boundary by rounding the cliffs at Terracina, where the sea broke on the rocks beneath the roadway, and they had entered a new terrain and almost a new season. Round Lake Fondi olives and myrtles, palm trees and prickly pear grew in the sun trapped by the south-facing hills. So abundant were the pomegranates and oranges, whose fruit still hung like lanterns amid the evergreen branches, that Goethe, forgetting Vicenza, recognized here the land for which Mignon had yearned. As the journey wore on, the hills and promontories, the lakes and bays and coastal plains combined and recombined in a kaleidoscope of enticing views, and with Tischbein's help Goethe began to try to catch on paper what would ever afterwards seem to him, with the addition only of a well-sited temple, the constituent elements of the perfect classical landscape. For the present, ancient remains were visible only as fragments in garden walls. After a night in a poor inn at Fondi they drove up into the hills again to come down to Gaeta, to the sea, and to their first view of the island of Ischia, straight ahead of them on the other side of the bay, and of Vesuvius, off to the left behind other peaks, and surmounted by a plume of smoke. On the beach Goethe found pebbles of blue and green glass, which he felt could date only from antiquity, mingling now indistinguishably in the shingle with the natural beauty of jasper and porphyry, starfish, urchins, and seaweed. Another night had still to be spent in the heights above Capua until, with a bitterly cold north wind blowing down from the snow-crowned Abruzzi, but bringing blue skies after some grey days, they reached Naples on 25 February, the First

Sunday in Lent, and took up their lodgings in a large inn overlooking the great square beside the Castel Nuovo.

'We eat crabs and eels and they do us no harm—neither will these delicate little creatures, and they may be nutritious', said Goethe in reply to Tischbein's warnings, drinking off a cloudy and densely-populated glass of water, and ordering another, drawn from the bottom of the cistern, so that he could study the fauna more closely. His first two days in Naples, however, he then had to spend in his unheated room, wrapped up and warming his hands at a chafing-dish, while he 'awaited the passing of a slight physical indisposition'. But once he was free to go out, the city and its surroundings had on him the intoxicating effect he had hoped for from Rome and not perhaps received. 'If Rome encourages study, here people are interested only in living.' 'It is disagreeable to have to remember Rome here; compared with this open location the world capital down in the Tiber valley seems like an ancient, ill-situated monastery.' He found himself thinking again of his father, as he had not done in Rome, and of the earlier Italian journey that had prefigured his own: 'My father . . . could never be wholly unhappy, because he could always think himself back to Naples'—that too perhaps, though Goethe did not yet know it, a prefiguration. The month now spent in the capital of the Kingdom of the Two Sicilies brought him one powerful new experience after another. Although in 1817 he destroyed the letters and diaries relating to this period after completing the relevant sections of his autobiography, the breathlessness of the original documents can be heard in his edited account of his first excursion, on Thursday 1 March, with a party assembled by an Austrian general, the Prince of Waldeck, to the volcanic hills and bays of the Phlegrean Fields and the famous Solfatara:

By water to Pozzuoli, gentle drives, happy walks through the most extraordinary country in the world. Beneath the purest sky the unsafest ground. Ruins of unimaginable prosperity, mangled and unappealing. Boiling waters, chasms exhaling sulphur, mounds of clinker hostile to plant life, bare repulsive spaces and yet in the end a vegetation always luxuriant, intruding wherever it can, rising above all the death, around lakes and streams, even firmly establishing the most magnificent oak forest on the walls of an ancient crater.

The caution with which, as a draughtsman, he had dutifully followed Tischbein's instruction is thrown to the winds and for a few days after the visit to the Solfatara he sketches in a furious, almost expressionist, manner as if, after an interval of eight years, he has rediscovered in drawing a vehicle comparable to poetry for the personal element in his experience. But it is not long before the bravado is followed by 'indisposition' and Goethe is writing, in tones reminiscent of Werther: 'When I want to put down words it is always pictures that come before my eyes, of the fertile land, the open sea, the shimmering islands, the smoking mountain, and I lack the organs to represent it all.'

First and foremost it was, as he had intended, the natural landscape that marked Goethe's Neapolitan Lent, especially its exotic components: the sea, the vegetation, and, above all, Vesuvius. The endlessly varied coastline between Pozzuoli and Salerno, the superb panorama of the Bay of Naples itself, the breakers running ashore after a few days of stiff wind, whose form he tried to fix in his sketch-book, all these had no precedent in his hitherto land-locked existence. Neither had the sight of the sails of the packet boat for Sicily dwindling away between Sorrento and Capri, nor the yearning that sight aroused. Naples could also show him species of plants he had not before seen growing in the wild, if at all: he had for some while been watching and drawing the germination of household beans—now he could study the leaf-forms of agaves and the branching process of the prickly pear. In this 'plenitude of an alien vegetation' he began to suspect he might be able to make more progress with botanical than he had with geological form, and one evening walking by the sea he had an 'inspiration in botanical matters' which made him think the Ariadne's thread within his grasp, the basic principle in accordance with which all plants grow and reproduce. Already the thought seems to have come to him that he too was now in Arcadia, (so adapting a motto that was originally coined to apply to Death) and he began to plan with Tischbein a set of 'Idylls' in verse and painting, a joint project of which nothing however was to come for another thirty years.

Yet in the garden of Paradise, to which, despite the earliness of the season, Goethe repeatedly compares the city and its environs, lay also the mouth of hell, and Goethe's geological interests were still sufficiently alive for him to want to inspect the fire and the brimstone at close quarters. Only the Prince of Waldeck's invitation to Pozzuoli had prevented him from climbing Vesuvius on the 1 March, and he made good the omission on the second. The ashen cone was enveloped in low cloud, however, and though quiescent was seeping large volumes of acrid smoke. After fifty paces up through the fumes, with his handkerchief to his face, Goethe had to turn back from his attempt to see the new crater, though he had now seen both old and recent lava and had reconnoitred the ground. He was particularly pleased at discovering stalactite-like formations in some of the smoking vents—distilled he thought from the volcanic gases and an indication perhaps of how other rocks might be formed 'without moisture and without melting'. A second ascent, four days later, in the company of Tischbein and two guides, was more successful: the weather was better and the mountain was now erupting showers of stones, clinker, and ash at regular intervals. During one of the pauses Goethe and his guide, their hats padded out with scarves and handkerchieves, mounted rapidly from the saddle below the Monte Somma to the rim of the crater and glimpsed the shattered interior of the smoke-filled basin, before being caught in the next eruption and being glad to get back to shelter with their lives and a liberal coating of ash. Finally, on 19 March, Goethe was tempted to make a third visit by the news that lava had begun to flow out of the side of the cone. With the same two guides he watched the fiery effluent, its glow dim in the brilliant sunshine, running out of a wall of smoke and down a channel perhaps ten feet wide that it had created for itself as it cooled, while the vents nearby whistled and bubbled like a kettle. He even made an attempt to come down from above on to the opening from which the lava was emerging, but the ground proved too spongy and hot, and the sulphurous fumes were intolerable. A bottle of wine refreshed him from his adventure, however, and as he drove back he was rewarded with a magnificent sunset over the bay.

Naples also offered Goethe a unique experience of the works of man—most notably of course of classical antiquities. Systematic excavation of Herculaneum had begun in 1738 and by 1750 so much had been uncovered, both there and at Pompeii, that the King, who maintained a jealous monopoly over both sites, decided to build a new museum to house it all in his palace at Portici. Security was so strict—sketching, and even note-taking, was prohibited—that knowledge of the collection, and even of the exceptionally rare and important specimens of Roman wall-painting, filtered out only slowly into the world of European learning. Indeed in 1787 King Ferdinand—whose main interest was hunting, but whose knowledge of art was adequate for possessiveness—was consolidating his holdings by transferring to Naples from Rome, to the outrage of the Roman artistic community, the Farnese treasures which he had acquired by inheritance. His agents in the transfer were his Keeper of Antiquities, the director of the Neapolitan porcelain manufactory, Domenico Venuti (1745-after 1799), and his Court Painter, Reiffenstein's former protégé in Rome, Philipp Hackert. Goethe and Tischbein visited Pompeii on 11 March in the company of Venuti and his wife, and Hackert's brother Georg, an engraver; Herculaneum and the museum at Portici were inspected a week later. Goethe paid particular attention to the Pompeian wall-paintings but, despite forewarning, he found himself surprised by the smallness of the rooms in the houses, and he was glad to shake off 'the strange, half disagreeable impression of this mummified town' in an *osteria* looking out over the sparkling sea at Torre Annunziata, where a frugal lunch was enlivened by the famous local Lacryma Christi. Characteristically he allowed his eye to be caught by the architectural similarity between the modern houses of the area and those preserved by the death-dealing volcanic ash, and a quick glance into the interiors confirmed his instinctive feeling that despite the vicissitudes of history the life of the ancient world continued, in its essentials, on the terrain, and under the skies, that had given it birth. He similarly turned from death and darkness to light and life after the visit to Herculaneum, which was if anything more certainly disagreeable, since the remains had to be inspected by torchlight sixty feet underground: his enthusiasm was aroused rather by the visit to the museum which, far better than the site itself, permitted the imagination to travel back into the distant age when not only had 'all these objects stood around their owners for living use and enjoyment' but the perfection of art they embodied had transfigured life and delighted the senses.

Goethe's interest in contemporary Italian life, suspended during his stay in Rome, revived in Naples. Naples in 1787 was after all, with half a million inhabitants, the third largest city in Europe, surpassed only by Paris and London. It did not have the sense of metropolitan identity of those national and nascently imperial capitals, or the intellectual self-assurance represented by such wholly urban figures as Diderot or Dr Johnson, but for a month or so it gave Goethe the experience, never to be repeated, of life in a big city. Through the endlessly crowded streets—'how everyone swirls along in confusion and yet each individual finds his way and his goal!'—he picked his way either on foot or, if he was making for the country, in the light, one-horse, two-wheeled and two-seater red and gold *calèches* that stood for hire in all the large squares, and as he went he observed with a perceptive and sympathetic curiosity. Any visitor would have noticed, and patronized, the porters waiting to be hired, the hucksters with lemons and iced water to make lemonade on the spot, the deep frying-pans at the street corners from which passers-by could take away fish and batter on a piece of paper. Not everyone though would have paid as much attention to the children, gleaning sawdust for sale as fuel, selling sweets or pumpkin to one another, or huddling on the paving stones still warm from the fire in which a blacksmith had worked an iron hoop, nor have watched closely enough to determine that the many men standing around were all actually engaged in some trade and that the Neapolitan loafer was a fiction of the guide-books, nor have realized the connection between the clean streets, the numerous pannier-loaded donkeys, and the cauliflowers, currants, nuts, figs, and oranges on so many little stalls: an essential part of the establishment of any of the innumerable market-gardeners was the donkey which carried out to his patch every scrap of dung or kitchen-waste that he could find, 'to accelerate the cycle of vegetation'. Goethe had enough experience of state affairs to recognize that, thanks to the climate, the plentiful fish-supply, and the fertile hinterland, the Neapolitans had no need to save and store for the morrow, let alone for the winter, that, as it would be put today, theirs was a subsistence economy. He saw that, despite great industriousness, there was no industry in the 'northern' sense, that there were no factories, technology was backward, and apart from lawyers and doctors there was little in the way of an educated middle class. For the few as for the many the watchword was enjoyable consumption: 'they want to be merry even when at work'.

Goethe however could not rid himself of his 'German disposition and the desire to learn and do more than simply enjoy' himself: 'it is a strange sensation for me to keep the company only of people who live for pleasure'. In the restricted circle that formed the great city's high society he sometimes felt as isolated as he did on the crowded streets, or as out of place as he had done playing cards at the court of Brunswick. Company of a familiar kind he found only in the earnest young jurist Gaetano Filangieri (1753–88), who reminded him of his brother-in-law Schlosser, and whom he 'never heard speak a trivial word'—not even at a grand dinner, at which, since it was Lent, all the meat dishes turned out to be dressed-up fish. Filangieri introduced him to the works of Vico (1688–1744), though Goethe seems to have thought them simply a worthy local product with nothing in them that would now be new to the wider world. Marchese Venuti, by contrast, for all his court responsibilities, took life less seriously: after showing Goethe Herculaneum and the Portici musuem, and no doubt under the influence of the Lacryma Christi, he (and his wife) took a full part in a battle royal on the beach at Torre Annunziata which began with sand-throwing and ended with a general ducking in the waves. Goethe meanwhile had sidled away to inspect the composition of the stone breakwaters. When Tischbein demonstrated at Venuti's house the art of portrait-sketching with brush and ink, the party clamoured to try their own hand and it was not long before they were painting beards and moustaches on each other's faces: Goethe could only shake his head and write 'they do not know what to make of me'. A more pretentious but essentially similar frivolity could be savoured at the magnificently situated residence, looking across the bay to Capri, of Naples' greatest Epicurean, the British plenipotentiary and foster-brother of George III, Sir William Hamilton (1730–1803). In Miss 'Emma Hart' (1761–1815), whose real name was Amy Lyon, the man whose studies of Vesuvius and Etna had brought him a Fellowship of the Royal Society, and whose collection of Greek vases had been bought by a special Parliamentary grant for the British Museum, had, Goethe wrote, 'found the summit of all the joys of nature and of art'. She had arrived in Naples in 1786, would become Hamilton's wife in 1791, and was already his mistress, as she would later be Lord Nelson's. Her speciality was precisely a fusion of 'nature' and 'art', even though these were the two realms that Goethe now believed a courtly culture had to keep apart: in a darkened room, beneath a light held by Hamilton, and in a specially made 'Grecian' gown, she would strike 'attitudes', as she called them, reminiscent of Greek statuary and expressive of different circumstances or emotions. Hamilton had also had a large golden frame constructed round an open booth lined with black velvet, in which she could represent famous paintings as a *tableau vivant*. Tischbein, whom Hamilton later commissioned to draw his second collection of Greek vases, painted a picture for the Prince of Waldeck of 'Miss Hart' as Iphigenia, in the moment of recognizing Orestes, who bears Goethe's features. Despite the ironical tone in which he writes of the old cavalier and his young beauty, Goethe clearly also enjoyed this particular society game—perhaps he envied the wealthy Hamilton, who had no need of courts, and who had 'constructed a beautiful existence for himself . . . and after traversing all the realms of creation has attained the great Artist's masterpiece, a beautiful woman'.

It was not just stuffiness or pomposity: Goethe was not quite at ease in Naples. If Rome was for those who 'study', Naples was for those who 'only live', and it had never been possible for Goethe 'only' to live, and to 'forget oneself and the world'. A strange sense of 'only'? Let alone of 'live'? Neither Johnson nor Diderot would have thought of study and life as mutually exclusive, or even opposed. Yet this opposition, not just of 'life' and 'study', but of 'life' and 'consciousness of self', was central to the culture which was becoming yearly more dominant in Germany, and with which Goethe, if he was to retain his German identity, had to find his *modus vivendi*, even when he was a thousand miles distant from Weimar. Whether it was the climate, the landscape, the great city, or the religion, there was something about Naples—which after all might have seemed in many ways the compensation for his disappointment with Rome —that was not compatible with his task in Germany:

Naples is a Paradise; everyone lives in a kind of intoxicated obliviousness of self. It is the same for me; I scarcely recognize myself; I seem to be a quite different person. Yesterday I thought: 'Either you used to be mad or you are so now'.

Naples was taking Goethe out of himself. The diary that he kept, and that he sent off, this time at frequent intervals, to Frau von Stein, was destroyed in 1817. But it was for general consumption—Fritz also saw it, and so did Goethe's mother—and as he wrote it he felt that it contained 'only a few things in detail—I cannot and do not wish to say anything about the whole, about my inmost self'. Rather than finding himself in the 'great mass of knowledge and new ideas' that his journey was giving him, he was finding it necessary to stand aside from his experience in order to retain some hold on the identity he would need when he was back in Germany. With Sicily, and some weeks in Naples, still before him, Goethe, writing to C. G. Voigt, his colleague in the Mining Commission, spoke of it all as already simply food for the memory:

Beautiful and splendid though this world is, one has nothing to do in or with it. To be sure there cannot easily be a more beautiful situation than that of Naples and the memory of such a sight gives savour to a whole life . . . I say nothing of the rest . . . It will eventually furnish us with a good conversation in the Thuringian forest, on walks, on some cosy evening.

Goethe's long reconciliation with the court world, to which he would 'eventually' have to return, continued even in Naples. When the local nobility drove out in their finery every Friday, 'my heart warmed to them, for the first time in my life'. Unlike the Roman theocracy, the Naples court had a familiar, secular, absolutist structure. And in Philipp Hackert it had a German artist who could demonstrate the art of successful adaptation to it. From 14 to 16 March Goethe stayed with Hackert, who had a comfortable suite of rooms in the royal palace out in the country at Caserta, and who there instructed the Princesses in painting and gave evening talks to the royal circle on aspects of the fine arts. Hackert used as his text-book Sulzer's *General Theory of the Fine Arts and Sciences*, whose Gottschedian assumption that art could be systematically taught and learned Goethe had fifteen years before so fiercely attacked as alien to the ambitions of the young generation. 'What a difference there is between a man who wants to build himself up from within [that is, such as Goethe] and one who wants to have an effect on society and give it instruction for domestic, practical purposes!' Yet now Goethe felt bound to approve: it seemed that after all Sulzer would do quite well for Hackert's domestic purposes, and that these were not to be spurned—for were they not the condition on which Hackert was able to practise the art that mattered to him? Was it not an acceptable compromise to teach Sulzer to the world—which 'only' had to live, and needed nothing further—if this left you able to build yourself up from within? 'Much knowledge was here imparted, the way of thinking was one in which so stout a fellow as Sulzer had found contentment, and should that not be adequate for people of society?'

Goethe's poetry was perhaps a robust enough growth to accept such a reduction to an esoteric process of self-cultivation, and still be able repeatedly to find new ways of breaking out into the public awareness. But his drawing was a frailer gift. He had drawn almost no original landscapes between 1779 and 1786. The stimulus of escape had set him sketching once more as he left Carlsbad, he had taken up sepia, and in Rome he had cautiously begun to try himself out in the new medium of water-colour. But only in his first few days in Naples is there any sign of a resurgence of the personal style of his best years, 1774 to 1777, and soon afterwards began the inner withdrawal from his Neapolitan experiences, the reluctance to find himself in the wealth of life around him. Hackert's blunt judgement—'You have got talent, but you are incompetent'—and his advice to study for eighteen months, if he wanted to 'produce something that will give pleasure to you and others', not only finally undermined Goethe's confidence but completed the reduction of this modest branch of his art to a mere accomplishment, a means of adaptation and integration into court society. Tischbein had been a more understanding mentor, encouraging his amateur pupil to follow his bent, but Tischbein, unlike Goethe, was not on holiday, and was now preoccupied with establishing himself in Naples, where from 1789 to 1799 he was to be Director of the Academy of Arts. He introduced Goethe to another local German landscape artist, C. H. Kniep (1748–1825), who had more time than he to act as Goethe's 'permanent companion'. Kniep, a year older than Goethe, three years older than Tischbein, always sat down to eat in a dinner-coat, the buttons of which were separately rewrapped in paper after use, and when beginning a sketch liked to mark out a frame, to draw his horizons with a ruler, and to sharpen his pencils frequently —just the man, in fact, to carry out Hackert's prescription. Goethe, however, was looking for more than a drawing master: no longer able to find in his own drawings an adequate embodiment of his experience of the landscape, he needed, in that pre-photographic era, a hired draughtsman to keep a visual record of his travels. Kniep did some satisfactory landscapes for him in the neighbourhood of Naples, the two got on well together, and Goethe quietly passed a personal milestone: he abandoned in his drawing any serious attempt at that fusion of the autobiographical and the representational which was the driving force in his poetry, and left to Kniep the task of being camera.

By 23 March, at the latest, after three weeks of unusual vacillation, Goethe had decided to extend his journey and be in Sicily at Easter, rather than back in Rome. He had been toying with the idea since December and perhaps he was inexplicitly postponing the moment when he would have to turn his face northward again and prepare himself, however gradually, for life in Germany once more. He was also rather curious to know what a sea-voyage was like. Perhaps too he hoped, by a tour on which he could receive no mail, to regain something of the freedom and even something of the inspiration, of his wandering days in the previous autumn. In his month in Naples his drawing had not developed as expected and he had made no progress with *Tasso* beyond resolving, despite the doubts about the new *Iphigenia* expressed by Frau von Stein (and later by Seidel), that its poetic prose should also give way to the new blank verse. Moreover he had not given up hope that his entire expedition might yet be given a satisfactory symbolic shape, perhaps by turning it from the flight to Rome, which it had originally been, into an Italian Journey. Rome had proved an unsatisfactory culmination, and despite the novelties of the last month he could not honestly share the local sentiment, 'See Naples and die!': possibly Sicily, 'which points me towards Asia and Africa', might still provide a convincing terminus, a symbolic node, like the St Gotthard in 1779. 'It is no small thing to stand on that strange and remarkable point towards which so many radii of world history are directed.' A circuit of the other half of the Neapolitan kingdom might turn what had so far been an erratic progress from one mild disappointment to another into a purposeful voyage of discovery. 'On this journey I am certainly learning how to journey—whether I am learning how to live I do not know. The people who do seem to know how to do that are too different from me in character and in manners for me to be able to lay claim to that talent.' It was agreed that Kniep should accompany him free and sketch as they went, and that Goethe would supply him with materials and become the owner of a certain number of the finished drawings. The next days were hectically busy and a violent thunderstorm prevented Goethe from fitting in, as he had hoped, a visit to the ancient temples at Paestum, south of Salerno, which therefore had to be postponed until after his return—assuming that the unsettled weather permitted the departure at all. At midday on Thursday 29 March, however, after much waiting in a harbourside coffee-house hoping that the adverse south-westerly wind would drop, the two travellers boarded the Sicilian packet, a neat little American-built corvette, lying at anchor before the Molo, and eventually, with the sunset, it moved slowly out to sea.

The Gardens of Alcinous

As a result of his destruction of the documents there is as little contemporary evidence for the details of Goethe's Sicilian tour as for his stay in Naples.

Furthermore, as he began, in 1816, to compose his own account of events, Goethe himself seems to have been faced with a shortage of material ('if possible, construct a diary' he adjured himself at the time in his working notes). The narrative in the *Italian Journey*, therefore, although it is almost our only source, must be treated with some caution. It is, for example, beyond doubt that bad weather extended the crossing from Naples to four and a half days and that for much of that time Goethe was seasick and lay on his back in the little cabin he shared with Kniep. It is rather less certain whether during this time he meditated as coherently and effectively as he claims to have done on the structure of *Torquato Tasso*, since Kniep, who himself remained stolidly unaffected by the sea, is independently reported to have said that Goethe was quite delirious and took the footsteps of the sailors on the deck above him for those of his grandmother. Goethe's recent reflection on Hackert's position at the Neapolitan court may have brought the subject-matter of his play closer to him, and he certainly set out with the intention of taking *Tasso* further while he was in Sicily, but nothing was committed to paper for another year. A different project had already claimed his attention by the time that he could stand on deck on the morning of 2 April, the Monday in Holy Week, watching the slow approach of the Sicilian coastline. The grey mass of Monte Pellegrino lay to the right, and a great curve of bays and promontories to the left, all under a brilliant sun, whose light spilled towards him from the south round white houses and through the first springtime green of the trees, while behind him rolled, unilluminated and unsparkling, the northerly waters of a wine-dark sea. As they docked in Palermo at three in the afternoon he felt like Odysseus cast up on the island of the Phaeacians. 'I wish you could see the flowers and the trees', he wrote a little later to Fritz von Stein, 'and could have joined in our surprise when after a troublesome crossing we found on the seashore the gardens of Alcinous.' The story of Nausicaa, the daughter of King Alcinous, who meets the shipwrecked Ulysses on the strand, had returned to his mind: as himself a wanderer, even a seafarer, in what seemed a Homeric world, Goethe felt drawn again to a theme in which there was so much 'that I could have drawn from nature, in accordance with my own experiences', phrases that could apply as much to his escape from Weimar, and its Calypso, as to the natural luxuriance now surrounding him.

I have no words to express how this queen of islands has received us: with freshly green mulberry trees, evergreen oleander and hedges of lemons etc. In a public garden lie broad beds of ranunculus and anemones. The air is mild, warm and scented, the wind balmy . . . and meanwhile I am preparing for all my loved ones another monument to these happy hours of mine.

The monument was to remain unfinished, but it preoccupied him throughout his stay in Palermo and was the constant background to his thoughts for the rest of his tour.

Goethe and Kniep stayed in Palermo, a town as large as Rome, for just over a fortnight, lodging in a large inn near the harbour. 'I doubt whether I have in my life been for sixteen days in succession so happy and contented as here.' They walked out around the bay, admiring the town's situation, though only occasionally venturing into the steep labyrinth of filthy streets that made up its centre; they made excursions to the local points of interest recommended by the guidebooks—a geological outing on to the Monte Pellegrino; a day at the bizarre palace of the Prince of Palagonia, decorated with statues of monsters, many standing awry or on their faces, and furnished with chairs with legs of unequal length or spikes concealed in the upholstery (Goethe was not amused); visits to the collections of antiquities of the Prince of Torremuzza and the Benedictine monks of Monreale—and time and again Goethe returned to the public garden (probably that of the Villa Giulia) that had first charmed him with its pergolas of lemon-trees and its wealth of exotic and to him unknown plants not yet in leaf, all within sight and scent of the hurrying waves of the blue-black sea. On Good Friday he travelled up a miraculously engineered road to visit the shrine of Saint Rosalia, a cave high among the cliffs of the Monte Pellegrino, surrounded by an enclosed courtyard that served as a church, but otherwise left in its natural condition, except for a remarkably life-like and beautiful statue of the reclining saint. The hostility to Christianity that had become increasingly marked when he was on the mainland seems to have retreated into the background during Goethe's stay in Sicily. He was entranced by the silence and simplicity of this spot and remained alone there with the statue long after the chants of vespers had died away.

Easter Day in Palermo began with a battery of fireworks exploding at dawn. Goethe then found himself invited to dine with the Viceroy, the only social engagement of his stay, but one which brought him a surprising meeting with a Count Statella, a Knight of Malta. The Count had studied in Erfurt, was anxious for news of the Thuringian families, and even enquired after the man in Weimar 'who in my time was young and lively and made rain and sunshine there—I forget the name, anyway, he was the author of *Werther*'. Two months later Statella was back in Weimar, bearing greetings from Goethe to Frau von Stein.

It may have been in the course of his further conversation with Statella that Goethe began to entertain the idea of a trip to Malta, easily reached by the coastal packets on their circuit from Naples, and included in his German guidebook. His interest might already have been attracted to the island, however, through his curiosity about a figure who, most notoriously in Europe at this time, claimed association with the chivalric Order that still ruled it. Count Cagliostro, as he called himself, had been imprisoned in the Bastille after playing a major, though not certainly criminal, role in the Diamond Necklace Affair of 1785, which had profoundly shaken Goethe's confidence in the stability of the existing social order. His claims to high birth and occult knowledge, and his connections with the most irrational wing of the Masonic movements, still surrounded him with an air of mystery in the world at large, but in Palermo there was a widespread belief that he was the ne'er-do-well son of a poverty-stricken local family. Anxious to locate the real basis of a structure of deceit that had cast a shadow over the French throne, Goethe made contact with Antonio Vivona, a Palerman lawyer reputed to know more of the matter. Vivona had provided the French authorities with documentary evidence of the genealogy of one Giuseppe Balsamo (1743–95), who on leaving Palermo had taken the married name of his, Balsamo's, godmother and great-aunt, Vincenza Cagliostro, and Vivona agreed to arrange for Goethe to meet Balsamo's family. Goethe was to masquerade as an Englishman, Mr Wilton, bringing news of Balsamo-Cagliostro's release from the Bastille and escape to England. In the company of one of Vivona's clerks, who was also able to act as an interpreter of the Sicilian dialect, Goethe, probably on the afternoon of 16 April, found his way into a little alley off Palermo's main street, up a rickety flight of stairs, through a kitchen and into a one-windowed, one-room dwelling, where Balsamo's mother and his widowed sister—still bringing up two of her three children—were sitting with an invalid woman whom, despite their poverty, they were looking after out of charity. The furniture was dilapidated, the saints' pictures on the wall were blackened with age, and the family's clothes were in tatters, but the room was clean. Delighted to hear good news of Giuseppe, the women were not backward with their reproaches to him for neglecting them in his good fortune and for not repaying the fourteen ducats he had borrowed from them on leaving. Goethe promised that if they wrote a letter to Balsamo he would return the following day to collect it, which he did. As he left, the children implored him to return for the feast of Saint Rosalia in July, the high point of their year and a festival whose splendour, they thought, he could not imagine.

Goethe knew that he had deceived some worthy people and he was touched by their need. He soon determined that he did not have enough money with him to repay Balsamo's debt himself, but on his return to Germany he recounted the episode in Weimar and Gotha, showed round the letter, which of course he had been unable to deliver, and so collected the sum of 100 dollars, which was transferred anonymously to the family in November 1788—in time for Christmas, as they wrote in their letter of thanks to Balsamo, whom they imagined to be the benefactor. But Goethe must have taken away from the Palermo tenement more than a debt of decency. He had seen in all their ordinariness and simplicity the circumstances from which—there could now be no doubt—'Count Cagliostro' had sprung, and there was nothing in them whatever to account for the entire fabric of mystery woven around that figure. Like the absurd decorations of the palace of the Prince of Palagonia, Cagliostro's claims and antics were 'a nothing that wants to be thought a something', and somehow, inexplicably, they had succeded in their aim. There was an eerie, even alarming, element in the power that such nothingness, such pure deceit, could acquire in human society: perhaps there were forces at work of which Goethe had hitherto taken no account. At any rate there was a theme here that would not let him go, and at some point in the next four weeks he began to think that the best way to treat this combination of farce, trickery, and ultimately empty menace was in a comic opera, initially to be called *The Mystified* (*Die Mystifizierten*), though eventually it became the prose drama *The Grand Kophta* (*Der Groß-Cophta*).

But with the days growing warmer and the gardens of Alcinous filling with blooms round about him, Goethe's attention was for the present fixed on Nausicaa, or rather on *Ulysses among the Phaeacians*, as he was still planning to call the play. In its first drafts the princess figures under the name, both ethically more significant and metrically more convenient, of her mother, Arete (= virtue). It is possible, though rather improbable, that Goethe had simply forgotten the name Homer gives her. In any event, it was not so much Nausicaa's fate that was dominating Goethe's thoughts as the landscape in which Homer had set her, and the figure in that landscape of Ulysses the wanderer. He drew up a schema for a five-act tragedy, but drafted only two vivid scenes that had caught his imagination from the start of book 6 of the *Odyssey* (of which he bought a Greek text with Latin translation on 15 April). The two episodes—the princess's maidens playing ball on the seashore, and the shipwrecked Ulysses rising from his bed of leaves under an olive grove— belong together, and with the delicacy of a silver-pencil drawing, they depict the dawning of consciousness after a terrible disaster, the awakening of a mind to natural surroundings that appear welcoming but in which some strange power is concealed—possibly, though the good fortune would be almost too much to hope for, a power that promises human companionship. After the maidens have flitted across the stage, like images in a dawn dream hinting at love, Ulysses comes out of a cave, as Faust comes out into the forest, and speaks a fine adaptation of lines from the *Odyssey*:

> Was rufen mich für Stimmen aus dem Schlaf?
> Wie ein Geschrei ein laut Gespräch der Frauen
> Erklang mir durch die Dämmrung des Erwachens?
> Hier seh ich niemand! Scherzen durchs Gebüsch
> Die Nymphen? oder ahmt der frische Wind
> Durchs hohe Rohr des Flusses sich bewegend
> Zu meiner Qual die Menschenstimme nach.

What are the voices calling me from my sleep? As it were a babble, a loud converse of women, sounded through the twilight of my awakening. Here I see no one. Are nymphs playing through the bushes? or is the fresh wind, moving through the tall reeds of the river, imitating to my torment the human voice?

One might think of Caliban's speech—'Be not afeard—the isle is full of voices'—but the sense of mystery in these words is created not by the presence of superhuman powers in a sordidly human world, but by the ghostly presence, or possible presence, of the human in the natural world. Their strangeness is related to that of 'Elf King', or of the elegiac lines Goethe addresed to the nymphs peopling the solitude of the Weimar park. There is here none of the confidence of possession in Faust's hymn of gratitude to the Earth Spirit—from that Goethe was now separated by the agonizing turn in his relations with Frau von Stein, by his frustrations in Rome and his unease in Naples. There is, rather , the uncertainty of a wounded man who cannot quite believe that he has been transported to a land of Faery where all his sufferings may be made good. The rest of Ulysses' speech is devoted to what he has just lost—above all to the companions he has just lost—in the jaws of the sea. Goethe thus con-flates the shipwreck which brings Ulysses to the Phaeacians (*Odyssey* v) with that which brought him to Calypso (*Odyssey* XII), and indeed the whole monologue would come more appositely from a castaway on Calypso's enchanted island of Ogygia, rather than on the blessed coast of the kingdom of Alcinous. Maybe the drama Goethe really wanted to write was a drama about Calypso, and not a drama about Nausicaa at all.

Critics have rightly observed that Homer's story of Nausicaa, the story of the princess who brings the traveller to her father, hoping secretly that she may have found her bridegroom, only to see him travel on, is readily equated with the story of Friederike Brion, of a woman used and abandoned by a man devoted to what he believes to be a higher purpose, a theme to which Goethe returned obsessively in the works he conceived in the period up to about 1776: *Götz, Clavigo, Urfaust, Egmont,* even (in the Mariana story) *Wilhelm Meister.* But the woman to whom Goethe came after a shipwreck in which he lost almost all his comrades, who embodied virtue for him in a land which was at once a haven and a desert island where he could converse only with nature-spirits, and to whom he could have said, in the last words which Homer's hero addresses to Nausicaa, 'through all my days your name on my lips will be like a god's because you gave me my life'—that woman was not Friederike but Frau von Stein. Goethe's relationship with Frau von Stein could not be assimilated to his relationship with Friederike, any more than the figure of Calypso could be assimilated to that of Nausicaa, and so Goethe left the drama unfinished. But his strange regression to the theme of his guilty detachment from a woman, which had last gripped him over ten years before, after the breaking of his engagement to Lili, and the subconscious association of his material with a flight from an enchantress, both suggest a hidden awareness of an impending crisis in relations with Frau von Stein, which he was not as yet able to face, or express, directly. On 18 April, possibly only two days after first drafting Ulysses' monologue, he wrote, for example, in the only letter from Sicily to Frau von Stein that he did not burn, that 'absence and the great distance separating us have so to speak purged away everything that latterly had stagnated between us' —had he ever recognized the true import of these lines he would certainly have burned them also. Again, in the same letter he told her of the tears of joy he had wept at the thought of the pleasure he could give her with the play that was progressing so well—even though in that play the outline of a parting was just beginning to become discernible to which he would give adequate artistic form only two years later, in *Tasso*. And, lastly, like Ulysses he found himself, as he wrote, in an 'inexpressibly beautiful' land in which the natural world, to which in his years with Frau von Stein he had increasingly devoted himself, seemed to have reached the classic perfection defined by the gardens of Homer's Alcinous. Yet Goethe had devoted himself to nature because, as he had told the nymphs in the Weimar park, it offered him a support and consolation that the world of men and women did not: the perfection of nature was at the same time the perfection of solitude, and so as Ulysses stepped out into paradise he was troubled by the ghostly voices of a human companionship that he had lost.

'How much joy I have every day from my little bit of knowledge of natural things', Goethe wrote in his letter of 18 April, 'and how much more I would have to know if my joy were to be complete.' The day before, if his later account is to be trusted, he had been in his beloved garden, revolving his play in his mind, and had been tempted to test the perfection of the natural world about him and to try for a decisive extension of his knowledge of it. He began to look in this wealth of vegetation for the primal plant ('Urpflanze'), to try to find exemplified in nature the basic pattern of plant growth that had come to him in that moment of illumination while walking along the sea-front in Naples three weeks before. The difficulty with that basic pattern—in which plant-form was probably reduced to a repeated sequence of such elements as sprout, leaf, and growth-point—had been that it was 'so abstract' that Goethe feared no one would be able to relate it to the complex variety of actual plants. But if Sicily really was the perfection of the natural world, populated by the full range of complex plant-forms, then surely it must also house, in classic simplicity and purity, and as visible as the rhomboidal granites he had seen in the Harz, the basic and unelaborated Plant on which all others were variations, the real case corresponding to his abstract intuition? The search, however, was unsuccessful and Goethe perhaps came to realize that if he was ever to have success he would have to look further afield than the gardens of Alcinous: the perfection of Sicily was personal and poetic, rather than botanical. It was in a sense 'the turning-point of the whole adventure': in the centre of his classical paradise he had discovered that he was still a fallen being, still not intellectually and instinctually at one with Nature.

Despite his disappointment over the 'Urpflanze' (if we assume that he did seriously look for it), the sensation remained with Goethe that in Sicily he was getting to the bottom of things—of the Cagliostro affair, of what the *Odyssey* was about, of botany, of the purpose of his entire journey. He still felt capable of fusing the world and his mind into a symbolical unity. 'My journey is now getting a shape', he told Frau von Stein, 'in Naples it would have been cut off too abruptly'; and to Fritz he wrote that only here could one really get to know Italy, and that he would soon have reached his journey's goal. Five days later he was in Agrigento on the south coast, the furthest point from Weimar he ever attained, and from then on he was, as he promised to Fritz, travelling homewards. But the goal of his journey was not to reach a particular place, nor was it even, as he tended to suggest in later life, and as the remark to Fritz might be thought to imply, to become acquainted, in the Agrigentine ruins, with the ancient Greek architecture of which he had as yet seen nothing. Had that been his purpose Goethe would certainly have followed the normal route for eighteenth-century tourists, from Palermo through the coastal classical sites of Trapani, Marsala, Mazara, and Selinunte, all of which he passed by in order, he told Fritz, to see something of Sicily's mountainous interior. Simply 'to have seen Sicily' he later repeatedly called 'an indestructible treasure for the whole of my life', and that estimate was unaffected by his judgement that the temples of Paestum on the mainland surpassed anything of the kind to be found on the queen of islands. In Sicily Goethe felt that he had got to the bottom of something by touching classical, indeed 'hyper-classical', soil, but he owed this feeling more to his reading of Homer, and to his preoccupation with *Nausicaa*, which was to be 'a dramatic concentration of the *Odyssey*', than to any archaeological observation. Indeed one might doubt whether he was capable of such observation, of the disinterested recognition of what is alien. His perception of Sicily—and even of Homer—was, for his time, distinctly conventional, already perhaps a little old-fashioned: he completely ignored the Norman and Byzantine, and largely overlooked the Moorish, elements in what he saw; he regularly, like Winckelmann, interpreted Hellenistic work as classical; and the same coastline that brought the *Odyssey* to life for him evoked the exclamation, 'Now at last I understand the Claude Lorrains'. Goethe evidently learned a certain amount while he was in Sicily, but he no more acquired any significant factual knowledge from going there than he did from climbing the Brocken in 1777 or crossing the Furka in 1779. His permanent acquisitions were, rather, a repertoire of remembered scenery, and a pattern to his journey that enabled him to claim success for his whole Italian adventure: he had traversed Italy from end to end and had at the last found, in a landscape transfigured by his 'poetic mood', that immediate access to the classical past that had escaped him amid the ruins of Rome and Pompeii.

Goethe, Kniep, and a Sicilian guide set out on muleback from Palermo on 18 April and were soon high up in white rocky hills, the blue sea away to the right and ahead of them a roadway lined and partly covered with bushes and herbage 'wild with blossom'—hawthorn and blazing yellow broom, sheets of red clover, borage, wild onions, and orchids. After two nights in the hill-town of Alcamo they rode out to the lonely remains of the Greek temple of Segesta (fifth century BC) which Goethe rightly recognized had been left unfinished:

The situation is remarkable. On an isolated hill at the uppermost end of a broad long valley the temple looks out into the far distance over much countryside but only a tip of the sea.

The area lies still, in a mournful fertility.

Everything cultivated and almost unpopulated.

On flowering thistles countless butterflies were swarming and wild fennel stood 8 or 9 feet high, it looked like a tree nursery.

There is no trace in the neighbourhood of where a town may have lain.

The wind soughed in the columns as in a wood and birds of prey hovered shrieking over the entablature. They must have had young in the crevices.

Goethe captures, unforgettably, the atmosphere of his visit; he takes extensive notes on the details of the temple's construction; but on its architectural and aesthetic qualities he is silent: 'Of the whole I say nothing . . .'. The silence, now and on the rest of his tour, is striking, given the strength of his interest in Palladio and Vitruvius only a few months before. On the rest of the journey, via Castelvetrano and Sciacca, to Agrigento (then known as Girgenti and numbering some 20,000 inhabitants), his main concern was identifying the minerals washed down from the hills in the river-beds or on the seashore, and delight at seeing his first cork-oak. For the four full days between 23 and 28 April which the party spent in Agrigento, lodging with a pasta-maker, since there were no inns, a local priest was their guide to the ruins of the ancient city. Most of these lay half-hidden among the trees and vineyards on the southerly seaward slopes below the modern hill-top town, though when he looked out of his window in the morning Goethe could see the remains of the temple of Concordia etched against the eastern sky. But after a day looking at such works of pagan art as were preserved in the cathedral, and a day pushing through vegetation to locate this or that felled and crumbling stone giant, he was really marking time while Kniep completed a suitably extensive set of views for the portfolio. His own few drawings give prominence to the landscape rather than the remains. The fact was that he disliked decay, and more important to him than any differences between the architecture of Greek Agrigento and that of Imperial Rome was the ruinous condition of both.

It cannot therefore have been difficult for Goethe to decide not to continue on the coast road to Syracuse, as had been his original plan, for he knew that there, after the destructive earthquake of 1693, only more rubble awaited him, however ancient. Instead he turned, eccentrically but characteristically, to life, to the modern life of Sicily, in which might lie a truth about the life of antiquity. He had not yet seen many signs of the cereal crops that had reputedly made Sicily into the granary of the ancient world. By taking the inland route to Catania he could see this agricultural marvel for himself, and incidentally also pass through the ancient township of Enna where the daughter of the corn-goddess, his own heroine Proserpina, was said to have been taken down to Hades. The decision to omit Syracuse was at the same time a decision about the shape and scope of the whole tour, for it implied abandoning also the project of a visit to Malta, to which, after Syracuse, the packets proceeded. As on the Gotthard in 1779, Goethe found himself in need of the assistance of Terminus, the god of boundaries: before leaving for Sicily he had recognized that the trip might divert him too much from his 'principal purpose', which lay in Rome, and now, on the south coast of Sicily, looking out across the sea towards distant cloudbanks, low on the horizon, that were said to betray the presence of Africa, he felt he had reached his purpose's utmost limit. For a similar reason, no doubt, he had not long hesitated before declining a proposal made in Naples by the Prince of Waldeck that they should visit Greece and Dalmatia together. Goethe's world was the Western, the Holy Roman, Empire: he had now lived at its centre and travelled to its periphery; what lay outside its limits concerned him only in so far as it was reflected within them. And so Agrigento became the turning-point. The guide was re-engaged to take them to Catania and Messina and a boat for Naples, and as the mules plodded out of Agrigento on the sunny morning of Saturday 28 April, Goethe was beginning the long homeward journey back to Weimar.

The eccentricity of Goethe's choice of route soon became evident: under the hot sunshine of midday and afternoon the treeless and sparsely inhabited rolling downland and its unending sea of wheat and barley seemed a 'fertile desert' and Caltanissetta, which they were relieved to reach by evening, proved to have no inn adapted to receive travellers of quality—indeed the inhabitants had still not heard the news of the death of Frederick the Great in the previous year. The following day it began to rain heavily, the country was more broken, and it was necessary to ford the swollen River Salso. After toiling up abominable roads to Enna (then Castro Giovanni) on its lonely height, the drenched little expedition had to put up in a garret with no glass in the windows and no hot food available. For the next two days, until they arrived in Catania on the Tuesday evening, 1 May, they were riding under cloudy skies through long empty valleys given over to rough pasture and thistles, and so, despite an occasional glimpse of the flanks of Mount Etna, there were not even any picturesque views for Kniep to draw as a reward for their privations.

Catania had at least a decent inn to offer, and private collections, both of ancient artefacts and of geological specimens (especially of volcanic origin), and also another helpful clergyman willing to put himself and his carriage at the disposal of the all too infrequent visitors, and to act as cicerone: Goethe and Kniep stayed till Sunday. But Catania's main attraction for Goethe was neither its museums nor its Graeco-Roman theatre and baths, 'so buried and submerged that there is pleasure and instruction in them only for one with the most specialist knowledge', but Mount Etna, and he had long promised himself some instructive comparisons with Vesuvius. Etna, however, is a serious mountain, and on being warned that it was too early in the year and there was too much snow lying for a safe ascent, Goethe settled for the secondary crater of the

Monti Rossi. On Friday 4 May he and the long-suffering Kniep made their way on their mules across the uneroded and lifeless lava fields towards the 6,000-foot double peak, while the white mass of Etna brooded in the background. Kniep stayed behind at the base of the red cone while Goethe mounted to the rim. A stiff east wind had got up, however; Goethe had to spend all his effort on keeping his footing and preventing his hat, his coat, and himself from being blown down into the crater. Stupefied by the wind he was hardly able to appreciate the magnificent coastal panorama 'from Messina to Syracuse' which Kniep, from a less advantageous but more sheltered position, was calmly sketching below.

Riding on the Sunday between the shoulders of the great mountain and the sea, Goethe reached Taormina in the early afternoon. Immediately he and Kniep climbed up out of the town to admire the Graeco-Roman amphitheatre, built into a natural bowl between two summits. Here at last was an ancient site which could be imagined in its heyday without requiring of the visitor 'a powerful talent for restoration':

If one takes one's seat where the uppermost spectators used to sit one has to confess that never can an audience have had such objects before it in the theatre. To the right castles rise up on the higher cliffs, further down lies the city . . . Then one looks along the whole length of the Etna ridge, on the left the coast down to Catania . . . and then the enormous, steaming volcano completes the broad and spacious canvas—not fearsomely however, for the moderating effect of atmosphere makes it appear more distant and more gentle than it is.

The whole of Monday was spent in Taormina to allow Kniep to draw this 'most extraordinary work of art and nature', while Goethe found himself a quiet spot to sit and think in a deserted and overgrown orange grove down by the sea. His Sicilian tour was nearly at an end; the warm and fertile coastline with its ever-changing harmony of sea and sky and strand and green-clad cliffs had returned to displace the bleaker impressions of the interior; away to the south rose the great white mountain, smoking a little; he had just been privileged to enjoy one of the most spectacular vistas of the ancient world; maybe the wind had dropped and it was calm and hot. His thoughts turned once again to *Nausicaa*, the vessel in which he hoped to capture the unique atmosphere of the last six weeks, during which an island spring had turned into early summer; resting in this moment of achievement, before the bustle of departure began, he perhaps felt briefly a stillness and immediacy of pure being that Winckelmann had attributed to the greatest sculptures of antiquity, and on a loose piece of paper noted down two lines from a speech of Ulysses' in praise of the land of the Phaeacians. No longer, however, does the wanderer find his own solitude reflected in the natural world, nor does he even, like Faust, claim it as his possession. For an instant Goethe's poetry attains an objectivity beyond the mere depersonalization that he had learned from Frau von Stein and Spinoza:

Ein weißer Glanz ruht über Land und Meer
Und duftend schwebt der Äther ohne Wolken

A white sheen rests upon land and sea and the hazy* aether floats without a cloud

This is the land of Mignon's desire, present and fulfilled and untroubled by the least motion of desire itself (for once Goethe's use of the verb 'schweben' suggests the pure stillness of 'floating', rather than the poised energy and multiple movement of 'hovering'). It is a landscape at once wholly impersonal, and wholly gratifying, an earthly counterpart to the home of the gods described by Homer in book VI of the *Odyssey.* These exquisitely melodious lines (there is scarcely a vowel in them that is not echoed at least once, and in the middle of the first line the rhythm itself seems to stand still) with their extreme chasteness, even abstractness, of means to a highly sensual end are the equal of anything in Valéry's *Cimetière marin,* which they inescapably recall. Of course any lines of Valéry's of which one might think are part of a great, completed, philosophical poem—Goethe's are a fragment of a fragment, a gem in the rubble. But this solitary unrhymed couplet should really be compared to a haiku, or to the second 'Night Song of the Wanderer' (which has only three more words to it), and the significance of the one moment of fulfilment it expresses, without parallel anywhere else in Goethe's writing, should not be understated. In particular, the momentariness of the insight these words represent is not at issue in them: they are innocently unreflective. Precisely because they are a fragment, and are not finally inserted into a dramatic context where they would become the utterance of a limited and active subjectivity, with its own purposes and desires, they remain uniquely uncontaminated by personality, and an image for us of the objective perfection Goethe would always attribute to ancient art. They may be the last words of *Nausicaa* that Goethe ever wrote—if so they are also the last, the culminating words of that poetry of desire in which the first half of his life's work consists—and it could be said that in making them possible the plan for the drama had served its turn. Whenever and wherever in Sicily they were written, they expressed Goethe's awareness of a perfect moment which he could not ask to tarry without corrupting it.

One more grand and terrible sight awaited Goethe before he left Sicily for ever. On the Tuesday afternoon, after a dramatic ride along the coastal cliffs, with the east wind driving the breaking waves over the roadway, the travellers came to a wasteland as desolate as the lava-fields of Etna: the site of the city of Messina, totally destroyed by the earthquake of four years before, with the loss of many thousands of lives. One night spent in eerie silence in the only building re-erected above the ruins was enough, and for the next two days they sought out the livelier surroundings of the shanty town that had sprung up to the north

---

* The modern sense of 'duftend'—'aromatic'—is not however necessarily absent from Goethe's phrase.

of the city area. Goethe and Kniep now parted from their faithful guide, whose name has not been recorded, and, unsettled by a meeting with the city's mad Irish governor, Don Michael, or Michele, Odea, decided to embark as soon as possible on a French merchantman making for Naples. Conditions again were adverse, the east wind was still blowing, the ship made slow progress, and Goethe soon felt the qualms of seasickness once more. He emerged from the communal cabin as they passed the volcanic island of Stromboli, 'a strange sight —such an ever-burning chimney in the middle of the sea, without any other shore or coastline' (the effect is completely lost in Goethe's own later drawing of the scene, in which some non-existent rocks make up a conventional foreground—in 1775 he would have used part of the ship for that purpose). As the Neapolitan coastline approached, the sight of a great fume over Vesuvius, glowing and flickering red at night, also brought him on to the deck. But otherwise he remained in 'the horizontal position', drawing up an understandably jaundiced balance for the whole Sicilian excursion, for which it is true that, apart from Kniep's sketches, he had remarkably little to show. He had drawn little himself, *Tasso* had made no progress at all, *Nausicaa* was but a few pages of notes, his numerous geological observations were amateurish and inconclusive. And as for antiquity—Agrigento, Catania, and above all Messina were only more evidence of the frailty and transience of what human beings build. He had not needed to leave Rome to learn that. The journey appeared likely to end in complete disaster when, on the third day out from Messina, the ship was becalmed within sight of Capri, was caught by the currents, and began to drift on to the Faraglioni Rocks. The lives and belongings of all on board were saved by a puff of wind, however (the experience was later reflected in two poems of about 1795, 'Sea Calm' and 'Prosperous Voyage' ('Meeresstille' and 'Glückliche Fahrt')), and in the middle of the morning of Monday 14 May, they disembarked, with much relief, and few thanks for the ship owner and his unskilful captain, in the bustling harbour of Naples.

Neapolitan life continued as happily unselfconscious as before—more happily, in fact, now that Lent was over and the butchers' shops were full again, flowers and vegetables were everywhere and there were saints to celebrate like St Philip Neri (1515–95) on 26 May. The founder of the Oratorians, his shrewd wisdom and good humour in the cause of 'conjoining the spiritual, even sacred, with the worldly' were a saintliness Goethe could understand. After some busy letter-writing to catch the post on 15 May he entered, almost with a sense of obligation, on a social round which before Sicily he had avoided, and he enjoyed it, even though he felt he was 'getting lazier and lazier here'. Visiting nobility were no longer shunned, Countess Lanthieri, gratified no doubt by her accurate guess at Goethe's destination in Carlsbad the previous summer, put in an agreeable appearance, there were evenings at the theatre and the opera ('though I am really too old for these frivolities'), and concerts at the Austrian ambassador's—Goethe was impressive with his knowledge of older Italian

music, drawn no doubt partly from Kayser and partly from his stay in Venice—
and there was dinner at Hamilton's, with Miss Hart this time showing off her
musical talents. Hamilton then allowed Goethe and (more remarkably) Hackert
an inspection of his store-rooms piled high with ancient sculptures, vases,
bronzes, and other *objets d'art*, some of which had clearly lost their way between
Pompeii and the royal museum at Portici. Goethe even consented, as a favour to
the studious Filangieri, to call on his wife so that a visiting Englishman could set
eyes on the author of *Werther*, and could pay him the understated compliment
—in the event delivered on the doormat, since he was leaving as Goethe arrived
—: 'the work did not have as violent an effect on me as on others but whenever
I think of what it must have taken to write it I am filled with new astonishment'.
He might have been even more astonished had he learned that the author of
that explosive little book had a few days before been in Pozzuoli inspecting
piddock-holes in the columns of a Roman temple, in the hope of settling a
fierce geological controversy. (The temple had clearly been submerged in water
at some time since it was built, but Goethe could not accept the widely
canvassed theory that the sea-level of the Mediterranean had fluctuated by
thirty feet in the last two thousand years. He thought that a lake must have
formed round the temple after a volcanic eruption, which was right in so far as
he appealed to local volcanic activity rather than a continental catastrophe, but
which ignored the possibility that local earth movements had first lowered the
temple and then raised it again.)

There was a serious side to Goethe's last three weeks in Naples. The return
journey to Weimar had already begun, but some things had still to be clarified.
What, for example, had been the purpose of his expedition, and had it been
attained? Was the classical world that he once thought he had come to see only
an abstraction, its monuments so gnawed by time that it could no more be
found in the real world than the primal plant that he had unsuccessfully sought
in Palermo? He paid another visit to Portici, 'the alpha and omega of all
collections of antiquities', where he could savour 'the joyous aesthetic sense' of
the ancient world, but not to Pompeii or Herculaneum, dominated by the sense
of destruction. One final attempt, however, to rescue from the ruins a tangible
image of the classical past succeeded beyond all expectation. At some point
during the ten days after his return from Sicily, possibly from 21 to 23 May, an
expedition with Kniep took him via Salerno to Paestum, where on a desolate
coastal plain lay the ruins of the ancient Greek settlement of Posidonia. In its
three excellently preserved Doric temples, two of them from the archaic period,
Goethe at last met with Greek architectural monuments on which he did not
have to exercise a talent for restoration but to which he could react as directly as
in the previous year to the Roman amphitheatre at Verona. The confrontation
however was not easy, particularly with the two older temples:

the first impression could only provoke astonishment. I found myself in an utterly alien
world. For . . . our eyes, and through them our whole inner being, are directed and

emphatically attuned to a more slender architecture, so that these squat, conical, crowded masses of columns seem to us oppresive, yes, terrible.

But the lessons of study in Rome, of Winckelmann and Hirt, were not forgotten, nor the painfully won understanding that works of art could not, like works of nature, be grasped by the simple concept of the innocent eye, even when they were relatively undamaged:

I soon took a hold on myself, however, remembered the history of art, considered the age whose spirit found such a fashion of building appropriate, recalled to my mind the severe style of its sculpture, and in less than an hour I felt myself on good terms . . . only when one moves around them, through them, does one really communicate life to them; one feels the life out of them again that the architect intended, yes, that he created into them.

It was a rare moment: not perhaps since his first acquaintance with Duchess Luise had Goethe met a presence so completely other, and been provoked to woo it so intensely with his own subjectivity, persuading it to abandon its otherness and become 'myself not myself'. The courtship was a success. Back in Naples he wrote jubilantly to Herder that Paestum—especially the temple of Poseidon, the most recent of the three—was 'the final, and I might almost say, most splendid image ['Idee'] that I can now take, complete and whole, back with me to the North'. He put it beside the experience of the Sicilian landscape which had opened his eyes to the sheer truthfulness of Homer's descriptions and similes in the *Odyssey*. By means of these 'images' he felt he had gained an understanding of the principles of ancient art that could well serve retro-spectively as the purpose of a journey which had begun as a desperate clutching at a private superstition, in which neither Paestum nor Sicily had figured at all. Far from its being a flight to Rome, to still the desire and hopes of thirty years, he now saw the journey as a tour of the homeland of classical culture, on which he had 'hastened from peak to peak', roughly surveying the terrain in preparation for a more detailed visit later, and from which he now took back a guiding 'image'—as it were, an Ariadne's clue—by which to direct his future investigations.

Goethe needed the sense of conviction with which he wrote to Herder. He was probably replying to a letter handed over to him on 24 May by a Viennese visitor who had collected the mail waiting for him in Rome. There were, in addition to several letters from Fritz, five from Frau von Stein and three from Carl August. Weimar, from which he had been mentally almost completely detached for the last two months—and very relaxing months they had been—was now demanding his attention again. He had to be definite about his achievements, and clear about his intentions. His plan was to leave Naples on 1 June, to be in Rome for the Corpus Christi ceremonies on the seventh and to stay there until after the Feast of Saints Peter and Paul on the twenty-ninth, aiming to reach Frankfurt for his birthday at the end of August, and then to stay with his mother for some months, working quietly on the last four volumes for

Göschen, on his travel-notes, and perhaps on *Wilhelm Meister* and *Nausicaa* before his leave terminated at Christmas. But what awaited him then? From 27 to 29 May he gradually composed a long letter to the Duke asking for a complete revision of his official commitments, which, he said, had grown out of his personal relationship with Carl August and were secondary to it. The Duke, after expressing his satisfaction with the audit of the Duchy's financial affairs under Goethe's administration, had proposed to end the interim arrangements made in 1782 by making Schmidt into Vice-President of the Chamber, with all the day-to-day responsibility, and Goethe into the President. This Goethe diplomatically declined, asking to be relieved not only of the Chamber but of all detailed administrative work, 'which is not in my nature', and to be used instead for the things which only he could do. These he did not specify, though he mentioned the possibility of a general advisory position, and his hope that he would be able in future 'to live for your delight and that of many others' suggests that he was thinking especially of a literary and cultural role, as does his appeal: 'Give me back to myself, to my fatherland, give me back to you, so that I may start a new life, and a new life with you.' He delicately alluded to his almost misanthropic seclusion of recent years by emphasizing his recent interest in the Neapolitan social world, and by identifying as one of the main benefits of his journey 'the cheer it has added to my existence'. He even more delicately hinted at his own ultimate freedom of action, not only by the proposal to spend so much time in Frankfurt, but in the words: 'I have seen a great and goodly slice of the world and the conclusion is: that I want only to live with you and in your domain.' To Frau von Stein a week later he was more direct: 'I would rather have death than my life of the last few years.'

In the event Goethe's departure from Naples was delayed until 3 June by the arrival of the Prussian special envoy, Marquis Lucchesini, fresh from his negotiations in Rome for Papal recognition of Dalberg's appointment as Coadjutor Bishop of Mainz. He and Goethe had much to discuss, especially since Lucchesini had last met with Carl August only weeks before, in March, to talk of this matter and of the related question of the League of Princes. The Imperial view that even on an Italian sketching-trip Goethe was a man worth watching was not wholly groundless. Goethe took to Lucchesini, who seemed to him a model of the state servant and man of the world that he was not and could not be: 'he seems to me to be one of those people who have a strong moral stomach and can always be good trenchermen at the world's table. Whereas our sort is like a ruminant animal, that from time to time gets too full and then can take no more until it has finished the repeated process of chewing and digestion.' The two days' delay gave Goethe a little more time to complete his round of leave-taking social calls—at Venuti's he also said farewell to an ancient bust of Ulysses that had particularly appealed to him, though he did not reveal what he was thinking as he did so—but he was ready to go, or at least said repeatedly that he was: 'another fortnight would have taken me further and further down and away from my purpose . . . And really I am glad to be leaving Naples, since fundamentally I have nothing to do here and the gay life is not my line . . . one cannot come to one's senses in the place.' Tischbein had gone back to Rome while he had been in Sicily. Kniep was making preparations for marriage: theirs had been a business arrangement at first, but it had proved a turning-point in Kniep's career and had grown into friendship. The parting was warm. Goethe's one regret was that on 1 June there was a particularly violent eruption of lava from Vesuvius, which he was no longer able to observe at close quarters. Instead, sitting in the carriage for three days, retracing his earlier route from Rome, and pondering all that had happened in the interim, he began that consolation with memories which was to last the rest of his life, and concluded his Neapolitan adventure with a remembered spectacle. He thought back over the last few evenings, leaving the theatre by the harbour and walking out along the Molo in the warm night air under a full moon and seeing

at a single glance the moon, the moonlight edging the clouds, the moonlight in the sea and on the edges of the nearest waves, the lamps of the lighthouse, the fire of Vesuvius, its reflection in the water and the lights on the ships.

Southern Italy had become what it would always remain to him—a memory of light.

In the High School of Art: 1787–1788

Goethe's plans for his return journey were soon modified. Back in Rome on the afternoon of Wednesday 6 June, he was by Friday writing to Frau von Stein that he could complete his work for Göschen as easily there as in Frankfurt, and so perhaps did not need to depart promptly at the end of the month—he merely had to be sure not to leave so late that the autumn or winter weather would make it difficult to cross the Alps. Not that his Neapolitan experiences had altered his opinion of the ill-sited monastery on the banks of the Tiber. On the contrary, on the Thursday he had felt completely alienated by the Corpus Christi celebrations. 'I am spoiled once and for all for these church ceremonies, all these labours to give currency to a lie seem vapid to me.' What did arouse his enthusiasm were the tapestries which on that day were used to adorn the colonnades in St Peter's Square, and which, like those he had once seen in Strasbourg, had been woven to designs by Raphael (themselves preserved, then as now, in the English royal collection at Hampton Court). 'Rome is the only place in the world for the artist and that in the end is what I am', he explained after seeing them—yet even these words need careful interpretation. On the Monday after writing them, Goethe left Rome again to spend a fortnight sketching in Tivoli and the Alban Hills with Philipp Hackert. It was hardly the act of a man desperate to see, or see again, the art-treasures of the ages, and for whom time was running out. For the next year Goethe was to associate Rome primarily with 'art', and to call himself, repeatedly, an 'artist', but what he meant by these terms is not as straightforward as it may seem.

Hackert, who had come up from Naples with Venuti to arrange for the removal of the Farnese collection, was the guest of Baron Reiffenstein and was happy to spend time among the woods and waterfalls and picturesque classical ruins of Tivoli, inducting Goethe into some of the secrets of his trade, of which he had a very precise conception. There were, he thought, for example, only three forms of leaf which the painter needed to know, the chestnut, the poplar, and the oak, and all other foliage could be constructed from these elements. Goethe reciprocated with glimpses into his own workshop: readings from his fragments of *Tasso* and *Faust*. Maybe he also contributed his own reflections on the unity of plant life. Hackert encouraged him to think well of his 'little talent', and the prospect of having time to improve it further, 'even if only as an amateur', made all the more welcome the amiable suggestion in a letter from the Duke that he was master of his time and need not hurry home. Conceivably Carl August sensed, rightly or wrongly, that four months in Frankfurt might in the long run prove a more dangerous counter-attraction to Weimar than a necessarily limited holiday in Italy. So on 29 June, when the fireworks had died away on the Castel Sant' Angelo and the myriad lights illuminating every rib and cornice of St Peter's had burned themselves out, Goethe did not after all leave Rome as in Naples he had intended. He had by then already written to his mother that she could no longer expect to see him on his birthday, 28 August, for that was the date to which he had now postponed his departure, as a year before he had intended it for his departure from Carlsbad. He evidently still hoped, by retaining such significant dates, to preserve a symbolical pattern to his journey, but that urge was now weak and overshadowed by a more pressing concern.

It was no light matter to stay in Rome for the hottest months of the year and Goethe proposed to use them only secondarily for the undemanding task of improving his drawing: his main purpose was to take up again a struggle he had abandoned five years before, to finish *Egmont*, intended to be the main item in the fifth volume of the Göschen edition, and the work in which, more directly than in any other, he had addressed the political issues of his time. A heat wave began in the last week of June of such intensity that it was said to surpass anything known in Spain or Portugal: an overstatement perhaps, since meteorological records show that even in August the midday temperature did not rise above 28°C, although the afternoons no doubt got rather hotter. The streets were empty during the day, only in the warm moonlit nights did they come to life, and the sound of zithers and merry-making could be heard until dawn. Fortunately, however, Goethe was able to take over the large and cool studio, which Tischbein vacated on 2 July on removing finally to Naples, and here he shut himself away for July and August. Every day he walked out at sunrise through the Porta del Popolo to the fountain of Acqua Acetosa, where he took the waters in Bernini's pump-room, and by 8 o'clock he was back in the Corso working on his play and drawing architectural subjects, probably mainly from books, and scorning the local habit of the siesta. On Sunday mornings Angelica's carriage would call for him and after a visit to one of the great collections or churches with her and Zucchi and Reiffenstein they would all dine together. During the week one evening was also spent at Angelica's, others were used for a visit to the comic opera, or simply for a cool walk—to see from the top of Trajan's Column the sun setting over the city, for example, or to enjoy the moonlight on the palaces and squares. Otherwise Goethe kept the company only of his fellow lodgers, Bury and Schütz, of J. H. Meyer, the 'trusty Swiss' whom he was now getting to know better, and of Moritz, who saw him 'for some hours almost every day' and was grateful to be educated in any aspect of Goethe's thinking, whether about ancient art and mythology, which Moritz was now studying closely himself, or about speculative botany. In the summer heat the seedlings, whose germination Goethe had earlier observed and drawn —a prickly pear, a Mediterranean pine, some date-palms and doubtless others too—were growing strongly: their fresh green must have contrasted well with the white plaster casts.

Around the end of the first week in August Goethe received Carl August's reply to the long letter about his future in Weimar which he had written at the end of May, shortly before leaving Naples. On 21 July the Duke had left his territories to take up the rank of major-general in the Prussian army, in which in December he would receive the command of the Sixth Regiment of Dragoons. His mind was full of the prospect of active service, since Prussia, in alliance with England, was intervening in the Netherlands to restore the royal house of Orange, tottering in the face of revolt. Immediately before leaving to take his part in the crushing of the latter-day Egmonts, he wrote to Goethe that he would be more than happy to retain him, even in the rather indefinite roles that were being proposed, but did not see the need for any detailed decisions, or indeed for Goethe's return, until he was himself back in Weimar early the following year. Goethe gratefully took the hint and now, on 11 August, formally applied for an extension of his leave that would enable him to stay in Italy until Easter 1788. In justification he offered not only his hope of completing *Tasso* by the new year, and *Faust* by Easter—*Egmont* being finished, he said, with only a slight exaggeration; he also outlined, as the principal attraction Rome now held for him, an intensive course in drawing that he was arranging for himself. A week before, Maximilian von Verschaffelt (1754–1818), the son of the Director of the Academy of Drawing in Mannheim, had begun a series of public lectures on perspective which Goethe was attending. In September and October he hoped to be able to make sketches in the open air which he could then work up into finished landscapes, before embarking in the first months of 1788 on the human figure. He knew that he had a certain talent and he wanted to see how far he could develop it, though he admitted only to aiming at a social accomplishment like that of the musician who can play a piece at sight for the pleasure of his friends. It was an odd programme for a finance minister, and no less odd for a poet, even one long given to surprising gestures. By no stretch of interpretation could Goethe's Italian journey now be accommodated to the pattern of his excursions to the Brocken and the Furka: a new period had begun, with its own rationale, and he was now living a settled existence outside Weimar as the duchy's most expensive student. But Carl August took it in his stride.

Weimar too had been tolerant, at first. Once the secret of Goethe's whereabouts had been revealed around the turn of the year, the bewilderment among those he had left behind had given way to interest in the latest doings of their respectable scapegrace. The Duchesses, Frau von Stein, Fritz, and Knebel, had read travel-books and studied maps to follow Goethe's movements and fill out his own reports. Carl August arranged for the erection in the park of a small stone altar, copied from a Herculanean model, and dedicated ambiguously to 'the genius of this place' ('genio huius loci'). The Dowager Duchess, who had long had a cultured interest in Italy, was so taken with Goethe's venture that she started planning for an Italian journey of her own and soon Herder was following suit. There had also been a general sympathy in Weimar society for the loyal servant of the duchy whose journey south was explained as necessary to restore his ravaged health. In Rome Goethe had concealed his birthday from his new friends, and spent it alone, but in Weimar 28 August 1787 was marked by a celebration at the cottage by the Ilm. The host was Knebel, who since the end of October 1786 had intermittently been living there, trying out the Wertherian existence for himself. Frau von Stein was away at Gross Kochberg but her sister and sister-in-law came, as did the elder Voigt, from the Mines Commission, and his wife, two of Herder's sons (Herder himself was ill), and some others—among them the 28-year-old Schiller. Schiller's completed blank-verse tragedy *Don Carlos* had appeared earlier in the year, at the same time as *Iphigenia,* marking, like Goethe's play, a considerable change from the style of the works by which he had first become known, and he had now been in Carl August's territories for a month, breaking a journey to Hamburg, like Goethe twelve years before him, but destined, like Goethe, to stay for a lifetime. He was naturally curious about this lion of the past, now returning with such force to the literary arena he seemed to have left, who was none the less also a courtier loved and revered by a circle that even in his absence he could bring together to celebrate him. Schiller did not entirely approve of their—as it seemed, too sensuous and unspiritual—enthusiasm for the empirical sciences of nature, and perhaps there was already in him a touch of envy, but he savoured the irony that, after a wreath of Ilmenau heather had been hung on the cottage door to welcome Goethe on his expected return at Christmas, and after a substantial dinner, the health of the minister, still lingering in the capital of the classical world, should be drunk in hock in his own house on the proposal of the author of *The Robbers*. There was a little firework display in the evening, and, before the moon rose, Goethe's garden was illuminated, with as the centrepiece the Altar of Good Fortune, the monument to the power that brings rolling stones to rest.

There were those, however, who doubted whether Goethe would be seen in Weimar again. After the departure of the Duke in July, and once it became clear that his favourite, hitherto far more the driving spirit in the administration than the dutiful Prime Minister von Fritsch, could not be expected back before the middle of 1788, the duchy entered a state of suspended animation, in which gradually murmurings began to be heard. Quite apart from his neglect of home affairs, Carl August's military ambitions were attended with risks and considerable expense for his subjects. Goethe was by the winter generally thought to have gone soft in the head, and Schiller reported:

Goethe's return is uncertain and his permanent withdrawal from affairs of state is regarded by many as practically definite. While he is painting in Italy the Voigts and Schmidts have to sweat like mules on his behalf. He consumes in Italy, for doing nothing, a salary of 1800 dollars, and they have to bear a double burden for half the money.

Certainly it must have been galling, even for the good-humoured and tireless Schnauss, who himself was an amateur artist, to receive from his colleague a letter, however well-intentioned, written in Frascati and complaining that his correspondence was suffering because he had so much sketching to do during the day and had to spend so much time in the evenings working the sketches up: 'and perhaps once in a while one dabbles in water-colours and so time passes as if it were all inevitable'. But Goethe himself knew, and told the Duke, that it was only the presence of thoroughly reliable deputies that kept his long furlough from being irresponsible: Schmidt in the Chamber, fending off the criticisms aroused by new accounting procedures that reduced the opportunities for corruption; Voigt, dealing with a fatal accident and two serious floods at Ilmenau, which exhausted the company's capital and made it necessary to raise more from the shareholders; and his personal standby, the faithful Seidel, who kept his accounts, acted as his intermediary with Göschen and Bertuch, and renewed the lease on his Weimar house for another year, and to whom he commended the education of the now adolescent Fritz von Stein. It was also Seidel's task to introduce and accommodate in Weimar young Filippo Collina, the son of Goethe's Roman landlord. Collina had agreed to accompany the Dowager Duchess as courier on her Italian expedition, which Goethe had persuaded her to defer until 1788, to give her time to prepare it properly, and perhaps to give himself time to execute his own plans without interruption.

As far as his art studies were concerned, Goethe kept to the timetable he had announced to the Duke, even advancing it a little. His studies of perspective in August were accompanied by an experiment in clay modelling—within days he was writing to Frau von Stein that in this new medium he had found an 'Ariadne's clue through the labyrinths of the human form' that had enabled him to shape a good head almost immediately. The practical experiment was not followed up, but Goethe continued throughout his stay in Rome to study the various traditional numerical formulae for the proportions of the head and body. His work on *Egmont* kept him in the city throughout August but with Bury and Lips, who were commissioned to make copies from Michelangelo's *Last Judgement,* he was able to spend time in the Sistine Chapel when it was officially closed, and even, against his usual practice, to nod off after lunch, sitting on the Papal throne. The manuscript of *Egmont*, however, was finally ready on 5 September and immediately afterwards he spent some days in Frascati with Reiffenstein, no doubt in order to recover from the 'inexpressibly difficult task' of 'completing without rewriting' a work begun twelve years before in very different circumstances. But it was still a little early in this long, hot summer for the *villeggiatura*, the retreat of Roman society to the country resorts in the surrounding hills, during which he had promised himself some intensive sketching. For that one had to wait until the first rains of autumn brought a second spring to the baked landscape, and during the middle of the month he began the rewriting of *Erwin and Elmira*, which had been so popular in Weimar as a 'Singspiel', and which he now wished to recast as a fully musical *opera buffa* in the Italian manner of *Jest, Craft, and Vengeance.* Kayser himself was planning to come to Rome to continue his work on early church music and Goethe saw the opportunity of a close collaboration and an immediately accessible professional commentary on the technical aspects of his libretto. Finally, on 25 September, he left with Reiffenstein for his villa in Frascati, from which on 4 October they travelled over to Albano and their ultimate goal, the former residence of the General of the Jesuits in Castel Gandolfo, then the property of Thomas Jenkins, the English art-dealer, where open house was kept throughout the season.

The Alban Hills, whose volcanic origins Goethe recognized, with their long fertile slopes rich in olive groves and vineyards and with their sudden picturesquely wooded descents to the crater lakes Albano and Nemi, were the scene at this time of year of a social traffic that reminded Goethe of the German spas. He made off quietly in the early mornings to sketch in peace and then gave the rest of his day to conversation and gentle walks, so that in his three weeks in 'one of the most beautiful spots on the face of the earth' he spoke to more Italians, and spoke more Italian, than he had done in the previous year. There was even a little theatre in Castel Gandolfo where Pulcinello provided amusement on those evenings that were not given over to copying, inking, colouring, or arranging the morning's work. A touch on the heart-strings was also a conventional accompaniment to resort life. One day, while helping an attractive young Milanese woman, Maddalena Riggi, the sister of one of Jenkins's employees, to read the English newpapers Goethe realized, or later claimed to have realized, on learning that she was engaged to be married, that he was once again in danger of 'a Werther-like fate'. He courteously avoided her in future, though the incident probably gave rise to the one independent lyrical poem that he wrote between leaving Carlsbad and returning to Weimar: 'Eros as Landscape-Painter' ('Amor als Landschaftsmaler'). But it was surely elsewhere that any danger for Goethe's feelings lay. Angelica Kauffmann was in Castel Gandolfo too, and encouraging him in his landscape drawing: eight years older than Goethe, and childless, there was enough of the maternal in her warm affection for him to maintain, if in a milder climate, that aura of sexually neutralized emotion in which he had chosen to live for the last twelve years. She and Zucchi and Reiffenstein took as close an interest in Goethe's literary progress as that other little circle in Weimar: he read *Egmont* to her as he worked on it, she drew an illustration for it, which Lips engraved, and Goethe presented her with the de luxe copy of the first four volumes of the Göschen edition (despite his reservations about misprints, and the poor quality of type and paper) when they reached him in September. Indeed without consulting Göschen, Goethe arranged for her and Lips to adorn the future volumes also and be well remunerated for it. During the summer Angelica had been painting his portrait and Goethe withdrew as instinctively from the feelings it contained as he did from Maddalena Riggi: 'a pretty lad', he wrote, 'but no trace of me'. Herder put the position more accurately: 'she has fixed Goethe's appearance with great delicacy, more than he himself possesses, which is why everyone declaims about unlikeness, but there is really none of that in the picture'. Certainly the portrait can hardly be less true to the sitter than the white marble bust which Trippel also completed during the summer, in a style modelled on a recently discovered antique Apollo (now in the British Museum), and with a mane of long curls. Goethe had the gracious irony to comment, probably many years later, that he had no objection to leaving behind in the world the idea that he had looked like that.

The neo-classical *goût grec* was continuing to make headway, particularly in France, still the home of the most influential courtly culture in Europe, as had been apparent from the summer exhibition of the French Academy in Rome which Goethe had seen before going to Frascati. There had been nothing displayed to compare in convulsive effect with David's *Oath of the Horatii* of three years before, but the historical painters, whose genre was traditionally regarded as the summit of art, were clearly following in David's footsteps. Goethe's attention was drawn equally by the representatives of the supposedly inferior art of landscape—a reconstructed view of Rome in the age of Diocletian, quasi-classical pastorals in the tradition of Gaspard Poussin—on which, after all, he was still concentrating himself. But the sketching with which he was now passing his fresh, sunny mornings in Castel Gandolfo (though the weather became more changeable towards the end of October) was of a very different kind from that which had faded away in Naples and Sicily. He was no longer drawing 'picturesque' views—pictures of real places that tried to grasp their appeal to a specific, individual observer. He was now, under the academic and courtly influence, not only of Hackert, but of the whole artistic community in Rome, consisting as it did almost entirely of students, teachers, and court pensioners, training himself in the drawing of 'ideal landscapes', in which elements taken from nature, but from disparate scenes and contexts, were combined into a perfect landscape of the imagination. Many of Goethe's drawings from his *villeggiatura* survive—few of them can be identified as direct representations of specific places, indeed it was only in the summer and autumn of 1787 that some of his landscapes using Sicilian motifs were worked up, including coastal scenes for which the Alban Hills certainly offered no originals. He was now preoccupied with technique, and had purged his art of its personal, autobiographical, element.

Back in Rome on 22 October, Goethe saw before him four or five clear months in which to carry through his plans for artistic self-improvement. It was like the start of a new semester, and indeed 'for one who is willing to apply himself Rome is a true high school; while to other visitors it inevitably soon seems gloomy and dead, for which reason most of them quickly hurry on to Naples, the place of life and activity'. For Goethe 'art' was 'a serious matter' and he was now 'going to school' to study it. He began to draw the human figure, 'of which outside Rome one can form only an imperfect conception', having hitherto 'averted my eyes from it as if from the light of the sun'. Alternately drawing for himself, and studying ancient sculptures, he spent November and December on the head, and in January proceeded to the rest of the body, paying particular attention to muscular anatomy, the feet, and the hands, with which he concluded his 'course' on 24 January. As it happened, G. A. Camper, the son of the illustrious Dutch anatomist who had pronounced a negative judgement on Goethe's manuscript treatise on the intermaxillary bone, was in Rome at the end of the year and—until he went down with syphilis— gave lectures to the German artists on his father's system, which Goethe was able to attend, while he continued four times a week to attend classes on perspective. In a letter of recommendation written in November he also claims close knowledge of the work of the young Hamburg architect J. A. Arens (1757–1806), who had studied in Copenhagen and worked in France and England before coming to Rome. But his principal instructors, apart from himself, were J. H. Meyer and young Bury, who watched over his progress at home in the Corso. When at the start of November Kayser too arrived to stay in the Casa Moscatelli, taking over the little room Goethe had first occupied, when his piano was installed in the studio, and tuned, and he could play through the score of *Jest, Craft, and Vengeance,* there was some justification for Goethe's description of these rooms, in which poetry, painting, art-history, and music (and even a little botany) were all being pursued at once, as 'our domestic academy'.

Kayser's presence was important in more ways than one. As one of Goethe's oldest acquaintances from his Frankfurt years he added an intimacy, and a sense of continuity with the past, to a life so far lived 'entirely among strangers, in a strange land without even a familiar servant to lean on'. As a musician he could guide Goethe through his art by playing and analysing at home the pieces they heard in concerts and at the great church festivals: St Cecilia's Day, Christmas, and, eventually, the climax of the year, which Goethe would after all be able to experience, despite having preferred to it in 1787 the opportunity of visiting Sicily, the celebration of Holy Week and Easter. But above all with Kayser Goethe was able to pursue his plans for building up a stock of libretti— maybe even full scores—for German-language operas in the Italian manner. These plans were already quite extensive. In the final version of *Egmont* Goethe had introduced a substantial musical element—songs, intermezzi, incidental music, and a finale—and once in Rome Kayser set about composing what was required. If the music was ready to be advertised in conjunction with the fifth volume of Göschen's edition, *Egmont*, appearing as the first product of Goethe and Kayser's collaboration, might pave the way in the public mind for future works. Göschen had expressed himself willing to take on *Jest, Craft, and Vengeance*, not originally included in the contents of the collected edition, and even before Kayser arrived Goethe had written excitedly to his musical friend of his intended Cagliostro opera, for which he had already drawn up a plan and written a couple of appropriately cynical songs. In the event, however, Goethe's literary efforts during the winter of 1787–8 were not spent on *The Mystified* nor, for that matter, on *Tasso,* as he had once announced to the Duke—but on the material required to fill out the *Egmont* volume, the honorarium for which he would need to help pay for his return journey to Weimar. In addition to *Erwin and Elmira, Claudina of Villa Bella* was also to be rewritten as a full Italian opera, and this proved a much lengthier task than expected. By the time he had finished, his new versions had 'only the title and a few songs' in common with the first. *Erwin* was not ready until December, the last act of *Claudina* was not ready until February, and Goethe gave up the idea of submitting *Lila, The Fisherwoman,* and *Jery and Bätely* to a similar operation. Easter was early in 1788—23 March—and the shadow of his departure was beginning to fall: 'Whoever leaves Rome has to renounce art—outside it everything is mere dabbling ['Pfuscherei].'

The strange episode of Goethe's year as a part-time art student in Rome calls for an explanation. It was not a natural continuation of the flight from Carlsbad: Goethe was mentally fully prepared for his return to Germany when he set out from Naples at the start of June with the two great 'images' within him of Paestum and Sicily. Neither his drawing, nor his study of the remains of the ancient world had developed so promisingly in the previous eight months that to break off then would have been manifest amputation. The mere attraction of idling in the warm South explains nothing, despite what Schiller and his informants thought: had that been uppermost in Goethe's mind he would have remained in Naples, and besides, apart from his *villeggiatura* and five days in mid-December spent walking in glorious autumnal weather through the Alban Hills with Kayser and Bury, Goethe was very busy. One might question the value of much of this busyness, especially in respect of Goethe's drawing, and Goethe himself in later life represented the year as a period in which he learned to abandon the illusion that he might be a great visual artist, and reached the insight that 'I am too old to do more than dabble from now on'. But there is no contemporary evidence that Goethe entertained any grandiose ambitions. From as early as July it is clear in his correspondence that he had set himself two limited goals: to improve his drawing technique, and to lay a sound foundation of practical understanding for future study and appreciation; and in both these projects he eventually felt himself to be fully successful. Moreover, to represent Goethe's second period in Rome as a story of the loss of illusions is not to give due weight to the active contentment that radiates from his letters (by contrast, it could be added, with the period before his visit to Naples), to his own later repeated and unqualified assertions of his happiness at this time, and of its importance in his development, or to the bitter anguish that he suffered on leaving. One might also question, though Goethe himself never did, the peculiar quality of the literary work done during this year and of the aesthetic ideas lying behind it—but that is all part of what stands in need of explanation.

In the letter to Carl August announcing his imminent departure from Rome, Goethe gave his own summary of what his Italian journey had achieved: 'I can truly say: in these one and a half years of solitude I have found myself again; but as what?—As an artist!' Whatever did he mean? Certainly not that he was now a skilled professional in the visual arts, the sense that the term 'artist' ('Künstler') usually bears in his earlier letters. But if he meant that he had now recognized himself as a literary 'artist' we must ask further what can have been of such decisive importance for his literary self-understanding in a period in which study of the plastic arts bulked so large. And we shall also have to ask how it was that the term 'artist' could come to have this special, literary sense.

A clue to the answer to the first question is given by the context of Goethe's remark. He is first and foremost an artist—by contrast with any other administrative tasks to which Carl August may choose, and is welcome, to direct him. Less politely: Goethe is willing to remain a servant of the state of Weimar, on condition that the primacy of his 'artistic' vocation is acknowledged. And it is his time in Italy which has clarified what that vocation is. While he has been in Italy, he has learned something about his relation to Germany. The year in Rome, indeed, was not remarkable for anything Goethe learned about Italy, present or even past. He remained resolutely uninterested in Roman administration and in the city's aristocratic establishment, even though he no longer had the excuse of a visit too short to be wasted in empty courtesies. He did not use the unexpected bonus of these extra months in what he now called 'paradise'—transferring to the ill-situated monastery the epithet once reserved for Naples—to observe the life and circumstances of the Roman people, or to speak their language. He did not even set about acquiring a systematically deeper knowledge of the antiquities he had cursorily surveyed the previous winter: not until his departure was approaching did he resume his study of ancient sculpture, and when he visited the Borghese gallery in February 1788 he noted that a whole year had passed since he was last there. His own return to active drawing in January 1787 had coincided with the end of his spell as a Roman tourist. He was not thereafter filling sketch-books with mementoes of striking scenes or much-loved buildings or works of ancient art: the 800 or more sheets which he took back to Weimar were overwhelmingly either ideal landscapes or studies in perspective and anatomy based on printed sources, on a small number of casts, and on some life-drawing in the studio.

After his return to Rome in the summer of 1787 Goethe lived in an essentially German world: German in its language, its personalities, and its interests. It was also a consciously professional and to that extent uncourtly world. Goethe strikes a false note in his correspondence with Carl August, as if he is countering a reproach, when he claims to be reading the newspapers assiduously and offers some reflections on the vulnerability of Italy to Imperial expansionism, now that war has broken out between Russia and Turkey, and France is preoccupied with its nobility's struggle with the Crown for economic power, a struggle whose long-term consequences no one as yet foresaw. Whatever he may have been in Naples, Goethe was not, in Rome, a travelling diplomat, not even on furlough. He resisted 'all the importunities of the so-called *beau monde* . . . who can can give me nothing, and to whom I have nothing to give'—and if Marquis Lucchesini, whose appetite for social and political life he had once admired, and who was now living in the Corso, only 500 yards away, chose never even to call on him, it was not only, as Goethe surmised, because there was now no way in which Goethe could be of use to him, but no doubt also because the King of Prussia's special envoy could not be certain in what attire, or in what company, he would find the Baron opposite the Rondanini. Goethe was living surrounded by artists, many of them young and learning their trade; he was himself busy with his own trade, as he came to see it, completing a commercially published edition of his works; true, he was at the same time pursuing what might seem a dilettantish course of instruction in the visual arts and a somewhat similar course in music with Kayser; but he had no wish, and no need, to associate with anyone who did not share the professional's single-minded commitment, whether as practitioner, like Angelica, as patron, like Reiffenstein, or as theorist (and modest pupil), like Moritz. Rome for Goethe in the summer and winter of 1787 was the uniquely suitable place for the collective, and emphatically 'serious', devotion to 'the arts' of a group of essentially middle-class people who wanted only to get on with their job without having to accommodate and justify it to a local court. 'What makes life in Rome so agreeable is that there are so many people here who spend their whole lives thinking about art and practising it.' This was the world that Goethe described to Kayser as 'the fulfilment of all my desires and dreams—how can I leave the one place on the face of this earth that can become paradise to me?' It was not really in Rome, or in Italy, that Goethe spent that transcendently happy year: Rome and Italy had, by June 1787, proved ultimately a source of disappointment. Goethe spent his uncovenanted year of grace in an ideal Germany, in Germany as it ought to be, 'the promised land' to which Möser had once pointed the way, the fulfilment—however conditioned and restricted—of the hopes of fifteen years before. In December 1787, Herder presented to Carl August and the Margrave of Baden, at their request, a (thoroughly practical) scheme for a German National Academy, to be supported by the League of Princes, and to bring together, once a year, in scholarly and literary activity, the intellectual class scattered through the civil services of the many German states. The scheme itself came to nothing, but a model for it was already operating in the 'domestic academy' on the Corso.

As much as Herder's project, however, the life of the German colony in Rome was reliant on princely patronage, even if the princes were far away in the north. To that extent, Goethe's decision to stay in Rome, rather than rejoin his mother in Frankfurt and the republican world of the old Empire, was a decision to stay with Weimar. He was not taking the mail-coach home, not even via a detour through Italy. Although a mental detachment from his earlier life had begun, he was in material terms committing himself more firmly than ever before to practising his art in dependence on a court. But distance so attenuated the political and social constraints of the Weimar connection that life in Rome could seem like a fulfilment of the dreams of the 1770s for a national culture without the courts at all. That illusion however exacted its price: if Rome was distant from the princes, it was distant also from the publishing houses, without whose creation of a national consciousness the 1770s could not have begun to dream in the first place. Goethe had remained fundamentally true to the ambitions of Herder, Möser, and the age of book-battling 'geniuses' when he had removed to Weimar and so to a vantage-point nearer than Frankfurt to the motor forces in the national life. What Weimar had not been able to provide was a culture for more than the solitary soul and the single friend on his bosom. When he signed his contract with Göschen Goethe had first begun to turn away from the private interiority which had held him bound for ten years. Now in Rome he had discovered instead a whole cultural matrix—a city, a tradition, and like-minded companions—but it was divorced from the German nation. In particular 'one feels the lack of a livelier literary commerce', especially of course with Germany: for nearly two years the only German books Goethe acquired were those sent or brought him by friends, and his detachment from the public mind was all but complete. Even of the poor reception of his *Literary Works*, the corner-stone of the new self-understanding that had begun with his gradual withdrawal from the Weimar administration, he learned only through occasional discouraging remarks in letters from Göschen. (Of the 1,000 subscribers expected, only 550 were found.) Here, in the very home of 'the alpha and omega of all art', where 'as an artist I could venture to live, if only I could transport some of my friends here', Goethe was certainly happy and contented and—his own word—'sensible' as never before or after, but Lessing, it will be remembered, had noted that if Goethe ever came to his senses he would not be much more than an ordinary person. Goethe needed Germany to drive him mad, he needed its frustrations and deprivations, its wayward and uncomprehending public, as he needed 'the seed of madness lying in every parting', and especially in his parting from Rome, if he was to be more than an ordinary German intellectual of his day. In the slightly unreal atmosphere of his Roman milieu—a German world without Germany's political and cultural recalcitrance—some unrealistic attitudes to his own, essentially public, art of literature were fostered. In so far as they implied a detachment from the courtly world in which Goethe had so long taken refuge from the public domain they were a necessary step towards a maturer self-understanding, but at no other time in his adult literary career did Goethe approach so closely to a merely conventional wisdom in aesthetic matters.

At the most superficial level, that represented by his collaboration with Kayser in the planning, writing, and rewriting, of opera libretti, Goethe's conception of literature during his Roman year was a continuation of the illusions associated from the beginning with his work on *Jest, Craft, and Vengeance.* He fully recognized the need to work for a public, to leave behind the increasingly sterile solitude into which until 1785 he had sought to barricade his soul. But as yet he obstinately refused to recognize the kind of public he was most suited to reach. He saw himself as a technician, working for the theatre with the task of 'subordinating the poetry to the music', and leaving it to others to tinkle on the heart-strings. When Seidel objected to the resulting mutilation of *Claudina of Villa Bella,* Goethe emphasized that what was now to be printed was merely the writer's contribution to a theatrical production in which many others would be involved. The new *Claudina,* like the new *Erwin and Elmira,* he told Frau von Stein, was intended to be sung to music, not to be read. The false premiss, which Goethe in his busy Roman isolation could not recognize, was the assumption that Germany, particularly northern, Protestant Germany possessed either a theatre to put on such works, or audiences to enjoy them. The public for which Goethe had originally written *Claudina* was real enough, but it was a reading-public, and he was still not reconciled to the notion, with which he had so successfully co-operated in the early 1770s, that the culture of modern Germany had to be a culture of the printed word. Briefly, in 1785 and 1786, as he began to redefine his new existence and to plan his collected edition, the reconciliation had seemed to be taking place, as Aurelia's development in *Wilhelm Meister* shows. But by December 1786, when Goethe abandoned his plans for dedicating the edition 'to the German public', this interlude of realism was over, though fortunately it had proved long enough to permit the rewriting of *Iphigenia*. The theatre audience for which *Claudina* was revised, and *Jest, Craft, and Vengeance* written, was a phantasm, conjured up in Goethe's mind to justify to himself the suppression of the uncourtly autobiographical art of the emotions which was his true genius and for which he had still not identified the appropriate public medium. Since his conversations in Naples with Philipp Hackert that autobiographical element had been largely expunged from his drawing, and it was in deference to the concept of impersonal technique that in his subsequent art-studies Goethe gave such prominence to ideal landscapes, and to drawing from text-books and sculptures. Even if the Roman domestic academy was a Weimar stripped of its courtly appearances and devoted to the ideals of the age of Storm and Stress, it was still, at the end of however long a string, an appendage of an absolutist regime and in it there still prevailed the courtly hostility to emotional individualism and to the bourgeois art of print, which created the medium in which the emotions of individuals could become those of a public. It was still an extension of that strange prison into which Goethe had been locked with his audience of one, the only one who really understood him. Only while Goethe was a genuinely solitary wanderer on his way to Rome, still, in spirit at least, engaged in a symbolical journey towards an object of desire, could he confront and accept the truth about himself and his audience—that he was writing *Iphigenia* to be read, not to be performed, and so writing it for the world, and not just for one 'pure soul'; and to that insight the play owes, in great part, its unique formal maturity. Once he was settled in Rome, in Weimar's extramural art school, the old puzzlements closed in and his revision of his musical comedies, and even in part of *Egmont*, was based on the pretence that they were destined for immediate production: 'I think *Egmont* will be performed straight away. Here and there, at any rate.'

The hesitation shows that Goethe knew his position was ambiguous. His Roman life might seem to recall the purely middle-class world of the Storm and Stress—the days of the *Frankfurter Gelehrte Anzeigen*—but in reality he was caught up in a crisis of loyalty to the court similar to that which had produced Sentimentalism, the official culture, dominated by values of interiority rather than publicity, in which the middle class had been able to collaborate with the absolutist regime. In his attempt to resolve that crisis he was assisted by a systematic revival of Sentimentalist aesthetics undertaken at this time by Carl Philipp Moritz. Moritz was not yet the protégé of a court, he was supported by publisher's loans, but he had been obsessed since his adolescence not only by the author of *Werther* but also by the Sentimentalist ideology to which Werther falls a victim. Goethe tells us that Moritz's little treatise *On the Plastic Imitation of the Beautiful* (*Über die bildende Nachahmung des Schönen*), written in the winter of 1787 and published in 1788), which discusses the processes by which works of art are produced and appreciated, resulted from their long conversations in the latter part of 1787, and acknowledges that the thoughts that then emerged between them were destined for a successful subsequent career in 'the mentality of the age'. What Goethe does not say, though he does not deny it either, is that those thoughts had a longer lineage than the hot summer of 1787 could provide—just as the second sojourn in Rome was itself not a rootless novelty but the fulfilment of the dreams of an earlier generation. Equally, though these thoughts for a limited period expressed his own fundamental belief about his literary calling, they had only a short future before them. Their influence on the revision of *Egmont* will concern us later. The concepts of 'Art' and 'the Artist' contained in Moritz's treatise, however, are important clues to the self-interpretation with which Goethe in 1788 explained his return to Weimar.

Goethe's sweltering and solitary birthday in 1787 was refreshed by mail from Germany, including a packet from Herder: his latest publication, his definitive word on the Spinoza controversy initiated by Jacobi, a modest little volume, with the title *God: Some Conversations* (*Gott. Einige Gespräche*, about, of course, not with). For all the warmth with which he welcomed them, the dialogues— between a learned reader of Spinoza and Leibniz (Herder?), an impassioned investigator of nature (Goethe?), and a worried lady (Frau von Stein?)—cannot have brought Goethe much that was new. The Spinozan scheme is here scarcely expounded, but immediately translated into Leibnizian terms—instead of two divine attributes, matter and mind, we must assume that we know an infinite number, the individual 'substantial forces' to which Leibniz gives the name of monads—and all of Spinoza that remains is his freedom from any Christian ballast which might be incompatible with the wholesome concept of necessity, of the overriding rational order of all things, far transcending any limited notion of particular providential purposes. There is little in the book, except perhaps, as we shall see, in its final section, that Goethe had not already sketched out, either for himself or on the basis of conversations, of which *God* too may be a kind of record, at the time of his 'Spinozan Study'. Moritz, however, to whom Goethe almost immediately lent it, devoured it with passionate enthusiasm, and the Leibnizian terminology of individual 'forces', all proceeding from and reflecting their one origin, provides the metaphysical foundation for his aesthetic treatise. In *On the Plastic Imitation of the Beautiful*, however, the name of that origin is not 'God' but the quasi-Spinozan 'great totality of Nature' and there is more than a passing similarity to Tobler's essay on Nature: Moritz too represents human activity as the mediate activity of Nature herself. Furthermore, whereas Leibniz had seen every soul as a living mirror, a 'small divinity in its own sphere', making, in its perceptions, a small-scale analogy of the universal order, Moritz reserves the privilege of creativity, of likeness to Nature, the original and first maker, for the plastic artist ('bildender Künstler'). As in the Sentimental theory of genius, Moritz's 'artist' possesses, pre-eminently among creatures, the characteristic of the creator. The 'artist's' great soul is unusual for its high degree of likeness to the soul of souls—what Leibniz called 'God'—of which others are but fainter and more derived reflections:

If Nature herself has impressed the sense of her creative power into a man's whole being, and the measure of beauty into his eye and soul, he does not remain content with observing her; he must imitate her, strive to emulate her, spy on her in her secret workshop and with the burning flame in his bosom must form and create even as she does.

And the work—'every beautiful whole'—that proceeds from the hands of such an artist is an analogy and bears the imprint of the supreme beauty of the 'great whole' that is Nature.

When Moritz's treatise had been published, and was being widely discussed in Weimar, the objection was made that it did not sufficiently distinguish the beauty of Nature, which is a real totality, from the beauty of a work of art, which is a merely apparent totality. The distinction between reality and appearance, which was to be of cardinal importance in the aesthetics of the next decade, is actually present in *On the Plastic Imitation of the Beautiful,* but is explicit only in what the essay says about tragedy. That Moritz's argument did not particularly emphasize the distinction between the work of art and the work of Nature must however have been a merit in Goethe's eyes, since accepting the disseverance of the two was the most painful condition attached to his 'reborn' understanding of courtly culture. That continuity with his own earlier attitudes was worth a certain collusion with the 'excessive demands of oneself' which many years before he had depicted, half humorously in 'Wanderer's Storm-Song', tragically in *Werther,* and which Moritz was now exploiting to suggest a new mode of life for him. The products of Nature and the products of human artifice may well be distinct (the treatise implies) but the powers which produce them are conjoined in the artist's soul, which is said to be equally intimate with both. Moritz's theory, though essentially Sentimentalist, is, however, novel in the sharpness with which it marks off the 'artist' from all other feeling souls, and in the prestige, and so potentially the social significance, it assigns to his activity. In his Roman enchantment, where the conflicts that had exhausted him in Germany were so muted that they seemed to have reached an ideal harmony, Goethe was tempted by a new and highly seductive form of the Leibnizian compromise between the middle-class and courtly worlds. For, having accepted the de-Christianized monadism, which in *God; Some Conversations* Herder passes off as Spinozism, Moritz links it, in *On the Plastic Imitation of the Beautiful,* with the quasi-religious interpretation of artistic activity and appreciation that had first become explicit in Winckelmann's writings. At the same time he extends the application of Winckelmann's pagan piety from the visual arts to literature, and so follows Lessing, in his *Laocoon,* who in turn was following Batteux, in positing a field of aesthetic activity in general, of which poetry and painting are sub-species. Unlike Lessing and Batteux, however, Moritz—or is it Goethe?—has a name for this general, and now religiosified, field—'Art'.

It is difficult for the modern English reader, well used to the terms 'art' and 'creative', and their various juxtapositions, to realize how recent are their present senses, and that in both cases they are importations from German usage. Not until the end of the nineteenth century did the English word 'creative' cease to be a purely theological term, and at about the same time the word 'art' took on its modern, general, and rather high-flown meaning (' . . . but is it Art?'). That meaning was unknown to Dr Johnson, who in his *Dictionary* refers in this context only to the many different 'arts', and chooses as an example 'the art of boiling sugar'. Even for Herder, in part III of his *Ideas*, which reached Goethe in October 1787, 'art' ('Kunst') often means something like 'technology', covering for example ship-building and navigation. The notions that literature, the visual arts, and music have something in common which sets them apart from mere technical crafts, that their products are self-contained little worlds, in something like the way in which the great world, the universe, is self-contained, that these products can no more be judged by standards (for example, moral standards) external to them than can the universe, and that their producers are thus analogous to the producer of the universe Himself and so are rightly called creators—these notions had all separately been developing in the matrix of eighteenth-century German aesthetics and Moritz brings them together within a brief compass and demonstrates their interrelation. In particular Moritz builds on an argument he had already expounded in a short article in 1785 to show how the products of 'Art'—whether seen, or heard, or, as in the case of literature, accessible only to the imagination—share a special characteristic of beauty in general which he is one of the first to identify: that it serves no purpose. When Oscar Wilde remarked that all art is quite useless, he was uttering not a paradox but a tautology: uselessness was one of the features which German thinkers, a hundred years earlier, had used to define the new activity of 'Art' in the first place. That uselessness is closely related, as Moritz shows, to the new activity's status as the nearest thing to a religious activity allowed by a completely secularized world-view. But this self-sufficiency and quasi-religious character of Moritz's 'works of Art' has a social and indeed an economic implication.

If Art is the successor to religion, then the priests of Art have as much right to be supported by society as the priests of religion, and if works of Art are to be produced and appreciated 'for their own sake', then those who produce them—and even those who simply know about and understand them—should be maintained for the sake of their Art and not because of any other utility that they may incidentally possess. Music, theatre, painting, and sculpture had been patronized by the German absolutist courts for purposes of entertainment and ostentation; in Moritz's treatise we see the first outlines of an aesthetic theory which implies that such patronage is desirable regardless of any purpose served and which associates literature, hitherto a largely private and middle-class affair, of little courtly value, and scholarship, hitherto confined to the universities, with the other adornments of court life. The assertion of the autonomy of the work of Art—and the identification of works of literature as works of Art, and so as autonomous—was the explicit form of an implicit assertion of the autonomy of the Artist—and particularly of the Poet—at an absolutist court. It was, that is, a claim for patronage (because Art is a state function, like religion) but at the same time a claim for the independence of the self-employed bourgeois craftsman, a priest in an (established) church of one. When Goethe told Carl August that in Italy he had rediscovered himself 'as an artist!', and added, 'anything else I may be is for you to assess and to utilize', he was making a doubly daring and advanced use of the term 'artist': as applicable not just to painters and sculptors but to a man of letters (and not applicable to lowly craftsmen), and as implying a court function dignified in itself, and not simply in virtue of its association with the person of an acting President of the Chamber. In that novel usage (acknowledged by the exclamation mark) lay a proposal for resolving Goethe's long-standing dilemma—to reconcile in himself the courtly and the middle-class elements in German culture by becoming a self-sufficient creator of literary beauty who was maintained by his patron but did not subserve any of his patron's purposes.

It was a proposal which was destined to have a widespread influence on German life and literature in general, but which for Goethe had a peculiarly personal significance. It was a sign that the interiorization—that is, the assimilation to a middle-class sensibility—of court values, to which he had committed himself in his relation with Frau von Stein ('let man be noble . . .') no longer satisfied his sense of his independent worth. He might in future be materially dependent on the court, but far from seeking to derive his inspiration from it, he was asserting the duty of the court to support whatever he might undertake. This involved of course only a partial, and to that extent a flawed, interpretation of the role he might now hope to play in establishing a culture for the German nation, but it was at least a step on the road back to recovering the ambitions of the 1770s in a more fulfillable form. What the proposal omitted was what Goethe's Roman existence omitted: any direct relation with his German public. Since its invention, Art after all has never mixed well with the public: 'You are just a prosaic German,' was Goethe's reply to Seidel's criticisms of the new *Claudina*, 'and think a work of art should slip down as easily as an oyster'. Moritz actually deduces from his definition of beauty that the only person who truly and fully experiences the beauty of a work of Art is the 'Artist' in the moment of the work's 'creation': the aesthetic theory that he records, and that Goethe must at the time at least partially have shared, gives a role to the recipient of a work of art, which is emphatically subordinate to that of its maker. This emphasis may to some extent reflect Goethe's thoughts about the manifestly different order of his talents as a poet (essentially a producer) and as a draughtsman (essentially a consumer), and it also certainly reflects Moritz's feeling about the difference between his own modest, and mainly critical or receptive, gifts and those of the productive poet whom he worshipped. In Moritz's elevation of the Artist to the status of a demi-god, in his equation of creativity with the masculine and receptivity with the feminine, in his self-abasement before the fulgurations of the creative power, it is not difficult to see the particular cast of his own sensibility. Goethe was perhaps no more averse to receiving Moritz's tribute—so similar to that paid by the ravished Winckelmann or Heinse to the arrogant features of the Belvederean Apollo—than he was to leaving in the world the idea that he had looked like Trippel's Apolline bust. But his literary art—as, surely, any literary art—was an art of communication, of dialogue, and if he was ever to overcome the spiritual solitude that had been intensifying for the last ten years and had become absolute when he had arrived in Rome and his inner dialogue with Frau von Stein had ceased, he would have to leave behind the flattering apotheoses of his Roman period and find a means of conducting a conversation with a real audience once more.

Personality apart, Moritz's idolatry, and his assumption that Goethe possessed some superhuman affinity for the natural order, is understandable if one reflects that not only was he now privy to Goethe's unpublished literary work, and a daily witness of his talent for drawing, but he was also receiving private lessons in Goethe's new system of botany, 'my *Harmonia plantarum*' intended to supplement Linnaeus. Yet a partial detachment from Leibnizian ways of thought is already discernible in Goethe's botanical studies after his return from Naples, though Moritz was in no position to notice any shift. On the one hand, Herder's *God* prompts Goethe to think he too has found a botanical *'hen kai pan'*, a single universal divine principle manifest in all things, and some comments on experiments Seidel was conducting in Weimar on crystallization show him still equally committed to the two principles of the *Monadology*: the unique 'manner of being' of each object and the universal intercommunication of all things. On the other hand, the failure to find the primal plant in Sicily has clearly left its mark, and the simple trust in a necessary harmony between his own innocent perceptions and the natural order—provided only he can maintain the purity of his vision—is shaken, just as it was shaken in relation to works of art when he realized the importance of understanding them historically. He remains confident that he is on the track of a principle of unity for the entire vegetable kingdom but he is uncertain what the status of that principle will be: he no longer expects to find the primal plant growing on an Italian hillside. In June he describes it simply as a 'model' which, if taken in conjunction with a 'key', will permit one to construct an infinite number of plants, all of which are possible, and some of which may actually exist. By October 1787, however, the primal plant is no longer even a model, but has faded into a 'formula' which 'explains' the shapes only of all real plants. He is particularly excited to discover a complete new plant, itself flowering, growing out of the flower of a carnation, which he regards as a vindication of his theory (whatever that may have been), but he does not describe the phenomenon as a model, let alone as the primal plant. At about the same time he takes up again his old interest in physiognomy, several drawings show an interest in the series of gradations by which one profile modulates into another, even an animal face into a human, and there is an attempt to reduce to a numbered sequence the possible shapes of the lips. The pattern already established in the cases of geology (in which he took no further interest while in Rome) and the intermaxillary bone is repeating itself: the moment of inspirational insight is followed by a long period of empirical investigation, in the course of which it becomes increasingly difficult to define the significance of the original 'discovery' and keep hold of Ariadne's clue. For all his assurances of rapid progress, it will take Goethe another three years to formulate his botanical principles, and by then the concept of the primal plant will be practically forgotten.

At a deeper level than either Goethe's toying with libretti or his self-stylization as an Artist specially intimate with the cosmos an altogether new theme can be discerned, however. Even those were attempts, however superficial, at the solution of a serious problem: what to do and be when his leave ran out. In his correspondence, however, and most plainly in his literary works, particulary *Egmont,* one can catch hints of a markedly more sober and more sombre train of thought, also set in motion by the steady lapse of time. Already in Naples, when he believed he was on the way home, he had been thinking of the very landscape about him as the material of future memories. Back in Rome the fear of loss grows: the 'beautiful sights . . . remain alive in my soul and can never be taken from me', he has 'collected the finest, most solid treasures', such as his visits to Sicily or to Paestum, which are now an 'indestructible' component of his life, to be carried home with and in him. The determination that his own happiness shall remain, through memory, ever-living and ever-fresh must have been reinforced by the change and decay that had disappointingly affected most of the monuments he had visited. Through-out the summer and autumn his almost obsessive redrawing and retouching of Sicilian scenes into ideal landscapes may reflect a need to make a permanent haven in the mind for the significance and value of experiences already passing beyond the reach of repetition. In Weimar for so long now one year had been much like another, the court was relatively young, the estates, the forests, the Harz, were indefinitely revisitable, even the Swiss journey of 1779 had been conceived as a recapitulation. Italy, in its every moment, was unique, and flowing irrevocably away. It was no longer part of some symbolical story—it was precious in itself, and so destined to pass. On 22 September there came a tangible reminder that in Weimar, and Frankfurt before it, the river of time, if quieter, had run no more slowly. 'Four slender little volumes, the outcome of half a life', the first instalment of Göschen's edition of his *Literary Works,* sought Goethe out in Rome—'not a letter in them that has not been lived, felt, enjoyed, suffered, thought', they too unique, and they too now past. By November the shadow of the homeward journey, which would again become possible once winter was over, began to deepen, but now Goethe felt his state of uncertainty, poised between past and future, south and north, to be typical of the whole of human life. 'Periamo noi, periano anche i bicchieri' ('We do not last, and wine glasses do not last either') was the bleak wisdom of a little Roman girl he met, and he made it his own.

Much of what Goethe had recently been reading must have inclined him to this view. The last dialogue in Herder's *God* discusses what the religion of Nature and Necessity can say to calm the fears aroused by the transience of human, and indeed all organic, life, and the concluding fifteenth book of the third part of the *Ideas,* which Goethe found 'splendid' and which 'quite transported' Moritz, raises the same question in relation to the entire course of history, seemingly a pointless chronicle of loss and destruction. In *God* Herder offers what are effectively the consolations of Leibniz: there is no death in the universe, individual souls do not die but shrink into imperceptibility, all that is composite is, however, by its nature destined to be dissolved into its parts for the sake of the continuing and 'permanently active life of the world-spirit'. The individual dies because it achieves the supreme moment of its own existence in giving birth to new life. In a similar section of *On the Plastic Imitation of the Beautiful* Moritz adds his own characteristically ruthless touch: 'Is it not through the permanent destruction of the individual that the species maintains itself in eternal youth and beauty?' Herder, in his *Ideas,* develops his appeal, first made in *God,* to the mathematical laws of harmonic motion, in order to argue that the human species as a whole is necessarily involved in an enormous process of self-improvement. As each flower has its moment of glory for the sake of the next generation, as each human individual is born to strive for a particular perfection which only he can achieve, so each historical human culture reaches its 'maximum', the moment of perfection of its own peculiar contribution to the store of human achievement, and though it then rapidly passes away, humanity in general is further advanced towards its collective 'maximum' of reason and justice. By means of this doctrine of 'maxima', Herder hoped to circumvent the arguments Kant had put forward, partly in criticism of the earlier books of the *Ideas,* to the effect that human perfection was something no individual human being, or group, could attain, but only the species as a whole.

Goethe was selective about these speculations on the 'cosmic order' and took from them what related most closely to his own circumstances. He welcomed *God* for its stark Protestant contrast with the Roman 'Babylon' that surrounded him, just as he welcomed the *Ideas* as the true 'gospel' with which to vanquish the apologists of Christianity, Lavater, Jacobi, and Matthias Claudius. But it was a gospel for one who was 'not expecting any Messiah', and the rational progressivism and eudemonism—the attempt at a historicized version of Leibnizian optimism—did not elicit any comment from him. Instead, almost perversely, in view of Herder's strenuous attempts to find a large-scale pattern in human history, Goethe seems to have seized on Herder's notion of the individual 'maximum' as the only purpose fully and perceptibly embodied within the historical process, and to have interpreted the conclusion of the book to be 'that the moment is everything and that the merit of a rational human being consists only in so conducting himself that his life . . . contains the greatest possible quantity of rational, happy moments'. It was not an acknowledgement of the power of death, but at least it was not—as Herder's thoughts in reality perhaps are—a repetition of the conceits with which Prometheus once sought to allay the curiosity of Pandora. It was a conclusion that fitted well the mood of a man the fulfilment of whose long-standing dreams had at last come but was even now passing away, who was uncertain with what dreams to replace them, and indeed was tempted to blame dream and desire itself for the pain inflicted by the passage of time—how he loathed the Christians that traded in such unrealities and wished them to their devil, with which they belonged, for the devil was 'from the beginning a friend of lies . . ., intimations, desires etc.!' (He even felt alienated from the 'indeterminate spirituality' of Rembrandt, whose prints he had for years been helping Carl August to collect.) These sentiments must have been uttered to Moritz, and were clearly not wasted on him, since his own farewell to Rome, some months after Goethe's, took place on the Capitol, where he stood with Herder watching the sun set, 'resolved "to enjoy every beautiful scene in life to its ultimate moment, with no complaint or grumbling that it must end"'.

Goethe's heart was heavy as January 1788 drew towards a close and with it his self-imposed art course. 'All this time I have had no sense of any of the ills that plagued me in the North . . . I have a number of indications that I shall leave this well-being behind me in Italy, along with various other good things.' Yet he knew he had to go back, and not simply because his leave terminated at Easter. At other times, when the prospect of return was not imminent, he could recognize the truth quite easily: Germany was his nation, and with Italy he really had 'nothing in common'; he needed his roots, 'man needs but little; he cannot do without love and the security of his relationship with those who have been elected and given to him'; especially, in his little academy, he felt how much he needed Jena: 'when I think back to Germany', he wrote to Knebel, asking him to greet his Jena friends 'I do not want to live anywhere other than among you all'. By this stage in his life there were possibly financial factors too: he was now used to a level of expenditure which could only be sustained if he had a salaried post. His Wilhelmiad—as he called his Italian adventure in allusion to Wilhelm Meister's readiness to surround himself with a subsidized entourage—had swallowed up all his current income including the fees from

Göschen, and he was reduced to calling in a loan made to the unfortunate Plessing. He was not negotiating from a position of abject dependency in seeking to arrange with the Duke a new basis for his continued presence in Weimar, but it was a little obsequious, and a little calculatedly so, to claim that 'the end of all my labours and wanderings is and always will be the desire to adorn your life', even if his, and Moritz's, new ideas about the nature of art made it easier to believe. But Goethe also knew that Rome was a temptation: he sensed the cost of exile—'in a foreign land, I notice, the heart readily becomes cold and brazen, because love and trust are seldom called for'—and it was not only in bitterness that he called Rome a 'magnetic mountain' and 'the city of the Muses, who sing more dangerously than the sirens'. Perhaps indeed only there could he be 'unconditionally happy'—but perhaps also he had other things to pursue, even at the cost of happiness. The completion of the Göschen edition, 'this *Summa Summarum* of my life', was in prospect, as indeed was his fortieth birthday, and it was not too early to be thinking of 'starting a new page'. Even so, at the turn of the year, Goethe could not but feel that 'Time, alas, which everywhere passes swiftly, seems here to hasten twice or thrice as much'.

Then, at the end of January, everything took on a new aspect. On the twenty-fourth a letter arrived from Carl August that sent Goethe rushing out on a long walk to master his joy and excitement. The Duke, now on his way back from the Netherlands, was in Mainz, partly to continue discussions about resistance by the League of Princes to any renewed military threat from Joseph II, and partly to undergo treatment for a venereal infection. In his letter, very warm and full of political and personal news (Goethe innocently took the medical details to refer to haemorrhoids), he asked Goethe to stay on in Italy to await the arrival of the Dowager Duchess later in the year. The length and promptness of Goethe's reply left no doubt about his reaction, but his words were measured: he had of course intended to set out from Rome at Easter at the latest, the Duke would need to confirm any arrangements from Weimar, until confirmation was received he would take no special steps but hold himself in readiness to leave at a moment's notice, but he could devote the whole of April to equipping himself with new lodgings, a servant, and respectable clothes, presenting himself to the French and Imperial ambassadors, to Cardinal Prince Buoncompagni (the head of the Curial administration), and to Prince Rezzonico (the 'Senator' or supreme magistrate of Rome), so smoothing the path for the illustrious visitor, for whom he and Angelica could undoubtedly find some suitable female companions, and in May he could make an excursion to Naples introducing himself at the royal court there, and he would have to spend some time preparing the ground in Florence also, as the Duchess would undoubtedly wish to visit the Habsburg court there and to call on Lady Hannah Cowper (one of the three daughters of Charles Gore, a rich Englishman whom Carl August was trying to entice to Weimar), and incidentally, for the Duke's information, his resources were fully adequate to his original plan of returning after Easter, but if he were to stay on he would of course be involved in substantial unbudgeted expenditure. Goethe knew this scheme meant 'a new life' in Rome, a metamorphosis into a courtier, and so as much an end to his student existence as if he had gone back to Germany—but the prospect of not going back to Germany was irresistible, intoxicating. The day after he wrote his long letter the Carnival began. The theatres had been open since Boxing Day—the rage of the season was Giuseppe Gazzaniga's opera *Don Giovanni*—but Goethe had seen, little of the, as it seemed to him, mediocre productions, 'for I usually spend my evenings in conversations about art'. This year, however, the Carnival was not to be the tedious affair that the outsider had found it in 1787. Goethe's mood had changed, he had anyway already agreed with Bertuch that he would write an account of the Carnival for one of his journals, for which Schütz would provide the illustrations, and this time he was not just surveying the scene from his window but was in among the throng, in Angelica's carriage, or pushing his way through on foot, with his long coat and round broad-brimmed artist's hat marking him out as a Northerner and so serving effectively as fancy dress. What with the 'far too noisy days', he observed, when Ash Wednesday (5 February) had passed, it had taken him the best part of a fortnight to complete one short letter. Also, he had found a better way to spend his evenings.

Carl August, busy with his own investigations in Holland, had long been pressing his old companion-at-arms for some barrack-room gossip about the female half of the Roman population. Goethe was unable to oblige: he had been greatly struck by the openness of homosexual friendships in Rome, he wrote on 29 December, but the women were another matter—either they were well under lock and key, or they were only available on a long-term and high-priced contract, or they were too public and too dangerous. Piffle, responded the Duke, nothing that a spot of mercury (then the only treatment for syphilis) wouldn't cure, all this self-absorption of Goethe's was bad for a man, and only a dry stick would not let himself be tempted into the garden. Goethe's feelings had certainly been stirred in the autumn of 1787, during the *villeggiatura*, and the poem 'Eros as Landscape-Painter' is a sophisticated return to the suggestive, anacreontic manner of his earliest verses. It even uses the Pygmalion motif again to make its concluding *pointe:* Eros paints a landscape for the poet and makes it come alive by introducing into the painting a female figure, dipping her foot into a purling stream, and 'with everything, everything moving', the poet asks, 'do you imagine I stayed sitting firmly on my rock, like a rock?' A little unrhymed song written at about the same time for the revision of *Claudina* complains, in unusual, agitated, quasi-classical metres, of the disturbance Cupid is bringing into the poet's household, but it is clear from the correspondence with Carl August that, for most of the winter, the disturbance was simply that Goethe could not get 'the damned second pillow' out of his mind, as is also suggested by some drawings of Hero and Leander from these months (when, too, as part of his 'course', he was making drawings from life).

No doubt the approach of the Easter deadline encouraged a decision—if Italy was in this respect too to bring him the fulfilment that would mark the end of the years of deprivation in Weimar, then time was now short. Something else, to which he certainly gave no conscious attention, must have made the decision easier. 'He has gone cold towards his friends', Fritz von Stein had written to Frau Goethe, doubtless reflecting the views of his own mother, and in 1787 the once significant date of 12 November seems to have passed unremarked. In the matter-of-fact reports that Goethe still regularly sent to Frau von Stein during his Roman studies nothing seems to have matched the agonized intimacy of the last letter before he left for Naples, in which he told of the destructive effect on him of the impossibility of possessing her. Without Goethe's noticing it, that crisis was now over; its final symptom had been his project of a drama on Nausicaa-Calypso; in that his interest had quietly died, and with it the ten-year-old unconsummated marriage of souls. Similarly perhaps he had never noticed that only if the friendship were chaste could it perform its symbolic function—bringing middle class and nobility together in an interior union from which power-relations were absent—and that only for that reason had it been begun. If what now followed was adultery, it came, as adultery often does, as the result, rather than the cause, of the end of a relationship. Interiority, Goethe thought, had had its day. Perhaps, after all, it was for this, more than anything else, that he had come to Italy in the first place.

Little is known of the episode. At some time in January, during the preparations for the Carnival, possibly only a few days before receiving the momentous letter from Carl August, Goethe made the acquaintance of a young widow, possibly the 24-year-old Faustina Antonini, née di Giovanni, the daughter of an innkeeper, with whom she and her 3-year-old son were now living. She was willing, available, and secure, and in a turbulent Carnival atmosphere Goethe quickly started to make assignations, probably under the name, in so far as a name was necessary, of Philipp Seidel. Maybe there is some truth in the story, as he later told it, of her writing a Roman 'IV' in spilt wine on a table in the inn, so evading the watchful eye of her guardian, to fix the hour of the night at which he would call; maybe, however, the family was complaisant enough to make subterfuge unnecessary and Goethe was suggesting either marriage or some such long-term relationship as he had outlined to Carl August on 29 December (possibly in allusion to these very negotiations). On 16 February he wrote to the Duke that he

could tell of some agreeable promenades . . . you, as *Doctor longe experientissimus*, are perfectly correct, that such moderate motion refreshes the spirits and puts the body into a delightful equilibrium. As I have discovered more than once in my life, noting on the other hand the inconvenience when, from the broad way, I have sought to introduce myself to the narrow path of abstinence and safety.

This is the first sexual encounter of Goethe's for which there is documentary evidence, and the stilted braggadocio of his remarks—not quite successfully repressing the triumph that reveals previous inexperience—makes it unlikely there were many predecessors.

From the middle of January to the middle of March Goethe's stay in Rome became, as a result of 'various measures', 'more and more sweet, useful and happy. Yes I can say that in these last eight weeks I have enjoyed the greatest contentment of my life and now at least know an extreme point against which I can in future calibrate the thermometer of my existence.' Physically and emotionally at peace at last, he was in his paradise and with good expectations of remaining there for months to come. After a wet winter the spring was lush: laurel, viburnum and box, peaches, almonds, and lemons came into bloom, and the gardens were full of anemones, hyacinths, and primroses. With his drawing course finished and the complete text for volume v of the edition (*Egmont* and the comic operas) finally despatched by 9 February, he was devoting himself to tasks which, though he might not have admitted it, were more congenial: colouring and retouching landscapes, and thinking about his next three volumes —*Tasso, Faust,* and the collection of his poems. With Kayser he was already studying the music for the great church ceremonies to come and, now that he was preparing himself for the fashionable circuit, it was a particular pleasure to see his old friend well received, and his playing admired, in the residence on the Capitol of the cultured Prince Rezzonico. Even Tischbein's announcement that he would shortly be returning to Rome from Naples, and would be requiring the studio once again, was no great inconvenience: it seemed likely that Goethe would be needing more space anyway, he had engaged a young German servant, Carl Pieck (also known as Federico Palatino), and at the Casa Moscatelli the whole floor above had just become free. Here in big light rooms, high enough to look down over roofs green from the winter rains, walls yellow with the new growth of ivy, and between them blossoming gardens, balconies, and terraces—a transfigured reminiscence, perhaps, of the view from his Frankfurt attic—Goethe's Roman year culminated in the setting up of his own gallery. Kniep's finished water-colours of the Sicilian journey were now arriving and could be studied in the portfolio or exhibited on the easel, a glowing pool of colour to catch the eye of the visitor as he came in, while round the walls stood the white plaster-casts that had accumulated over the months, some of them Goethe's, some of them Tischbein's: impressions of bas-reliefs, including sphinxes and other animals from one of the Egyptian obelisks, some genuine marble fragments, a Hercules, Junos and Jupiters, the monumental Juno Ludovisi, and that strangely fascinating antique *memento mori* from the palazzo opposite, the Medusa Rondanini.

She was not out of place. One way or another an end was imminent to his eighteen months of privacy, so much of which had been spent reliving the emotions of his youth and recasting their literary expression, and even though he now seemed suddenly to be sharing the table of the gods, like his old heroes Prometheus and Tantalus, the recollection of mortality would not let him forget entirely that their fall also had been swift. The path from Olympus to the underworld, he later wrote, led past the pyramid of Cestius, and one evening after the Carnival was over, yielding once more to the feelings with which his Sentimental contemporaries had sorrowed at the graves of Werther, and of Gray's elegist, he drew his own moonlit tomb, beside the pyramid, on the edge of the Protestant cemetery, where Keats and Shelley, and also another, far dearer to Goethe, would one day be buried. 'The most beautiful and solemn cemetery I ever beheld . . .', Shelley would write, on seeing 'the tombs mostly of . . . young people who were buried there' and he would go on: 'one might, if one were to die, desire the sleep they seem to sleep.' It was to be the destiny of the coming Romantic generation to live through to the last the possibilities —the temptations—that Goethe had known and expressed and resolutely put behind him: the unspoken meaning of his drawing is that he can remain in his beloved Rome and continue to eat the food of the gods only at the cost of an exile from Germany that would spell emotional death.

The end, when it came, was quick. On or about 15 March, the Saturday before Holy Week, letters arrived from Weimar, to which Carl August had finally returned a month before. Here consultations with Anna Amalia, and no doubt with Fritsch, brought a conclusion to the courteous negotiations about his future which Goethe had begun nine months previously with his letter from Naples. The Duke accepted that Schmidt should now become President of the Chamber, insisting only that Goethe should retain the right to take the chair at any of its meetings; he was however to have no specific duties—opportunities for these would doubtless present themselves in time: it was anyway better for Goethe to devote himself to what he thought important—even the management of the burgeoning park in Weimar was to be taken over by Bertuch, and, like Schmidt, Goethe was to receive a salary increase of 200 dollars a year. The Duke's old friend Wedel and the elder Voigt were to take up posts in the Chamber, and Herder—who was perhaps dreaming himself into the role of president or secretary of his projected National Academy—was also to have his salary raised. In warm and friendly tones the Duke put all these proposals to Goethe, hoping that he would find them acceptable as the basis on which he might return to Weimar in the not too distant future. It was an offer of outstanding generosity, which gave Goethe all he had asked for, and he knew it, but the letter that contained it was—however absurd and irrational this might be—one of the most desperately disappointing he had ever received. For Goethe, the Duke continued, had better things to do than act as courier for the Dowager Duchess; he need not feel that for that reason he had to tarry in Italy. It was the expulsion from paradise. To be sure, it was only a hint. But there was also a letter from Herder, whose views on his own letter the Duke had sought before despatching it, which spelled out plainly that this was too good an opportunity to pass up, that the Duke had to make his final arrangements soon, and the voices in Weimar questioning whether Goethe would ever return were becoming more insistent. It was, in a sense, Herder's retaliation for the letter which Goethe, on the eve of departing for Italy, had written to urge him not to move to Hamburg. There is no record of Goethe's attending the Palm Sunday service, and presumably his mood did not permit it, but by the Monday he was sufficiently master of his emotions to begin to write back to the Duke in a vein of strangulated levity: 'To your kind and heartfelt letter I reply at once with a joyful: "I come!"' On 11 April, almost immediately after receiving Goethe's assent, Carl August promulgated the new appointments in Weimar.

In the absence of most of the correspondence, the Duke's reasons for not pursuing his proposal that Goethe should take responsibility for his mother's Italian tour are not clear. He was later to assure Goethe that he would have been happy for him to stay longer in Italy, for a year or more if necessary, and Anna Amalia was reputed to be desperately anxious that he should accompany her there. There can be little doubt that no one else in Weimar was willing to be so indulgent to the favourite—not his colleagues, not Frau von Stein (and so not Duchess Luise), and, alas, probably not Herder either. (And to one or other of these sources must be attributed the malicious rumour reported by Schiller that 'the Duke has refused him an extension of his leave'.) It may have been in deference to these murmurings that, without himself setting a date for Goethe's return, Carl August terminated a scheme which would certainly have doubled the length of Goethe's absence, and Herder, in his account of the calumnies of envy and the dangers of delay, may have glossed a little inaccurately the intentions of the Duke. Certainly Goethe himself described his return as a response to 'a hint from our most gracious Lord, and a summons from my friends', and later attributed to Herder, or some 'evil spirit' motivating him, the responsibility for 'summoning me back', indeed 'driving me out', from 'my pretty little lodging' in Rome. To the Duke Goethe replied with a letter that after its opening is a masterpiece of self-control and proper gratitude for what he is being offered, remarkable for the easy briskness with which he discusses the arrangements for his departure. His grim mood is glimpsed only when he firmly clarifies that he will not be required for particular administrative tasks and that the War Commission will remain attached to the Presidency of the Chamber (so ensuring a continuing financial control over military matters, without himself being responsible for it), and when he unhesitatingly asks the further favour that—given his 'strange and unmanageable constitution' which has caused him 'to suffer much even though in complete freedom and while enjoying a happiness most earnestly prayed for'—his leave may continue in Weimar so that he may complete the last three volumes of his *Literary Works*. And there is a fleeting moment of reproach when he refers to the time it will take to unravel certain 'threads . . . that have been drawn tighter since your letter from Mainz'.

Goethe showed his usual ruthlessness in following through a decision once taken. Letters went immediately to Frau von Stein, Herder, and Seidel announcing his return, to a transport firm in Hamburg, and to his mother, telling her that he would not after all be coming back via Frankfurt. Time was not sufficient for the long visit originally planned, what little was at his disposal he would rather spend in Italy, and besides he did not wish to return across the Gotthard, from which the natural route ran down to Zurich where he could not avoid 'touching the circuit of the prophet'. With his new-found, and now painfully costly, commitment to a court life, to a purportedly Spinozist aesthetics, and to the art of pagan antiquity and the humanist Renaissance, Goethe did not wish once again to be exposed to Lavater's missionary enthusiasm for the scalps of converts—he had just been the target of the embarrassingly familiar and unctuous dedication of Lavater's most recent study of the conversion of the doubter, *Nathanael* (1786). But Goethe was still prepared to take an artist's and a historian's interest in Christianity, indeed, after the initial shock caused by the letters from Weimar, he did not deviate from his original purpose of observing all of the liturgy of Holy Week, completing his correspondence by Tuesday, and then putting all thoughts of departure out of his mind until the festival was over.

Was there any balm for him in what he saw? Certainly not in the press and the struggle necessary to get a good view of the washing of the feet and feeding of twelve pilgrim priests on Maundy Thursday, but perhaps in the 'unimaginably beautiful' singing in the slowly darkening Sistine Chapel that followed, the old office of Tenebrae and the Miserere of Tommaso Bai. The mediation of friends had secured him a place in the Chapel, where on Good Friday morning Allegri's Miserere and Palestrina's setting of the Improperia, the Reproaches, were sung: 'The moment when the Pope, divested of all his splendour, steps down from his throne to venerate the cross and everyone else stays in his place, all are silent, and the choir begins, "Populus meus, quid feci tibi?", is one of the most beautiful of all remarkable functions.' At 11 o'clock on Easter Saturday night Goethe was alone in his room, iron in his soul, writing to Frau von Stein of what he had seen: 'nothing has exactly impressed me, but I have admired everything'. Suddenly, from the battlements of the Castel Sant' Angelo a cannonade resounded across the city, the church-bells all began to ring, and the streets were full of the sound of muskets and mortars being fired in celebration of the Resurrection. The following morning, from a specially constructed gallery high up under the dome of St Peter's, he looked down on to the High Mass of Easter Sunday: 'at that moment I should have liked to be a child, or a believer, in order to see everything in its intensest light'. Even if they were not exactly an impression, the moment, and its associated emotions—of despair, of personal exclusion from a brilliant image of hope, and a resolve to pursue his own path, trusting, however faintly, that it might reach the same goal in the end —these could not be forgotten and ten years later formed the kernel round which was to grow his revision of the opening scenes of *Faust*.

The last month was, of course, hectic in the usual way for such departures:

saying and writing farewells, packing and despatching belongings, arranging, selecting, and discarding papers, buying presents for those at home, procuring a plaster-cast of what was believed to be Raphael's skull, in which Carl August had expressed a special interest, and Goethe had also to fit in last visits to particular collections and works, and in some cases—notably a number of Claudes—first visits too. It proved too difficult and risky to transport his plaster casts to Germany and so the pantheon amid which he had briefly lived had to be distributed among friends who also were being left behind. Angelica received the Juno Ludovisi. Tischbein did not in the event return from Naples, and took his leave by letter, and Kayser was travelling back with Goethe to be introduced at Weimar, but Angelica, Moritz, and Bury—dubbed 'Fritz the Second', as the honorary Fritz von Stein in Goethe's Roman household—had to watch tearfully the daily preparations for the end. 'The departure from Rome cost me more than was right and proper for my years', Goethe told the Duke—indeed he said he 'wept like a child' every day for the last fortnight. Of one parting Goethe said nothing, at the time or later, and it may be that in Faustina's case, as in Friederike's, the essential information was communicated in writing. On Saturday 19 April, five days before he left, he arranged for the transfer to a special account with Reiffenstein, in the name of Philipp Seidel, of 400 scudi: this substantial sum (well over 500 dollars) could only have been collected after he had left and was presumably compensation paid for the early ending of the relationship. It was generous, but no doubt Goethe did not want to be pursued, and he at least had probably not regarded the liaison as anything other than a business arrangement. On the Monday night a full moon illuminated the city and Goethe walked alone down the Corso to the Capitol and the Colosseum, no doubt promising himself inwardly, as he passed the familar monuments made strange and ghostly in the silver light, that he would return to enjoy them in the sunshine once more. On Tuesday he bought a little present for the Duke—a *Cupid and Psyche* with illustrations after the Raphael frescoes in the Farnesina, old friends since he had a similar series at home in Weimar. On Wednesday he said goodbye to Angelica: 'it pierced my heart and soul', she said. Early on the morning of Thursday 24 April, Moritz, Bury, Carl Pieck, and possibly others gathered round the coach in which Goethe and Kayser installed themselves. Moritz was at his best, talking freely, wittily, and easily, even after the postilion had blown his horn. Bury could not restrain himself and had to turn his face away since Goethe had told him not to weep. The coach began to move and, passing quickly through the Porta del Popolo, down the Via Flaminia and over the old, pre-Christian Ponte Milvio, took Goethe out of Rome for ever.

'At the end of the time in Rome I had nothing more to say', Goethe wrote to Knebel a month later, excusing his long silence, 'it was a tough business, taking my leave'. It was certainly the most painful crisis in his life since the death of his sister, and instinctively he realized that the prime necessity for him was the management of his state of mind: 'I have at least seven different moods an hour', he told Carl August in a letter written on the road, 'and I am cordially pleased that this scrap of a letter has fallen into the jolly seventh'. Sightseeing on the two-month journey that now began was undertaken more as a deliberate distraction than for its own sake: a haze lies for us over the details. Goethe kept no diary and in later life made only limited use of what he saw. Kayser's journal gives us the itinerary, but Goethe's soul was withdrawn from the objects around it, preoccupied with itself. Kayser, more of a musicologist than a musician, was more interested in libraries than art and architecture, and the two often left each other to their own devices. Passing through Viterbo and leaving Orvieto to the right, the travellers on 27 April reached Siena, where they spent a day climbing the Torre del Mangia and visiting the cathedral and, among other places of artistic pilgrimage, the church of San Domenico, where Goethe found a great deal of interest, for 'the more practised eye', in the earliest known dated painting, Guido da Siena's *Madonna and Child* of 1221. From 29 April until 11 May, Whitsunday, Goethe was in Florence, then a city two-thirds the size of Rome, which it equalled in its art collections and which, in the eyes of a man of the Enlightenment, it surpassed through being the seat of the government, not of an obscurantist Church, but of Emperor Joseph II's more successful brother, the benevolent, rationalist, and frugal Grand Duke Leopold. Writing to Carl August (whose attention he directed to the merits of the Grand Duke's administration) Goethe claimed to have seen 'almost everything Florence contains in the way of art', which may be taken as evidence of the degree of his abstraction from his surroundings. He later recalled having spent most of his time in Florence in the public gardens, for the weather, which in April had been changeable, was now set magnificently fair. He certainly, however, saw the nearly one hundred ancient masterpieces in the Uffizi galleries, particularly the Niobe and the Medici Venus, but also much early Florentine painting, both there and in the remarkable private collection assembled in the wake of Leopold's dissolution of monastic houses by the pseudonymous painter Berczy. Berczy's collection was chronologically arranged and it seems likely that here at last Goethe made the beginnings of an acquaintance with Giotto, Cimabue, Masaccio, and Fra Angelico.

The stage from Florence to Bologna must have been particularly painful for Goethe: it was the one part of his route which he had travelled before, in the opposite direction. This time he spent a little longer in Bologna—four days, while Kayser pursued his musical researches—and then, skirting the Apennines, went on through Ferraran Modena to Parma, where two days were devoted to a long awaited study of the frescoes and paintings of Correggio. After crossing the Po at Piacenza they reached Milan on the evening of 22 May. Here they stayed for a week, and Goethe, desperately seeking still a pattern in his experiences and a culmination to the great adventure even as it slipped away, drew what sustenance he could from Leonardo's *Last Supper*, 'a real keystone in the vault of one's notions of art'. For Milan was the last great Italian city in the slow torture of the return journey and here Goethe's mood got the better of him: the Alps were growing clearer, the Apennines were but a suggestion on the horizon, the cathedral could not be more Gothic. He allowed his imagination to escape into self-pity: how he would have rejoiced seeing the city for the first time, coming down from the Alps, with all the promise of the Apennines, the blue skies, 'the fruit, etc', before him! On 24 May, a month out from Rome, he wrote his letter to Knebel: scenting the air 'of mountains and the fatherland' he felt 'if not better, then different' and told how a local mineralogical collection had returned him to the mountain pastime of geology, completely neglected in Rome, 'where no stone was looked at if it was not formed. Form had displaced all interest in matter. Now even a crystal is important again, and a shapeless stone is something. Thus does human nature help itself, when there is no help to be had.' He resolved to buy himself a hammer for the passage into Switzerland, and to 'knock away at the rocks, to dispel the bitterness of death'.

The mood of self-preoccupation, Goethe later wrote, enveloped him day and night throughout the journey. The agony of leaving had somehow to be made his own, accommodated to his personality, so that it might eventually issue from him in poetic form. 'But I shied away from writing a single line, for fear that the delicate aura of inmost pain might evaporate. I almost wished not to look at anything in order not to be disturbed in this sweet torment. But soon enough it came home to me how beautiful the world looks when we contemplate it with a mind that has been touched by emotion.' The return from Italy was releasing the power of subjectivity, the well-spring of his poetry, held in abeyance since his arrival in Rome and the completion of his symbolical journey, and of *Iphigenia*. The work that was shaping itself, crystallizing out of this jealously protected emotional suspension, was another verse drama, *Torquato Tasso*, the monument to leaving Italy, as *Iphigenia* had been the monument to the arrival in it. Goethe tells us that he wrote some new material for the play—the first for seven years—in the gardens in Florence, and he bought a copy of Tasso's pastoral *Aminta* in Milan. But the earliest identifiable documentary evidence of the shift in Goethe's sensibility is provided not by his play at all but by the drawings he was making from the very day of his departure from Rome. Although these are all more educated, and skilful, than their counterparts on the journey down from Carlsbad, and although the interest in the 'picturesque corner', or special perspective, shows no sign of returning, there are indications from as early as the second day, in the neighbourhood of Viterbo, that the influence of Kniep and Hackert is being shaken off. Whereas in the (not very numerous) real landscapes from the second Roman period geometrical—even symmetrical—composition and firmness of line are the dominant values, and light in its very evenness seems almost to disappear, the sketches of April and May 1788 increasingly serve a personal, autobiographical purpose, and show a more fluid composition and a breakdown of line into minute, even pointillist, strokes, with an at times quite striking recovery of light (even of reflections) and of plasticity. Strong emotions are sometimes suggested by some rapid sketches of mountain landscapes, perhaps from the road between Florence and Bologna, but in a delightful group of three minute brush studies of trees, only four inches by six, force and shimmering light combine with assurance and control, and one of them surely belongs among the best of Goethe's drawings. Mountains, eventually identifiable as the Alps, bulk ever larger in these scenes. After leaving Milan on 28 May (despite the later importance to him of the locality, there is no evidence, and indeed it is highly improbable, that Goethe visited Lake Maggiore) Goethe and Kayser continued northwards towards Lake Como, where they embarked to sail the full length of the lake on the twenty-ninth. As he moved gently but irresistibly into the embrace of the mountains, Goethe drew like a man possessed: at least twenty views of the lake-shore and the towering heights above, in which the interest in a painterly 'impression' seems sometimes like a faint anticipation of Turner. Goethe was not now drawing with a view to future compositions in accordance with anybody's orthodoxy. He was trying to grasp and fix the last moments of happiness as they slipped away. On 30 May, having spent the night at Riva, he and Kayser drove up the valley to Chiavenna and then on up the tortuous road to the Splügen Pass, where they entered German-speaking lands once more.

The Great Soul: Works, 1787–1788

Goethe's literary work while he was in Rome was, like the Italian journey itself, an arduous collaboration with a long-standing inner necessity. In completing, or rewriting, sketches and fragments first conceived many years before, the inspired expressions of now dead moments, he was, quite as much as in coming to Italy in the first place, fulfilling the plans of the father who had once made it a lesson in diligence to complete his son's drawings for him.

Finishing my older things is being astonishingly useful. It is a recapitulation of my life and my art and by being forced to go back and adapt myself and my present way of thinking, my more recent manner, to my first one . . . I get to know myself . . . Had I dropped the old things and left them I would never have got as far as I now hope to.

This is a truer insight than the claim to Göschen that had he not spent 'priceless moments' on the work of revision he could have presented the publisher with four volumes of completely new material: posterity may question how far Goethe was really likely to progress with *Nausicaa, The Mystified*, a botanical treatise perhaps, and a Wilhelm Meister prosecuting his theatrical mission among the Italians, and no doubt putting on productions of *Jest, Craft, and Vengeance*—and posterity may reasonably feel that it would rather have the versified *Iphigenia*. For all the superficial appearance of new worlds and new starts, Goethe's time in Italy was, at its profoundest level, a 'recapitulation', like the Swiss journey of 1779, but more successful. It was an attempt to make sense of his life so far, to integrate the man he had been, and the art he had practised in Frankfurt, with what he had been and had done for twelve years in Weimar, to accommodate his Storm and Stress origin to the reality of a courtly existence rather than be reduced to a sterile silence by their incompatibility.

Caspar Goethe would assuredly have been particularly pleased to see his son take up *Egmont* again, if only because, of all his earlier works, it offered fewest prospects of easy integration into Weimar. Its exceptionally explicit opposition to courtly absolutism had been the fundamental reason why Goethe had eventually abandoned the project in 1782, even though it was so well advanced. It was politically impossible to complete it in its original form, and Goethe's attempts at an adaptation had only created new difficulties. The distinction introduced between the public and the private Egmont, parallel to that between the poet's public and private self, threatened the unity of the entire play. Clärchen was given a new role as the focus of Egmont's 'real' life, but was this personal relation between two characters enough to outweigh the principal story of revolt, invasion, and power politics? Was that to be reduced to unreality? How to reintegrate the divided self, and how to enter once more into a productive relation with the public world were questions for Goethe as well as for his drama. During his Roman summer, and with the aid of Sentimentalist aesthetics, Goethe, in his conversations with Moritz, came to see for himself a possible new future at Weimar, with what had in the past eight years been his private role as poet elevated now to a potentially public function, 'as an artist'. To be a creative imitator of the ultimate reality, of the purposeless beauty of the great totality of Nature, was to belong to the most exclusive nobility of all, on which the rest of merely receptive humanity was gratefully dependent. Similarly, and simultaneously, in the rewriting of *Egmont*, a public and national function is rediscovered for the hero, but in virtue not of his political action but of his exceptional private personality, the 'real' self which is revealed to Clärchen, which is the secret of his attractive power, and which—displacing both politics and love—will now form the unifying centre of the work.

Only six months before taking up *Egmont* once more, Goethe had completed *Iphigenia.* In that play he had already successfully integrated a Storm and Stress content with a courtly form by concentrating both form and content on the perspective and preoccupations of a single 'great soul'. Everything pointed towards the question Iphigenia posed herself in the last act: would the real order of things in the world outside herself confirm and fulfil the moral requirements of her heart? Would the world fall into place around the divine voice within her? Would Thoas say yes? If there was in the end something arbitrary about Thoas' saying yes—if the poet finds no way of accounting for the fortunate harmony between the moral order and the way of the world which provides his play with its happy conclusion—the story of his *Egmont* gave Goethe almost immediately the possibility of treating the opposite case. When

Egmont rides carefree into the forecourt of Alba's Dark Tower, he is, like Iphigenia, entrusting himself to the goodness of another human heart, trusting that, if it is only permitted to do so, that heart will show itself in instinctive harmony with his. But Alba is a Thoas who says no. The play ends, not in a glimpse of fulfilment, but in death, the ultimate denial of the moral pretensions of our hearts. The grim undercurrent to Goethe's reflections after his return from Naples was the passage of time, the necessity of death, the importance of the moment as the measure of human happiness and achievement. These themes are central to his final revision of *Egmont,* for they are used to define the extraordinary nature of its hero's personality.

Egmont is from the start distinguished by his attitude to death. He is first introduced to us in association with a motif that will remain peculiarly his own (the leitmotif seems a more deliberate element in the structure of this play than of *Iphigenia)*—the horse, a figure of nobility and boundless energy. But it is a horse that is shot dead beneath him. In the first scene, a soldier in the Brussels crowd tells of the near-suicidal cavalry charge which Egmont led across the sea-shore at Gravelines, head-on into the French artillery, which was overwhelmed. The incident is later recalled by Clärchen, who has seen it depicted in a wood cut, and by Egmont himself in the very last scene of the play: 'Now life is coming to an end, as it could have come to an end earlier, much earlier, even on the sands at Gravelines.' Egmont's attitude to the business of living is in essence a refusal at any time to be daunted by, or even to consider, the possibility that the next moment might be his last: 'Do I live only in order to think about living? Shall I fail to enjoy the present moment, in order to be certain of the next? And then consume that one with cares and notions?' To this first principle of his being he remains as true as Iphigenia to hers. He accepts Alba's invitation and enters his palace in the same spirit in which he charged the French positions—to do otherwise would be to betray his own self—but this time the fire strikes down the rider, not the horse. He has refused to allow the present moment to be corrupted by care for the next, knowing, as he did at Gravelines, that the next moment might be deadly, and this time it is. This principle, by which he lives and dies, is the secret not only of his courage, and of his energy, but also of his attractive power over others. Whatever is said of him in his absence, in the moment of meeting—whether with the turbulent crowd, or with Clärchen's mother or with his own secretary—he disarms criticism and gains hearts by his direct and complete engagement with the moment's task: he immediately understands the position and needs of the others—perhaps better than they do themselves—and asks (almost as if he wanted to teach a lesson) for a similar understanding of his own needs. It is an infectious egocentrism, as if he were appealing to the best in others in the belief that it is identical with what he knows to be the best in himself. Carefree, and happy with who he is, he finds it a strange folly to live a life gnawed by care or dissatisfaction. But the energy flowing so uninhibitedly through him makes him a dangerous man to be near.

Carl Philipp Moritz saw each individual as a centre of 'active force', reflecting and derived from the force that drives the universe; but in those in whom that force is pre-eminently present, but is not transformed into the 'plastic force' of the representational artist, it manifests itself as destruction: lesser centres of force must either subserve its interests or be swept away. It is Apollo, Moritz says, the image of youth and beauty deified, as in the Belvederean sculpture, who fires death-dealing arrows upon men. Brackenburg, Egmont's secretary, and Clärchen are all inconvenienced by Egmont in his life, and the last two are themselves destroyed in the catastrophe of his death; even the Regent, Margaret of Parma, is in a sense one of Egmont's victims: she too has been touched by affection for him, but knows that his plain speaking and untrammelled living are among the chief sources of the unrest that brings her political career to an end. Egmont has something of the demiurge as Moritz imagines him; like the Belvederean Apollo he is attractive, masterful, and negligently lethal. Indeed Egmont compares himself, if not to Apollo, then to Phaethon, when, in a passage Goethe almost certainly wrote in Rome, he links his leitmotif of the horse with his characteristic understanding of time as a cataract of moments so discrete that even pronouns scarcely retain their identity from one sentence to the next:

As if lashed by invisible spirits the sun-horses of time are bolting away with the light carriage of our fate; and we can only keep calm and valiant and grasp the reins and steer the wheels, now right, now left, here from a stone, there from a fall. Whither the race, who knows? He can hardly remember whence he came.

In December 1788 Moritz told Caroline Herder that, as in a perspective drawing (and, we might add, as in Leibnizian metaphysics), so in a play one should look for the central point from which all the structural lines radiate outwards, and he went on to identify the central point of *Egmont* as the scene at the formal centre, the only scene in which Egmont and Clärchen are both alive, and both on the stage together (III. ii). The scene begins with a bitter-sweet song from Clärchen on the pangs of love, which is subjected to a commentary full of foreboding:

MOTHER . . . Youth and fine love, it all comes to an end; and a time comes when you thank God for a place to creep in and take shelter.
CLÄRCHEN [shivers, is silent, then starts]. Mother, leave time to come of its own accord, like death. Thinking of it in advance is terrible—So what if it comes! If we have to— then—we will behave as we can.

There follows at once what is effectively a theophany: Egmont enters dressed as Clärchen has requested in the full glory of his court attire, rather as Jove appeared to Semele in the full majesty of divinity. The scene concludes with Clärchen kneeling before Egmont, hearing of the distinction between his public and private identities, receiving the confession of the private Egmont's love and exclaiming: 'So let me die! The world has no joys but these!'—unaware as yet that by inviting into her house not only the private but also the public Egmont she has indeed brought upon herself a ruin as complete as Semele's, if less immediate. In choosing as the kernel of the play this moment when Clärchen, in a state and posture of total dependence and 'feminine' receptivity, is irradiated by the fully revealed 'masculine' active power of Egmont's semi-divine personality, and voices her presentiment of her own destruction, Moritz was no doubt reflecting Goethe's own analysis. When Egmont says in Act II , 'I have risen high and can and must climb higher; I feel in me hope, courage and strength ['Kraft']. I have not yet reached the summit of my growth', he is not only saying that his life must contain what Herder would call a 'maximum', a moment in which his individuality is supremely expressed, but is also unconsciously identifying for us that maximal point as the next scene in which he will appear, his meeting with Clärchen. And it is in the nature of such maxima, it will be remembered, to be the harbingers of the mortality through which the universe remains ever young.

According to Moritz, the centrality of Egmont's relation with Clärchen, and its close association with the death of both characters (an association which only the semi-divine view of Egmont's personality makes natural), relegated the political issues to the margins of the play. Goethe probably thought the same, and felt as a result that his drama was less embarrassing and needed less rewriting than he had once imagined—in particular it is likely that Act IV, which in 1781 he had 'hated', was left much as it was. His principal task in 1787 was, he wrote, 'ending' *Egmont*, and that probably meant writing most of Act V, except for the greater part of Egmont's dialogue with Alba's son, Ferdinand. In the opening scenes of that act there is a preponderance of that highly literary language in which, for example, in Act II Egmont invokes 'the sun-horses of time', and which assorts so ill with the more naturalistic idiom of the early crowd-scenes that it is likely to belong to the last stage of composition. These are the scenes which show Egmont as a political failure: freeing him from prison is not an undertaking in which the citizens of Brussels are prepared to risk their skins. Since the grounds for Egmont's popular appeal now lie not in the cause that he represents (as must at least partly have been the case in the original version), but in the dynamically egocentric individual that he is, their attitude is not inconsistent with their earlier adulation. Clärchen's rallying cry is not 'Liberty or Death', but 'Egmont's liberty, or death!', and that is not a political cry, and it does not get a political response. Similarly, in the second scene of the act, Egmont consoles himself with hopes of an uprising by the people which will secure not their freedom but his, and these hopes remain illusory. But it is not of course Goethe's purpose to show us either a love or a way of life which, being purely—or even selfishly—personal in their scope, fall victim to large-scale public events which they are neither willing nor able to influence. His aim is precisely to find a mechanism by which Egmont's way of life and love, though detached now from any immediate political efficacy, can none the less be restored to a general, and even a political, significance.

For in his own life, too, Goethe is trying to find his way back to a public role after years of imprisonment in a solitude from which no one in Germany showed any interest in releasing him. Before going to Naples he had written to Frau von Stein of what he hoped to learn from the Italians, whose 'incredibly carefree nature' protected them 'from growing old':

my situation will be [the very happiest] as soon as I can think *only* of myself, if I can banish from my mind what I have so long regarded as my duty and properly convince myself: that man should snatch whatever good comes his way as a fortunate prize, and look neither right nor left, let alone concern himself how well or ill some *totality* is faring.

To take thought to achieve such an ambition is of course self-defeating, and in Naples Goethe soon learned that he was not really interested in going native. Back in Rome however—and partly no doubt in conversation with Moritz—he saw a way by which 'thinking only of himself' could become a service to the totality to which he had decided he belonged. The artist could have no higher duty than to be himself: in Moritz's treatise we learn that the 'active force' that animates the world-order manifests itself pre-eminently in the creative soul of the artist, but also, to a lesser degree, in the receptive souls of his audience, who, by his mediation (that is, by seeing the world order in his works), also take part in the universal life. The artist's self-fulfilment is at once the fulfilment of a function for all.

Egmont is not an artist; but he is a pure manifestation of that active force, standing, in Moritz's words, 'at the highest point of his activity; war, wrath, the cry of battle, supremest life, is present at the very brink of his destruction'. Egmont lives for others by living through to the end, to death itself, the principle that he will allow death no dominion over his thoughts. In his cell, to which he has come only through the fearlessness which the anxious call folly, he passes on to Ferdinand not—as was probably the case in Goethe's original conception—the standard of revolt, but an example for all of the highest form of life: 'I am ceasing to live; but I have lived. Live thus too, my friend, gladly, and with pleasure, and do not flinch from death!'

This principle, the heart of Egmont's being, is manifested with most explicitness when Egmont, at the 'summit of his growth' and the centre of the play, reveals himself, in splendid intimacy, to Clärchen. When his hero comes to die for remaining true to the principle, Goethe gives him, in his last minutes before execution, a vision of what he is remaining true to: it bears the features of Clärchen, and so for only the second time in the play the principal male and female characters are on the stage together and the moment of Egmont's death parallels the supreme moment of his life. If in Frankfurt Goethe's first conception of the play was essentially political, and in Weimar the division in Egmont's personality became central, in Rome the relation between Egmont and Clärchen has become the dominant motif. Although Egmont's freedom from the fear of death, and so the vigour of his commitment to life, may be appropriately symbolized by Clärchen, however, we might none the less ask why the symbolic vision should also bestow on him the crown of victory in the coming war of liberation, as if the purely personal freedom for which he is dying were a political freedom. Perhaps, indeed, it is not political—though it seems to be an implication of the play that political freedoms, institutions, and traditions are valueless if men are not also freed, as Egmont is, from care—but the point of the final scene is to show us that it is certainly not selfish. Egmont's way of life is a human ideal, he fulfils completely the potential of the 'active force' in which all share, if to a lesser degree. By maintaining that way of life, even at the cost of personal death, he gives to all an image of what they are living, fighting, and dying for, which can inspire them as they could not be inspired by the task of prolonging the years of Clärchen's lover:

I am dying for the freedom for which I lived and fought [namely, at Gravelines], and to which I now sacrifice myself in suffering . . . Friends, better cheer! . . . to save what is most dear to you, fall joyously, as I now give you an example!

In some cryptic sections of *On the Plastic Imitation of the Beautiful,* which almost certainly are his reflections, in the light of his reading of Herder, on what he had heard and seen of the manuscript of *Egmont,* Moritz tells us

Humanity cannot . . . raise itself higher than to the point at which, through nobility in action, it draws the individual out of his individuality, and perfects itself in the beautiful souls which are capable of stepping over from their limited selfhood into the interests of humanity and losing themselves in the species.

In particular, Moritz belives that in tragedy the real suffering of an individual's death is turned into a representation, or 'appearance', of death which, as a symbol of the process by which life is maintained, is no longer supremely terrible (for the individual) but rather supremely beautiful (for the species). Since in German the one word 'Erscheinung' means both 'appearance' and 'apparition', we may take it that for Moritz the dream vision of Clärchen is the moment when Egmont's death ceases to be a terrible reality and becomes a beautiful appearance. In his last minutes Egmont is being transformed, and he knows it, from a suffering individual into a public symbol. 'I now live for you', he says to Ferdinand, who has already received Egmont's horse, the figure of the ever-youthful life of the universe, 'and have lived enough for myself'. He has from the start of the play—in Clärchen's woodcut or the conversations of the townspeople—been a significant image as well as a physical person, and now he is becoming one of the most efficacious of political images—a martyr. Perhaps the fullest and most direct of Goethe's treatments of the theme of death issues therefore in the conscious transformation of its hero into an immortal ideal. Egmont's recovered equanimity in the face of death accompanies a willing submersion of his limited identity in the operatic paraphernalia of mass-politics: strident rhetoric, rousing music, slogans, legends, simplifed images and emotions. Goethe is of course principally concerned with the overcoming of the division in Egmont's personality by the discovery that a private and personal identity can have a public function. But if he has thereby coincidentally anticipated the mechanisms of ideological politics that were to dominate Europe in the coming revolutionary age, it is by one of the coincidences that are somehow the lot of genius.

Reactions in Weimar to the new version of the play were mixed. Carl August, though of course long familiar with its theme, must have been slightly irritated by the close and unflattering analogy with his recent campaign, and wrote a long critique of those scenes which betrayed to him the author's inexperience of high affairs of state. In general it would seem to have been the recent additions to the last act that were most criticized. Frau von Stein appears to have objected to the apotheosis, in the play's closing moments, of a loose-living woman. Schiller, in a most acute review, published in the *Allgemeine Literaturzeitung* in September 1788, described the conclusion as 'a *salto mortale* into a world of opera', and analysed 'the ethical part of the play quite well', Goethe thought, but showed little understanding of its 'poetical' part. Certainly Schiller fails to see the underlying connection between the manner of Egmont's life and the manner, and transfiguration, of his death. In Goethe's later, and not wholly good-tempered, words: 'the unpoetical amateur of art, comfortable in his self-satisfied philistinism, usually takes offence at the point where the poet has sought to resolve, adorn, or conceal, a problem. The comfortable reader wants everything to continue on its natural course, but even the unusual can be natural . . . ' Probably he felt that there was the same justification for the two planes of reality in his final act as for the two centres of action, one earthly and one heavenly, in Raphael's *Transfiguration*: if one wishes to depict 'a relation between the world of ideas and the world of real events' ('ein ideeller Bezug aufs Wirkliche') both must be present in one's depiction. But in the art which Goethe had mastered in 1774 and 1775, in the first version of *Werther* and the *Urfaust*, for example, that relation was present without the need for a depiction of two separate worlds. 'The dream had rather a frigid effect', wrote an early reader of *Egmont*, 'as did Clärchen's suicide: there is something in them of the new letters in *Werther*', a remark which not only precisely identifies that 'cold and brazen' element which Goethe knew had entered his sensibility while he was in Rome, but which also diagnoses the very problem in Goethe's development, and the development of his play, which he was endeavouring to conceal: that in twelve years at Weimar the needs of the heart and the realities of social life had drifted so far apart that in the new *Werther* they are completely divorced, and only by an act of aesthetic violence can they now be brought together in *Egmont*.

Of course *Egmont* also demonstrates that, in Rome, Goethe had found the will, and made a first attempt at finding the means, to bridge the gap. The task was so difficult, however, that, he acknowledged, 'the play is there, more as what it could be than what it should be', and this perhaps explains why with his little operas he did not even make the attempt. In both *Erwin and Elmira* and *Claudina of Villa Bella* the stratum containing the problems of the 1770s, and particularly all trace of Storm and Stress loyalties, is obliterated. The introduction of a new pair of lovers, parallel to Erwin and Elmira, and the elimination of Olimpia and Bernardo, had been decided before Goethe went to Italy, in order to provide a more conventionally constructed cast and the opportunity for more trios and a concluding quartet. But with its two representatives of the older generation the entertainment loses also its authoritative middle-class voices, whose realism was sometimes old-fashioned and sometimes subversive, and was in tension throughout with the Sentiment-alism of the two youngsters. The engaging element of buffoonery, which mediated between these two forces, is reduced to theatrical farce, and similarly that other intermediate emotion, Erwin's moment of nostalgia for the hermitage, the place of deprivation and longing which he is leaving for ideal bliss with Elmira, is emphatically deleted:

> Ohne Thräne kann ich lassen
> Diese Hütte, dieses Grab.

Without a tear I can leave this hut, this grave.

Emotional complexities are also ironed out in the rewritten *Claudina*—and with them, and more surprisingly, dramatic complexities too: gone are the jealous nieces and Crugantino's (now, probably on the model of a young Roman singer, Rugantino's) competition with his brother for the love of Claudina, gone also, with the elimination of the family friend Don Sebastian, is the depth of perspective provided by the older generation and by the opposition between Crugantino's Storm and Stress and the socially acceptable Sentimentalism of Claudina's milieu. Rugantino, tired of outlawry at the beginning of the piece, is paired off with Lucinde, as Claudina with Pedro, and all villainy is left to Rugantino's companion Basco. The action is simplified to the point where it threatens to run out in the middle of the last of the three acts into which the work is now divided, and a new episode has to be invented to prolong it: Goethe's attention was clearly waning and an error in the naming of the dramatis personae escaped him even in proof. The new *Erwin and Elmira*, though reduced to the parallel tiffs and reconciliations of two sets of lovers, conducted now in rather functional blank verse, is at least a competent libretto. The new *Claudina*, which has too much of the iambic, and occasionally trochaic, recitative, is not even that. Both revisions are evidence for the ferocity of Goethe's determination to establish his credentials as an 'artist'. Only that can explain his publishing to a readership, and as poetic works, texts which purportedly were to be regarded only as subordinate elements in musical entertainments They succeeded as neither, for in them Goethe had not accommodated his past—as he did in *Egmont,* which, though not written as an opera, attracted the attention of Beethoven—he had simply suppressed it.

In *Faust* Goethe knew he faced a challenge of an altogether different order. Of all his fragments it was the one that had longest lain untouched, and in which most was invested. In February 1788, in confident mood, with his operas despatched, and in the middle of his eight weeks of unconditional happiness, he wrote to Carl August that nothing now faced him 'but the hill *Tasso*, and the mountain *Faustus*'. Since only two acts of *Tasso* were at this stage completed, while the *Urfaust* had long been regarded as nearly finished, this assessement clearly implied a new conception of the play, and of what remained to be done to it. The *Urfaust* had shown us a magician summoning spirits of nature out of the night in the hope that by turning his back on religious and moral orthodoxy he could satisfy the endless longings of his heart. The play had not shown us precisely how this figure came to be associated with his sinister shadow, Mephistopheles, but the story of his disastrous attempt to live out his new freedom through his love for Gretchen, and of their being overtaken in the end by the cruel realities of sin and judgement which he had repudiated, had otherwise been coherent and complete in itself. Imaginatively it was wholly true to the spirit and moral trajectory of its folk-story original: it did not need this or that episode or named character—the Pope, or the Duke of Parma, or Helena—to be immediately recognizable as the parable of the man who sells his soul to the devil. Only someone who no longer understood the symbolic relation between the *Urfaust* and the puppet-plays which inspired it could imagine that Goethe's draft was a defective rather than a modern *Faust*. Perhaps Goethe in Italy had temporarily lost that understanding, but it is more likely that in revising his old scheme he was motivated not by a desire to complete something that needed little completing and already had its own poetic integrity, but by a driving need to change its purport altogether. What he may have had in mind is suggested by a passage in Moritz's *On the Plastic Imitation of the Beautiful*, which is evidently a reflection on the blank-verse monologue of Faust, now in the scene 'Forest. Cavern', which Goethe in all probability wrote in Rome, or on the way to it, in the winter of 1786–7. Moritz uses a paraphrase of the monologue to describe the 'calm feeling of self' through which a soul passes in its transition from the extreme manifestation of the 'destructive active force' to the operation of the 'gentle, creative plastic force' which is responsible for artistic representation. Moritz therefore seems to have imagined the monologue as coming at the end of a period when Faust's activity has been at its most destructive—no doubt at the end of the Gretchen story—and as marking the transition to a period when Faust would become an artist, as Goethe and Moritz now understood the term. There is no reason to assume that this impression of Moritz's did not derive from what Goethe had told him. By the winter of 1787–8 at the latest Goethe must have resolved, then, that if *Faust* was to be completed, it was not to be as the tragic drama of the collision between a modern sensibility and the world from which it had emancipated itself. At the very least it was to be continued, and continuation was hardly possible without a change of mode, scale, and conclusion, and in particular without a change in the status and implications of the Gretchen episode. Goethe's move to Weimar had, after all, been a commitment to search for a place where the emancipated sensibility could be at home, and to search for it in a social and political context on which the *Urfaust* had not even touched. In the world of the court and of officialdom the only potential tragedy for the isolated heart was that of self-destruction. Over the years that threat too had receded, and the enemies of feeling were now tedium, pointlessness, transience, and death. The soul had to assert itself and its innate noblity: as a creative, artistic force, the soul was the equal of any courtly power and was elevated as far above the domesticity and religiosity of Gretchen's narrow, German, burgher world as any baron on the grand tour.

There had always been too intimate a bond between Goethe and his Faust for the poet simply to fill the gaps in the narrative and publish it as over and done with, 'more as it could be than as it should be'. Even though he now hoped to complete the play by Easter 1789, Goethe still felt the need to make Faust's life incorporate the more important of the new possibilities he had discovered in his own. He needed to prove to himself and to others that the yearning heart of a Crugantino did not simply have to be forgotten, but could continue to grow and house new forms of activity. But how were these new possibilities to be accommodated within the existing structure of the play, and of Faust's career? The traditional format, it was true, permitted any number of episodes to follow on Faust's contract with the Devil, as magical fulfilments of his wishes, bought at the price of his soul. But Goethe had not been able to formulate an agreement in which he could believe between Mephistopheles and his modern Faust, and the one major episode in his tragic tale, Faust's involvement with Gretchen, sprang directly from Faust's religious and moral emancipation, depicted in the play's opening scene, and required no diabolical intervention to explain it. Moreover the moral catastrophe which had terminated Faust's grand experiment in living beyond good and evil had left Goethe with a concluding scene, 'Prison', after which it had been next to impossible to imagine any continuation. During his blessed eight weeks in the early spring of 1788, however, Goethe drew up a plan (now lost) and wrote one, or more probably two scenes which established a new framework for the drama and opened up the serious possibility of a replacement for the tragic conclusion of the *Urfaust*.

Of course Goethe's mind must already have been turning to *Faust* throughout the winter of 1787–8. *Egmont* was finished, but its concern with the exceptional individual, with a life lived to the full, with the high points of human experience and their necessary passing, was prolonged in Goethe's conversations with Moritz and in his reading of Herder's *God* and the third part of his *Ideas*. As the prospect of return to Weimar drew closer, he had finally resolved to arrange his first sexual liaison, and thoughts of time and pleasure and the moment preoccupied him. What shape was he to give to the life that stretched uncertainly before him but that could no longer be ordered around symbolical gestures of fulfilment, a life in which no Messiah was to be expected and desire was a devilish power? In a world as ruthless and godless as that of Moritz,

Herder's was the only gospel: the worth of human life lay in the accumulation by the species of unique and unrepeatable, 'maximal', moments of achievement. But of what interest was such collective accumulation to the individual? Having reached the one 'summit of his growth' was he, like Egmont, to be content to perish? Was Italy to be Goethe's summit, and such life as he had thereafter to be spent in the condemned cell? The old obstreperousness of Hanswurst bubbled up again, a recalcitrant resistance to all this deadly secular piety: 'it always comes down to the same point', said Egmont, 'I am to live in a way I don't want to live'. Perhaps Goethe recalled Humanus, the central, Herder-like figure in the *The Mysteries,* in whom all the 'maxima' represented by the religions of mankind were to be united. Why should not he too, like Moritz's Leibnizian 'artist', reflect within his own compass the wholeness of the universe? In February 1788, after a twelve-year silence, Goethe's Faust, in mid-sentence, exploded into speech once more:

> Und was der ganzen Menschheit zugeteilt ist,
> Will ich in meinem innern Selbst genießen,
> Mit meinem Geist das Höchst' und Tiefste greifen,
> Ihr Wohl und Weh auf meinen Busen häufen,
> Und so mein eigen Selbst zu ihrem Selbst erweitern,
> Und, wie sie selbst, am End' auch ich zerscheitern.

And I wish to enjoy in my inner self what is allotted to the whole of humanity, to grasp with my spirit the height and the depth, to heap on my bosom its weal and woe, and so expand my own self to its own self and like itself at last be shattered too.

It was, in thought and in vocabulary, a continuation of the overweening grandiloquence with which Faust had summoned the Earth Spirit. But it contained a wholly new programme for the the development of the drama, after that first night of visions. Unmoved by the remote and abstract symbols of universal order in the sign of the Macrocosm, rejected by the Earth Spirit as too puny to share in the mighty energies by which the natural world is kept in being, Faust was to turn instead to the field of human activity—but of *all* human activity. *Faust* was to become the story, not of a man seeking to live free from moral, natural, and theological constraints, but of a man seeking to encompass within himself all the possibilities and achievements of the human race, to incorporate humanity with all its 'maxima', its 'crowns' as he here calls them, into a single individual soul. If his play was to represent even a fraction of Faust's attempts to realize this encyclopaedic ambition, it is no wonder that Goethe now regarded the task of completing it as mountainous.

However, the dialogue between Faust and Mephistopheles that followed on this would-be titanic opening at least held out the promise of resolving one of the major uncertainties left in the *Urfaust:* the manner of Mephistopheles' introduction into the plot, and the nature of his relationship with Faust. The scene was intended to follow almost immediately on the great first scene of the play: whatever the precise form in which he first manifested himself (that was still unclear) Mephistopheles would be, after the Macrocosm and the Earth Spirit, the third and most prosaic of the supernatural forces conjured up by Faust the magician. As for a formal agreement, a selling of Faust's soul to the devil, in all probability there would have been no such thing, had Goethe carried out the plan he at this time envisaged—Faust might even have expressed Goethe's own now vehement disbelief in the Christian scheme, which alone could give meaning to such a sale. Instead Faust would simply agree to allow Mephistopheles to be his companion, and Mephistopheles would agree to give Faust whatever assistance he could in realizing his grandiose ambitions. Such a loose, and theologically inexplicit, association would have been in keeping with the partially modernized form of the legend which the *Urfaust* had always offered. Goethe's ninety-eight lines of dialogue do more than simply establish this association, however: they also sketch out a new version of the fundamental tension between the two partners. In the *Urfaust*, Faust's belief in the possibility of a life beyond good and evil, and in the supremacy of human feelings, becomes the instrument of his and Gretchen's tragedy: Mephistopheles, for his own good reasons, does not share that belief, but is able to use it to extend his power over Faust and Gretchen. In the new scene a difference of opinion remains, and a threat of destruction should Mephistopheles be proved right; but the destruction threatened is not significantly moral, and Gretchen is not centrally involved. Faust wishes it to be possible for him to *be* all that is open to humanity: Mephistopheles mocks this ambition—how will Faust unite in one man bravery and cowardice, naïvety and calculation?—but he thinks it fully possible to *have* all the satisfactions human life affords, and he is a specialist in providing them. So let them plunge 'straight into the world', and while Faust thinks he is pursuing by the only means available to him, his ideal of expanding his self, Mephistopheles will in fact be supplying him with empty pleasures (for as a nihilist the devil thinks pleasure both to be the only reality, and to be ultimately empty). In this hectic course in human experience Faust will, in Mephistopheles' view, certainly not satisfy his nobler—or at any rate more metaphysical—ambition, and may well lose it altogether in a random quest for sensual gratification, caught eventually like a bird in the lime.

By the spring of 1788 Goethe was probably rather embarrassed by Gretchen. Her story was elevated neither in its subject nor in its setting—his mind was on Junos, not German wenches—and there was too much Christianity, and perhaps even too much tragedy, about it for his present taste. *Faust* was altogether too grand a work to peter out in a peccadillo—the great souls were necessarily destructive of those near them, and if Egmont needed his Clärchen, or Goethe his Faustina, this was, at least in one respect, a biological matter of no eternal significance. The new theme for the play scaled down the Gretchen intrigue in two ways. First, it implied that the action of the play as a whole would now consist of a series of episodes, each of which would recount, in a different context, Faust's grasping after 'the crowns of humanity': among these the Gretchen story would figure simply as one of several, instead of being the single crucial test of Faust's magical way of life. Maybe it would lose its devastating conclusion. Almost certainly it would not be the last episode in Faust's career. Quite possibly it would be followed by scenes traditional in Faust plays: a visit to the Imperial court, for example, which would take the action into the political world, or the conjuration of Helen, the quintessence of ancient beauty, which might have been the occasion for turning Faust into an artist, if Moritz's report is correct. The story of Faust and Helen had always appealed to Goethe but had been displaced in the *Urfaust* by that of Faust and Gretchen: now perhaps Goethe saw the possibility of making amends for his discourtesy to a classical queen. Second, the dramatic tension common to these different episodes would no longer be the moral and religious tension which sustains the *Urfaust*—which vision will prove more adequate to the human reality? that of Faust? or that shared in their so different ways by Gretchen and Mephistopheles? Instead, the dramatic interest will repose on a question about Faust's spiritual stamina—can he maintain his original vision and purpose, or will he in the end accept the Mephistophelean assessment of what is worth having in life—'the joys of earth' —and lose himself in the vanities of sense? Faust's involvement with Gretchen will no longer raise the question whether he has sinned, but whether he has become a sensualist.

Much more overtly than the dialogue between Faust and Mephistopheles, the second scene that Goethe wrote, or at least began, in the spring of 1788— many years later he recalled composing it in the gardens of the Villa Borghese — was concerned with modifying the status of the Gretchen story. But 'Witch's Kitchen' also met an important formal requirement. Since Goethe was not at this stage planning a contract between Faust and Mephistopheles (and the signing of the pact—in blood—was one of the longest and most colourful scenes in the Faust puppet-plays), something more striking and ceremonious was needed to inaugurate their joint venture into 'the world' than a conversation in which one party inquired 'How do we set about it?' and the other replied 'We just leave.' Goethe had recourse to a theme previously hardly connected with the Faust story, and almost entirely absent from the *Urfaust* (it may have come to him through Fuseli and Lavater)—that of the witch and witchcraft. Mephistopheles, as an intimate of witches, introduces Faust to one of his acquaintance who, amidst her steaming apparatus and cavorting monkey familiars, prepares a magical potion which rejuvenates Faust by thirty years and additionally, or thereby, has an aphrodisiac effect. The motif may derive from Annibale Carracci's fresco, in the Palazzo Farnese, of Circe offering the cup to Ulysses, but Goethe when he took up *Faust* again was anxious to recover the tone and manner of his original scenes, which was very far indeed from any kind of neo-classicism. With some justification he felt that he had achieved this aim in his dialogue between Faust and Mephistopheles. However, the farcical, if unfunny, hocus-pocus of 'Witch's Kitchen' has nothing in common with the profound nature-magic that opens the *Urfaust*, nor with the true and sinister devilishness of Mephistopheles in the Gretchen scenes, while even the versification is at times flat and padded, or reminiscent of the jingles in the recently rewritten operas. The artificiality is partly deliberate, for it belongs with a new clarity in Mephistopheles' identification of himself as the Satan of Christian mythology or at least an emanation of him—in the *Urfaust* he is unsettlingly evasive about the matter. As a more supernatural figure he thus becomes less credibly serious, especially for an Enlightened eighteenth-century audience, and the threat that he poses to Faust in the later scenes, as a reminder of the real possibility of moral destruction, is correspondingly reduced.

Drinking the witch's potion, in Goethe's plan of 1788, sealed the bond between Faust and Mephistopheles: it launched Faust on a magical existence of indefinite duration without requiring an agreement about the eternal future of his soul, and it had the further advantage that, if Goethe wished eventually to redeem his hero, it could always be repudiated, provided Faust were willing to forgo the rejuvenating effect and return to the natural processes of age. Making Faust thirty years younger was a motif probably suggested to Goethe in the first place by his own feeling that in Rome he had experienced a personal, and sexual, 'rebirth'. It was certainly not a dramatic necessity: Goethe had not added anything to his text that unambiguously made his hero older, and the Faust who made love to Gretchen was clearly not a 50-year-old. Indeed the rejuvenation introduced new problems of consistency, for the Faust of the opening monologue is, as was noticed on its publication, still rather too turbulent for a man of 50. Much more seriously—and unfortunately one cannot be certain that Goethe at the time did not intend this—the new scene degraded Gretchen's love for Faust, for its object was no longer a passionate young genius but an animated waxwork. Goethe at this stage, however, was interested much less in Gretchen's perspective than in Faust's. And in two complementary ways 'Witch's Kitchen' modifies Faust's relation with Gretchen, aligning it with the new concepts in the Faust-Mephistopheles dialogue Goethe had just written, but subordinating it to the larger, more encyclopaedic scheme of the play, and subordinating Gretchen herself to Faust's concerns. On the one hand, Faust discovers among the witch's equipment a magic mirror which gives him a vision of perfect female beauty: this ideal inspires him to try to leave at once and set out to seek it even before he has drunk the potion. His life thus acquires an overriding purpose, analogous to his original aim of attaining the crown of humanity and largely independent both of the programme Mephistopheles has in mind, and of the assistance he can give: a quest for a feminine ideal; and Gretchen and Helena, we may take it, are intended by Goethe simply as stations on the way. On the other hand, the potion itself is emphatically physical in its effects, suffusing every pore of Faust's body, and causing 'Cupid to stir and spring up and down'. It is an instrument of Mephistopheles' intention that Faust should not 'overleap the joys of earth':

> Du siehst, mit diesem Trank im Leibe
> Bald Helenen in jedem Weibe.

With this drink in your belly you will soon see Helen in any woman.

Any female body will do for 'enjoyment' as Mephistopheles understands it, and the belief in a higher beauty and a higher form of union with it is, for him, an illusion. The difference of opinion between Faust and Mephistopheles on love in this scene reflects their difference in their earlier dialogue. Whichever is right, however, Gretchen's status is reduced: in the intrigue which follows more or less immediately on 'Witch's Kitchen' she is either one of several passing embodiments of a more lasting ideal, or the purely physical object of Faust's artificially stimulated lust. At best she may hope to be an earthly Venus, whom Faust embraces while his eyes are fixed on nobler and more classical forms.

After the arrival of Carl August's letter on 14 or 15 March, Goethe can have had no more stomach for his work on *Faust*. The mood of confidence left him— in a sense, never to return—and he now spoke of the play, due to appear in volume vii of the edition, as something to occupy him in the winter months in Weimar. On the way back he would give his attention to *Tasso*, intended for publication in volume vi at Michaelmas, 'and so on a journey there will be dramatized the fortunes of a man whose whole life was a journeying to and fro'. From Carl August Goethe scarcely concealed the affinity he felt with the Italian court poet or the appropriateness of the subject to his current state of mind: 'as the attraction that led me to this theme proceeded from my inmost nature, so it is quite extraordinary how the work I am undertaking to finish [the play] connects with the finish of my Italian career, and I cannot wish it to be otherwise'. True, the Duke had once advised him against the idea, and, as with *Egmont*, had he not already started it he would not have chosen to take it up now. But he had not been put off completing *Egmont* by the knowledge of the military intervention in Holland and the possible embarrassment that play might therefore cause his patron, and he was certainly not now in a mood to spare the Duke's feelings, however merry his letters might seem. In turning to *Tasso* he was not, as had been the case with *Egmont* and *Faust*, attempting to adapt to a new atmosphere and new phase of his life works conceived in quite different circumstances; nor was he trying to define for himself a possible future life-style: he was trying to think through to the end a work that had from the start been conceived as an explanation of the life he had come to live and to which he was now returning. It required him to formulate what he thought and felt about his present condition. It was a challenge to resume, in circumstances that were emotionally more demanding than any he had known for a decade, that art of the emotions by which he had first made his name: in his old age he would take up the words of a French critic and call *Tasso* 'an intensified

Werther'. The 'elegiac mood' so carefully nurtured and protected on the slow journey back came to dominate the work: by the time he reached Milan he knew that 'the first acts will have to be almost completely sacrificed' to it. As he passed through the landscapes of central and northern Italy in the springtime he looked on them with the eyes of a man about to lose them. Infused with his emotions, what he saw entered the texture of the play, which thus became the most poetic of the 'glimpses of the promised land' that he took back with him, 'as evidence that I have been in Paradise'. From now on—the paradox is striking—the promised land lay behind him: he had enjoyed the fulfilment, or what he was prepared to regard as the fulfilment, of the hopes of fifteen years before, and what lay ahead of him was bereavement, or at best the struggle for recovery. Italy would in future, in his poems as in his drawings, be the image of a happiness remembered, real, but remote. Even though Goethe would always recall his second sojourn in Rome as the moment in his life that most clearly approached to the ideal, and to true happiness, the evidence of what he wrote during that time—the operas, the one poem, the revisions of *Faust* and *Egmont* —is that it was not in moments of happiness, but in banishment from them— hoping for them in the future, or looking back on them as past—that he could practise the art that was truly his own. *Tasso* took new shape in his mind as the Alps grew larger before him—'like a coast on which I am to land after a strange voyage'—and as the vein of poetry was opened within him once again. But the new theme he had discovered for his redrafted play, which it would take him another year to complete, was the suffering in which poetry had its origin:

I took heart and found the courage for freer poetic activity; the thought of Tasso was taken up and . . . I wrote the passages that even now directly bring back that time, those feelings . . . In respect of my fate I could compare myself to Tasso. The anguished yearning of a passionate soul, irresistibly drawn into an irrevocable exile, runs through the whole drama.

8

The Watershed
1788–1790

Old and New Faces: June–December 1788

THE WEATHER deteriorated. 1788 was to be a bad summer in northern Europe and in May hot spells alternated with thunderstorms. Through the dramatically narrow gorge of the Via Mala, of which he made an accomplished sketch, Goethe reached Chur with Kayser in the pouring rain. Their route then lay down the more open valley of the upper Rhine, through Liechtenstein to the southern shore of Lake Constance, along which they drove to arrive, on the evening of 3 June, in the Golden Eagle in Constance itself. Here Goethe made an unpleasant discovery. He read in the newspaper that Herder, whom he knew to be planning his own Italian journey, had already set out and so would not be in Weimar to meet him. It looked as though three years would have passed by the time the two friends saw each other again, and Goethe wrote a letter to await Herder's arrival in Rome, to welcome him enviously in that happy place and introduce him to his artist acquaintances.

I do not know whether I am waking or dreaming as I write this to you. It is a hard test I am having to endure . . . When you get to Castel Gandolfo ask your way to a pine-tree that stands not far from Mr. Jenkins' house, not far from the little theatre. That is the one I had before my eyes when I so longed for you to be there with me. Farewell.

Goethe had last stayed in the Golden Eagle on his return from Switzerland in 1779, as he braced himself to resume life in Weimar after the restorative days in Lavater's house in Zurich. Now he arranged for a reprise of that happy interlude, though it had this time to be without the company of Lavater. From Rome he had suggested that another member of the Zurich circle, Barbara Schulthess, enthusiastic, motherly, and with no missionary ambitions, should spend a few days with him as he skirted the domains of 'the prophet'. At 9 o'clock on the following morning she arrived at the inn with a nephew of hers and with Döde (Dorothy), the second of her three teen-aged daughters, whom the privilege of this outing made the envy of her sisters, consumed as they were with curiosity not only about their mother's famous friend but also about any news of the quiet Mr Kayser, their former piano teacher. Slim and sunburnt, in his brown suit and fine shoes with semi-precious stones on the buckles, Goethe —'oh, I cannot see enough of him he is so friendly and nice' wrote Döde— entertained the young people with tales of the Carnival in Rome and Frau Schulthess with talk of Italian art and artists, which the two found so absorbing that they did not notice the others falling asleep. For a week the party stayed together, walking out in Constance and the lakeside meadows, talking endlessly over long meals at the shared table where, in Barbara Schulthess's words, 'everything was so domestic and I was mother', and where on one occasion Döde was at last able to abstract a memento for her elder sister, a piece of bread

Goethe had kneaded in his fingers throughout the dinner. This was no doubt the return to Germany that Goethe would have wished to have: a happy reunion of family and friends, a wholly middle-class affair untouched by the formalities of rank or public life, a direct and energetic interest in the things he had seen and in his personal reactions to them. It was what he had valued in the atmosphere of Lavater's household, in which his 'withered' heart had once been able to expand again, but from which he was now cut off by anti-Christian feelings, fostered, in part at least, by the court to which instead he was proceeding, and where his welcome would be rather different.

As the carriage moved on through Ulm into Franconia, Goethe prepared himself for his home-coming. He noted down a number of resolutions for his future conduct: 'concealment of present condition . . . not to make comparisons with Italy . . . understand *everyone's* existence . . . not hard or brusque'. After another week's travelling three days of which were spent seeing the sights of Nuremberg, he and Kayser drove up from Coburg into the Thuringian Forest, spending their last night in Saalfeld, and stopping off for the next afternoon in Jena. At 10 o'clock in the evening of Wednesday 18 June they arrived in Weimar by the light of the full moon. At 6 o'clock the following morning Fritz von Stein, who had not been expecting the return of 'the Privy Councillor' until the twentieth, was surprised by a summons to attend again at the house on the Frauenplan which had been his home for three and a half years: he was now nearly 16 and was so overcome by the reunion that he could hardly utter a word. At 8 o'clock, on his way no doubt to pay his respects to the Duke and his official colleagues, Goethe called in at the offices of the Chamber and embraced a new arrival, Dr C. J. F. Ridel (1759–1821), who had been appointed in his absence, but with his support, as tutor to the little Crown Prince, Carl Friedrich. It was a welcome not just to a new face, but to his own past, for Ridel's fiancée, who would come to Weimar in 1791, was Amalia Buff, one of the crowd of brothers and sisters who had clustered for their supper round Lotte Kestner in Wetzlar, in the summer of 1772. But this day was devoted to the present, to the court— 'everyone is fighting to get at him', Ridel wrote to Schiller, himself eager for a meeting—and at dinner that afternoon he was the object of 'general rejoicing and curiosity', particularly from Carl August who monopolized his conversation, now and for days to come. On this occasion at least Goethe kept his resolutions. To the other diners, who included Prince August of Gotha and a retired Danish ambassador who knew Rome and Senator Rezzonico well, he seemed courteous, lively, good-humoured, and more than usually talkative, even if he diplomatically evaded serious issues and spoke always from his markedly personal point of view. The face at the table that must most have gladdened Goethe, however, was that of Herder, for Canon J. H. F. von Dalberg, younger brother of the Statthalter of Erfurt, who was taking Herder with him and had offered to pay his expenses, was not after all leaving for another six weeks, only a little before the Dowager Duchess. She too was as excited by the return of her 'friend' as by the prospect of repeating a journey he had rendered peculiarly enticing—'we have heard nothing but Italy and Italy, it is unbearable' complained the court ladies—and 'they say she will move heaven and earth to prevail on him to go back with her'.

Weimar was to be in turmoil until August, with the preparations for all these departures, and with a stream of visitors, mainly English, presenting themselves at court and requiring to be politely entertained, and Goethe, who now dined at court practically every day, was of course one of the chief attractions at the table. In July the English widower Charles Gore (1729–1807) arrived, with two daughters, Eliza and Emily—Anna Amalia would be visiting a third, Lady Hannah Cowper, in Florence. Gore had made his money in shipbuilding; in retirement he had taken to travel, and as a true compatriot of Sterne's Uncle Toby he had his hobby-horse: he employed his amateur talent as a water-colourist to build up a large collection of views of sailing-ships and other nautical scenes, all drawn by him at first-hand. For an Englishman to be wealthy and well-informed is tantamount to being cultured and would have been grounds enough for Carl August to want the Gores to settle in Weimar, even had there not been the matter of the romantic and unrequited passion he had conceived for Emily, in whom he claimed to see an ideal companion for his stay-at-home wife. Although, however, Eliza immediately discovered a tenderness for Goethe and began to translate passages from *Werther* into Italian, Goethe himself found the family 'fine enough if one lives with them after their own fashion, but in moral and artistic matters they are so limited that to a certain extent I am unable to converse with them'.

With whom was he to converse? It seemed to many that 'Goethe has acquired by his journey, by the freedom he enjoys and no doubt by the hope of returning to Italy, a gaiety he has not had for a long time'. Knebel, who had been in Ilmenau on the eighteenth, had the impression, after their first meeting at the end of June, that 'Goethe is not that unhappy here. He knows how things are and recognises that one must regard those that have passed as a dream. But if the dream was good, memories remain which can make rich and happy the moment in which we find ourselves.' But with whom were those choice memories, those rich and happy moments, to be shared? Frau von Stein had stayed on in Weimar, rather than retiring to Gross Kochberg, in order to welcome Goethe and the Gore family. So greedy, however, had the Duke been for Goethe's company that the Saturday afternoon when Knebel arrived was probably her own first meeting with him of any length, and her verdict was less sanguine than Knebel's: 'I was with Duchess Luise and the Herders at Goethe's. He showed us some engravings of Claude Lorrain and some ancient cut stones. We had not been with him long when Knebel came in too, and so our old group was together; whether with the old spirit I rather doubt.' She had her own reasons for thinking that since 1786 an epoch had ended: it was just a year since her invalid son Ernst had finally succumbed, at the age of 20, to what was probably bone-cancer; the health of her husband, who in October would suffer the first of a series of crippling strokes, had already begun to decline; her sister, Luise von Imhoff, had come back to Weimar to live separately from her husband, who had now settled in the artist colony in Munich and was only a month away from death; and she herself at 46 now faced the change of life. Old intimacies, even if continued and untroubled, would have been a frail enough consolation. But for her there was no glossing over the act of deception that had begun the two years' infidelity whose end was now being celebrated. 'In my heart too there is a premonition that our circle of amity will be sundered and our little group scattered. Goethe has on his conscience that he took the first step . . . ' Yet only in the dialogue with her, which had so long enabled him to distinguish his *dehors* and his *dedans,* could Goethe hope to express or discuss the full truth about his condition: 'To you I think I may say', he wrote, asking her to be gentle with his 'lacerated being', 'that my interior is not like my exterior'. But to tell her what the cheerful and sociable exterior concealed, to tell her what he had lost in leaving Italy, was only to confirm the implication of his original departure—that Weimar, and she, had become a burden to him, and that he would rather be elsewhere. Small wonder that she 'gave this confidence an unfriendly enough reception'. 'Pleasure of travellers', Goethe later noted, 'to talk of what they have seen and experienced. Tic of those who have stayed at home to show less interest, which only makes keener the feeling of what one is deprived of.' To Goethe's attempts to draw richness and happiness from the memory of the great things that had passed, Frau von Stein responded with accounts of the antics of her dog Lulu. 'We exchanged nothing but tedium with one another', she commented, with determined disillusion. 'She is out of temper, and there do not seem to be any prospects', was Goethe's assessment.

Frau von Stein returned to Gross Kochberg on 22 July, as a grey, wet summer closed in on Thuringia. 'I shall carry on my life as best I can, strange a task though it is. Kayser is leaving with the Duchess, which puts an end to all my hopes for the beauties of music. The gloomy sky swallows up any colour. Herder is leaving too now . . . '. Herder's departure for Switzerland and Italy on 5 August was followed by that of Anna Amalia's little caravan on the fifteenth. In her two coaches, taking, in addition to the Dowager Duchess and her lady-in-waiting, Fräulein von Göchhausen, Einsiedel, Kayser, and Filippo Collina, there would have been ample room for Goethe too. But to have fled back to Italy so soon would have been true cowardice and would have solved nothing: Goethe had taken his decision, Weimar was his future, and sooner or later, and whatever the cost, he had to learn to live there, so it might as well be sooner. Moreover, he had resolved to complete his collected edition before he was 40, and of that there could be no question if he now consented to waste a year as a travelling courtier. The Rome he had loved and left had been a Rome centred on his own literary and artistic development: it was not to that Rome that he would return in the train of Anna Amalia. Perhaps also there was something of the less noble feeling that if the Dowager Duchess wanted his company she should have said so more forcefully to her son in February and spared him the agonies of the last five months. So he stayed, as Weimar emptied. The day after the Dowager Duchess, the Duke himself left for the garrison at Aschersleben, and the Gores, who at least brought with them a touch of the exotic, also moved on for a while. It must have been one of the most despondent moments in his life and thirty years later the bitter isolation of the period was still fresh in his mind, but the courage should not be ignored with which he resisted the easy palliative that had been offered him:

From Italy, rich in forms, I was directed back to a shapeless Germany, exchanging a serene for a sombre sky; my friends, instead of consoling and embracing me, brought me to despair. My delight in the remotest objects, practically unknown to them, my suffering, my laments for what was lost, seemed to offend them, I felt no sympathy, no one understood my language. I could not reconcile myself to this painful condition . . .

Goethe found a new friend, even though he may not at first have realized that she was destined to be, as Frau von Stein once had been, the inheritor of all the others, and the means of reconciling him for a generation and more to a situation that did not cease to be painful. Shortly before 11 July, less than a fortnight after the first uneasy meeting of the 'old group', while he was walking in the park beside the Ilm, a 23-year-old woman, Christiana Vulpius (1765–1816), approached him with a petition for assistance from her brother August (1762–1827), an inexperienced young author in whom Goethe had shown some interest before he left for Italy. The Vulpius family of lawyers and pastors (the name is a sixteenth-century humanist's Latinization of 'Fuchs') had long been established in Southern Saxony and in Weimar itself. Christiana's grandfather had been a successful Weimar advocate, but his son Johann Friedrich had to leave the University of Jena without a degree and achieved only a miserably paid post as a clerk in the ducal administration, rising eventually to Court Archivist, which was a better title, though not much better a job. August and Christiana were the only survivors of four children, but their mother died when they were young, their father remarried, and another four children were born. Despite the family's straitened circumstances August was able to study law (also heraldry, numismatics, and history) at Jena and Erlangen and started to write for publication. His career as one of Germany's most successful authors of pulp romance was far in the future, however. In 1782 Johann Friedrich Vulpius was charged with malversation in his office and peremptorily dismissed by the Privy Council (of which, of course, Goethe was a member). Clemency was shown, nevertheless, in view of his large family, and perhaps because the charge proved less well-grounded than originally thought, and Goethe found him another post in his own Highways Commission. It was doubtless partly in the aftermath of this affair that in 1786 he died, aged about 60, having reputedly—though there is no evidence—taken to the bottle. Christiana, together with her aunt Juliana Vulpius, was left to look after the family and on four days a week earned herself something in the manufactory of artificial flowers that Bertuch's wife had set up five years before, to provide pin-money and an occupation for 'girls of the middle class'. It is possible that Goethe first saw her when inspecting the workshop in the summer of 1786. August meanwhile, who had become the private secretary of the miserly Baron von Soden in Nuremberg, was in a desperate position once his employer found someone willing, as he was not, to combine his duties with those of a lackey, and dismissed him. Herder, passing through Nuremberg, took pity on Vulpius and asked Goethe to try to do something for him, but Christiana had by then already presented her own petition. Goethe had immediately written to August on 11 July, promising support, and exerted himself repeatedly over the next eighteen months to find him a position. Jacobi was at the time looking for a private secretary and tutor to his children, but Vulpius turned out not to know enough French, and Goethe was uncertain how warmly he could recommend him to so close a friend. When Vulpius moved to Erlangen Goethe sent him money and wrote a letter of recommendation to the professor of theology there; this too proved fruitless and in April 1789 he sent him ten dollars so that he could try his luck in Leipzig, where Göschen was asked to try to find him some niche in publishing work, Goethe expressing willingness to give him further occasional support. Göschen thought him a discontented young man, too unreliable for such tedious work as book-selling or proof-correcting, insufficiently educated to take on translations, too unproven as an author, and unlikely to find a tutorship because of his lack of French and music. None the less he did what he could and Goethe sought further help from Breitkopf, the music publisher, who had printed his own first poems twenty years before. A visit to Leipzig in October 1789 probably had as its main purpose the attempt to find Vulpius something more secure than occasional piecework—although this was for the present unsuccessful, more money was advanced, and Vulpius continued to write.

Christiana was a spirited, energetic, practical, and straightforward character. Her round and cheerful face with its strong, almost masculine, nose, framed at this time by thick waves of brown hair, already showed a tendency to plumpness. She had had some formal education, though, like most women of her time, only to a primary level; her spelling was even worse than that of Goethe's mother, whom in many respects she resembled and with whom she eventually had warm and happy relations, but she was not illiterate. She was used to looking up, with a rather childlike trust, to older men with difficult lives —her brother and her father—but when she made her appeal to Goethe he can have been stirred only briefly by that strongest of aphrodisiacs, the feeling of power: as his continual efforts on August's behalf show, he kept conscientiously his side of whatever bargain was spoken or unspoken between them. He was alone, and nearly 39, and had for half his life allowed others, and especially one other, to enjoy in his poetical works and his communication of his ceaseless mental activity, the fruits of his unfulfilled and diverted desires. He had sacrificed himself to the court as completely as Kilian Brustfleck had ever expected Hanswurst to sacrifice himself to the German public, and he did not at the moment think he was seeing much gratitude. His experiences in Rome had emboldened him to take a sexual initiative, without a commitment to the marriage that he still feared, and perhaps he had already resolved to seize any opportunity that might arise in the fluid and uncertain days after his return, when old habits were not yet re-established and the shape of his new existence was still unclear. In later years Goethe and Christiana celebrated 12 July, the day after he had written to August, as the day on which their relationship was sealed. Perhaps Christiana was summoned to the Frauenplan—for Knebel was still living in the cottage in the park, which would otherwise have provided a more convenient rendezvous—to hear what the Privy Councillor had done or was proposing to do for her brother. But she was old enough—and, malicious tongues would later say, experienced enough—to know what he wanted her to do for him. She could no doubt see her own advantage, and that of her family; Goethe will certainly have deployed all his incomparable charm; she was not by nature unwilling; and perhaps there was already the germ of what is called love.

Another assignation was probably made for a night in the coming week, but Christiana could not make herself free, or, living on the other side of the little town, in what is now the Luthergasse near Herder's church of St James, she felt unable to come unobserved. The result for Goethe was a wakeful night and one of the last of his poems of unachieved fulfilment, 'Morning Complaint' ('Morgenklagen'), which, in the unrhymed trochaic pentameter he had used in Italy for 'Eros as Landscape-Painter', lists all the sounds the sleepless poet hears as night goes over into early morning in a market-town until the sun has risen and there is still no sign of his beloved. But there is still something pubescent and anacreontic about the narrator of 'Morning Complaint', who before going to bed tests the hinges of his door to make sure they do not creak. 'The Visit' ('Der Besuch'), written in the same metre, but a little later, for its setting is the cottage in the park, which Knebel vacated on 19 July, is both erotically and personally more mature. Stealing quietly on his beloved in the little house the poet finds her asleep—Goethe made a pencil-sketch of the scene—and his desire to waken her is restrained by his fascination with the clarity of vision which her state permits him. Awake, she enchants him with her body, asleep, she allows him to reflect on and recognize his love:

> Wärs ein Irrtum, wie ich von dir denke,
> Wär' es Selbstbetrug, wie ich dich liebe,
> Müßt' ich's jetzt entdecken . . .

If it were an error, the way I think of you, if it were self-deception, the way I love you, I should surely now discover it . . .

And so, after sitting long in the quiet assurance of his love, and leaving two oranges and two roses as a sign of his visit, he steals away again, unfulfilled for the present, but looking forward to a double reward for his delicacy that very evening. In the confidence of past and future physical satisfaction, and in a certain reminiscence of Propertius, but also in the direct but questioning relationship to another person, absent from Goethe's writing since his extraordinary verse-letter of 1776 to Frau von Stein ('Why did you give us the deeper vision'), 'The Visit' anticipates the eruption of an altogether new kind of poetry, beginning in the autumn of 1788, Goethe's most original new venture for six or seven years to come.

The collection and revision of his shorter poems for volume viii of his *Works,* and some painfully slow progress on *Tasso,* to be the main item in volume vi, were Goethe's principal tasks during the cold wet summer. 'The weather . . . deadens my spirit', he wrote to Herder, on his way into the Italian heat, 'how can one live when the barometer is low and the landscape has no colours? . . . I could bear anything, if only it was always fine.' His answer was to 'live like a snail, withdrawn into its shell'. The rain, however, must actually have made it easier to conceal the idyll developing in the cottage by the Ilm, which was revealed to no one, not even the Duke: the only consolation Goethe admitted to Frau von Stein was the open fire he had laid for him at the end of August. Carl August had returned to Weimar on the eighteenth, after only three days away, having injured his foot in a fall from his horse, but Goethe's almost daily dinners *à trois* with the convalescent Duke and the long-suffering Duchess hardly interrupted his seclusion and certainly did nothing to dispel his conviction that in the duchy 'in political matters nothing is to be done'. 'I put as good a face on things as possible and inwardly am in despair, not about this particular case but because this case represents his, and our, entire fate.' The Duke's passion for European politics and soldiering, which had led to his injury, was, like his passion for Emily Gore, a pure self-indulgence and his impatient inability to subordinate his fancies to a long-term and consistent plan for the improvement of his own restricted territories was reflected in his impatience with the doctor's orders through which alone he could hope to recover. Goethe at times felt like Epimenides, the classical Rip van Winkle, awakening from supernatural sleep into a world to which he has become a stranger. His inward despair did not go unnoticed: 'for all his fortunate circumstances and brilliant talents he actually is unhappy, for he is pursued by a tormenting disquiet, a sombre melancholy'. The sharp eyes of the ladies saw more: 'it is a pity that he always has his armour on, but sometimes I see through it'; 'he is living now without nourishing his heart. Frau Stein thinks he has grown sensual, and she is not wholly wrong. Going to court and eating at court now means something for him'; 'before he went to Italy I liked him better; just the look in his face, he has lost something of his refinement'. Frau von Stein, naturally, had understood best: her refusal to be reconciled might not meet with general approval, but she could tell that Goethe was no longer willing to pay the price for reconciliation. But because theirs was not a personal but a symbolic relationship, because he valued not what she was, but what she stood for—a kind of relationship in which she had long acquiesced and which she had perhaps actually inaugurated—they had neither of them the resources, or the incentive, to reach a compromise.

In the first half of September the weather improved and on the fifth, a day of uninterrupted sunshine, Goethe set off with Fritz, Sophie von Schardt, Frau von Stein's sister-in-law, and Caroline Herder, of whom he was taking special care during her husband's absence, on the four-and-a-half-hour drive up rough roads to Gross Kochberg. The first to come out from the tall manor-house to meet the party was the 22-year-old Charlotte von Lengefeld (1766–1826) who had spent much of her childhood in the Stein household in Weimar as Frau von Stein's honorary daugher. Then Frau von Stein herself appeared and greeted everyone cordially except Goethe, who spent the rest of the day in a bad mood as a result. None the less, Italy was the main subject of conversation, with Goethe's drawings being inspected in what remained of the morning, while Herder's letters were read out in the evening and provided a topic in which everyone could agree to find a warm interest. The following day the weather was equally beautiful and Goethe read some of the little essays he was writing up from his travel-notes for publication in the *Teutscher Merkur*, including his meditative description of the shrine of Saint Rosalia in Palermo, which must have accorded well with the mood of the evening walk, by the light of a golden moon. But Goethe's devotion to the moon was coming to an end: after two years in the sun he was losing the taste for that cool, reflected, infertile light, and it would be well over a decade before he could recover in his poetry the symbolic power it had once had for him. During this summer of 1788, perhaps around this very time, he redrafted, for volume viii of the Göschen edition, the poem which ten years earlier had invoked that symbol upon his new life with Frau von Stein—shut off unhating from the world—and in doing so he made of 'To the Moon' an epitaph on their friendship, eloquent, tender, and understandingly responsive as neither of them, in their behaviour and conversation, was capable of being.

While Goethe was in Italy Frau von Stein had herself rewritten the original version of the poem—which, it will be recalled, contrasted the solitary death of Christel von Lassberg with the mutual love of Goethe and Frau von Stein—and turned it into an address from her to the moon, lamenting the faithlessness of her departed 'friend'. She eliminated the references to the river as the place of death and made of it instead the receptacle of her quiet tears, writing two new strophes which must surely count as the supreme achievement of her otherwise modest poetic talent and which give voice, as does nothing else, to the silent anguish of the frigid months and years after the betrayal of 1786:

Mischet euch in diesen Fluß!
Nimmer werd' ich froh,
So verrauschte Scherz und Kuß
Und die Treue so.

Jeden Nachklang in der Brust
Froh– und trüber Zeit
Wandle ich nun unbewußt
In der Einsamkeit.

Mingle with this river! Never shall I grow glad again, thus did merriment and kisses die away, thus did faithfulness too. With every echo in my breast of glad and sorry times, I now pace mindlessly in solitude.

When Goethe saw these new lines, whether on the September visit to Gross Kochberg or earlier, he saw also the possibility of a new poem which was true to his and Frau von Stein's present pain as the older version had been true to the mood of mysterious renunciation in which they had sought refuge in each other. He accepted that the speaker should now be a woman—indeed, should be Frau von Stein—and accordingly maintained the new references to himself as the speaker's 'friend', who had once had the power to transform her life. The direct mention of his 'departure' was too unjust for him, and was dropped, but he allowed the accusation of unfaithfulnesss to stand, perhaps in the penitent spirit in which in 1771 the faithless lover of Friederike Brion had had Weislingen poisoned before the eyes of the world. But Goethe also incorporated into this newly complex and many-layered poem his own emotions of the time, of which Frau von Stein, and indeed everyone else around him, seemed oblivious—the very emotions that were driving them apart and were the source of her pain. After her plaintive lines on the loss of happiness and the passage of time he inserted a strophe which, as much as it referred to Frau von Stein's loss of him referred to his loss of Italy and to the memories which for all their choiceness and richness were reminders of deprivation:

Ich besaß es doch einmal,
Was so köstlich ist!
Daß man doch zu seiner Qual
Nimmer es vergißt!

I once possessed it, the thing that is so precious. Oh that to one's torment one never forgets it!

The one disadvantage of the new version is that, with the excision of the themes of Christel von Lassberg, of death and ghosts, the tight logic is lost which linked together the moon, the poet's heart, and the river in its different moods, lethal or life-giving. The difference between the river in winter and the river in spring is now merely decorative, while in the earlier version it was the turning-point of the poem. Goethe disguised the new looseness of the connection between the moon and the river by a beautiful ritornello device so that the repeated mentions of the river become, like a musical accompaniment, independent of the main theme but linked to it, now more and now less audible:

> Rausche, Fluß, das Tal entlang,
> Ohne Rast und Ruh,
> Rausche, flüstre meinem Sang,
> Melodien zu,

Rush, river, down the valley, without rest or stay, rush, and whisper melodies to my song,

The moment of pausing reflection in this strophe, of awareness that this is a song, a product of art, not simply an emotional effusion, is comparable, on the smaller scale, to the operatic moment of deliberate artifice in the conclusion of *Egmont*. The final strophes which follow may textually be little changed from the first version, but in their new context they are no longer the culminating wisdom of the poem, rather they are the self-conscious citation of a past or possible attitude. Whether the citation is in hope that friendship and devotion to the labyrinthine mysteries of the heart will soothe the pain of loss, or whether it is in painful recollection of a happiness which neither Frau von Stein nor Goethe will know again is left uncertain. What is clear is that the poet who has organized this poem stands above, and detached from, both possibilities. When Goethe spoke these words in his own person they expressed his intention of adapting to court existence by continuing to live, in the monochrome world of Frau von Stein and before her audience of one, the life of feeling that for the Werther generation had been a collective and public affair. Now that that peculiar compromise has proved physically unsustainable Goethe has either to live whole-heartedly for the court and eliminate from his writing the confessional element of shared feeling that had been its strength, or to find some new compromise that will distance him again from the court and give him access to another public. For the present he is following the first course, as the sovereign artist who puts words into, and indeed takes words out of, the mouth of Frau von Stein, in order, not to express a feeling, but to create a song. The result however is something extremely unusual in Goethe's early lyrical writing, something which anticipates his achievement in the completed *Tasso*. For the poem is characterized not by an absence of personal feeling, but by a balance of the feelings of two different persons. Two voices are heard which for twelve years have carried on a mental and emotional dialogue. Neither dominates the other, both are accorded equal respect, indeed at times both seem to utter the same words though with different intonations, for the controlling poet has achieved a detachment which can allow both parties to speak for themselves.

On Sunday 7 September, the whole group from Gross Kochberg drove down for the day to Rudolstadt, a little town of 3,000 inhabitants nestling between the hills in the upper valley of the Saale, and residence of the Princes of Schwarzburg-Rudolstadt. The Lengefelds—Charlotte, her widowed mother, and her married sister Caroline von Beulwitz (later von Wolzogen) (1763–1847) with her Privy Councillor husband—lived together in their family home in Rudolstadt, and this summer an almost daily visitor there, walking in from the village of Volkstädt, where he was writing industriously in rural seclusion, was Friedrich Schiller. Schiller had led a wandering and penurious life since *The Robbers* had taken Mannheim and Germany by storm in 1782. Fleeing the tyranny of his native Württemberg he had briefly been resident playwright at the theatre in Mannheim, managed by W. H. von Dalberg, another younger brother of the Statthalter of Erfurt, but, despite a further great success with his third play, the socially critical *Cabal and Love*, (*Kabale und Liebe*) (1784), his only work of contemporary realism in the manner of Lenz, his contract was not renewed. He was rescued from total impoverishment by some unknown admirers in Leipzig, C. G. Körner (1756–1831) and L. F. Huber (1764–1804), both officials in the Electoral Saxon administration, and about to marry the two sisters Minna and Dora Stock, whose mother had combed the down out of Goethe's hair when he was a student and whose father had taught him the rudiments of etching. Körner and Huber invited Schiller to stay with them in Leipzig for as long as he wished, and thanks to their support he was able, between 1785 and 1787, to complete his blank-verse drama *Don Carlos* and the first volume of a *History of the Revolt of the United Netherlands,* as well as some prose fiction written for money. Already in 1784 Schiller had become emotionally entangled with Charlotte von Kalb (1761–1843), a gushing and fanciful lady married to a brother of the dismissed President of the Weimar Chamber, who had conceived the notion of divorcing her husband and securing Schiller's future by bringing him her private fortune in marriage. In 1786, with the lengthy court proceedings begun, she moved to Weimar, and when Schiller came in the summer of 1787 he was introduced by her and accepted on the Weimar circuit—for example, at the celebration of Goethe's birthday—as the heir-apparent. During the winter, however, which he also spent in Weimar, Schiller made the acquaintance of Charlotte von Lengefeld, who was staying with Frau von Stein's sister, Luise von Imhoff at the time, and once he had moved out to Volkstädt in May a complex relationship with both Lengefeld sisters began. Caroline was the more fiery, emotional, and intellectual, but she was married; Charlotte was a lover of nature, reading, and drawing, quiet, practical, and single. Schiller understood his own mind better than his own feelings and wrote repeatedly to his Leipzig friends that there was no question of his falling in love with anyone, or of any intimacy that could not be equally shared with both Charlotte and Caroline: 'I have weakened my feelings by dividing them and so the relationship lies within the bounds of warm and rational friendship.'

Although ten years younger than Goethe, Schiller was then, though he did not realize it, at a somewhat similar point in his personal development: he too was weary of years of solitary toil and self-denial and looking for the security and support of a lasting partnership. There were other parallels. Having written the most overtly political, and theatrically most successful, dramas of the Storm and Stress, Schiller had entered a more Sentimentalist phase, and while working on the more formalized *Don Carlos* had made his relationship with Charlotte von Kalb into the symbol of a spiritual revolution much as Goethe had done with his own Charlotte. Encouraged by Wieland, ever generous to young talent, he had published in the *Teutscher Merkur* some anonymous but severely self-critical *Letters on Don Carlos*, had striven to regularize his versification, and in March 1788 had published in the same journal the first of his mature philosophical poems, 'The Gods of Greece' ('Die Götter Griechenlands'), which could well pass for a manifesto of Hellenism. In fact Schiller had little antiquarian interest in the art and culture of the classical world, but 'The Gods of Greece', a hymn of hate against the Christianity which —especially in its rational, eighteenth-century form—has destroyed the humanism and sensuous polytheism of antiquity, showed that in their religious, or anti-religious, views Schiller and Goethe were again converging. Schiller, who since the days of his medical training had an unusual sympathy for materialist conceptions of the human animal, emphatically rejected the notion of an after-life, and in his attempt to define some alternative absolutes was, in the late summer of 1788, already beginning to sketch out another philosophical poem on the unique importance of 'art' and 'artists'. He differed from Goethe, of course, in his want of office and patronage, and of private means to give him independence and freedom of choice should office or patronage be offered him, but his conception of the dignity of his calling was, if anything, all the higher. He was consumed by an ambition warranted by his abilities—after his first visit to Weimar he wrote to Huber 'the upshot of all my experiences here is that I recognize my poverty but set a higher value on my mind than had previously been the case'—but Goethe stood before him as one who had already realized that ambition, at once an admired ideal and a crushingly successful rival: '*Iphigenia* has given me another really good day—although I have to pay for the pleasure that it gives me with the depressing feeling of never being able to produce anything similar.'

Knowing that Frau von Stein was at Gross Kochberg, only a few miles away, Schiller had been looking forward to the possibility of Goethe's visiting Rudolstadt for over a month. The Lengefelds themselves were anxious to bring about a meeting between 'their' two poets and were a little disappointed that, though Goethe was perfectly friendly, there was not the electric charge to the occasion that they had hoped for.

The first sight of him [Schiller wrote] considerably lowered the high estimation of this attractive and handsome figure that had been instilled in me. He is of medium height, his bearing is stiff and so is his walk; his face is sealed, but his eye very expressive, lively, and one lingers with pleasure in his glance. For all its gravity his countenance has much benevolence and kindness. He is *brunette* and seemed to me to look older than by my calculation he can be. His voice is altogether agreeable, his speech fluent, intelligent and vivacious; it is altogether a great pleasure to listen to him; and when he is in a good humour, as on this occcasion he was, more or less, he speaks readily and with interest. We made each other's acquaintance quickly and without the least constraint . . . Of Italy he speaks willingly and with passionate recollections; but what he told me of it gave me the most accurate and immediate impression of the land and the people. Above all he can bring vividly home how this nation more than all others in Europe lives amid *present pleasures* . . . In Rome there is no debauchery with single women, but it is all the more usual with those who are married . . . For a foreigner the dirt is almost totally intolerable.

He is full of praise for Angelica Kaufmann; in respect both of her talent and of her heart. Her circumstances are apparently extremely fortunate; but he speaks with delight of the noble use she makes of her wealth . . . He seems to have spent much time in her household and to feel his separation from it with regret.

. . . All in all the great notion of him that I did indeed have has not been reduced by this personal acquaintance, but I doubt whether we shall ever come very close to each other. Much that is still of interest to me, that I still have to wish and to hope for, has had its day for him; he is so far ahead of me (less in years than in experience of life and self-development), that we shall never again come together on the way; and his whole being is from the start differently constituted from mine, his world is not my world, our ways of thought seem essentially distinct. On the other hand no full and firm conclusions can be drawn from such a meeting. Time will tell the rest.

Perhaps few would warm to such cat-like attention. The Lengefelds attributed the relatively unemotional nature of this meeting to Goethe's preoccupation with lost Italy. Goethe's own later explanation for his subsequent distance from Schiller was that he disliked *The Robbers*, which in its violence and extremism belonged to an era he wished to put behind him. It must, however, have been perfectly obvious to him at the time that the man he met was detached from the emotional and aesthetic world of *The Robbers*, and also that he had something in common with the author of 'The Gods of Greece', which he took with him from Rudolstadt and read with approval. Yet Schiller was not misjudging the differences between them either, for in one respect he was still the author of *The Robbers*, the play that had had a greater *retentissement* than anything Goethe had written since *Werther*: he still submitted his work, as he wrote in announcing his journal the *Rhenish Thalia* (*Rheinische Thalia*), to one tribunal only, the public. 'The public is now everything for me, my study, my sovereign, my trusted friend. To it alone do I now belong.' In 1788 it was still Schiller's intention to live by his pen, indeed he had no alternative. As long as he remained remote from the notion of a self-justifying Art, altruistically supported by court patronage, and as long as Goethe, diverted by his Italian journey from the original purpose of the Göschen edition, remained distrustful of the contemporary German literary public, they would both continue to live in separate worlds. Only when both had shifted their ground would a rapprochement become possible.

The meeting with Schiller provoked Goethe to reveal his hand, indeed, he swore to secrecy everyone who on 8 September heard his discussion of 'The

Gods of Greece' in the carriage back from Gross Kochberg down the Saale valley to Jena. But it can hardly have been his scheme for an investigation of the concrete evidence of the ancient world's theology—its sculpture—that he was so anxious to conceal. The plan was not obviously a dangerous secret: first to identify and eliminate from study all those images of Greek gods known to be modelled on historical characters, and then to use physiognomical principles on the remainder to determine the physical features by which the ancients represented their moral and aesthetic ideals. (No human characteristic, Goethe argued, was ever purely represented by a single real individual: the perfect humanity found in a Minerva or Apollo must have been drawn not from one model, but from many.) But while envisaging the means for putting the plan into effect, Goethe must have realized he had let his fancy run away with him and had exposed more of his true wishes than was prudent. 'Finally he said that if Louis XIV were still alive he thought that with his support he would be able to carry the whole thing out: he had had a sense for the grand design; with 10 to 12,000 dollars a year he could carry it out in ten years, it would have to be in Rome, of course.' It is harmless enough to wish for ten years in Rome when one has had two, and for 10,000 a year when one has nigh on 2,000 (Schiller in 1789 would have been very happy to scrape together 1,000), but to wish for Louis XIV when one has Carl August could look like ingratitude.

Goethe swiftly and wittily redeemed his *faux pas* over the next week which, after a brief halt in Weimar to collect Prince August, he spent in Gotha, his first visit to the familiar old palace since his return from Italy. He was anxious to repair relations between Duke Ernst and Tischbein, who seemed on the point of disappearing into Neapolitan service; he had drawings of Kniep to show round in the hope of securing new commissions; and there was the collection he had resolved to make for the benefit of Cagliostro's family. But he spent much of his time writing one of the last new pieces for his eighth volume, which he intended to conclude with a series of longer poems relating to poets and artists. In 1774 he had briefly thought *The Artist's Earthly Pilgrimage* might be completed with *The Artist's Deification*, showing the reward, in the form of posthumous fame, for a painter's struggle in his lifetime to serve his art and feed his family. In now writing *Artist's Apotheosis* (*Künstlers Apotheose*) Goethe not only completed what had before been merely sketched, but presented, in semi-farcical form, his new programme for the practice, and support, of Art—a rather different affair from what had been envisaged in 1774. The artist of the *Earthly Pilgrimage* is now dead, but a student admirer is assiduously copying his works, as the means to acquiring facility and understanding of the craft. A dilettante pooh-poohs this uninspired imitation and demands 'Nature' instead, but the student prefers to stay with the professionals, from whom he learns the necessity of exercising his own judgement and recognizing the shortcomings of even the greatest masters: 'One must love one's art and not one's hero . . . Learn to recognize what he achieved and then to recognize what he wanted to achieve'. To these echoes of the lessons Goethe learned in Rome—the distinction between nature and art, and the need for study and scholarship to understand the ideal perfection that is only rarely expressed in the work itself— is now added a new factor, unthought of in 1774: a princely patron, who buys the masterpiece which mattered more to the artist in the *Earthly Pilgrimage* than all the commissions he undertook to earn his daily bread. Looking down from heaven, the artist, while glad to see the prices his work is now fetching, concludes with the wish that future painters may receive such benefits of patronage while still alive: 'alas, I lacked a prince who valued talents'. Such princes, who save the *homme moyen sensuel* from prostituting himself in the market simply to secure his creature comforts, and who free him to devote himself to achievements of Art which a future generation of connoisseurs will acknowledge, were fortunately to be found both in Gotha and in Weimar, to which Goethe returned on 8 September, well satisfied with a week that had been 'fruitful in more senses than one'.

The arrangements Goethe had made with his prince before returning from Rome had been contained in their personal correspondence and it was only gradually that Weimar at large divined the new order concealed behind the reshuffle. In the middle of November Schiller still thought it news that 'it is pretty well decided that he will stay here, but in a private capacity . . . he is now only in the Mines Commission, purely as a hobby'. Carl August accepted his role as Maecenas with good grace and was happy, for the present, for Goethe to 'continue living as *homme de lettre[s]*', just as Knebel continued 'as student of man and nature', but there was evidently far from general agreement with Sophie von Schardt's view that 'he is useful enough to the Duke by being his friend; I cannot see the injustice that the world finds in his keeping his salary'. The mine at Ilmenau, where the task of drainage was making slow progress, was firmly in the hands of the Voigt brothers, the elder in the administration in Weimar, the younger, the geologist, on the spot up in the hills: Goethe's responsibilities were more or less confined to keeping up an emollient correspondence with shareholders known to him personally, such as Jacobi. In Jena the old palace, with Büttner's library on the ground floor, the scientific instrument collection on the second, and the first conveniently free for visitors, was a continuing attraction; partly because Knebel was still living there, until the end of November, when he moved over to Weimar for a change; and partly because there was still much anatomy to learn from Loder, whose lecture- course on the muscles he attended from 9 to 21 November. Once there, however, he could not restrain himself from involvement in the little tasks that presented themselves: purchase of land to complete the flood-protection works he had begun in 1784, a rebinding programme for the books in Büttner's library, collecting publicly and privately expressed suggestions for the improvement of the university, some quickly organized crowd-control when the students gathered to say farewell to Eichhorn, the influential biblical scholar, leaving for a chair in Göttingen. One task at least, though, was of the first importance: possibly at the suggestion of the elder Voigt Goethe resolved at the end of November to promote the appointment of Schiller as professor of history, nominally on the grounds of the success of his history of the Dutch Revolt, but more probably because of his obvious appeal to the younger generation, which might help to counteract the loss of Eichhorn, whose replacement, the equally influential H. E. G. Paulus (1761–1851), was as yet little known. There was initally no stipend attached to the post—although fees were paid by those attending the lectures—but it opened the way to secure an official career if Schiller wanted it. During a stay in Gotha in early December Goethe secured the agreement of Carl August and Duke Ernst to the plan, and although the approval of the other Saxon duchies had also to be given, the matter was as good as certain by the end of the year. Schiller appreciated Goethe's energetic involvement on his behalf—which certainly indicated esteem, if not necessarily for his literary work—but would have preferred more time to prepare himself. 'The best way to learn is to teach', Goethe told him patronizingly, if truthfully.

Carl August was clearly acceding to Goethe's request to be left free of official duties until the edition of the *Works* was completed, and it is noteworthy that even 30 January 1789 passed without a contribution of any kind from Goethe to the celebration of the Duchess's birthday. One service, however, Goethe must have agreed to perform fairly soon after his return, and probably very willingly: the informal overseeing of the development and education of the Crown Prince, now 5 years old. After an initial consultation with Ridel, and a period of gaining the child's confidence, he began in the autumn of 1788 a series of occasional trips away with the Prince, usually to Jena (he encouraged the inhabitants to think their Prince might one day study there, as the sons of George III were studying at Göttingen), and often in the company of August Herder, one of his godsons. 'I am so happy that he is concerned for my children', said Duchess Luise, 'for he has an excellent way with children', and Carl August thought the effect on his son 'exceptionally good'. But this was essentially a service of friendship, not a duty of office, and similarly it was as a friend of Carl August's —certainly not as a member, past or present, of the ducal Chamber—that Goethe undertook some financial negotiations to rescue their mutual friend Merck, who had set up a cotton-mill to employ widows and orphans just as the prices of cotton goods slumped. Goethe arranged for Carl August to underwrite a loan to Merck of 2,000 dollars from the Frankfurt banker J. J. Willemer, even though it was clear to both of them from Merck's letters that he was on the verge of insanity, a prey to irrational terrors (on his death his financial affairs were found to be in good order), and aware that his emotions were destroying him.

How far this confidence extended in respect of Christiana Vulpius it is difficult to say. In a letter to the Duke in October there was a first hint that Goethe was living a 'natural' rather than an 'unnatural' life, and a month later, after borrowing a Propertius from Knebel, he began including with his correspondence what he called *'Erotica'*, love poems, at first of a fairly harmless kind, usually in classical metres and usually epigrammatic. If by then he had let the Duke into his secret, and wished to show off to him, there would be an explanation for the embarrassing *thé dansant* of 7 November:

Frau Schardt told me . . . that . . . he scarcely spoke a word to any sensible woman but kissed the hands of the little misses one after another, said nice things to them and danced a great deal. Frau Kalb finds it abominable that he excites the young girls so etc. In short, he has no further interest in being anything to his friends. He is no use in Weimar any more; on the contrary, I think the slavering over the young girls did not make the best of impressions on the Duke, who was present.

A week later Goethe was writing from Jena to the Duke and boasting of a visit to the country seat of the Gotha minister A. F. C. von Ziegesar (1746–1813) and his three daughters—Goethe does not trouble to mention the fourth, Silvie, who was 3 at the time—which he described as 'inspecting the Ziegesar bloodstock'. The 'two bachelors'—he and Knebel—were looking forward to a return visit, at which 'we shall endeavour to commend ourselves. In front of you I am not ashamed of the student vein which is beginning to revive in me.' The implication of these episodes is that Goethe did not as yet consider his liaison with Christiana to be permanent and that he was beginning to look around for a titled young lady to marry. Already in September he had suggested to Wieland that he should contribute more regularly to the *Teutscher Merkur* since 'even though I have no children to feed' some more money would be welcome, and Caroline Herder had recorded (and rather tactlessly communicated to her husband) Goethe's view that the three volumes so far published of Herder's *Ideas* all came down in the end to the principle that a 'man needs a comfortable home of his own'. An intuition of this development, and a pronounced jealousy of Goethe's continuing careful and gentlemanly attention to Caroline Herder, is sufficient to explain Frau von Stein's unchanging frigidity: whatever Goethe may have told the Duke, there is no reason to assume that by the end of 1788 the name of Christiana Vulpius meant anything special to Frau von Stein.

Rome in Weimar: December 1788–May 1789

'Purgatorial fire' was Goethe's description of the gossip directed by the Weimar ladies at his student-like behaviour. His offence was that after well over a decade as a celibate pet he had become sexually dangerous once more, a 39-year-old adolescent. The safely unmarrying Major von Knebel was distinctly preferred: 'Knebel . . . has many female friends', wrote Charlotte von Kalb, 'and with genuine uprightness is their affectionate admirer. All of us here are satisfied with his conduct, and he is now the most frequent object of praise.' On 3 December however a rival to Knebel appeared, and, when Goethe returned on the following day from discussing Schiller's professorship in Gotha, took up residence with him. His ugly little face, with its turned-up nose and monkey features, his chirpy, boyish manner, radiating sexual non-participation, his enjoyment of female company, particularly when he could entertain and instruct it, his skill as a raconteur and 'interestingly comical' conversation (Schiller's phrase) soon made Carl Philipp Moritz into the new favourite. Within a fortnight of his arrival he had been introduced into every circle in Weimar, including that of Frau von Stein, and he had spent much time with the Duke. Even his breath-taking Goethe-worship seemed delightfully selfless, and he was after all telling the ladies that they were privileged to live in the neighbourhood of a poet comparable only with Shakespeare, Homer, or the Greek tragedians (though these last were 'for the present' a little 'too elevated for us'). There was, he claimed, no point in reading anything but the best— Schiller's works were mercilessly dissected, and not a shred of poetry or drama left to them—even to what Schiller regarded as Goethe's 'mediocre' productions, including of course *Egmont*, Moritz gave this 'canonical' status. Herder, who had met Moritz in Rome in September and October, was irritated by what his wife told him of the 'fuss' the ladies were making of him, though he thought Goethe's 'dreadful enthusiasm for . . . a crushed and sickly creature' was understandable, given that Moritz was 'infatuated with him and has designed his whole philosophy to deify him as the summit of the human race'.

An exceptionally hard winter for all Europe had begun—in Paris the temperatures would fall to $-28°C$ and the authorities would keep public bonfires burning on the street corners—when Moritz arrived in Weimar, penniless and in a thin and threadbare overcoat. Goethe bought him some clothes and probably advanced him money (and subsequently reclaimed his expenses from the Duke). He had left Rome in October since there was a prospect of his being appointed to a professorship at the Berlin Academy of Arts, and he hoped to force the issue by being present in Berlin, and also to secure some influential support by stopping off in Weimar on the way. For Goethe, who had already sought to commend Moritz to Carl August in June, the visitor could not have come more conveniently, to help him survive his first northern winter for three years by bringing new intellectual stimulation, memories of Italy, and significant help with his own literary projects. *Tasso* in particular was growing, if 'slowly, like an orange-tree', and Goethe hoped that Moritz would stay until it was finished; quite what the 'part' was that Goethe later acknowledged Moritz had had in the play must be uncertain, but it is unlikely to have been limited to the questions of prosody on which he had already given Goethe advice in Rome. Moritz was toying with some ideas of his own—a systematic aesthetics to replace Sulzer's *Dictionary*, and, more modestly, an essay on *Werther*, which he committed to paper shortly after leaving Weimar—but his only real work was giving English lessons to Carl August, who was anxious to be able to deliver compliments to Miss Gore in her native tongue, and who paid the exceptionally generous fee—probably a tactful subsidy—of 175 dollars. The Duke was impressed by Moritz's powers of observation and analysis, and by his idiosyncratic and entertaining turn of expression, and allowed him to dine frequently at the princely table. Schiller too, who was spending the winter only two doors away from Goethe, in the vain hope that Goethe might call on him, had several meetings with Weimar's new star (whom he had briefly met before in 1785) and found him 'a deep thinker' with whom conversation was very agreeable 'because we lit upon my favourite ideas'. These may have been some of the ideas that are common to *On the Plastic Imitation of the Beautiful* and the long philosophical poem *The Artists* (*Die Künstler*) on which Schiller was working at the time, though a first draft had been completed before Moritz's arrival: such ideas as that a work of art is 'sufficient to itself', that human artists are analogous to the one great Artist, or that death, 'gently' dealt out by Necessity, is transformed by art (however, the notion, prominent in *The Artists,* that ancient religion was also a form of art is absent from the treatise and anticipates rather Moritz's later work on mythology). *On the Plastic Imitation of the Beautiful* was already published, but seems to have been available in Weimar at this time only in manuscript form. None the less it circulated widely. Schiller had his copy from Frau von Stein, studied it closely and had a high opinion of its wealth of ideas, though he thought 'Herderian ways of thinking very visible' in it. Herder however, who had heard Moritz read the essay in Rome, and who through his correspondence with Caroline contributed, after a month's interval, to all the Weimar discussions, thought it 'utterly Goethean' and repudiated 'this whole philosophy . . . it is selfish, idolatrous, uncaring and desolating in its effect on my heart'. But Schiller was surely right: all that separated Herder's views on death, necessity, and the supreme reality, from those of Moritz was a rationally ungrounded ethical benevolence which had only a nostalgic relation to the Christian religion. Moritz and Schiller were both in their own ways too ruthless to have much patience with nostalgia.

Herder, however, had Duchess Luise, Frau von Stein, and Knebel on his side. Knebel quarrelled twice with Goethe in January 1789; in both cases Moritz was involved, though the root of Knebel's irritability probably lay deeper, in some such feeling of usurpation by Italy as had distempered Frau von Stein. Knebel told Herder that Goethe had

brought back from Italy a lot of narrow-minded ideas such as . . . that our being is too limited even to form any concept of the existence and essence of things; that everything is *absolutissime* limited to the individual existence; and that therefore nothing remains for us to think or understand but single cases and investigations or the scope of art etc.

Most of these 'ideas' are not in fact at all related to what are known to have been Goethe's preoccupations in Italy: the emphasis on the limits of knowledge looks much more like a first reaction to Kant's critical writings, which Goethe—·

possibly feeling he could no longer ignore someone so highly regarded in Jena, and so vigorously promoted by the *Allgemeine Literaturzeitung*—began to study this winter. The concentration on individual existences, and on art, as the only fully knowable things suggests an attempt to meet Kant's sceptical onslaught by taking refuge in Moritzian concepts. Knebel says that these 'propositions' were 'microscopically' discussed with Moritz and gives his own hostility to Moritz's treatise as a reason for the increasingly personal nature of the argument. Open 'war' began at the end of January when Goethe published a fictitious letter, entitled 'The Study of Nature ' ('Naturlehre'), in the February issue of the *Teutscher Merkur,* which haughtily dismissed speculations Knebel was known to have entertained about the relation between the ice-crystals forming on the window-panes of snowbound Weimar and the plant life which their shapes resembled. Goethe's rather uncharacteristic assertions that it is more important to find distinctions between natural phenomena than similarities—he admits that he has himself frequently offended against this principle—have a particularly Kantian flavour, but Knebel simply found it insulting that such a salvo should be discharged in public without prior consultation with the target. Moritz acted as intermediary between the two contestants, and it was arranged that a conciliatory reply from Knebel should appear in the next issue of the *Merkur,* but almost immediately the differences over *On the Plastic Imitation of the Beautiful* led to a second argument, which was resolved only after Moritz had left. It was now Knebel's turn to complain that crucial distinctions were being blurred: it was he who identified the principal weakness of the treatise as its failure to distinguish between the real whole that is Nature and the apparent whole that is a work of Art. Goethe readily took the point, once it was agreed that this was the source of Knebel's hostility, and that the problem lay simply in Moritz's confused expression: the courtly distinction between nature and art, which had first been borne in on Goethe during his first months in Rome, was actually implicit in the later parts of Moritz's argument. Something was here being glossed over, however: namely, that if in the earlier parts of the treatise the two kinds of wholeness were too radically diffferentiated, the entire analogy in those sections between the Artist and the Creator would be undermined, and on that analogy reposed the claim for the metaphysical significance of Art and the social importance of the Artist as something more than a specialized craftsman. Knebel had isolated the precise point at which Moritz was endeavouring to modify the traditional courtly conception of the arts, as an ornament of power, in order to make them the professional province of a new, secular, clerisy. Such acuity is only ever propelled by personal animosity, however well Knebel, the courtier, and Moritz, the vagabond apprentice, may have professed to get on. Once Moritz had left, Goethe produced a summary of the main theses in Moritz's booklet, which was eventually published as a review in the *Teutscher Merkur,* to which Knebel now found he could fully assent. As Goethe's summary was largely direct quotation from Moritz's original, it was difficult, as Caroline Herder remarked, to see what the fuss had been about. Knebel and Goethe could now both agree that the beauty of nature was seen in the regular laws by which the totality of things is ordered (a sentiment with which Leibniz would have happily concurred), and that the beauty of a work of art derives from the artist's personal intuition of the wholeness of things, translated into a sensuously appealing 'figure' or 'form' ('Gestalt'). This was a definition *ad hominem,* which, since it no longer contained an account of the relation between the order of a work of art and the order of the universe, beyond an appeal to the artist's private and incommunicable experience, effectively abandoned the metaphysical ambitions of *On the Plastic Imitation of the Beautiful.* But it was enough between friends, and Knebel thought the correct relation had been found between the 'philosopher' (himself) and the 'artist' (Goethe).

As far as Moritz was concerned, his plan in coming to Weimar succeeded beyond all expectation. He left for Berlin on 1 February in the company of Carl August, who was visiting the Gores, and who intervened personally with the King of Prussia to secure his appointment to the professorship. His two months at Carl August's court had been a personal triumph, but they also permanently affected Weimar's intellectual atmosphere, regardless of any systematic weaknesses in his writing. A system does not have to be coherent for its individual components to be influential, and the ideas of Art, and of Goethe as an Artist were powerful medicine. 'Moritz's essay has given me a fully comprehensive concept in relation to Art' ('Totalbegriff für die Kunst'), Caroline Herder told her husband, and her correspondence is suddenly full of talk about 'art' and 'artists', 'poetry' and 'poets', the terms being sometimes used as if they were nearly synonymous. Herder himself, outside Moritz's and Goethe's sphere of influence in Rome, was still using 'art' exclusively of the visual arts when, writing to Goethe, he savaged the Renaissance and Baroque painting which had provided the two aestheticians with their inspiration, and contrasted it with the classical sculpture which was all he personally wanted to see: 'I could almost say that I have never felt so coolly about art as here . . . once it was a fair blossoming of human striving, but now it is a flower-factory'. 'For me these are all puddles left by a dead sea', was his exasperated response to his wife's account of the current Weimar obsession, 'however loudly Goethe proclaims their sweetness'. He recognized that art in the ancient world was not at all the same thing as art in the modern world, and could appreciate artefacts —such as representations of Greek divinities—which originally had a religious function, while sensing with distaste the ambition of modern works, at any rate in the Moritzian interpretation of them, not to serve religion, but to supplant it. He had already identified the 'idolatrous' element in the stylization of Goethe as 'the Artist' *par excellence,* which Goethe had begun with his letter of self-discovery to Carl August, which Knebel had accepted, and which Moritz was doing all he could to promulgate. It was certainly a trial of his scanty patience when his wife announced her conversion to the new sect. Caroline had asked

Goethe whether the unimpressive character of Leonora in *A Farce of Father Porridge,* now published in volume viii of the *Works,* was really a representation, as rumour had it (and as was indeed largely the case), of herself. Goethe replied by skirting the question and appealing to the principle he had explained in the carriage back from Gross Kochberg: 'the poet took only as much from one individual as was necessary to give life and truth to his subject; the rest he took from himself, of course, from the impress on him of the living world'. The same was true of *Tasso,* he said, for all that it contained much of significance about his own person. Caroline professed herself fully satisfied, 'now that I think of him so completely as a poet. He absorbs and digests material from the "totality of the universe" (as Moritz calls it), into which I too belong, and all other circumstances are subordinate to the poet. I now see that clearly and daily see him more in his proper light. He is simply a fortunate favourite of Nature.' Not merely had Caroline learned that she had to expect to be absorbed and digested by a superior monad for the sake of the representation of a totality in a work of Art, she also had understood that in daily life the Artist stands in an abnormal relation to the human material that surrounds him and ordinary worshippers must be modest in their expectations:

I have really had a great revelation about Goethe. He just does live like the poet *with the totality,* or *with the totality in him,* and so we single individuals must not expect more from him than he can give. He does think of himself as a higher being, that is true, but he is still the best and most steadfast of them all. Since understanding what a poet and artist is, I do not ask for a closer relationship, and yet when he comes here I feel there is a very good spirit round and within him.

She could now give instruction to a more recent arrival in the congregation, Charlotte von Kalb, who was too crassly equating the figures in *Tasso* with Carl August, Duchess Luise, Goethe, and Frau von Stein.

I corrected her a little on the point. Goethe does not at all want it interpreted that way either. The poet depicts a *whole character,* as it has manifested itself to him in his soul; but a single human being alone does not possess such a whole character . . . That he borrows traits from his friends, from those living around him, is quite right and necessary; that is how his figures become true without having, or being able, to be a whole character in life.

Herder can be forgiven for overlooking the sound sense in some of these remarks (which is essentially Goethe's) and reacting instead to the Moritzian absurdities.

Old Harry take the god round which everything is supposed to be a masquerade that he uses as he pleases. Or to put it more gently: I creep away from the great Artist, the unique reflecting totality in the totality of Nature, who regards even his friends and what happens to him as paper for him to write on, or colour on his palette for him to paint with. Panegyrics of the kind Moritz is giving must enervate if they are not treated with contempt.

Called to order by her husband, Caroline deeply regretted her aberration:

Dearest angel, you are completely and absolutely right about him. You judge him as man to man . . . We know friend and foe are aware of his monarchical attitude and hundred little vanities, and my idolatry has not gone so far that I can actually regard them as divine attributes . . . I would not for anything in the world exchange you for a vain poet. That I have made so much of it comes from my not having had any very concrete notion of the poet and poetry, the artist and art, and I was just like a child that has learned a new letter . . . *Tasso* . . ., in what it shows and what it says, confirms all this deification of the poet . . . Oh, I wish I could unwrite all my letters about him and Art! What is the poet and poetry to me!

Moritz's theory, in its later adaptation by Schiller, was to have a profound influence on German aesthetics in general: his establishment of an autonomous realm of Art proved important for many literary figures in determining their relationship to the political powers that were to be. The effect of Moritz's interpretation of Goethe himself, whom his treatise was, according to Herder 'tailor-made to fit', was if anything even longer-lasting, and even more certainly undesirable. But, for all the huff and puff about 'the Artist', which in the end Goethe could do without, since his position in Weimar, and in the Duke's esteem, did not depend on ideology, there was a close and genuine affinity at this time between the views of Goethe and Moritz and of Herder too, had he had the courage to admit it. For all three human cultural achievement was a possible source of meaningful order in a world from which providential divine guidance was absent. In the first months of 1789 Goethe saw through the press —Moritz, inaccurately, reading the proofs—the description of the Carnival which he had begun in Rome: of all his writings it most closely, and perhaps even deliberately, fulfils the Moritzian requirements for a work of art. Of all his writings it is also the most godless. Beautifully printed in Berlin by the typographical specialist J. F. Unger (1753–1804), with Schütz's illustrations engraved and coloured in Weimar by G. M. Kraus, *The Roman Carnival* (*Das Römische Karneval*) was first published by Bertuch in an edition of 250 copies which almost immediately went out of print, and then republished in 1790, without the plates, in the successfully up-market 'journal of luxury and fashions', the *Journal des Luxus und der Moden*, which Bertuch had launched in 1786. Within this exquisite physical framework Goethe presents a microcosm: the Carnival, which he describes in a series of separately titled paragraphs of varying length, written in a coolly detached and unvarying present tense, is a world of its own. No attempt is made to incorporate an account of normal Roman life: the subject is exclusively the saturnalian behaviour of the masked and costumed citizenry and the brief culmination of the day's excitement in the horse-race down the Corso, and when Ash Wednesday is reached the Carnival is over and its description is over too. The material setting, in the Corso and its neighbouring streets, is almost as restricted as the stage of the neo-classical theatre. But there are no individual actors: there is only the shifting, jostling, multi-coloured crowd of carriages, horses, and human beings crammed into this narrow compass, from which scenes, faces, episodes, names, emerge for a phrase or a line or two, and are then swallowed up in a collective, self-steering, order or disorder, as the mass takes on a shape to permit the passage of dignitaries or the running of the race, and in the next moment loses it again. Sexuality and murderous violence are everywhere mimed, invoked, alluded to, but the city watch and the impromptu gallows threatening instant retribution secure the crowd from everything but unavoidable traffic and racing accidents, which it is large and indifferent enough to absorb and ignore. The narrrator himself has no individual part in the events, except as a typical and anonymous body in the throng: he merely observes this world and reports on it for an alien audience to whom everything he says is strange. Until, that is, the concluding 'Ash Wednesday meditation' is reached and it is made explicit, precisely as Moritz's theory of tragic conclusions requires, that what we have seen is a representation, not just of a world, but of the world, of the life of the human species itself: love and birth and crowded activity, tense expectation and momentary fulfilment and death have all passed before us, and while the milling and nameless actors in this pageant have, now that it is over, forgotten it as if it had never been, the narrator and his audience have an ordered picture of it before them and can reflect on its meaning for themselves and their entire race. That meaning is not found, however, in the Lenten and Easter mysteries which, for the Roman population, the Carnival ushers in and to which Goethe makes scarcely an allusion, except to say that the Carnival is the surviving form of a pre-Christian festivity. Nor is it found, as Moritz, and even Herder, might have found it, in the perpetual springtime of resurgent life, expressed and celebrated, for Moritz at least, in the beauty of the work of art that depicts the tragedy of individual death. Goethe's conclusion is distinctly more sombre and, assisted now by an awareness of Kant's arguments for the limitations on our possible knowledge, gives definitive expression to the mood that had first come over him a year and a half previously, when he had returned from Naples to Rome, had begun his revision of *Egmont* and had studied and discussed Herder's historical speculations with Moritz. The most intense pleasures of life, he says, should be compared to the horses that flash past in the race on the Corso, and, 'since life as a whole is ungraspable, unenjoyable, even questionable', we should remember 'the importance of every momentary and often seemingly insignificant enjoyment that life has to offer.'

The passion which Schiller had noted in Goethe's evocation of Italy as the home of present pleasures was deeply rooted. Withdrawn from the 'element' that had there nourished him with enjoyment at every turn, but was in Weimar reduced to a memory, he felt, with a metaphysical keenness rivalling that of his own Egmont, the value of any and every rich and happy moment. His hostility to 'lies . . . intimations, desires etc.', to anything that distracted from the tangible satisfactions of present experience, became fully explicit. The curiosity of Frau von Stein and Caroline Herder about their dreams infuriated him. 'One becomes a dream, a blank, oneself if one seriously concerns oneself with these phantoms', wrote the author of 'Why did you give us the deeper vision?' and in a little verse expostulation, 'Bright and Brazen' ('Frech und froh') he announced that his heart now spurned the 'sweet pains' of yearning love, which made only for a dog's life. He now wanted no admixture of pain, absence, or longing at all:

> Nur vom Tücht'gen will ich wissen,
> Heißem Äuglen, derben Küssen.

I am only interested in the real thing, eager eyes and smacking kisses.

There is a close relation betweeen sexual abstinence and certain kinds of religious emotion. Goethe had discovered bodies, his own and a woman's, and for a while there seemed to him to be nothing else. In his first sustained experience of sexual contentment, the sense of an unfulfillable desire, that might in this material life be the token of transcendent otherness, and so might furnish the ground for a never-terminating symbolic quest, abandoned him. Three days before his birthday in 1788 Caroline Herder reminded him of the birthday present he had received the previous year, Herder's book of conversations on Spinoza and Leibniz; ' "Then I was given God", he said smiling, "and this year I don't believe in one" '. 'Disagreeable things must be going through his mind', Caroline commented. As it happened, Goethe at the time was reading her husband: the fourth part of his *Ideas*, largely complete in manuscript, though not published until 1791, and in particular the seventeenth book, on the early history of the Christian Church. Herder had by now largely given up the struggle of his early years to accommodate historical Christianity and rational Enlightenment and had settled for the latter. Christ was a figure about whom little was known: as a human teacher of pure humanism he had nothing much to do with the religious structure that bore his name. The 'religion about Christ', Herder wrote, adopting the distinction already propagated by Lessing, the mindless worship of Christ's person and his cross, was a 'turbid efflux' from 'Christ's religion', his pure ethical teaching. While Goethe could read with approval Herder's disillusioned account of the spiritual power-politics by which the priestly hierarchy established its sway over the minds of Europe, he was no longer interested in such cautious distinctions, which for another two centuries would enable liberal Protestants to proclaim the essence of Christianity from pulpit and rostrum. Christianity now seemed to him a quite accidental, if thoroughly effective, impediment to straight thinking about the one solid world that there was, and when he wrote to Herder a week after his birthday his phrasing left open the possibility that both forms of Herder's religion were equally fabulous:

Christianity you have treated as it deserves; for my part I am grateful to you. I am now getting the opportunity to look at it from the artistic side as well and there things really do start to look wretched . . . The fable about Christ [das Märchen von Christus] means that the world can last for another 10,000 years and no one will come properly to their senses because you need as many resources of knowledge, understanding and logic to defend it as to dispute it. Then the generations get confused, the individual is a sorry thing, whichever party it declares for, the *whole* never comes together as a *whole*, and so the human race totters hither and thither in a paltry botch, which would none of it signify if only it did not have so great an influence on matters that are so important to human beings.

Matters of life and death, in fact. What was one to say to a Merck, desperately conscious of the onset of madness, and longing for a 'peasant's faith' so that he might pray? Goethe wrote back, sensibly enough, that he was sure writing the letter had been a relief for Merck and promised a willing ear in future. For Duchess Luise, deeply depressed after another stillbirth, Goethe had recourse to a sterner and quainter remedy, though a tried one, and read her the account of the origins of the Roman hierarchy from the manuscript of Herder's *Ideas*. When he wrote in December to F. L. Stolberg on the death of his wife Agnes at the age of 27, his consolations were judiciously chosen to avoid any of the indeterminately spiritual effusions that had once been the staple of their relationship. Stolberg, however, took up the implicit challenge and wrote back, telling of his wife's growing certainty of the Christian afterlife in the last weeks of her illness. Goethe's reply the following February, immediately after Moritz's departure, was a masterpiece of inapposite diplomacy, but it contained his most open declaration of loyalty to Epicurean atheism and materialism:

Even if I for my part adhere more or less to the teaching of Lucretius and restrict all my ambitions within the circumference of life, it is still a source of rejoicing and refreshment to me to see the universal mother Nature also letting gentler tones and echoes resonate for tender souls in the undulations of her harmonies, and in so many ways granting finite man a sense of belonging with what is eternal and infinite.

'From the artistic side' Goethe's adoption of the views that the only things we can know are individual material objects of the senses, and that the only things of value are individual happy moments, was to prove of deeper and more lasting effect than Moritz's theories about the processes of aesthetic production. It was Goethe's attitude to the visual arts that longest felt this influence, partly because that inevitably continued to be determined by the memory of his Italian experiences, and partly because in the coming decade those arts were affected by the Kantian turn in German intellectual life to a much lesser extent than literature or the natural sciences. A series of essays in the *Teutscher Merkur* in late 1788 and early 1789 was the principal medium in which Goethe elaborated his new understanding of works of art as 'present pleasures'. 'It is my conviction', he wrote to J. H. Meyer, now in Naples 'that the highest purpose of art is to show human forms that are sensuously and aesthetically ['sinnlich'] as significant and beautiful as possible. As for psychological and ethical ['sittlich'] subjects, only those should be chosen which are most intimately connected with the sensuous-aesthetic and can be indicated by form and gesture.' A first point, then, in the new aesthetic is that the ambitions of the visual arts are restricted within the circumference of life, they are essentially material, and in the most literal sense: their matter is matter. Shortly after arriving back in Weimar, Goethe wrote to the great classical philologist, C. G. Heyne (1729–1812), professor and librarian in Göttingen, that on the basis of his studies in Italy he would like to investigate

how far the material that was being shaped determined the artist to shape the work thus and not otherwise . . . You can see that I am starting very much with my feet on the ground and it might seem to many that I am treating the most spiritual of affairs too earthly; but may I be permitted the remark that the gods of the Greeks were not enthroned in the seventh or tenth heaven, but on Mount Olympus and took their giant steps not from sun to sun but at most from peak to peak.

An outline of this project duly appeared as the essay 'On the Theory of the Plastic Arts' ('Zur Theorie der bildenden Künste') in the October issue of the *Teutscher Merkur:* Goethe argued that the fracture patterns of Egyptian granite made it a suitable material for obelisks, that the change from the use of limestone to the use of marble in the building of Greek temples had determined the change from the Doric to the Ionic order, and that the Gothic style resulted from an illegitimate transfer into large-scale stone structures of techniques appropriate only to small-scale wood-carving (rather, one might add, as the middle-class literary movement of the 1770's, which had identified itself with that style, had tried to apply to German history and society the realistic and novelistic techniques that were appropriately, or at any rate permissibly, applied only to domestic miniatures). What was prettily elaborate as a reliquary was grotesque as Milan Cathedral. A similar line of thought underlay the scheme for investigating the expression of ideal characteristics in Greek sculpture. Just as the Greek thought of his gods as living in a material and earthly heaven, on a mountain-top he could see, so the perfections he imagined them to possess must lie innate in human beings—and that to Goethe now meant: in visible human bodies. His anatomy course at Jena in the autumn of 1788 was, he confided to Caroline Herder, part of his preparation for this investigation. Physiognomy was the scientific study of the human material of art and he wrote confidently to Caroline's husband of 'physiognomical discoveries relating to the formation of ideal characters' that would astonish him on his return. Little trace of these 'discoveries' remains however, beyond a rather odd series of modulated studies of the balding and choleric Knebel, who at one point is given the ears of a satyr, and at another is idealized, à la Trippel, by the addition of flowing Apolline locks.

    To the requirement that art should concern itself with material things that give substantial satisfaction to the senses, Goethe added, secondly, the requirement that it should be purged of those elements of feeling, interpretation, and, ultimately, morality, which he called 'psychological and ethical', or 'sittlich'. The expression of the artist's personality, of his desires, beliefs, or other affections, has no place in the business of art, as that is defined in the essay 'On Simple Imitation of Nature; Manner; Style' ('Über einfache Nachahmung der Natur, Manier, Stil'), which appeared in the *Teutscher Merkur* in February 1789, and which Caroline Herder found an 'incomparable' adjunct to Moritz's treatise in ordering her new thoughts about art, although the issue it is concerned with lies outside Moritz's scope altogether. When an artist, for example a flower-painter, is engaged in the 'simple imitation of Nature', he is, Goethe tells us, unselfconsciously concentrating on the obedient reproduction of the immediate and superficial appearances of things: though limited—it is most suited for miniature work—such an attitude can produce art of the highest quality. In order, however, to broaden the scope of art to encompass, say, a whole landscape, a different approach is needed: such an extensive and complex object can only be grasped by an art of abbreviation and suggestion which is symbolic, like a language, and, like a linguistic utterance, bears strongly the impress of the individuality of the speaker. This level of artistry is called by Goethe 'manner', and though he reiterates that the term is not pejorative, it is none the less the only one of the three levels of art which is said to be in any danger of producing inartistic results and becoming 'empty and insignificant', it is given little praise in its own right, and it is defined as no more than the 'means' to achieving the highest level, 'style'—which however can also be reached directly from the first level, 'simple imitation', without the intervention of 'manner'. In 'style' is found the true, or at any rate the perfect, integration of the artist's personality with the natural objects he is depicting. Goethe returns to the example of flower-painting. If the artist is also a botanist, if behind his reproduction of superficial appearances lies long study of the objects themselves, or the factors that cause them to become as they are, their processes of growth, the influence of the environment, their specific and generic affiliations, then what is truly 'characteristic' of them—their particular relation to the general order—will become apparent in his work. The supreme artist, in short, appears in his own productions not as an emotion expressed, but as an intellect applied. In writing all this about 'style' Goethe no doubt felt very grown-up, but the true import of his essay obviously lies in what it says about 'manner'—the level of art which Goethe characterizes by reference to language and landscape, the only two media in which he had himself up to now been productive, and whose fate under a regime of 'style' remains unclear. In subordinating 'manner' to 'style'—itself characterized by reference to his new interest, botany, in which he had as yet achieved nothing—Goethe was endeavouring to mark a turning-point in his own development. He had once accepted a distinction between his inner and his outer lives in order to accommodate himself to the court world but the poetry that his inner life had produced had been intended for the eyes of a public consisting of one woman only. Now that he sought not just accommodation to the court but complete assimilation to it, and saw himself as writing for the sake of art rather than of any public, even a public of one, poetry itself was not allowed to give asylum to the inner life: intimations and desires were of the devil and art, like life, was all matter, and externality. He was putting behind him, or putting in its place, the art of feelings, to which he would soon have devoted the greater part of eight volumes, and substituting for it an art of things, sensuously perceived and immediately enjoyed. It remained to be seen how many volumes the new art could fill and whether it would rise to more than the literary equivalent of flower-painting.

The essay 'On Simple Imitation of Nature . . .' does not refer to the point under close discussion with Knebel at the time of its publication: the insufficiently definite distinction in Moritz's treatise between the order of nature and the order of art. Goethe, however, had already, and loudly, proclaimed this distinction, the third and final component in his new attitude, the previous November, in another essay drawn from his Italian experiences and published in the *Teutscher Merkur*, 'Female Roles Played by Men on the Roman Stage' ('Frauenrollen auf dem Römischen Theater durch Männer gespielt'). The practice referred to in the title is seen by Goethe as a survival from classical drama and an important clue to its character. Excellent though female impersonators may be, their performance always has about it an element of artificiality, of deliberate pretence. That now seems to Goethe a defining feature of art itself:

one experienced here the pleasure of seeing not the thing itself but its imitation, of being entertained not by nature but by art, of watching not a particular individual [eine Individualität] but a quintessence of observation and reflection [ein Resultat].

That sense of the difference between art and life, between the real thing and its representation is so valuable to Goethe at this time because it is so opposite to the sense which the then burgeoning art of realistic literature, particulary the novel, sought to arouse: that there is no difference between fact and fiction, that we are reading about the same things whether we read about 'Germany', 'ten thousand a year', and 'a good match', in *Werther* and *Mansfield Park,* or whether we read about them in the newspapers. Art, as Goethe now wants to conceive of it, is not part of a process of communication at all—it does not tell us things, or tell us about things—and it certainly is not part of that middle-class process of communication which is the publishing industry. That too Goethe has put behind him. Art is a matter for the closed society of the court, and it is a matter of enjoyment. The artist neither communicates nor expresses: he creates a thing that gives pleasure to his patron.

One may doubt whether such a view can consistently be applied to literature. But by making not literature but the visual and performative arts—traditionally, and for various quite straightforward reasons, the objects of courtly patronage —into his paradigm of Art itself, Goethe reverses the—essentially middle-class —priorities Lessing had established in his *Laocoon.* There the visual arts are treated as a channel of communication, an alternative form of literature. Goethe's new, courtly, and materialist aesthetic is most fully expressed in his correspondence with J. H. Meyer about a fresco of Annibale Carracci's in the Farnese Palace, which revealed to him what now seemed this crucial deficiency of the *Laocoon*. Carracci's picture shows Circe offering Ulysses the magic potion—a prostrate companion with swine's head reminds us of its effect— while behind him Mercury approaches on winged feet to drop into the cup the moly that will protect him. Meyer pointed out to Goethe that two separate episodes from the *Odyssey*—in which Mercury gives Ulysses the plant well before he enters Circe's palace—were here combined into one image. Goethe saw the significance of this direct contravention of Lessing's principle that painting should deal only with simultaneous and poetry only with successive events. Lessing's principle assumed that painting and poetry depicted different selections of information from the same continuous story: the poet told of the battle and the painter represented the costumes, the poet should not describe, and the painter should not narrate; both, however, were partial representations of the same full reality, within particular formal constraints. Carracci's picture suggested to Goethe that 'the painter should and can narrate', because the perfection of a painting was not after all to depict something true about the real world outside itself, but to be a wholly satisfying—and therefore self-contained —material object: the symmetrical patterning of figures on ancient vases was not a deficiency in verisimilitude but an artistic merit, because symmetry is sensuously pleasing, and art is a 'decoration'. In Lessing's account our minds are never on the picture, but on the story of which it represents a part, as in the Christian art Goethe had repudiated in Italy there was 'never a matter of present interest, always something awaited in fantasy'.

The Ancients regarded the picture as a *complete* and *self-contained whole*, within its space they wanted to *show* everything, the picture was not to make one think *of* something, one was to think *the picture*, and *see* everything *within* it. They compressed together the different phases of the poem, of the traditional story, and in this way brought *succession* before our eyes, because the picture is to be seen and enjoyed by our *bodily eyes*.

Carracci's picture enabled Goethe to link Moritz's emphasis on the autonomy and amorality, the 'uselessness', of the complete whole that is a work of art, with the various new, materialist, strands in his own thinking. *Ulysses and Circe* became for him an emblem of the attitudes he now attributed to the classical and pagan world as the source of a perfection in art which in the modern era, save for the work of Raphael, was rarely to be found: material, sensuously satisfying, self-contained, unemotional, unrealistic, and courtly. Such a work of art was the occasion for a happy moment of present enjoyment. It was an aesthetics to go with his Lucretian metaphysics.

It was also an aesthetics that in two respects was extremely difficult to sustain, unless it was completely divorced from any application to Goethe's own art of poetry. First, its hostility to the values of the middle-class reading public was hardly compatible with the attempt to address that public in an eight-volume publishing venture. A better solution than this had to be found to the problem of how the poet was to live with the court to which, by his return from Italy, he had newly committed himself, and eventually Goethe would recover a relation with his public and confine his materialist aesthetics to arts which he no longer seriously practised himself. Secondly, though, there was an intrinsic defect in the cult of the fulfilled and autonomous moment of present pleasure: it had nothing to say about the most insistent reality in Goethe's experience since he set out from Rome, the fact that happy moments pass, that the pleasures of memory are not those of present satisfaction, and that when desire has been fulfilled, it returns. The symbolic journey to Rome had ended in disappoint-ment, but by an extraordinary effort of reinterpretation Goethe had arranged for that disappointment to be followed by something that could count as the fulfilment of twenty years' waiting, found, not in Rome's ruinous physical reality, but in its German artist's colony—his domestic academy—and in the simplest of sensual satisfactions. That fulfilment he had begun to transplant to Weimar, but he was thereby only postponing the question: what next? How could a symbolic existence be maintained together with the pretence that the reality to which all symbols referred was the completeness of sensual fulfilment in the material here and now? It was a pretence, of course, and in that lay the eventual answer to the question. When the Carnival festivities were over Goethe dismissed the Roman populace, for which there were only present pleasures, into obliviousness—but he was left with his own mind, and the silent company of his reader's mind, facing ungraspable, unenjoyable, and question-able 'life as a whole'. The reflective, feeling self, always thinking of, and desiring, something other than what lay before its bodily eyes, could not be excluded permanently from Goethe's experience. Too much memory and loss lay behind even the satisfactions he had brought back from Italy, and in his literary work of this time, if not in his aesthetic theory, memory and loss continue a subterranean existence. Despite Goethe's conscious and confessed belief in the importance of the earthly and his embarrassment with what he calls 'manner', an awareness of the fragility of the precious moment keeps alive the confessional, autobiographical, strand in his major literary projects of 1788 and 1789: *Tasso, Faust,* and what came to be known as the *Roman Elegies.*

Through the lonely summer of 1788 and the long winter that followed—the snows did not melt until the first week in April—Goethe maintained within his mind, and to some extent in his Weimar environment, a certain extra-territorial Roman enclave. He was seeking, not to conceal from himself how much he had lost, he wrote to Herder, but to settle in again 'as far as possible'. 'Hibernating like a hamster', and heating his rooms to what others found an intolerable temperature, in despite of snow and grey skies, he was 'feasting on the best of art and nature'—which latter may have been a reference either to his botanical investigations or to his concubine. With Moritz as well, whose presence helped him 'survive the most burdensome part of the winter' and gave him a 'crammed' and fruitful January, much of his studious life on the Corso was indeed reconstituted, either in his town-house or the secluded cottage, and his work on the edition, especially on *Tasso,* seemed like a 'conversation with distant friends'. Goethe's own drawings of Italian scenery had to be sorted through, shown off, or even added to by impromptu illustrative sketches from memory, and these, or a relevant book such as Hamilton's work on Vesuvius and Etna, *Campi Phlegroei,* provided occasions for the 'good conversation . . . on some cosy evening' that he had once promised Voigt from Naples. Gifts to friends and colleagues spread the golden memories—Voigt himself received a Sicilian, Fritsch a Terracina, landscape; and when the spring came August Herder, whose mother under Goethe's tuition had acquired a taste for Mediterranean fruits (figs were grown under glass in Jena), was delighted to report to his father that the Italian seeds and the pine-cone Goethe had given him had all germinated: would father please bring back some more 'since Privy Councillor Goethe says that there all the flowers have a nice smell'. But it was above all through correspondence that 'a breath of the South' reached Goethe under his northern sky, especially from Meyer, an assiduous and voluminous letter-writer, who as the most literate of the German artist colony tended to act as their secretary. Meyer was not demonstrative about his feelings but the warmth of the friendship in the other letters—the passionate devotion from 'child' Bury, the long laments at their parting from Angelica, even from Hirt a proposal to dedicate a book to him—must have contrasted painfully with the frigid gossip that surrounded him in Weimar. Perhaps he wondered who most deserved his loyalty. Certainly he devoted much time to commending his old companions. Kniep he prevailed on to produce a price-list of his work according to size, then he showed off specimens, particularly in Gotha, and finally in February he was able to send a substantial order (six large, six medium, and six small) to be despatched, and paid for, through Hackert. Bury, who flattered himself that he had an eye for a bargain on the Roman art-market, and needed funds, got away with borrowing money in Goethe's name to purchase some indifferent paintings with optimistic attributions, which were also eventually sold to the Duke of Gotha. Tischbein was a more difficult case: after Goethe had soothed Duke Ernst in Gotha in September, so that Tischbein might have a line of retreat if he was not appointed to the Academy in Naples, Reiffenstein alerted Tischbein's paymaster to his continuing involvement with the Neapolitan court, and Goethe feared that Tischbein, whom he no longer felt able to trust, would fall between two stools. In the event, though, the fears proved groundless, and Tischbein got his appointment.

For as long as Herder and the Dowager Duchess were in Italy, treading in his footsteps, it was only natural, and even proper, that half of Goethe's mind should linger in Rome and Naples. Herder's trip, unfortunately, had early run into difficulties. Dalberg had chivalrously offered a place in the carriage to the widow of Siegmund von Seckendorff, the talented Weimar gentleman-in-waiting, who had died in 1785 at the age of 41. Mme Seckendorff was travelling to Italy for her health, but it was not long before her further designs on the Canon were apparent, and indeed successful. Herder found himself *de trop*, and —more seriously for a man with a wife and six children at home—was asked to contribute to the costs of the journey by Dalberg, whose purse was suddenly under unexpected strain. (Goethe's estimate was that an Italian tour could not be done by someone of his class for less than 2,500 dollars.) After a journey with the love-birds of exquisite discomfort, Herder reached Rome on 19 September and was greatly relieved when Anna Amalia arrived a fortnight later and he could attach himself to her party instead. Goethe followed and discussed it all from afar with Caroline, who found his advice helpful, particularly on financial arrangements, although her husband, irritated at discovering that he did after all need the formal clerical garb which Goethe had persuaded him to leave behind, and which he would now have to have made for him at considerable expense, was less sure of its value: 'All Goethe's suggestions about Rome are worthless . . . I cannot, will not, and do not want to live as Goethe lived here . . . nor do I have much from Goethe's companions. They are young painters with whom in the end there is not much to be done, let alone living with them for years . . . They are all well-meaning people, but too remote from my circle . . .'. The Duchess, however, had been well conducted by Filippo Collina along a route that had already taken her through Verona and Milan, and her first month in Rome was spent, according to Goethe's prescription, on art and antiquities, in the company of Angelica Kauffmann, Bury, Schütz, and Verschaffelt, with guided tours from Reiffenstein, and later Hirt. In November she allowed herself to be the visiting celebrity, although pleading her health to excuse her from the grander dinners, had a private audience with the Pope, who presented her with a fine mosaic of the Arch of Constantine, was received into the Arcadian Society, and made a good friend in Senator Rezzonico. She, Einsiedel, and Fräulein von Göchhausen all wrote delightedly back to Goethe of their success in what was effectively a de luxe and stylized repetition of his original spontaneous exploits, and even Herder's temper improved under the influence of an excursion to Tivoli and the Roman habit of addressing him as 'Archbishop'. The winter weather proved not to be good however—which further convinced Herder of Goethe's unreliability—and on 1 January 1789 Anna Amalia decamped with her entourage for Naples, and the company of Venuti, Hackert and Hamilton, and of Goethe's favoured trio, Tischbein, Kniep, and Meyer, returning to Rome for the last days of the Carnival at the end of February (which like Goethe in 1787 she found a disagreeable experience). Here she installed herself in a villa in the Trinità dei Monti area, not far from Angelica, until 19 May, when she returned to Naples for the rest of the year, despite Goethe's 'ex-Chamber-Presidential' warnings of the cost, and so parted from Herder, who gladly set out on his homeward journey at last.

Not merely was Goethe kept informed throughout by a dozen different correspondents, he himself arranged to contribute to Anna Amalia's enjoyment of the tour by ordering paintings for presentation to her in Rome and Naples by Verschaffelt and Kniep, and by sending her volume viii, the poetry volume, of his *Works*, as soon as it was published in early 1789, as well as, through Caroline Herder, a copy of the opening scenes of *Tasso*, read by Herder to the whole party in Tivoli in May (Schütz depicted the occasion in water-colour). Anna Amalia was asked to play her part in building up Weimar's Italian collections by buying books, music, coins, engravings of antiquities, impressions of gems, and even a collection of Sicilian mineralogical specimens. But this vicarious prolongation of his past was fundamentally an illusion, and when the illusion failed the recollection of loss was all the more painful: 'Holy Week', he told the Dowager Duchess, when it was over, 'which was always in my thoughts, brought me almost to despair and I had to do everything I could to direct my thoughts away from those happy regions.' The music he had heard in St Peter's was for Goethe, in that pre-technological age, as unrenewable a pleasure as the colours of the Claudes or the Raphaels he had seen. Kayser, his great musical resource, had from the start of the journey felt ill at ease with his courtly company, and had parted from the Duchess in Bolzano and returned to Zurich, so severing relations with Weimar—and, incidentally, ensuring that, in Goethe's words 'all the immense labour' on *Jest, Craft, and Vengeance* was wasted. True, in these months a new musical relationship was beginning—the greatly more gifted J. F. Reichardt (1752–1814) had set the new *Claudina* and visited Weimar to discuss a forthcoming production. But Reichardt was Kapellmeister at the Prussian court in Berlin and there was no prospect of any resumption for Goethe of the lessons in musical analysis with which in 1788 Kayser had prepared him in Rome for the liturgy of the Easter Triduum.

While Goethe had had a particular interest in Kayser's return to Italy, hoping to see him come back as a favourite, ready to settle in Weimar, there was no particular consolation for him when Reichardt set out on his own Italian journey, no more than when in the same year both Sophie von la Roche and Knebel's youngest brother Max, travelling-companion to the Margrave of Ansbach, also followed the example he had set. The same was true of the tour of Anna Amalia and Herder. These replications of a gesture that in his own case had had hidden motives, unguessed, to some extent, even by him, could bring him no more reward than, years before, the re-enactment of Werther's anguish in the hearts of so many readers had brought relief to his pain. It was a measure of Goethe's myth-making power that the Dowager Duchess of Weimar could spend 40–45,000 dollars on repeating and reinforcing, in sometimes remarkable detail, the pattern he had established—as in 1779, when he took Carl August to Switzerland, the resources of the entire duchy were indirectly at his personal disposal in his quest for a meaningful shape to life. But once the journey itself had been displaced from the centre of Goethe's attention by the emotional trauma of its passing, and of his return, Anna Amalia's or anyone else's repetition bore much the same relation to it as a *Werther*-imitation to *Werther*, or as Tischbein's portrait of Anna Amalia in the ruins of Pompeii to his portrait of Goethe beside the ruins of the Appian Way: a stiffly self-conscious and superficial analogue, in which, not surprisingly perhaps, the symbolic paraphernalia is not integrated with the central figure, and is reduced to clutter. Only Herder had the independence of mind to grasp, at least partially, what he had allowed himself to be persuaded to—Moritz, writing to the departed Goethe from Rome, welcomed Herder as the Holy Ghost, 'the Comforter, who brings us your peace'—and he resisted it, if only with petulance: 'The day after tomorrow', he told his wife on his return to Rome from Naples, 'Trippel is starting on the bust of me which the Duke has ordered, and which is to be a pendant to his one of Goethe. Oh these wretched pendants! Goethe had himself idealized as an Apollo, how will I, poor thing, look by contrast, with my bald pate!'

It was generally assumed that Goethe would return to Italy and that only the date was unsure. Although 'I can and must not say . . . how painful it was for me to leave the fair country', he did not conceal from Meyer that 'my most eager wish is to meet you there once more'. Frau von Stein of course knew his state of mind: 'you are right', she wrote acidly to Knebel, 'there is much resemblance between the character of Caesar and that of our friend, and for that reason he feels himself more at home in Rome'. Originally he was expected to join Anna Amalia in the winter of 1788–9, 'and this would be a very good thing'. 'Goethe flourishes best in Rome': a sentiment which Herder endorsed and passed back to Goethe adding that it was certainly not true of himself. Winter passed, however, and Goethe stayed in snowbound Weimar. The following March he spoke of going to Rome in September and returning with the Duchess in the summer of 1790. The plans were probably changing with the lengthening timetable for the completion of the *Works,* and specifically of *Tasso,* which was becoming 'like a dream in which you are so near to your object and yet cannot get hold of it'. The underlying plan however does not seem to change: Goethe's principal task is the work on the Göschen edition, which explains both his remaining in Weimar and his holding aloof from most official duties, and while he is engaged on this task he is in a parenthesis between visits to Italy—as soon as it is finished he will be free to join the Dowager Duchess in the happy regions again. Goethe, however, was already looking beyond even a second visit, which, after all, could only conclude in a second painful return, and—on the assumption that no Louis XIV would appear, giving him ten years in Rome on 10,000 a year—was making more permanent arrangements for the transfer of the Roman spirit to Weimar, and events came to meet him.

At the beginning of 1789 Carl August announced his intention of rebuilding, at long last, the burnt-out ducal palace, the Wilhelmsburg. If only for financial reasons, it would be a long process, and the first step was to set up a

Commission to appoint an architect, draw up his brief, and eventually superintend the work. In asking Goethe to be his personal representative on this Commission, of which the other members were Voigt, Wedel, and Schmidt, the Duke was acting wholly within the spirit of their agreement of the previous year. Even though the task would gradually become onerous—from an early stage the Commission met weekly and at its first meeting on 25 March the Duke presented it with forty points to consider—it was a new, and in principle a limited, undertaking of a kind particularly suited, and even flattering, to someone who had just spent a year in the semi-professional study of the arts and architecture. Goethe had discovered himself 'as an artist'—'artifex', Herder called him, using the Latin term without irony, for he did not yet know what was afoot—and the Duke was taking Goethe at his word (if in its older acceptance). 'None of us, of course, understands the job', Goethe was careful to tell him, all the same, and the most the Commission could do, he went on, was to choose its experts well. Although his colleagues might not be convinced by his arguments for shallower pitched, Italianate, roofs, there was no dissent from his choice of an acquaintance from his Rome days, J. A. Arens, to be the scheme's architect. Here, then, was the opportunity to persuade a talented and promising artist from the right stable, and from whom Goethe hoped to learn much, to settle in Weimar, and he might even build something that would make Goethe's physical environment more Roman: Goethe started to sketch plans for Italian bridges in Tiefurt or an idealized Weimar park, where in 1788 three classical pillars were erected to suggest a ruined temple. But Arens was busy, and slow both in answering letters and in coming to view the site, and the Commission contented itself for the present with resolving to retain as much of the shell of the palace as the builder thought possible, and to drain and fill the old moats. Arens finally arrived in June—'and I once again enjoy the presence of an artist'—but stayed only for three weeks.

Goethe had more immediate success with another member of his Roman circle, who, however, was an acquaintance of much longer standing, the Swiss engraver Lips, already collaborating with Angelica on the frontispieces and title-page vignettes for the later volumes of the Göschen edition. At the end of March Goethe wrote to Lips, now living in his own old lodgings on the Corso, asking if he would consider moving to Weimar: the Duke was prepared to pay a retainer of 150 dollars a year (enough to cover food and accommodation) if he would give a little instruction in engraving in the Drawing Academy, and Bertuch could guarantee commissions from publishers of 500 dollars for some years to come, though Goethe quite understood that he would not wish to devote all his time to such commercial work 'of lesser significance for art', in Lips' own words. Although the proximity of publishing houses in Gotha and, especially, Leipzig was one of the features Goethe mentioned as likely to make Weimar attractive to Lips, along with the wide diffusion of the *ALZ* from its base in Jena, he also pointed to the art-collections in Dresden, Berlin, Cassel and Gotha and held out the possibility both of collaboration with himself and of the growth of a school of pupils who could undertake many of the routine commissions of a workshop. In this prospectus we can already see the outlines of a vision of Weimar as combining the benefits of princely patronage with an intellectually liberal and commercially (though not of course politically) active middle-class society, in which a 'thinking and free-born [that is Swiss] artist' would not feel out of place. When Lips arrrived in December 1789 he specifically asked to be listed in the Weimar almanac not as a ducal functionary, but as a freelance attached to the Drawing Academy. Goethe, for his part, had no wish to go so far, but the 'artists' he wanted to encourage, and of whom he saw himself as one, had in his eyes precisely that noble autonomy, even if, without advertising the point, they depended for food and accommodation on their state pension. When Lips readily accepted the offer at the end of April, Goethe was delighted. 'Now I am covered for next winter', he wrote, for *Tasso* was still not finished, so he had to give up hope of leaving for Rome in September and was again in need of someone like Moritz to help him through.

Shortly after writing to Lips Goethe was able to make another 'acquisition' and to hold out to Meyer the prospect that he too might eventually receive a similar offer. It was not until August, however, when Herder had returned from Rome and given him a full account of Meyer's intentions that he could confirm the proposal. Meyer was to receive a strictly secret retainer of about 150 dollars for the two further years that he wished to devote to study, and at the end of that time, when he would be looking for a 'quiet nook', he was to come to Weimar, where Goethe would ensure that he found a position. In the meanwhile Meyer was to continue to collect, if not on the same scale as Anna Amalia, the materials necessary for survival in 'a small-town existence in the North'. Meyer's support was to come from the Duke, but Goethe also turned to Duchess Luise to further such local talent as there was: a young artisan, F. W. Facius (1764–1843), was showing an unusual gift for steel-engraving, and money was found for his education in the hope that he would eventually be able to cut gems in the classical style. The Dowager Duchess too was kept informed of these arrangements, partly because Goethe hoped she might take enough of an interest in Bury to bring him back to Weimar with her, 'and then we can set up a nice little Academy'. Had Goethe's hint been taken, and had Kayser and Tischbein not let him down, he would indeed have transferred the domestic academy of the Casa Moscatelli, practically in its entirety, to Weimar and Gotha. 'One cannot live without artists, either in the North or in the South', and Goethe was mobilizing the court to provide them.

In one other respect Weimar in 1789 was being remodelled on the pattern of Goethe's Rome, and this, though less expensive than buying artists, was very much more controversial. By the beginning of March Goethe's secret was out. Fritz von Stein was probably the first person, after the Duke, to know, though he had so far received so strictly moral an upbringing that he could not at first work out who the 'corpulent little female' could be whom he met one day by accident at Goethe's cottage, and who claimed to belong there herself. On 8 March Fritz's mother told Caroline Herder that 'he has got the young Vulpius girl as his Clärchen and frequently has her come to him etc.' 'She holds it very much against him', Caroline told her husband, and she too was disappointed. 'Since he is so excellent a man, and is 40 years old too, he should not do something by which he so demeans himself to the level of the rest. What do you think about it?' The comparison of Goethe with the 'careless' Egmont was already current in Weimar (the English word being used, probably in the sense of 'carefree'), but Egmont's relation with Clärchen was a symbol of a somewhat earlier stage in Goethe's adaptation to court existence than his liaison with Christiana. Herder in Rome could recognize the truth: 'What you write about Goethe's Clärchen I find more displeasing than surprising. A poor girl! I couldn't permit myself that for anything.—But people have different ideas and the way he lived here with uncultivated, though goodhearted, people could not have produced any other result.' Christiana's role in Goethe's life was soon to change—she would see to that—but for the present she was precisely what Herder's reaction implied: a German Faustina. She too had her part in the programme by which Goethe was still trying to construct meaning out of the end of meaning that he had experienced on his arrival in Rome in 1786. The atheist and materialist priest of pure Art was closing his eyes to his real circumstances and striving to recreate Weimar as a lasting image of the place, and indeed the time, in which desire had terminated; and along with the importation of artists, and the redefinition of his relation to the court, and of art itself, that involved also a head, even if now a German head, on the 'damned second pillow'.

Goethe's first portrait sketches of Christiana are homely and personal, but after he had begun a study of the heads on ancient coins in November 1788 he drew a series of profiles of her which assimilate her increasingly to these classical models, eventually absurdly so. A similar process occurred in the *Erotica*, as he called the poems which Christiana inspired. 'Morning Complaint' and 'The Visit', from the summer of 1788, are still in a modern, trochaic, metre and in a Weimar setting, despite their anacreontic overtones. In the autumn of that year, however, Goethe began to be attracted again by the idea of writing epigrams in German forms of classical metres, specifically elegiac couplets, a genre in which he had already experimented when composing his inscriptions for the Weimar Park. He studied the form carefully with Knebel who was translating Propertius at the time, investigated an early eighteenth-century German venture in rhymed elegiacs, and made an unsuccessful attempt at hendecasyllables. The insignificant little poems that resulted, some of which he used as scarcely more than space-fillers in volume viii of his *Works*, which at the end of 1788 was being prepared for the press, he also classified in virtue of their subject-matter as *Erotica*, and it is no surprise that he should have been reading, and no doubt looking for models in, Catullus and Propertius, and probably Tibullus and Ovid's amorous poetry too. But in December another work of Ovid's appears in his correspondence which suggests a more serious aspect to this game, the *Tristia,* Ovid's lamentations over his banishment from Rome. 'That my Roman friends should think of me is very reasonable', he wrote to Herder, 'I too cannot expunge from my heart a passionate recollection of those times. I cannot tell you with what emotion I often repeat to myself Ovid's lines . . .' and he then cited the description from the *Tristia* of Ovid's last night in Rome, which forty years later he would use to conclude the *Italian Journey.* This is the first hint that Goethe's reading of the Latin elegists might lead to poems in which Rome provides not only the verse-form but also the scene, and even the subject, and in which a link is established between his sexual fulfilment and the great deprivation which was dominating his emotional and intellectual life. There is no evidence when exactly he began to write longer, less epigrammatic pieces which conflated the figure of Christiana with the figure of his Roman mistress, nor when he conceived the idea of collecting these together into a semi-narrative cycle set in Rome, nor can the date or order of composition of individual poems be determined. None the less the likelihood is that most or all of the twenty-two poems of this kind were written between the autumn of 1788 and the winter of 1789–90: some of them were probably already written into the notebook in which Goethe was collecting all his *Erotica* when in December 1788 or January 1789 Moritz was amused to find it stowed under the plaster-cast of Raphael's skull.

Goethe, too, often speaks of these poems as 'jokes', but if there is humour in them it is just the good humour of people enjoying themselves in a simple and innocent way. The love depicted is, as he had demanded in 'Bright and Brazen', quite devoid of any pain, or of any longing beyond physical impatience, but also of anacreontic coyness or prurience. Gone are the interrupted climaxes, and the mock modesty with which Cupids avert their gaze in the bridal chamber. Goethe now leaves to others the fiddling and fumbling with brocade and jewels, stays and furbelows, which in another kind of poetry might have served to heighten desire, or even stand proxy for its fulfilment, but are here only a delay and a vexation:

> Näher haben wir das! Schon fällt dein wollenes Kleidchen,
>> So wie der Freund es gelöst, faltig zum Boden hinab.
> Eilig trägt er das Kind, in leichter linnener Hülle
>> Wie es der Amme geziemt, scherzend aufs Lager hinan.
> Ohne das seidne Gehäng und ohne gestickte Matrazzen,
>> Stehet es, zweyen bequem, frey in dem weiten Gemach.
> Nehme dann Jupiter mehr von seiner Juno, es lasse
>> Wohler sich, wenn er es kann, irgend ein Sterblicher seyn.
> Uns ergötzen die Freuden des ächten nacketen Amors
>> Und des geschaukelten Betts lieblicher knarrender Ton.

We have that nearer to hand! Your little woollen dress falls, as soon as your friend has loosed it, in folds down to the ground. Quickly he bears the child in her light linen covering, as befits a nurse, laughing up on to the bed. Without silken hangings and without embroidered mattresses it stands free in the spacious room, comfortable for two. Then let Jupiter have more of his Juno, let any mortal, if he can, have a better time. We are delighted by the joys of true naked Eros and the sweet creak of the rocking bed.

At the heart of the *Elegies,* as they will eventually be called (the term is not used for any of the *Erotica* until November 1789), there lies a completely happy experience such as we find in no other of Goethe's works: two exceptionally different people—a poet plagued by introspection and by public obsession with his *Werther,* and an uneducated but temperamental and far from subservient middle-class woman—meet as any man and woman may meet and find that they make up a 'we', in a fuller and more untroubled sense than that—anyway relatively rare—pronoun has in any of his previous poems, even those to Frau von Stein (it does not occur in 'To the Moon'). The literary complexities in which that central experience is enveloped should not lead us to overlook the biographical core, which in the event makes up much of the charm of the future cycle: the rapid start to the liaison, the eager waiting for darkness, the lovers in bed, secure from the wind and rain outside, conversations about childhood, the woman rising in the morning to waken the fire from the ashes (material for an erotic metaphor), the quarrel and reconciliation, the gradual diffusion among the local citizens of knowledge of the relationship—it would be a mistake to see these simply as established commonplaces in a particular kind of poetry, just as it would be a mistake to derive the features in Goethe's idealizing portraits of Christiana simply from goddesses and coins. The very stylization which these elements have undergone tells us something of the far from happy personal circumstances amid which Goethe's relationship with Christiana gave him a delight and a support which he was not, and never would be, capable of fully articulating. Because in the winter of 1788–9 it was becoming a real relationship (and not just a business arrangement like the Roman affair) it could not be exhaustively translated into its symbolical, would-be classical, analogues, and, after the *Elegies,* Goethe did not again make the attempt. The lengths he goes to in these poems in trying to make the relationship stand for other things (such as the recovery of ancient Roman attitudes to love and beauty) is a measure of the power of this new presence in his life. The unusual metrical form—German had at this time volumes of hexameters, but little in the way of elegiacs—and the stilted diction and word order it often requires are a first and immediate alienation; by contrast, say, with the relaxed and realistic manner of 'The Visit' they are a clear assertion that we are reading art, and not an imitation of nature. The setting in contemporary Rome, not Weimar, permits the (partial) transfer of the happiness radiating from the sexual encounter to the encounter with the relics of ancient art and culture (which had in reality been so ambiguous). The concealment of Christiana behind the mask of Faustina

introduces a contrast between the German poet and his supposedly Mediterranean lover, so that their relationship reflects Goethe's own divided life and
loyalties at the time he was writing. The more and less overt allusions to the
works of the Latin elegists suggest that the poet is deliberately allowing his
experience to be moulded by a literary pattern—a parallel (and of course a
contribution) to Goethe's construction of a Roman enclave in Weimar. Even an
episode of physical love-making can be made to stand for the new aesthetic,
according to which total and present sensuous satisfaction should be provided
by a material that dictates its own characteristic form:

> Und belehr' ich mich nicht, indem ich des lieblichen Busens
>   Formen spähe, die Hand leite die Hüften hinab?
> Dann versteh' ich den Marmor erst recht: ich denk' und vergleiche,
>   Sehe mit fühlendem Aug', fühle mit sehender Hand.

And am I not instructing myself by observing the forms of her delightful bosom, by
directing my hand down her hips? Then at last I truly understand marble: I think and
compare, see with a feeling eye, feel with a seeing hand.

In this, one of the best of his 'erotic jokes', the poet effectively boasts of the
subordination of his personal experience to the exigencies of Art:

> Oftmals hab' ich auch schon in ihren Armen gedichtet
>   Und des Hexameters Maß leise mit fingernder Hand
> Ihr auf den Rücken gezählt.

Often too before now I have composed while in her arms and with fingering hand have
gently counted out on her back the measures of the hexameter.

But the last laugh is on him, for he is permitted the leisure for such little male
vanities only when his beloved is asleep. As in 'The Visit', the sleeping woman
is the image of the new presence in his life, the symbol of the irreducible and
inexpressible other identity, that cannot be controlled or interpreted away in
symbols, and for whom there is only a name, Christiana.

   Carl August might be surprised, Goethe told him, in May 1789, to find
among the *Erotica*—though not in the semi-fictional group that became the
Roman Elegies—a panegyric of himself. Yet the eighteen-line 'epigram' that
he wrote on the tenth of that month could stand as a summary of the new
attitudes that had made the *Erotica* possible, and at the same time it states with
great clarity their dependence on the social and political order found in Weimar
and on the personal régime of the Duke. The materialism that had led Goethe
to reduce Herder's philosophy of history to man's need for a comfortable home
is expressed, both here and in an associated poem, with deliberate *naïveté*—he
needs 'five natural things', he says, food and, especially, drink, a home, clothes,
and a sweetheart, and it might therefore be thought that his praise of Carl
August has been purchased with a bribe:

Denn mir hat er gegeben, was Große selten gewähren
Neigung, Muße, Vertraun, Felder und Garten und Haus.

for he has given me what the great rarely grant: affection, leisure, trust, fields, a garden, a house.

But as an artist, with manifold needs but a poor grasp of money matters, he needs a patron. The bourgeois world of commercial publishing gave him a success he has often cursed, but nothing more material, at home or abroad:

Hat mich Europa gelobt, was hat mir Europa gegeben?
Nichts! Ich habe, wie schwer! meine Gedichte bezahlt.

If Europe has praised me, what has Europe given me? Nothing! I have paid, how dearly, for my writings.

Only a court, then, can support the poet as he needs to be supported, but in the German-speaking world neither Vienna nor Berlin, only Weimar, limited in territories and resources, has taken on the function for which ancient Rome offers such illustrious models:

Niemals frug ein Kaiser nach mir, es hat sich kein König
Um mich bekümmert, und Er war mir August und Mäcen

No emperor ever asked after me, no king concerned himself for me, and *he* has been my Augustus and my Maecenas.

The 'panegyric' of 10 May 1789 was Goethe's definitive declaration that a German national culture, the promised land to which the intellectuals of the 1770s had set out, and in which he thought he had lived during his year in Rome, could be permanently established, in Weimar and perhaps in other similar centres, through the patronage of artists by the absolute rulers of feudal principalities. Five days before the poem was written, the Estates General of France had convened in Versailles for the first time since before the Thirty Years' War and the death-agony of feudal Europe had begun.

In the crisis that was to come the universities, of all the institutions of eighteenth-century absolutist Germany, were to prove the best survivors, and in the long term Goethe was to find his niche as much in the world of learning as in the court and the administration. Even in 1789, indeed, Berlin did not totally neglect Goethe, for in February he received an invitation to become a member of the Berlin Academy of Arts, an offer surely more to his bent than the membership of the Commission for rebuilding the ducal palace he had just accepted. 'Goethe feels more comfortable in Jena; he feels himself at home there, and a stranger here', Caroline Herder wrote from Weimar, and Goethe himself in retrospect gave Jena great importance in reconciling him to Germany after his return from Rome, yet at the time this role developed only gradually. After attending Loder's anatomy lectures in November 1788 Goethe did not visit Jena again until July of the following year. He remained alert to possibilities of development, particularly in the sciences: when the professor of mathematics and physics, J. E. B. Wiedeburg, died in January he arranged for the purchase of his apparatus, though the charitable motive of assisting the widow and children was as important as the scientific. Otherwise his involvement was for the present limited to personnel matters, some of which were of course of great moment. Schiller's appointment was not, however, a sign that Jena was about to become a haven for former geniuses, as Goethe showed firmly when the poet Bürger—with whom he had once had a correspondence on intimate 'Du' terms —called on him during Reichardt's visit in April. Bürger had hopes of being asked to transfer from Göttingen to Jena, to some such position as Schiller had, whom he was also visiting, but Goethe kept the conversation brief and at a sufficiently formal level to prevent discussion of the issue, and quickly returned from the ante-room to hearing Reichardt play through his score of *Claudina*. Bürger was incensed at the 'ministerial' manner of his former comrade-in-arms, perhaps with some reason, but he had no inkling of the complex changes wrought by the passage of fifteen years during which his own attitudes had remained largely unaltered. Goethe was more concerned at this time to keep Schütz satisfied, the editor of the enormously successsful *ALZ*, who had asked for a rise and for whom, by juggling funds, he was able to find an extra 200 dollars a year. It is a measure of the importance Goethe attached to the figure of Schiller that in mid-May he reported to Carl August on Schiller's removal to Jena to take up his post in the same breath in which he told of Schütz's temporary departure for Paris to observe events there. However, the promise of the meeting in Rudolstadt in September, and of Goethe's work on Schiller's behalf at the time of his appointment, was not borne out by what seemed to Schiller to be Goethe's avoidance of him during the winter. Perhaps Goethe was not entirely pleased at the continuing presence of the other poet's Storm and Stress plays in Bellomo's regular repertoire. Schiller consoled himself by writing to Körner:

To be near Goethe frequently would not make me happy. Even towards his closest friends he shows not a moment's effusion . . . I do believe he is an egoist to an unusual degree. He has the gift of captivating people . . . but he is always able for himself to avoid attachment. He benevolently makes his existence known, but only like a god, without giving himself . . . Men should not allow such a being to grow up around them. As a result I find him hateful, although I love his mind with all my heart and think highly of him. I regard him as a proud old prude who needs to be got with child to be brought low in the eyes of the world. It is a quite strange mixture of hate and love that he has awoken in me, a feeling not dissimilar to that which Brutus and Cassius must have had for Caesar . . . Goethe also has much influence on my wishing to see my poem [*The Artists*] properly finished and perfect . . . I set great store by his judgement . . . I will surround him with spies, for I will never myself interrogate him about me . . .

Much more lies behind this extraordinary outburst than personalities. Schiller's triangular relation with the Lengefeld sisters continued its mysterious course, but beneath the surface thoughts of marriage and the need for an income may have been crossing his mind. His Jena office gave him a position but not a stipend, it was costing him money through the necessity of taking a degree and fitting himself out, and the preparation of lectures took time away from the writing which at present alone supported him. He felt exploited by Weimar, and impatient with its inaccessible presiding genius, whose apologists could not account for their devotion: 'one really has too little sheer life to spare to spend time and trouble deciphering people who are indecipherable'; 'this person, this Goethe, is simply in my way, and he so often reminds me that Fate has dealt hardly by me. How easily was *his* genius borne along by his fate, and how *I* have had to struggle up to this very moment!' It is the complaint of a whole generation, indeed of a whole class over several generations, the secularized German intelligentsia, without office, but without economic independence, looking with envious fascination on the great exception who had both. Schiller felt he breathed more easily when he moved over to Jena, but the university was no more his natural medium than the court. His inaugural lecture on 26 May, 'What is universal history, and to what end is it studied?', attracted an audience of three to four hundred (Goethe did not attend), and in the evening he was serenaded and applauded for it, according to the German custom, but it began with a scathing satire of 'professional scholars', who were said to be unlikely to derive much from Schiller's course and were contrasted with 'philosophical minds', with those, that is, who did not expect their education to bring them a livelihood, and whose curiosity about the point of it all had survived the collapse of a dogmatic theological structure. The lecture identified a public, and played up to it, but was not calculated to commend Schiller to his colleagues. In its opening remarks, in its emphasis on the 'middle class' as the sole vehicle of culture, in its deliberately moralistic account of a historical process culminating in the present (a rather primitive fusion of Kant and Herder), and in its consequent scorn for a merely nostalgic or antiquarian interest in the past, it suggests no willingness on Schiller's part to compromise with the institutions of the absolute State or with the aristocratic culture of its officials, uninterested in the public conscience and inclining to a belief in the authority of past models, to all of which Goethe had now, if privately, professed a loyalty. Schiller soon came to feel his intellectual isolation in Jena and perhaps it was partly in reaction to it that in August he secretly became engaged to Charlotte von Lengefeld. With that, however, compromise became necessary at last: to marry he reckoned he needed an income of 1,000 dollars a year. From his writings he could earn perhaps 600; his wife would bring an income of 200, but 200 more still had to be found. Could Carl August be brought to provide a stipend after all for his new professor? Perhaps Goethe or Frau von Stein could use their influence. In the event it was Frau von Stein who swayed the Duke, who was anyway glad to help provide for the younger Lengefeld, and at a private meeting in the Residence in January 1790 he offered Schiller a pension, adding in a low voice and with an embarrassed expression that 200 dollars was all he could afford. For him it was some guarantee that Schiller would not leave at the first opportunity, and for Schiller marriage on 22 February 1790 was worth 20 per cent of his freedom.

Jena lost a figure who more than any other had attached Goethe to both town and university when the chancellorship of the university fell vacant and in October 1789 Knebel was passed over for the succession by the Duke. Knebel's disappointment was so great that at first he wished to leave the duchy altogether. Goethe, however, persuaded him to move to Weimar for the winter, and only in April 1790 to start a year's leave with his family in Ansbach. However, the most important personnel question Goethe had to deal with in 1789, though it directly threatened the future of Jena, did not concern a university appointment at all. Goethe in fact advised Herder against considering the chancellorship of Jena, on the grounds of the hostility he would meet from entrenched academic interests. But Herder was now one of the greatest names in German theology and philosophy, his preaching was a magnet, and any university that could capture him could count on luring away a substantial proportion of its competitors' students. In early 1789 Herder, while in Italy, was offered a chair at Jena's arch-rival, Göttingen, and asked to name his terms. The offer was communicated to him through his wife, and Goethe, who foresaw both the ruin of Jena and his own loss of his only intellectual companion should Herder accept, was well-placed to influence the outcome. His advice both to Caroline and to her husband was at once solidly practical and nicely calculated. Herder had already confessed himself 'tired of the connection with princes and princesses', who were all 'foolish children in the end', and had even told Caroline that he doubted whether Goethe could mean much more to him personally. He had never been a support to him in official matters and 'my journey [to Italy] has unfortunately made his selfish existence, so completely and essentially uninterested in others, clearer to me than I could wish . . . It is painful to feel that one has lost a pleasant dream . . .'. Göttingen, with its liege-lord safely absent in England, was the most professionally academic and least courtly of the German universities, and its library was already so rich that Herder reckoned the books he could write from its resources would provide all the capital he needed to bequeath to his children: he was powerfully attracted. Goethe's first reaction, however, was disarming—he warmly welcomed the offer. Now at last, whether he stayed or went, Herder could secure for himself a satisfactory settlement. Goethe advised Herder to take no decision until he returned, and then to resolve the question purely by the balance of financial advantage; emotional considerations should be left out entirely. 'His temperament he will take with him wherever he goes', he said to Caroline, and he was surely right. 'In an academic senate there are the most stupid and infuriating scenes and decisions—And in a university it is the charlatan of a professor who counts for most. All the professors will be against you, because you are their intellectual superior. As man and as professor you will lead a much more dissatisfied life than in Weimar . . . '. Goethe proclaimed himself confident that Carl August would do whatever was necessary to retain Herder, but he cannot have been certain—Herder's distinction, and his closeness to Duchess Luise, did not necessarily outweigh the difficulties caused by his personality. Goethe had incurred Herder's anger by not pursuing an offer from the Duke to help out with his Italian expenses, on the grounds that it was Dalberg's duty to pay and the Duke's goodwill should be reserved for a better occasion. Carl August seems to have been well-intentioned but to have needed a little coaxing, and the terms on which Herder finally agreed to stay were not outstandingly generous; he was promoted to Vice-President of the Consistory, and so in effect administrative as well as spiritual head of the duchy's church and schools system, while being relieved of routine ecclesiastical duties; his stipend went up by 200 dollars to 1,800 dollars and parity therefore with Goethe's; the Duke secretly undertook to pay all his outstanding debts of about 2,000 dollars, and a pension was to be paid to Caroline should she be widowed. The one problem was the future of Herder's children, espcially the university education of his boys, and Goethe sought to arrange individual guarantees of sponsorship from the privy purses of the ducal family. Carl August and Luise did their bit, but an elaborate scheme put to Anna Amalia in Italy was turned down flat. Goethe returned to the charge and the Dowager Duchess agreed to provide an annual sum for one child when needed, though the agreement seems to have been forgotten. On 9 July Herder was back in Weimar 'in good humour' and his appointment was promulgated on 24 August. While he was packing to leave Rome he had looked through all the letters he had received and 'found so many real proofs of Goethe's manly loyalty, friendship and love towards me—I will hold on to that', and Caroline too found her disenchantment with the vain poet dispelled by the affair: 'Knebel is an unsteady, uncertain reed . . . Goethe does not change; he stands on firm ground . . . he is the only purely good person here.'

'I am a different man': June–December 1789

The humiliation of the old prude was nearer than Schiller thought. At some time between the beginning of May and the middle of June Christiana told her lover that she would be bearing a child of his in the coming December. Goethe was immediately presented with a series of private decisions of the utmost consequence, though he cannot have failed to consider them hypothetically in the preceding months. Medical opinion of the day now required a long period of sexual abstinence from him and so the question had to be resolved what store he set by the person, rather than the body, of his mistress. There is no evidence that he ever considered simply paying her off, as he had done in Rome: the *Erotica* show that for months there had been too much mutuality in the relationship for that course, even if it had been practical in so small a society as Weimar. It was not now, if it ever had been, a short-term affair. Was he then to set her up in her own establishment, bringing up her bastard(s) alone, while he continued to look out for a wife from the noble rank to which he had belonged since 1782? That would have been the courtly solution, scandalous, but within a familiar and to that extent acceptable framework, and though the comings and goings between her house and the Frauenplan would have been the target of ceaseless observation and gossip, it would have preserved between him and Christiana a distinction, both of lives and of station, which would have made her purely physical function clear to all. This possibility must early have been excluded—assuming, that is, that Goethe was ever emotionally capable of contemplating so complete a surrender of his staid burgher identity to the ethos of the aristocracy his father had despised. If Christiana told him she was pregnant as soon as she knew or suspected it, the news may have influenced Goethe's decision, which was certainly taken by 8 May, not to go to Italy in September. Unlike Carl August, who was usually absent during his wife's confinements, Goethe intended to stay at home; and that intention alone indicates how he now thought of Christiana. But where was home to be? And if there were not to be two separate establishments, had the long-feared and long-deferred moment of marriage come at last? A marriage to a commoner, without education or money, who even in Frankfurt and before his ennoblement would, because of her father's failure to graduate, have been his social inferior, could not but be deeply shocking in Weimar, although here as elsewhere marriage between male commoners and titled women was admissible (for example, Schiller and Charlotte von Lengefeld, or Goethe's great-uncle, J. M. Loen). Eventually, however, the von Steins and von Schardts and von Kalbs, and the duchesses too, would have consented to sit at the table of a legitimate Frau von Goethe, however much they despised her conversation or the unmentionable arts by which she had betrayed the solidarity of her sex and lured her husband away from his pure devotion to them. Goethe must really have been held back from marriage in 1789 by the same fear of 'being' rather than 'becoming' that had terminated his relations with Friederike Brion and Lili Schönemann, the fear of losing the freedom of endless self-reinterpretation that is manifest in his verse of this time as a fear of the attractive power of a woman whom he dares to depict only when she is asleep. But there was a further reason, and a persuasive one, that he could give to himself, and to others. Marriage was possible in eighteenth-century Germany only through a Christian ceremony. Goethe claimed belief neither in Christ nor in God; the *Erotica* that he was writing he cited in these very days of decision as evidence of his 'paganism'; he could not without hypocrisy seek, for the most intimate undertakings of his life, the blessing of a Church whose mythology and hierarchy had in his view reduced the rational conduct of human affairs to a paltry botch. Carl August would certainly have been sympathetic to such an argument, and may also have been relieved that Goethe was not proposing to try the patience of his court by a formal marriage; and informal arrangements were not necessaarily permanent.

Goethe will have consulted the Duke, and have received an assurance that whatever he did would be tolerated, and indeed supported, either in the first week of May or after 6 June, when the Duke was back in Weimar. It may be that the burst of gratitude that expressed itself in the 'panegyric' of 10 May is to be associated with some such conversations. On 12 May he wrote to Carl August: 'And what is it in the end that distinguishes the powerful man, but that he makes the fortunes of those who belong to him, and can do so easily, with variety, and on a large scale? while a private individual has to grind away for his whole life to set up a couple of children or relatives in any sort of comfort.' Maybe he was thinking of Herder, but at the same time he was urging himself to finish *Tasso*, 'at whatever cost', and the 250 dollars that he would receive for volume vi, and that would materially assist him in setting up as a private individual with children and relatives, cannot have been far from his mind. But by resolving neither to discard Christiana nor to marry her Goethe had posed for himself a peculiarly difficult further problem. She could not simply move in with him as if she were his wife, receiving the Weimar nobility and entertaining Goethe's sometimes illustrious guests without the protection of his title: Goethe's lady-friends, at least, would just not have come. In 1788 the rumour that Sir William Hamilton had at last married 'Miss Hart' was a great relief to Baron Reiffenstein, since it meant that the Dowager Duchess would not now be dishonoured by a visit to Hamilton's house: the rumour was untrue, and Anna Amalia went all the same, but Naples was a long way from Weimar. Nor, in the house on the Frauenplan, could Christiana and her child be kept in the background like servants. Goethe rented only one floor: there was neither enough space nor were the rooms disposed so that embarrassment could be avoided. He would have to move. Fortunately at this time two adjacent apartments could be made vacant in a grace and favour dwelling of the Duke's, the Great Huntsmen's Lodge, or Grosses Jägerhaus, a few yards round the corner from the Frauenplan, but outside the town gates on the road to the summer residence of Belvedere, and looking out towards the Latin Garden. Until a suitable house could be found to buy Goethe could live in one of these as the single man that in law he was and that to some extent he still wished to be, and next door Christiana could live with her entourage, since she for her part did not wish to be separated from her aunt Juliana and her 13-year-old half-sister Ernestina. Goethe meant what he said when he talked of providing for children and relatives, and in taking on Christiana he was taking on responsibilities as heavy as any that would have been imposed by a legal marriage. The apartments were refitted during the summer and in November the move began. In the difficult negotiations, probably during June that eventually led to this far from foregone conclusion,—'*in re incerta*' as he said— he was loyally and discreetly supported by the elder Voigt, now his closest friend in the ducal administration. At the end of June a letter went to August Vulpius in Leipzig, probably telling him of the new arrangements, so we may assume that the essential decisions had been taken by then. 'I find myself thinking more and more about a domestic existence', he told Carl August on 5 July.

At the beginning of this critical period, on 5 May, Charlotte von Stein, her health shaken by her husband's decline and by an emotionally difficult winter, set out to take the waters at the western spas of Wiesbaden and Bad Ems. She was burdened, 'as with an illness', she told Charlotte von Lengefeld, with the thought of her 'former friend of fourteen years', who seemed to her like a beautiful star she had seen fall from heaven. Caroline Herder had originally thought with Duchess Luise that the source of the dissension lay simply in the Italian journey, in Frau von Stein's refusal to forgive and Goethe's refusal to ask forgiveness. Once she knew of the Vulpius affair, it was clear that if Frau von Stein was 'very, very unhappy' it was because Goethe 'has, she thinks, turned his heart completely away from her and given himself completely to the girl, who used to be a common whore'. Frau von Stein told herself, evidently, that her sufferings were not simply those of jealousy, or wounded pride, or anger at the betrayal of those countless reiterations of exclusive love, but the pain of one who sees a dear friend become untrue to himself, his duty, and his highest ideals, and slip down into the mire. Immediately before leaving Weimar she wrote to Goethe, for the first time speaking, with her usual immovable certainty in matters of propriety and moral judgement, of the relationship that threatened to ruin him. As if to emphasize the charitable and personal nature of her concern she took her way to Bad Ems through Frankfurt, where for the first time she met Goethe's mother, and in Offenbach she spoke to Sophie von la Roche, for her anxieties were such as might be shared by other women. She was also perhaps testing the roots of the strange phenomenon that had dominated her life for so long. Goethe did not reply to her letter at once, and perhaps she had written when she did because she did not intend him to. It was probably a time when he was particularly uncertain of his plans anyway, and he was determined to finish the three scenes of *Tasso* still outstanding, so on 20 May he left with the Crown Prince and Ridel to spend a fortnight in the quiet and elegant Italianate surroundings of the Belvedere Palace. Here he made much of the necessary progress with his play, he felt satisfied with what he had done, and perhaps he found some clarity and confidence about the future direction of his life. From here, on 1 June he was able to send Frau von Stein a letter in which he gravely, directly, and with overwhelming self-assurance, identifies their emotional impasse, accepts no moralizing glosses and remorselessly lays bare unnameable feelings, while leaving them unnamed. Without malice, though without total candour in respect of his motives, it must surely, for all the absurdity of its conclusion, in which God speaks with the voice of a periwigged herbalist, count among the most terrible love-letters ever written:

I thank you for your letter which you left for me, although it saddened me in more than one way. I hesitated to answer it, because in such a case it is difficult to be honest and not to wound.

How much I love you, how well I know my duty towards you and Fritz, I showed by my return from Italy. If it had been a matter of the Duke's decision, I would be there still. Herder was going, and since I did not foresee being able to be anything for the Crown Prince I can scarcely have had anything else in mind but you and Fritz.

I do not like to repeat what I have left behind in Italy, you have given my confidences about that a sufficiently unfriendly reception.

Unfortunately when I arrived you were in a strange mood and I frankly confess that the way I was welcomed by you, and received by others, was exceedingly hurtful to me. I saw Herder, the Duchess, depart, the seat in the carriage empty that had been urged upon me; I stayed for the sake of my friends, as for their sake I had come, and at the same moment had to hear obstinately repeated to me that I might as well have stayed away, I had no interest in people etc. And all that before there could be any question of a relationship that seems to offend you so much.

And what sort of a relationship is it? Whose rights are curtailed by it? Who lays claim to the feelings I bestow on the poor creature? To the hours I spend with her?

Ask Fritz, Frau Herder, anyone who is reasonably close to me, whether I am less interested, less communicative, less active on behalf of my friends than before. Or whether rather I do not now for the first time belong properly to them and to our society.

And it would have to be a miracle that only with you should I have lost the best, most intimate relationship.

How keenly I have felt it still to exist whenever I have found you attuned to talking to me about interesting subjects.

But I readily confess that I cannot tolerate the way you have treated me hitherto. When I was talkative, you sealed my lips, when I was communicative you accused me of indifference, when I was active for friends, of coldness and negligence. You have scrutinized all my looks, found fault with my movements, my manner of being, and always put me *mal à mon aise*. Where was trust and openness to flourish when you repelled me with your deliberate moods?

I should like to add much more, if I did not fear that in your state of mind it would be more likely to insult than placate you.

Regrettably you have long spurned my advice in respect of coffee and introduced a régime which is extremely damaging to your health. As if it were not enough that it is difficult morally to get the better of certain impressions, you increase the tormenting and hypochondriac power of sorrowful notions by a physical means the harmfulness of which you for a while clearly perceived and which out of love for me you avoided for a time and felt better for it. May the waters and the journey prove beneficial. I do not completely give up hope that you may recognize me once more. Fare well. Fritz is content and visits me assiduously. The Prince is lively and in good spirits.

Frau von Stein's stay in Wiesbaden—then, before its transformation at the start of the nineteenth century, a wretched, bug-ridden, unfrequented resort—had been lonely and miserable enough. Her only reply to the long-awaited letter was to write on it, in anger, pain, and incredulity, the exclamation: 'O!!!'. A week after his cannonade, and back in town, Goethe feared that his self-defence had been too forceful and wrote again in more conciliatory terms:

There can have been few pages harder for me to write than my last letter to you and probably it was for you as disagreeable to read as for me to write. However at least our lips are now opened and it is my wish that we may never again close them towards each other. I have known no greater happiness than my confiding trust in you, which was unlimited from the very beginning; as soon as I can exercise it no longer I am a different man, and am bound subsequently to change even more.

But he can hardly have seriously believed in the tentative proposal that follows for a new collaborative venture, with Frau von Stein as a guardian angel preserving his relationship with Christiana from degeneration. Thirteen years before, in *Brother and Sister,* he had imagined a heavenly Charlotte endorsing and encouraging her lover Wilhelm's union with an earthly Mariana. Then and now such a solution was 'impossible'—then, because one person, Frau von Stein, could not combine both roles, now, because, if the roles were distributed to two people, they could only be rivals.

Goethe knew that the end had come, that he stood at the threshold of a greater change than any his adult life had so far brought him. He cannot have been surprised that on those letters there followed the silence rehearsed before when he had just arrived in Rome. It was apparent from their converse of the last year that Frau von Stein had shut herself off from him, and when he left a note in her house to welcome her back on 6 July—'from an old former friend' was her description if it—it would seem to have met either with no reply or with a request to terminate the correspondence. If she had not already reclaimed her letters, she did so then, and Goethe was left to his domestic existence. It was a cruel punishment she imposed on herself: even in October she was still tormented by 'bad memories', though she then 'sank into quiet mourning for her relationship with G. and in that she seemed truer and more harmonious than in the unnatural state of indifference or contempt . . . '. However it is striking, if understandable, that even at this stage, in his letters and personal utterances—*Tasso,* effectively completed during the crucial weeks of May and June, is a rather different matter—Goethe did not acknowledge that more was at issue than jealousy and the deadlocked emotions on which he concentrated when his lips were first opened. Frau von Stein could not be expected to tolerate or understand his sensual union with a social inferior, for her whole life, and such identity as she had, reposed on a denial of sensuality and an unquestioning assumption of the centrality of the court world. She was incurious about what might lie beyond the bounds of that world, whether elsewhere or in Weimar, and the middle classes acquired significance for her only as they entered it, possibly as intellectuals, but more particularly as officials with a potential for ennoblement. Christiana Vulpius was not merely low, she was alien; and for the existence as court artist that Goethe was now gradually defining, in substitution for his career as a loyal, industrious and self-sacrificing minister, there were also no precedents in Frau von Stein's experience or that of her family. But Goethe could not bring himself to recognize that he was asking the impossible of her in asking her to accept the necessity of his life with Miss Vulpius, because to recognize that would have been to recognize that the nearly fourteen years of his confiding trust in her had equally excluded a necessity of his existence. The trust had been unlimited only because, at great cost, he had first limited himself in order to find acceptance. What had been excluded was not simply his sexuality but his personal origins in the German middle classes, and his literary and intellectual relation with them. Perhaps it was an obscure understanding of this that had taken Frau von Stein, in her extremity, on her first visit to Goethe's mother.

Goethe, though he as yet hardly knew it, was returning to his roots. In the years dominated by Frau von Stein he had learned that simply to deny and suppress his origins, and to seek fulfilment of his needs and ambitions within the world in which she was content to move, led only to silence. His final commitment to life at the Weimar court made indispensable the definition of a domain within which he was not of the court, within which the world and the attitudes from which he came, and which it remained his task to represent on the edge of the centre of political power, could maintain, in however strange a form, an autonomous, and even to some extent a defiant, existence. Carl August used a deliberately telling metaphor when he described Goethe's seclusion during the winter of 1788–9 as 'living on the income of his great capital, which seems so secure that no external events or shortages can make him fear its dwindling'. This was, for a number of crucial years, one function of his relation with Christiana: not so much the assertion of bourgeois values, for however well-ordered and private the ménage it could hardly be called respectable, but the assertion that Goethe's life and significance were not exhausted by his court role. For the year and a half during which he was writing his *Elegies* a secret, pseudo-Italian love-affair provided him with that necessary independence—for it is of the essence of the love in those poems to be a secret, and the cycle into which they were eventually gathered terminates with the affair's becoming public—but thereafter he derived it simply from the opposition between the expectations of Weimar's ruling stratum and the class and legal standing of the woman he chose to manage his household. In later life Goethe indirectly admitted that these were years of great suffering, caused by a conflict between his decision to live 'outwith or beside the law, or perhaps even in contravention of law and custom', and his sense none the less of 'the necessity of remaining in equilibrium with ourselves, with others, and with the moral order of the world'. But he did not admit—for the point by then no longer mattered—that at the time it was essential to his spiritual and emotional survival that that conflict should be maintained, and with it his distance from the political and social institutions through which alone none the less, he realized after 1788, his mission could be fulfilled.

'Freye Liebe sie läßt frey uns die Zunge, den Muth' ('Free love leaves us with a free tongue and a free spirit'), he wrote in the *Erotica*, in defence of his refusal to accept the marriage bond. Only once he had established a new relation with the reading public as a counterweight to his court role was he in a position to regularize his relation with Christiana. This is not of course to say that he did not love her or that he simply manipulated her (her interests and those of the little Vulpius tribe were after all not neglected). On the contrary, if love was the vehicle of his freedom it was not for that any the less love, as he showed finally when freedom was secured by other means and he and Christiana became legally what physically and socially they had been long since, two in one flesh. In his erotic poems of the later part of 1789 a certain personal directness begins to break through the idealizing mould, in association with the event that precipitated the great change:

> 'Ach, mein Hals ist ein wenig geschwollen!' so sagte die Beste
> Ängstlich.—'Stille, mein Kind! still! und vernehme das Wort:
> Dich hat die Hand der Venus berührt; sie deutet dir leise,
> Daß sie das Körperchen bald, ach! unaufhaltsam verstellt.
> Bald verdirbt sie die schlanke Gestalt, die zierlichen Brüstchen,
> Alles schwillt nun, es paßt nirgends das neuste Gewand.
> Sei nur ruhig! es deutet die fallende Blüte dem Gärtner,
> Daß die liebliche Frucht schwellend im Herbste gedeiht.'

'Alas, my throat is a little swollen!' said my dear friend anxiously.—'Quiet, my child, quiet, and hear the pronouncement: the hand of Venus has touched you; she is telling you gently that she will soon, alas, irresistibly deform your little body. Soon she will ruin your slender figure, your pretty little breasts, everything is swelling now, your newest dress will not fit anywhere. Just be calm! the falling blossom tells the gardener that in the autumn the lovely fruit will swell and grow.'

From 23 July to 17 August Goethe was away in the Eisenach territories of the duchy, showing the Crown Prince the palace in Wilhelmsthal, and drawing up proposals for the development of the little town of Ruhla as a spa, testing the waters and designing a promenade to lead up into the surrounding woods. In a letter from Ruhla back to Herder, now in Weimar, we hear for the first time a new note, which will recur throughout his correspondence with Christiana (the first surviving letters to her date from 1792). Goethe is homesick, and not for Italy: 'Here we are in the land of the famed mountain-nymphs, and yet I can assure you that with all my heart I long to get home and to be once more with my friends and a certain little *eroticon* whose existence your wife will no doubt have confided to you.'

A significant date was now fast approaching. On 28 August Goethe would be 40. He could not have foreseen when he went to Italy that he would reach this landmark without the company of Frau von Stein and on the point of becoming paterfamilias, but he had long intended that the edition which was to present and round off the 'first . . . period of my writing' should by then be complete. In June 1789 it was too late to meet the deadline precisely, but Goethe concluded that month of many decisions with a step that made it possible for him to fulfil his obligations, to Göschen and to himself, and to be 'a free man at last', by the end of the calendar year. Volume viii, the anthology of *Miscellaneous Poems* and shorter pieces such as *Artist's Apotheosis*, had appeared in February. *Tasso* was 'on the point of being finished', needing only 'retouching', and had become so long that it needed to be accompanied only by *Lila* to make up volume vi. The one outstanding task, the only fragmentary work announced in 1786 which remained to be completed, other than *Elpenor*, was *Faust*. But *Faust* was a mountain, and it had not shrunk since Goethe's work on it in Rome. On 5 July, after drawing up a list of projects for the coming year, Goethe announced to Carl August that he had solved the problem by a *coup de main:* 'I will publish *Faust* as a fragment, for more than one reason. More on that when we meet.' Although he thus compromised the principle he had adopted after his success in rewriting *Iphigenia*, 'not to publish anything piecemeal or unfinished', he was able to leave *Faust* as the sole exception to the principle (apart from *The Mysteries*) by dropping *Elpenor* altogether and drawing into the edition the already finished *Jest, Craft, and Vengeance*, which as a separate musical project was now dead. With *Jery and Bätely* and *Jest, Craft, and Vengeance, Faust: A Fragment* would now make a substantial volume vii to complete the edition—'for this time', as he remarked in correspondence. We can only speculate what the more than one reason may have been that Goethe invoked in his conversation with the Duke: the desire to finish by the significant date, before the birth of his child, in time to visit Anna Amalia in Italy, by hook or by crook anyway and so earn the fee on the final volume, these may all have played a part, but they come down to a recognition that *Faust*, as newly conceived in the scenes written in Rome, is a special case, too big and too important to be brought to a rough and ready conclusion like *Egmont* and with an indeterminate potential for development. Simply in order to make his manuscript publishable, and bring it to a state which was open to future change but did not pre-empt it, Goethe had to 'bury himself' for most of October and the first part of November and he did not in the event despatch *Faust* to Göschen until 10 January 1790. But after his decision of early July this could be regarded as mopping up. The last major operation needed to reduce his youth to eight neat small octavo volumes proved after all to have been the work on *Tasso*. The 'retouching' continued throughout the summer and only on 27 August did Goethe send his publisher the definitive text of the last two acts. That was a greater celebration than anything on the following day: at midday Goethe dined at court and in the evening he entertained a few friends at home on the Frauenplan—Carl August, Prince August of Gotha, Knebel, grumbling at how much he was having to socialize these days, Charlotte von Kalb, and one of the ladies-in-waiting. No attempt seems to have been made to mark the occasion as unusual. Perhaps there was an uncomfortable awareness of the absence of any communication from Frau von Stein.

The summer of 1789 is not, however, nowadays remembered solely because Goethe reached the age of 40. At the very moment when he was packing up the first half of his life and preparing to settle into domesticity, patronage, and a vocation to Art, events were in train that were to demonstrate and bring home —one day, in 1806, to his own kitchen—that human beings are not only the makers but also the victims of their fate and that the world is a larger place than our symbolic interpretations of it. It might of course be regarded as yet another finesse of the genius for coincidence that accompanied him that at the point when he had by his own efforts established a caesura in his development, a caesura of a different and unwelcome kind should be forced on him from without. Yet that it is now possible to see 1789 as a turning-point for Goethe is the result of half a lifetime's subsequent work by Goethe himself to understand and express what in that year began to happen to France, to Europe, and to him.

In June the Estates General of France, summoned as a result of the failure of the nobility over the previous two years to agree with the Crown on economic and social reform, fell under the domination of the Third Estate through the defection to the bourgeois cause of a significant number of the clergy. In Count Mirabeau (1749–91), J. J. Mounier, and the Abbé Sieyès (1748–1836), the Third Estate found powerful defenders of its right to proclaim itself first the National and then, after the Tennis Court Oath of 20 June, and with the participation of the other Estates, the Constituent Assembly, making it their task to provide France with a written constitution. The middle classes throughout the country had already been politically activated by the discussions and elections that prepared for the meeting of the Estates General, and they set up the National Guard to protect themselves and their property when the King menacingly withdrew the army from Paris and laid plans for the forcible dissolution of the Assembly. The fear either of military intervention or of politically motivated riot was well founded, for the poorer classes too, being those who suffered most from the calamitous state of the national economy, now looked to reform to provide alleviation for the rapid rise in food prices: long-term inflation and the bad harvest consequent on the recent series of wet summers combined to produce price increases of 150 per cent and more for wheat and rye in 1789. After the dismissal of Necker, the popular Finance Minister, which led directly to the storming of the Bastille, disorder, lynchings, and the establishment of local revolutionary committees spread from Paris to all parts of France during July and August ('la Grande Peur'). Meanwhile the Constituent Assembly continued its theoretical deliberations, resolving in an all-night sitting on 4 August to abolish all feudal dues and services (a decision of more apparent than real benefit to the poor, since only those already wealthy could afford the compensation payments required to buy out noble interests) and on 26 August promulgating the Declaration of the Rights of Man and of the Citizen. In October the King, reluctantly, and the Assembly, more willingly, moved from Versailles to Paris, which was thus identified as the centre of government of a modern nation-state, rather than the largest city in a feudal kingdom, and in the next two months the advance into modernity continued with a time-honoured step, foreshadowed alike by the Reformation and by the Enlightenment: the nationalization of Church property, ready for its distribution to the makers of the new order. The issue of paper money, *assignats*, secured on this nationalized land, only contributed to inflation.

The immediate reaction of the German population at large to events in France in 1789 was imperceptible. Only the reputed abolition of tithes and feudal dues appears, understandably enough, to have met with some echo in the peasantry. In the smaller German states, whose poorest inhabitants were anyway frequently better off than the French urban proletariat, there was sufficient intimacy between rulers and ruled, and a sufficiently controlled, mercantilist economy, to blunt the revolutionary effect of the food shortages. In Weimar the Apolda weavers, on the verge of starvation again, received bread supplies organized by the elder Voigt. In Bonn, four days before the fall of the Bastille, the Archbishop-Elector of Cologne was mobbed by an enthusiastic populace grateful for the government measures that had averted famine. The birth-pangs of the modern state in a mass society and a capitalist economy were not felt in the fragmented German nation with its essentially medieval constitution. Only the intelligentsia responded at once, recognizing, whether in friendship or in hostility, that the long process of Enlightenment had suddenly become practically effective in a grand attempt to give a purely rational pattern to the political and social order. (In this they probably showed a clearer understanding of the ideological and historical significance of the Revolution than did the enthusiasts of the English-speaking world, with the exception of course of Tom Paine and, *mutatis mutandis,* of Burke.) Kant's acquaintances noticed how assiduously he now read the newspapers, and how much the Revolution came to dominate his previously multifarious conversation. Klopstock, in an ode, welcomed the summoning of the Estates General as 'the greatest action of this century' and one which reminded him that the 'Franks' were after all blood-brothers of the freedom-loving Germans (the historian J. von Müller (1752–1809) carried this Germanophile interpretation even further and declared 14 July 1789 'the best day since the fall of the Roman empire'), and the other poets associated with Klopstock, notably Voss, F. L. Stolberg, and Bürger, were similarly carried away. Only a few, such as Matthias Claudius, were hostile from the beginning to the triumph of Enlightenment, though Jacobi early had doubts. In Weimar the most emphatic friends of the Revolution, though in conversation rather than in print, were Herder, Knebel (both of whom may have been predisposed by a chronic personal disaffection from their surroundings), and Prince August of Gotha. Wieland published his first thoughts on the matter in the *Teutscher Merkur* in October 1789, from the point of view of a 'citizen of the world', and was initially deprecating, though he warmed to the Revolutionary cause the following year. Frau von Stein and Duchess Luise never wavered in their opposition. Goethe did not share the intoxication either, though he seems largely to have avoided the subject since there is practically no evidence of his views for six months after the fall of the Bastille, apart from a disagreement in November with the musician Reichardt, who had been in Paris in 1785 and was temperamentally inclined to the popular cause. Schiller, who from the start was apprehensive about the course events might take, called Reichardt 'impertinent', and Caroline Herder thought him a 'bad man'. There was anyway little in the Parisian drama, or its actors, to appeal to Goethe. He was still enough of a Frankfurt burgher, with painful memories of the attitudes of the Görtz faction when he arrived in Weimar, to have scant sympathy with a nobility concerned only to preserve its privileges and neglectful of its duty to govern for the common good. He had spent ten years struggling with the inconsistencies of those born to rule, and Carl August or Ernst II were at least well-intentioned: a class so corrupted by self-interest as to flinch from the task presented to the French 'Notables' in 1787 must have seemed to have brought its fate upon itself as the dominoes began to fall in the latter months of 1789. Goethe had feared for the solidity of the French ruling caste in 1785, when he learned of the Diamond Necklace Affair, and his reading of Saint-Simon's *Mémoires* in May 1789 must have confirmed his fears. Yet he had no time either for those who were working the downfall of the old rulers. Mirabeau's publication of some scandalous revelations about his sojourn at the Prussian court, including allegations about the Duke of Weimar, had discredited him in Goethe's eyes long before the papers found on his death revealed him as a double-dealer who sold himself for money. The crucial role of the all but secularized clergy, personified by Sieyès, in the first stages of the upheaval was particularly provoking to one who not only had just pledged his loyalty to the First Estate but who saw himself as an honest atheist. It was one thing to betray your own class in the name of Christian principles when you did not believe in Christ, it was a triple treachery to bring the whole house down round the ears of those who shared neither your initial faith nor your subsequent modifications of it. The extraordinary power wielded in these months by men of straw reminded Goethe of the equally uncanny Cagliostro who, supported by nothing, had risen from a room in Palermo to his part in the Necklace Affair and had so begun, it now seemed, the destruction of the greatest monarchy and most populous nation in Europe. Certainly there was at least the possibility that the network of political associations that had sprung up in France, both during the elections for the Estates General and in the aftermath of 14 July, was connected in some obscure way with the Masonic and Illuminist movement that had harboured Cagliostro, so when the proposal was made to open a new Masonic lodge in Jena in April 1789 Goethe's vigorous opposition was dictated by more than a dislike of obscurantism: both the *ALZ*

and the university itself—by means of lectures on the history of Masonry—
were to be enrolled in a campaign of publicity intended to break the spell of
secrecy which was the movement's strongest weapon. Perhaps all the Masonic
talk of inner circles and occult forces was as banal a deception as Frau von
Ziegesar's assertions, which excited Weimar society in February, that she had
'magnetic' powers and in a catatonic trance could see and hear through her
fingers, yet Goethe regarded it as a most serious libel that Carl August was said
to have taken part in demonstrations of 'magnetism' in Carlsruhe: these were
'nothings' that could become a 'something' that could overturn a throne. What
Goethe was in fact witnessing in France were the first manifestations of the
power of organized, or partially organized, public opinion to channel the
energies of a mass—whether 600 deputies in an assembly, or as many local
branches of some national movement—so as to produce effects inexplicable in
terms of the personal qualities of the apparent leaders. Without, in 1789
(though they would soon come), such concepts as political parties, slogans,
public relations, the media of communication, or even 'organization' itself in
any but a biological sense, Goethe none the less understood the phenomenon,
even if he had only the terms 'magic', 'deception', or 'nothing' to describe it.
What he most abominated about the Revolution in its early days was its
obsessional hold over men's thoughts and words, written and spoken, its
usurpation of the medium of interpretation that he as a literary man had
gradually developed for the symbolic understanding of his own life and
purposes. At a date that cannot be precisely determined he rewrote what is now
the second of the *Roman Elegies* so that the poet finds in his love-affair a refuge
from pursuit not, as originally, by his own contribution to European print-borne
mythology, the figure of Werther, but by a 'modern fable' the endlessly and
pointlessly repeated discussion of the 'raging Gauls', of the 'storm threatening
us from without', and the people of Europe and their kings, about which
everyone in society is expected to read and have an 'opinion'.

On the list of tasks for the year 1789–90 which Goethe drew up for himself at
the start of July only two literary projects figured, apart from the *Erotica*, after
such work as was still outstanding on the Göschen edition. Both of them were
opera libretti, and both of them were old: *The Mismatched Household* and *The
Mystified*. Only the latter held Goethe's interest, for the story of Cagliostro
suddenly had a sinister topicality, and he actively pursued it with Reichardt,
who began his setting, well before the text was complete, with the songs Goethe
had written in Italy. But most of the items on Goethe's list, other than such
mundane but in all ages time-consuming matters as 'furniture' and 'house
purchase', and the perennial problems of Peter im Baumgarten, related not to
writing but to study: of the principles of art, of ancient literature, now Greek as
well as Roman, and of the Greek language itself, and above all of scientific
topics, mineralogy, chemistry (probably for mineralogical purposes), microscopy, osteology, and especially botany. The new epoch Goethe saw beginning
for him would not be devoted to literature, or only to a literature which had

'style' rather than 'manner' and was founded on solid study of the objective world. He would find small support for such an undertaking from a German public obsessed with 'opinions' on matters which did not concern them and of which they were largely ignorant. The autumn and winter of 1789–90, when Goethe, as in the previous year, was in his Italian enclave 'striving to dissipate the mists of the atmosphere by the light of the mind', had a decisive effect on his work in geology, botany, and optics, even though this did little to reconcile him to his readers and were of little profit to his poetry.

The visit of A. G. Werner (1750–1817), the famous professor of mineralogy at the Saxon mining academy in Freiberg, on 16 and 17 September, was not only of scientific importance but also an agreeable prolongation of Goethe's Italian experiences. Throughout the two days the main subject discussed, both in Weimar and in the scientific collections in Jena, was volcanoes, of which Goethe, who had seen Etna and Stromboli, and trodden Vesuvius and the Phlegraean Fields, had a range of direct knowledge Werner could only envy. Werner was the principal protagonist of the 'Neptunist' theory of the origin of most modern rocks in a primal watery fluid and Goethe, who had originally wanted to avoid making hypotheses about processes beyond all possibility of immediate observation, found himself swayed by his authority. Already in Italy he had inclined to the view that volcanoes were geologically superficial and characteristically maritime phenomena, originating not far below the sea-bed, perhaps through the intrusion of water into burning coal-seams. Werner now confirmed this view for him and thus persuasively removed the major visible and contemporary evidence for his opponents' belief in a fiery origin of the earth as we know it. Werner's pupil, the younger Voigt, up in Ilmenau, had apostatized from his master's teaching, and had proclaimed himself a 'Vulcanist', but Goethe thought even he would now have to return to the fold, and began to write an essay intended to bridge the gap between the two schools and their conflicting bodies of evidence. He was now prepared to make a general defence of the necessity after all of hypotheses in natural science, and to put forward a specific one of his own. His suggestion that basalts had indeed crystallized out of the sea, but that the sea at the time was boiling, took elements from both theories but is in implication more Vulcanist than Goethe realized, once its defective chemistry is ignored. He could not have chosen a worse example than basalt, but if he now thought he inclined to Neptunism it was less because of a prejudice in favour of water than because of an inarticulate awareness of the enormous time-scale of geological change. At the moment it was probably no more than a coincidence that his growing conviction that the earth had been formed by a process of slow development, rather than violent upheaval, should have been clarified and formalized at a time when political Europe seemed bent on demonstrating that an opposite rule prevailed in human affairs. One day, however, the connection would become symbolical.

Goethe's botanical work had continued in a desultory fashion since his return from Italy. Already in August and September 1788 he told Caroline Herder that he was writing up his 'System of Plants' and asked Göschen to secure him copies of certain works of Linnaeus, but a year later the 'System' was still no more than a subject of supper-time conversation with Herder and Knebel. It was perhaps Goethe's desire not to appear personally responsible for every new initiative in the duchy that caused him in January 1789 to ask Knebel to take on the task of promoting his protégé Batsch's scheme for a Botanical Institute in the ducal garden in Jena. In October Carl August approved the scheme, appointing Batsch as the Institute's director and commissioning Goethe to set it up. Almost immediately a dramatic turn in events forced Goethe into close consultations of a scientific nature with Batsch and the plantsman Dietrich. On 11 November the *ALZ* carried a publisher's announcement of a forthcoming treatise by the botanist C. K. Sprengel to be entitled *Essay in Elucidation of the Construction of Flowers* (*Versuch die Konstruction der Blumen zu erklären*) which had a doubly galvanizing effect on Goethe. First, he determined not to be anticipated and to reduce his own thoughts to systematic form for publication at the same time as Sprengel's work, the following Easter. By 18 December he had a draft in a series of paragraphs, numbered as in contemporary scientific textbooks for easy reference, which he asked Batsch to read and comment on. On 20 December he went to Jena, staying in Knebel's old quarters, conferred about his draft with Batsch and resolved to expound the main theses more fully and to include more examples and also illustrations (though in the end it was published without these). A publisher had still to be found, but Goethe was hopeful that Göschen would bring it out, as a favour to his most distinguished author, if nothing else, and may have asked Bertuch to put the proposal. Secondly, however, Sprengel provided Goethe with a title, and so in a sense with his main thesis. The notion of the 'construction of flowers' would have been repugnant to Goethe in more than one way. The word 'construction' had overtones of that theoretical deduction of the real world from first principles which was becoming increasingly popular in a Kantian, perhaps one should already say post-Kantian, intellectual atmosphere, and which was incompatible with Goethe's own current conviction of the primacy of the senses. It was also reminiscent of the word 'composition', which Moritz, perhaps already influenced by Goethe's botanical thinking, had rejected as inapplicable to works of art, 'because such a work is not *put together* from *without*, it is *unfolded* from *within. One* thought embodied in several forms ['Figuren']'. Goethe's acceptance of the distinction between nature and art was not so wholehearted that he would let it stand in the way of his long and deeply held conviction of the unity of nature and his hostility to its mechanical analysis: if works of art grew from a single principle, *a fortiori* so did works of nature. And finally Goethe wanted to explain more than the shapes of flowers: he was seeking an explanation for the structure of the whole organism, and so ultimately of the whole vegetable kingdom. The title of his own treatise therefore now became, in direct competition with Sprengel, *Essay in Elucidation of the Metamorphosis of Plants* (*Versuch, die Metamorphose der Pflanzen zu erklären*). The term 'metamorphosis'

contains the kernel of the entire argument, and perhaps for this reason Goethe
was extremely secretive about it, not using it in conversation or correspondence
until the type was set in March 1790; but it is anyway quite possible that it was
only by seeking the *mot juste* in opposition to Sprengel's title that Goethe hit
upon the central term round which his thoughts of the last five years finally
crystallized. 'Metamorphosis' was of course already an established zoological
concept, applying to the different stages of insect growth from egg, through
larva and pupa, to imago. Goethe's use of it in his title, though it had precedents
in Linnaeus, was deliberately provocative, suggesting that the same, or a
similar, process could be seen at work in the different stages of plant growth,
which he took it upon himself to define. This revelation of a regular sequence
of different forms in which, in each individual plant, a single initial principle is
embodied, or 'developed from within', would provide also an explanation of the
unity in variety of the whole plant kingdom, or at any rate of the dicotyledonous
angiosperms to which Goethe confines his investigation. (Goethe does not deal
with plants that produce only one seed-leaf, or cotyledon, on germination, such
as grasses or lilies, nor with gymnosperms, those plants, such as conifers,
which, unlike the angiosperms, do not bear their seeds in an ovary, nor for that
matter with plants that do not reproduce by seed at all, such as ferns.) What,
then, is common to all the plants he treats is no longer a particular structure,
which might be found instantiated somewhere in its pristine form, but a
particular formula for their growth through different shapes over time. At an
earlier stage Goethe had already found it necessary to couple the notion of the
primal plant as a 'model' with the notion of a 'key' or 'formula' for the model's
application: now he has dispensed with the model altogether, and the concept
of metamorphosis has completely displaced that of the primal plant, which plays
no part in his *Essay*.

The single thought from which Goethe's account of plant development
begins, the single principle which a plant variously embodies, is that of the leaf.
'Hypothesis. Everything is leaf,' he had written while in Italy, ' . . . a leaf that
merely draws in moisture under the earth we call a root; a leaf expanded by
water etc. we call bulbs'. As a plant grows, a single organ—to which for
convenience' sake we give the general name of 'leaf'—is 'metamorphosed' into
different specific shapes. The process of metamorphosis can itself be reduced
to two opposite possibilities, which alternate in the life-history of the plant:
expansion and contraction. If we begin with the growing (not dormant) plant in
its most contracted state, we find a germinated seed with two cotyledons, two
simple, internally and externally undifferentiated, leaves. There follows a first
stage of expansion. More leaves grow, larger, and with a more distinctive shape,
than the cotyledons, and with a more complex internal structure, and as the
plant grows from one leaf to the next so between them the stem is formed.
Along its length the leaves, which as cotyledons appeared in pairs or a ring, now
arrange themselves alternately or in a spiral. This stage of growth Goethe calls

the 'successive' stage, and the plant could in principle carry on indefinitely sprouting from leaf to leaf, and indeed will do so if its diet is too rich. But if the diet is restricted the plant will stop expanding and a new phase of contraction will inaugurate the flowering stage in its life-cycle. The leaves of the stem cease to appear successively, a new stem, leafless and of finer texture, emerges, and at its end the leaves reappear in a new form, growing now not alternately but simultaneously and gathered into a ring. This circular arrangement of leaves, which seems to Goethe to be a contraction of the lineally expanded organs that made up the stem, is the calyx, the ring of sepals, usually green, which surrounds the flower proper, enclosing it initially in a green bud. A new phase of expansion brings about the formation of the flower itself, the corona of leaves or petals which usually indicate by their brilliant colour that they are the culmination of the growth of the individual plant. But the splendid flower is not the end of the plant's story, for a new phase of contraction draws the petals together to form the stamens and pistils, the reproductive organs at the flower's centre. Goethe's somewhat confused account of the process of fertilization in plants is intended to minimize the distinction between sexual and vegetative reproduction so that he may represent his final stages of expansion and contraction as, like their predecessors, steps in the development of the same individual plant with which he began. For the final and in some species (such as gourds) most spectacular expansion of the leaf is said to be the formation of the fruit, the container of the seed, while its final and most complete contraction is the formation within the fruit of the seed-capsule itself.

Goethe's metamorphic botany can be seen as a continuation of his work on the intermaxillary bone: it almost immediately led to a revival of his interest in comparative anatomy and shares many of the preoccupations, strengths, and weaknesses of his earlier, and in 1789 still unpublished, scientific treatise. It shows at first sight little trace of the principle rather superciliously advanced against Knebel at the start of the year, that in the natural sciences it is more important to define differences than similarities. On the contrary, it seems rather to be written out of a spirit of opposition to Linnaeus' great achievement in marking off species one from another by precisely defined, and often countable, features: not only, a Linnaean might object, does it suggest that every part of a plant is, save for the process of metamorphosis, identical with every other part, it also claims that to produce for this process a formula in respect of which every species is identical with every other species is actually an 'elucidation', rather than an obfuscation, of the order obtaining in the vegetable kingdom. Now this concern with the unity to be found in the multiplicity of natural phenomena, with processes of transition, and with the continuity of seemingly discrete identities, is part of the Leibnizian inheritance which had already conditioned much of Goethe's work in geology and anatomy and which he shared with many, if not most, of his contemporaries—notably Charles Bonnet (1720–93), whom he had met in 1779 and whose compendious and

informative *Contemplation de la nature* of 1764 he by now possessed in an edition
of 1783. At the same time Goethe's essay on the *Metamorphosis of Plants* does
something more than simply rehearse the commonplaces of eighteenth-century
natural historians. As important as any assertion of structural identity or analogy
—of organ with organ or plant with plant—is the emphasis on dynamic change,
on the plant as an individual identity that in the course of time adopts different
conformations in accordance with an innate principle. That is of course a more
truly Leibnizian view of things than any static hierarchy of beings and it had a
particular appositeness for Goethe at a time when he was aware that his own life
was changing radically, fruiting, yes, but whether expanding or contracting it
was difficult to say. The individualism of the essay—its concentration on the
single plant to the exclusion of such other botanical units as the species, or the
local population, or the sexes—and its materialism—no divine or indeed any
other purpose is served by the development of the plant, which is determined by
the interaction of internal and external material forces—also reflect Goethe's
contemporary attitudes. In two respects, however, the essay also hints at the
future. The suggestion that the instrument by which successive different forms
of the plant are brought into being is a power which manifests itself in two equal
and opposite ways, in this case expansion and contraction, is destined for much
elaboration and has no obvious Leibnizian antecedents. And the conception in
the treatise of 'the leaf', which has much irritated the commentators, is a first
and unreflected form of a notion which within a few years will be one of the
fixed points in Goethe's new understanding of natural science. What precisely
is 'the leaf'? To say that, suitably modified, the leaf 'is' the root, or 'is' the calyx,
or 'is' the stamens, or indeed 'is' the leaf, seems to the unsympathetic observer
just a form of higher nonsense, which yields no clearly verifiable or falsifiable
assertions. Yet more than mystical vagueness is at work in the concept which
reveals to Goethe the unity in the variety of the shapes we see in the world of
plants. It would have been open to Goethe to attribute that unity to the
operation of some hidden principle, of a monad, or a conative or formative
force, or to the existence of a generalized organ prior, in some metaphysical
sense, to all specific organs. A later age has added the possibility of reference to
the common building-block, the plant-cell, or to descent from a common
ancestor. But all of these explanations of the manifestly but mysteriously unified
phenomena that are the shapes of plants involve an appeal to something other
than the phenomena themselves, which are thereby turned into the mere
consequences of some invisible and shapeless cause. Goethe instead seeks to
elucidate the phenomena by appeal to one of their own number, by choosing
one of them—the leaf—as basic or central, and inviting us to see its organized
relations with its fellows: 'we can equally well say . . . a petal-leaf of the calyx is a
contracted . . . stem-leaf . . . as we can pronounce a stem-leaf to be an . . .
expanded petal'. 'Elucidation' of vegetable growth is not a matter of inventing
occult mechanisms that will explain *why* plants have the shapes they do, but of

understanding clearly *what* the shapes are. Goethe is remaining true to his task of finding order in the world as it presents itself to his five senses without grounding the order in processes that are, or seem to him to be, inaccessible to his senses. As yet, however, the procedure is philosophically unselfconscious, and the essay on the *Metamorphosis of Plants* contains no reflection on the problems it implicitly raises about the nature of scientific method or the foundations of knowledge. Despite these features that point to later developments in Goethe's scientific work it is therefore best regarded as the culmination of a period which began with his mineralogical and anatomical studies of 1780 and 1781 and which was dominated by the attempt to extract from the empirical detail of natural history the rational and analogical structure of the Leibnizian ladder of being.

It was optics, and in particular the theory of colour, that was to transform Goethe's scientific studies, but the second half of 1789 saw only the first foreshadowings of the new epoch. In July Goethe listed colour, which he had discussed more curiously than seriously with Angelica Kauffmann in Rome, as one of the aspects of painting he wished to investigate further. In the summer or autumn he borrowed from Büttner in Jena the prisms, mirrors, cards, and so on necessary to make the standard experiments in demonstration of Newton's main propositions. But he also needed a room into which the sun shone directly and which could easily be blacked out. This the house on the Frauenplan could not provide, though in the Jägerhaus to which he had moved by the beginning of December there was a long narrow room with a southwesterly aspect ideal for conversion into a camera obscura. The turmoil incident to the move and to Christiana's confinement, and his preoccupation with his botanical treatise, prevented him, however, from setting up Büttner's equipment until well into the new year. The loan had been long and Büttner was, perhaps through Loder, pressing for its return, but he, like the dawn of Goethe's new epoch, would have to wait.

The move itself was a protracted affair. For ten days at the end of September and beginning of October Goethe was at Aschersleben in the foothills of the Harz watching Carl August's manœuvres, and showing a degree of interest in his prince's little follies which he had displayed only for the first time in May when writing his panegyric of the man who had provided him with his house. He returned to Weimar to 'bury himself' in the work on *Faust,* from which his only distractions were his short trip to Leipzig to further August Vulpius' affairs, and the installation in the Jägerhaus of a large new heating stove of experimental design, which took some three weeks. There were frequent visits to the new apartments during the fitting-out and Goethe enjoyed their open airy situation. By 2 November *Faust* was 'fragmented', and the study of Greek could begin, but it had to be dropped again after a fortnight, when botany had become a matter of urgency, and Goethe was at last 'gently manœuvring into my new quarters. The heavy artillery has gone on ahead, the corps is in motion, and I am covering the rearguard', a facetious description, to the Duke of course, of Christiana, Aunt Juliana, Ernestina, Goethe's two manservants, and Lips, who had just arrived and was lodging for the present with his patron. At the beginning of December Goethe himself finally moved across, just before the first false alarms that Christiana was going into labour, and became to all intents a married man. 'I am gradually getting my possessions into order', he wrote to Anna Amalia on the fourteenth, 'and remembering my fair days beyond the Alps.' The new stove ensured an Italian temperature, but the welcome in the *Erotica* to the unknown arrival who would complete the family circle cast off all, or most, Roman affectations:

> Wonniglich ist's, die Geliebte verlangend im Arme zu halten,
>   Wenn ihr klopfendes Herz Liebe zuerst dir gesteht.
> Wonniglicher das Pochen des Neulebendigen fühlen,
>   Das in dem lieblichen Schoß immer sich nährend bewegt.
> Schon versucht es die Sprünge der raschen Jugend; es klopfet
>   Ungeduldig schon an, sehnt sich nach himmlischem Licht.
> Harre noch wenige Tage! Auf allen Pfaden des Lebens
>   Führen die Horen dich streng, wie es das Schicksal gebeut.
> Widerfahre dir, was dir auch will, du wachsender Liebling—
>   Liebe bildete dich; werde dir Liebe zuteil!

Blissful it is to hold the beloved, longing, in your arms when her beating heart first confesses love to you. More blissful still to feel the pounding of the novice to life moving and continuously feeding itself in the sweet womb. Already it is attempting the leaps of impetuous youth; it is already knocking impatiently, yearning for heavenly light. Tarry a few days yet! On all the paths of life the Horae [goddesses of the seasons] will be your strict guides in accordance with the bidding of Fate. Whatever may befall you, you growing darling—love formed you; may love be your lot.

December, however, was an anxious time. It seemed as if there might be complications—Weimar talked with satisfied indifference of the possibility that Christiana might die and noted a 'gentler', 'more spiritual' look in Goethe's face—and Goethe stayed at home for most of the first half of the month, except when he had to consult with Carl August on some state matters of the highest secrecy and danger. The Duke was being encouraged by Prussia to accept the kingship of Hungary offered him by some dissident nobles who were hoping to break away from Austria: fortunately both he and his counsellor could recognize a wild adventure and were not disposed to pursue it, but negotiations continued. Goethe was probably difficult company for the women of his household during these weeks and it was no doubt with relief that they saw him off to Jena on the eighteenth for his botanical discussions with Batsch. When he returned on Christmas Day it was to the news that he had that day become the father of a son. August Walther Vulpius, named after his uncle but no doubt also after the Duke, was christened in the vestry of St James' Church on the twenty-seventh with Aunt Juliana entered in the register as his only godparent and with no mention of his father. The end of the year was busy. Goethe was dining at court every day—the final refusal of the Hungarian crown was despatched on the twenty-eighth—and he gave two large receptions in his new apartments. In the press of guests he hardly had time to exchange more than a few words with a 22-year-old nobleman from Berlin, Wilhelm von Humboldt (1767–1835), who had come over from Göttingen, where he was studying, to visit Weimar with his future wife, Caroline von Dacheröden. Goethe later excused himself on the grounds that he had been 'in a very unfortunate [or; unhappy] mood': perhaps it was just an excuse, perhaps it was still not clear that both mother and child would survive and prosper, perhaps it was a certain jealousy of the novice to life—ordinary though the emotion would have been, Goethe was capable of giving to his discovery of ordinary emotions a more than ordinary significance.

Goethe had once hoped that a symbolic monument by Fuseli would set the seal on his thirtieth year, 1779, in which he had written *Iphigenia,* started a new administrative career, and recapitulated his earlier life by introducing the Duke to its landmarks. What failed at thirty more than succeeded at forty: the year was superabundant with symbolic terminations, and of these the birth of his first child surely, if not obviously, put most demands on his emotional resources. The painful severing of his strange bond with Frau von Stein or the change in his relation with the court, manifest in his laborious removal from the Frauenplan, where he had first resided as the duchy's triumphant ministerial factotum in 1782, to the Jägerhaus beyond the city walls, these were certainly accompanied by some feelings of relief. The same was also true of the completion of his *Literary Works,* the monument to twenty years of autobiographical literature that had begun in 1769 with *Partners in Guilt:* 'now we can get on with other things' he wrote, as he pushed aside the fragmented manuscript of *Faust.* What those other things might be worth, and whether botany, Greek, *The Mystified,* the Romanization of Weimar, or even the unpublished and slumbering *Wilhelm Meister* could even in twenty years amount to a match for those eight volumes, were questions buried deep within him which at this stage it was unprofitable to ask. The violence in France, the increasingly bizarre and apocalyptic political pronouncements emanating from the Assembly, were a—rather distant—source of alarm, but could not yet be seen by all to mark the beginning of a new era. To a dispassionate observer the most likely future course of events had to be that France would fall into anarchy and that order would be restored by force, possibly by foreign intervention as two years previously in the Low Countries. In the meantime the Revolution's demands on a thinking man's attention were more vexatious than emotionally perplexing. But the birth of little August was a watershed: in one moment Goethe was reassigned to a new generation, and to a new condition of life. Other soulmates might be found after Frau von Stein; yet further changes in his relationship with the Weimar, or another, court were always possible; it was not yet apparent that the events in France were irreversible and destined to have an irreversible effect on every individual life in Europe. But once August was born Goethe had as certainly completed a course he would never run again as has the leaf when it is metamorphosed into the fruit and the seed. Like all children, August retrospectively furnished the goal of his parents' previous life. *This* was what lay at the end of all the aimless, seductive, anguished questing of unattached or half-acknowledged Eros, that had tormented Goethe for twenty years, fascinating all who met him, and expressing itself in the ever-juvescent symbolic novelties of his literary works, and in the twists and turns of his unpredictable but mysteriously logical career. August was the inescapable evidence of the terminus to desire, which, Goethe confided to his collection of *Erotica*, was also the terminus to the symbolic existence that had from the start been a substitute for the religious interpretation of life:

> Ob erfüllt sey was Moses und was die Propheten gesprochen
>   An dem heiligen Christ, Freunde, das weiß ich nicht recht.
> Aber das weiß ich: erfüllt sind Wünsche, Sehnsucht und Träume,
>   Wenn das liebliche Kind süß mir am Busen entschläft.

Whether what Moses and the prophets said has been fulfilled in sacred Christ, I do not, friends, rightly know. But this I do know: wishes, desire and dreams are all fulfilled when my little sweetheart drifts deliciously into sleep on my bosom.

Fulfilment is an end too, and it makes the end sensible by being the beginning of a new and other life. Fatherhood demanded of Goethe the recognition that he had entered on an age, not of possession of what had been desired, but of new labour on behalf of what had been given him. To free himself for that task he would have to put behind him any thought that he could still linger in the age of desire, clinging in a Roman Weimar to the symbolic fulfilments he had once created for himself. At the end of 1789 Goethe still had to liberate himself from the spell of Italy.

Summa Summarum: *The Edition Completed*

'All these recapitulations of old ideas, these reworkings of subjects from which I thought myself for ever detached, to which scarcely an inkling of mine could extend, cause me great pleasure', Goethe wrote to Carl August from Rome in February 1788. 'This *Summa Summarum* of my life makes me confident and glad in turning a new page again.' It was always clear to him that the projected edition of his *Literary Works* meant more than the opportunity of printing unpublished manuscripts or republishing scattered pieces in an authorized or more accessible form. It meant more even than a means of disciplining himself and meeting the criticism passed by Anna Amalia to Herder, with a shrug of the shoulders, that 'it was a pity [Goethe] could never actually get anything done'. It was also a decisive and—until Goethe began writing his own commentaries on his work—a definitive act of self-interpretation, which subordinated to a grand design the myths which he had created in order to understand his experience, and so liberated him from them once more. The Göschen edition is the first of several such liberations, but it already contains all but a handful of the works posterity would associate with Goethe's name, and already most of these have undergone an editorial process which leaves them representative of two, or even more, quite different self-understandings. In some cases the two editorial levels are close together in time, but while in *Götz* this does not materially add to the complexity of the work, the two stages of *Partners in Guilt* are separated by nothing less than the discovery of autobiographical literature, and between the writing of *Proserpina* and its insertion in *The Triumph of Sensibility* lies a gulf not of time but of attitude. Usually it is the work for Göschen from 1786 to 1789 that creates the duality in the final text: the Werther who is so recognizably a creature of his time that a generation identified with him, and the Werther who is a windowless monad and whose self-destruction leaves his surroundings largely untouched; the Iphigenia who, sustained by benevolent gods, is a healing principle of moral devotion in the life of a sick Orestes, and the Iphigenia who is the opposing counterpart of the later Werther, the figure of a self-authenticating moral autonomy which has only to be willed in order to be possible; the Faust who is both a young Titan destined for a tragic collision with the moral order he rejects, and a victim of disillusioned middle age who embarks on an odyssey through experience of unspecified duration towards an uncertain goal. In Egmont no less than three strata are discernible: a political martyr, a divided self, and a superhuman centre of active force that consumes those who come too close to it. The many-layered work that in each case results positively invites multiple partial interpretations, yet resists any attempt to reduce it to a single overall pattern. What the unity of its parts may be remains at once fascinating and mysterious like the sphinx, or any human personality which combines both determinate character and the capacity to surprise. The different layers of meaning, like the different phases of Goethe's life in which they have their origin, may not be equally attractive, but the whole into which they are more and less successfully integrated presents us, not with his personality, either *in toto* or as it was at a particular time, but with something like the principle of personality in general: a fusion, which cannot be further interpreted, of the necessary and the contingent, the present and the past, being and becoming. 'Incommensurable' Goethe later called those works of his that have this quality, though at the time of the Göschen edition, under the sway of a more limited theory of Art, he only half understood it. Incommensurable may of course be a euphemism for incoherent: sometimes there is little or no integration, only a jarring of incompatibles, or an act of violent subjugation, like that which rewrote *Claudina*; but sometimes there emerges a 'beautiful monster', a 'myself not myself', which embodies all the internal dialogue, or dialectic, that belongs to being a modern person, caught between religion and secularity, subjectivity and objectivity, bourgeois society and the autocratic State. Only minutes lie between the different layers of 'On the Lake', years between those of 'To the Moon'; the contingencies of a whole lifetime were necessary for Goethe to be able to write, and were made necessary by him in writing, his longest and pre-eminently incommensurable poem, *Faust.*

Of all literary genres it is in fact the lyric poem that is most suited to the representation of internal dialogue. To Goethe the eighth volume of his *Works,* that devoted to his miscellaneous poems, put together during 1788 and published early in 1789, was a microcosm of the edition itself, 'a Summa Summarum of the feelings of a whole lifetime . . . a strange affair, and it could have been an even more motley show. I had to leave out too much.' The volume consisted of *Lumberville Fair* and several associated farcical pieces—Herder disapproved since that merely offered a target to the critics and 'these youthful scribbles and jokes are never really suitable for print'—followed by two 'collections' of poems: the first nearly fifty love-poems ranging in length from the eight pages of 'Lili's Park' (Goethe thought the volume a little thin and urged Göschen to be generous in the lay-out) to the briefest and most anodyne of the recent *Erotica.* The second was made up of a dozen long reflective poems, such as 'Song of Mahomet' or 'Winter Journey in the Harz', then a group of recent epigrams in elegiac distichs, and finally a sequence of fourteen poems on art and the artist culminating in 'The Poetical Mission of Hans Sachs', 'On the Death of Mieding', and the dramolets *The Artist's Earthly Pilgrimage* and *Artist's Apotheosis.* The volume, numerically the last, concluded with the fragmentary *Mysteries,* the *Dedication* to which had opened volume i, so that the whole edition is framed in stately stanzas, yet by implication is left open to being extended or completed in the future.

Goethe spent the summer and autumn of 1788 not merely collecting and arranging the poems but also 'polishing' them. He clearly could not publish 'Limitation' ('Einschränkung'), for example, in its original form as a puzzled question to Fate about what he and 'my Carl' could be doing in the 'narrow little world' of Weimar in August 1776. The revised version is not only less specific in its reference, but also clearer and simpler in its effects, eliminating such characteristics of Goethe's earlier period as the address to Fate or the favoured adjective, 'dear' ('lieb'), or noun, 'obscurity' ('Dumpfheit'). The price paid is a loss of psychological complexity, as in the revision of the electrically irregular rhythm of 'Needed Love' ('Liebebedürfnis') into the metre and manner of the trochaic *Erotica* of 1788. Two hundred years later the uniqueness of those not quite fathomable moments in the first summers and winters in Weimar seems more valuable than regularity and purity of diction. Goethe knew, however, when he had done something really well, and his revision of 'On the Lake' is cautious: 'soft mists' ('weiche Nebel') is more objective, has less 'manner', than 'dear mists' ('liebe Nebel') but is not obviously inferior, and the new version of the opening lines retains both the all-important rhythm and even perhaps the metaphor of the umbilical cord, though in a less crass form than at first, and firmly subordinated to the metaphor of suckling:

> Und frische Nahrung, neues Blut
> Saug' ich aus freier Welt;
> Wie ist Natur so hold und gut,
> Die mich am Busen hält!*

And fresh nourishment, new blood, I suck from the free world; how fair and kind is Nature that holds me on her bosom!

The original variety of physical relations between the poet and nature has however been rationalized away, while the sense that the poet is at the centre of a vortex is lost when a centrifugal movement is introduced by the (otherwise vigorous) change of 'cloud-girt' ('wolkenangetan') to 'cloudily heavenwards' ('wolkig himmelan') and by the substitution of the more proper 'meet' ('begegnen') for the more mannered 'counter' ('entgegnen'):

> Die Welle wieget unsern Kahn
> Im Rudertakt hinauf,
> Und Berge, wolkig himmelan,
> Begegnen unserm Lauf.

The wave lifts and rocks our boat in the rhythm of the oars and mountains, cloudily heavenwards, meet our course.

In these various little changes, none perhaps of outstanding importance, we can see signs of the priority of art over personality which Goethe asserted in the essay *On the Simple Imitation of Nature* . . . and which led to the introduction of a self-conscious reference to 'my song' into the new version of 'To the Moon'. It is but a step from the priority of art to the priority of the artist, who in the Moritzian view is as commandingly arrogant, and indeed erotic, a figure as any Egmont. 'Just as love is the highest perfection of our feeling nature, so the production of beauty is the highest perfection of our active force', we read in *On the Plastic Imitation of the Beautiful,* and just as Caroline Herder had to expect to be paint on Goethe's palette, so Clärchen had to find the supreme moment of her life in kneeling before her lover. This Egmont-like view of the artist, which permeates *Artist's Apotheosis* (and recurs in some of the as yet unpublished *Elegies*) has an unfortunate effect on a number of the love-poems in the 'First Collection', in which the woman is newly cast by the revision into a conventional victim's role, most notably 'My heart pounded', retitled 'Welcome and Parting' ('Willkommen und Abschied'), of which the concluding strophe now runs:

> Doch ach, schon mit der Morgensonne,
> Verengt der Abschied mir das Herz:
> In deinen Küssen welche Wonne!

* For the original version of these lines, see above p. 204.

In deinem Auge welcher Schmerz!
Ich ging, du standst und sahst zur Erden
Und sahst mir nach mit nassem Blick:
Und doch, welch Glück, geliebt zu werden!
Und lieben, Götter, welch ein Glück!*

But alas, already with the morning sun parting constricts my heart: in your kisses what ecstasy! in your eye what pain! I went, you stood, your eyes on the ground, and followed me with moist eyes, and yet what bliss to be loved! and to love, ye gods, what bliss!

In the version of 1771 the parting is an awkward adolescent affair out of which the confidence of mutual love shines as the one certainty. It is not clear how long the meeting has lasted, or why the woman has to go, or why the poet keeps his eyes on the ground—a posture suggestive of a host of suppressed feelings from guilt and sorrow to a reluctance to face the truth. In 1788 the situation is clarified and stylized into a parting at dawn after a night of love, with the man going and the woman staying, something at once more sensual, more dramatic, and more conventional. The roles of the lovers are more simply differentiated: he enjoys the 'bliss' of sexual victory, while 'pain' is reserved for her. The mysterious posture in the fifth line has become largely functionless, suggesting perhaps the woman's shame at past intimacies or at her feelings of regret at the departure. Since it is now only she who is tearful, the poet, in the jubilant counter-assertion of the last two lines, is no longer consoling himself with the thought of a shared affection: instead, he is assuring the woman that, although she is sorrowful, either he can rejoice in being able to love and be loved by her or she should count herself privileged in being able to love and be loved by him. A similar reduction of confused but shared emotion into a sterotyped differentiation of roles occurs in the new versions of 'Rose upon the Heath', in which the rose's resistance becomes completely ineffectual and the boy suffers no pain at all, and 'Evening Song of the Huntsman', in which the man becomes a sexually dominant figure who leaves the woman behind in order to range the world alone.

Outside the erotic context Goethe's polishing of his poems can lead to an enrichment rather than a reduction of meaning. At the last moment before printing he changed one word in the contrast of gods and men which concludes 'Bounds of Humanity', so that the final strophe now reads:

Ein kleiner Ring
Begrenzt unser Leben,
Und viele Geschlechter
Reihen sich dauernd
An ihres Daseins
Unendliche Kette.†

* For the original version see above p. 112.
† Original version above p. 321.

A little ring rounds our life and many generations string themselves on to the endless chain of their existence.

By the summer of 1788, as he told Caroline Herder, Goethe had discarded any belief in divine powers external to human life or determinant of it, and certainly any fear of their predatoriness. It is only by their own efforts, and not through the operation of some inscrutable destiny, that human beings come to share in some structure of meaning, or chain of culture, that transcends the individual existence. A similar train of thought led Goethe to delete from the invocation, in 'Divinity',* of the 'higher unknown beings that we sense' the line 'Let man resemble them' ('Ihnen gleiche der Mensch!') which alone in the poem assured the higher beings of an independent existence: the divinity of human beings was primary, the Lucretian Goethe thought in 1788, not derivative from an external model. In 'Bounds of Humanity', however, the revision created a suggestive ambiguity, where none had existed before: originally 'the endless chain of their existence' which the many little lives of individuals make up had to belong to the gods; in the printed version of the poem, although the grammar of the sentence now requires that the many generations make up their *own* existence by stringing individual lives together, the possibility that they may be making up the existence of the *gods* is kept open by the continuing opposition, in the poem's argument, between gods and man. The question whether the development of humanity is part of a larger process or is self-moving and self-sufficient is left unresolved; or, to be precise, a solution is suggested according to which it is true both that humanity has its bounds, and that outside those bounds lies not as in the earlier version a mysterious power, but nothing. This deliberately ambiguous solution brings Goethe to the threshold of Kant's critical philosophy.

As the prospect of completing his *Summa* came closer Goethe became more aware of its potentiality to be his own *monumentum aere perennius*. In his eighth volume he paid particular attention to the sequence of his poems and their layout on the left and right of each opening, 'because of certain relations' between them. 'Divinity', for example, had to end on a left-hand page in order that it should be complemented by two epigrams in classical metres, 'Duke Leopold of Brunswick' ('Herzog Leopold von Braunschweig'), originally composed as an inscription for the Tiefurt Park, and 'To the Ploughman' ('Dem Ackermann'). These attempt to establish an attitude to death consonant with the ethical humanism of the rules of life formulated in 'Divinity'. The epitaph for Anna Amalia's brother, who in 1785 was drowned taking part in rescue operations after a flood, suggests that—like Egmont in the final version of the play—he will be divinized by death and, as a supernatural force, carry through to completion the virtuous task in which, as a living man he failed. 'To the Ploughman' is a version of the biblical comparison of life to the grain of

---

* Above p. 351.

wheat which falls into the ground, rewritten so as to avoid any hint of a resurrection and offering instead only an indeterminate 'hope'. The two epigrams clearly indicate that the preceding poem is to be given as anti-theistic a reading as it will bear. Such deliberate interpretative hints are, however, a rarity. In general Goethe's purpose in the arrangement seems to be to point up similarities, or occasionally oppositions, in mood or subject-matter, or, in the case of his classicizing epigrams, to build up three or four minute poems into larger units—to suggest, in fact, that each of the two 'collections' is composed, or, as Moritz would have said, unfolded, in accordance with an underlying principle. The volume is therefore not simply a miscellany, but a work of art in its own right, and Goethe evidently felt similarly about the next volume to which he turned, the sixth in the series devoted to *Tasso* and *Lila*. In this case he was particularly concerned that Göschen should avoid any of the misprints that had disfigured Unger's printing of *The Roman Carnival*: 'With the extreme care I have applied to this drama I also wish it to arrive uncorrupted in the hands of the public.' Already as he passed the manuscript to Göschen plans were laid, which seem never to have come to fulfilment, for a de luxe edition on fine paper and in roman type (rather than the Gothic used for the *Works*). Goethe himself did not shun the word rejected by Moritz to indicate the degree of artistic deliberation that had gone into the work which more than any other was the fruit of his Italian journey: 'Your approval', he wrote to Herder, 'is ample reward for me for the quite improper care with which I have worked up this piece. Now we are free of any desire to undertake so strict a composition again.'

Torquato Tasso is a work of detached and conscious artistry, yet its subject is the intense suffering of an artist; it was intended by Goethe to embody the Italian virtue of purely impersonal style, yet its setting in Italy was of the greatest significance to him personally; it is rigorously courtly in its form and in its subject, yet it contains some of the most withering denunciations of a court Goethe ever wrote. It is thematically and structurally by far the most complex play Goethe had yet completed and though intended as a final reckoning with the 'first period of my writing' it anticipates in several respects the art of his middle and old age. Superficially, at least, it is a regular drama in the French neo-classical court style: five acts in verse—though the modern pentameter blank verse of *Iphigenia*—representing the events of a single spring day in about 1577, and confined to one locale, Belriguardo, the summer residence of the dukes of Ferrara, though with a change of scene for each act. There are only five characters, all of them noble, though Tasso himself belongs only to the minor bureaucratic nobility of recent creation, and the action is concentrated on a single event: the emergence of Tasso's madness. Announcements, from the first lines of the play onwards, that some or all of the characters intend to leave by the evening lend to the action the urgency that tradition requires. Within the acts the stage is never left empty ('liaison des scènes'), and each new entrance is usually announced in the last lines of the preceding scene. Unlike *Iphigenia, Tasso* has no variation in verse-form though some irregularities in metre, these nearly always of dramatic or psychological significance. The diction is elevated and decorous, at times allusive and stiff; the discussion of personal matters is often conducted obliquely through generalizations, or *sententiae,* whose alternation (stichomythia) is the nearest approach in the play to animated dialogue. Otherwise speeches tend to be long, and of the nineteen scenes in Acts II to v eight are monologues and only two involve more than two speakers. At the same time there are certain elements of opera, or masque, alien to the strictest French traditions: in Acts I and v there are moments when each character in turn utters briefly on a significant topic, so contributing to a choral effect; costumes and stage properties—two busts, a laurel crown, a sword—have an important role; and at three or four points the physical action on the stage develops into a symbolic tableau. The exclusive concern with the affairs and state of mind of a single central character, the richness of reference to political and literary figures of the Italian Renaissance, and the discursive sensuality of some descriptions of the landscape, are also contrary to the principles of French neo-classical tragedy. Above all, though dramatic in form, *Tasso* is composed as a book: unperformable, through sheer length and stasis, on any stage, as Goethe himself came to recognize, it is a grand meditation on the art of poetry, the true conclusion and coping-stone to the edition, and to the half of Goethe's life which the edition summarized. In writing it Goethe concerned himself far more with its poetic structure than with its dramatic effect, and thereby he unwittingly anticipated the pre-eminence that in the second half of his life he would, eventually and reluctantly, give to the written over the spoken word. It is in many ways the axis round which his entire literary career revolves.

Act I takes place in the gardens of Belriguardo, decorated with busts of the poets. Princess Leonora d'Este, sister of Duke Alfonso II of Ferrara, and Countess Leonora Sanvitale, known as the object of some of Tasso's love-poems, both of them dressed as shepherdesses, are weaving wreaths—the Princess one of laurel, with which she crowns the bust of Virgil, the Countess, to whom the name Leonora is reserved throughout the play, one of many flowers, with which she crowns the bust of the author of her own preferred epic, Ariosto. Their conversation soon turns to Tasso, the distracted young poet, who we learn is really in love with the Princess but who, in order to avoid impropriety, addresses her only in poems apparently written for her namesake, the Countess. Alfonso joins them, looking for Tasso, whose extreme suspicions that he is being spied on or plotted against are a cause of concern to this measured and benevolent manager of men. Having expressed the hope that Tasso's great *opus, Jerusalem Deliver'd,* will soon be complete, Alfonso is surprised and delighted when Tasso arrives and presents the manuscript himself. To mark the occasion, Alfonso takes the laurel wreath from the bust of Virgil and, despite his reluctance, uses it to crown his court poet, so anticipating, in the intimacy of a private circle, the formal coronation on the Capitol arranged for the historical Tasso in 1595 but prevented by his death. The almost mystical ecstasy of Tasso's response to the impromptu ceremony is interrupted by the arrival of the play's man of action, Alfonso's minister Antonio Montecatino, just returned from a difficult but successful diplomatic mission in Rome, who clearly thinks the honour excessive for the practitioner of a useless art, the poverty-stricken young orphan who has become the pampered plaything of the ladies. If poets are to be praised at all, his vote would go the colourful and entertaining imaginary world of Ariosto, so aptly symbolized in Leonora's variegated wreath.

We follow the characters indoors for the second act, and in its first scene the Princess, still in her pastoral costume, recalls her first meeting with Tasso and tries to encourage him to live on better terms with her brother, with Antonio and with Leonora. In language reminiscent of 'Winter Journey in the Harz' and the seventh of the *Elegies,* both of which occupied Goethe in 1788–9, she warns Tasso against the dangers of solitude in which the spirit

> strebt
> Die goldne Zeit, die ihm von außen mangelt,
> In seinem Innern wieder herzustellen,
> So wenig der Versuch gelingen will.

strives to reconstruct within itself the golden age it lacks outside it, however little the success it meets with in the attempt.

Tasso reacts with passion to the mention of the Golden Age which he has depicted in his pastoral poetry, and of which the Princess's costume is a reminiscence, but to his evocation of that Age the Princess responds with a reminder that, if it ever existed, it is now past, and the only Golden Age that matters is that which human beings constitute among themselves by moral and seemly behaviour. This highest form of civilization is in the care particularly of women, and Tasso is thus provided with a cue to declare that all the female virtue shown in his epic derives from the single model now before him. At first the Princess seems not disinclined to let him unveil

> das Geheimnis einer edlen Liebe
> Dem holden Lied bescheiden anvertraut

the secret of a noble love modestly entrusted to fair song

but as he becomes more emphatic she breaks off the conversation with a warning against excess. Tasso resolves to abandon solitude: as Antonio approaches he falls upon him and offers him his hand, asking him for initiation into his special art, 'the moderate use of life'. The older and higher-ranking man is at first distant and cautious, then patronizing, sarcastic, and openly hostile, belittling the value of poetic achievement. Provoked beyond endurance, and in defiance of all courtly—and theatrical—convention, Tasso draws his

sword and challenges his tormentor. Alfonso's sudden entrance cools the
tempers but Antonio, somewhat perfidiously, suggests that Tasso must now be
severely punished. Alfonso, however, mildly asks Tasso to give him his word to
stay in his room until told he may leave it. Tasso takes this merely symbolic
punishment with the full gravity with which he is used to treating symbols, and
lays down both his sword and his laurel crown as equally forfeit, now that he is
dishonoured and cast out from 'the hall of the gods'. Left alone with Antonio,
Alfonso makes it gently clear that in such an incident the more experienced
man is to blame. He suggests the damage may be repaired if Leonora Sanvitale
speaks to Tasso first and if Antonio then brings him the news of the Duke's
pardon and himself seeks Tasso's friendship.

In the third act the Princess and Leonora form their own plans for restoring
Tasso. Leonora thinks that Tasso needs to spend some time away from the
court, for example in Florence, where she will shortly be staying and so will be
able to win him over while they are on neutral ground. The Princess reluctantly
agrees to losing the man whom she confesses to be the light of her life. We
learned in Act II that a sickly childhood kept her secluded from a noisy town and
the chivalrous pomp of the court, in this conversation we hear of the consuming
power of the beauty that came into her enclosed little gardens when Tasso
entered them like the sun. Now he is to be withdrawn and her world is wrapped
in mist. Leonora reveals in a monologue that her motive in wishing to remove
Tasso temporarily from the court is her rivalry with the Princess, of whose pale,
'moonlight', passions she speaks pityingly: if Tasso is firmly bound to her in
Florence the world will always identify the Leonora of his poetry with her, not
with the Princess. Antonio, however, with whom Leonora flirts in the next
scene, proves unwilling to co-operate with her scheme, but in accordance with
the Duke's plan he sends her to Tasso to prepare the ground for his own visit.

Tasso is on stage for the whole of Act IV, which is set in the room to which he
has been briefly banished, and three monologues of his alternate with two
dialogues, with Leonora and Antonio respectively, and show his gradual
descent into dementia. Leonora's attempts to persuade him that Antonio is not
as hostile as he appeared remind him how he loathes the courtier's pretensions
of superiority—'I recognize only one lord, the lord who feeds me'—and how
bitter, in his view, is Antonio's jealousy of the favour the Muses have shown
him. Like a child Tasso refuses to be enticed away from the delights of fury and
hatred and in this mood Tasso can see Leonora's proposal that he should go
to Florence only as a veiled indication that he is no longer wanted at the court.
Leonora assures him that no one in the court is secretly scheming to harm him:
these notions are all products of his over-active poetic imagination. Alone once
more Tasso explodes into sarcastic indignation at Leonora's wiles. We naturally
have some sympathy with him, for he is indeed the victim of intrigue, if not of
the intrigue he imagines. Deception is all about him, he feels, and the plan that
he should go to Florence, where the Medici are bound to honour him, seems an

obvious trap, intended to provide evidence for rumours of his disloyalty to Ferrara. Tasso, however, will be more cunning than the cunning world that is persecuting him: since everyone wishes him to go, go he will, but not to Florence. When Antonio comes to make his peace Tasso asks him to secure from the Duke permission for a journey to Rome, where he can discuss his epic with the great city's many scholars, and revise and polish it, for despite having presented it today in a provisionally finished form he cannot regard it as perfect yet. Antonio advises him against going away anywhere: he should instead exploit the moment of the Duke's favour to the full. Tasso insists, however, and Antonio eventually agrees, with deep misgivings, to present his petition. The perverse logic of madness has driven Tasso beyond the point where he might see in this reluctance anything but further evidence of Antonio's treachery. In his final monologue Tasso begins to doubt the reliability even of the Duke and his sister. Not having heard a word yet from the Princess he convinces himself that 'she too' has joined the ranks of his enemies and at the crest of a new wave of self-pity glimpses clearly for the first time his own impending mental disintegration:

> Und eh nun die Verzweiflung deine Sinnen
> Mit ehrnen Klauen auseinander reißt,
> Ja klage nur das bittre Schicksal an,
> Und wiederhole nur, auch *sie!* auch *sie!*

And now, before despair tears your senses apart with brazen claws, yes, just denounce bitter destiny and just repeat: *she* too!, *she* too!

For the final act we return outside, to another part of the gardens. Alfonso is somewhat vexed by Tasso's request: he tells Antonio he does not want any rival patrons in Rome or Florence tempting his favourite away from Ferrara.

> Ich bin auf ihn als meinen Diener stolz,
> Und da ich schon für ihn so viel getan,
> So möcht ich ihn nicht ohne Not verlieren.

I am proud to have him as my servant and since I have already done so much for him I should not like to lose him without necessity.

Antonio expresses his own irritation at Tasso's disordered life, his indulgence in spicy foods and drink, and his irrational suspicions: whatever his restless spirit is looking for in Rome, Naples, or elsewhere can be found, if at all, only at home in Ferrara. Alfonso, however, treats all these idiosyncrasies with generous tolerance—'not everyone serves us in the same way'—and decides to give Tasso the leave he asks for. However, his patience is tried to the limit when the poet asks also for the return of the manuscript with which he is so dissatisfied. Fearing that revision will only spoil the poem, or perhaps even that if he once parts with the manuscript he may not see it again, Alfonso agrees only to let Tasso have a copy of his original. He urges Tasso to tear himself away from the self-preoccupation that is threatening to destroy him, conducive though Tasso thinks it is to poetic activity. 'The human being will gain what the poet loses.' Tasso compares himself, in response, to the silk-worm that creates by an irresistible urge, and out of its inmost being. True, what it creates is its grave, the cocoon in which it dies to its present existence, but perhaps there is a new life as a butterfly beyond that grave too, though what it might be Tasso cannot guess. Left alone he dismisses all he has just heard as a parroting of Antonio, and then discovers, as the Princess approaches, that he is still not confident of his emotional detachment from this hateful court. Tasso's second and final dialogue with the Princess is the most moving and powerful depiction, in a play rich in understanding of the different degrees of pathological delusion, of a mind slipping out of contact with those most desperately trying to reach it, and not entirely by its own fault. To the harmless, awkward, enquiry, 'So you are going to Rome?' Tasso responds at such length that he is drawn into expressing his greatest fear, that he will not after all complete his poem as he wishes. Quickly he breaks off and talks of going on to Naples, and Sorrento, his home town, where his sister Cornelia still lives. His description of the journey becomes ever more elaborately microscopic, and even as the Princess interrupts him he is taking two lines of verse to cross his sister's threshold. The Princess has recognized the full degree of his absorption in a potentially endless fantasy and tries to draw him back into the present by upbraiding him. That of course is without effect, but a word of kindly sorrow can still catch his attention. It leads, however, to another excess of fantasy on Tasso's part: this time he paints in detail a life of devotion to her as gardener and caretaker of her remotest country residence. So evident is the Princess's distress that Tasso abandons his mistrust and asks for help; but all the Princess can do is assure him of her affection and tell him he has to help himself. It is a superbly dramatic picture of a good intention too generous for the limited sensibility that has formed it, and which therefore fades away into the vacuum between two different personalities:

> Gar wenig ist's was wir von dir verlangen,
> Und dennoch scheint es allzu viel zu sein
>
> .   .   .   .
>
> Du machst uns Freude wenn du Freude hast
>
> .   .   .   .
>
> Und wenn du uns auch ungeduldig machst,
> So ist es nur, daß wir dir helfen möchten
> Und, leider! sehn daß nicht zu helfen ist;
> Wenn du nicht selbst des Freundes Hand ergreifst,
> Die, sehnlich ausgereckt, dich nicht erreicht.

It is very little that we are asking of you, and yet it seems to be all too much . . . you make us happy when you are happy yourself . . . and even if you make us impatient it is only that we should like to help you and see, alas, that there is no help unless you yourself seize your friend's hand which, stretched out in yearning, cannot reach you.

Of this archetypally unappealing appeal Tasso hears only the expressions of impotent affection and interprets them as the open declaration of the love that on both sides has hitherto been formally inexplicit. The other three characters are now seen in the background gradually approaching, closing in on the pair, but carried away by 'that feeling which alone can make me happy on this earth', and which he has to his own misfortune too long resisted, Tasso falls into the Princess's arms and embraces her. This unpardonable act of *lèse-majesté* brings the Duke's brief and definitive condemnation of his unmanageable subject, 'Er kommt von Sinnen, halt ihn fest' ('He is taking leave of his senses, hold him fast'), and after its momentary convergence on the play's central figure the court disperses in horror as swiftly as it came, leaving only Antonio standing beside the demented Tasso.

The catastrophe is complete but, after a long silence, a coda follows, a dying cadence that lays the harrowed emotions to rest. At first Tasso raves against each of his friends in turn: Antonio, however, does not react to Tasso's charges, but affirms that he will stand by him in his need and urges him to take such hold on himself as he can and rescue what identity is left to him. Tasso's mood switches from rage to tearfulness as he sees the dust rising from the departing carriages and realizes it is too late to restore what is lost for good:

> bin ich *nichts,*
> Ganz *nichts* geworden?
> Nein, es ist alles da und ich bin nichts;
> Ich bin mir selbst entwandt, sie ist es mir!

have I become *nothing*, wholly *nothing?* No, everything still is, and I am nothing; I am wrung from myself, and she from me!

In the last lines of the play Antonio's admonitions take effect; Tasso recognizes that one thing 'remains' to him, the cry of pain and the ability to give it poetic expression,

> Und wenn der Mensch in seiner Qual verstummt,
> Gab mir ein Gott zu sagen, wie ich leide.

And when the human being falls silent in his torment, a god gave me the gift to say how I suffer.

At this recognition by Tasso of his only real and lasting role, as the disembodied voice of human pain, Antonio clasps his hand as he earlier refused to do when Tasso was seeking to join him as a colleague, and the curtain falls on the symbolic picture of their embrace, while Tasso utters a final metaphorical interpretation of their ambiguous relationship:

> So klammert sich der Schiffer endlich noch
> Am Felsen fest, an dem er scheitern sollte.

Thus at the last the mariner clings fast to the rock on which he was destined to founder.

In 1827 Goethe said to Eckermann about the subject-matter of the play:

I had the life of Tasso, I had my own life, and in putting together two such strange figures with their individual characteristics the picture of Tasso formed itself within me, to which as a prosaic contrast I opposed Antonio, for which also I did not lack models. And then the other circumstances of the court, of life and love, were the same in Weimar as in Ferrara and I can justifiably say of my depiction: it is bone of my bone and flesh of my flesh.

Goethe was not thereby retracting his earlier denial of Charlotte von Kalb's theory that Alfonso is to be identified with Carl August, Tasso with Goethe, the Princess with Duchess Luise, and Leonora with Frau von Stein. Nothing said in 1827 contradicts Goethe's view of 1789 that characters in art do not have simple or single originals in nature but are put together by the artist from features that in reality may be widely scattered. Similarly a single real model may leave traces in more than one fictitious character. Because Alfonso so obviously shares the rank and function and judicious generosity of Carl August (but not his military adventurism), and because there are in his case no competing possible models, apart perhaps from Duke Ernst of Gotha, it is tempting to go on and to equate Goethe with his Tasso because both achieved literary success at a young age, both received court patronage as a result, and both attracted the enmity of an older generation of courtiers; for both, furthermore, Rome and Naples were places of special importance, Rome because in it they hoped to see the completion of a major literary project, and for both perhaps it was true that what they sought abroad they would only truly find at home; neither could tolerate being addressed in a fatherly tone; and both, for that matter, as Goethe troubles to inform us, had a sister called Cornelia. When Tasso is reduced to despair by the sight of the carriages drawing away down the dusty road it is natural to think of Goethe watching Anna Amalia depart with opposite her an empty seat. Once the equation of the two poets has been made, misleading parallels (or excessively simple contrasts) may be drawn between them in respect of issues that are obviously of the first importance to both: their poetry, their self-absorption, their relation with the court, their love of a woman of the highest rank. But, quite apart from elements he could draw from biographies of the historical Tasso, Goethe's own experience offered him more models than himself for Tasso's derangement: just as he had once incorporated in *Werther* details from the life and death of young Jerusalem, so he now had before him the examples of the Plessing of 'Winter Journey in the Harz', with whom he had remained in contact since 1777 and to whom he still sent complimentary copies of his works; the struggling young Schiller, perhaps, asserting the rights of his art against a court existence; and particularly at this time the sad letters of Merck, in which reports of vivid fears, uncontrollable despondency, and a physically disordered life, combine with a terrible lucidity about his condition. More significant still, though, are the traces of Goethe to be found in characters other than Tasso. This is most obvious in the case of Antonio. Although Antonio's initial arrogance recalls the contempt Count Görtz showed for a bourgeois *arriviste* in

1776, and his indifference to artists the pragmatism of Marquis Lucchesini, the great trencherman at the table of life, Goethe too by 1789 was enough of a courtier to know from first-hand experience, in Brunswick for example, the diplomacy that costs 'many days, some spent in impatient waiting, some purposefully wasted'. And if Merck was Tasso, it was Goethe who was Antonio to him, not merely speaking of 'help', but arranging large loans, and in his letters offering what he could: a willing ear, and encouragement to Merck to use his intellect in sorting out which of his troubles could be remedied, and by what means. When the Princess excuses what seems to her, as to Alfonso, Antonio's uncharacteristic brusqueness at his first meeting with Tasso, she does so in words Goethe no doubt longed, or thought he ought, to hear about himself from the lips of Frau von Stein:

> Es ist unmöglich, daß ein alter Freund
> Der lang entfernt ein fremdes Leben führte,
> Im Augenblick da er uns wiedersieht
> Sich wieder gleich wie ehmals finden soll.

It is impossible that an old friend who has long led a strange life in distant parts should, in the moment when he sees us again, be exactly what he always was.

In a less self-pitying mood, Goethe presumably modelled Antonio's impatience with Tasso's harmful diet on his own impatience with Frau von Stein's coffee-drinking. The Princess too at times reflects Goethe's emotions. In the last desperate interview between her and Tasso we can perhaps catch, in her unavailing attempt to stretch out to Tasso a hand he will not grasp, a shadow of Goethe's own struggle to establish communication with a resolutely un-listening Frau von Stein, in a crisis in which she would inevitably suffer more than he. The Princess's lament for the passing of her old relationship with Tasso has the triple ambiguity of the lament in the final version of 'To the Moon'—not only the woman's loss of the man, but also the man's loss of the woman, and the other loss which symbolizes them, and of which too they are symbols, that of Italy:

> So selten ist es, daß die Menschen finden,
> Was ihnen doch bestimmt gewesen schien,
> So selten, daß sie das erhalten, was
> Auch einmal die beglückte Hand ergriff!
> Es reißt sich los, was erst sich uns ergab,
> Wir lassen los, was wir begierig faßten.
> Es gibt ein Glück, allein wir kennen's nicht:
> Wir kennen's wohl, und wissen's nicht zu schätzen.

It is so rare that human beings find the very thing that seemed destined for them, so rare that they retain what the fortunate hand happens to have seized. What had just yielded to us tears itself free, what we had eagerly grasped we let go free ourselves. There is such a thing as happiness, but we do not know what it is: we know what it is, and do not know its worth.

Where, then, there are traits of Goethe in Antonio and the Princess, there are corresponding traits of Frau von Stein in Tasso—her diet, her personal loss, her refusal to be converted from a course that could only bring her suffering. She too has left her mark on more than one character. Apart from her similar court status, Charlotte von Kalb's identification of her with Leonora is rather improbable, though her sister-in-law, Sophie von Schardt, would be a possible model for that figure, or even Frau von Kalb herself, with her transparent designs on Schiller ('the ladies here are of quite remarkable sensibility', Schiller wrote from Weimar, '. . . they would all like to make a conquest. E.g. there is a Frau von S[chardt] whom in any other society you would label a fully qualified *fille de joie,* a fine face, not ugly, eyes vivacious but very greedy'). But Frau von Stein is principally present in the play, of course, in the figure of the Princess. We must certainly not underestimate the contribution to this character from Duchess Luise: her rank, her closeness to Duke Alfonso as co-patron of the arts, the restricted regime of her life, and her constantly elegiac mood, all suggest that most silent and unchronicled of Goethe's attachments, and the companion-piece to Tasso in the sixth volume of the *Works* was *Lila,* the playlet most transparently devoted to the Duchess, at any rate in its revised form which Goethe was now printing. The function of the Princess in Tasso's life, however, is mainly defined in terms learnt by Goethe in his relation with the Duchess's lady-in-waiting. Tasso asks the Princess:

> wo ist der Mann?
> Die Frau? mit der ich wie mit dir
> Aus freiem Busen wagen darf zu reden.

Where is the man, the woman, to whom as to you I can dare to unbosom myself freely?

and the love that grew in their daily conversations was marked, she says, by a growth in mutual knowledge and understanding, by purity, harmony, moderation, and virtue. 'I had a friend', she says, in almost the words and tone of 'To the Moon' or of Frau von Stein's remarks about Goethe after early 1789. She sees herself as the representative to him of the society of 'noble women' and of the quality of 'seemliness' which it is their task to maintain—indeed, so 'unselfish' is she that a friend has told her she cannot quite appreciate the selfish needs of her friends. At a less conscious level, her sensually deprived childhood, and her combination of erotic encouragement to Tasso with an abrupt retreat into propriety, amount to a pattern which also marks the personality of Frau von Stein. Through Tasso her deprivation is indirectly remedied and the active life that in itself is too violent a stimulus for her fragile constitution is presented by him to her at one remove:

> Ich mußt ihn lieben, weil mit ihm mein Leben
> Zum Leben ward, wie ich es nie gekannt.

I had to love him, for with him my life became life as I had never known it before.

To Tasso in turn she is the inspiration of his poetry and she is the woman who has 'healed' the false tendencies of his heart and to whom he entrusts himself as teacher and guide. We can no more wholly identify the Princess with Frau von Stein than we can identify Tasso with Goethe, but the parallels are not superficial or irrelevant.

Although there can be no doubt that *Torquato Tasso* is bone of Goethe's bone and flesh of his flesh, we shall only misrepresent its relation to his personal development if we make direct and exclusive equations of characters in the play with figures in his life. If we want to understand the play's significance we must in the first instance look at its internal structure. Tasso's final cry of annihilation —'I am wrung from myself, and she from me' —is far more than a hyperbolical epitaph on Goethe's peculiar love-affair with Frau von Stein. For one thing, we must remember the Princess's affinities with Duchess Luise. To embrace Alfonso's sister is as impossible in Ferrara as to embrace Carl August's wife in Weimar or Jove's queen on Olympus, yet this impossible act of hubris, which Goethe believed to have been the crime of Tantalus, is being said by Tasso to be so intrinsic to his sense of personal identity that the frustration of the act has as its direct consequence his final descent into madness. Tasso desires the impossible, and in his first private conversation with the Princess, in Act II Scene I, he reveals—to us, if not to her—why he can be what he is only if he does desire what to the world, or at any rate to the Ferraran Court, is morally, practically, and politically impossible. Where, he there asks, is the Golden Age now, the Age 'for which every heart yearns in vain'? His verbal picture of that state of perfection in which 'all that is pleasing is permitted' is not merely an evocation of an ideal Italian landscape with classical figures, nor is it simply an ecphrasis, a putting into words, of an Arcadian scene painted by Poussin or Claude—perhaps such a painting as Poussin's *Landscape with two nymphs and a snake* of 1659— it is also a richly sensual exercise of the poetic imagination. Indeed when the Princess refuses to join in Tasso's yearning for that lost perfection, and changes his motto to 'all that is seemly is permitted', she does so in a speech devoid of all sensual ornament or visual appeal, denying that what Tasso has just so vividly evoked has ever really existed, and dismissing it as a flattering illusion fostered by 'the poets'. In rejecting Tasso's vision of a world in which there is a perfect harmony between morality and sensuality, in rejecting the pre-established affinity he claims between the yearning of his heart and the object that necessarily—he believes—corresponds to that yearning, the Princess is rejecting poetry itself. If then the Princess so explicitly refuses to share the ambition which is central to the poetry which in turn is central to Tasso's life, why does Tasso love her? Tasso gives the answer himself later in the scene: in her he has 'seen it with my own eyes, the original patttern ['Urbild'] of all virtue and beauty'. She is the manifestation in the flesh, the model—and so the guarantee of the reality—of all the qualities of his imaginary heroines,

Es sind nicht Schatten, die der Wahn erzeugte,
Ich weiß es, sie sind ewig, denn sie sind.

They are not shadows engendered by illusion, they are immortal, I know, because they are.

Tasso's art of the imagination does not detach him from reality, as we see most clearly when it runs away with him in the final act, and he imagines every concrete detail of his journey to Naples or of the Princess's country residence. As the Princess recognizes in the first scene of the play, 'the real attracts him forcefully and holds him fast'. It is essential to the fulfilment he envisages, most potently and lucidly in his depiction of the Golden Age, that it should be physically possible and should take place in the world accessible to the senses. The beautiful as he imagines it is simply the perfection of the real. In loving the Princess Tasso is loving the object for any (unalienated) subject, the reality which the perfection imagined in his art ought by its nature to possess. 'Oh teach me how to do the possible', he beseeches her, for it is of the essence of his ideal that its realization should be possible, even if in practice he is not prepared to observe the conditions that she lays down—conditions of 'moderation and abstinence' and the artificiality of court life.

Tasso's art is the quintessence of the poetry of desire—rooted in the real world which it longs to see transformed into the sacrament, the realized and realizing symbol, of the perfect happiness of which it carries the image within itself. It comes from life and returns to life and to those who enjoy art it gives double enjoyment of life. Tasso is, in respect of his art, as closely, but as incompletely, identical with the Goethe of 1788–9, as was Werther with the Goethe of 1774 in respect of his sensibility. If Goethe called *Tasso* 'an intensified Werther', it was partly because in both the novel and the play he was depicting an obsession from which he was already partially detached, and from which he wished to be detached still more, although in both cases the obsession ran too deep, and was too intimately involved with his spiritual and artistic identity, to be exorcized by an act of will, or by its expression in a single work of art. Goethe never left Werther completely behind him, as *Tasso* itself shows, and, fortunately for his poetry, he never left Tasso completely behind him either, although the play which bears the poet's name is an attempt to do just that. For *Torquato Tasso* shows us the destruction, and indeed the self-destruction, of the poetry of desire, and endeavours to show us also its replacement by a more limited, impersonal, and objective poetry.

The action of the play begins and ends with a moment of hubris, and there is a sense in which the two moments are identical, like leaves and flower, the one the unfolding and explication of the other. In both it is possible to discern an element of provocation, or complicity, by a court which does not understand, or does not wish to understand, Tasso's poetry. To the embrace of the Princess in Act V corresponds the crowning with bay in Act I. To be precise, the transfer to

Tasso of the wreath properly belonging to his master Virgil is only the second stage, or symbolic manifestation, of the true act of hubris itself, which prompts Alfonso's gesture, the declaration by Tasso that his poem is complete. For a poem which embodies perfection at once unconditioned and supposedly real can be completed only by an impossibility, only if the Golden Age can be restored round about it, so that its description of what is desirable is at the same time a description of what is. In an imperfect world such a poem can be complete, that is both completely beautiful and completely true, only at the end of time. This view of course is not shared by the court, whose lower opinion of the nature of poetry is shown by their belief that completion of a poem is a straightforward matter and by the frivolous readiness with which they embark on the presumptuous ceremony of coronation, which in a sense they force on Tasso. Unlike them, he knows and believes in the real meaning of the symbols they are invoking, and he quails. Symbols to him are not, as they are to Leonora and Antonio, airy nothings, 'mere' signs: meanings, he says, have more effect on him than events. Hardly has he declared the poem complete when he realizes his mistake: he shrinks from the laurel crown, which belongs properly only to 'heroes', to those who achieve a real and limited goal within the real and limited world of time, and asks the gods to preserve it for him in the heavens as an ideal, luring him for ever onwards:

> daß er hoch und höher
> Und unerreichbar schwebe! daß mein Leben
> Nach diesem Ziel ein ewig Wandeln sei!

May it float higher and higher, unattainable. May my life be an eternal pilgrimage towards this goal.

This is what his struggle with the poem up to now has been, a striving for an unattainable goal, for his aim has been nothing less than that his spirit should 'draw together the ultimate bounds of all things' and so 'round his poem to a whole'—thus recreating, as Moritz required, the great totality of Nature. No wonder Alfonso has hitherto had to grumble impatiently (like Anna Amalia?) that Tasso 'can never finish' and permanently postpones the moment when his audience can enjoy the completed work. Such a demand could only issue from a complete misunderstanding of the task in which Tasso believes himself engaged. For Tasso that moment must always be too early, and yielding to the importunities of Alfonso and the court and telling them that it has at last arrived must, once the confusion has passed, seem to him an act of impiety towards his own inspiration. The same belief in the real power, and the real meaning, of poetry which leads him into what the court regard as an excessive reluctance to go through with their playful ceremony also leads him to take what seems to them an excessively meticulous view of Alfonso's sentence on him in Act II. He regards his arrest as the suitable punishment for his earlier transgression in accepting the laurel crown and lays the wreath down again although it 'seemed to have been given me for eternity':

Zu früh war mir das schönste Glück verliehen,
Und wird, als hätt ich sein mich überhoben,
Mir nur zu bald geraubt.

The fairest happiness was bestowed on me too early and, as if I had been overweening in
it, is reft from me only too soon.

His brief presence in the 'halls of the gods' is, like that of Tantalus, followed by
a sudden fall. He lays plans for going to Rome to revise his manuscript, now
that the illusion of its completion is past, and so resumes his endless pilgrimage
towards the unattainable laurels. Antonio agrees with this interpretation of
events:

So jung hat er zu vieles schon erreicht
Als daß genügsam er genießen könnte.

So young he has achieved too much to be capable of contented enjoyment.

But the enjoyment Antonio is here envisaging, like that of whose continual
postponement Alfonso complains, is not that absolute and unconditioned
enjoyment demanded by the poem as Tasso understands it, realizable only
through the restoration of the world itself in a new Golden Age, and clutched at
once more when he falls into the arms of the Princess: it is a limited and relative
enjoyment of 'the goods of this life' whose value is appreciated only by those
who have undergone 'the toil of this life' in order to achieve them.

Such incomprehension of Tasso's true motives is general. When Tasso
makes his announcement to the court and hands over his manuscript to
Alfonso, each of the characters on stage in turn expresses in one line the
significance of the action as he or she understands it. Not one of them,
however, agrees with the poet himself. The pleasure of a finite 'work'
completed, of a service done to the public, of personal fame: Tasso rejects all
these partial satisfactions with words at once of supreme modesty and of
overweening pride: 'Mir ist an diesem Augenblick genug' ('This present
moment is enough for me').

If it really is the case, as the court before him wishes to believe, that he has
completed his poem, in the sense in which he understands that task, then this
indeed is for him the consummation of all desire. Such a consummation is of
greater value than any work of art, for no work of art, not even that which he has
just supposedly finished, can have a higher goal than to represent (and so
recreate) a completely fulfilled moment.

A moment which someone with Tasso's understanding of reality and his
desire for perfection can declare totally satisfying must lie outside the normal
sequences of life and history. After his coronation Tasso appears transported in
an ecstatic vision to Elysium, a timeless paradise, and the words 'present' and
'presence' recur emphatically in the text. Alfonso, however, in lines that enact
the precipitancy of events, brings about the passing of Tasso's supreme moment

as he announces the arrival, no, the presence, of Tasso's supreme adversary:

> Er ist gekommen! recht zur guten Stunde.
> Antonio!—Bring ihn her—Da kommt er schon!

He has come! Just at the right moment. Antonio!—Bring him hither—Here he comes.

Antonio and Tasso are both men of the present moment. But whereas Tasso sees eternity in a single fulfilled moment, and does not waver in his loyalities and attitudes throughout the play, Antonio sees the moment as the only reality in the flux of time and is, as befits a diplomat, a virtuoso performer in the art of mutability. His inital rudeness to Tasso is held by all to be out of character, and is explained, as we have seen, by the effect on him of the set of circumstances momentarily prevailing at the time of his arrival. But he is no prisoner of that passing weakness. He can subsequently learn to see his error, and can apologize for it, to the Duke and also to the man he has offended. At the end of the play he can respond to the new need created by the deterioration of Tasso's condition, can sympathize and assist, and even appear to be as firm as a rock, because he knows how to manage time. He presses Tasso to exploit the moment of the Duke's favour, insisting on the importance of being 'present' in Ferrara to do so, while Tasso disdains to make poetry, and the matters of eternal importance with which it deals, dependent on such passing contingencies. Antonio knows that time passes and has long since reconciled himself to that cruel fact. For Tasso the discovery that 'the present' is no more is the ultimate pain, as agonizing as the supreme moment was fulfilling: in the play's last lines he sees the court that crowned him, and that his poem addressed, dispersing and departing without even the possibility of saying farewell,

> Laßt mich nur Abschied nehmen,
> Nur Abschied nehmen! Gebt, o gebt mir nur
> Auf einen Augenblick die Gegenwart
> Zurück!

Just let me take my leave, just take my leave! Give, oh, just for one moment give me back the present!

Now Tasso's only companion and support is the man who took no part in his coronation, and whose arrival interrupted his 'present' forever. The collaborator with time is his only protection against a loss so total that it threatens to encompass even the loss of his conscious mind.

Goethe told Caroline Herder that the theme of his play was 'the disproportion between talent and life'. In the most general, conceptual, terms, there is a mismatch between the art that sees an endless importance, the capacity for an endless fulfilment of desire, in the things of the real world, and the law of life by which with the passage of time those things lapse into nonentity. In dramatic terms the 'disproportion' is between the man who practises the art and the circumstances which make it possible for him to do so, and also, within the man, between his art and himself.

The cultural circumstances of Ferrara are closely parallel to those of Weimar. In a nation—Italy, but it might equally be Germany—politically fragmented but aware of its nationhood, full of competing patrons, there are, the play tells us, popular and urban forces striving for expression; these however are not represented by the courts, which are the centres of a culture based on absolute autocratic rule, and, if more aristocratic than that of the towns, also more intimate and personal. Leonora is, as usual, the unthinking voice of the *ancien régime* when she contrasts Ferrara with republican Florence:

> Das Volk hat jene Stadt zur Stadt gemacht,
> Ferrara ward durch seine Fürsten groß.

The people made that city into a city; Ferrara grew to greatness through its princes.

The Princess adds a significant qualification which accommodates the self-esteem of the intelligentsia:

> Mehr durch die guten Menschen, die sich hier
> Durch Zufall trafen und zum Glück verbanden.

More through the good people who met here by chance and, happily, combined.

Chance or no, Tasso is clear and emphatic that Ferrara has supplied him with everything. His 'lord who feeds' him rescued him from poverty, and gives this 'artist' ('Künstler') the patronage during his lifetime for which the protagonist of *Artist's Apotheosis* prayed, rather than fruitless acknowledgement after he is dead. True, Tasso's first abrasive meetings with Antonio show that he accepts only the absolutism of the patron, not the aristocratic order of social privilege that supports it. But he also recognizes that to the court he owes both the subject-matter of his poem and its audience. Could he have drawn the knowledge of war and government shown in his epic from his own resources, out of himself? If his art is to be based in reality the artist must have access to the real world he is depicting. That perhaps was the single most important consideration in Goethe's own transfer from Frankfurt to Weimar. There is, however, no trace or equivalent in *Tasso* of the reading public that was the complicating factor in that decision of Goethe's, no mention even of the art of printing. Instead there is Tasso's declaration of exclusive loyalty to the court, his theme, and also his only audience: 'An euch nur dacht ich wenn ich sann und schrieb' ('When I was pondering and writing I was thinking only of you all'). For him as an 'artist' the little circle of 'friends' displaces the 'world', as it did for Goethe in his early Weimar years and in the first version of 'To the Moon':

> Wer nicht die Welt in seinen Freunden sieht
> Verdient nicht daß die Welt von ihm erfahre.
> Hier ist mein Vaterland,
>
> .   .   .   .
>
> Hier spricht Erfahrung, Wissenschaft, Geschmack,
> Ja Welt und Nachwelt seh ich vor mir stehn.

Die Menge macht den Künstler irr und scheu:
Nur wer *euch* ähnlich ist, versteht und fühlt,
Nur der allein soll richten und belohnen!

He who does not see the world in his friends does not deserve to be known to the world.
Here is my fatherland . . . Here there speaks the voice of experience, knowledge, taste,
yes I see the world and posterity stand before me. The masses puzzle and startle the
artist: only he who is like you, understands and feels as you do, he alone shall judge and
shall reward

If this is Tasso's initial commitment the depth of the subsequent tragedy is
clear. He is at the end of the play abandoned and rejected by those of whom and
for whom he wrote, and who have carried off with them a poem which bears his
name but which he no longer recognizes as his own. We cannot say that for this
disaster his own unstable temperament is entirely to blame. In the scene of his
coronation there are already hints that the court is not all that it appears to be in
his flattering account of it. None of its three representatives shares Tasso's
understanding of the completion of the poem as a moment in which life itself is
perfected and subject and object are fully harmonized, none of them
understands the pre-eminence for Tasso of the *truth* of what he has written.
Alfonso has already told us in the previous scene that he looks forward to the
completion of the epic for the fame it will bring to himself and his duchy, once it
is in his hands he regards it as a commodity he possesses, and calls 'it in a
certain sense my own', and as the play develops he shows himself concerned
above all to retain Tasso in order not to lose to his rivals a valued if eccentric
servant. Leonora's preoccupation with the personal fame brought by literary
success becomes her motive for the intrigue which crystallizes Tasso's
suspicions of those around him, and she never shows any insight into, or even
interest in, Tasso's poetry, except in so far as it relates to a woman called
Leonora:

Wie reizend ist's, in seinem schönen Geiste
Sich selber zu bespiegeln!

How attractive it is to mirror oneself in his fair mind!

Her favoured poet is Ariosto, for reasons similar, her colourful wreath suggests,
to those given by Antonio: precisely not the truth of the poetic vision, which is
so important to Tasso, but the brilliantly illusionistic quality of his fantastical
inventions. Ariosto's poetry is represented by Antonio as an entertaining
pretence, and all art, in Antonio's view, is but a luxury, of value only as an
ornament of the state power, or perhaps, as Leonora suggests to him, as a
relaxation after the business of the day is over. Even the Princess, though she
has more insight into Tasso than the others, and certainly more love for him, is
using him, if in a highly refined sense, for her own purposes. It is he who 'like a
flame' brings heat and light and colour into her pale and delicate world, faded
by sickness and abstinence. Indeed the kind of love she has for Tasso, and

wishes him to have for her, is not wholly, or even greatly, compatible with his genius. The sensuous richness of his speeches contrasts with the pallor of hers, and if his flame-like personality, expressing itself in his fiery tastes in food and drink, is associated by her with the sun and sunshine, Leonora uses the correctly complementary image for the derivative and reflected energy by which the Princess lives:

> Denn ihre Neigung zu dem werten Manne
> Ist ihren andern Leidenschaften gleich.
> Sie leuchten, wie der stille Schein des Monds
> Dem Wandrer spärlich auf dem Pfad zu Nacht;
> Sie wärmen nicht und gießen keine Lust
> Noch Lebensfreud umher.

For her affection for the worthy man is like her other passions. Like the quiet light of the moon they gleam sparsely on the wanderer's path at night; they do not warm, and do not spill around them pleasure or joy in life.

Tasso, whose art is to imagine the full, glowing detail of other peoples' lives and deeds and feelings, cannot be for ever deprived of 'joy in life'. In his dialogue with the Princess about the Golden Age, in their different assessments of the moment of fulfilment in Act I, in their different approaches to the declaration of love, we see a conflict not just between unsocial excess and moral decorum, but between two imaginations, one of which is, however gently and understandably, parasitic upon the other and restrictive of its freedom. (The rewritten *Lila* was perhaps coupled with *Tasso* in volume vi as a farcical attempt at the cure of such a deprived sensibility.) When Tasso, standing on the brink of catastrophe, concludes Act IV with the repeated words 'she too! she too!', it is not just his delusion speaking. It is a recognition, however madly exaggerated, that there is no one in Ferrara who understands him or his art, that, despite the kindness he has received there, he has in Ferrara no public, and that this absence of an objective human world to support his imagination means the disintegration of his poetry. The solitude in which Tasso ends contrasts grimly with the concluding vision in *Egmont*, which was the vehicle for a confidence that its hero's personality could even in death inspire a whole nation. Between the two conclusions lay two years in which Goethe's ardent resolve to find a new role in Weimar had had time to cool in a prosaic reality and in the anguish of loss and personal upheaval. By the summer of 1789 Goethe had no difficulty in recapturing the sense of constriction and isolation, and the fear of sterility, that in 1786 had driven him, like Tasso, to a flight to Rome. Deprived of his own Italian Golden Age, and of his own moon-like companion, surrounded by a trivial and gossiping court with no interest in what he meant by 'Art', with even the edition of his *Literary Works* attracting less attention than had once been hoped, he might well feel that in leaving Frankfurt all those years before he had fled the fate of Werther to meet one that was little better. By Tasso's repetition and 'intensification' of Werther's tragedy we may measure the significance for Goethe of the turning-point in his life and art marked by the two years after his return from Italy.

Of course it is possible to see much of Tasso's suffering, like that of Werther, as self-inflicted, and 'the disproportion between talent and life' is often taken to be some congenital unpracticality of poets which makes them unsuitable courtiers and embarrassing dinner-guests. It is certainly important to Goethe's artistic purpose that we should see Tasso's behaviour as unreasonable, his suspicions as excessive, his demands as at times infantile, and his self-pity as at times unjustified. But it is equally part of Goethe's design that we recognize the grounds the court has given for Tasso's suspicions, and the provocations it has offered, above all by its misunderstanding and even contempt for that which is of the first importance to him—the nature of poetry. There cannot be a simple opposition between Tasso's art and the demands of 'life', because the perfection of Tasso's art is its truth to life—life transfigured by the desire for its perfection, no doubt; but it is still profoundly offensive for Tasso to hear from Antonio that poetry is an agreeable tissue of illusions. There is therefore nothing adventitiously personal about Tasso's descent into madness. Like the suicide of Werther-Jerusalem it stands in a symbolic relation to Goethe's own concerns—even more so, in fact, than the earlier case, for madness involves the degeneracy of the very organ which makes it possible for Tasso's poetry to be true to life in the first place, the imagination. If by the end of the play Tasso has lost his reason, the poet has destroyed himself as surely, if not as bloodily, as did Werther with Albert's pistols. It is not simply, as Leonora believes, that Tasso's poetic imagination takes over his life and deludes him into seeing around him the intrigues and deceptions of which he writes in his epic and which are no part of ordinary experience. There is a deeper madness which corrupts the intellectual powers themselves, the very sources of poetry, when poetry is most profoundly understood, not as a medium of illusion, but as an instrument for the apprehension of truth. The fading grasp of organizing reason on the wealth of sensory detail Tasso's imagination presents is horribly apparent in his final dialogue—or rather, accompanied monologue—with the Princess. The essence of the debility is an ever-narrower focus on the immediate datum of consciousness, which in the evocation of the caretaker's tasks on the Princess's country estate is even detached from the narrative sequence that gives a vestigial appearance of order to Tasso's account of his imaginary journey to Naples. This restriction of Tasso's attention to his single central consciousness has been becoming ever more pronounced since the start of the play. A presentation of the real world as capable of perfection, of transformation into a Golden Age, is necessarily, if silently, always accompanied by reference to the heart that yearns for that perfection. (That is why the poetry of desire is always at least implicitly autobiographical.) Like Werther in his last stages of collapse Tasso's mind at the last races away in detachment from its objects. In the fifth act he slips rapidly from deliberate deceit, through mental disaggregation, to the abstractions of his final address to the Princess—before his attempt to assault her—in which reality has been reduced to a ragged set of reiterated pronouns:

> Nichts gehöret mir
> Von meinem ganzen Ich mir künftig an
>
> .   .   .   .
>
> Du hast mich ganz auf ewig dir gewonnen,
> So nimm denn auch mein ganzes Wesen hin.

In future nothing of my whole me belongs to me . . . you have won me for yourself for ever, so take then my whole being too.

With these words the poetry of desire reaches its empty culmination, the only one that circumstance allows.

One last metamorphosis of Tasso's poetry is still possible, however, for as long as total insanity is held at bay. Madness is the worst pain possible for a poet, for it deranges his ability to lament or even to express his sufferings. Goethe leaves it an open question whether the last words of his play are to be understood as the last sane words Tasso is capable of uttering and whether the silence that follows is that of spiritual night, though much speaks for that view, not least the historical Tasso's career and posthumous reputation. But before the shadows close in, Tasso, supported by Antonio, recovers enough composure, in the coda that is appended to the play's catastrophe, to be a poet, as he understands the term, for a little longer. Antonio's firm and compassionate advice enables him to come to his senses, recognize that he is still a human personality, and that he still, as a poet, has a unique vocation. He is not—or not yet—annihilated, but he is surrounded by a dead unspeaking world which he no longer has the power to enliven with his desire into a potential Golden Age. The separation of subject and object is total. But pure personality, unmodelled on any classical or other literary predecessors, may still find a voice in poetry: his suffering may still provide the material of song, to which other, prosaic, men may not rise. His autobiographical art can, briefly at least, continue: no longer as a poetry of desire for his Princess and the Golden Age, for they are definitively and tragically denied him, but as a lament for them, a poetry of loss, 'the tear . . . the cry of pain', the voice of tormented nature. And though Antonio, who is precisely not the man of words, can at this point say nothing, he can make the symbolic gesture of friendship and solidarity, for at last his attitude and Tasso's coincide. No longer for Tasso is poetry directed in desire towards a moment of real fulfilment, instead poetry in so far as it speaks of fulfilment speaks of it as lost and unreal: the happy moment passes, never to return, Tasso's 'present' can never be restored to him, and in the knowledge of the irreversible flux of time Tasso is now at one with Antonio, the man of the moment, whom every moment frees for new existence.

Torquato Tasso is subtitled *A Drama*. At a time when it was practically finished, however, Goethe referred to it in correspondence as 'a tragedy', and the comment on the nearly completed manuscript by Carl Philipp Moritz, with whom Goethe had discussed the play intensively at a time when the last act was already in final draft, also deserves to be given full weight. It uses the terms Moritz had applied to the conclusions of classical tragedies and also, probably, of *Egmont*, in *On the Plastic Imitation of the Beautiful*:

I see a point at which what is most painful and oppressive in human circumstances perfects itself into the most benign Appearance . . . Tasso is . . . the supremely spiritual, the most delicate humanity, oppressed by even the most soft and gentle environment, approaching its dissolution; having lost the centre of gravity which attached it to reality and so only able to reach its true perfection in Appearance. The tragic depiction of this gentle and spiritual element at the moment when it detaches itself in lamentation and sinks in upon itself is certainly the supreme point of poetry . . .

Moritz clearly sees Tasso's final speeches as the last stages of a tragic introversion, the last ripples of an alienated spirit subsiding into non-being. We too should at least attempt to see the play in this tragic light and not assume that Tasso's claim to the divine power to say how he is suffering brings about a last-minute redemption of the catastrophe. The one suffering to which that divine power is certainly inapplicable is the suffering of true madness (rather than the feigned madness of poetic tradition'), and in 1789 Goethe certainly did not think of poetry as a purely personal cry of pain or as the voice of nature. The discovery that the poetry of desire could lead a new and metamorphosed existence as the poetry of loss was of immense importance for his future development, but it was not a discovery that in 1789 he was in a position to exploit. At this time he thought that 'manner' had to give way to 'style' and that what had to take the place of the poetry which in Tasso had compassed its own destruction was not a tear, or an animal cry, but Art. Moritz thought that the final impression left by *Torquato Tasso* was that of a perfected Appearance, and by 'Appearance' he meant something like 'self-conscious art'. Tasso's personality, his personal suffering, and the personal art into which his suffering leads him, are all absorbed into a final, harmonious, symbolic tableau, beautiful in itself, of which personality is only a subordinate component. More is at issue, in fact, than the concluding ambiguous image of Tasso clasping Antonio, the mariner clasping the rock. The Art which for Goethe has displaced Tasso's autobiographical poetry of desire, including its last and most attenuated phase as the poetry of loss, is the Art demonstrated in the 'composition', as he dared to call it, of the whole play, an Art which is sovereign and impersonal, though it may well draw upon personal data as its material. Tasso's final moment of illumination draws our attention to the role of art in transfiguring suffering such as that in which he founders, but it is not an art that he knows or can envisage which performs this role for him. The poem taken from him and given to the world by Alfonso he disowns—no limited and concrete 'work', as the Princess called it, can satisfy the demand for completeness made by poetry as he understands it. Tasso's poetry is spun out of him as the cocoon is spun out of the silkworm of which it will eventually be the grave, and in succumbing to madness Tasso is the victim of his own poetry. Tasso does not know what brilliant creature will emerge from the grave in which his own art has destroyed itself, but we do. The butterfly which Tasso becomes is a play called *Torquato Tasso*, and the art which takes the place of his is Goethe's Art, which is responsible for the poetry of Tasso and the prose of Antonio, the anti-poetry of the Princess and the melodramatic calculations of Leonora, just as it has put those figures together from the colours Weimar laid out on its palette. No character in the play is to be identified with any character in Goethe's life, though features of individual lives may have been rearranged to create new individuals within the work of art. No emotions within the play, not even the emotions of a poet driven to distraction by the court to which he has committted his life, can be assumed without further ado or qualification to belong in Weimar to the contexts in which they are set in Ferrara, for unlike Tasso Goethe does have a literary predecessor on whom to model and by whom to express his sufferings, namely Torquato Tasso. There is much paradox and self-deception here on Goethe's part, and later in his life he was able to recognize that the play was in reality as confessional as anything else in the *Works* from the period it was supposedly terminating. He had in fact used the excuse of Art to provide himself with a reserved area within which his older manner could undergo what would eventually prove its most important transformation. But Goethe could be very obstinate in a bad cause and it was not emotional confessions, but *Tasso's* partly operatic, partly French-classical form, its stylized dialogue, regular verse and stately visual effects, and even its freedom from misprints and the fine roman type in which he hoped eventually to see it appear, that provided Goethe with his model of 'Art' in the new epoch of writing that began for him as he sent off the last sheets of his manuscript to Göschen on the day before his fortieth birthday.

After the carefully composed volumes vi and viii the seventh volume of the *Works*, the only one still outstanding by 28 August, was disposed of with something very close to haste. However, unlike *Jery and Bätely* and *Jest, Craft and Vengeance, Faust* still required revision before it could appear, even as a fragment. While working on *Tasso*, Goethe had evidently continued to think about the new plan for *Faust* that he had conceived in Rome. Antonio's reflections on Tasso's unsuitability to enjoy 'the goods of this life' may echo Mephistopheles' monologue, probably written in Italy, in which Faust is characterized by a boundless desire that 'overleaps the joys of earth', and the parallel between the two pairs of characters is sufficiently close for it to be likely that through moulding the one Goethe was preparing his understanding of the other. There are even moments in Antonio's complaints to Alfonso about

Tasso's irrational and self-indulgent behaviour, and in the Duke's tolerant responses, which sound like anticipations of the discussion between Mephistopheles and the Lord in the 'Prologue in Heaven' prefaced many years later to *Faust: Part One*. Prologues in hell already existed in the puppet-plays about Faust, and in the attempts at Faust-plays by literary men, and in March 1789 prologues were on Goethe's mind—he described Act 1 Scene 1 of *Tasso* as a 'prologue' to the whole work—so he may already have been thinking of setting his *Faust* in a semi-comical cosmological framework (a Lucretian atheist could of course write nothing else), and noting down pieces of dialogue towards it. If so, he had passed a milestone in the work's composition, for he had begun to consider the metaphysical terms to which Faust's relation with Mephistopheles could be reduced, so both sketching a definition of the play's philosophical scope, with more at issue than the mere sequence of experiences envisaged in Italy as making up Faust's career, and taking the first steps towards a more formal wording of Faust's agreement with the devil than anything he had so far attempted. His task in preparing his still somewhat provisional drafts for publication was, as well as polishing up their language to the standard of the other volumes in the edition, to produce from them as coherent a whole as he could, while leaving it open to future development and not committing himself to a theme or to episodes on which he had not yet finally decided. In the event, the only substantial body of finished manuscript material he witheld from publication were the last three scenes of the *Urfaust*, all mainly in prose, which show Faust learning of Gretchen's crime and condemnation, and riding to release her from prison, only to be rejected by her and apparently claimed by Mephistopheles. The scenes were too conclusive, and the emotions they contained too tragic, to be evidently compatible with the new scheme of a Faust bent on accumulating the high points of human experience, and either Goethe was quite uncertain how they could be included in a final version or he was considering suppressing them altogether. Perhaps also they seemed too close to the Christian moral perspective of the original story: for some similar reason Goethe at this point deleted the little scene 'High Road' ('Landstraße'), in which Mephistopheles averts his gaze from a wayside Calvary. Such indirectly suggestive indications of Mephistopheles' diabolical nature were out of place in a drama which now contained the gross and unserious Satanistic japes of 'Witch's Kitchen' (a scene which, though begun in Rome, may have been completed in Weimar).

With no defined terminus to the action, and no commitment to the role or dimensions of the Gretchen story in the final version of the play, it was now open to Goethe to reconsider the guiding principle that was to structure Faust's career. Possibly he had already felt in Italy that the loose association with Mephistopheles, inaugurated in the dialogue he had written in March 1788, was insufficiently compelling as a central theme, even when accompanied by the question: would Faust remain true to his high ambition or would he succumb to mere sensuality? In *Tasso* he had now written a drama of nearly 3,500 lines in which the question how life was to be lived after a moment of complete fulfilment was posed in the first act, and in which the two central male characters were, *inter alia*, differentiated through the ability of one to see every moment in human life as instinct with the possibility of perfection, and the ability of the other to exploit each moment for what it offers and to pass uncomplaining, and perhaps unthinking, to the next. The nature of the fulfilled moment and of the passage of time, and the shape to be given to human life as a whole, were all issues preoccupying Goethe in 1788–9. When he began his last spell of concentrated work on *Faust*, in October 1789, Goethe was clear that an essential feature of the play was to be, not merely Faust's craving for the 'crowns of humanity', but his expression of dissatisfaction with them or their substitutes, in the form in which Mephistopheles would offer them in each, possibly momentary, episode. Only thus can we explain why Goethe, in recasting the scene 'Auerbach's Cellar' from prose into verse, to match all the other scenes he was publishing, should have reduced Faust to the role of a silent, passive, and indeed bored, spectator, while Mephistopheles conducts the conversation with the drinkers and performs the conjuring tricks with the wine which in the original version were reserved to Faust. It is now evidently Mephistopheles' role to present Faust with discrete experiences, and Faust's to pass judgement on them, probably negative. They will make up a 'course', we are told in lines almost certainly written in 1788–9, a 'new career' in which Faust will pass first through low life (the Gretchen story no doubt) and then high life (perhaps scenes at court and with Helena, yet to be written). This is a completely new dynamic for the play, displacing the moral testing of Faust the emancipated genius, which sustains the *Urfaust*. Since it is in principle capable of indefinite extension to any number of episodes it is well suited to the encyclopaedic scheme for the plot on which Goethe had resolved in 1788 but which in 1789 he saw no prospect of completing immediately. Furthermore, it is possible that when he adopted this new pattern for his play's development he already thought of its natural dramatic embodiment as being a wager, or at any rate a promise by Faust not to rest in his pell-mell pursuit of whatever passing objects of desire Mephistopheles put before him. In the scene 'Forest. Cavern', ('Wald und Höhle') which makes its first appearance in volume vii of the *Works*, Mephistopheles replies to Faust's complaint that by telling him to move on to 'something new' he is interrupting a moment of peaceful contemplation, with the words:

> Nun, nun! ich lass' dich gerne ruhn,
> Du darfst mir's nicht im Ernste sagen.

Well now, I am happy to let you rest, you do not have to be [or possibly: you should take care not to be; or even: our agreement does not allow you to be] too serious in telling me about it.

However we interpret the second line, the implication is clear that if Faust does 'seriously' ask for rest when Mephistopheles offers him novelty the devil will get his reward. A conditional agreement of this kind is very close to one at least of the formulae by which Goethe, when he was completing *Faust: Part One*, eventually decided his partners should bind themselves.

The scene 'Forest. Cavern' was almost certainly the last major new addition to Goethe's manuscript before he set one of the clerks in the Weimar Chancery to write out a fair copy of *Faust: A Fragment* ready for despatch to Göschen. The work on 'Auerbach's Cellar', important though its implications are, was essentially editorial, as was the revision of the satirical scene between Mephistopheles and the student—it was abbreviated, its internal organization was clarified, and a few weighty lines linked it with the rest of the action—and the stylistic grooming of those Gretchen scenes which were to be published (everything as far as the moment in 'Cathedral' when Gretchen swoons, calling for her neighbour's smelling-salts). But 'Forest. Cavern' involved a significant new structural decision: on the placing and function of the great blank-verse monologue written in Italy, in which Faust gave thanks to the Spirit of Earth for bestowing on him all he had asked for. It had perhaps never been clear, from the time of its conception, where this hymn of gratitude belonged, with its verbal echoes suggesting fulfilment of the yearnings expressed in Faust's very first speeches. In Rome Goethe would seem to have thought of it as lying altogether outside that part of Faust's career covered by what he had so far written of the play. In October 1789 he took two steps in order to tie it into the dramatic argument of his *Fragment*, and made it the occasion for expounding what he now saw as the play's principal tension. He associated Faust's hymn of fulfilment with his sexual conquest of Gretchen by placing the monologue after the scene 'At the Well', when Faust and Gretchen have already spent the night together and Gretchen is beginning to feel qualms at having yielded to a love which she none the less believes to be essentially good. The assumption that all forms of ambition, desire, and the quest for meaning found their terminus in sexual union, and nowhere else, was one which underlay the decorous frustrations of *Tasso* no less than the plain talking of the *Erotica.* Second, however, Goethe also arranged for the hymn to be followed by an immediate reversal of mood and plot, so that the fulfilment of which it speaks becomes only momentary. He decided to build into one scene both the monologue and a section of dialogue from the *Urfaust*, in which Faust acknowledges that he is bound to bring ruin on Gretchen and in a mood of savage and self-hating despair calls on Mephistopheles to assist him in a last brief and devilish venture. It was a considerable challenge Goethe set to his own ingenuity to invent a transition between these two so nearly opposite emotional states and perhaps he did not wholly succeed. He unsettled the apparent calm of the monologue by adding the second blank-verse section in which Mephistopheles is blamed for Faust's awareness that no human being experiences perfection, so preparing us for the interruption of this idyll too. In this second half of the monologue, Mephistopheles' role of denying value to all things is quickly sketched; we are told that he is filling Faust with desire for the image that Faust saw in the magic mirror in the Witch's Kitchen—a suggestion of a feminine ideal lying beyond and behind the figure of Gretchen; and Faust's own new career, from high point to high point of experience, is characterized in the words:

> So tauml' ich von Begierde zu Genuß
> Und im Genuß verschmacht' ich nach Begierde.

So I stagger from desire to enjoyment and in enjoyment I pant for desire.

The lines are densely functional, a programme in a nutshell for the whole play, but they prepare admirably Mephistopheles' entrance and Faust's ill-tempered contributions to the dialogue that follows. Mephistopheles defends his view that there is nothing to human motivation beyond sexual desire, not even in so apparently ethereal a matter as the contemplation of nature, and proceeds to an experimental demonstration by whipping up again Faust's previously quiescent lust for Gretchen. A whispered aside reveals to us that it is by this means that he intends to entrap Faust, and the concluding section from the *Urfaust*, into which we are now transferred, seems to suggest that he is likely to succeed. Dramatically speaking, it is not obvious what more Faust can do than he has already done to bring about Gretchen's ruin, though the masculine attitudiniz- ing which treats this wilful crime as only a passing episode in a great man's career fits well with the sensualist notions that prevail in the scenes Goethe wrote in Rome in 1788. Conceptually, however, and regarded as a self- interpretation by Faust the new scene is subtly different from the Roman dialogue between Faust and Mephistopheles, which in turn is of course very different from Faust's self-interpretations in the *Urfaust*, notably in the amoral opening monologue and the remorseful scene 'Dark Day. Countryside', not included in the *Fragment*. 'Forest. Cavern' seems in fact to be inclining a little more in the direction of the *Urfaust* conception than the material of 1788: not only the incorporation of Faust's self-lacerating speech from the *Urfaust* suggests this—as he himself approached domesticity in November 1789 Goethe perhaps felt less indulgent towards Faust's crime than he had done as an unattached bachelor the previous year—but also the hint of a new conditional agreement, a proto-wager, with the Devil. Such an agreement would imply that Faust's career in partnership with Mephistopheles is to be seen no longer as an attempt to fulfil humanity's proper and highest ambitions but as an unnatural exertion of the will which can be maintained at fever-pitch only by some artificial stimulus. The coexistence of several different layers in the play, each with a more or less different account of the motives of Faust and Mephistopheles, and the relations between them, gives the *Fragment* a structure not unlike Goethe's life and works as a whole: a series of overlapping but partially independent interpretations of a central, and possibly changing, mystery. It is in this closeness of relationship between the writer and the play, rather than between the writer and the play's principal character, that the special status of *Faust* among Goethe's works is grounded. While the *Urfaust* layer of the *Fragment,* and particularly the character of Gretchen, made for its popularity, once volume vii had appeared, with the female and more literary public, the new generation of more philosophically inclined young men were particularly fascinated by the more recent scenes and by the motifs they contained of the conflict between the spiritual and the sensual, and of the ambition to embrace all human experience. However, although these themes touched a nerve of the age, and although they successfully wrested Goethe's play away from its original design and opened it up, potentially, to much new material, they seem not to have captivated or inspired Goethe's imagination, and he made no further progress with *Faust* for another eight years. Perhaps it is difficult to have tragic sympathy with someone languishing for a little desire, to alleviate the boredom of enjoyment.

Faust was not the only fragment to concern Goethe in the autumn of 1789. When he disavowed any intention of repeating so elaborate a 'composition' as *Tasso* he added, 'I feel easier with the fragmentary manner of erotic jests'. Yet the *Erotica,* particularly those poems which became the *Roman Elegies,* resemble *Tasso* in that a strict and artificial external form, and an outer shell of cultural and historical allusion, enclose, and even foster, a highly personal and confessional content. This is absent from the contemporary work on *Faust,* where the external constraints are lacking, and perhaps it was the belief in his devotion to impersonal Art that kept Goethe free at this time to develop, in *Tasso* and the *Elegies,* elements of his poetry that he would not be able to acknowledge, to himself or to others, for many years to come. The *Elegies* were not made into a separate collection until the winter of 1790–1 and, with one exception, were not published until 1795, but they are an essential component of any full picture of Goethe's literary production in the year in which so much in his life and work that had hitherto been fragmentary was reduced to an appearance of order and at least temporary completion.

The *Roman Elegies* are sometimes called Goethe's first poem-cycle, although that title properly belongs to the First and Second Collections of miscellaneous poems in volume viii of his *Works,* or to some of the subsidiary groupings within them, and they were neither composed as a sequence nor, when they were eventually arranged, did Goethe achieve a clearly significant thematic or narrative order apart from the appropriate placing of three or four poems at the beginning and end of the series. None the less they were written by Goethe in the midst of various attempts between 1788 and 1791 to group his smaller poems into larger units: the older ones into volume viii, the newer ones, mainly, into the *Erotica,* from which several subsequent selections were made, one of them being the *Elegies.* The concern from the start for the literary context into which a newly-written poem might be inserted is only one aspect of a self-consciousness which is among the *Elegies'* most striking features, if we compare them with Goethe's earlier lyrical work. Repeatedly we find the poem, the poet, the process of writing, or the poetic form of the elegiac distich being referred to in the poem itself: the fifteenth elegy is actually about the time which it takes the poet to write it, and which he has to see pass before the hour of his appointment with the beloved comes. Such manifest self-reference is no part of 'My heart pounded', 'On the Lake', 'Elf King', or even 'Dedication', though we have seen it make a first appearance in the final version of 'To the Moon', and it has a decisive role in the last scene of *Tasso*. It is a more fundamental feature of the *Elegies* even than their classical heritage, for whether it is a matter of verse-form, literary reminiscence, or mythological allusion, that heritage is always present not as an authority to be obeyed or imitated but as the material of a conscious analogy with the present:

> Und so gleichen wir euch, o römische Sieger! . . .
> Amor schüret die Lamp' indes und denket der Zeiten,
> Da er den nämlichen Dienst seinen Triumvirn getan.

And so we are like you, o Roman victors! . . . Eros meanwhile tends the lamp and thinks back to the times when he performed the same service for his triumvirate [i.e. Catullus, Tibullus and Propertius].

Contrary to what is often asserted, the *Elegies* are quite as much poems about thinking as they are poems about sensual experience (Goethe writes his eighth elegy, not about Faustina as a child, but about his—and Faustina's—thinking about Faustina as a child). The making of parallels with a past culture, known largely through books, quite as much as the explicit telling of 'old stories' (Elegy XIX) or reference to the future publication of the *Elegies* as the story of this modern love-affair (Elegy XX), is an acknowledgement that others have thought and been poets before us, that experience comes to us not pure, but already shot through with literary and other pre-formed symbolic patterns, and that the poet cannot, and perhaps should not attempt to create *de novo* and from his own resources the 'symbolic' component of his existence. The *Elegies* thus resemble *Tasso* in making past literature a part of of their subject-matter, to an extent not previously seen in Goethe's work (in *Werther* and *The Triumph of Sensibility* the literary presence was essentially contemporary and satirical), and they anticipate future attempts by Goethe and by later poets from Browning to Pound to make use of, often ironical, effects of similarity and contrast with material drawn from a past or otherwise alien literary context. (Goethe would eventually have particular recourse to the literatures not only of Rome and Greece, but of England, Persia, and, if to a lesser extent, India and China.) It was a recovery of poetry after years of sterility and the crisis marked by *Tasso*, but it was not a return to those simple moments of direct contact between the self and the object of experience that create the unique moods of his earlier poems up to the second 'Night Song of the Wanderer'.

For the first time in Goethe's verse the experiencing self in the *Elegies* is regularly identified as a poet. No longer simply an 'I', nor the metaphorical artist or sculptor who had stood in for the writer in the more introspective works from the *Prometheus* drama to *Artist's Apotheosis*, the narrator is specifically a 'poet' ('Dichter'), indeed in the first version of Elegy II specifically the author of *Werther*, and the professional business of writing—metre, publication, writer's block—is at issue in Elegies V, XIII, and XX. This partial objectification of the self into a character within the poems contributes to a tone and effect quite new in Goethe's verse writing, though there were signs of it in *Wilhelm Meister's Theatrical Mission*: a humorous irony at the expense, not of a confident, self-regarding central 'I', for that feature of, for example, *Satyros* or *The Birds* is now obsolete—but of the poetic form itself, an awareness of its artificiality and its limitations.

> Einem Dichter zuliebe verkürze die herrlichen Stunden,
>   Die mit begierigem Blick selig der Maler genießt

For the sake of a poet abbreviate the splendid hours that with greedy gaze the blissful painter enjoys

is the prayer of the poet waiting for the sun to set and the time of his assignation to come, and at the end of this elegy the time ('die Länge . . . dieser Weile') it has taken to compose the brief history of Rome which makes up its second half is, by a pun, labelled mere tedium ('die Langeweile') separating him from the beloved, to whom, once we reach the final lines, it is at last time for him to set off. Writing poems is very far from being the same thing as making love, and there seems to be a deliberate contrast between the learned ballast of some of the elegies (notably III and XII) and the simple pleasures in among the bushes to which they are supposedly intended to invite the quite uneducated woman to whom they are addressed. A similarly ironical change of tone concludes Elegy VII, when the voice of Jove is heard calling the poet down to earth from the fancy in which he has identified himself with Tantalus, walking with the gods on the summit of Olympus, and deliberate ambiguities of meaning and effect draw to our attention that these poems are pondered constructions of words, not spontaneous communications of thought or feeling. Double entendre is a natural humorous resource of erotic poetry and there can be little doubt about the concealed point of a poem (Elegy XI) which concludes with the suggestion that a statue of Priapus, the ithyphallic god of gardens, should 'stand' in the poet's studio beside the images of his parent, Bacchus and Venus. A much more complex effect is achieved in the thirteenth elegy when the poet, after complaining that the dawn hours, which used to bring inspiration for his writing, are now otherwise occupied in bed, exclaims

> Welch ein freudig Erwachen, erhieltet ihr, ruhige Stunden,
>   Mir das Denkmal der Lust, die in den Schlaf uns gewiegt!—

What a joyful awakening, you quiet hours, if you have maintained [or: if you were to receive] the monument to the pleasure that rocked us to sleep!—

Perhaps Goethe is praying for the Muse to bestow on him the gift of a literary monument to love, another *eroticon* like the (suppressed) elegy in praise of linen shifts; perhaps also, in the circumstances, his mind is on an obelisk of a different sort. The duplicity of the wording reflects the dual role of the speaker, as poet and as lover, which is the poem's theme.

Self-consciousness, ambivalence, irony, literary indirection, these are the features of the *Elegies* that are truly novel in Goethe's writing, rather than the 'limitation . . . to a sensually graspable present' of which the critics prefer to speak. On the contrary, if there is a major theme in the *Elegies*, beside that of physical love, which marks them out as contemporary with *Tasso, The Roman Carnival,* and the reflective additions to *Faust,* it is that of time, its mysterious nature and passage. Near the centre of the collection we find in the ninth elegy a superb evocation of fulfilment in flux, a fusion of the excitement of sensual love with the untroubled resting in a timeless moment that overcame Goethe almost by accident when he was working on his fragment of *Nausicaa.* The whole poem, written throughout, until its final reflective lines, in a present tense that eliminates the distinction between present and future (and even future perfect), is devoted to the single moment of 'uprushing flame' (D. Luke's translation) which never ceases through the whole course of the poem. Throughout the poem, too, the flame is both the open wood fire which is needed on autumnal nights in a country that is spared the German stove (and which was a treasured feature of Goethe's cottage) and also an image of the upsurge of physical love:

> Herbstlich leuchtet die Flamme vom ländlich geselligen Herde,
>> Knistert und glänzet, wie rasch! sausend vom Reisig empor.
> Diesen Abend erfreut sie mich mehr; denn eh' noch zur Kohle
>> Sich das Bündel verzehrt, unter die Asche sich neigt,
> Kommt mein liebliches Mädchen. Dann flammen Reisig und Scheite,
>> Und die erwärmete Nacht wird uns ein glänzendes Fest.
> Morgen frühe geschäftig verläßt sie das Lager der Liebe,
>> Weckt aus der Asche behend Flammen aufs neue hervor.
> Denn vor andern verlieh der Schmeichlerin Amor die Gabe,
>> Freude zu wecken, die kaum still wie zu Asche versank.

The flame glows autumnally from the companionable rustic hearth, crackles and gleams —how quickly!—rushing up from the kindling. This evening I rejoice in it more; for before the faggot is consumed into charcoal and declines beneath the ashes my sweet girl comes [or: is coming]. Then kindling and firewood flare, and the heated night becomes for us an illuminated festival. Tomorrow morning she leaves the couch of love and busies herself swiftly waking flames again from the ashes. For Eros gave to the little flatterer more than to others the gift of awakening joy that had hardly yet sunk back quietly as into ashes.

Although the reflective nature of the conclusion reminds us that we are still with the poet whose sweet girl has not yet come, this love is so confident, so devoid of anxiety and sure of its object, that the 'not yet' is of no significance and does not even need to be expressed. The poet's emotional state does not change, just as the image of uprushing flame does not change, whether his mind is on the present, the future, or the past.

Yet immediately on the timelessly active present of the ninth elegy follows the *carpe diem* of the tenth, which begins playfully but ends in the sombre mood of *The Roman Carnival*:

> Freue dich also, Lebend'ger, der lieberwärmeten Stätte,
> Ehe den fliehenden Fuß schauerlich Lethe dir netzt.

Rejoice then, you who are alive, in the nest warmed by love, before chilling Lethe wets your fleeing foot.

The appearance of merely playing a variation on an ancient literary commonplace allows Goethe to stand briefly in the shadow of a death more final than any that threatened Egmont, whose obsession with the moment was really an attempt to live without an awareness of time at all. If moments are precious in the *Elegies* it is because here poetry is ceaselessly striving, and usually failing, to grasp the transition from 'still not yet' to 'forever past' in which life consists: the goddess invoked in the fourth elegy as the patron of the poems is Opportunity, who can be seized only once. The first elegy speaks of the transfiguring effect of love on the appearance of Rome to the visitor—but the love has still to come, and the transfiguration is 'not yet'. The 'so rapid' moment of opportunity, is, in the third elegy, a recovery for the lovers of a mythical Golden Age of instantaneous fulfilment:

> In der heroischen Zeit, da Götter und Göttinnen liebten,
> Folgte Begierde dem Blick, folgte Genuß der Begier

In the heroic age when gods and goddesses loved, desire followed the glance, and enjoyment followed desire

but by the end of the fourth elegy 'that time is over'. In Elegy VI the liaison is already under threat and Elegy VII, although it begins with the exclamation 'how happy I feel in Rome!', turns into a prayer, associated with some of Goethe's most intimate memories of his last months in the city, that he may not be cast out from paradise but allowed to linger there until death takes him into the underworld through a tomb standing close to the pyramid of Cestius. The moment of love-making is usually outside or on the margins of the poem's scope, anticipated and remembered at once in the seventeenth elegy, frustrated altogether in the sixteenth. The fourteenth elegy appears to deal with the same expectant moment with which the ninth elegy begins, and which there is happily indistinguishable from the moment of fulfilment when the sweet girl comes, but instead it proves to be entirely, impatiently, and interminably caught in the 'not yet'. The poet brusquely orders his servant to light the lamp even though it is not yet dark and the time of the assignation has not come:

Unglückseliger! geh und gehorch'! Mein Mädchen erwart' ich.
　　Tröste mich, Lämpchen, indes, lieblicher Bote der Nacht!

Wretch, go and obey! I am waiting for my girl. Meanwhile console me, little lamp, sweet harbinger of night.

There is no 'meanwhile ' in Elegy IX—there the expressions of time are 'before' ('eh noch') and 'hardly' ('kaum')—but, in the later elegies in particular, it is the 'meanwhile' that predominates. For Egmont too there was no empty 'meanwhile' between moments of fulness, only the continuous onward rush of the sun-horses of Time. The poet of the fifteenth elegy, so much of which is purportedly written to while away the time before he is due to meet his beloved, shows himself aware of an inner contradictoriness of his feelings quite alien to Egmont's remorseless self-consistency, when he prays both that he may long continue to see the sun shine on the great buildings of Rome, and that the evening assignation may quickly come:

Spinne die Parze mir klug langsam den Faden herab.
　　Aber sie eile herbei, die schön bezeichnete Stunde!

May Fate be cunningly slow in spinning out my thread, but may the moment so beautifully indicated to me [namely, by the Roman IV written in spilt wine] hasten to come!

But the period when Goethe wished, and thought it possible, to construct literary images of self-consistent and self-contained personalities such as Iphigenia and Egmont, is over, the ambition shattered in Tasso's experience of the reality of time and limitation, as earlier in the experience of Werther. Poetry now seems something to occupy the time in between fulfilments, the time of contradictory and self-conscious states such as expectation and recollection, nostalgia and irony.

A particularly prominent theme, in fact, in a number of the elegies is a conflict or incompatibility between poetry and fulfilled love. The fifteenth elegy concludes by apologizing to the Muses for treating them so cavalierly but reminds them that, for all their pride, they have always given precedence to Eros, as if they cannot in the end express the feelings and value of the momentary events in which, it is held, human happiness really consists, and between which the territory of poetry lies. The superficial aesthetic to which Goethe consciously subscribed in 1789, and which for a good seven years was to divert him from any further development of what the *Elegies* achieved, held of course that the work of art, particularly the ancient work of art, offered a moment of complete sensual satisfaction. The deceptively ingratiating arguments of that aesthetic are put forward by Eros in the first part of the thirteenth elegy when he approaches the poet with various encouragements to write in an amorous vein: the great works of antiquity, whose ruins lie all about them were inspired by sensual love; the 'school of the Greeks' is still taking pupils and it is time Goethe produced something in that spirit; he has only to fall in love as the classical poets and artists did and their spirit will revive in him —if he imitates the ancients in his life he will find that his art has classical qualities too:

> War das Antike doch neu, da jene Glücklichen lebten!
> Lebe glücklich, und so lebe die Vorzeit in dir!

After all the ancient was modern when those happy ones were alive. Live happily, and so let antiquity live in you.

But, the poet continues, in giving him the subject-matter for classicistic poetry —that is, in giving him a love-affair—Eros prevents him from writing it. He has the material, but no longer has the time, the energy, or the peace of mind. Dawn no longer finds him writing, but watching his sleeping companion and fearing that she will awake, and that her opening eyes will interrupt his 'quiet enjoyment of pure contemplation'. This is precisely what happens in the last line of the poem: 'Blick' ihr ins Auge! Sie wacht!—Ewig nun hält sie dich fest' ('Look into her eyes. She is awake!—for ever she now holds you fast').

The line is too breathless and excited to be interpreted merely as a rueful or humorous acknowledgement that the poet has finally yielded to the domination of women and is no longer able to write another word: it is at once much more erotic than that, and much more poetical. Given that a number of other elegies end in, or on the verge of, a moment of physical love, the last word, 'fast', probably has the erotic overtones of 'firm', like the 'monument of pleasure' referred to some lines earlier. The verb 'to hold fast' ('festhalten') in German also has the literary and representational sense present in the word 'monument', that of 'seizing' or recording a moment or a likeness for the benefit of posterity, and echoes of this sense may be present too. The poem, by ending with a perspective on eternity, itself becomes statuesque: a moment, literally the 'twinkling of an eye' ('Augenblick'), is eternalized in the literary equivalent of marble. The very fact that the poet, distracted by passion, can no longer continue to address us, is used, with great ingenuity, to invoke, and perhaps to claim for the poem, the permanence of that art from which the poet says love is excluding him. There is a strong resemblance to the last scene of *Tasso* and to the manifestation there of an impersonal art through the depiction of the limits to the poetry of personal experience. The last words suggest at one and the same time an eternal captivation of the man by the woman, an eternal embrace and an eternal sexual arousal, and also a freezing, as by Medusa's stare, into a monument of love, *aere perennius,* for future ages to admire. Cleverly though Goethe has made this poem out of his inability to write a poem, the achievement is only possible because the poem breaks off where it does, at the point where love and life take over from poetry, where the woman herself becomes active, and where we as readers, standing on the threshold of the relationship between the lovers, are politely refused admission to their secret. Fame and love are long-standing enemies, the nineteenth elegy tells us, and one implication of that is the fear, expressed long before in *The Artist's Earthly Pilgrimage,* and now in the thirteenth elegy, and perhaps in *Tasso,* that the poet who achieves sexual fulfilment can lay down his laurels. But another implication is that love deserves protection from ill-fame—not merely from gossip, but from the publicity attendant on the celebration of love in poetry. The traveller of the second elegy finds in love a refuge from newspaper-fed chatter, whether political or literary, the twelfth elegy rejoices in a private satisfaction cut off from 'the world', and the eighteenth pleads with the public of Rome, the 'Quirites', to allow the lovers to enjoy undisturbed the security of the bed where they lie listening to wind and rain beating on the windows. Is there not then a paradox in the poet's broadcasting this private happiness in verse?

Indeed there is, says the twentieth, and concluding, elegy: taciturnity, the ability to keep secrets, is one of the characteristics of a mature man, yet when the heart is full the lips will speak, and the poet finds himself longing to tell the world of his 'fair secret', like the barber who longed to tell of his discovery that King Midas had ass's ears, though he feared to break his oath of silence. (The unflattering comparison reflects the literary convention that falling in love is a comical weakness—bringing even Hercules, as the previous elegy reminded us, to wear women's clothes and wield the distaff.) The barber's solution was to tell the reeds by the river, which themselves whispered the story to the world, and the poet too spreads abroad the story of his love. But like the barber he does so indirectly. He does not directly confess his feelings to his friends, not even through the medium Goethe had made so peculiarly his own, that of the symbolism of nature:

> Mein Entzücken dem Hain, dem schallenden Felsen zu sagen,
> Bin ich endlich nicht jung, bin ich nicht einsam genug.

And, finally, for telling my delight to the grove and to the reverberant rock I am not young enough and not lonely enough either.

That period of his youth, a period when his poetry and his life were equally full of symbols of his own desires, is over: he is no longer a solitary Narcissus. His life now contains someone else, someone not reducible to his desires and their symbolic representation, and in the poetry into which the heart of a mature man overflows there will always be a 'secret' (the word occurs five times in Elegy xx), an awareness that the poem leaves some things unsaid, and that what it does say is subject to certain limitations, to which it may well point in irony, humour, or paradox. To the new love, and the new awareness of her otherness, belongs a new, more formal, and more self-conscious manner:

> Dir, Hexameter, dir, Pentameter, sei es vertrauet,
> Wie sie des Tags mich erfreut, wie sie des Nachts mich beglückt.

To you, hexameter, to you, pentameter, be it confided how she makes me joyful by day, how she makes me happy by night.

The *Elegies*, with their classical metres and allusions and deliberate stylizations, are art, not nature, a literary game, not confessional poetry, and yet they serve in the end a confessional purpose; like the reeds by the river they are the medium of indirection and speak the barber's words, if not with the barber's voice. They contain the beloved, but do not depict her, speak of her, but do not tell us about her:

> Zaudre, Luna, sie kommt! damit sie der Nachbar nicht sehe;
> Rausche, Lüftchen im Laub! niemand vernehme den Tritt.

She is coming, O moon, delay your appearance, so that my neighbour does not see her; rustle, breezes, in the leaves, let no one hear her step.

True, a mature poet's verses take his life and feelings to the public, but what they take contains a secret, something left unspoken in the very moment of its revelation; and that is because, unlike what he may have said in his youth, the 'fair secret' no longer concerns solely the poet himself, but is a joint possession of him and someone else who is not publishing her feelings to the world. The last lines of the twentieth elegy, and so eventually of the whole collection, are:

> Und ihr, wachset und blüht, geliebte Lieder
>
> .   .   .   .
>
> Und entdeckt den Quiriten, wie jene Rohre geschwätzig,
> Eines glücklichen Paars schönes Geheimnis zuletzt.

And you, beloved songs, grow and bloom . . . and, talkative like those reeds, reveal to the Quirites a happy couple's fair secret at the last.

As it happened, some years were to pass before Goethe found himself in the relation of indirect intimacy with the German public which these lines foreshadow. The *Elegies* are an immediate poetic response, of astonishing energy and novelty, to the most important event in his adult life after his move to Weimar in 1775, the establishment of a permanent relationship with Christiana Vulpius. That event necessitated a complete reordering of his understanding of his life and art. The completion, in these circumstances, of the edition of his *Works*, the monument to what was now a past era—three volumes since 12 July 1788, all of them containing significant revisions and substantial new writing— would have been remarkable enough for any ordinary genius. The capacity for simultaneous renewal and rebirth demonstrated by the *Elegies* is without parallel. However, what was achieved momentarily, and in a 'fragmentary manner', in the crucible of poetry took longer to permeate all the organs and contexts of a uniquely complex existence. The process was also retarded by some more and less contingent obstacles, both external and internal. The Göschen edition was not so successful that of itself it re-established Goethe's relationship with German readers. Permanent though the liaison with Christiana was in Goethe's eyes, it was provisional in the eyes of everyone else, and relations with the Weimar court were to be in a state of strain and confusion for the better part of two decades. The turmoil of the Revolution soon touched

Goethe's personal affairs, and his attitudes, both political and cultural, became more defensive and inflexible as a result. The memory of Italy, the continued intercourse with mediocre artists, and the associated aesthetic theory, remained something of an incubus. Only as an extraordinary surge of intellectual creativity gathered momentum in Germany at large in the middle years of the new decade, under the double impact of political events in France and the, as it seemed, equally revolutionary philosophy of Kant, did the subtle art of the *Roman Elegies* come into its own as the appropriate manner for the new epoch in Goethe's life.

Farewell to Italy: *January–June 1790*

Goethe dined at court almost daily during the first weeks of the new year. Christiana could not feed the child herself and, since he was slow to put on weight, and 'the women' foresaw little progress until the first twelve weeks were over, we may deduce that young August suffered from colic. There may not have been much to attract Goethe at home and he must anyway at the time have felt particularly close to the Duke. Carl August could not officially take notice of his Privy Councillor's illegitimate offspring, but privately he had indicated that he regarded himself as a godparent and, strange though the ménage in the Jägerhaus might be, he was no doubt pleased that his great protégé was at last leading a natural life. Now that Goethe had declared his duty to the Göschen edition discharged—and Carl August must have felt just a little like Alfonso, who could call Tasso's poem 'in a certain sense my own'—he was free, as had been agreed, to take on the role of personal adviser, private secretary, and minister without portfolio, for which he was more suited than for that of executive civil servant. The Duke had a new scheme for the division of responsibilities between himself and the Privy Council, which was given final form in an edict of 6 April. Precisely those areas in which he could expect most assistance from Goethe—foreign affairs, the university of Jena, new building and highways projects and new taxation arrangements with the Estates—were particularly reserved to the Duke's personal decision, while the Privy Council was given greater autonomy in those matters—notably legal administration and Church affairs—in which the Duke (and Goethe) had little interest. Already in January, immediately after the affair of the Hungarian Crown, Goethe was discussing with Carl August the—remote—possibility of Weimar's having a claim to the succession in Lusatia, east of the Elbe, beyond Leipzig, and advising him on the diplomatic mission Prussia wished the Duke to undertake in order to gain Electoral Saxony for a joint war against Austria, currently weakened by its conflict with Turkey, by the continuing turbulence in the Netherlands, and by the illness of Joseph II. When in February Goethe expressed the hope that he might be able to accompany Carl August on the manœuvres planned for later in the year as part of the pressure on Austria he was simply demonstrating that close personal association with his ruler which had become the essence of his official role and which now outweighed his long-standing but still unconcealed hostility to military adventures. 'Bring your affairs to a happy conclusion', he wrote to Carl August after he had left for Berlin on the first stage of his mission on 17 January, 'and bring back to us confirmation of our beloved peace. For since the purpose of war can really only be peace it is very proper for a warrior to make and maintain peace without war.'

The prospects for peace improved dramatically on 20 February when another era came to an end for Goethe with the death of Joseph II. Maria Theresa and her son had between them ruled the Holy Roman Empire since well before Goethe was born. Joseph died a disappointed man who, alarmed by events in France and the disorder in his own territories, had already begun dismantling his earlier reforms, but his successor, his brother Leopold the Archduke of Tuscany, promised as Emperor a continuation of the enlightened but prudent policies Goethe had already admired in Italy. Moreover, there was not yet any indication that France was entering on a period of military expansion. In the first half of 1790 the Revolution was introspectively devoted to consolidating itself, to the federation of local National Guards into a countrywide political structure, and to the elaboration, and dilution, of the proposals for abolishing feudal dues. In May the National Assembly formally renounced all wars of aggression. But it was clear to Goethe where his loyalties lay, clearer than ever perhaps, and this may be one of the meanings of his cryptic remark in a letter to Jacobi at the beginning of March: 'You can well imagine that the French Revolution was a revolution for me too'. When the commanding officer of Marie Antoinette's Swiss Guards, himself a minor German poet, visited Weimar on 8 February, he, like so many, found Goethe 'decorous and frigid' at their introduction, but then noted that at dinner he was 'full of merriment, parodied the tone of the members of the National Assembly —defended sophisms wittily, Germany warmly'. Evidently Goethe had already decided that the old Imperial German constitution, even with the artificialities of its courts, was a surer guarantee of such freedom as was practically attainable than the flights of rhetorical and theoretical fancy to be heard across the Rhine, a view in which loyal subjects throughout Germany were increasingly willing to concur. Provided the rulers continued to devote themselves to the well-being of their people, an era of peaceful prosperity could lie ahead—so at any rate Goethe judiciously glossed Carl August's interest in the other matters of Goethe's special concern in January and February. 'For these last few days we have occupied ourselves as seriously with the plans for rebuilding the palace as if we were looking forward to the peaceful reign of Solomon', he wrote at the end of a long visit from Arens. Much of January in fact was spent in discussion with the architect, and in introducing him in Erfurt and Gotha, where some small commissions were forthcoming, though not enough to secure Goethe's long-term goal of persuading his old Roman acquaintance to settle in the Weimar area. This time, however, the Weimar committee made sufficient progress for the rebuilding of the Wilhelmsburg to be officially inaugurated on 9 April.

More surprising, for Goethe, was a new sign of Carl August's interest in the arts of peace: 'That in present circumstances you still wish to concern yourself with the most mechanical of sciences, the German theatre, gives grounds for hope to us worshippers of Pax that this quiet beauty will continue to reign for a while.' The Bellomo troupe's contract was not due to expire until 1792 but Weimar was tiring of them and Carl August was beginning to think that he should see to the re-establishment of a permanent professional court theatre. While in Berlin he discussed the question with Kapellmeister Reichardt; out of the discussions emerged the idea that Weimar might become the home of a company devoted to the highest standards of theatrical and musical art; and Reichardt wrote in excitement to Germany's most respected actor-manager, Schröder in Hamburg, to ask his views. Schröder's response, an open letter to Reichardt, which was circulated also to Goethe and the Duke, has not been preserved, but its general tone of commercial common sense is apparent enough from Goethe's comments to Reichardt:

I knew in advance that he would answer thus, for I know his circumstances. A *German* theatre-manager would be foolish to think otherwise. Our public has no notion of Art . . . The Germans are on average decent worthy people, but they have not the least notion of what constitutes originality, invention, character, unity or execution in a work of art. In a word, that is, they have no taste. On average, of course. The coarser part is taken in by variety and exaggeration, the more educated by a kind of *honnêteté*. Knights, robbers, philanthropists and their grateful beneficiaries, an honest and worthy Third Estate, a perfidious nobility etc. and everywhere a carefully maintained mediocrity, from which at best a few steps are ventured downwards into vulgarity or upwards into nonsense, those have for ten years been the ingredients and the character of our novels and plays. What in these respects my hopes are for your theatre, whoever may be its director, you can imagine.

Goethe's devotion to his state was still in essence a personal loyalty to its prince. As the artist which he now believed himself principally to be he felt quite as isolated, as cut off from any public, or even from any like-minded spirits as his Tasso. He was grateful to Reichardt for his efforts in setting *Claudina*, and now *Erwin and Elmira*, and he was evidently more disappointed by Schröder's reaction than he liked to admit, but it was a disappointment to which he had steeled himself since his return from Italy. The court could perhaps provide the artist with patronage, but it was doubtful whether even the court could provide him with an audience, certainly not for a work of the subtlety of *Tasso*. 'It is not for production', was Knebel's judgement on the proofs of that play, brutal, but correct, for it was not performed until 1807. As wounding in its way was Göschen's rejection of the manuscript of the *Metamorphosis of Plants*, completed in January, with its implication that Goethe was not a name on which he wished to risk any more capital. A local publisher, however, was found in Gotha, willing to bring the treatise out at Easter, and Goethe felt able to tell both Reichardt and Jacobi that he was now embarking on a new career as a natural historian, as if, having completed his eight volumes, he was washing his hands, if not of literature, then at least of the theatre and the general reading public.

Goethe was already planning a second botanical treatise, to complement the first, and an essay 'on the *Form of Animals*' to carry over, or back, into zoology the metamorphic approach, of which a first anticipation had been his work on the intermaxillary bone. But February 1790 brought a quite new development in his scientific interests and one of such overwhelming importance—its consequences would occupy him until, quite literally, his dying day—that it can properly be called a turning-point. The weather was frequently fine during the winter of 1789–90, and from Italy it was reported to be 'heavenly': there was still, after two postponements, an opportunity for him to go and meet Anna Amalia on her slow—and, like his, reluctant—return journey, and in January, in a letter to Einsiedel he dropped a hint, which he knew was almost certain to be taken up, that he was willing to be invited to do so. It therefore seemed likely that in the near future he would be travelling again, and so unable to make serious use of the optical equipment which he had borrowed from Büttner and which he therefore decided to return. But before packing it up he thought he would have a quick look through the prism in order to see again the phenomena of refraction which he had not consciously seen since he was a student:

I was standing at the time in a completely whitewashed room; mindful of the Newtonian theory, I expected, as I put the prism before my eyes, to see the whole white wall bathed in different gradations of colour, to see the light returning from it to the eye broken up into so many coloured lights.

But how astounded I was that, while the white wall seen through the prism remained white, only where something dark impinged on it did a more or less definite colour manifest itself, and that finally the window-mullions appeared most vividly coloured while there was no trace of colouration on the luminous grey sky outside. I did not need to reflect long in order to recognize that a boundary was necessary to produce colours and as if by instinct I said aloud to myself that the Newtonian doctrine was false.

What Goethe saw is perfectly explicable on Newtonian principles. Had the light passing through his prism come from a point-source, as in a camera obscura, and so been effectively a single ray, he would have seen the phenomenon he expected, a ray of white light dispersed into a sheet of coloured bands. When many rays pass through the prism together, as when it is held up to a window, the spectra produced by the dispersal of the individual rays overlap and reconstitute the original white effect. Only when the source is limited or interrupted—by the window-frame or mullions, for example—will the rays along its edge be able to produce a spectrum that is not overlaid by any others and so remains separately visible to the observer. Goethe tells us that he soon came to hear this explanation from 'a nearby physicist'—possibly J. H. Voigt, Wiedeburg's successor as professor of mathematics in Jena—whom he felt he ought to consult in view of his own lack of experience in these matters, but 'whatever objections I made . . . whatever display I made of my experiments and convictions, I heard nothing but his initial credo and had to tolerate being told that experiments in the camera obscura were much more suitable for acquiring a true view of the phenomena'. Goethe felt himself faced once again with professional savants willing to repudiate the evidence of their five senses and sacrifice the living concept of the thing to what people had said about it. He seems not, however, to have noticed that obstinacy, dogmatism, and an invincible adherence to an initial credo were more truly characteristic of his own attitude than of the surprisingly tolerant professionals against whom he gradually began a forty-year long campaign, which can fairly be called obsessive, to establish a non-Newtonian theory of colour. He thought he had seen with his own eyes that the experiment which purported to show that white light could be broken up into constituent colours was a fraud. Having had that experience he could see, and thought others must also see, that it was a patent absurdity to assert that any set of colours, which by their nature are darker than white, could possibly be recombined into something brighter than any of them were to begin with. The falsity of Newton's account of colour was incomparably more evident, if only prejudice were discarded, than human possession of an intermaxillary bone. We shall later be concerned with the intellectual journey on which Goethe now set out, and with the truths that he discovered on his way, but for the present we need concentrate only on the moment of the journey's inception. Once again, as with geology, comparative anatomy, and his botanical insight while walking beside the sea at Naples, the revelation of an instant provided the Ariadne's thread to guide Goethe through a boundless labyrinth of subsequent empirical investigation. But this moment was different. Not simply was it a more extreme case than any of its predecessors: Goethe's insight was achieved on this occasion without any prior study at all; and, in a more fundamental, even abstract, way than before, what was at issue was the evidence of his senses and nothing else. The occasion also had an unusually personal charge: it has more than once been compared to a moment of religious conversion such as was the normal opening to a life in Christ for a member of the Pietist conventicles which Goethe briefly frequented between 1768 and 1770. A recent exegesis, of great brilliance, has shown that the underlying structure of Goethe's unwearying argument for a new chromatics is that of a defence of Arianism—the heretical belief that Christ was not divine—against the tyrannical sophistries of the established Trinitarian Christology. Light is tortured, indeed crucified, with the instruments of the scientists, who, like a churchful of theologians parroting inherited dogmas, endeavour to split up the pure simplicity of divinity into seven colours or three persons or some other magical number in which they would rather put their trust than in what their eyes and their reason tell them. It may have been within a couple of months of

his moment of conversion that Goethe began to make the comparison of Light
suffering at the hands of Newton and his followers with Christ suffering at the
hands of the orthodox:

> 'Alles erklärt sich wohl', so sagt mir ein Schüler, 'aus jenen
>   Theorien, die uns weislich der Meister gelehrt.'
> Habt ihr einmal das Kreuz von Holze tüchtig gezimmert,
>   Paßt ein lebendiger Leib freilich zur Strafe daran.

'Everything is readily explained', a pupil says to me, 'by the theories our master wisely
taught us.' If you have once properly carpentered the wooden cross, a living body can of
course be punished by being made to fit.

The comparison is of profound significance though it may at first seem difficult
to understand why Goethe should wish to make it at this point in his life. Early
1790, however, was the time of a metaphysical crisis for Goethe, a crisis of
identity. The symbolical life which he had led for twenty years as a secular Son
of Man, after casting off the orthodox requirement to live his life in imitation of
the Son of God, was by then at an end. The power of natural desire, on which
he had relied to give purpose to his life and a seductive fascination to his poetry,
had been tamed. The law and the prophets had been fulfilled and the end of it
all was social and intellectual isolation and a crying infant next door. The view
Mephistopheles expresses in 'Forest. Cavern' that human beings are simply
animals, and all their high-flown emotions concealed lust, must have seemed
very plausible. At the same time, naturally enough, Goethe's faith in any sort of
transcendent divine providence and his residual respect for the historical Christ
had finally evaporated and his long-standing suspicion of ecclesiastical
institutions had turned into hatred and contempt. Christiana, after all, was an
outcast, and he an eccentric, because of the absence of a church ceremony, and
his son had been welcomed into the world at a hasty and secretive function
which made of his very existence an embarrassment. Yet if Christendom's great
structures of meaning were an illusion, if even love and the sense of union with
Nature were but ripples on the surface of the animal reproductive process, what
was he? and what was he for? —a question he rejected for natural phenomena,
but could not reject for himself. The tawdry consolations of Art made little
sense in the crowded solitude of Germany, especially since his own art, that of
literature, was dependent on the mass of shared meanings and values embodied
in the language, not of one man but of a whole society:

> Was mit mir das Schicksal gewollt? Es wäre verwegen,
>   Das zu fragen; denn meist will es mit Vielen nicht viel.
> Einen Dichter zu bilden, die Absicht wär' ihm gelungen,
>   Hätte die Sprache sich nicht unüberwindlich gezeigt.

What destiny intended with me? That would be an audacious question for mostly it does
not intend much with many. To make a poet—that intention might have succeeded, had
the language not proved an insuperable obstacle.

Goethe in early 1790 was strongly tempted to see his life as a meaningless failure: perhaps the doctrine of original sin was true, he mused bitterly, and he was himself a botched being.

In the moment of revelation, when Goethe looked through the prism, there were two elements: an awareness of what he, personally and immediately, was seeing; and an awareness of oppostion to a universal orthodoxy, 'the Newtonian doctrine was false'. Here suddenly there was a new church for him to hate, with its popes and inquisitors, its dogmatic theologians and its obstinately simple faithful, the dupes of the hierarchy. Its dogmas related to the material and sensual world, not to the tissue of illusions in which the Christian Church specialized, so to devote one's life to assailing it was not to commit oneself to a battle with phantoms, in which 'no one will come properly to their senses' and 'whichever party [the individual] declares for, the *whole* never comes together as a *whole*'. Here was a battle in which the cause of a worthwhile truth did stand to be won. And of what was this Newtonian scientific church the enemy? What was the heresy it had persecuted as remorselessly as the Christians persecuted the plain and original truth about their own founder, in heretic after heretic recrucifying their own Christ? Nothing other than what Goethe saw before him now, the evidence of *his* own senses, the light flooding pure and unbroken into *his* eyes. The Christ whom the Newtonians recrucified was the pureness and simplicity of white light as it manifests itself to any observer, but above all to this one—the integrity and reliability therefore of the individual's perceptions, of his sense of his own being, and of his intimacy with the world he perceives. Goethe could no longer live a symbolical life, with himself as the reinterpreted Christ-figure at its centre, but he had discovered, in a moment of insight, how to transfer that pattern to the world of scientific activity, with the role of the suffering Son given to the observing consciousness that comes innocent into the world, and to the pure light that enlightens it. If there are paranoid, even faintly psychotic, features of Goethe's later obsession with the theory of colour, it is because that subject became for him a refuge for a providential, religious, understanding of things which he was not able to maintain in the rest of his complex existence. That Goethe's concern with chromatics had its origin in a moment of quasi-religious conversion during a dark night of the soul, and that it subsequently took on, at times, a pathological tinge, does not, however, mean that what he thought and discovered in this area of knowledge is worthless, or reducible to the psychological needs which the activity fulfilled for him. On the contrary, if Goethe's theory of colour is not, as he was provocatively to claim at the end of his life, more important than all his 'poetical' works, it is certainly more important and more original than all of his other scientific works, and the different stages of its development will deserve our consideration quite as much as those of other, more literary, projects which also took many years to come to completion.

In the February and early March of 1790, however, it was not clear that

Goethe had already passed down the road to Damascus—if we reapply the terms of a little poem from these months in which Goethe prays to Christ to come again to engineer human sensuous fulfilment, and proclaims himself ready, in such a mission, to act the part of St Paul. Optics was evidently his new enthusiasm, but at first its demands were limited and predictable: Büttner's equipment was retained and there was an intensive series of experiments, but the dimensions of the project and its capacity for absorbing Goethe's energies and attention were not yet apparent. What could absorb Goethe, however, at this time, and it is a measure of his spiritual desolation that it did so, was sex. Not the physical love, the happiness of which had inspired the *Elegies*, but the impersonal processes that fascinate the schoolboy and the pornographer. To Carl August, whose visit to Berlin had a medical as well as a political purpose, he announced at the start of February his resumption of marital relations, after the birth of August, though the Duke probably found less jolly than Goethe did his further comment that as a humbly monogamous poet, without a prince's obligations to the female sex at large, he could keep his *membrum virile* 'purissimum'. His member, perhaps, but not his mind. For the unmarried Prince August of Gotha he wrote in February a free exposition in Latin of the functions of the various Roman deities supposed, according to St Augustine in the *City of God*, to preside over the various stages of coition, and at the same time, and in the same salacious obscurity of the learned tongue, a number of comments on the text of a seventeenth-century collection of obscene Latin verse, Schoppius' *Priapeia*. He was particularly interested in establishing the correct form of a schematic representation, or hieroglyph, of the male genitalia, to which the *Priapeia* allude, and which consisted, reading from left to right, of a capital letter E, a long bar, and a capital D. This symbol recurs among Goethe's other papers and drawings of early 1790—there even seems to be an attempt to find the same structure in the reproductive organs of plants—and two prominently phallic scenes, complete with symbolic snake and tortoise, were probably drawn at the same time. A contemporary sketch of a monstrously erect garden herm can be associated with two *Erotica* devoted to Priapus, which ask the god to bless the poet's organ and keep it from flagging. It was probably to his scholarly—and experimental—work on the *Priapeia* that Goethe was referring when he told Jacobi that in response to the French Revolution, which had had so powerful an effect on him too, 'I am studying the ancients and following their example as far as that is possible in Thuringia'. Although Goethe had long had, and always retained, a healthy plainness of thought and speech about sexual matters, the exclusively scatological fantasies of early 1790, which continued for some months, were an isolated episode and are evidence at least of an unusual state of mind if not of emotional and imaginative impoverishment. One would hardly guess from his writings, drawings, or letters that he had a son who must just have begun to smile.

That Goethe should return to Italy was a plan at least as old as Anna Amalia's departure without him in August 1788 and he had taken the first steps towards carrying it out in January 1790, by dropping his hint to Einsiedel, the day after he posted off the manuscript of *Faust: A Fragment.* To be sure, so much had changed in the previous eighteen months that his motives now for making the trip cannot have been clear to him. There was, however, no new and tangible reason at the end of February why the plan should be altered; Carl August in Berlin cannot have been surprised to receive Goethe's formal request for leave, written on the twenty-eighth; and Schiller was no doubt accurately reflecting the general opinion when he speculated that once back in Italy Goethe was unlikely to return to Weimar. 'He is yearning once more for his Italy', Knebel wrote, 'and wants to be in Venice by Palm Sunday.' Goethe certainly needed a rebirth, but he must at times have doubted whether Italy could again provide it. Having made his arrangements he was anxious to be off—after Arens's visit there was plenty to occupy his subordinates in the programme for the Wilhelmsburg; he had conducted an expert recommended by Werner through the Ilmenau workings and the advice received on drainage and the access road could be acted on in his absence; some fiscal matters had equally been set in motion and did not require him to stay—but the only pleasure he seems definitely to have promised himself was that of hearing the Holy Week music in the churches of Venice: 'In order myself, as a pagan, to have some profit from the good man's passion I have to hear the singers of the conservatoria and see the Doge in his solemn procession.' Perhaps he intended to leave himself open to the unpredictable—'I have my own way in which I have to live or be utterly miserable' was his justification to Herder for not taking a friend with him, such as Knebel—but this time he was being accompanied by his servant J. G. P. Götze whom he encouraged from the start to keep a diary and to take an interest in the works of art they saw, as if he expected no more than to repeat a journey over familiar territory which could be made interesting only by a pedagogical purpose. Having left Weimar early on 10 March he was on the twelfth no further than Jena, and from there wrote to Herder, to whom he commended Christiana and little August in case of an emergency: 'This time I am leaving home unwillingly and this standstill nearby makes my desire to return even keener.'

The reason for the delay was an urgent commission from Goethe's colleagues on the Privy Council. In Jena on 4 March some Weimar soldiers, under one Sergeant Wachtel, on being taunted by some students about an incident earlier in the day, had failed to follow their orders and arrest the students but had attacked them instead. On learning of the relatively light punishments imposed on the offenders—fourteen days imprisonment for Wachtel, partly on bread and water—the entire student body, and some of the professors, had protested to the military authorities and had demanded, and been promised, more substantial 'satisfaction'. Passions were running high, and loyalties were switching by the hour. The danger to be feared was not so much a general riot as a mass emigration of students to another university, which would have been a disaster for the local economy and could have done permanent damage to Jena's academic future. Goethe's task, as an emissary of the highest authority in the land, in the absence of the Duke, was to investigate and if necessary either confirm the sentences already passed or impose new punishments. He hoped initially to be able to settle the matter without passing the decision back to Duke, but after long consultations in a heated atmosphere, to which old grievances among the professors, and between them and the military, contributed, he had to abandon that hope. The students were pacified by being given three days of a minister's attention but could not be deflected from their purpose of appealing to the Duke, as was their right. Goethe in the end counted himself fortunate to have averted the calamity which threatened at least twice: when the soldiers effectively mutinied, on hearing a rumour they were to be severely punished; and when the students proposed they should set up their own patrols to keep order. To the first crisis Goethe's answer was to ignore it, and indeed it went away; the students' proposal he refused to countenance. If that pass was sold, he thought, mindful of events in France the previous July, Jena could take over from Leipzig the title of 'little Paris'. The whole affair, however, was completely unpolitical and bears out the pattern of innumerable minor disturbances in Germany in 1789 and 1790: the sovereign power was so widely dispersed through Germany's small territories that order was easily kept locally, grievances could be heard personally, and the organs of national representation and publicity were lacking to give symbolic significance to an individual event or to allow many incidents to cumulate into an identifiable general cause. For Goethe himself, though, the complex animosities he was asked to keep in balance were disagreeable: 'if I could ever come to hate this place,' he wrote to Herder, 'it would have to have been during these last few days', and if he was taking advantage of the opportunity to consult with Professor Voigt about Newton's colour theory the results cannot have endeared Jena to him either. He left at a quarter to seven on the morning of the thirteenth, having sent long reports to Fritsch and Carl August, and when in April the Duke returned and terminated the matter by having Wachtel flogged and cashiered Goethe was long since in Italy.

In a chaise put at his disposal by the Duke, Goethe drove continuously for two days and two nights via Coburg and Bamberg to Nuremberg. The fine weather had turned to rain and snow, the city was dirty and cold. Stopping only for a few hours to see some Dürers and buy some books, he resolved not to make a detour to Ansbach to visit Knebel's mother and sister, to whom Knebel had asked him to deliver a letter, but to press on to Augsburg which, after another overnight drive, he reached at 9 o'clock in the evening of 16 March. He clearly wished to put Jena's troubles and any doubts about his project firmly behind him. Although the days of its great prosperity were past and it could not compare as a trade centre with Frankfurt, the free city of Augsburg appealed to Goethe in the way Regensburg had done in 1786: the Holy Roman Empire, Catholic, but tolerant, and German rather than Habsburg, at its apogee in the burgher culture of the fifteenth and sixteenth centuries, was something like his natural element, and he valued it all the more now that it and everything like it seemed to be under question from the rationalist constitutional theoreticians of France. 'I shall stay a few days more in Augsburg for here the sweet odour of freedom greets me, that is, of the greatest constitutional limitation', he scribbled in a fragmentary draft of a letter, probably never sent. His feelings were precipitated by the Requiem Mass for the dead Emperor held in the Cathedral on 18 March:

But the veiled crown on velvet cushions, the arms of the kingdoms and provinces painted on card, the many candles, candelabra, silver and other circumstances gave me in a single moment a more profound feeling of his dignity, his rank, his destiny, his unfortunate arbitrariness and power than any words could have impressed on me. Altogether I feel here once more that there is no better condition than that of a pagan living among Catholics.

Catholicism here, needless to say, stood for more than religion: it stood for the federalist, internationally minded bourgeois Germany of the cities of the south and west, which had given him birth and to which socially and politically he by instinct belonged; the 'paganism' marked his Protestant and individualist inheritance, his need for mental freedom from the society that surrounded him. It is a striking confession of loyalty to a political constellation which was on the point of eclipse and which would not be ascendant in Germany again for 160 years. That ideal 'condition' of a free man in a bourgeois Germany was what Goethe had really sought and found when he had been happy in Italy, and his confessional note shows what really made for the difference between life in Rome and life in Weimar, where he could indeed be a pagan, but not 'among Catholics'. As he wrote it, he must surely have known that he was not on his present journey going to recapitulate even his recent past: his commitment to Weimar, however alien to his origins that place and its political order might be, was now as complete as it was inevitable.

Leaving Augsburg on the nineteenth, the day after the Requiem, the chaise climbed through Füssen into the snow-capped Tyrol and at 2 a.m. on the twenty-first Goethe rejoined his old route of 1786 at Innsbruck. He allowed himself a little longer than on his first visit—a day to inspect the treasures of the Ambras Palace—but on the twenty-second he was on the road again, pausing only for half an hour in the middle of the afternoon on the summit of the Brenner, no longer the mysterious frontier to fulfilment, reaching Bressanone well before midnight, and driving on into the following day to arrive in Trent at 5 o'clock in the afternoon. In the Tyrol only the spurge had been beginning to bloom, but in Trent the fruit-trees were coming out and there were even lizards to be seen. After a night in Rovereto Goethe arrived in Verona on 25 March: the peaches and cherries were in full bloom but there was a cold wind from the north-east and the plains did 'not yet have an Italian look about them'. Had he been prepared to press on he might still have been able to reach Venice by the twenty-eighth, but the attractions of a Christian liturgy were outweighed by the only substantial antiquities his route had to offer, the Veronese amphitheatre and the Museo Lapidario with its moving funerary reliefs, and he stayed on in Verona until Palm Sunday.

Leaving their chaise at the inn, Goethe and Götze hired transport to Vicenza and Padua, where they embarked for Venice, coming out into the lagoon on the afternoon of 31 March, the Wednesday in Holy Week. Lodgings were found on the Rialto, the hotel Einsiedel had recommended proving not to exist, and on Maundy Thursday, despite the cold, Goethe and his manservant watched the Doge and the Papal Nuncio bringing the Easter indulgences in solemn procession with the Senators, clad in black and violet, across St Mark's Square, while the cannon fired a salute. During the night it snowed and the going became more difficult even than usual in that city of water and mud, but the travellers struggled to St Mark's Square both morning and afternoon on Good Friday to see the Confraternity of St Roch bury the image of the Saviour and then process to the cathedral to worship the phial of His still liquid blood. On Easter Saturday things improved a little: the wind dropped at last, the school of the Confraternity was visited, and Götze was introduced to Goethe's favourite Palladian building in the Carità. True, there was as yet no indication from Anna Amalia when she might be expected, but it was not just irritation at the uncertain arrangements, at the weather, or at superstition triumphant, that spoke in the letter Goethe wrote that day to Carl August:

I must by the way confess in confidence that this journey has delivered a fatal blow to my love for Italy. Not that in any sense things have gone badly for me, how could they? but the first bloom of affection and curiosity has fallen off and one way or another I have just become a little more like Smelfungus [in Sterne's *Sentimental Journey*]. And then there is my affection for the *Erotion* [the name of a slave-girl celebrated by Martial] I have left behind, and the little creature in swaddling clothes . . .

If Goethe's love-affair with Italy could come to an end so quickly—within two years of his return to Weimar, and within two weeks of his second crossing of the Brenner—a certain kind of interpretation of it is clearly impossible. It cannot have been a sudden discovery of the qualities of classical art and of the possibility of a sensual sun-drenched modern existence in imitation of the ancient world which between them gave Goethe's life a new direction for ever —or if it was that, then it was little more than a coincidence that it took place in Italy. The truth is that, as we have seen, the realities of contemporary, or even ancient, existence in Italy had little part in the experience Goethe called by that country's name. Even the reality of the surviving ancient artefacts had only a superficial effect on him. Not a single work, not even the middle temple at Paestum, so carried him away that he had to endeavour to recreate it in stone or in verse when he returned to Weimar, or had repeatedly to draw and describe it, as he obsessively redrew the imaginary landscapes of Claude and Hackert. Scarcely a month after he was first back he confessed as much to Heyne, the presiding genius of German classical philology in Göttingen:

I had a strange enough experience and yet at bottom a quite natural one. I myself only recognize it now, after my return, from the letters I wrote from there to my friends and which I am now able to see again.

At the beginning I still had the confidence, and it still gave me pleasure, to observe details, to treat and judge them after my own fashion; however, the further I advanced into things and the more I came to grasp the sheer range of art, the less I dared to say and my last letters are a kind of lapse into silence or, as Herder puts it, dishes in which one misses the food. Only when I have recollected myself will I realize what I have gained . . . What I could offer the public are fragments that signify little and satisfy no one . . . If I were to be setting down a general credo about my observations on ancient and modern art I would say: that it is true one cannot feel enough respect for what remains to us from the ancient and more modern period, but a whole life is necessary to qualify this respect properly, to recognize the peculiar value of each work of art and neither to demand too much of it, as a [merely] human creation, nor on the other hand to be too easily satisfied.

Goethe did not spend the years from 1786 to 1788 in Italy, as he much later admitted when searching for a title-motto for what is now known as *The Italian Journey;* he spent them in Arcadia, in a creation of his mind and heart, his needs and longings, into which as much of the real Italy was mixed as was necessary to convince him that the object of his desires had a place and habitation on this earth. When he wrote to Heyne he had already begun to understand this, to learn that what he had really seen there, like what he could really see anywhere, was the work not of gods but of men. From Carlsbad he had set out with a different expectation, that of gaining access at last, as a latter-day Tantalus, to the dwelling of the gods, to the home of perfection, and the journal he had kept on his way to Rome seemed to him, after only a month back in Weimar, a shameful private secret, 'pudenda . . . very stupid stuff that now stinks in my nostrils'. When, at much the same time, he asked Frau von Stein to let him see whether his letters to her from Italy contained material suitable for publication in the *Teutscher Merkur,* he was clear that 'without such an intention I should not at all have wanted to look at the old papers again'. For the two years after June 1788 Goethe was gradually detaching himself from the myth of Italy he had himself constructed in the years before he went there. As long as he was still in Weimar the pain and bewilderment caused by this loss of Arcadia, already decorously manifest in his letter to Heyne and most powerfully symbolized in the sufferings of Tasso, could be attributed to his physical departure from Italy: 'forget me not', he inscribed on one of his earliest landscapes drawn from memory of the Campi Phlegraei, and those around him certainly thought that was where his heart lay. But Goethe already knew or suspected, in one of the many secret drawers of his heart, that just as his physical presence in Italy was not the sole occasion for his hymns to fulfilment in *Iphigenia,* or in Faust's stately monologue, so his absence from Italy was not the sole cause of the sense of loss to which Tasso gave anguished expression. In the spring of 1790 it required only a few days beyond the Alps for the misgivings with which he had set out to become certainties and for it to become clear to him that there had always been another factor in his love for Italy besides the fascination exerted by a particular place:

> Das ist Italien, das ich verließ. Noch stäuben die Wege,
>   Noch ist der Fremde geprellt, stell' er sich, wie er auch will
>
> .   .   .   .
>
> Schön ist das Land! doch ach, Faustinen find' ich nicht wieder.
> Das ist Italien nicht mehr, das ich mit Schmerzen verließ.

This is the Italy I left. The ways are still dusty; the foreigner is still bilked, whatever he may do . . . Fair is the land! but alas, I cannot find Faustina again. This is no longer the Italy I left in anguish.

Not of course that Goethe was suddenly seized by memories of his Roman mistress. Rather, when he had first come to Italy the journey had been leading him towards a fulfilment, possibly an ultimate fulfilment, of desire, and that he had eventually named Faustina; now fulfilment, in so far as it was possible for men rather than for gods, lay behind him in Weimar and the journey was only taking him further away from it. Christiana and August's absence had taught him to love them; 'they are very close to me and I happily confess that I love the girl passionately. How much I am attached to her I have felt for the first time on this journey.' They now were his Faustina, the ultimate object and terminus of desire, and without that vivifying presence Italy was empty and drained of significance. It was not just that he had become 'a little more intolerant than last time of the swinish ways of this nation'; 'among other laudable things I have learned on this journey is that I absolutely cannot be alone again in future and cannot live outside our fatherland'. 'My attitudes are more domestic than you imagine', 'I long very much to be home'. It had once seemed to Goethe that Rome was the great lodestone, the magnet towards which all desire aligned itself, and for five years or more before he finally set out he had been sick with yearning for the treasures of the south—even his headlong journey down to Verona in 1790 seems marked by a memory of that great attractive force. But suddenly, in a matter of days, the realization dawned that the force was in Rome no longer; it had shifted, as abruptly and mysteriously as the magnetic pole; the first bloom of affection and curiosity had fallen away and he confided to the notebook which served him as a travel diary that scenes which in Naples had lured him on had no power over him now:

> Glänzen sah ich das Meer, und blinken die liebliche Welle,
>   Frisch mit günstigem Wind zogen die Segel dahin.
> Keine Sehnsucht fühlte mein Herz; es wendete rückwärts,

Nach dem Schnee des Gebirgs, bald sich der schmachtende Blick.
Südwärts liegen der Schätze wie viel! Doch einer im Norden
Zieht, ein großer Magnet, unwiderstehlich zurück.

I saw the sea gleam and the sweet wave sparkle, with a favourable wind the sails drew briskly away. My heart felt no desire; the yearning gaze soon turned backwards to the snow of the mountains. Southwards lie so many treasures! But a treasure in the North, a great magnet, draws me back, irresistibly.

What then was Italy now? Now that it was no longer home to the great lodestone, no longer Arcadia, but a place of dusty roads and dishonest hotel keepers? 'For a year one is submerged in sensuous enjoyment', Goethe told Professor Heyne, when he visited Weimar in July 1790 and enquired further about Goethe's Italian experiences, 'but eventually one feels that we need something more, and that spiritual enjoyment is not possible surrounded by *that* nation.' Goethe recognized that in the end, to be the man and the poet that he was, he needed Germany, Weimar, Christiana, and all the strange tense and partial fulfilments they brought. It was, in the end, to the 'spirit' that he had given his loyalty. But was Italy then a mistake, a flattering illusion, now discarded, as the Princess discarded Tasso's Golden Age? Not necessarily—to have lived in Naples, Goethe told Herder, recalling no doubt his own father, 'will cast a ray of sunlight through the whole of your life'. The place that has been transfigured, indeed hallowed, by appearing to us, even if transiently, the sensuous vehicle of spiritual fulfilment, stays with us for ever as an image of that which we now recognize to be beyond our reach. The Italy Goethe inhabited between 1786 and 1788 is no more,—not because Italy has changed, but because Goethe has changed—but what remains, be it Italy as it now is, be it memories of the past, is a pledge of the happiness he will always continue to seek, however much his desires may have been transformed, transferred, constricted, or even fulfilled:

Emsig wallet der Pilger! Und wird er den Heiligen finden?
Hören und sehen den Mann, welcher die Wunder getan?
Nein, es führte die Zeit ihn hinweg, du findest nur Reste,
Seinen Schädel, ein paar seiner Gebeine verwahrt.
Pilgrime sind wir alle, die wir Italien suchen:
Nur ein zerstreutes Gebein ehren wir gläubig und froh.

The pilgrim strides out busily. And will he find the saint? Hear and see the man who performed the miracles? No, time has borne him away, you will find only remains, his skull, a few of his bones preserved. We are all pilgrims, we who seek Italy: we honour only scattered bones, full of faith and happiness.

The poem can also of course be read differently: to *seek* Italy, which is quintessentially a place of happiness *found*, is to give honour to dry bones rather than to a living human being. In 1790 Goethe knew that time had borne away what once had mattered most to him, but was unsure how to respond to the loss, precisely because what now mattered most to him was something else. A similar uncertainty attaches to a series of remembered Italian landscapes drawn shortly before or shortly after this trip in a roundel form: does the circular format indicate the assimilation of the landscape to a happily decorative, even symbolic, ornament, or does it emphasize the painful distance between memory and present reality?

Goethe spent over a month in suspension in Venice, coming to uncertain terms with the shift in the balance of power in his life. Not until a fortnight had passed did he hear from Einsiedel of Anna Amalia's planned arrival at the end of April, later postponed until 7 May, and not until 23 April did he receive a letter from Weimar. Even then the news was not good: little August was seriously ill for two weeks, and though by 3 May Goethe had heard from Christiana that he had recovered, he told Caroline Herder that he was 'very worried—I am not yet used to that'. (What he wrote to Christiana, however, or she to him, about the illness we do not know). The weather continued cold, and at times very wet, until the end of April; whole days had to be spent indoors. On Easter Sunday, after a cannonade beginning at 4 a.m., there was Mass to attend in Santa Maria della Pietà, with its female choristers, and two processions of the Doge to watch, and for a day or two more Goethe took Götze to hear the singing in San Lazzaro dei Mendicanti. But the main occupation, once the Easter ceremonies were over, was an intensive study of the development of Venetian art, mainly as preserved in the city's churches, from the earliest mosaics, which did not appeal to Goethe, and the Byzantine paintings in the iconostasis of San Giorgio dei Greci, in which he thought he saw survivals of classical elements, to the great masterpieces in colour of Titian, Tintoretto, and Veronese. Goethe's pleasure was as usual spoiled by his objection to the subject-matter of Christian painting; that 'we are all pilgrims' was not an insight he would ever apply extensively to the visual arts:

Nowhere does one see better than in the best paintings of these masters the poverty of the Christian and Catholic mythology and the wretchedness of the whole history of the martyrs. Cause and effect are always an invisible nothing, only the accessories occupied the artist and in turn attract the eye.

But he was able to admire the muscles of the executioners, and the charms of the female sinners, and to take a particularly close interest in the techniques of tempera and oil-painting and in the changes of colours over time. There was material here, and in Venice's unique land- and sea-scape, for his new theoretical interest in colour—at present he was concentrating on blue. Accompanied everywhere by Götze, who dutifully noted down—sometimes effectively to dictation—all that he was learning, Goethe also visited the workshop of the picture-restorers in the monastery of SS Giovanni e Paolo and made the acquaintance of the craftsmen. By the end of the month, 'I have almost made myself ill looking at paintings', he told Caroline Herder, 'and really had to break off for a week', and besides the weather was improving. There had already been an excursion to the glass-blowers on the island of Murano on 15 April; on the twenty-first there was a visit to the Arsenal to admire the ancient marble lions from the Piraeus and inspect two merchantmen under construction, and a troupe of acrobats provided some unexpected entertainment, especially an extraordinarily lithe and confident little girl. The following day was spent on the Lido. Venice was a city of stone and water, there was scarcely a tree whose greening would mark the progress of spring, so it was hardly suited to botanical studies. But there were plenty of sea-creatures to observe, fish and crabs, the rapidly multiplying piddocks he had last seen in the neighbourhood of the temple in Pozzuoli, whose pillars they had bored, and other molluscs which when swimming freely in the sea reminded him of the boneless agility of the young acrobat. His thoughts returned to zoology, and his projected complement to the botanical theory of metamorphosis. As they walked through the Jewish cemetery on the Lido Götze picked up from the ground the skull of a sheep and, in a crude anti-Semitic jest, presented it to his master as if it were that of a human being. It was another moment of insight for Goethe—into the similarity of structure, not of human and animal skulls, for that was familiar territory, but of the constituent bones of the skull, which had suddenly been presented to him in an unfamiliar light, and the vertebrae from which they proceed. His scribbled note, 'Characteristic of every vertebra to have an extension. Woodlouse', is the beginning of an attempt to reduce animal form to the repetition and transformation of a single basic unit, as plant form had been reduced to variations on the basic leaf. There were as yet no social distractions in Venice: those could wait until the Dowager Duchess arrived, and Goethe's only sustained personal contact in his first month seems to have been with the banker Zucchi, brother-in-law of Angelica Kauffmann, who treated him to long, and of course topical, expositions of the city's ancient republican constitution. There was therefore ample time for those scientific reveries in which Goethe naturally now found a refuge—objective, as it seemed to him— from troublesome emotions or from intellectual irritants, such as religion or politics. 'I cannot deny that sometimes this last month impatience has threatened to get the better of me, but I have also *seen, read, thought, written* more than otherwise I would do in a year, when the proximity of friends and my dear sweetheart make me fully comfortable and content.'

Immediately after announcing to Carl August the death of his love for Italy, Goethe went on to state his suspicion that the sum of his *Elegies*, to which he had probably not added anything since December, was now complete: 'there is not even a trace of that vein left in me'. If Italy was no longer the land of heart's desire, there was clearly no longer any reason to translate Weimar into Rome, either ancient or modern, and since the fateful moment when he became a family man, Goethe's grasp on his own identity had grown too insecure for the subtle play which the *Elegies* make with it. But he promised his correspondents a compensation, a booklet of epigrams, little satirical poems sprouting and multiplying in Venice as rapidly as the piddocks in the sea—by 4 May there were 100 of them, though some of these may have been taken over from the existing collection of *Erotica*, of which the whole project was but an extension. If the patron of the *Elegies* was Propertius, that of the *Epigrams* was to be Martial: 'these are fruits that do well in a big city: there is subject-matter everywhere and it does not take much time to do them'. They were to be as varied as the occasions of urban life, witty, scathing, obscene. In fact, however, the range of themes is considerably narrower than the model, and there is very little in the way of personal satire; Venetian society Goethe neither knew nor cared about, and Weimar society could hardly accommodate a Martial (or a Juvenal, whose works he bought on 12 April). There are reflections on Italy in general, and on his own relation to it and to Christiana, some of which have already been cited, and also a number of poems which are clearly located in the tourist's Venice: complaints about the rain, the mud, the lack of trees and grass; sarcasms at the expense of the priests and the Papal Nuncio during the Easter ceremonies; allegorizations of coffee-shops and snuff-sellers (Goethe disliked both commodities) and of a yo-yo player (one of the earliest references to the toy); glimpses of the lions at the Arsenal (a poor substitue for the Piraeus, the lions think), of beggars and prostitutes, and gondolas, and the caged singers at the Mendicanti; and well over a dozen little pieces devoted to the troupe of acrobats Goethe saw on 21 April, particularly the young girl, to whom he gives (possibly correctly) the name Bettina. Three other more general themes reflect Goethe's preoccupations at a distempered time, although severe censorship has long obscured their decisive contribution to the collective character of what Goethe wrote in Venice. First, there is a small but significant group of poems relating to the French Revolution and to the political issues it raises. Superficially they might seem to express no more than the moralizing smugness of an uninvolved and reactionary foreigner: 'The sad fate of France . . . the masses became tyrant to the masses', 'if you would free many, dare to serve many', princes and demagogues are equally well-intentioned, and sometimes equally deceitful— and so on. Yet the even-handedness of these comments, which do not simply dismiss the Revolution as absurd or disgusting folly, suggests that in one respect at least the Revolutionaries have the poet's sympathy: yes it is true, he says, that the ranters on every French street-corner are mad

> Mir auch scheinen sie toll; doch redet ein Toller in Freiheit
> Weise Sprüche, wenn, ach! Weisheit im Sklaven verstummt.

To me too they seem mad; but a madman at liberty can utter wise sayings while in the slave, alas, wisdom falls silent.

Freedom of speech is one revolutionary freedom whose value this German literary intellectual can recognize, despite his preference otherwise for the old German constitution (the old Venetian constitution is also praised). That he knows how revolutionary it is, is shown by an epigram that was left unpublished throughout his lifetime:

> Dich betrügt der Staatsmann, der Pfaffe, der Lehrer der Sitten,
>   Und dieß Kleeblatt wie tief betest du Pöbel es an.
> Leider läßt sich noch kaum was rechtes denken und sagen
>   Das nicht grimmig den Staat, Götter und Sitten verlezt.

You are deceived by the statesman, the priest, the teacher of morals, and how profoundly, O rabble, you worship this trinity. Unfortunately it is now scarcely possible to think or say anything right that is not savagely wounding to the state, the gods and morals.

Deception is the universal corrupting feature of the prevailing social and intellectual order, and the poet is alone in seeing through it all. Essentially this is a religious rather than a political opinion, and Goethe got away with publishing the following exceptionally blasphemous epigram only by situating it in a context where it could seem to apply to 'apostles of freedom', the peddlers of political nostrums:

> Jeglichen Schwärmer schlagt mir ans Kreuz im dreißigsten Jahre;
>   Kennt er nur einmal die Welt, wird der Betrogne der Schelm.

Let me have every ranter nailed to the cross in his thirtieth year [that is, Christ should have been crucified three years earlier, before his preaching began]; if he once gets familiar with the world the deceived becomes a deceiver.

Second, then, the *Venetian Epigrams,* as they eventually came to be called, are Goethe's first explicitly and violently anti-Christian work and none of his later outbursts matches the directness of this first assault. Christianity is presented as a series of illusions, as indeed is moral 'stoicism', that is, deism—neither is a religion 'for free men'. We may deduce that only Epicurean materialism offers us the plain truth 'about God, man, and the world', a truth which, though not difficult to attain, is profoundly unpopular.

> Warum treibt sich das Volk so, und schreit? Es will sich ernähren,
>   Kinder zeugen und die nähren, so gut es vermag.
> Merke dir, Reisender, das und tue zu Hause desgleichen!
>   Weiter bringt es kein Mensch, stell' er sich, wie er auch will.

Why are the populace so busy and noisy? They want to feed themselves, beget children and feed those as well as they can. Note that, traveller, and do the same at home. No man achieves any more, whatever he may do.

There is no deeper secret to life than living, and if the poet had hundreds upon hundreds of years he would simply wish tomorrow to be as today. Christ was a foolish enthusiast who lost what life might have given him and prevailed on others to do the same,

> Folgen mag ich dir nicht; ich möchte dem Ende der Tage
>   Als ein vernünftiger Mann, als ein vergnügter mich nahn . . .

I do not wish to follow you; I should like to approach the end of my days as a reasonable and contented man. . .

His villainous disciples stole his body from the tomb to give credence to the fable of the resurrection, and Christians, Jews, and Moslems are all equally intolerant and persecuting fools. Goethe's venom is particularly reserved, however, for Lavater, the Christian apologist, who 'mingles sense and nonsense', who endeavours, that is, like the hermit-crab to conceal beneath a borrowed covering of human reason the 'naked pudenda' of his absurd belief. Goethe too loved his toys when he was a child, but discarded them when he grew up:

> So griff Lavater iung nach der gekreuzigten Puppe.
> Herz' er betrogen sie noch wenn ihm der Athem entgeht!

and similarly Lavater when young clutched at the crucified doll. Let him continue cuddling it in his deception when breath fails him.

Asked what is perhaps the ultimately religious question—'what can I hope?'— Goethe gives a reply that is uncompromisingly materialist, that is, erotic:

> Welche Hoffnung ich habe? Nur eine die heut mich beschäftigt,
> Morgen mein Liebchen zu sehn das ich acht Tage nicht sah.

What hope do I have? Only one that concerns me today, to see my sweetheart tomorrow, whom I have not seen for a week.

For the third thematic area that gives the *Epigrams* their peculiar character is the sexual, and it is closely related to the religious. The *Epigrams* are a direct continuation of that priapic phase of Goethe's sensibility which seems to have begun with the completion of the *Works* and the birth of his son, and can be read as further evidence of a breakdown of the meanings that had previously sustained his life. Hope and desire are reduced to their most elemental form as the sexual pruritus, the reality behind all the deceptions whether of priests or politicians. Sexual language or allusion gives rhetorical force to a number of the poems whose overt content is religious or political, and we have already seen some examples. Many epigrams, however, remained unpublished for a century because of their direct treatment of such unacceptable subjects as nudity, erection, masturbation, both male and female (the latter also oral), prostitution, sodomy, venereal disease, and the disappointments of a loose vagina, compared to a Venetian canal. The poet unsurprisingly associates Christianity with the suppression of sexuality, and it is suggested that Priapus has a better cure for religious hysteria than Christ.

The strange and unprecedented feature of the *Epigrams* is therefore that they are full of Goethe's opinions. That those opinions are rather jaundiced does not make the work more agreeable but is not a relevant consideration. Never before in his writing have views been expressed in so undramatized a form, so unattached to any *persona* other than that of Goethe at a particular time and in a particular place. The *Elegies* make an instructive contrast. There is no semi-fictional narrative, whether in Italy or in Weimar, in the background to the *Epigrams;* and there is no self-conscious pretence within the poems that what is being said is part of a revival of an ancient form, or an ancient way of life. There are remarkably few mythological, literary, or other classical references, in fact; the modern world of coffee, gondolas, and the French Revolution, not to mention yo-yos, provides the material of the *Epigrams,* not the relatively timeless symbolic world of wine, and the lamp, and the bed, and the fable of King Midas, on which the *Elegies* largely rely. Goethe speaks in his own voice, as in a journal or commonplace-book, with date and place over the entries, and that voice is oddly strained. But, of course, the one literary genre of which Goethe never made anything was the journal, the pure monologue. His forte was the letter, the monologue with an addressee, which thus became an inner dialogue. By 1790, however, the interlocutor in Goethe's inner dialogue was gone:

> Eine Liebe hatt' ich, sie war mir lieber als alles!
> Aber ich hab' sie nicht mehr! Schweig', und ertrag' den Verlust!

I once had a love, I loved it more than everything. But I have it no longer. Be silent, and bear with the loss!

In the *Elegies* the interlocutor is still there, but as a 'secret' whom poetry cannot touch. In the *Epigrams* at the point where in all Goethe's previous poetry there is a partner to be addressed and wooed, or an object (usually natural) to be enlivened with feeling, there is, apart from this brief exclamation, and one or two others like it, simply silence. There is no 'you', let alone a 'we', of any substance, and as a result the 'I', though it is in one sense more fixed and specified than ever before, has lost its unique power to fascinate. The irony at the expense of the form used, the paradoxical awareness of time passing, and yet never passing, within the poem, the sense of the fragility of the moment - indeed the moment itself, as a poetic theme—these features of the Elegies, in which an alert and reactive subjectivity was expressed and preserved against a day more favourable to its flowering, are all absent from the *Epigrams.*

   This would all matter less if these imitations of Martial achieved the formal precision and ruthlessness of their original. But to do that in a modern tongue they would need the resources of a Pope or a Rochester, of the rhyming couplet. The nearest English approach to the form of the *Epigrams,* however, is that of the distichs which Clough uses to open and close each canto of *Amours de Voyage*:

> So go forth to the world, to the good report and the evil!
> Go, little book! thy tale, is it not evil and good?

Some verbal patterning is possible (more in German, whose inflections make word order more malleable than in English), but the symmetry can never achieve the finality and inevitability of rhyme. Similarly the rhythmic effect is always, and necessarily, diffuse and tentative. In German as in English, though to a lesser extent, a main obstacle to the reproduction of classical metres is a shortage of spondees—that is, in an accentual system of versification such as modern English and German largely possess, of feet consisting of two successive stressed syllables. In German the difficulty can to some extent be overcome since the language contains many syllables which have an intermediate stress and so may be counted as either stressed or unstressed ('long' or 'short', in quantitative terms), as happens to be metrically convenient. An indeterminacy of metre is often the result, however, especially in Goethe's writing, in which the reproduction of ancient measures tends to be impressionistic rather than scholarly. So the telling and brutal line, 'So griff Lavater *iung nach der ge*kreuzigten Puppe', ('so young Lavater *reached out for the* crucified puppet'), can be read as the hexameter it is required to be only if the italicized syllables are read as two spondees. The natural way to read them, though, is as a stressed ('long') syllable followed by a dactyl, which turns the line into a hypermetric pentameter. The uncertainty does not matter much in longer poems such as the *Elegies* (or indeed *Amours de Voyage*), where it contributes to a conversational and reflective tone, but it dissolves the intended lapidary effect of the *Epigrams*. If a man is going to get on a high horse he must know how to keep his seat. In the *Epigrams* there is too much padding (the reader will have noticed the repetition of the meaningless half-line 'stell' er sich, wie er auch will', 'whatever he may try to do') and too much blurring of contours. We see emerging the dubious consequences of Goethe's adoption of classical metres which, when the collection of *Erotica* was begun, was little more than a poet's diversion, being overshadowed by his work in modern forms, on *Tasso, Faust,* and the edition of his miscellaneous poems. The *Elegies,* if they may by 1790 be regarded as a single work, were already Goethe's longest non-dramatic venture in a regular verse-form. It was, however, precisely in rhythmic variety, in changing effects and unique forms that Goethe's poetic personality had most memorably expressed itself. After the completion of the *Works* an iron curtain of hexameters and elegiacs rolls down, behind which that rhythmic inventiveness is hidden for nearly eight years. The new manner is perhaps at its least impressive in the rather unepigrammatic *Venetian Epigrams,* whose only poetic resource too often is a prescribed metre, uncertainly manipulated.

Yet it is possible to look at the *Epigrams,* and Goethe's flight into classical measures, rather differently. The lyrical moments in which he had discovered a new rhythm and a new poem had gradually been occurring less frequently since about 1780—only once or twice since 'Divinity'—and with 'Dedication' and *The Mysteries* he had shown the first signs of looking for regular forms which could take over the burden of invention. At the same time, and the two developments are no doubt closely conected, there was also a decline in Goethe's ability, or willingness, to create symbolic figures central to large-scale works of literary art, images at once of the self and of a tendency of the age, in responsive relation to a world which springs into being around them—Götz-Weislingen, Werther, Faust, Egmont, Wilhelm Meister, Iphigenia, Tasso. After the conception of Tasso in 1780 Goethe created no new figures of this kind, unless we include, as a transitional case, the unnamed and far from integrated pseudo-Propertius who is the 'I' of the *Elegies*. The Göschen edition effectively marked the end of works constructed in this way, and the survival of Wilhelm Meister, and even perhaps of Faust, into the new era was only apparent. A change of such magnitude in the creative process can hardly be attributed to a lesser factor—there is perhaps in the end none greater—than age. In poets of less weight and tenacity—and since we are talking of superlatives here we may as well cite Wordsworth or Gerhart Hauptmann—such a change has concluded the best part of the *œuvre*. Goethe's particular genius was not to persist in a mode whose time was past but to make a virtue of this necessity too, as of others before and after it: not to deny the change, but to find a way of writing in order to express it. The *Venetian Epigrams* may indeed be evidence of the final passing of Goethe's youthful sensibility, but they are more remarkable as evidence of his determination to keep writing, and to keep renewing the sources of writing, despite the effects of time. He would find more satisfactory ways of writing without a central symbolical figure, of writing a literature of opinions, of writing epigrammatically, and even for a while of writing in classical metres, but in the *Epigrams* he at least made a first attempt at divining water in the desert, and he knew that was what he was doing. The *Elegies* foreshadow much better the nature of the future achievement, but for that very reason they give less sense of the deprivation in which, and the odds against which, the achievement was won. The *Epigrams* were explicitly announced as compensation for the drying up of a poetic vein. They describe themselves as 'a disconnected conversation' with the Muses, of a kind which is 'congenial to the wanderer'. The image of the traveller, of the man who is not at home, is fundamental to the collection, explaining both the mood of discontent with Venice and the yearning to be back across the Alps. The experience of travelling, of being absent from the place of true satisfaction, forces the would-be sensualist and materialist poet, in the third epigram, into an acknowledgement that there is after all a spiritual world, for his mind is full of the delights of being with his lover, while the carriage bears his body further and further away from her. The poetry that he can now write is therefore not like the poetry of the *Elegies*, which were inspired by a fulfilment so absorbing that they could not express it; instead it is a poetry from which fulfilment is also absent, but for a rather different reason:

> Alle Neun, sie winkten mir oft, ich meine die Musen;
>   Doch ich achtet' es nicht, hatte das Mädchen im Schoos.
> Nun verließ ich mein Liebchen; mich haben die Musen verlassen
>
>   .   .   .   .
>
> Doch . . . du kamst mich zu retten
>   Langeweile! du bist Mutter der Musen gegrüßt.

All nine of them beckoned me, I mean the Muses; but I took no notice, had my girl in my lap [as in Elegy xiii]. Now I have left my sweetheart; the Muses have left me . . . But . . . you came to rescue me, o Tedium; welcome, Mother of the Muses.

The *Epigrams* are the fruit of the tedium of absence,

> Wißt ihr, wie ich gewiß zu hunderten euch Epigramme
> Fertige? Führet mich nur weit von der Liebsten hinweg!

Do you know, o Epigrams, how I can manufacture you by the hundred? Simply take me far away from my beloved,

and they are therefore naturally spiced with 'the sweetest spices of the world', hope and remembrance—remembrance of happiness with the beloved and hope of return to her. While there is in this remark some wishful thinking about the true character of the collection, it is also possible to see here, in the otherwise improbable context of the *Venetian Epigrams*, the first outlines of Goethe's future concept of renunciation.

It is unlikely that many of the *Epigrams* were written after the first week in May. Götze was sent to meet Anna Amalia in Padua on the second; on the fifth Meyer, who, together with Bury, was accompanying the Dowager Duchess out of Italy, arrived in Venice. Goethe's solitude was over, there was every prospect that the homeward journey would begin by Whitsun, the twenty-third, and his mood was transformed. Anna Amalia benefited: arriving in the early evening of the sixth she noted in her diary, 'I cannot say how I felt, I grew sad and oppressed. I did not find the grand notion that I suppose one has, coming into the city, into the canals, my sadness increased, all seemed melancholy to me. Getting down at the inn I met Goethe and was merry once more.' The sightseeing began the following day with St Mark's Square and the cathedral, but rain forced the party to return for the afternoon and evening to the Duchess's lodgings on the Canal Grande, where Zucchi visited and Goethe read from his epigrams, and Knebel's translations of Propertius. With familiar faces around him, his two artist-friends and Kapellmeister Reichardt as company, and the Duchess's name to open doors, Goethe spent a fortnight reviewing the pleasures of Venice—the paintings, the buildings of Palladio, the singers of the Mendicanti, the Doge's palace, the Arsenal, even the Prison (Goethe stayed outside with the ladies), and trips to the glaziers of Murano and down the Lido to Pellestrina to see the sea-wall Goethe had admired in 1786— and all no longer the disconnected occasions for a lone traveller's misanthropy but transfigured into the last chapter of a grand tour by their sharing in the evening splendour of the *ancien régime*. 'Everyone' of course had to call on the Dowager Duchess: her nephew the Prince of Brunswick was travelling back too, in his own party; the Duke of Sussex, one of the English princes, was in Venice with his Göttingen tutor and did not disdain to eat an ice with them all in a café on St Mark's Square; Prince Rezzonico, the Senator of Rome, visited; the Prussian and Imperial residents paid their respects; the French ambassador, the Marquis de Bombelles, his mind already far away in his native land and on darker things, arranged a reception at his majestic residence on the Canal Grande, to which all of course arrived by gondola. The weather was windy on Ascension Day, 13 May, and the ceremony of the Doge's marriage with the sea had to be postponed, but on the following Sunday the sea was calm and the sun shone for the most splendid occasion of Venice's year, with the Doge's barge, the Bucentaur, escorted out by vessels of the Republic's navy and by hundreds, if not thousands, of gondolas, St Mark's Square 'packed full', the cannons firing and all the church bells ringing. On Thursday 20 May the Spanish ambassador gave a ball for the entire diplomatic corps, from which the Weimar party did not return until two in the morning, and early on Saturday the twenty-second they left to sail up the Brenta to Padua.

Whit Sunday was spent visiting the university institutions of Padua, the observatory, the natural history collection, and the botanical garden, and eating strawberries in milk, but Goethe, who had evidently taken over from Einsiedel the direction of the Duchess's tour, knew what he thought would appeal to her most and arranged that they should move on to Vicenza the following day in order to admire the work of Palladio and make an excursion to the Rotonda. A day and a half was also spent in Verona, where Goethe and Götze were reunited with the ducal chaise, and there was a visit to a mediocre production in the amphitheatre. From Friday to Sunday the thirtieth Goethe was, for the only time on the journey, in new territory, when the Duchess moved to Mantua, then an Austrian fortress and much declined, being less than twice the size of Weimar, but reached through magnificent avenues of poplars and containing a good collection of antiquities and a fine Mantegna altarpiece. In Mantua Anna Amalia's Italian Journey effectively came to an end; she was apprehensive about her return to Weimar and old age after the greatest adventure of her life, but Goethe perhaps found it easy to keep up her spirits because his own feelings were very different. 'I yearn longingly to be home', he wrote to the Herders, 'I have left the Italian orbit of life completely.' True, he was parting here from Meyer, who was returning to his family in Switzerland, but Meyer was now secured for a future in Weimar, and though the tearful Bury was staying behind in Mantua and then going back to Rome, he had so commended himself to Anna Amalia that he could have good hopes of finding a place 'in the future artists' republic' with which the Duchess was evidently planning to console herself once she was back in Thuringia. One more day was spend in Verona and then the return began in earnest. After stops in Rovereto, Bolzano, and the Alto Adige the Brenner was passed on 4 June, and the party then spent three nights in Innsbruck being entertained by Archduchess Elizabeth of Austria, one of Joseph II's many sisters. Augsburg was reached on 9 June and here came a grim reminder that they were once more in the land which had driven Werther to distraction: among the letters waiting for them was one with the news that Knebel, who in April had begun his leave in Ansbach, had a month previously been out walking with his brother Max, heard a shot behind him, and had discovered that Max, a depressive young man, had taken his own life. But Goethe could now face even Werther. A meeting with Knebel had already been arranged in Nuremberg and he and his sister, to whom Goethe now delivered the letter with which he had set out in March, seemed to be bearing their bereavement well. The reunion with the Duchess, Einsiedel, and Fräulein von Göchhausen must have been a useful distraction for the Knebels, and the stay in Nuremberg from 12 to 15 June seems to have been merry, certainly more so than Goethe's few hours there in March, and to have helped Anna Amalia too to feel at home in Germany once more. Via Bayreuth and Hof the carriages, somewhat delayed by two broken axles, reached Jena on 18 June, arriving in Weimar at 11 o'clock in the evening, two years to the hour since Goethe's return from Arcadia. Whatever the coincidence symbolizes, it is not that time had stood still: even if you have not left the high ground, the landscape looks very different when you are coming down on the other side of the watershed. 'The Duchess is well and contented', Goethe wrote, 'as one is when one returns from Paradise. I am now used to it, and this time I was quite happy to leave Italy.'

Works Cited in the Notes

in alphabetical order of the abbreviations used

Ackerman, *Palladio*: J. S. Ackerman, *Palladio* (Harmondsworth, 1966).

Andreas, 'Vorabend': W. Andreas, 'Goethes Abschied von Carl August am Vorabend der Italienreise', *Goethe* 21 (1959), 54–69.

Anecdotes: J. Sutherland (ed.), *The Oxford Book of Literary Anecdotes* (London, 1977).

AS: J. W. Goethe, *Amtliche Schriften*, ed. W. Flach and H. Dahl (Weimar, 1950–72).

Bauer, *Kopf*: K. Bauer, *Goethes Kopf und Gestalt* (Berlin, 1908).

Behrends, *Einwohner*: J. A. Behrends, *Der Einwohner in Frankfurt am Mayn in Absicht auf seine Fruchtbarkeit, Mortalität und Gesundheit geschildert* (Frankfurt-on-Main, 1771).

Beutler, *Essays*: E. Beutler, *Essays um Goethe*, ed. C. Beutler (Zurich, Munich, 1980).

Bicknell, *Beauty*: P. Bicknell, *Beauty, Horror and Immensity: Picturesque Landscape in Britain 1750–1850* (Cambridge, 1981).

Biedermann-Herwig: *Goethes Gespräche*, ed. F. von Biedermann and W. Herwig (Zurich and Stuttgart, 1965–84).

Binder, *Faust*: W. Binder, *Goethes Faust: Die Szene 'Und was der ganzen Menschheit zugeteilt ist'* (Gießener Beiträge zur deutschen Philologie 82; Gießen, 1944; repr. Amsterdam, 1968).

Blackall, *Emergence*: E. A. Blackall, *The Emergence of German as a Literary Language 1700–1775* (Cambridge, 1959).

Blanning, *Reform*: T. C. W. Blanning, *Reform and Revolution in Mainz 1743–1803* (Cambridge, 1974).

Blanning, *Revolution*: T. C. W. Blanning, *The French Revolution in Germany: Occupation and Resistance in the Rhineland 1792–1802* (Oxford, 1983).

Blunden, 'Lenz': A. Blunden, 'J. M. R. Lenz', in A. Natan and B. Keith-Smith (eds.), *German Men of Letters*, vi (London, 1972), 207–40.

Blunt, *Poussin*: A. Blunt, *The Paintings of Nicolas Poussin: A Critical Catalogue* (London, 1966).

Bode: *Goethe in vertraulichen Briefen seiner Zeitgenossen*, ed. W. Bode, rev. R. Otto and P.-G. Wenzlaff (Berlin and Weimar, 1979).

Bode, *Bau*: W. Bode, *Goethes Leben: Am Bau der Pyramide seines Daseins. 1776–1780* (Berlin, 1925).

Bode, *Flucht*: W. Bode, *Goethes Leben: Die Flucht nach dem Süden. 1786–1787* (Berlin, 1923).

Bode, *Garten*: W. Bode, *Goethes Leben im Garten am Stern* (Berlin, 1922).

Bode, *Geniezeit*: W. Bode, *Goethes Leben: Die Geniezeit. 1774–1776* (Berlin, 1922).

Bode, *Lehrjahre*: W. Bode, *Goethes Leben: Lehrjahre. 1749–1771* (Berlin, 1919).

Bode, *Pegasus*: W. Bode, *Goethes Leben: Pegasus im Joche. 1781–1786* (Berlin, 1925).

Bode, *Rom*: W. Bode, *Goethes Leben: Rom und Weimar. 1787–1790* (Berlin, 1923).

Bode, *Ruhm*: W. Bode, *Goethes Leben: Der erste Ruhm. 1771–1774* (Berlin, 1920).

Bode, *Stein*: W. Bode, *Charlotte von Stein* (Berlin, 1912).

Boenigk, *Urbild*: O. von Boenigk, *Das Urbild von Goethes Gretchen* (Greifswald, 1914).

Böttiger, 'Bode's Leben': K. A. Böttiger, 'J. J. C. Bode's literarisches Leben', in *Michael Montaigne's Gedanken und Meinungen . . .* , trans. J. J. C. Bode, vi (Berlin, 1795), pp. iii–cxliv.

Bougeant, *Femme docteur*: G. H. Bougeant, *La Femme docteur . . .* , ed. A. Vulliod (Lyon, 1912).

Boulby, *Moritz*: M. Boulby, *Karl Philipp Moritz: At the Fringe of Genius* (Toronto and Buffalo, London, 1979).

Boyd, *Notes*: J. Boyd, *Notes to Goethe's Poems* (Oxford, 1944).

Boyle, 'Lessing': N. Boyle, 'Lessing, Biblical Criticism and the Origins of German Classical Culture', *German Life and Letters* 34 (1981), 196–213.

Boyle, 'Pascal': N. Boyle, 'Pascal, Montaigne, and "J.-C.": the Centre of the *Pensées*', *Journal of European Studies* 12 (1982), 1–29.

Braun, *Zeitgenossen*: J. W. Braun, *Goethe im Urteil seiner Zeitgenossen* (Berlin, 1883–5).

Bruford, *Culture*: W. H. Bruford, *Culture and Society in Classical Weimar: 1775–1806* (Cambridge, 1962).

Bruford, *Germany*: W. H. Bruford, *Germany in the Eighteenth Century: The Social Background of the Literary Revival* (Cambridge, 1935).

Bruford, *Theatre*: W. H. Bruford, *Theatre, Drama, and Audience in Goethe's Germany* (London, 1950).

Brüggemann, *Gottsched: Gottscheds Lebens- und Kunstreform . . .* (DLER Reihe Aufklärung 3), ed. F. Brüggemann (Leipzig, 1935).

Butler, *Fortunes*: E. M. Butler, *The Fortunes of Faust* (Cambridge, 1952).

Carlson, *Theatre*: M. Carlson, *Goethe and the Weimar Theatre* (Ithaca and London, 1978).

Carlyle, *Correspondence: Correspondence between Goethe and Carlyle*, ed. C. E. Norton (London and New York, 1887).

CGZ: Corpus der Goethezeichnungen, ed. G. Femmel (Leipzig, 1958–73).

Clough, *Poems*: A. H. Clough, *Poems* (London, 1895).

Dechent, *Kirchengeschichte*: H. Dechent, *Kirchengeschichte von Frankfurt am Main seit der Reformation*, ii (Leipzig, Frankfurt-on-Main, 1921).

Dietrich, *Faust*: *Faust* (Theater der Jahrhunderte. Faust. 1. Bd.), ed. M. Dietrich (Munich and Vienna, 1970).

DJG: H. Fischer-Lamberg (ed.), *Der junge Goethe* (Berlin, 1963–74).

DKV: Goethe, *Sämtliche Werke*, ed. D. Borchmeyer (Frankfurt, Deutscher Klassiker Verlag, 1985–).

Durrani, 'Love': O. Durrani, 'Love and Money in Lessing's *Minna von Barnhelm*', *Modern Language Review* 84 (1989), 638–51.

Ehrlich, *Wittumspalais*: W. Ehrlich, *Das Wittumspalais in Weimar* (Weimar, 1976).

Eliot, 'Goethe': T. S. Eliot, 'Goethe as the Sage (1955)', in *On Poetry and Poets* (London, 1957), 207–27.

Ephemerides: Ephemerides Societatis Meteorologicae Palatinae, Observationes Anni 1787 (Mannheim, 1789).

Etwas über Frankfurt: Anon., *Etwas über Frankfurt: Aus der Brieftasche eines Reisenden* (n.p., 1791).

Faber, *Frankfurt*: J. H. Faber, *Topographische, politische und historische Beschreibung der Reichs- Wahl- und Handelsstadt Frankfurt am Mayn* (Frankfurt-on-Main, 1788).

Fairley, *Goethe*: B. Fairley, *A Study of Goethe* (Oxford, 1947).

Federn, *Christiane*: E. Federn, *Christiane von Goethe* (Munich, 1917).

Fichte, *Werke*: J. G. Fichte, *Sämmtliche Werke*, ed. I. H. Fichte (Berlin, 1971 = Berlin, 1845–6).

Ford, *Europe*: F. L. Ford, *Europe 1780–1830* (London and New York, 1970).

Fowler, 'Orest': F. M. Fowler, 'The Problem of Goethe's Orest: New Light on *Iphigenie auf Tauris'*, *Publications of the English Goethe Society*, NS 51 (1980–1), 1–26.

Genton, *Promotion*: E. Genton, *Goethes Straßburger Promotion* (Basle, 1971).

Gerth, *Intelligenz*: H. Gerth, *Bürgerliche Intelligenz um 1800* (Göttingen, 1976).

Goldsmith, *Works*: *The Poetical Works of Oliver Goldsmith*, ed. A. Dobson (London, 1906).

Gooch, *Germany*: G. P. Gooch, *Germany and the French Revolution* (London, 1920; new imp. 1965).

Göttinger Hain: *Der Göttinger Hain*, ed. A. Kelletat (Stuttgart, 1967).

Gottsched, *Werke*: J. C. Gottsched, *Ausgewählte Werke*, ed. J. and B. Birke and P. M. Mitchell (Berlin and New York, 1968–87).

Gräf: H. G. Gräf, *Goethe über seine Dichtungen* (Frankfurt-on-Main, 1901–14).

Grumach: *Goethe. Begegnungen und Gespräche*, ed. E. and R. Grumach (Berlin, 1965–).

Grumach, 'Prolog': E. Grumach, 'Prolog und Epilog im Faustplan von 1797', *Goethe* 14/15 (1952/3) (Weimar, 1953), 63–107.

HA: J. W. Goethe, *Werke*, Hamburger Ausgabe, ed. E. Trunz (Munich, 1988). References without the date are to page numbers of the text, which are identical in all issues of the edition. References with the date are to the apparatus and may be correct only for the 1988 issue.

HABr: J. W. Goethe, *Briefe*, Hamburger Ausgabe, ed. K. R. Mandelkow (Munich, 1988).

HABraG: *Briefe an Goethe*, Hamburger Ausgabe, ed. K. R. Mandelkow (Munich, 1988).

Hagen–Nahler: W. Hagen, I. Jensen, and E. and H. Nahler (eds.), *Quellen und Zeugnisse zur Druckgeschichte von Goethes Werken* (Berlin, 1966–86).

Haller, *Alpen*: A. Haller, *Die Alpen und andere Gedichte*, ed. A. Elschenbroich (Stuttgart, 1965).

Harnack, *Nachgeschichte*: O. Harnack (ed.), *Zur Nachgeschichte der italienischen Reise* (Schriften der Goethe-Gesellschaft 5; Weimar, 1890).

Haskell and Penny, *Taste*: F. Haskell and N. Penny, *Taste and the Antique* (New Haven and London, 1981).

Heer, *Leibniz*: F. Heer (ed.), *Gottfried Wilhelm Leibniz* (Frankfurt-on-Main and Hamburg, 1958).

Hegel, *Werke*: G. W. F. Hegel, *Werke in zwanzig Bänden* (Theorie Werkausgabe), ed. E. Moldenhauer and K. M. Michel (Frankfurt, 1970).

Henderson, *Novels*: P. Henderson (ed.), *Shorter Novels: Seventeenth Century* (Everyman's Library 841; London, 1930).

Herder, *Briefe*: J. G. Herder, *Briefe*, ed. W. Dobbek and G. Arnold (Weimar, 1984–8).

Herder, *Werke*: *Herders Sämtliche Werke*, ed. B. Suphan (Berlin, 1877–1913).

Hinrichs, *Preußentum*: C. Hinrichs, *Preußentum und Pietismus* (Göttingen, 1971).

Homer, *Odyssey*: Homer, *The Odyssey*, trans. W. Shewring (Oxford, 1980).

Istel, *Rousseau*: E. Istel, *Jean-Jacques Rousseau als Komponist seiner lyrischen Szene 'Pygmalion'* (Publikationen der Internationalen Musikgesellschaft, Beihefte, 1; Leipzig, 1901).

Italienische Reise, ed. Golz: J. W. Goethe, *Italienische Reise*, ed. Jochen Golz (Berlin, 1976).

Italienische Reise, ed. von Graevenitz: J. W. Goethe, *Italienische Reise*, ed. G. von Graevenitz (Leipzig, 1912).

Jackson, 'Air': P. H. Jackson, '"Air and Angels": The Origenist Compromise in Haller's *Über den Ursprung des Übels*', *German Life and Letters*, NS 32 (1979), 273–92.

Jellicoe, *Gardens*: G. and S. Jellicoe, P. Goode, and M. Lancaster (eds.), *The Oxford Companion to Gardens* (Oxford, 1986).

Jugler, *Leipzig*: J. H. Jugler, *Leipzig und seine Universität vor hundert Jahren* (Leipzig, 1879).

Kalnein, 'Architecture': W. von Kalnein, 'Architecture in the Age of Neo-Classicism', in *The Age of Neo-Classicism* (Catalogue of the Fourteenth Exhibition of the Council of Europe) (London, 1972), pp. liii–lxvi.

Kant, *Werke*: I. Kant, *Werke*, ed. W. Weischedel (Frankfurt-on-Main, 1964). (The pagination for both the twelve-volume and the six-volume impression is identical. Volume numbers are given for both impressions.)

Kington, 'Mapping': J. A. Kington, 'Daily Weather Mapping from 1781', *Climatic Change*, 3 (1980–1), 7–36.

Klopstock, *Oden*: F. G. Klopstock, *Oden*, ed. H. Düntzer (Leipzig, 1887).

Klopstock, *Werke*: F. G. Klopstock, *Ausgewählte Werke*, ed. K. A. Schleiden (Munich, 1962).

Krogmann, *Friederikenmotiv*: W. Krogmann, *Das Friederikenmotiv in den Dichtungen Goethes: Eine Motivanalyse* (Germanische Studien, Heft 113; Berlin, 1932).

Krockow, *Warnung*: Christian Graf von Krockow, *Warnung vor Preußen* (Berlin, 1981).

LA: J. W. Goethe, *Die Schriften zur Naturwissenschaft*, Leopoldina-Ausgabe (Weimar, 1947–).

Leavis, *Pursuit*: F. R. Leavis, *The Common Pursuit* (London, 1952).

Leibniz, *Discours*: G. W. Leibniz, *Discours de métaphysique (1686)*, ed. H. Herring (Hamburg, 1958).

Leibniz, *Monadology*: G. W. Leibniz, *The Monadology and Other Philosophical Writings*, ed. R. Latta (London, 1898).

Leibniz, *Théodicée*: G. W. Leibniz, *Essais de Théodicée...*, ed. J. Jalabert (Paris, 1962).

Lessing, *Schriften*: G. E. Lessing, *Schriften*, ed. K. Lachmann and F. Muncker (Stuttgart, 1886–1924).

Lichtenberg, *Aphorismen*: G. C. Lichtenberg, *Aphorismen*, ed. A. Leitzmann (Deutsche Litteraturdenkmale des 18. und 19. Jahrhunderts, 123, 131, 136, 140, 141; Berlin and Leipzig, 1902–28).

Liljegren, *Sources*: S. B. Liljegren, *The English Sources of Goethe's Gretchen Tragedy: A Study on the Life and Fate of Literary Motives* (Acta Reg. Soc. Hum. Litt. Lundensis, 24; Lund, 1937).

Loram, *Publishers*: I. C. Loram, *Goethe and his Publishers* (Lawrence, Kan., 1963).

Luke, *Elegies*: D. Luke, *Goethe's Roman Elegies* (London, 1977).

MA: J. W. Goethe, *Sämtliche Werke nach Epochen seines Schaffens*, Münchner Ausgabe, ed. K. Richter (Munich, 1985–).

Mandelkow, *Kritiker*: K. R. Mandelkow, *Goethe im Urteil seiner Kritiker* (Munich, 1975–84).

Metz, *Friederike*: A. Metz, *Friederike Brion* (Munich, 1911).

Meyer, 'Haus': H. Meyer, 'Kennst du das Haus?', *Euphorion*, 47 (1953), 281–94.

Mollberg, *Kulturstätten*: A. Mollberg (ed.), *Weimars klassische Kulturstätten* (Weimar, 1926).

Monatsblätter: Göttinger Monatsblätter, 80 (October 1980).

Moritz, *Bildende Nachahmung*: K. P. Moritz, *Über die bildende Nachahmung des Schönen* (Deutsche Litteraturdenkmale des 18. und 19. Jahrhunderts 31), ed. S. Auerbach (Heilbronn, 1888).

Nautical Almanac: The Nautical Almanac and Astronomical Ephemeris for the year 1788, published by order of the Commissioners of Longitude (London, 178[8]).

Nettleton, 'Books': G. H. Nettleton 'The Books of Lydia Languish's Circulating Library', *Journal of English and Germanic Philology* 5 (1903–5), 492–500.

Nietzsche, *Werke*: F. Nietzsche, *Werke*, ed. K. Schlechta (Munich, 1954).

Nisbet, *Tradition*: H. B. Nisbet, *Goethe and the Scientific Tradition* (London, 1972).

Nohl, *Möller*: J. Nohl, *Goethe als Maler Möller in Rom* (Weimar, 1962).

Nollendorfs, *Urfaust*: V. Nollendorfs, *Der Streit um den Urfaust* (Paris and The Hague, 1967).

Parth, *Christiane*: W. W. Parth, *Goethes Christiane* (Munich, 1980).

Pascal, *Sturm und Drang*: R. Pascal, *The German Sturm und Drang* (Manchester, 1953).

Petriconi, *Unschuld*: H. Petriconi, *Die verführte Unschuld* (Hamburger Romanistische Studien, Reihe A, xxxviii; Hamburg, 1935).

Pound, *Poems*: E. Pound, *Selected Poems*, ed. T. S. Eliot (London, 1959).

Rabelais, *Œuvres*: F. Rabelais, *Œuvres complètes*, ed. P. Jourda (Paris, 1962).

Rabener, *Schriften*: G. W. Rabener, *Sämmtliche Schriften: Dritter Theil* (Leipzig, 1777).

Reed, 'Paths': T. J. Reed, 'Paths through the Labyrinth: Finding your Way in the Eighteenth Century', *Publications of the English Goethe Society*, NS 51 (1981), 81–113.

Rosenberg, *Bureaucracy*: H. Rosenberg, *Bureaucracy, Aristocracy and Autocracy: The Prussian Experience (1660–1815)* (Cambridge, Mass., 1958).

Rost, *Selbstmord*: H. Rost, *Bibliographie des Selbstmords* (Augsburg, 1927).

Rousseau, *Œuvres*: J.-J. Rousseau, *Œuvres complètes* (n.p., 1792).

Sagarra, *Social History*: E. Sagarra, *A Social History of Germany 1648–1914* (London, 1977).

Sauder, 'Aspekte': G. Sauder, 'Sozialgeschichtliche Aspekte der Literatur im 18. Jahrhundert', *Internationales Archiv für Sozialgeschichte der deutschen Literatur* 4 (1979), 196–241.

Schiller, *Briefe: Schillers Briefe*, ed. F. Jonas (Stuttgart, Leipzig, Berlin and Vienna, n. d.).

Schiller, *Werke*: F. Schiller, *Sämtliche Werke*, ed. G. Fricke and H. G. Göpfert (Munich, 1965).

Schöffler, 'Anruf': H. Schöffler, 'Anruf der Schweizer' in *Deutscher Geist im 18. Jahrhundert: Essays zur Geistes- und Religionsgeschichte*, 2nd edn. (Göttingen, 1967), 7–60.

Schöne, 'Auguralsymbolik': A. Schöne, 'Auguralsymbolik', *Goethe-Jahrbuch* 96 (1979), 22–53.

Schöne, *Farbentheologie*: A. Schöne, *Goethes Farbentheologie* (Munich, 1987).

Schöne, *Götterzeichen*: A. Schöne, *Götterzeichen, Liebeszauber, Satanskult: Neue Einblicke in alte Goethetexte* (Munich, 1982).

Schütz, *Illusions*: M. Schütz, *Academic Illusions in the Field of Letters and the Arts* (Chicago, 1933).

Shaftesbury, *Characteristics*: Anthony Ashley Cooper, Third Earl of Shaftesbury, *Characteristics* . . . , ed. J. M. Robertson (Gloucester, Mass., 1963; rep. of 1900 edn.)

Shelley, *Letters*: *The Letters of Percy Bysshe Shelley*, ed. F. L. Jones (Oxford, 1964).

Silk and Stern, *Tragedy*: M. S. Silk and J. P. Stern, *Nietzsche on Tragedy* (Cambridge, 1981).

Smith, *Theory*: A. Smith, *The Theory of Moral Sentiments* (London, 1759).

Soliday, *Community*: G. Soliday, *A Community in Conflict: Frankfurt Society in the Seventeenth and Early Eighteenth Centuries* (Hanover, NH 1974).

Stadt Goethes: *Die Stadt Goethes: Frankfurt am Main im XVIII. Jahrhundert*, ed. H. Voelcker (Frankfurt-on-Main, 1932).

Staiger, *Goethe*: E. Staiger, *Goethe* (Zurich and Freiburg, 1952–9).

Steiger: *Goethes Leben von Tag zu Tag: Eine dokumentarische Chronik*, ed. R. Steiger (Zurich and Munich, 1982–).

Stern, *Nietzsche*: J. P. Stern, *A Study of Nietzsche* (Cambridge, 1979).

Sterne, *Journey*: L. Sterne, *A Sentimental Journey* . . . , ed. I. Jack (London, 1968).

Strohschneider-Kohrs, 'Künstlerthematik': I. Strohschneider-Kohrs, 'Künstlerthematik und monodramatische Form in Rousseaus *Pygmalion*', *Poetica* 7 (1975), 45–73.

Sturm und Drang: K. Freye (ed.), *Sturm und Drang* (Berlin, Leipzig, Vienna, and Stuttgart, n.d.).

Sturm und Drang. Dramatische Schriften: E. Lowenthal and L. Schneider (eds.), *Sturm und Drang. Dramatische Schriften* (Heidelberg, 1972).

Sulzer, *Schriften*: J. G. Sulzer, *Vermischte philosophische Schriften* (Leipzig, 1773).

Theorie und Technik: D. Kimpel and C. Wiedemann (eds.), *Theorie und Technik des Romans im 17. und 18. Jahrhundert*, ii (Tübingen, 1970).

Ullrich, *Robinson*: H. Ullrich, *Robinson und Robinsonaden*, i. *Bibliographie* (Litterarhistorische Forschungen 7; Weimar, 1898).

Veil, *Patient*: W. H. Veil, *Goethe als Patient* (Jena, 1946).

Vierhaus, *Absolutismus*: R. Vierhaus, *Deutschland im Zeitalter des Absolutismus (1648–1763)* (Göttingen, 1978).

Vulpius, *Goethepark*: W. Vulpius, *Der Goethepark in Weimar* (Weimar, 1975).

WA: *Goethes Werke*, Weimarer Ausgabe (Weimar, 1887–1919).

Walker, *Home Towns*: M. Walker, *German Home Towns* (Ithaca, NY, and London, 1971).

Ward, *Fiction*: A. Ward, *Book Production, Fiction, and the German Reading Public 1740–1800* (Oxford, 1974).

Wells, *Development*: G. A. Wells, *Goethe and the Development of Science: 1750–1900* (Alphen, 1978).

Wolff, *Humanität*: H. M. Wolff, *Goethes Weg zur Humanität* (Munich, 1951).

Zimmermann, *Weltbild*: R. C. Zimmermann, *Das Weltbild des jungen Goethe* (Munich, 1979).

Zorn, 'Führungsschichten': W. Zorn, 'Deutsche Führungsschichten des 17. und 18. Jahrhunderts: Forschungsergebnisse seit 1945', *Internationales Archiv für Sozialgeschichte der deutschen Literatur* 6 (1981), 176–97.

Notes

Conversations with Eckermann are cited by date only. Other authorities are cited by the abbreviated forms listed above.

CHAPTER I

The Age of Goethe?

4 *German philosophy was being taught*: Carlyle to Goethe, 22 Dec. 1829, Carlyle, *Correspondence*, p. 162.

4 *last person with her own adult memories*: Sophie Bettmann, see *Monatsblätter*, p. 9.

6 *monarchical into a bureaucratic absolutism*: see Rosenberg, *Bureaucracy*; Zorn, 'Führungsschichten'.

6 *138 volumes*: in WA, counting part-volumes separately but excluding volumes entirely devoted to indexes.

6 *'People are always'*: Biedermann-Herwig, 3. 2, p. 792, No. 6878.

6 *'Goethe . . . lived and lives'*: *Menschliches, Allzumenschliches*, II. 2, 'Der Wanderer und sein Schatten', para. 125, Nietzsche, *Werke*, i. 928.

7 *'is about as unrepresentative'*: Eliot, 'Goethe', p. 219, my italics.

7 *'after all, the artist'*: HA, xii. 47.

7 *'The more I have learnt'*: Eliot, 'Goethe', p. 218.

7 *'If someone were to ask'*: Biedermann-Herwig, 3. 1, p. 730, No. 5539.

Princes, Pietists, and Professors

8 *20 and 27 million*: Bruford, *Germany*, pp. 158–9.

8 *24 million*: Ford, *Europe*, p. 38.

9 *94 spiritual and temporal*: for this and following information about the Empire, Bruford, *Germany*, p. 7.

9 *1,500 guilders*: Stadt Goethes, p. 66.

9 *16,000 cases*: Bode, *Ruhm*, p. 70; cp. HA, ix. 530.

11 *average loss was around one-third*: Sagarra, *Social History*, p. 5.

11 *not by the entrepreneurs but by the princes*: Blanning, *Reform*, pp. 8–10.

12 *affinity for state absolutism*: see Krockow, *Warnung*, pp. 133–5; Blanning, *Reform*, pp. 26–9, both based on Hinrichs, *Preußentum*.

13 *Spener left little mark*: see Dechent, *Kirchengeschichte*.

13 *two major problems of philosophy*: Leibniz, *Théodicée*, Préface, p. 30.

13 *implications for the German social order*: 'Von dem Verhängnisse', Heer, *Leibniz*, pp. 199–203.

14 *the first after Pascal*: see Boyle, 'Pascal'.

15 *'soul should often imagine'*: Leibniz, *Discours*, para. 32, p. 82.

15 *nine separate translations*: Ullrich, *Robinson*, pp. 43–50, 102–39.

15 *Neville's* Isle of Pines: conveniently accessible in Henderson, *Novels*, pp. 225–35.

18 *as a linguistic innovator*: Blackall, *Emergence*, pp. 26–48.

18 *'Geschäfte' and 'Geschäftsmann'*: HA (1988), ix. 805.

19 *regulation of the German Third Estate*: Gerth, *Intelligenz*, pp. 33–8, 44.

19 *'more than anything else'*: Blanning, *Reform*, p. 14; cp. Vierhaus, *Absolutismus*, pp. 51, 72.

The Literary Context, to 1770

20 *England mattered most*: Schöffler, 'Anruf', p. 23.

20 *382 clergymen*: ibid.

20 *in 1692 displaced Latin*: Ward, *Fiction*, p. 31.

21 *'First select'*: Versuch einer Critischen Dichtkunst, I, 4, para. 21, Gottsched, *Werke*, 6. 1, p. 215.

21 *passion for poetry*: Lob- und Gedächtnißrede auf . . . Martin Opitzen, Gottsched, *Werke*, 9. 1, p. 161.

21 *translated a French satire*: L. A. Gottsched, *Die Pietisterey im Fischbeinrocke*, e.g. in Brüggemann, *Gottsched*, pp. 137–215, taken from Bougeant, *Femme docteur* (a comparison of the versions shows that no claim for Frau Gottsched's originality can be sustained; the same is largely true of her husband's *Der sterbende Cato*, pieced together from translations from French and English (Addison)).

24 *movement of biblical criticism*: see Boyle, 'Lessing'.

25 *'expression in new symbols'*: Bruford, *Theatre*, p. 115.

25 *'from 1740 . . .'*: ibid. 114.

25 *saw the modern philosopher as the true Protestant:* Fichte, *Werke*, vii. 609.

25 *'the fear of the creeds'*: quoted Gerth, *Intelligenz*, p. 122, n. 291.

26 *after explaining*: Biedermann-Herwig, 3. 1, pp. 759–63, No. 5619, p. 760.

26 *'not omnisentient'*: ibid. 761.

26 *'The German philosophy'*: ibid. 722, No. 5524.

26 *'Amerika, du hast'*: HA, i. 333.

27 *'Souls in general'*: Leibniz, *Monadology*, p. 266 ('ensamples' is Latta's translation of 'échantillons').

27 *'second Maker'*: Shaftesbury, *Characteristics*, i. 136, Treatise III, 'Advice to an Author', I. iii.

29 *'sphere of association'*: Stern, *Nietzsche*, p. 127.

30 *'empfindsam' invented or revived by J. J. C. Bode*: Böttiger, 'Bode's Leben', p. iii.

30 *Lydia Languish borrowed*: Nettleton, 'Books'.

30 *'It had ever . . . been'*: Sterne, *Journey,* section 'Amiens', p. 43.

30 *'I sat down close'*: ibid., 'Maria—Moulines', 'The Bourbonnois', pp. 114, 117.

31 *'He watch'd and wept'*: *The Deserted Village,* l. 166, Goldsmith, *Works,* p. 28.

31 *'As we have'*: Smith, *Theory,* p. 2.

32 *'Nothing can happen'*: Leibniz, *Discours,* para. 14, p. 36.

32 *'absolutely sheltered'*: ibid., para. 32, p. 82.

34 *military and financial policies of Frederick*: Durrani, 'Love'.

34 *'courtly praise of . . . half-men'*: 'Fürstenlob', ll. 3–9, Klopstock, *Oden,* pp. 125–6.

34 *'dignity of the poet's subject-matter'*: HA, ix. 399.

34 *17 per cent of life's earnings*: Sauder, 'Aspekte', p. 214.

35 *'What is the history'*: Klopstock, *Werke,* p. 1052 (*Eine Beurteilung der Winckelmannischen Gedanken über die Nachahmung der griechischen Werke in den schönen Künsten*).

35 *'sacred history'*: ibid.

36 *intensity of his feelings*: ibid. 209 (*Messias* I, ll. 444–5).

36 *Divine self, which floats away*: ibid. 445 (*Messias* XI, ll. 40–3, 53).

36 *Goethe himself called it bizarre*: HA, ix. 270.

CHAPTER 2

Frankfurt and the Goethes

43 *a prosperous free city*: the following account principally from *Stadt Goethes,* only a selection of points being individually annotated.

44 *'Tage' with 'Sprache'*: *Faust,* ll. 6876–8.

44 *'genug' with 'Besuch'*: ibid., ll. 2901–2.

44 *'Das wäre mir'*: ibid., l. 10311.

44 *eleven honorific adjectives*: *Stadt Goethes,* p. 56.

44 *'peppersacks' or 'barrelsquires'*: ibid. 131.

44 *quotations from the Bible*: HA, ix. 251.

44 *not French but Italian*: Bode, *Lehrjahre,* p. 73.

44 *congratulatory poem in Yiddish*: Dechent, *Kirchengeschichte,* ii. 196.

44 *200 houses and a synagogue*: *Stadt Goethes,* p. 27.

45 *Catholics had the churches*: Soliday, *Community,* p. 5.

45 *twelve Lutheran pastors*: Dechent, *Kirchengeschichte,* ii. 265.

45 *private conventicles*: ibid. 224.

45 *three substantial local foundations*: *Stadt Goethes,* p. 19.

45 *'external potentates'*: ibid. 87.

46 *a town child*: cp. HA, xiii. 149.

46 *3,000 houses: Stadt Goethes*, p. 17.

46 *the average was around twenty-five*: ibid. 21.

47 *standing voluntarily in the bucket-chain*: HA, x. 83–4.

47 *the 15-year-old*: ibid. ix. 222.

47 *8.30 when the town-gates closed*: Faber, *Frankfurt*, i. 25–8.

47 *first thing that struck them*: see *Etwas über Frankfurt*.

48 *Hole of Pestilence*: 'Pestilenzloch', Behrends, *Einwohner*, pp. 110–19.

48 *same long-case clock*: Steiger, i. 20.

49 *'With the fresh young mind'*: WA, II. 12, pp. 5–6.

49 *42,500 guilders: Stadt Goethes*, p. 387.

49 *salary of his father-in-law, the Schultheiss Textor*: ibid. 395.

49 *country pastor at this time could expect*: Gerth, *Intelligenz*, p. 28.

50 *nominally about 7½ new pence*: Sagarra, *Social History*, p. 456.

50 *three days' wages: Stadt Goethes*, p. 396.

50 *Caspar Goethe's meticulously kept accounts*: ibid. 398–9.

50 *changed the spelling*: ibid. 367.

50 *saw the Doge wedded*: Steiger i. 27.

50 *model of a Venetian gondola*: WA, III. 1, p. 241.

51 *'had Prussian sympathies'*: HA, ix. 47.

52 *recounted the affair to his son*: ibid. 75–6; cp. ibid. x. 53–6.

52 *newspapers that he took were violently anti-Prussian: Stadt Goethes*, p. 403.

52 *respect for its institutions and traditions*: HA, ix. 182.

52 *'could not resign himself'*: ibid. 86.

53 *'inhuman prejudice': Stadt Goethes*, p. 410.

53 *Custine's advancing troops*: ibid. 14.

'More chatterbox than substance': 1749–1765

53 *'dilettantism'*: HA, ix. 32.

53 *thirty schoolteachers: Stadt Goethes*, p. 104.

53 *Maria Magdalena Hoff*: Dechent, *Kirchengeschichte*, ii. 194–7.

54 *Johann Schellhaffer*: Bode, *Lehrjahre*, p. 55.

54 *private tutors: Stadt Goethes*, pp. 167, 415–16.

54 *Latin teacher was a Turk*: Scherbius, see Dechent, *Kirchengeschichte*, ii. 196.

54 *a former Dominican from Naples*: Giovinazzi, see ibid.

54 *French by an émigré*: Roland, see ibid.

54 *formerly Jewish clerk*: Christfreund (Christamicus), see ibid. 195.

54 *J. G. Albrecht: Stadt Goethes*, pp. 156–8.

54 *'The Consistory gives'*: Dechent, *Kirchengeschichte*, ii. 195.

55 *first preserved poem*: HA, i. 7–8.

55 Miscellaneous Poems: ibid. ix. 142.

55 *Goethe saw playing the piano*: Grumach, i. 32; *Stadt Goethes*, pp. 333–4; Bode, *Lehrjahre*, pp. 148–9.

55 *little understanding between the two brothers*: HA, ix. 37.

55 *irritation at the grief*: Grumach, i. 16.

55 *H. P. Moritz*: HA, ix. 114.

55 *J. D. B. Clauer: Stadt Goethes*, p. 385.

56 *portrait—heavily corrected by a drawing master—of Clauer*: CGZ, I, No. 74.

56 *preferred to be referee*: Grumach, i. 20.

56 *three separate sets of clothes*: ibid. 41–2.

56 *'communicable'*: HA, x. 131.

56 *'read between the lines'*: ibid. 133; cp. ix. 227–30.

56 *played the part of Nero*: HA, ix. 109.

57 *read their works,—and to consume refreshments*: *Stadt Goethes*, p. 412.

57 *independent evidence that an official investigation*: Steiger, i. 138–9.

57 *possibly at the instigation of his father*: ibid. 141–2.

58 *'Arcadian Society of Phylandria'*: Bode, *Lehrjahre*, p. 165.

58 *'addiction to dissolute living'*: Grumach, i. 54.

58 *'treacherous' iridescence*: ibid. 56.

58 *'a good chatterbox'*: ibid. 55.

58 *basic legal text-books*: HA, ix. 145–6.

58 *interest in philosophy*: Steiger, i. 144–5.

58 *ill during the spring*: ibid. 154.

58 *riding lessons were not particularly successful*; HA, ix. 147–8.

58 *positively refused to dance*: Steiger, i. 159.

58 *earliest landscape drawings*: CGZ, I, Nos. 1, 2.

58 *'in Saxony . . . the land'*: Steiger, i. 158.

59 *'something of the character of a flight'*: ibid. 165.

59 *may have refused to lend money*: *Stadt Goethes*, p. 420.

59 *crippled young Prince of Saxony*: ibid. 368.

59 *present his boy with a puppet-theatre*: ibid. 412.

59 *public reading from Klopstock's* Messiah: ibid. 401.

60 *'in the matter of completing things'*: HA, ix. 145.

60 *'We went to Goethe's house'*: Grumach, i. 223.

60 *'Ah, if you had seen'*: Bode, i. 153, No. 228.

60 *'Vom Vater hab' ich'*: HA, i. 320.

60 *'Now I really understand'*: *Stadt Goethes*, p. 437.

61 *'stuffed full of* beaux esprits': ibid. 435.

61 *'the infuriating fatherly tone'*: Grumach, i. 252.

61 *very few expressions*: the only other serious case is a letter to Kestner of 10 Nov. 1772 (*DJG*, iii. 7–8) in which the mood is compounded partly out of Goethe's own ill-temper and partly out of a clear identification with his father.

61 *'The course of my father's'*: HA, ix. 32.

62 *own desire was to study history*: ibid. 241.

62 *'an academic teaching position'*: ibid.

A burnt-out case?: 1765–1770

62 *complaints of the poorer students*: Jugler, *Leipzig*, p. 50.

62 *red, pale green, and yellow*: ibid. 6.

62 *one-third of the size*: ibid. 5.

63 *'gospel of beauty'*: HA, ix. 314.

63 *troupe's annual subsidy of 12,000 guilders, and could attract students*: Jugler, *Leipzig*, p. 80.

63 *Rabener satirically listed*: Rabener, *Schriften*, p. 28.

63 *between three and six lectures*: HABr, i. 14, 26.

64 *water-colour relatively little*: CGZ, I, p. 46, ad No. 97.

64 *portrait heads*: ibid., Nos. 3–10, 17–27, 67–71, VIa, No. 36.

64 *theatrical scenes*: CGZ, I, Nos. 75–8.

64 *Landscape*: ibid., Nos. 45–8, 63; cp. 36, VIb, Nos. 1–7.

64 *few quick sketches*: ibid., Nos. 37, 74, 86, 89, 90–1.

64 *'fine thing a Professor'*: HABr, i. 13.

64 *getting used to the beer*: DJG, i. 83; Steiger, i. 173.

64 *seen Gellert*: HABr, i. 16.

64 *Gottsched who at 65 has married a 19-year-old*: ibid. 14.

64 *goose, snipe, or trout*: ibid. 15.

64 *Corona Schröter*: WA, i, p. 228.

64 *the 'Singspiel' or melodrama*: Steiger, i. 200, 275.

64 *any intention of becoming a 'dandy'*: HABr, i. 14.

64 *set of more fashionable outfits*: HA, ix. 250.

64 *particularly those to Horn*: DJG, i. 484.

64 *Böhmes were kind to him*: HABr, i. 39.

64 *a poem to be presented to his grandparents*: DJG, i. 94.

65 *composing an epithalamium*: HA, ix. 301.

65 *fifth act in the blank verse*: HABr, i. 21.

65 Beauties of Shakespeare: Steiger, i. 190.

65 *epithalamium was severely criticized*: HABr, i. 44; cp. HA, ix. 300.

65 *Behrisch and 'Käthchen' dominated*: HABr, i. 33.

66 *'he was in Leipzig'*: Grumach, i. 86.

66 *'the last thing I expected'*: ibid. 111–12.

66 *'drunk as a beast'*: DJG, i. 144.

66 *one quite unserious duel*: Grumach, i. 91–2.

66 *from a bolting horse*: HABr, i. 61.

66 *'my life here'*: ibid. 40.

66 *deformed their operators*: HA, ix. 300.

66 *he was soon so disgusted*: Grumach, i. 81.

67 *'We have not lost a friend'*: ibid. 82.

67 *'I shall divide'*: HABr, i. 33.

67 *sweets for their father's little dog*: Grumach, i. 108–9.

67 *label for the Schönkopf firm*: CGZ, VIb, No. 271.

67 *on Sundays he would go on after dinner*: HABr, i. 53.

67 *helped put on an amateur production*: Steiger, i. 255, 257.

68 *ruins of Dresden*: ibid. 262.

68 *'no place so godless'*: HABr, i. 39.

68 *only fifteen between September 1766 and May 1767*: ibid. 44.

68 Die Laune des Verliebten: Steiger, i. 216.

68 *conclave, consisting probably of Behrisch, Horn*: ibid. 227.

68 Poetische Gedanken: HA, i. 9–13.

68 *'Joseph . . . was condemned'*: HABr, i. 52.

69 *earthquake in Lisbon*: HA, ix. 29–31.

69 *altar of mineralogical specimens*: ibid. 43–5.

69 *'peculiar attitudes in religion'*: Grumach, i. 111.

69 *'Now it once occurred'*: ibid. 110.

69 *deliberately shunned Lessing*: HA, ix. 327.

69 *Oeser turned away all visitors*: ibid. 329.

70 *'can never be his wife'*: Grumach, i. 82.

70 *'almost without a girl'*: HABr, i. 55.

70 *'good friendship'*: DJG, i. 265; cp. HABr, i. 71.

70 *sexually inexperienced*: Bode, *Lehrjahre*, p. 302.

70 *His religious position*: HA, ix. 334–5.

71 *contracted some form of tuberculosis*: Veil, *Patient*, pp. 55–6.

71 *'Odes to my Friend'*: 'Oden an meinen Freund, 1767', HA, i. 21–4.

71 *a number of short poems*: *DJG*, i. 475–7, four of these are printed in HA, i. 18–20.

71 *a little collection of his most recent poems*: *DJG*, i. 193–7.

71 *'You are so merry'*: Grumach, i. 112–13.

72 *J. F. Metz*: Steiger, i. 286–7.

72 *a portrait of her English friend*: ibid. 292–3.

72 *Goethe compared his relation*: HABr, i. 79; cp. *DJG*, i. 505.

72 *not yet reformed*: Steiger, i. 303.

72 *drafting a new play*: HABr, i. 79; cp. *DJG*, i. 505.

72 *a chronic latent tonsillitis*: Veil, *Patient*, p. 27.

72 *oxalic kidney stones*: ibid. 39–40.

72 *'"Where is he now?"'*: *DJG*, i. 273.

72 *mother had turned to her Bible*: Grumach i. 120.

72 *'Saviour has at last caught up'*: HABr, i. 84.

73 *'too much the antithesis'*: Grumach, i. 114.

73 *courtly virtue of 'good taste'*: ibid. 118.

73 *'Goethe . . . still looks unhealthy'*: ibid. 124.

73 *'However healthy and strong'*: *DJG*, i. 280.

74 *'could not tolerate a brother-in-law'*: HABr, i. 102.

74 *Lessing lacked practical experience*: ibid. 98; cp. HABraG, i. 9–10.

74 *overstated the claims of poetry*: HABr, i. 91.

74 *only true teachers*: ibid. 104.

74 *wildness of his republican spirit*: *DJG*, i. 275.

74 *they are merely verbal gestures*: HABr, i. 90.

74 *'frigid, dry and far too superficial'*: HA, ix. 349.

75 *250 'awakened souls'*: Dechent, *Kirchengeschichte*, ii. 211.

75 *showed Wolfgang round*: Grumach, i. 124–5.

75 *'able to indicate the right path'*: HA, x. 57.

75 *Fräulein von Klettenberg*: Dechent, *Kirchengeschichte*, ii. 187–91, 211–13.

76 *creative role of Lucifer*: cp. HA, ix. 351–3.

76 *Voltaire's dismissal, quoted at length, of religious dogmas*: *DJG*, i. 428.

76 *theories about original sin*: ibid. 435.

76 *God and Nature*: ibid. 431.

77 *sign of Virgo*: ibid. 427.

77 *Moses Mendelssohn*: ibid. 437–9.

77 *'my efforts to become'*: HABr, i. 79.

First writings

77 *At the age of 17*: HABr, i. 43.

78 *energy of sexual desire*: ibid. 25.

78 *'A Song over The Unconfidence'*: ibid. 29–30 (Goethe's spelling and punctuation, except for the addition of quotation marks in strophes 5, 8, and 10).

80 *'first and most genuine of all poetic genres'*: HA, ix. 397.

80 *ballad 'Pygmalion'*: DJG, i. 183–4.

81 The Lover's Spleen: *Die Laune des Verliebten*: HA, iv. 7–27.

82 *'Du junger Mann'*: DJG, i. 292.

82 *'Gern verlass' ich'*: HA, i. 18.

83 *self-consciously 'solcher'*: DJG, i. 476.

84 *'My letter has all the makings'*: HABr, i. 63.

84 *'Your servant, Herr Schönkopf'*: ibid. 68.

84 *letter of 30 October 1765*: ibid. 15–17.

85 *'Louis Quatorze' style and therefore 'nowadays contraband'*: ibid. 89.

85 Partners in Guilt: *Die Mitschuldigen*, HA, iv. 28–72.

85 *second draft was probably complete by spring 1769*: DJG, i. 505, but see HA (1988), iv. 478, for a tenuous argument for Aug.–Dec. 1769.

86 *'I shall have a house'*: HABr, i. 103.

86 *'Denn zwischen Mann und Frau'*: HA, iv. 32, l. 106.

87 *'Dies ist nun alle Lust'*: ibid. 34, l. 165.

87 *projected into the winter of 1770-1*: it is Carnival time, perhaps Jan., a year after the marriage of Sophie and Söller (l. 65). Goethe in 1769 was expecting the wedding of Käthchen and Kanne in early 1770: if the action of the play is situated in Dec. 1770 or Jan. 1771, a year after that catastrophe, then Sophie will have been born, like Käthchen, in 1746 (l. 184), and Alcest will have left Leipzig, like Goethe, in the late summer of 1768 (l. 103).

87 *Sophie is relieved of blame*: cp. first version l. 188, second version l. 520, DJG, i. 324, 389.

88 *as a satirist of modern ways of* thinking: HABr, i. 104.

88 *'Rettung'*: DJG, i. 309–10.

88 *entire contents of Hell*: Grumach, 'Prolog'; cp. Zimmermann, *Weltbild*, ii. 279–80, though to speak in this connection of Goethe's 'metaphysisch-kosmologische Überzeugungen' (p. 281, my italics) is to confuse what Goethe *thought* with what he may have *known about*, and what he made use of as the material of his poetry.

88 *respectable representatives of the Enlightenment*: Leibniz, *Théodicée*, I, paras. 17–20, pp.

116–21 (also paras. 86, 268–9, 272); Haller, *Über den Ursprung des Übels* III, ll. 185–90 (Haller, *Alpen*, p. 73); Lessing, *Leibnitz von den ewigen Strafen* (Lessing, *Schriften*, xi. 461–87); contrast Jackson, 'Air'.

88 *Catharina Maria Flindt*: Boenigk, *Urbild*; Liljegren, *Sources*, p. 31.

CHAPTER 3

The Awakening: 1770–1771

91 *'To Italy, Langer!'*: HABr, i. 107.

91 *tapestries woven to the patterns of Raphael's cartoons*: HA, ix. 362.

91 *lights upon Isaiah*: DJG, ii. 318–19.

91 *'we must not seek to be'*: HABr, i. 114.

92 *Strasbourg, with a population of some 43,000*: Steiger, i. 357.

92 *his motley French*: HA, ix. 479–80.

92 *'I heard nothing I did not know'*: Steiger, i. 385.

92 *set out with Weyland*: ibid. 372.

93 *'Yesterday we had ridden'*: HABr, i. 109–10.

93 *'I live rather from one day'*: ibid. 112.

93 *'so profoundly boring'*: ibid. 115.

93 *'young man with an inkling'*: DJG, ii. 319.

94 *he passed with distinction*: ibid. 289.

94 *dissertation which would usually be of some 20–40 pages*: Steiger, i. 430.

94 *'excellent man' with his 'big bright eyes'*: Grumach, i. 148–9.

94 *asymmetry of Goethe's face*: Bauer, *Kopf*, p. 12.

94 *'sparrow-like' Goethe*: Bode, i. 20, No. 21.

94 *wrestle with him as Jacob*: HABr, i. 128.

95 *'strike sparks for a new spirit'*: *Journal meiner Reise im Jahre 1769*, Herder, *Werke*, iv. 435.

95 *'disciples who regarded me'*: to Caroline Flachsland, 22 Sept. 1770, Herder, *Briefe*, i. 229.

96 *idioms and turns of phrase*: *Über die neuere deutsche Literatur* I. 6, Herder, *Werke*, i. 162–6.

96 *'Will it be soon that'*: ibid. I. 17. 217.

97 *'German version of such an exercise'*: ibid. II. 3. 266.

97 *'an original writer'*: ibid. III. 7. 402.

97 *'can someone be a Pindar'*: ibid. 406.

97 *'inspiring with the warmth'*: HABr i. 133.

97 *'must remain true to his roots'*: Über die neuere deutsche Literatur III. 7, Herder, *Werke*, i. 405.

97 *personal dedication*: HA, ix. 478.

97 *'the age demanded'*: Pound, *Poems*, pp. 173, 184.

98 *'No Mercury or Apollo'*: Abhandlung über den Ursprung der Sprache, Herder, *Werke*, v. 51–3.

98 *'with every original author'*: ibid. 121.

98 *'a sum total of the operation'*: ibid. 136.

98 *'haben ihm nacherfunden'*: ibid.

98 *under pressure from the ruling French*: HA, ix. 367.

99 *'the oldest of old dears'*: HABr, i. 127.

99 *'they were intended for you'*: ibid.

100 *Herder's development of Robert Lowth*: HA, ix. 408.

100 *'irony' which Goethe found*: ibid. 429–30; cp. HABr, iv. 360.

100 *'poetic world', express an 'attitude'*: HA, ix. 429–30; cp. HABr, iv. 360.

100 *'Sessenheim' is more correct*: Bode, *Lehrjahre*. p. 376.

101 *Goethe's entire account in* Poetry and Truth: Metz, *Friederike*, pp. 22–3.

101 *'From time to time in November'*: Grumach, i. 167.

102 *'It is raining'*: HABr, i. 122.

102 *'the* conscia mens': ibid. 121.

102 *'all the dreams of your childhood'*: ibid. 122.

102 *refused to have any thing more to do*: DJG, ii. 285.

102 *exertions of the Brion family*: ibid. 286.

103 *'the second daughter'*: HABr, i. 272–3.

103 *letter of condolence*: ibid. 119–20.

103 *only 'enthusiast' for religion*: Steiger, i. 433.

103 *'no more than sound politics'*: Bode, i. 17, No. 15.

103 *printed Latin theses*: DJG, ii. 54–8.

104 *'with great good humour'*: HA, ix. 474.

104 *knowall 'with something missing'*: Genton, *Promotion*, pp. 17, 19.

104 *talk of an academic future*: ibid. 119.

105 *letter of application to the bar*: HABr, i. 124–5.

105 *by December he was writing to Salzmann*: DJG, ii. 70.

105 *issuing a formal reprimand*: Steiger, i. 494.

105 *'learning and the arts'*: HABr, i. 154.

105 *14 October, in the Protestant calendar the name-day*: Bode, *Ruhm*, p. 5.

105 *a cost of over six guilders*: DJG, ii. 67, 328.

106 *'now for the first time'*: HA, ix. 520.

106 *Susanna Margareta Brandt*: Beutler, *Essays*, pp. 85–92.

106 *'sombre remorse'; 'here for the first time I was guilty'*: HA, ix. 520.

107 *'I did it in order to ask you about it'*: HABr, i. 130.

Life and Literature: Works, 1770–1771

107 *nearly all the major literary works*: Krogmann, *Friederikenmotiv*, p. 120.

107 *raises the question in* Poetry and Truth: HA, ix. 468–71.

108 *probably an invention*: Metz, *Friederike*, pp. 82–6.

108 *generated not in Sesenheim*: HA, ix. 499.

108 *thirty or more letters*: DJG, ii. 285.

109 *'symbolical existence'*: HABr, i. 246.

109 *'poetic anticipation'*: HA, x. 431, 433.

110 *copied (and interpolated) by Lenz*: DJG, ii. 290–6.

110 *'Erwache, Friederike'*: HA, i. 29–30.

111 *'Es schlug mein Herz'*: ibid. 27–8, but for the absence of strophic division see DJG, ii. 32, and for the textual history, pp. 293–5.

113 *'Es sah' ein Knab''*: HA (1988), i. 509–10.

114 The History of Gottfried: *Geschichte Gottfriedens von Berlichingen mit der eisernen Hand, dramatisirt*, DJG, ii. 88–227 (the second version: *Götz von Berlichingen mit der eisernen Hand. Ein Schauspiel*, HA, iv. 73–175).

114 *Herder's opinion*: HABr, i. 133.

114 *'radical rebirth'*: ibid. 130.

115 *Duke of Marlborough*: *Anecdotes*, p. 18, No. 16.

115 *'characteristic' culture*: HA, xii. 13.

115 *'individualized' Germany*: Walker, *Home Towns*, p. 1.

115 *address to the Shakespeare festival*: HA, xii. 224–7.

116 Von deutscher Baukunst: ibid. 7–15.

116 *'that mysterious point'*: ibid. 226.

116 *Emil Staiger calls its lack of frame*: Staiger, *Goethe*, i. 95.

116 *no modern drama to compare with it*: Bode, i. 48, No. 59.

117 *'great man'*: HA, iv. 82.

117 *'Tell your captain'*: ibid. 139 (full text DJG, ii. 170, iii. 253).

118 *'Poor Friederike will find'*: DJG, iii. 46.

118 *fallen in love with Adelheid*: HA, ix. 571.

119 *'Du hast viel Arbeit'*: HA, iv. 139.

120 *'Scheltet die Weiber!'*: ibid. 116–17 (first version *DJG*, ii. 141).

Between Sentimentalism and Storm and Stress: 1772–1774

125 *Georg Schlosser*: Beutler, *Essays*, pp. 99–107.

125 *Merck, 'a singular man'*: HA, ix. 505.

126 *'man of leather'*: Grumach, i. 317.

126 *its development away from the realistic novel*: J. H. Merck, *Über den Mangel des epischen Geistes in unserm lieben Vaterlande*, in *Theorie und Technik*, pp. 5–10.

126 *difficult for Herder to give unambiguous encouragement*: Bode, i. 41, No. 48.

126 *Merck's abrasive instruction*: HA, ix. 572.

126 *Leuchsenring saw himself*: ibid. 558.

127 *J. W. L. Gleim*: ibid. 400.

127 *Thurn and Taxis*: ibid. 558.

127 *Gleim visited Darmstadt*: Bode, *Ruhm*, p. 52.

127 *'communion of saints'*: HABr, i. 133.

127 *Deinet the publisher called the 'invisible church'*: Genton, *Promotion*, p. 81.

127 *Shakespeare's 'Under the greenwood tree'*: Steiger, i. 491.

127 *'a certain similarity'*: ibid. 485.

128 *Luise von Ziegler*: Bode, *Ruhm*, p. 50.

128 *'I am beginning'*: Bode, i. 23, No. 22.

128 *'our heaven-sent friend'*: ibid. 27, No. 25.

128 *'If Goethe were of noble birth'*: ibid.

128 *obscure allusion in* Poetry and Truth: HA, ix. 521.

128 *hostile reviewer of the aesthetic writings*: *DJG*, iii. 93–7.

129 *His first review was probably*: Steiger, i. 488.

129 *last was a scurrilous attack on Georg Jacobi*: *DJG*, iii. 97.

129 *'patriotic emotion'*: HA, ix. 535.

129 *'the man in his family'*: *DJG*, iii. 88.

129 *'philosophers' could not 'grasp'*: ibid. 92.

129 *in his* Emilia Galotti: HA, ix. 569.

129 *the only and ultimate tribunal*: ibid. 535–6.

129 *'harmless but also fruitless'*: ibid.

130 *middens, some said to have been growing*: Bode, *Ruhm*, pp. 69–72.

130 *population of Wetzlar was a mere 5,000*: ibid.

130 *('Praktikanten'), of whom there were eighteen*: ibid.

131 *He and Goethe had known each other*: Grumach, i. 86.

131 *'He possesses what is called'*: Bode, i. 36–7, No. 44.

132 *Every Saturday afternoon after dinner*: Steiger, i. 514.

132 *'meinem Mädchen'*: Grumach, i. 201.

132 *a 'little quarrel' with Lotte*: Steiger, i. 525.

132 *'for what is love'*: Bode, *Ruhm*, p. 102.

132 *'with indifference'*: Grumach, i. 202.

132 *'daß er nichts als Freundschaft'*: ibid. 203.

132 *few days in Giessen*: Steiger, i. 528–35.

133 *had he not been in Wetzlar*: HA, ix. 553.

133 *midnight on 27 August*: Grumach, i. 210.

133 *warned, however, that his departure*: Steiger, i. 535.

133 *'and today, and tomorrow'*: Steiger, i. 534; HABr, i. 135, (1988) 592.

133 *'Wenn einst nach'*: *DJG*, iii. 73.

133 *'He is gone'*: HABr, i. 134.

134 *started to give her English lessons*: Steiger, i. 595.

134 *'has poured much well-being'*: *DJG*, iii. 11.

135 *thoughts 'of hanging myself'*: ibid. 8.

135 *'Poor young fellow!'*: HABr, i. 136.

135 *'But the devils'*: ibid.

135 *'You complain of solitude'*: ibid. 137.

135 *account in writing of Jerusalem's personality*: HA (1988), vi. 521–4; *DJG*, iv. 351–6.

136 *'Since he lacked all virtues'*: Bode, i. 41, No. 47.

136 *'an academy at Merck's'*: ibid. 40, No. 46.

136 *Goethe drew and engraved a vignette*: CGZ, VIb, No. 272.

136 *'a really proper bit of work'*: *DJG*, iii. 13.

136 *'lost myself in aesthetic speculations'*: HA, ix. 539.

136 *'I am revolving new plans'*: *DJG*, iii. 11.

137 *Cornelia's dowry was settled*: Bode, *Ruhm*, p. 210.

137 *'look forward to a deadly solitude'*: *DJG*, iii. 47.

137 *'and I am telling you'*: HABr, i. 147.

138 Ein Fastnachtsspiel vom Pater Brey: *DJG*, iii. 161–74.

138 *did not appeal to Herder*: ibid. 456; cp. Steiger, i. 602.

138 *'My poor existence'*: *DJG*, iii. 33.

138 *wrote asking for letters*: ibid. 50–1, accepting Steiger's dating, Steiger, i. 601.

138 *'I am alone, alone'*: HABr, i. 148.

138 *2 February he joined the College*: Bode, *Ruhm*, p. 174.

138 *once is he recorded as attending*: Steiger, i. 608.

139 *society game*: HA, ix. 232–6.

139 *marry in earnest*: ibid. x. 71.

139 *'it was easy to see'*: ibid. ix. 234.

139 *'If another and later species'*: Lichtenberg, *Aphorismen*, F342.

139 *second silhouette over his bed*: HABr, i. 138.

139 *'as sterile as a sand-dune'*: ibid. 145–6.

139 *a second Lotte*: ibid. 145.

139 *'cold misogynist'*: Bode, i. 43, No. 51.

139 Concerto dramatico: *DJG*, iii. 63–8.

139 *Rabelais's Panurge*: Le Tiers Livre, chap. 9, 26, 28, Rabelais, *Œuvres*, i. 437–40, 512–14, 521–3, also 517, 524.

140 Des Künstlers Erdewallen: HA, i. 63–7.

140 Des Künstlers Vergötterung: ibid. 67–8.

140 *civic affairs and Goethe 'did not resist'*: HABr, i. 154, 151.

140 *Horn who had now also obtained administrative office*: *DJG*, iii. 49.

140 *'I need too much for my own use'*: HABr, i. 156.

140 Zwo wichtige: *DJG*, iii. 117–24.

141 *'No mortal can maintain'*: ibid. 123.

141 Brief des Pastors . . . : HA, xii. 228–39.

141 *'divine love so many hundred years ago'*: ibid. 231.

142 *'with Lessing in mind'*: *DJG*, iv. 11.

142 *concerned for the salvation*: HA, xii. 237.

142 *Goethe's* Mahomet *did not progress*: *DJG*, iii. 128–33.

142 *poetic potential*: HA, x. 122.

142 Jahrmarktsfest zu Plundersweilern: *DJG*, iii. 134–47.

143 *'Was hilfts daß wir'*: ibid. 141.

143 *by June at the latest*: HABr, i. 149; cp. Steiger, i. 607, *DJG*, iii. 421.

143 *Merck had prevailed*: HA, ix. 572.

143 *April the printing began*: Steiger, i. 599.

143 *copies which arrived in the second week of June*: HABr, i. 149.

143 *150 copies were sent for sale*: *DJG*, iv. 4.

143 *borrow money before long to cover*: ibid. 263.

143 *'Boie! Boie!'*: Bode, i. 44–5, No. 56.

144 *'wholly repudiated the* compliment': ibid. 49, No. 62.

144 *'here and there'*: ibid., No. 60.

144 *'the approval you find everywhere'*: HABraG, i. 20.

144 *'My highest wish'*: HABr, i. 153.

144 'the most beautiful, most interesting monster': HA (1988), iv. 489.

145 *Caspar Goethe proved a willing and generous host*: Steiger, i. 613; cp. HA, x. 92.

145 'He is a thin young man': Grumach, i. 239–40.

146 'throw up at [the Jacobis]': Steiger, i. 621.

146 Satyros: HA, iv. 188–202.

146 Erwin und Elmire: *DJG*, iii. 36–61.

147 'my feelings, my ideas': ibid. 39.

147 'Götter, Helden und Wieland': HA, iv. 203–15.

148 'Have you ever died?': ibid. 208–9.

149 *copied it from an early date*: CGZ, I, Nos. 301–2, 305, VIb, Nos. 239–40.

149 'either an atheist—or a Christian!': HABraG, i. 26.

149 *Letter of the Pastor of ✳ ✳ ✳ a work of genius*: Bode, i. 48, No. 57.

150 'I am not a Christian': HABraG, i. 17.

150 *told Lavater in January 1774*: ibid. 19.

150 'If Jesus Christ is not my God': ibid. 25.

150 'curse of Cain': HABr, i. 149.

150 'the word of men': ibid. 159.

150 'we are symbols': HABraG, i. 16.

151 'Continue to love': Bode, i. 51, No. 64.

151 *Rubensian Dutchwoman*: HA, x. 31.

151 'friendship I do not want': HABr, i. 152.

151 'mad with joy': Steiger, i. 632.

151 'barrels of herrings': Bode, i. 54, No. 68.

151 'I am far more to her': *DJG*, iv. 6.

151 'la petite Madame Brentano': Bode, i. 54, No. 69.

151 'a compensation' fate had sent him: *DJG*, iv. 6.

151 *benches on duckboards*: Bode, *Ruhm*, p. 228.

151 'No branch of my existence is lonely': *DJG*, iv. 6.

Detonating the Bomb: Works, 1772–1774

153 'Ich bin ein Deutscher': F. Stolberg, 'Mein Vaterland', in *Göttinger Hain*, p. 188.

153 'If it is good character': *Sturm und Drang*, vol. i, p. lxi.

154 'At the auction in Frankfurt': ibid., p. lxxxi.

154 'as if our tongues were confused': Lichtenberg, *Aphorismen*, F499 (Apr.–May 1777).

154 'Alas I had resolved': *Sturm und Drang. Dramatische Schriften*, i. 417.

154 *recent criticism has shown*: Blunden, 'Lenz'.

155 *3,599 subscribers*: DJG, iv. 335.

155 *'who absolutely desires to transcend'*: Sturm und Drang. *Dramatische Schriften*, ii. 362, 363.

156 Enquiry into the Origin: 'Untersuchung über den Ursprung der angenehmen und unangenehmen Empfindungen', Sulzer, *Schriften*, pp. 1–98.

156 *'Stärke der Seele'*: 'Entwickelung des Begriffes vom Genie', Sulzer, *Schriften*, pp. 307–23, see p. 317.

156 *'All capacities of the soul'*: ibid. 309.

157 *'Ihr Edleren'*: Klopstock, *Werke*, p. 108.

157 *'Schön ist'*: ibid. 53.

158 *'Maifest'*: HA, i. 30–1.

158 *'Wandrers Sturmlied'*: ibid. 33–6.

159 *lines have been found difficult*: e.g. Staiger, *Goethe*, i. 70, also DKV, I. 1, p. 864, MA 1. 1, p. 849, which however come closer to the interpretation proposed here than does Zimmermann, *Weltbild*, ii. 108–9. Like Zimmermann though, these most modern editors do not consider the possibility that the word 'neidgetroffen' refers to the god's envy of the cedar, whose independence of his warming power is precisely what attracts his envious admiration (a possibility rendered more likely by the syntactic parallelism: 'kalt . . . vorübergleiten . . . neidgetroffen . . . verweilen').

160 *'Ganymed'*: HA, i. 46–7.

160 *described as pantheistic*: e.g. DJG, iv. 337, Boyd, *Notes*, i. 63, cp. pp. 39–40, Pascal, *Sturm und Drang*, p. 113.

160 *read Spinoza so methodically*: HABr, i. 476.

162 *'everything fits together'*: Bode, i. 53, No. 67.

162 *a mental audience*: HA, ix. 576–8; cp. HABr, i. 150.

162 *'meine Ideale'*: HABr, i. 152.

162 *H. M. Wolff*: Wolff, *Humanität*, pp. 17–28. A date as early as 1772 seems improbable, however.

163 *subject-matter of the ode*: HA, i. 44–6.

164 *'working up my situation'*: DJG, iii. 41, but contrast Zimmermann, *Weltbild*, ii. 122–6, for some good, though not decisive, arguments for the priority of the drama.

165 *the drama* Prometheus: HA, iv. 176–87.

165 *Rousseau's 'scène lyrique'* Pygmalion: DJG, iii. 19.

166 *'Und du bist'*: HA, iv. 178–9, ll. 100–9.

166 *Rousseau's monodrama ends*: Rousseau, *Œuvres*, xviii. 358–9.

166 Galatea (*it is Rousseau who invents*: Istel, *Rousseau*, p. 9; Strohschneider-Kohrs, 'Künstlerthematik', p. 54.

168 *'What it costs to dig'*: DJG, iii. 41.

168 *'Man kann nicht immer'*: Grumach, i. 257.

168 *'it will be some time'*: HABr, i. 150.

168 *'I'd like to tell'*: ibid. 147.

168 *my feelings to [Jerusalem's] story*: ibid. 159.

169 *'Germanic drama'*: DJG, iv. 11 (for the comparison of the subject-matter of *Götz* and *Werther*, HA, ix. 540).

169 *half-way house between monologue and dialogue*: HA, ix. 576–7.

169 *advising Frau von La Roche*: DJG, iv. 7.

169 *'breaking down the paper wall'*: HABr, i. 157.

170 *'spoken with awe'*: ibid. 173.

170 in, cum et sub: ibid. 157–8.

170 *'The effect of this little'*: HA, ix. 589–90.

171 *'We are not here talking'*: ibid. 583.

171 *The story of Werther*: we follow here the first edn. of 1774 (*DJG*, iv. 105–87) (the second edn. of 1787 will be found in HA, vi. 7–124).

172 *'white-hot expression of pain'*: HABr, i. 379.

173 *'Wenn das liebe Thal'*: DJG, iv. 107, letter of 10 May 1771; cp. HA, vi. 9.

173 *'Mußte denn das'*: DJG, iv. 138–9, letter of 18 Aug. 1771; cp. HA, vi. 51–3.

174 *'was ich weis'*: DJG, iv. 155, letter of 9 May 1772; cp. HA, vi. 74.

174 *sick child*: DJG, iv. 108, letter of 13 May 1771; cp. HA, vi. 10.

174 *'Und das Herz'*: DJG, iv. 161, letter of 3 Nov. 1772; cp. HA, vi. 84–5.

175 *Whether Goethe also knew*: Dr Boyd Hilton tells me 'buff and blue would seem to have become "Whig"—or, to be more precise, "Rockingham Whig"—colours round about the time of Fox's rise to prominence in that party, circa 1780', although the combination already had political significance as the chosen badge of Washington and the American insurgents. It seems difficult to determine who, if anyone, influenced whom.

175 *the novel was too widely known*: Braun, *Zeitgenossen*, i. 148.

175 *'Criticize, should I?'*: ibid. 64.

175 *'the fate of Sodom and Gomorrah'*: ibid. 104.

175 *suicide associated with a reading of* Werther: Rost, *Selbstmord*, pp. 316–26.

176 *'myself was in this case'*: HA, ix. 583–4.

176 The Robbers: *Die Räuber*, Schiller, *Werke*, i. 481–618.

176 *paternal authority and the compensating cult of blood-brotherly friendship*: Gerth, *Intelligenz*, pp. 46–7.

177 *Matthias Claudius's suggestion*: Mandelkow, *Kritiker*, i. 20.

The Messiah and his Nation: 1774–1775

178 *'a town where nobody reads'*: Bode, i. 115, No. 163.

178 *Princes of Saxe-Meiningen*: ibid. 104, No. 143.

179 *'man of few words'*: Grumach, i. 342.

179 *'too arrogant and dogmatic'*: Bode, i. 131, No. 191.

179 *'chameleon' or 'Proteus'*: see the treatment of this theme in Fairley, *Goethe*, pp. 4–11, 230–1, 269.

179 *'entrancing company'*: Bode, i. 132, No. 192.

179 *'in the most vigorous conversation'*: ibid. 108, No. 152.

179 *'wish I had noted down his table-talk'*: ibid. 61, No. 87.

179 *wrote a five-act tragedy*, Clavigo: HA, x. 71.

179 The Artist's Earthly Pilgrimage: *DJG*, iv. 231–5.

179 *'take the key from your heart'*: Grumach, i. 259.

179 *'handkerchief always ready'*: ibid. 270.

179 *'burning down his summerhouse'*: HABr, i. 160.

179 *'give milk without having given birth'*: ibid. 174.

179 *'rolling on the moral snowball'*: *DJG*, v. 4.

179 *he reached Frankfurt*: Steiger, i. 656.

180 *Lavater's notes of their first day*: Grumach, i. 263–5.

180 *Bürger was glad*: Bode, i. 56, No. 75.

180 *feared a loss of their dignity*: ibid.

181 *'still an unbridled, immoderate man'*: ibid. 57, No. 77.

181 *With Wieland relations remained difficult*: HABr, i. 160.

181 *initiated a direct correspondence*: ibid. 161.

181 *'drops of independent spirit'*: ibid. 163.

181 *'If You exist'*: HABraG, i. 27.

181 *'Scripture gives not a single example'*: ibid.

181 *distinctness of the human osseous structure*: *DJG*, v. 366–7.

182 *collected some 30,000 guilders*: Bode, *Ruhm*, p. 305.

182 *Passing over his legal work to Schlosser's brother*: Steiger, i. 663.

182 *'Hoch auf dem alten Turne steht'*: HA, i. 81.

183 *'and he and I'*: *DJG*, iv. 222.

183 *Sunday 24 July*: Steiger, i. 668–70.

183 *'Der König von Thule'*: HA, i. 79–80.

183 *'Es war ein Buhle frech genung'*: ibid. 81–2.

183 *midnight conversation hopes*: cp. HA, v. 25 (*Iphigenie*, ll. 673–9).

183 *'What Goethe and I'*: Bode, i. 67, No. 93.

184 *'dam is gone'*: HABr, i. 165.

184 *remembered the sight of their bodies*: HA, viii. 274–6.

184 *'Only in such moments'*: *DJG*, iv. 222.

184 *'What is the heart of man?'*: ibid. 245.

184 *'Herr Klopstock, the favourite'*: Grumach, i. 296.

184 *'short, plump, neat'*: Steiger, i. 682.

184 *diplomat's ability to talk*: HA, x. 62.

184 *'how far is Goethe our friend?'*: Bode, i. 84, No. 119.

184 *'whose heart is as great'*: ibid. 71, No. 104.

185 *three advance copies to the Kestners*: HABr, i. 170; *DJG*, iv. 249.

185 *Kestner's response*: HABraG, i. 36–7.

185 *'il est dangereux'*: Bode, i. 103, No. 141.

186 *'I am inclined to forgive him'*: ibid. 76, No. 115.

186 *'Werther must—must'*: HABr, i. 173.

186 *Lessing, embittered at what he saw as a libelling*: Bode, i. 107, No. 151.

186 *'does not need a key'*: ibid. 86, No. 122.

186 *'truly national book'*: ibid. 95, No. 137.

186 *'work for* everyone': ibid. 158, No. 236.

186 *'Even if Jerusalem is not Werther'*: Bode, i. 110, No. 156.

187 *'Do you imagine a Roman'*: ibid. 74, No. 111.

187 *'You know my love for English'*: ibid. 108–9, No. 153.

187 *historically quite appropriate*: HABr, i. 163.

187 Clavigo *selling, in Deinet's words, like hot cakes*: Bode, i. 68, No. 98.

187 *'Everything I have read of yours'*: HABraG, i. 55.

188 *'most terrible and loveable'*: Grumach, i. 258.

188 *'The more I consider it'*: ibid. 285.

188 *'the greatest man'*: HABraG, i. 38–9, 40.

188 *'splendid figure acting at a prince's side'*: Bode, i. 72, No. 107.

188 *'Goethe has divine force'*: ibid. 135, No. 199.

188 *'Goethe is a god'*: Steiger, i. 736.

188 *'This Goethe of whom'*: Grumach, i. 297.

188 *'come by the name of a Shakespeare'*: Lichtenberg, *Aphorismen*, E69, written July 1775.

189 *'Poet—dramatic god!'*: Shakespear, Herder, *Werke*, v. 227.

189 *'world of dramatic history'*: ibid. 221.

189 *'Mein Busen war'*: HA, i. 61.

189 *'whole teaching about Christ'*: HABr, i. 182.

189 *'Zwischen Lavater und Basedow'*: HA, i. 90.

190 The Wandering Jew: *Der Ewige Jude*, *DJG*, iv. 95–103.

190 *'O mein Geschlecht'*: ibid., l. 132.

190 *'O Welt voll wunderbaarer'*: ibid., ll. 136–41.

191 *'Now there's a man'*: *Sturm und Drang. Dramatische Schriften*, ii. 40.

192 *'Wer sein Herz'*: *Satyros*, ll. 374–5.

192 *'Erklärung eines alten Holzschnittes'*: HA, i. 135–9.

192 *'Er fühlt'*: ibid., ll. 11–14.

192 *'Die spricht'*: ibid. 39–46.

193 *felt rather distant*: HA, ix. 590.

193 *'Generalbeichte'*: ibid. 588.

193 *'atmosphere in Goethe's home'*: Grumach, i. 368.

193 *'Everything is in a ferment'*: ibid. 299.

193 *'Another thing that makes me happy'*: HABr, i. 177.

193 *The public, he said, was the echo*: ibid. 175.

193 *hostile to Wieland as the editor of the* Teutscher Merkur: Bode, i. 103, No. 142; 105, No. 146.

193 *'break up' first*: Bode, i. 57, No. 78.

193 *'People fear his own fire'*: ibid. 68, No. 96.

193 *'An Schwager Kronos'*: HA, i. 47–8.

194 *'I have been lying curled'*: HABr, i. 171.

194 *talk of a career for him*: HA, x, 69, 112.

194 *'common people' who 'are the best'*: HABr, i. 161.

194 *middle classes of whom, he noted, his current literary circle*: HA, x. 117.

194 *nobility whose favour he had gained*: ibid. 116.

195 *'most extraordinary phenomena'*: Bode, i. 91–3, No. 133.

195 *'enjoy the best of men'*: Grumach, i. 306.

195 *'radical' cure of any 'displeasure'*: Bode, i. 97, No. 139.

195 *'You do not consider how dangerous'*: ibid. 101, No. 140.

195 *refused to believe in the seriousness*: Steiger, i. 693.

196 *'Say goodbye'*: Bode, i. 94, No. 135.

196 *Fritz achieved his own version of the Goethean fusion*: HABraG, i. 49–50; on *Eduard Allwill* cp. *DJG*, v. 250.

196 *Knebel, to whom Goethe had lent*: Bode, i. 93, No. 133.

196 *Kraus thought well of Goethe's portraits*: ibid. 108, No. 152.

196 *Kraus's drawings and etchings of Weimar*: HA, x. 171–5.

196 *wrote to Wieland to congratulate him*: Bode, i. 112, No. 161.

196 *thought the furniture too luxurious*: Grumach, i. 317.

196 *embarrassed when a new satire on Wieland*: Bode, i. 112, No. 160.

196 *not well received*: ibid. 116, No. 165.

196 *Görtz thought it an 'obscenity'*: Bode, i. 111, No. 159.

196 *'write a lot to me'*: HABr, i. 181.

197 *Wieland was after all on the wrong side*: DJG, v. 12.

197 *'I will come to a bad end'*: ibid. 16–17.

197 *'dear old thing they call God'*: ibid. 22.

197 *herself noticed the similarity*: HABr, i. 615.

197 *'Lili', as Goethe called her*: Beutler, *Essays*, pp. 180–308.

197 *final bankruptcy not coming until 1784*: ibid. 188.

198 *as the carnival season of 1775 began*: HABr, i. 176–7.

198 *'thoroughly brilliant period'*: HA, x. 106.

198 *20 April an informal engagement*: Steiger, i. 716–17.

198 *'Lili's Park'*: HA, i. 98–101.

198 *'first woman whom I deeply and truly loved'*: to Eckermann, 5 Mar. 1830.

199 *ready 'to surrender to him her virtue'*: HABraG, ii. 564–7 (Beutler seems not to grasp the plain meaning of this avowal).

200 Hanswursts Hochzeit: *DJG*, v. 183–95.

200 *his hair was beginning to fall out*: Grumach, i. 318; cp. *DJG*, v. 18.

200 *'Die Welt nimmt'*: DJG, v. 185–6.

201 *'Mahomets-Gesang'*: HA, i. 42–4.

201 Salomons Königs von Israel: *DJG*, v. 357–9.

201 *'the four of us'*: Bode, i. 124, No. 179.

202 *'Werther's uniform'*: ibid.

202 *test the strength of the relationship*: HA, x. 127.

202 *'thought I was approaching the port'*: HABr, i. 182–3.

202 *'You will not stay with them long'*: HA, x. 128.

202 *'old place now so new'*: HABr, i. 183–4.

202 *in two minds whether to continue*: Bode, 128, No. 182.

203 *no agreement with his father*: Steiger, i. 721, 'Zur Emmendinger Reise'.

203 *not borne out by the contemporary evidence*: especially *DJG*, v. 249.

203 *Margrave of Baden's highest-paid official*: Beutler, *Essays*, p. 102.

203 *advise him to break with Lili?*: HA, x. 133.

203 *sent to Knebel and so to Carl August*: DJG, v. 230.

203 *'disinterred' in April*: ibid. 20.

203 *'still feel the main purpose'*: HABr, i. 184.

203 *Jakob Gujer, called 'Kleinjogg'*: Bode, *Geniezeit*, pp. 116–21.

204 *'It is strange that a German'*: Bode, i. 129, No. 185.

204 *'Ich saug' an meiner Nabelschnur'*: HA, i. 102.

206 *under some external pressure*: Steiger, i. 736; Bode, i. 129, No. 186.

206 *'splendid high sunshine'*: *DJG*, v. 236.

206 *'omnipotently terrible'*: ibid.

206 *'bare rock and moss'*: ibid.

207 *sketch he made of his garret-room*: CGZ, VIb, No. 174.

207 *seven sketches have survived*: CGZ, I, Nos. 106–12.

207 *of motion in a landscape*: ibid. 117–18, 126, 130.

207 *reduced almost to abstraction*: ibid. 117.

207 *'point of view'*: ibid. 108–9; cp. 105.

207 *'would look well in a picture'*: Bicknell, *Beauty*, pp. ix, xiii.

207 *behind a stove or the back of a seated figure*: CGZ, I, Nos. 97, 110; cp. 85, 88, VIa, No. 37, VIb, No. 21.

207 *in landscapes the eye is in an emphatically high*: ibid. I, Nos. 106, 108, 113–4, 120, VIb, No. 22.

207 *hospice on the Gotthard*: ibid. I, No. 121.

207 *footpath winding down*: ibid., No. 120.

207 *cornering downward track*: ibid., No. 130. See plate 13.

207 *random portrait heads*: ibid, No. 111; cp. VIb, No. 259.

207 *a collaborator of Lavater*: ibid. VIb, Nos. 241–6.

207 *Swiss were a 'noble race'*: HABr, i. 185.

207 *'Holy Switzerland of the German Nation'*: *DJG*, v. 246.

208 *'glad that I know a country like Switzerland'*: ibid. 243.

208 Dritte Wallfahrt nach Erwins Grabe: HA, xii. 28–30.

208 *'She sees the world as it is'*: *DJG*, v. 232.

208 *'he could not sleep'*: Bode, i. 141, No. 208.

209 *writing to Augusta Stolberg*: HABr, i. 187–9.

209 *'I am stranded'*: *DJG*, v. 249.

209 *chorale he wrote for the wedding*: 'Bundeslied einem jungen Paar gesungen von vieren', HA, i. 93–4.

209 *'the most cruel most solemn*: HABr, i. 193–4.

209 *'burning tears of love'*: ibid.

209 *'I tremble at the thought'*: ibid.

209 *'state of continual inner warring'*: Bode, i. 92, No. 133.

209 *'if God does not work'*: ibid. 185–6, No. 279.

209 *'as I saw the sun'*: HABr, i. 192–3.

210 *'What a life'*: ibid. 195.

210 *translation of the* Song of Songs: *DJG*, v. 360–5.

210 *'become pious again'*: *DJG*, v. 249.

210 *wrote to Princess Luise*: ibid. 503.

210 *to Knebel a note*: ibid. 246.

210 *'if I were to go to Italy'*: ibid. 249.

210 *'invisible lash of the Furies'*: HABr, i. 190.

210 *'I am again watching'*: *DJG*, v. 249.

210 *'Duke had quite fallen in love'*: Bode, i. 143, No. 212.

210 *portrait-sketch that Goethe drew of the young ruler*: CGZ, VIb. No. 232.

211 *Zimmermann himself, 'an achieved character!'*: *DJG*, v. 252.

211 *'father and mother came'*: ibid. 260.

211 *to whom Goethe wrote*: ibid. 505.

211 *'I packed for the North'*: ibid. 402–3.

211 *'innocent guilt'*: ibid.

212 *'For eight to ten days'*: Bode, i. 142, No. 210.

212 *'The divine man'*: ibid. 145, No. 216.

'Innocent guilt': Works, 1774–1775

212 *'if I were not now writing dramas'*: HABr, i. 179.

213 Clavigo: HA, iv. 260–306.

213 *'Don't write me such rubbish'*: HA, x. 72.

213 *'Möge deine Seele'*: HA, iv. 294.

214 *'pendant to Weislingen'*: HABr, i. 162.

214 Claudine von Villa Bella: HA, iv. 216–59.

216 *'Do you know the needs'*: ibid. 256.

216 *marriage* à trois *in the court at Carlsruhe*: *DJG*, v. 433.

216 Stella: HA, iv. 307–47.

217 *'Aber wie's geht'*: ibid. 312.

218 Urfaust: HA, iii. 365–420.

218 *not as incomplete as has from time to time been asserted*: e.g. Petriconi, *Unschuld*, p. 112, and see Nollendorfs, *Urfaust*.

218 *'nearly finished', 'finished'*: Grumach, i. 298.

218 *'half-finished'*: ibid. 392.

218 *'A work for* everyone': Bode, i. 158, No. 236.

218 *'Usually people have only one soul'*: Ibid. 153, No. 228.

218 *'profound impersonality'*: Leavis, *Pursuit*, p. 130, quoted and discussed in Silk and Stern, *Tragedy*, p. 274.

219 *death which, according to Hegel*: *Phänomenologie des Geistes*, B. IV. A. 'Selbständigkeit

und Unselbständigkeit des Selbstbewußtseins; Herrschaft und Knechtschaft',
Hegel, *Werke*, iii. 148–9.

219 *famous issue of the periodical*: Lessing, *Werke*, viii. 41–4.

220 *'the godhead has not given man'*: ibid. iii. 386.

220 *'lest God's enemies'*: Dietrich, *Faust*, p. 289.

221 *'the one supremely great tragic drama'*: Schütz, *Illusions*, p. 65.

221 *'end of such fantasies'*: Grumach, i. 292.

222 'Faust *came into being'*: to Eckermann, 10 Feb. 1829.

222 *'Zwar bin ich'*: HA, iii. 367, ll. 13–16.

222 *'Welch Schauspiel!'*: ibid. 369, ll. 101–3.

223 *drawing dated between 1773 and 1775*: CGZ, I, No. 87.

223 *'In Lebensfluten'*: HA, iii. 371, ll. 149–56.

223 *'Du gleichst dem Geist'*: ibid. ll. 159–60.

224 *'Mißhör mich nicht'*: ibid. 406–7, ll. 1123–53.

225 *'es steht ihm'*: ibid. 408, ll. 1181–2, 1190.

226 *'Ich kann die Bande'*: ibid. 416, ll. 31–3.

227 *'Er sieht sein Schätzel'*: ibid. i. 82.

228 *'unhoused . . . unman'*: ibid. iii. 415, ll. 1414–15.

CHAPTER 4

Weimar in 1775

233 *carriages regularly made detours*: MA, 2. 2, p. 710.

233 *at 5 a.m. on 7 November*: details of Goethe's arrival in Steiger, i. 760–1.

233 *a large château*: Bruford, *Culture*, p. 57.

233 *'Redoutenhaus'*: ibid. 121.

233 *Town Church*: ibid. 310.

234 *'Wittumspalais'*: Ehrlich, *Wittumspalais*.

234 *quarter of its 6,000 inhabitants was made up by the court*: Bruford, *Culture*, p. 67.

234 *rate of exchange*: Sagarra, *Social History*, p. 456.

234 *upper class, in W. H. Bruford's words, 'stood out'*: Bruford, *Culture*, p. 62.

235 *Lavater visited, his Swiss German*: Grumach, iii. 28, 30.

235 *great-great-uncle of the historian Lord Acton*: Bruford, *Culture*, p. 55.

236 *dreamed of living as a private gentleman*: Bode, *Pegasus*, p. 84.

236 *University of Jena*: Bruford, *Culture*, pp. 369–70.

236 *an army of around 500 infantry*: ibid. 110–11.

236 *next to no industry*: Bode, *Bau*, pp. 6–9.

237 *von Stein family drew from all sources*: Bode, *Stein*, p. 316.

237 *'So I have been mounting'*: to Knebel, 17 Apr. 1782, HABr, i. 395.

237 Triumph of Sensibility: WA, I. 17, p. 5.

237 *style of Chancery documents*: *AS*, i. 420–1.

237 *notes for the continuation of his autobiography*: WA, I. 53, pp. 383–8, esp. 384.

238 *ducal household*: Bruford, *Culture*, pp. 69–70.

238 *there were also six pages*: ibid. 60.

238 *Baron von Imhoff*: Bode, *Pegasus*, pp. 296–8; Grumach, ii. 542.

238 *no great church-goers*: Grumach, ii. 434.

239 *in a glass carriage*: Ehrlich, *Wittumspalais*, p. 6.

239 *guests selected for their intellect*: Grumach, ii. 397.

239 *Frau von Stein was reported to be skating*: ibid. 64.

239 *one of the pages recalls being dressed as a demon*: ibid. 64–5, cp. iii. 50.

Why Goethe Stayed

240 *Carl August spent 600 dollars on buying*: Bode, *Garten*, pp. 23, 76.

240 *spent considerably more than his official salary*: Beutler, *Essays*, pp. 399, 401; *Stadt Goethes*, p. 419.

241 *subscribers to the* Teutscher Merkur: Bruford, *Culture*, p. 295.

241 *regret the prospect of a shadowy retirement*: ibid. 40.

241 *Görtz, within the year*: ibid. 69.

241 *Countess Görtz and her friend Countess Gianini*: extracts from letters in Grumach ii and iii *passim*.

242 *play cards with the Duchess's party*: this seems to be the true sense of Böttiger's anecdote, retailed by Bruford, *Culture*, p. 82; cp. the dating by Grumach, ii. 363, 366, and Caroline Herder's remark, p. 378.

242 *close to the von Stein family*: Bruford, *Culture*, p. 63.

242 *'You know his good qualities'*: Bode, i. 188, No. 286.

242 *hunting, riding, camping out*: Grumach, i. 478 (cp. iii. 39), 446, 419, 480.

243 *wedding-night of F. J. Bertuch*: ibid. i. 419.

243 *authoritative plain-speaking*: e.g. WA, III. 1, p. 79.

243 *refused any confrontation*: cp. Grumach, ii. 412.

244 *keystone of the ducal family*: cp. Bode, i. 206, No. 321.

244 *In 1782 he wrote to his mother*: on 11 Aug., HABr, i. 368–70.

244 *'Everything pointed'*: HA, x. 174.

244 *court of geniuses*: cp. Hufeland's recollections, Grumach, i. 473–4.

244   *'If it is possible'*: ibid. 386.

244   *great engraver Chodowiecki*: ibid. 412.

245   *amateur theatre has been somewhat exaggerated*: cp. Bruford, *Culture*, pp. 120–1.

245   *'Goethe will doubtless'*: Bode, i. 173, No. 260.

245   *wild behaviour of 1776*: Grumach, i. 446, 409, 414.

245   *Carl August was said to have remarked*: Bode, i. 210, No. 327.

245   *Duke so 'strong and healthy'*: ibid. 229, No. 370.

245   *'his incredible services'*: Grumach, iii. 37.

245   *Duke the only person there showing signs of development*: WA, III. 1, p. 87.

245   *maturing personality*: ibid. 86.

246   *Merck—'the only man'*: ibid. 87.

246   *'he has now done'*: HABr (1988), i. 707.

246   *'all the business'*: ibid. 205.

246   *'Even if it is only'*: ibid. 207.

246   *neither satisfaction nor peace of mind*: ibid. 230.

246   *the lure of the 'world*: HABr, i. 209; WA, IV. 3, pp. 21, 37; WA, III. 1, p. 94, and cp. p. 51 ('Regieren!!').

247   *a 'hole'*: HABr, i. 289.

247   *'Fate, the object of my worship'*: ibid. 221, 239 (7 Nov. 1777, the second anniversary of his arrival in Weimar).

247   *'wholly concealed'*: ibid. 277. Next quote: ibid. 381.

247   *'help him on the way*: WA, III. 1, p. 94.

247   *'not my will is being done'*: HABr, i. 400.

247   *'I have less and less idea'*: ibid. 414.

247   *'the safest, trustiest'*: ibid. 238.

247   *'evil genius'*: ibid. 367.

247   *'My position'*: ibid. 368.

247   *'I could not think'*: ibid. 432.

248   *contemporaries who were already looking to Berlin*: cp. Goethe's own remarks in Book 7 of *Poetry and Truth*, HA, ix. 279–81.

248   *'the disproportion'*: HABr, i. 369.

249   *'under the republican pressure'*: ibid. 312.

249   *its own candidate for the bishopric*: Fürstenberg, see Bode, *Pegasus*, p. 301.

250   *'dégoûtantes platitudes'*: extracts from *De la littérature allemande* e.g. in HA (1988), iv. 492–3.

250   *Goethe was passed over in silence*: certainly not forgotten, as Bode has it, Bode, *Pegasus*, p. 4.

250   *Möser took up his cause*: extracts from Möser e.g. in HA (1988), iv. 493–4.

250 *letter to Möser's daughter*: HABr, i. 361–3.

251 *thirty-one hours' travelling*: WA, IV. 6, p. 296.

The Minister

252 *never missed a session*: WA, IV. 6, p. 218.

252 *records of the well over 600 sessions*: AS, i, pp. lxx–lxxviii; the following details all from *AS*, i.

252 *minute of Goethe's oral intervention*: *AS*, i. 49.

253 *silver-mines at Ilmenau*: the following details mainly from Bruford, *Culture*, pp. 104–7.

253 *on Goethe's birthday in 1783, to issue 1,000 shares*: MA, 2. 2, p. 954.

253 *February 1786*: Grumach, iii. 4–5.

253 *special committee on which he sat*: Bode, *Bau*, p. 10.

253 *Highways Commission*: Bruford, *Culture*, pp. 107–9.

254 *loan from the city of Berne*: Bode, *Pegasus*, p. 149, also for the following details.

254 *'the economics are on a sound basis'*: HABr, i. 434.

254 *attention of Frederick the Great*: Grumach, ii. 542–3.

255 *Goethe had himself proposed such a League*: *AS*, i. 54–5; cp. p. 50.

255 *recognized that he did not have the suavity*: see his 'Self-Portrait' of 1797, HA, x. 529.

255 *not a single case had he ever deliberately*: Grumach, ii. 379, iii. 53.

255 *'The Duke finds his existence'*: HABr, i. 416.

255 *'anyone who takes up his time'*: ibid. 514.

Frau von Stein

256 *9 January 1782*: WA, III. 1, p. 135.

256 *after about seven*: WA, IV. 4, p. 235.

256 *'Mme Stein is the cork-jacket'*: HABr, i. 381.

256 *probably on 12 November*: Grumach, i. 386, suggests 11 Nov., but note the importance to Goethe of the twelfth in the period 1779–85.

256 *silhouette which Goethe now felt did not do her justice*: Bode, i. 152–3, No. 228.

256 *'sees the world as it is'*: ibid. 141, No. 208.

256 *horses and bulls*: HABr, i. 318.

257 *had spent little time at home*: Bode, *Stein*, pp. 11–12, 19–20, 22.

257 *distant Scottish ancestry*: ibid. 7.

257 *two black pages*: ibid. 69.

257 *Knebel, who knew her well, described her*: HABr (1988), i. 641.

257 *Carl August thought her*: Grumach, iii. 45.

257 '*A woman of the* beau monde': Bode, i. 101, No. 140.

257 '*dear Angel*'—'*dear Lady*': HABr, i. 206.

257 '*missing' her at balls and receptions*: Grumach, i. 405.

257 '*because no one can understand*': Bode, i. 169, No. 254. It is not so much Goethe's failure to find his stick that should be psychoanalysed as Frau von Stein's thinking it worthy of mention.

258 *some new apartments*: Bode, *Bau*, p. 95, *Garten*, p. 127.

259 *talking exclusively about himself*: HABr, i. 307, 321.

259 *bread from the military bakery*: Bode, *Garten*, p. 259.

259 *studying Spinoza together*: HABr, i. 459.

259 *the relation between them as a marriage*: ibid. 350.

259 *the gloves he received on his initiation*: ibid. 307.

259 *receiving from her a gold ring*: ibid. 306.

259 *Fritz later said were the happiest*: Grumach, iii. 49.

259 '*only Goethe's* confidante': Grumach, ii. 46.

259 *paper left in a pair of trousers*: ibid. 301.

260 *act as postman*: Bode, *Pegasus*, p. 240; see e.g. WA, IV. 6, p. 349.

260 *friendly terms*: HABr, i. 318, 377, 445.

260 *re-establishing the 'Du'*: ibid. 350.

260 *satisfied that she does indeed believe*: ibid. 371, 390.

260 *resisting an attempt*: ibid. 378.

260 *more of 'Liebe'*: e.g. ibid. 355.

260 *emotional force of the letters is quieter*: e.g. ibid. 420, 430.

260 *possessiveness, even jealousy*: Branconi, ibid. 428; Lavater, ibid. 516; rocks, ibid. 429.

260 '*in the midst of good fortune*': ibid. 400.

260 '*I am a poor slave of duty*': ibid. 435.

260 *promise Frau von Stein later felt he had broken*: Grumach, iii. 48.

260 *small his circle of friends is*: HABr, i. 434.

260 *lacks the consolations of a home*: ibid. 508.

260 *pleads with her not to reawaken*: Bode, *Pegasus*, p. 240, citing also WA, I. 2, p. 125, 'Warnung'.

260 *his physical condition*: Grumach, ii. 293, 302, 305, 348, 412, 417, 431, 519.

260 *shortness of life*: HABr, i. 324, 435, 489.

261 *at court in Berlin*: ibid. 253.

261 *those who have 'Welt'*: ibid. 348.

261 *consciously striving to improve*: ibid. 420.

261 *embodiment of court values*: Bode, *Stein*, p. 1.

261 *object of Goethe's passion was the wife*: Grumach, iii. 46–7.

262 *'I can assure you'*: HABr, i. 449.

262 *dug frantically*: ibid. 248.

262 *'An den Mond'*: HA, i. 128–9.

263 *'My soul used to be'*: HABr, i. 250.

263 *'Apart from the sun'*: ibid. 279.

263 *'my soul is like an everlasting firework display'*: ibid. 310.

263 *'too ministerial'*: Grumach, ii. 318.

263 *'I fully understand'*: Bode, i. 308, No. 496.

263 *'gradually entered on the inheritance'*: HABr, i. 324.

263 *'how little I can subsist'*: ibid. 484.

264 *'be my substitute'*: ibid. 504.

264 *'I had intended'*: ibid. 271.

264 *'Adieu, sweet conversation'*: ibid. 349.

264 *'medium of love'*: Bode, i. 141, No. 208.

264 *'you have been transubstantiated'*: HABr, i. 391.

264 *'everything that has once passed'*: Bode, i. 301, No. 486.

265 *'what a man notices'*: HABr, i. 408.

265 *'the pyramid of my existence'*: ibid. 324.

265 *'Goethe can be kind'*: Bode, i. 244, No. 393.

265 *'She did not want to say'*: ibid. 249, No. 400.

265 *'Goethe always read too much'*: Grumach, iii. 45.

265 *a Tartuffe, as Jacobi saw him*: Bode, i. 247, No. 397, 'le pauvre homme'.

265 *'the greatest gift'*: HABr, i. 319.

265 *'If only the pen'*: ibid. 320.

266 *'if only my thoughts of today'*: ibid. 314.

266 *'Über allen Gipfeln'*: HA, i. 142.

Literary Difficulties of a Statesman

266 *'no poetical production'*: to Eckermann, 10 Feb. 1829: HABr (1988), i. 634.

267 *In 1780 he noted*: ibid. 303.

267 *it was produced in this form four times*: 6 and 12 Apr. and 12 Aug. 1779, and 30 Jan. 1781, DKV, I. 5, pp. 1009–11.

267 *'quarrelled a bit with the Muses'*: Bode, i. 289, No. 466.

267 *'unpoetical condition'*: HABr, i. 443.

267 *'my writing is being subordinated to life'*: ibid. 303; cp. Grumach, ii. 417.

268 *'He will do much good'*: Bode, i. 190, No. 289.

268 *as good as anything*: ibid. 262, No. 421.

268 *'always been a disagreeable feeling'*: HABr, i. 515; cp. Grumach, i. 470.

268 *ten unauthorized predecessors*: HA, x. 613.

268 *savage epigram*: WA, I. 5. 2, p. 161.

269 *Peter im Baumgarten*: Beutler, *Essays*, pp. 444–58.

270 *'Das Göttliche'*: HA, i. 147.

271 *ignorant of the public reactions*: WA, IV. 7, p. 173.

272 *'brought me almost to despair'*: HABr, i. 503.

272 *after October 1776 Goethe had nominal responsibility*: Bruford, *Culture*, p. 84.

272 *novelty of the expensive Greek costumes*: Grumach, ii. 115; Bruford, *Culture*, p. 120.

272 *'Iph. performed'*: WA, III. 1, p. 84.

273 *'the supreme masterpiece'*: Grumach, ii. 235.

274 *'They say you can best'*: HABr, i. 422.

274 *'Fortunate he'*: WA, I. 17, p. 179.

275 *shortage of dinner parties*: Bode. i. 480 (diary of Sophie Becker, 16 Dec. 1784).

275 *Tiefurt Journal*: Bruford, *Culture*, pp. 136–42.

275 *'You have isolated me'*: HABr, i. 451.

277 *'Einsamkeit'*: HA, i. 204.

278 *'false tendencies'*: HA, x. 529.

278 *'dumb as a fish'*: HABr, i. 504.

278 *'blessed me with natural science'*: ibid. 508.

278 *'passionless vegetable kingdom'*: ibid. 354.

278 *disciple of Spinoza*: ibid. 508.

278 *'evening prayer'*: ibid. 504.

279 *'circle of my existence'*: ibid. 434.

279 *'stones and plants'*: WA, IV. 3, p. 249.

CHAPTER 5

Confidence and Indeterminacy: 1775–1777

283 *Werther costume*: Grumach, i. 473, 474.

283 *student life*: ibid. 467.

283 *'geniuses' who visited*: ibid. 391.

283 *like the robbers in* Claudina: ibid. 398, 477.

283 *on 'Du' terms*: ibid. 390.

283 *Goethe, in Lenz's words, as their 'captain'*: ibid. 425.

283 *'You wouldn't believe'*: HABr, i. 208.

283 *blind man's buff*: Grumach, i. 390.

283 *Goethe rolled on the floor*: ibid. 403.

283 *dubbed the Stolbergs knights*: ibid. 390.

283 *Christmas Eve*: HABr, i. 202.

283 *charade representing the deadly sins*: Grumach, i. 406.

283 *less imaginative hooliganism*: ibid. 398, 393, 445–7, 478.

283 *'art to trample on the conventions'*: ibid. 402–3.

283 *'Things are frightful'*: Bode, i. 191, No. 292.

283 *crippled with rheumatism*: Bruford, *Culture*, p.87.

284 *fire-fighting*: e.g. WA, III. 1, p. 11; cp. *CGZ*, I, Nos. 135, 142, 155, VIb, No. 175.

284 *pencil-sketches of comrades*: *CGZ*, I, Nos. 153, 287, 295–9, and cp. 293.

284 *continuous with the manner of the drawings from Switzerland*: ibid., No. 296, p. 103, and cp. No. 152.

284 *charcoal-burners, highflown conversations on poetry*: WA, III. 1, p. 17.

284 *inscriptions cut*: Grumach, i. 449, 478.

284 *'Ilmenau'*: HA, i. 107.

284 *'union of souls'*: HABr, i. 220.

284 *'It's a strange thing, the government'*: ibid. 231.

285 *second dwellings outside the city walls*: *Stadt Goethes*, p. 40.

285 *guided tours*: Bode, *Garten*, pp. 27–8.

285 *restoration work*: ibid. 72.

285 *rented apartment*: ibid. 85–6.

285 *'Everything is so still'*: HABr, i. 216–17.

285 *fumbling around in the dark*: ibid. 217.

285 *attempts to express in pencil, charcoal, and chalk*: *CGZ*, I, Nos. 155–160, VIb, No. 27.

285 *pancakes with him and Seidel, and at Easter*: Grumach, i. 481–2.

285 *oaks and beeches*: for work in the garden, Bode, *Garten*, pp. 88–9, 239–40, 137.

286 *hope to look on the growth*: HABr, i. 230.

286 *need for 'resignation'*: ibid. 213.

286 *'all that had had to pass'*: ibid. 210.

286 *'journey was intended to draw up the balance*: ibid. 212.

287 *poets, he told Goethe*: HABraG, i. 58–9.

287 *Goethe's perhaps excessively condescending reply*: HABr, i. 215.

287 *excommunication of Weimar*: HABraG, i. 59; cp. Bode, i. 185–6, No. 279, i. 191, No. 292.

287 *the departures of Klinger*: Bode, *Garten*, p. 81.

287 *'asinine trick'*: WA, III. 1, p. 28; cp. Bode, *Bau*, p. 51.

287 *'What is man'*: WA, III. 1, pp. 26–7.

287 *'Goethe says he has handed over'*: Grumach, i. 413.

Works, 1775–1777

288 *'Seefahrt'*: HA, i. 49–50.

288 *added to it an epigraph*: Stadt Goethes, p. 419.

288 *as it seemed to the poet Bürger*: Bode, i. 218–19, Nos. 347, 350.

288 *support for Bürger's plan*: Grumach, i. 415–16, 462–3.

289 *began dictating*: WA, III. 1, p. 34, but cp. Bode, i. 146, No. 220; Wolff, *Humanität*, pp. 28–39, 63–5, followed by *DJG*, v, 511–15, seems to me to attribute too much to the Frankfurt period.

289 *'mysteries of the difference of the sexes'*: WA, I. 51, p. 14.

290 Lila: WA, I. 12 pp. 39–86, but all three of the principal stages of the text (1777, 1778, and 1788) are printed in DKV, I. 5, pp. 29–34, 35–62, 835–69, see also pp. 933–4 and MA, 2.1, pp. 616–17.

290 *'Jägers Abendlied'*: HA, i. 121.

290 *'Rastlose Liebe'*: ibid. 124.

291 *'Wandrers Nachtlied'*: ibid. 142.

291 *'Warum gabst du uns die tiefen Blicke'*: ibid. 122–3.

291 *'I cannot account'*: HABr, i. 212.

293 Die Geschwister: HA, iv. 352–69.

293 *'I can now stay'*: ibid. 354, 369.

294 *'Wilhelm, it is not possible'*: ibid. 369.

Tragedy and Symbolism: 1777–1780

294 *'such happy times'*: HABr, i. 233.

294 *on 8 June*: ibid. (1988) 659.

294 *'Dark, lacerated day'*: WA, III. 1, p. 40.

294 *'With my sister'*: HABr, i. 240.

294 *'without one's being able'*: HA, x. 131.

294 *'Alles gaben'*: HABr, i. 234; HA, i. 142.

295 *'we are together'*: HABr, i. 234.

295 *in Ilmenau with the Duke*: WA, III. 1, pp. 44–6.

295 *on 4 July*: Grumach, ii. 23–5.

296 *Peter im Baumgarten who arrived with his pipe*: Steiger, ii. 116; Beutler, *Essays*, p. 449.

296 *bust of Lavater*: Grumach, ii. 72.

296 *in August 1778*: ibid. 71, but cp. Bode, *Bau*, pp. 121, 223, and Steiger, ii. 153.

296 *Goethe laughed*: WA, IV. 3, p. 168.

296 *'I am very altered'*: ibid. 188.

297 *avoided a meeting with the uncrowned king*: WA, III. 1, p. 50.

297 *'a thoughtful, serious, frigid Englishman'*: Grumach, ii. 34.

297 *'Yesterday coming away'*: HABr, i. 239.

297 *'happened on my diary'*: WA, III. 1, pp. 26–7.

297 *on 29 November*: details of the journey from HABr, i. 241–6.

297 *obscurely hinted*: ibid. 246.

297 *as has been clearly shown*: Schöne, 'Auguralsymbolik'.

297 *diluting his poetic inspiration*: HABr, i. 243.

298 *young Plessing who had written to him*: Grumach, ii. 45–51.

298 *in thirteen years the forester had never climbed*: Schöne, *Götterzeichen*, p. 44.

298 *night of the ninth*: Grumach, ii. 53.

298 *'So I sat there'*: HABr, i. 247.

298 *'At a quarter past one'*: WA, III. 1, pp. 56–7.

298 *'The goal of my longing'*: HABr, i. 246.

299 *Devil's Pulpit and Witch's Altar*: Bode, *Pegasus*, p. 214, illustration after p. 216.

299 *'marvellous opening and melting of the heart'*: Grumach, ii. 54.

299 *completing the first book of* Wilhelm Meister: WA, III. 1, p. 59.

300 *to take supper with him*: Grumach, ii. 73.

300 *'This inviting sorrow'*: HABr, i. 249.

300 *Weimar's little boat*: WA, IV. 3, p. 215.

300 *mechanism of a great state being prepared for war*: HABr, i. 250–1.

301 *'how the great of the world'*: ibid. 249.

301 *proud and taciturn*: ibid. 253.

301 *'gave such general offence'*: Grumach, ii. 81.

301 *'I feel more comfortable'*: HABr, i. 251.

301 *'in a quite different situation'*: ibid. 250.

301 *'all Germany below me'*: ibid. 253.

301 *'dream' that 'the gods' had here allowed*: ibid. 249.

301 *drawing of the lower storeys*: CGZ, I, No. 197. See plate 28.

301 *classicizing tempietto*: ibid., No. 199.

302 *valley of the Ilm*: Mollberg, *Kulturstätten*, pp. 16–18.

302 *'Goethe's new poems'*: Grumach, ii. 84.

302 *open-air entertainment*: ibid. 86–92.

302 *'Actually I am not necessary here'*: HABr, i. 254.

303 *more 'husbandly'*: ibid. 241.

303 *delicately linear charcoal portrait*: CGZ, I, No. 292.

303 *burly chaperone*: ibid., No. 288.

303 *'radical clarification'*: WA, III. 1, p. 77.

303 *stiffness of his movements*: Grumach, i. 395, 468, ii. 506, iii. 40.

303 *rode from Leipzig to Weimar*: Bode, *Garten*, pp. 92–3.

303 *control of his diet*: WA, III. 1, p. 78.

303 *5 feet 9¼ inches*: i.e. 176 cm, Bauer, *Kopf*, p. 26.

303 *'idea of purity'*: WA, III. 1, p. 94.

304 *roused the other with a slipper*: ibid. IV. 3, p. 228.

304 *'nasty' position*: Grumach, ii. 118.

304 *Council forbade any discussion*: ibid. 119.

304 *closing of the Austrian borders*: WA, III. 1, p. 82.

304 *'frozen off from everyone else'*: ibid. 72.

304 *'living with the people of this world'*: HABr, i. 263–4.

304 *'solitude is a fine thing'*: ibid.

304 *'pressure of affairs'*: WA, III. 1, p. 77.

304 *one of his few satirical drawings*: CGZ, I, No. 307.

304 *fifty-six yards of white linen*: Bruford, *Culture*, p. 120.

304 *'took him for an Apollo'*: Grumach, ii. 115.

305 *dropping the Picturesque interest in the 'point of view'*: CGZ, I, Nos. 142, 145, 151, 172.

305 *by a bridge in the Weimar park*: ibid., No. 195.

305 *in the estate at Wörlitz*: ibid., No. 197. See plate 28.

305 *beside his garden fence*: ibid., No. 221.

305 *combination of geometrical*: cp. also ibid., Nos. 183, 205.

305 *'proto-Impressionistic'*: ibid. ad No. 221, p. 83, ad Nos. 151, 152, p. 63.

305 *'a little too cavalierly'*: HABr, i. 265.

305 *'old husks'*: WA, III. 1, pp. 93–4.

305 *'crucifixion' of Jacobi's novel* Woldemar: Grumach, ii. 126–8.

306 *highly charged discussions*: WA, III. 1, pp. 86, 92–3, 94–5.

306 *'For the first time I would be returning'*: HABr, i. 267.

306 *had recalled, to give him confidence*: ibid. 245.

306 *'he is still very inexperienced'*: WA, III. 1, p. 92.

306 *provided Frau Goethe could be prevented*: HABr, i. 268–9.

307 *court hissed*: Grumach, ii. 135; Bode, i. 239, No. 385.

307 *'by inspiration from the angel Gabriel'*: Grumach, ii. 140.

307 *soon rumoured*: ibid. 143.

307 *'the old mother'*: ibid. 142.

307 *'The 18th September'*: ibid. 141.

307 *'I am recapitulating'*: HABr, i. 270.

308 *Friederike, who later wrote a 'good' letter*: WA, III. 1, p. 111.

308 *'think with contentment'*: HABr, i. p. 273.

308   *called on Lili*: HABV, i. p. 273.

308   *at Emmendingen*: Grumach, ii. 147–9.

308   *spent a month simply touring*: ibid. 149–71, Bode, *Bau*, pp. 266–337, Steiger ii. 220–38.

309   *unable to draw 'a single line'*: WA, IV. 4, p. 138; cp. *CGZ*, VIb, Nos. 179, 180.

310   *'journey into Hell'*: HABr, i. 282.

310   *on 3 November*: the journey to Zurich from Grumach, ii. 171–93, Bode, *Bau*, pp. 337–89, Steiger, ii. 239–56.

311   *'the toughest thing I have ever done'*: Grumach, ii. 186.

311   *'summit of our journey'*: HABr, i. 283.

311   *'not guessing my future fate'*: WA, I. 19, p. 300.

311   *recapitulation resumed*: Zurich to Weimar from Grumach, ii. 193–217; Bode, *Bau*, pp. 390–430; Steiger, ii. 258–69.

311   *'seal and highest peak'*: HABr, i. 284.

311   *'moral deadness'*: ibid. 287.

312   *first set pencil to paper in Frankfurt*: *CGZ*, VIb, Nos. 37, 38, cp. 34.

312   *total cost of nearly 9,000 dollars*: Bode, *Bau*, p. 430.

312   *written to Lavater*: HABr, i. 287–90.

Works, 1777–1780

313   *'Wer nie sein Brot'*: HA, vii. 136.

313   *'Am I then in the world'*: WA, III. 1, p. 9.

314   Proserpina: HA, iv. 455–62.

314   *Hope floods her heart*: *Proserpina*, l. 166.

314   *prove refreshing*: ibid., l. 197.

314   *confirming expression*: ibid., ll. 216–17.

314   *more firmly than ever bound*: ibid., l. 210.

314   *queen of a realm of shadows*: ibid., ll. 221 ff; cp. ll. 47–9.

314   *'Leer und immer leer!'*: ibid., ll. 75–6; cp. ll. 78–86.

314   *the eating of the forbidden apple*: ibid., l. 227.

314   *fear of a double condemnation*: ibid., l. 230.

315   *chorus of Fates invisible*: the stage direction is possibly a later addition, WA, I. 17. p. 345.

315   The Triumph of Sensibility: *Der Triumph der Empfindsamkeit*, WA, I. 17, pp. 1–73.

315   *'sludge at the bottom'*: ibid. 56.

315   *'stupid interpretations'*: ibid. III. 1, p. 62.

316   *'fate of Tantalus'*: ibid. I. 17, p. 66.

316   *confused Tantalus with Ixion*: *Proserpina*, l. 62; cp. HA (1988), iv. 621–2.

316   *moment of 'hope'*: WA, I. 17, p. 71.

316   *'Winter Journey in the Harz'*: 'Harzreise im Winter', HA, i. 50–2.

318   *drawing of the Brocken by moonlight*: *CGZ*, I, No. 190. See plate 18.

318 *love, he assured Frau von Stein, of her*: HABr, i. 449.

320 *national theatre*: WA, I. 51, p. 69.

320 *'I won't utter it'*: ibid. 87.

320 *'the good cause'*: cp. ibid. 36.

320 *'Two years ago'*: Grumach, ii. 84.

321 *'Grenzen der Menschheit'*: HA, i. 146–7.

321 *'chain of culture'*: Herder, *Werke*, v. 134, 142, 586; ibid., xvi. 229.

321 *'I worship the gods'*: HABr, i. 251.

321 Iphigenia in Tauris: *Iphigenie in Tauris*, the first, prose, version of *Iphigenie auf Tauris*, WA, I. 39, pp. 321–404.

322 *'What pride'*: HA (1988), v. 411.

322 *Wieland, conversely, if for the same reason, welcomed*: *Briefe an einen jungen Dichter* 3, Mandelkow, *Kritiker*, i. 97.

323 *'I think I hear'*: WA, I. 39, p. 372.

323 *How can she bring back blessings*: ibid. 386.

324 *considering a play on this subject since 1776*: HA (1988), v. 430–1.

324 *identification between Iphigenia and her mother Clytemnestra*: cp. Fowler, 'Orest'.

324 *'grant blessed deliverance'*: WA, I. 39, p. 371.

325 *'If you are the truthful ones'*: ibid. 393.

325 *'You know that you are speaking'*: ibid. 394.

325 *'so often calmed me'*: ibid. 396.

326 *Publication would have been directly contradictory of its purpose*: Bode, i. 291, No. 470.

326 *inscriptions he was to compose for the Weimar park*: Grumach, ii. 460.

327 *Jery and Bätely*: *Jery und Bätely*, WA, I. 12, pp. 1–38 (the names are said to be Swiss abbreviations of Jeremias and Bathilde, DKV, I. 5, p. 1456).

327 *'I can still sniff'*: WA, I. 35, p. 7.

327 *'Song of the Spirits'*: 'Gesang der Geister über den Wassern', HA, i. 143.

327 *'not quite your and my religion'*: Bode, i. 245, No. 395.

327 Letters from Switzerland: *Briefe aus der Schwiez*, WA, I. 19, pp. 223–306, the denomination 'Zweite Abtheilung' dating from Goethe's later work on the collection in 1796 and 1808.

328 *which he compared to Xenophon's* Anabasis: Grumach, ii. 232.

328 *Wieland's 'incredible eye'*: WA, III. 1, p. 115.

328 *'if I succeed'*: ibid. IV, 4, p. 202.

328 *'little mention of people'*: ibid. I, 19, p. 273.

329 *first letter of all*: ibid. 224–5.

329 *'inner growth'*: ibid. 225, cp. ibid, 272.

329 *'My eye and my soul'*: ibid. 224.

329 *the setting sun*: ibid. 228.

329   *'one then gladly gives up'*: WA, I, 238.

329   *'the fashion', seek out the recommended perspectives*: ibid. 240, 259, 261, 263.

330   *predicting the weather*: ibid. 236.

330   *'one is enveloped'*: ibid. 271–2.

330   *'most desolate region'*: ibid. 290.

330   *'I am convinced'*: ibid. 291.

330   *'one to see it'*: ibid. 247.

330   *Carl August's diary*: Grumach, ii. 182.

331   *village of Fiesch*: Bode, *Bau*, p. 369.

CHAPTER 6

Between Society and Nature: 1780–1784

335   *Duke cut his hair*: HABr, i. 297.

335   *Goethe now appeared at court*: Grumach, ii. 220, 222.

335   *'well-intentioned bird'*: HABr, i. 321.

335   *full-length bathing-costume*: Bode, *Garten*, p. 267.

335   *natural sequel*: cp. HABr, i. 332.

335   *'very prosaic'*: WA, III. 1, p. 105.

335   *Major von Volgstedt*: ibid. 114, 126; ibid. IV. 5, p. 29.

335   *By August 1781*: WA, III. 1, p. 130.

335   *record the rhythms of his mood*: ibid. 112.

335   *give up drinking wine*: ibid. 114.

335   *join the Masonic lodge*: HABr, i. 294.

336   *'some work of genius'*: WA, I. 17, p. 359.

336   *'Cold Shower'*: 'Kalte Dusche', CGZ, I, No. 310.

336   *'Since I have had to do*: HABr, i. 314.

336   *despatching Voigt*: HA, xiii. 251–2.

337   *from F. W. von Trebra*: WA, III. 1, p. 121.

337   *Merck in Darmstadt*: HABr, i. 328–30.

337   *from amateurs in Berne and Zurich*: Bode, *Bau*, p. 481.

337   *professors in Freiberg*: WA, iv. 4, p. 256.

337   *his old friend Sophie von la Roche*: HABr, i. 314.

337   *Leisewitz talked*: Grumach, ii. 255.

337   *interim report*: HABr, i. 337.

337   *natural history collection and a large private library*: AS, i. 190, 195; cp. Bode, *Bau*, p. 229.

337 *Loder, who was also a Freemason*: Grumach, ii. 304, 351.

337 *Academy of Drawing*: Bode, *Pegasus*, p. 62.

338 *remained in his memory*: HABr, i. 390, 439.

338 *tendency to revert*: ibid. 316.

338 *in his deepest nature*: ibid. 347.

338 *not to undertake another journey*: WA, IV. 5, p. 117.

338 *serious quarrel*: HABr, i. 325.

338 *denied to Lavater any interest either in marriage*: ibid. 324.

338 *ecstatic though softly spoken intimacy*: WA, IV. 5, p. 92.

338 *she is now 'new'*: ibid. 94.

338 *'conversion' in his innermost self*: ibid. 92.

338 *a new openness*: ibid. 97

338 *purification is said*: ibid. 87–8.

338 *spirit of benevolence*: ibid. 97.

338 *past readiness to disregard*: ibid. 122–3.

338 *daily acknowledges more deeply*: HABr, i. 356.

338 *'God and Satan'*: ibid. 355.

338 *'concepts which man has'*: ibid. 353.

338 *'love of the universe'*: ibid. 355.

338 *essay called* Nature: *Die Natur*, HA, xiii. 45–7.

339 *based, as Goethe himself confirmed*: WA, IV. 6, p. 134.

339 *captures well the atmosphere*: HA, xiii. 48.

339 *'She makes chasms'*: ibid. 47.

339 *'The Goblet'*: 'Der Becher', WA, I. 2, pp. 106–7.

339 *'withering away'*: Grumach, ii. 293.

339 *'growing uglier'*: ibid. 305.

339 *before August 1781*: WA, III. 1, p. 129; Bode, *Bau*, pp. 467–8.

340 *'Erlkönig'*: HA, i. 154–5.

340 *started his anatomical lectures*: Grumach, ii. 329.

340 *signed an agreement*: ibid. 330.

340 *at an interview with Anna Amalia*: ibid. 330–1.

341 Der Geist der Jugend: ibid. 344–5.

342 *'I do of Syriac'*: Bode, i. 281, No. 454.

342 *'So he is now'*: ibid. 283. No 456.

342 *'I can have nothing to fear'*: Grumach, ii. 375.

342 *'Dating from midsummer'*: HABr, i. 401.

342 *J. C. Schmidt*: *AS*, i, pp. xxxvii–xxxviii.

342 *'How much happier I should be'*: HABr, i. 397.

343 *'I was truly created'*: HABr, i. 406, cp. 405.

343 *'I am settling down'*: ibid. 375.

343 *'politico-moral-dramatical'*: ibid. 380.

343 *'born to be a writer'*: ibid. 406.

343 *revision of* Werther: ibid. 426.

343 *'ghostly anticipation'*: ibid. 427.

343 *'wedded' to the duties of his office*: ibid. 435.

343 Die Fischerin: WA, I. 12, pp. 87–115.

344 *a congratulatory poem*: ibid. 4, p. 222.

344 *'Venetian Carnival'*: Grumach, ii. 406–8.

344 *'I have completely separated'*: HABr, i. 416.

344 *'I see no one'*: WA, IV. 6, p. 298.

344 *'most agreeable order'*: HABr, i. 398.

344 *'I am wholly yours'*: ibid. 371.

345 *'I too am friendly'*: ibid. 420.

345 *put in order*: ibid. 415–16, 398.

345 *'went pale'*: Grumach, ii. 491.

345 *'pale as whitewash'*: ibid. 460.

345 *regards to Klopstock*: ibid. 462.

345 *reconciliation of all was that with Herder*: HABr, i. 433.

345 *arts were better than dogs*: ibid. 424.

345 *French, and then an English mistress*: Bode, *Pegasus*, pp. 156–9.

345 *no longer have any confidence in Carl August*: Bode, i. 289, No. 465.

345 *painted eggs*: Grumach, ii. 418.

345 *celebrate his birthday*: ibid. 423.

346 *meeting roughly once a week*: ibid. 441.

346 *'world and natural history'*: WA, IV. 6, p. 224.

346 *'Goethe suffers too'*: Grumach, ii. 441.

346 *'l'honnête-homme à la cour'*: ibid. 440, the allusion is not only to the two senses of 'honnête homme' (gentleman, and honest man), but also to the novel by Goethe's great-uncle, J. M. von Loen.

346 *'Onwards, onwards'*: ibid. 429; Wells, *Development*, pp. 50–1.

346 *reminisced about their ascent*: Grumach, ii. 427–8.

346 *Professor Lichtenberg*: ibid. 429–30.

346 *'monstrous' misrule*: WA, IV. 6, p. 213.

347 *'most mobile part of creation'*: HA, xiii, 255.

347 *'not in a picturesque fashion'*: WA, IV. 6, p. 402.

347 *'Nothing here is in its original position'*: HA, xiii. 257.

347 *Herder's formulation: Ideen*, 15. Buch, *Werke*, xiv. 204–7; cp. *Werke*, v. 511–13, 584–6.

347 *'Leitfaden'*: HA, xiii. 257.

347 *metaphor of Ariadne's thread*: Reed, *'Paths'*.

347 *same metaphor determined Goethe's own formulation*: HABr, i. 450; cp. ibid. i. 464, 511, 514, 515, and HA, i. 199, ll. 1–8.

348 *mechanisms of rock-formation*: Wells, *Development*, pp. 47–9.

348 *we do not know*: HA, xiii. 254, but LA, I. 1, p. 96 says it was both.

348 *'The simple thread'*: WA, IV. 6, pp. 297–8.

348 *'a simple principle'*: ibid. 308.

348 *'harmony in the effects'*: ibid. IV. 9, p. 346.

348 *rocks, the massive slabs*: Wells, *Development*, pp. 47–9, Bode, *Pegasus*, pp. 212–3, LA I. 1, p. 98.

348 *divisions of the rock mass*: ibid. 100.

348 *overlapping parallelepipeds*: ibid.

349 *this underlying grid*: CGZ, I, Nos. 276, 280, Vb, Nos. 183, 184, and p. 98, VIb, No. N32.

349 *another specific natural phenomenon*: HABr, i. 464.

349 *bringing relief to loyal subjects*: Grumach, ii. 447–8, 454–5.

349 *ordering the demolition*: ibid. 456–7.

349 *'not thought to be crazy'*: HABr, i. 440.

349 *By the end of October*: Grumach, ii. 498.

349 *little treatise*: HA, xiii, 184–96.

349 *'quite believe that a professional* savant': HABr, i. 475.

350 *Professor G. A. Wells*: Wells, *Development*, p. 16.

350 *Herder had said in the second book of* Ideas: Herder, *Ideen*, 2. Buch, 4. Kap., *Werke*, xiii. 66, although Herder thought the intermaxillary bone separates man and apes 'for ever', p. 118.

350 *'like the coping stone'*: HABr, i. 436.

350 *very variations which plague*: WA IV, 6, p. 354.

350 *poem 'Divinity'*: HA, i. 147–9.

351 *Goethe's 'taciturnity'*: Grumach, ii. 415.

352 *'Goethe lives in his thoughts'*: ibid. iii. 47.

352 *'best . . . of all mortals'*: HABr, i. 285.

352 *'anoint' the Duke*: ibid. 279.

352 *'moral deadness'*: ibid. 287.

352 *roused him to parody*: ibid. 402.

352 *'quackery'*: HABr, i. 431.

352 *'Not one heartfelt'*: ibid. 517.

352 *'you are in a worse case'*: ibid. 231.

352 *'I am a very earthly man'*: ibid. 279.

352 *'you can transfer your all'*: ibid. 364; cp. 389.

353 *no more than a 'we'*: ibid.

353 *'bird of paradise'*: ibid.

353 *'Wieland has philosophized'*: Grumach, ii. 434.

353 *'You continue . . . to extend'*: HABr, i. 364–5.

353 *'decidedly non-Christian'*: ibid. 402.

353 *'not be persuaded by an audible'*: ibid. 403.

353 *prayed little*: Grumach, ii. 490.

353 *Herder's sermon at the baptism*: HABr, i. 424.

353 *consoling Duchess Luise*: WA, IV. 6, p. 257.

353 *revive Goethe's old interest in the Jewish philosopher*: Grumach, ii. 504.

353 *he told Jacobi*: Bode, *Pegasus*, p. 223.

353 *'our saint'*: WA, IV. 6, p. 420.

354 *goût grec*: Kalnein, 'Architecture', p. lv.

354 *catalogue of the works of Palladio*: Meyer, 'Haus', pp. 282–3.

354 *'I am not skilful'*: WA, IV. 6, p. 347.

354 *Countess Görtz was certainly not deceived*: Grumach, ii. 482.

354 *'not seem as picturesque'*: HABr, i. 439, 452.

354 *'beginning to bore me'*: ibid. 452.

355 *in 1782 or, more probably, 1783*: possibly 12 Nov. 1782; WA, IV. 6, p. 210, but the poem looks like a later insertion.

355 *'Kennst du das Land'*: HA, vii. 145.

Works, 1780–1784

356 Egmont: HA, iv. 370–454.

356 *the autumn of 1775*: ibid. x. 181.

356 *not composed methodically*: ibid. 170–1.

356 *'impossible fourth act'*: HABr, i. 380.

356 *scenes could be left untouched*: HA, xi. 366.

356 *'too unbuttoned'*: HABr, i. 385.

360 *trait was added to Ferdinand's characterization on 5 December 1778*: WA, III. 1, p. 72.

361 *reporting to Knebel on the new order*: HABr, i. 416.

361 *hybrid prose of Iphigenia*: HA (1988), v. 497 (Tobler).

361 *set himself as a deadline 12 November*: WA, IV. 5, p. 4.

361 *same date in 1781*: ibid. 216.

362 *how does the poet live* after: see final version, ll. 501–7.

362 *intended by nature to be one*: ll. 1704–6.

363 *not until 1787*: WA, IV. 7, p. 236 (contrast entry for *Egmont*).

363 *he advised Goethe against*: HABr, ii. 90.

363 Elpenor: HA, v. 309–331.

364 *vengeance on whoever may have been responsible*: there is no evidence for the view in ibid. (1988) 643–4, that Elpenor swears to avenge his dead father.

365 *all discord is harmony not understood*: Pope, *Essay on Man*, I, ll. 289–90.

366 *drawing of either this scene or its original*: *CGZ*, I, No. 304.

367 *'beginnings of theatre'*: WA, I. 51, p. 194.

368 *if Goethe 'ever came to his senses'*: Bode, i. 262, No. 421.

368 *'Heiß mich nicht reden'*: WA, I. 51, p. 260; cp. HA, vii. 356–7.

368 *'irresistible yearning'*: WA, I. 52, pp. 3–4.

369 *like that of the leader of a Moravian congregation*: ibid. 76.

370 *'It is rare that'*: ibid. 52, p. 95.

370 *'a merry young actress'*: ibid. 49.

370 *'Many of our readers'*: ibid. 53.

372 *performing dogs*: ibid. 144.

373 *'It has often caused me disgust'*: ibid. 163.

373 *'stout vagabonds, noble robbers'*: ibid. 189.

375 *'On the Death of Mieding'*: 'Auf Miedings Tod', HA, i. 114–20.

376 *'O Weimar'*: ibid. 115, ll. 39–42; cp. Bode, i. 313–14, No. 509.

376 *'Ilmenau'*: HA, i. 107–12.

376 *'Unschuldig und gestraft'*: ibid. 110, l. 119.

376 *unable to explain*: ibid. ll. 112–7.

376 *'unable to tell you'*: ibid. l. 97.

376 *journey to Switzerland*: ibid. l. 166.

377 *'old rhymes'*: ibid. l. 22.

377 *diplomatic conference in Brunswick*: Grumach, ii. 478.

377 *'Zueignung'*: HA, i. 149–52.

377 *geological expedition*: Grumach, ii. 472–3.

379 *allegorical mechanisms*: e.g. ll. 96–104.

'I thought I was dead': 1785–1786

379 *attended only two*: *AS*, vol. i, pp. lxxviii–lxxix.

380 *'I am patching the beggar's cloak'*: WA, IV. 7, p. 51.

380 *new shaft*: Bode, *Pegasus*, p. 219.

380 *lifting engine*: ibid. 271.

380 *taxation of the Ilmenau district*: ibid. 30–1.

380 *'itch' for war*: HABr, i. 474.

380 *Goethe was reading Count Necker's*: ibid. 484.

380 *'indescribable impression'*: HA, x. 433.

380 *'maturing sense of what is human'*: Grumach, ii. 518.

381 *'nothing that can be changed'*: ibid.

381 *'Duke . . . has sent off the courtiers'*: HABr, i. 482.

381 *excisemen and some travelling Jews*: AS, i. 395–9.

381 Scherz, List und Rache: WA, I. 12, pp. 117–80.

381 *'When I wrote the piece'*: HABr, i. 477.

382 *'lack of feeling'*: WA, I. 35, p. 9.

382 *'tinkle on those strings'*: HABr, i. 495.

382 *'comprehend the difference'*: ibid. 493.

382 *attribute to the success of the* Seraglio: HA, xi. 437.

382 The Mismatched Household: *Die ungleichen Hausgenossen*, WA, I. 12, pp. 223–51.

383 *professed to Jacobi complete ignorance*: WA, IV. 7, p. 173.

384 *'disinclination for all literary quarrels'*: HABr, i. 507–8.

384 *'Studie nach Spinoza'*: HA, xiii. 7–10.

384 *confessed to reading only desultorily*: HABr, i. 476.

384 *'abstruse generalities'*: HA, xi. 98.

385 *'science' and 'art' can perform the function of 'religion'*: HA, i. 367; cp. HABr, i. 508.

385 *'in single things', in 'herbis et lapidibus'*: HABr, i. 476.

385 *'one can only* believe *in God'*: ibid. 508.

385 *'not an atheist'*: ibid. 475.

385 *little further progress in mineralogy without a knowledge of chemistry*: WA, IV. 8, p. 4.

385 *dissection of seeds, including a coconut*: ibid. 7, p. 24.

385 *F. G. Dietrich*: HA, xiii. 154.

386 *native orchids and large colonies of sundew*: Grumach, ii. 530–1.

386 *hitherto rather repelled him*: HA, xiii. 160.

386 *'I am agog'*: HABr, i. 514.

386 *'first great notions'*: ibid. 496.

386 *'no dream, no phantasy'*: ibid. 514.

386 *sentence of Spinoza's*: ibid. 508 and note (1988) p. 758.

386 *gave him 40 louis*: WA, IV. 7, pp. 55–6.

386 *in total to 530 dollars*: Bode, *Pegasus*, p. 295.

386 *'whole caravan'*: Grumach, ii. 517.

387 *'only by his beautiful eyes'*: ibid. 533.

387 *'fainéantise'*: HABr, i. 480.

387 *'The necessity of always being among people'*: ibid.

387 *son of Count Brühl*: Grumach, ii. 538.

387 *informal little groups*: ibid. 533–6.

387 *'completely different existence'*: HABr, i. 498.

387 *publisher's announcement*: Grumach, iii. 23.

387 *'invite all good men'*: ibid. ii. 547.

387 *'great gift for social intercourse'*: ibid. iii. 72.

388 *society of observers of Nature*: ibid. 62.

388 *'favourite scheme, Jena'*: ibid. 59.

388 *'living encyclopaedic dictionary'*: WA, IV. 7, p. 24.

388 *willingly discussed*: HA, xiii. 156; Grumach, ii. 508.

388 *at least twenty times*: Grumach, vols. ii, iii.

388 *'if only a beginning'*: HABr, i. 481.

388 *J. F. A. Göttling*: AS, i. 362–3.

389 *equipment of the eccentric August von Einsiedel*: WA, IV. 7, pp. 15–17.

389 *A. J. G. C. Batsch*: ibid. 176–7.

389 *professors' salaries*: AS, i. 364–75.

389 *effectively a visitation*: ibid. 427–36.

389 *atmosphere of the university*: Bruford, *Culture*, p. 373.

389 Allgemeine Literatur-Zeitung: ibid. 300–2.

389 *'Areopagus of the public'*: Grumach, iii. 284.

390 *stimulated C. L. Reinhold*: Bruford, *Culture*, pp. 295, 372.

390 *indirectly contributed much*: Grumach, iii. 55.

390 *shared Kant's view*: ibid. ii. 513, 516–7.

390 *'You have to take the place'*: HABr, i. 504.

390 *'without you and him'*: ibid. 517.

390 *'as dumb as a fish'*: ibid. 504.

390 *devote his 'whole life'*: ibid. 508.

390 *seemed 'ever silent'*: Bode, i. 318, No. 518.

390 *'something fearfully stiff'*: Grumach, ii. 506.

390 *'is now mine too'*: WA, IV. 7, p. 67.

390 *'Nur wer die Sehnsucht kennt'*: ibid. I. 52, pp. 225–6; HA, vii 240–1.

391 *'once a very famous poet'*: Bode, *Pegasus*, p. 263.

391 *'the desert sand'*: Bode, *Garten*, p. 271.

391 *schema for his* Literary Works: WA, IV. 7, p. 235.

391 *'thought I was dead'*: ibid. 8, p. 83.

391 *told Eckermann*: 3 May 1827.

392 *Polish count's proposal*: Grumach, ii. 537.

392 *Franz Kobell*: WA, IV. 5, p. 12; cp. *CGZ*, VIb, No. 45.

392 *persuade the Duke of Gotha*: HABr, i. 408.

392 *'For some years now'*: WA, III. 1, p. 290.

392 *'you should enter upon'*: HABr, i. 443.

392 *'in the Italian manner'*: Grumach, ii. 505.

393 *'if I hold out'*: HABr, i. 489.

393 *picture by Tischbein*: ibid. 491.

393 *mastering the Italian language*: ibid. 503.

393 *writes on 5 May to Kayser*: ibid. 510.

393 *Italian lesson in Jena*: WA, IV. 7, p. 220.

393 *Princess Gallitzin*: Grumach, ii. 548–50.

393 *'we were quite natural'*: WA, IV. 7, p. 109.

393 *'a certain cooling off'*: Bruford, *Culture*, pp. 165–6.

394 *'once again' bringing her joy*: WA, IV. 8, p. 10.

394 *would not be returning*: Bode, i. 319, No. 524, presumably based also on conversation since the period is not stated in WA, IV. 8, p. 7.

394 *she walked across*: ibid.

394 *'incognito in forests and mountains'*: ibid.

394 *found it too lonely*: Grumach, iii. 49.

394 *Jacobi received a hint*: HABr, i. 515.

394 *'element of the* absolute': Grumach, iii. 45.

395 *set out from Carlsbad on 28 August*: WA, III. 1, p. 147.

395 *'the wishes and hopes of thirty years'*: ibid. 326.

395 *by his fireside eating stewed cherries*: ibid. IV. 7, p. 247.

395 *'almost as overripe'*: HABr, i. 512.

395 *'take as an oracle'*: ibid. 516.

395 *Caroline and Lavater arrived*: ibid. 517.

395 *originally thinking of eight months*: e.g. WA, III. 1, pp. 296, 327; ibid. IV. 8, p. 119.

396 *Bertuch came to him with a warning*: Grumach, iii. 18–19.

396 *Unknown to Goethe*: Loram, *Publishers*, p. 14.

396 *Göschen was taken aback*: Grumach, iii. 23.

396 *fee should be paid in advance*: Loram, *Publishers*, p. 18.

396 *6,000 copies at up to 8 dollars*: Bode, *Pegasus*, p. 323.

397 *'I am going'*: HABr, i. 518.

Works, 1785–1786

397 The Mysteries: *Die Geheimnisse*, HA, ii. 271–81.

397 *fifty stanzas were written*: WA, IV. 7, pp. 37, 38.

397 *Brother Mark*: Grumach, iii. 67.

398 *'bade him renounce'*: l. 116.

398 *general account of what his intentions*: HA, ii. 281–4.

399 *'Von der Gewalt'*: ll. 191–2.

399 *drew up a plan*: WA, IV. 7, p. 138.

400 *Wilhelm's going to Italy*: cp. also HA, xi. 217.

400 *'I myself cannot take pleasure'*: HABr, i. 427.

400 *small potential audience and three performances*: Carlson, *Theatre*, pp. 48–9.

401 *'canonical'*: WA, I. 52, p. 247.

402 *'how dull, insistent'*: ibid. 246.

402 *'highest conception of my nation'*: ibid. 255.

402 *'I began to despise'*: ibid. 257.

403 *'delivering up to the public'*: HABr, i. 515.

404 *'a great deed'*: WA, I. 52, p. 234.

404 *'The hour of judgement'*: ibid. 249.

404 *'poet and artist'*: ibid. 255.

405 *repeated reflections in this book*: WA, I. 52, pp. 226, 240, 281.

405 *'the hero has no plan'*: ibid. 248.

405 *Everything which he had promised himself*: ibid. 282.

405 *'gently guiding'*: ibid.

405 *'a something that has no name'*: ibid. 277.

406 *'since I cannot offer'*: WA, IV. 7, p. 235.

407 *took as his standard*: ibid. 8, p. 388.

407 *'more producible'*: HABr, i. 511.

407 *obligation to Kestner and Lotte*: HABraG, i. 83–4.

407 *weaknesses 'in the composition'*: WA, IV. 7, p. 237.

408 *sand with which she blotted a note*: HA, vi. 41.

408 *'how I worship myself'*: ibid. 38.

408 *'how wittingly'*: ibid. 44.

408 *'As Nature declines'*: ibid. 76.

408 *distance from the court*: HA, vi. 73–4, 81, ll. 34–5.

408 *drawing based on his own experience*: CGZ, I, No. 308.

409 *'we may almost say'*: HA, vi. 107.

409 *'she would have liked'*: ibid.

409 *'it was too late'*: DJG, iv. 174.

409 *memory of Werther's fiery embraces or to displeasure at his audacity*: ibid. 182; HA, vi. 118.

409 *because the two cannot abide each other*: DJG, iv. 169.

409 *his presence makes Werther uncomfortable*: HA, vi. 93.

409 *'might perhaps still have been rescued'*: ibid. 119.

410 *evening readings from his unpublished works*: Grumach, iii. 66, 65, 70.

410 *elaborate ceremony in celebration*: ibid. 67–9.

410 *asked Herder for some urgent assistance*: HABr, ii. 8.

410 *Aloysia von Lanthieri*: Grumach, iii. 69, 71; WA, III. 1, p. 147.

411 *'The ten years in Weimar'*: HABr, ii. 10.

CHAPTER 7

The Road to Rome

415 *Bavarian roads*: WA, III. 1, p. 148.

415 *'How glad I am'*: ibid. 149.

415 *'leaving by the wayside'*: ibid. 154.

415 *not let him name his destination*: HABr, ii. 16.

415 *'my guardian spirit'*: WA, III. 1, p. 155.

415 *another good omen*: ibid. 158.

415 *'meeting my characters'*: ibid. 160.

416 *'inner self left me no peace'*: ibid.

416 *'From here the waters flow'*: ibid. 162.

416 *addressee of his inner conversation*: HABr, ii. 12.

417 *'concerned with sensuous impressions'*: WA, III. 1, p. 175.

417 *fresh fruit*: ibid. 150, 154–5.

417 *valley seemed crammed with vines*: ibid. 176.

417 *sprayed with chalk*: ibid. 180.

417 *'born and brought up here'*: ibid. 177.

417 *overcoat was soon packed away*: ibid. 185.

418 *The inn*: ibid. 183.

418 *he ate figs all day*: ibid. 184.

418 *moved his table to the doorway*: ibid. 181.

418 *'it went well'*: ibid. 182.

418 Georgics: 2, ll. 159–60.

418 *'greatest natural phenomena'*: WA, III. 1, p. 182.

418 *scene from his own* Birds: ibid. 185.

418 *'one great garden'*: ibid. 186.

418 *old-fashioned Italian carriages*: ibid. 321–2.

418 *great amphitheatre*: ibid. 194–5.

419 *(tennis-like) game of balloon*: ibid. 196.

419 *promenading on the Piazza Brà*: ibid. 212.

419 *shouting and singing*: ibid. 201–3.

419 *allusions to the chorus*: ibid. 215, 217, and cp. HABr, ii. 13.

420 *'What was there in them'*: ibid. 201.

420 *'Here a man beside his wife'*: ibid. 199–200.

420 *'like a ripe apple'*: ibid. IV. 8, p. 29.

420 *vintage was beginning*: ibid. III. 1, pp. 215–16.

420 *'lively public'*: ibid. 222.

420 *'one enjoys with the Vicentines'*: ibid. 217.

421 *debate organized by the local academy*: ibid. 222–3.

421 *went into a bookshop*: ibid. 230.

421 *'I must stay here'*: ibid. 224.

421 *'see these works physically'*: ibid. 213.

421 *Teatro Olimpico*: ibid. 214.

421 *'Palladio was a truly great man'*: ibid. 213–14.

421 *Villa Rotonda*: ibid. 218.

421 *giant columns on the Loggia*: ibid.

422 *unclassical features of the Loggia del Capitaniato*: Ackerman, *Palladio*, p. 123, and cp. CGZ, III, No. 2.

422 *'The greatest difficulty'*: WA, III. 1, p. 214.

422 *'earns little thanks'*: ibid. 214–15.

422 *Catholic religion which seems so intrinsic*: ibid. 228.

422 *'Towards evening'*: ibid. 220–1.

422 *'green sea'*: ibid. 234.

423 *sharp, sure presence*: ibid. 240.

423 *coronation of Mary*: ibid. 207

423 *in the botanical garden*: ibid. 237.

423 *leaves of the Palmyra palm*: HA, xiii. 163.

423 *'as is proper'*: WA, III. 1, p. 220.

423 *Villa Valmarana*: ibid. 227. For 'il Burchiello' see Plate 20.

423 *'my page in the book of Fate'*: ibid. 240.

424 *stay until* Iphigenia *was finished*: ibid. IV. 8, p. 29; ibid. III. 1, p. 242.

424 *as the housemaid remarked*: ibid. 248.

424 *to the ghetto*: ibid. 249, 284.

424 *the drains*: ibid. 287, 252.

424 *bargaining in the fish-market*: for these Venetian scenes, ibid. 248, 278–92, 249–50, 264, 256, 276–7.

424 *the theatre, which Goethe assiduously attended*: ibid. 255–6, 259–60, 269–71, 275, 295, 291–2.

424 *verses of Tasso and Ariosto*: ibid. 279–81.

424 *'empty and null'*: ibid. 276.

425 *not being written for stage*: ibid. 275.

425 *church music*: ibid. 252–3, 284.

425 *tastelessness of the subjects*: ibid. 282–3, 246, 257.

425 *Palladio's master, Vitruvius*: ibid. 290.

425 *'the scales fall'*: ibid. 250.

425 *Il Redentore*: ibid. 257.

425 *Santa Maria della Carità*: ibid. 254–5, 268, 292–4.

425 *'how backward I am'*: ibid. 261.

425 *adaptation of old-established*: ibid. 268–9.

425 *'Palladio was permeated'*: ibid. 257.

425 *from the tower of St Mark's*: ibid. 249.

425 *to the Arsenal*: ibid. 265–6.

426 *'Alas, alas'*: ibid. 266.

426 *manservant he had hired*: ibid. 262.

426 *to the Lido*: ibid. 271–2.

426 *whole day on the Lido*: ibid. 285–8.

426 *'What a delightfully splendid thing'*: ibid. 288.

426 *artificial world of Venice*: ibid. 285.

426 *higher ingenuity of life*: cp. the emphasis on *'lebendiges'* in the original, deleted in HA, xi. 93.

426 *modelled on a giant crab*: WA, III. 1, p. 246.

426 *suspicion of republican constitutions*: ibid. 278; cp. 268.

426 *quality of 'completeness'*: ibid. 323, cp. 320.

427 *exposure to the 'southern climate'*: ibid. 251, 289.

427 *'first stage of my voyage is over'*: ibid. 296, promising Frau von Stein the journal of the second stage by the end of Nov.

427 *promised Seidel*: ibid. IV. 8, p. 35.

427 *still hoped to finish* Iphigenia: ibid. III. 1. p. 315.

427 *'hegira' from the Duke's birthday*: HABr, ii. 15.

427 *by the mailboat*: WA, III. 1, pp. 298–9.

427 *'obey my own impatience'*: ibid. 302.

427 *'the world rushes away'*: ibid. 305.

427 *'I cannot enjoy anything'*: ibid. 303.

427 *'the yearning of thirty years'*: ibid. 302.

427 *aiming now to reach Rome*: ibid. 303.

428 *'What can one say'*: ibid. 306–7.

428 *places hallowed by ancient civilizations*: cp. ibid. 182, 241.

429 *glad the next day to be leaving*: ibid. 300, 311.

429 *Loiano and Montecarelli*: ibid. 313.

430 *Count Cesarei*: Grumach, iii. 84.

430 *'Why do you think'*: WA, III. 1, p. 319.

430 *three hours in Florence*: ibid. 316; cp. ibid. IV. 8. p. 37.

430 *'the different currencies'*: ibid. III. 1, p. 321.

430 *'sacred gallows-hill'*: ibid. 323.

430 *who refused to read the Bible*: cp. ibid. 235.

430 *Santa Maria Sopra Minerva*: ibid. 323–5.

430 *'Homerical style'*: ibid. 322.

431 *'after thirty years'*: ibid. 326.

431 *sleeping in his clothes*: ibid. 329.

431 *'the third work'*: ibid. 327.

431 *'two more nights!'*: ibid. 328.

431 *transition from the limestone hills*: ibid. 330.

431 *Porta del Popolo*: HABr, ii. 16.

431 *'capital of ancient'*: WA, IV. 8, p. 37.

431 *Dante, Rabelais, and Montaigne*: HA (1988), xi. 614.

431 *gratitude 'to heaven'*: WA, III. 1, p. 331.

'Quite as much toil as enjoyment': October 1786–February 1787

431 *Tischbein came*: Grumach, iii. 87.

431 *four weeks*: WA, III. 1, p. 327.

432 *The mile-long Corso*: HA, xi. 485–6.

432 *coachman of Cardinal Carafa's*: Nohl, *Möller*, p. 10; WA, IV. 8, p. 352.

432 *'law and the prophets'*: ibid. 42.

432 *'day of initiation'*: ibid. III. 1, p. 331.

432 *Pygmalion's statue*: ibid. IV. 8, p. 38.

432 *over 160,000 inhabitants*: Bode, *Flucht*, p. 185.

432 *280 monasteries*: ibid. 195.

432 *7,000 priests*: ibid.

432 *'not to corrupt my imagination'*: WA, IV. 8. p. 139.

433 *spoke German*: Bode, i. 364, No. 615.

433 *Shelley's correspondence*: Shelley, *Letters*, pp. 58–60, 84–90.

433 *'sad and laborious task'*: WA, IV. 8, p. 46.

433 *'wash my soul clean'*: ibid. 83.

433 *'food for thought'*: ibid. 123.

433 *at 'the goal'*: HABr, ii. 20.

433 *'not so much enjoying things'*: WA, IV. 8. p. 177.

433 *until 9 o'clock*: Bode, i. 322, No. 529.

433 *300 churches and over 80 palazzi*: Bode, *Flucht*, p. 218.

433 *Goethe was himself acting as guide*: WA, IV. 8, p. 135.

434 *entry charge*: Bode, *Flucht*, p. 249.

434 *All Souls' Day*: HA, xi. 126–30.

434 *'honest Swiss fellow'*: WA, IV. 8, p. 83.

434 *'earnest, blunt and brisk'*: Nohl, *Möller*, pp. 15–16.

435 *Italian sobriquets*: WA, I. 32, p. 451.

435 *'Baron opposite the Rondanini'*: HA, xi. 134.

435 *'not wasted a moment'*: WA, IV. 8, p. 188.

435 *'aqueducts, baths'*: HABr, ii. 22.

435 *walking with Tischbein*: Grumach, iii. 167; for the date WA, IV. 8, pp. 49, 51.

435 *'greatest work of genius'*: ibid. 45.

436 *a Minerva*: ibid. 130, 145.

436 *frescoes of the Carracci*: HA, xi. 352; WA, IV. 8, p. 54.

436 *Galatea in the Villa Farnesina*: ibid. 55; HA, xi. 368.

436 *'pleasure of the first impression'*: WA, IV. 8, p. 45.

436 *'obliterated manuscript'*: ibid.

436 *St Cecilia's Day*: ibid. 63.

436 *'infatuated with Michelangelo'*: ibid. 71.

436 *merits of Raphael and Michelangelo*: Nohl, *Möller*, pp. 32–4.

436 *take Michelangelo's part*: HA, xi. 373.

436 *avoided the debate*: ibid. 390.

436 *'like Nature, always right'*: ibid. 454.

437 *'fearful stare'*: WA, IV. 8, p. 100.

437 *Christmas Eve*: Grumach, iii. 101; WA, IV. 8, p. 104.

437 *'ceremonies . . . seem to me'*: HABr, ii. 41.

437 *euphony of the Greek*: WA, IV. 8, p. 131.

437 *'spoilt for this hocus-pocus'*: ibid. 158.

437 *Pope simply the best*: HABr, ii. 48.

437 *'I read Vitruvius'*: ibid. 22–3.

438 *along with Bury and Schütz*: Nohl, *Möller*, p. 22.

438 *Moritz was already known to him*: WA, IV. 8, p. 68.

438 Anton Reiser: HABr, ii. 34.

438 *'fairest dreams'*: Grumach, iii. 91–2.

438 *a touch of homosexuality*: Boulby, *Moritz*, pp. 140, 264.

438 *man he called 'God'*: Grumach, iii. 268.

438 *treatise on German prosody*: Boulby, *Moritz*, pp. 147–50.

438 *day by the sea at Fiumicino*: Grumach, iii. 96–9.

438 *Goethe arranged for a roster*: ibid. 98.

438 *bandage was removed*: WA, IV. 8, p. 115.

438 *'as confessor'*: ibid.

438 *'when I was with him'*: HABr, ii. 28–9.

439 *'like a mirror'*: ibid. 45.

439 *'Tischbein and Moritz'*: ibid. 33.

439 *'How I long'*: WA, IV. 8, p. 74.

439 *She would appear to have asked*: WA, IV. 8. pp. 79, 93; Bode, *Stein*, p. 261; Bode, *Flucht*, p. 281.

439 *'So that was all'*: WA, IV. 8. p. 79.

439 *'in Rome he sometimes'*: Grumach, iii. 242.

439 *bewildering surroundings*: HABr, ii. 29.

439 *'it is seeming ever more'*: ibid. 31–2.

439 *apprehension of loss*: ibid. 33–4.

439 *'life-and-death struggle'*: ibid. 34.

439 *'bitter-sweet' letters*: ibid. 39.

440 *'Since the death'*: ibid. 43.

440 *a 'new man'*: ibid. 18.

440 *being 'born again'*: ibid. 20, 25, 26, 27.

440 *sense of the 'solidity'*: ibid. 19.

440   *remains for the 'greatness'*: HABr, ii. 22.

440   *'someone who has truly seen'*: WA, IV. 8, p. 125.

441   *'have to* un*learn'*: HABr, ii. 33.

441   *'like an architect'*: ibid. 34.

441   *'monstrous passion'*: WA, IV. 8. p. 119.

441   not *'say in what specifically'*: HABr, ii. 43.

441   *'You know my old'*: WA, IV. 8, p. 66.

441   *'nature is easier'*: HABr, ii. 31.

441   *'The least product'*: ibid.

442   *Minerva Giustiniani*: WA, IV. 8, p. 130.

443   *'not only the artistic'*: HABr, ii. 33.

443   *'did not the French influence'*: ibid. 48.

443   *read* Fanny Hill: WA, I. 32, p. 436.

443   *'It is terrible'*: HABr, ii. 50.

443   *'Here as everywhere'*: ibid. 47.

443   *'many solid objects'*: WA, IV. 8, p. 190.

443   *'exercising my eye'*: ibid. 137.

444   *'different divinities'*: ibid. 153.

444   *'out from behind my barricades'*: ibid. 153–4.

444   *Arcadian Society*: HABr, ii. 39.

444   *evening gatherings*: Grumach, iii. 112–13.

444   *'Efigenia'*: e.g. Bode, i. 322, No. 529.

444   Egmont, Tasso, *and* Faust: WA, IV. 8, p. 143.

444   *'entry into my fortieth year'*: ibid. 178.

444   *edition of ten volumes*: ibid. 198.

444   *too time-consuming*: ibid. 157.

444   *'small-minded German manner'*: HABr, ii. 48.

444   *rapid progress*: WA, IV. 1, p. 184.

444   *ten coloured views to send to his mother*: ibid. 202; see *CGZ*, II, Nos. 54–8, VIb, Nos. 55–6, 186; cp. also *CGZ*, II, No. 47.

444   *through its spies*: Bode, i. 333–4, Nos. 550–1.

444   *appointment as coadjutor*: ibid.

444   *Carl August was actively intriguing*: Steiger, ii. 614.

445   *Hirt was portrayed*: by F. G. Weitsch, Nohl, *Möller*, pl. 8, opp. p. 65.

445   *'reflecting on the fate'*: Bode, i. 322, No. 529.

445   *'The first period'*: WA, IV. 8, pp. 196–7.

446   *'time for me to leave Rome'*: ibid. 159.

446 *Vesuvius had recently begun to erupt*: HABr, ii. 23.

446 *Naples in the new year*: ibid. 28; cp. WA, IV. 8, p. 83.

446 *Rome for the Easter ceremonies*: HABr, ii. 35.

446 *generous and understanding letter*: Andreas, 'Vorabend', p. 69.

446 *series of possible itineraries*: WA, IV. 8, p. 120.

446 *a visit to Etna*: ibid. 160.

446 *not be required back before Christmas*: HABr, ii. 45; cp. WA, IV. 8, p. 177.

446 *squeeze it in in April and May*: WA, IV. 8, p. 163.

446 *declined to advise him*: ibid. 202.

446 *'dismembered world'*: ibid. 194.

446 *'happier world'*: ibid. 188.

446 *'unspeakably beautiful'*: ibid. 204.

446 *'rather provincial'*: HABr, ii. 44.

446 *'glad' to be leaving it*: WA, IV. 8, p. 188.

446 *'not really a feast'*: HA, xi. 484.

446 *'life of Rome'*: WA, IV. 8, p. 197.

446 *coarse and plebeian*: Boulby, *Moritz*, p. 153.

447 *slaughter of 1,000 pigs*: HABr, ii. 38.

447 *riderless horse-race*: Bode, *Flucht*, p. 307.

447 *'tedious fun'*: WA, IV. 8, p. 183.

447 *had had more than enough*: ibid. 188.

447 *'more noteworthy natural spectacle'*: ibid. 197.

447 *'recover the desire'*: ibid. 204.

447 *'unbelievably and inexpressibly beautiful'*: ibid. 185.

447 *almond-trees were in full bloom*: ibid. 187.

447 *crocuses and pheasants' eyes*: ibid. 159.

447 *fragmentary manuscript of* Tasso: ibid. 204.

A Glimpse of Fulfilment: *Iphigenia* and 'Forest. Cavern'

448 *'Der mißversteht'*: Iphigenie auf Tauris (HA, v. 7–67), ll. 523–4.

449 *'er dichtet'*: ll. 524–5.

449 *said in prose*: WA, I. 39, p. 371.

449 *'unsre Rückkehr'*: ll. 1339–40.

449 *'Rettet mich'*: ll. 1716–17.

449 *'hunt down the joy'*: l. 1364.

450 *Thoas explicitly asks*: l. 327.

450 *'brazen bond'*: l. 331.

450 *personal moral defects*: ll. 332–4.

450 *'axial' scene*: HA, xi. 205.

450 *prayer that he should be released*: l. 1330.

450 *'Es löset'*: l. 1358.

450 *'release' from 'the curse'*: l. 2115; cp. ll. 2117–18.

450 *'speak to us only'*: l. 494.

452 *'So steigst'*: ll. 1094–1124.

453 *'denn mein Verlangen'*: WA, I. 39, p. 323.

453 *'Denn ach!'*: ll. 10–14.

454 *'Den gelben matten'*: WA, I. 39, p. 365.

454 *'Durch Rauch'*: ll. 1142–3.

454 *'let some half-verses'*: HABr, ii. 42.

455 *'Wie gärend'*: ll. 1052–5.

455 *'Doppelt wird'*: ll. 1525–31, Goethe's punctuation leaves the grammatical structure highly ambiguous.

456 *'inspired by none other'*: HA (1988), v. 412–13.

456 *perfectly 'Greek'*: ibid.

456 *'one cannot read'*: ibid. v. 412–13.

456 *monologue by Faust*: ibid. iii. 103–4.

457 *attributed some of his work on Faust to January*: WA, I. 32, p. 483.

457 *completing his fragments*: WA, IV. 8, p. 83.

457 *'Erhabner Geist'*: ll. 3217–18.

457 *phrase echoed in Goethe's letters*: WA, IV. 8, pp. 83, 159.

457 *pallid concluding section*: ll. 3240–50.

Uneasy Paradise

458 *last scenes of* Faust: Part Two: l. 11559.

458 *For Tischbein*: HA, xi. 355.

458 *passed for a post-station*: reproduced in *Italienische Reise*, ed. Golz, No. 58, p. 185.

458 *Mignon had yearned*: HA, xi. 181.

458 *try to catch on paper*: CGZ, II, Nos. 67–80.

459 *'We eat crabs'*: Grumach, iii. 117.

459 *'awaited the passing'*: HA, xi. 185.

459 *'If Rome encourages study'*: ibid. 208.

459 *'It is disagreeable'*: ibid. 189.

459 *'My father . . . could'*: ibid. 186.

459 *'By water to Pozzuoli'*: ibid. 187.

459 *furious, almost expressionist*: CGZ, II, Nos. 87–91, 97–105. See plate 25.

459 *reminiscent of Werther*: HA, xi. 209, not certainly contemporary.

460 *superb panorama*: ibid. 220.

460 *tried to fix in his sketch-book*: ibid. 196; cp. CGZ, II, Nos. 126–7.

460 *packet boat for Sicily*: HA, xi. 189.

460 *germination of beans*: CGZ, Vb, Nos. 51–7.

460 *agaves and the branching process*: ibid. Nos. 63–6.

460 *'plenitude of an alien'*: HA, xiii. 162.

460 *'inspiration in botanical'*: ibid. xi. 222.

460 *basic principle*: WA, I. 31, p. 338, 'Urpflanze'; cp. HA, xi. 205.

460 *now in Arcadia*: Harnack, *Nachgeschichte*, pp. 232–4.

460 *garden of paradise*: HA, xi. 184, 207, 211, 216.

460 *earliness of the season*: ibid. 223.

460 *mouth of hell*: ibid. 211, 216.

460 *climbing Vesuvius*: ibid. 188–9.

460 *second ascent, four days later*: ibid. 192–5.

460 *third visit*: WA, III. 1, pp. 332–3; HA, xi. 214–16; Steiger, ii. 591.

461 *excavation of Herculaneum*: Haskell and Penny, *Taste*, p. 74.

461 *agents in the transfer*: ibid. 76.

461 *surprised by the smallness*: HA, xi. 198–9; WA, I. 47, p. 235.

461 *'this mummified town'*: HA, xi. 199.

461 *Torre Annunziata*: Grumach, iii. 121; HA, xi. 204.

461 *visit to Herculaneum*: ibid. 211–12.

462 *'how everyone swirls'*: ibid. 211.

462 calèches *that stood for hire*: ibid. 333.

462 *hucksters with lemons*: ibid. 335.

462 *deep frying-pans*: ibid. 341–2.

462 *children, gleaning sawdust*: ibid. 333–4.

462 *huddling on the paving stones*: ibid. 200.

462 *cauliflowers, currants, nuts, figs*: WA, IV. 8, p. 208.

462 *'cycle of vegetation'*: HA, xi. 334.

462 *despite great industriousness*: ibid. 336.

462 *no factories*: ibid. 338.

462 *'want to be merry'*: ibid.

462 *'German disposition'*: ibid. 216.

462 *'it is a strange sensation'*: ibid. 209.

463　*'never heard speak'*: HA, xi. 203.

463　*the works of Vico*: ibid. 192.

463　*battle royal on the beach*: Grumach, iii. 120–1; cp. *Italienische Reise*, ed. von Graevenitz, No. 51, opposite p. 120.

463　*When Tischbein demonstrated*: HA, xi. 205–6.

463　*'summit of all the joys of nature and of art'*: ibid. 209.

463　*as a* tableau vivant: ibid. 331.

463　*'Miss Hart' as Iphigenia*: *Italienische Reise*, ed. von Graevenitz, No. 56 before p. 125.

463　*ironical tone*: HA, xi. 209, 330.

463　*'constructed a beautiful existence'*: ibid. 217.

463　*those who 'study'*: ibid. 208.

463　*those who 'only live'*: ibid.

464　*'Naples is a Paradise'*: ibid. 207.

464　*sent off, this time at frequent intervals*: WA, IV. 8, p. 419.

464　*for general consumption*: Bode, i. 336, No. 557.

464　*'only a few things'*: WA, IV. 8, p. 212.

464　*'great mass of knowledge'*: ibid. 208.

464　*food for the memory*: HA, xi. 204, 'Bild in der Seele'.

464　*'Beautiful and splendid'*: WA, IV. 8, p. 210.

464　*'my heart warmed'*: HA, xi. 190.

464　*'What a difference'*: ibid. 207.

465　*taken up sepia*: CGZ, II, No. 3.

465　*'You have got talent'*: HA, xi. 206–7.

465　*C. H. Kniep*: Grumach, iii. 125–6; in *Italienische Reise* Kniep is misrepresented as a youthful beginner.

465　*'permanent companion'*: HA, xi. 213.

465　*a dinner-coat*: Grumach, iii. 155.

465　*mark out a frame*: HA, xi. 218; cp. Grumach, iii. 156.

465　*horizons with a ruler*: CGZ, II, No. 146.

465　*sharpen his pencils frequently*: HA, xi. 218; cp. Grumach, iii. 156.

466　*By 23 March*: WA, IV. 8, p. 210; cp. HA, xi. 217.

466　*doubts about the new* Iphigenia: HA, xi. 208; WA, IV. 8, p. 213.

466　*symbolic shape*: ibid. p. 212.

466　*'See Naples and die!'*: HA, xi. 189.

466　*'points me towards Asia'*: ibid. 222.

466　*'no small thing'*: ibid.

466　*'On this journey'*: ibid. 223.

466　*agreed that Kniep*: ibid. 218.

466 *hectically busy*: ibid. 223.

466 *visit to the ancient temples at Paestum*: WA, IV. 8, p. 210; in *Italienische Reise* Goethe attempts to place the visit to Paestum before his departure for Sicily: traces of the actual visit on 15–17 or 21–3 May remain, however, and create the illusion that there were two visits.

466 *had to be postponed*: HA, xi. 222, 'soll manches nachgeholt werden'.

The Gardens of Alcinous

467 *'construct a diary'*: WA, I. 31, p. 340.

467 *the structure of* Torquato Tasso: HA, xi. 226–8.

467 *Goethe was quite delirious*: Grumach, iii. 129–30.

467 *Sicilian coastline*: HA, xi. 228.

467 *northerly waters of a wine-dark sea*: ibid. 230.

467 *'I wish you could see'*: WA, IV. 8, p. 211.

467 *'that I could have drawn'*: HA, xi. 300.

467 *'I have no words'*: ibid. 231–2.

467 *Palermo, a town as large as Rome*: Bode, *Flucht*, p. 384.

468 *'so happy and contented'*: WA, IV. 8, p. 211.

468 *Villa Giulia*: HA (1988), xi. 646.

468 *shrine of Saint Rosalia*: HA, xi. 237–40.

468 *Statella was back in Weimar*: Grumach, iii. 132.

468 *trip to Malta*: HA, xi. 281.

468 *easily reached by the coastal packets*: I am indebted for these points to Professor R. C. Paulin.

469 *Antonio Vivona*: Grumach, iii. 135.

469 *afternoon of 16 April*: Steiger, ii. 600.

469 *knew that he had deceived*: HA, xi. 264.

469 *recounted the episode in Weimar*: WA, I. 31, p. 301.

469 *'a nothing that wants'*: HA, xi. 242.

470 *he bought a Greek text*: WA, I. 31, p. 339, not necessarily reliable.

470 *'Was rufen mich'*: HA, v. 69, ll. 23–9.

471 *story of Friederike Brion*: HA (1988), v. 482; Steiger, *Goethe*, ii. 46.

471 *'through all my days'*: Homer, *Odyssey*, p. 96.

471 *'absence and the great distance*: HABr, ii. 51.

472 *'How much joy'*: ibid.

472 *'Urpflanze'*: HA, xi. 266.

472 *repeated sequence of such elements*: CGZ, Vb, No. 58.

472 *'so abstract'*: HA, xi. 222.

472 *gardens of Alcinous*: HA, xi. 267.

472 *'turning-point of the whole adventure'*: HABr, iii. 419.

472 *'journey is now getting a shape'*: HABr, ii. 51.

473 *Selinunte, all of which he passed by*: there is no contemporary evidence of a visit to Selinunte, *pace* Bode, *Flucht*, p. 409.

473 *'an indestructible treasure'*: HABr, ii. 52, 54.

473 *'hyper-classical'*: HA, xi. 300.

473 *'dramatic concentration'*: ibid. 298.

473 *ignored the Norman and Byzantine*: ibid. 241, 247, 249, 273 and footnotes.

473 *'understand the Claude Lorrains'*: HA, xi. 231.

473 *remembered scenery*: ibid.

473 *'poetic mood'*: ibid. 300.

474 *'The situation is remarkable'*: WA, III. 1, p. 341.

474 *Girgenti and numbering some 20,000 inhabitants*: Bode, *Flucht*, p. 412.

474 *own few drawings*: CGZ, II, Nos. 160–1; III, No. 31 (VIb, No. 188 is probably later), and cp. HA, xi. 278.

474 *original plan*: HABr, ii. 51, 'Küsten'.

475 *'principal purpose'*: HA, xi. 222.

475 *visit Greece and Dalmatia*: ibid. 223.

475 *'fertile desert'*: ibid. 282.

475 *'so buried and submerged'*: ibid. 296.

476 *Friday 4 May*: for the dates see WA, I. 32, p. 476; Steiger, ii. 605–7.

476 *'talent for restoration'*: HA, xi. 296.

476 *'if one takes one's seat'*: ibid. 297.

476 *Monday was spent in Taormina*: Kniep's drawing in *Italienische Reise*, ed. von Graevenitz, No. 87, after p. 176.

476 *thoughts turned once again to* Nausicaa: HA, xi. 298.

476 *on a loose piece of paper noted down*: WA, I. 10, pp. 414–15, 422–3, contrast HA (1988), xi. 481. The dating however is not vital.

477 *'Ein weißer Glanz'*: HA, v. 72, ll. 135–6.

477 *book VI of the* Odyssey: l. 44.

477 *any lines of Valéry's*: e.g. 'Fermé, sacré, plein d'un feu sans matière/Fragment terrestre offert à la lumière'.

478 *parted from their faithful guide*: WA, I. 32, p. 476, dates as in Steiger, ii. 608.

478 *'a strange sight'*: HABr, ii. 55.

478 *own later drawing*: CGZ, II, No. 176.

478 *'horizontal position'*: HA, xi. 314.

478 *Faraglioni Rocks*: Steiger, ii. 610.

478 *'Meeresstille' and 'Glückliche Fahrt'*: HA, i. 242.

478 *butchers' shops were full*: HA, xi. 341.

478 *'conjoining the spiritual'*: ibid. 327.

478 *he enjoyed it*: WA, IV. 8, p. 219.

478 *'getting lazier'*: HABr, ii. 57.

478 *nobility were no longer shunned*: WA, IV. 8. p. 216.

478 *'really too old'*: HABr, ii. 54.

478 *concerts at the Austrian ambassador's*: Grumach, iii. 163–4.

479 *dinner at Hamilton's*: HA, xi. 330–1.

479 *'not have as violent an effect'*: ibid. 325.

479 *been in Pozzuoli*: ibid. xiii. 287–95; Wells, *Development*, pp. 68–9.

479 *'alpha and omega'*: HABr, ii. 57.

479 *via Salerno to Paestum*: HA, xi. 218–20.

480 *'most splendid image'*: ibid. 323.

480 *'hastened from peak to peak'*: HABr, ii. 54.

480 *letter handed over to him on 24 May*: Steiger, ii. 611.

480 *in addition to several letters from Fritz*: WA, IV. 8, pp. 219, 215; HABr, ii. 53.

480 *reach Frankfurt for his birthday*: WA, IV. 8, p. 217.

480 *stay with his mother*: HABr, ii. 55.

481 *long letter to the Duke*: ibid. 53–7.

481 *'rather have death'*: ibid. 59.

481 *Prussian special envoy, Marquis Lucchesini*: Steiger, ii. 614.

481 *man of the world*: HABr, ii. 57–8.

481 *'strong moral stomach'*: ibid. 57.

481 *'another fortnight'*: ibid.

482 *'at a single glance the moon'*: ibid. 58.

In the High School of Art: 1787–1788

482 *writing to Frau von Stein*: HABr, ii. 59.

482 *'I am spoiled'*: ibid. 58.

482 *Tivoli and the Alban Hills*: Steiger, ii. 618–19.

483 *three forms of leaf*: Italienische Reise, ed. von Graevenitz, No. 53, between pp. 122, 123.

483 Tasso *and* Faust: Grumach, iii. 165.

483 *'little talent'*: WA, IV. 8, p. 236.

483 *already written to his mother*: ibid. 420.

483 *surpass anything known in Spain*: HA, xi. 369.

483 *meteorological records show*: Ephemerides, p. 432.

483 *zithers and merry-making*: WA, IV. 8, p. 234.

484 *Bernini's pump-room*: Steiger, ii. 620.

484 *scorning the local habit of the siesta*: Grumach, iii. 236.

484 *Trajan's Column*: HA, xi. 371.

484 *'some hours almost every day'*: Grumach, iii. 174.

484 *Carl August's reply*: Bode, Rom, p. 49.

484 *rather indefinite roles*: HABr, ii. 62; cp. ibid. 86.

484 *Maximilian von Verschaffelt*: Steiger, ii. 624.

485 *read travel-books*: HA, xi. 348.

485 *Herculanean model*: Bode, Flucht, p. 367.

485 *ravaged health*: Bode, i. 339, No. 560.

485 *celebration at the cottage*: ibid. 339–40, Nos. 561–2.

485 *Wertherian existence*: Bode, Garten, p. 298, 'mein Öfchen', p. 299.

486 *those, however, who doubted*: Bode, i. 330, No. 543.

486 *soft in the head*: WA, IV. 8, p. 308, 264.

486 *'Goethe's return is uncertain'*: Bode, i. 344, No. 574.

486 *written in Frascati*: WA, IV. 8, p. 264.

486 *reliable deputies*: HABr, ii. 86.

486 *new accounting procedures*: WA, IV. 8, pp. 289–91.

486 *Voigt, dealing with a fatal accident*: ibid. 165, 274, 339; Bode, Rom, pp. 103–4.

486 *education of the now adolescent Fritz*: WA, IV. 8, pp. 327, 335.

486 *Filippo Collina*: ibid. 288.

486 *persuaded her to defer*: HABr, ii. 70.

486 *give himself time*: HA, xi. 429–30.

487 *'Ariadne's clue'*: HABr, ii. 65.

487 *numerical formulae for the proportions*: cp. CGZ, III, Nos. 121–2, 138, and pp. 54–5; HA, xi. 386.

487 *Papal throne*: HA, xi. 390.

487 *Egmont, however, was finally ready*: ibid. 394.

487 *'inexpressibly difficult'*: ibid. 431–2.

487 *first rains of autumn*: Bode, Rom, p. 64.

487 *Kayser himself was planning to come to Rome*: WA, IV. 8, p. 255.

487 *Alban Hills, whose volcanic origins*: HABr, ii. 68.

487 *'most beautiful spots'*: WA, IV. 8, p. 273.

487 *spoke more Italian*: Bode, Rom, p. 67.

487 *little theatre in Castel Gandolfo*: HA, xi. 415.

488 '*a Werther-like fate*': ibid. 427.

488 '*Amor als Landschaftsmaler*': ibid. i. 235–7.

488 *Angelica Kauffmann*: ibid. xi. 419.

488 *Angelica had been painting his portrait*: *Italienische Reise*, ed. von Graevenitz, No. 109, between pp. 210, 211, Herder's comment, p. 353.

488 *bust which Trippel also completed*: HA, xi. 397.

488 *summer exhibition of the French Academy*: HA, xi. 387–8, 391. The reconstruction has not been traced and may be a later invention of Goethe's.

489 *his* villeggiatura: CGZ, II, Nos. 250–325, VIb, Nos. 69–89. See also plate 29.

489 *like the start of a new semester*: HABr, ii. 79.

489 '*true high school*': WA, IV. 8, p. 286.

489 '*art' was 'a serious matter*': ibid. 261; cp. HABr, ii. 66.

489 '*going to school*': HABr, ii. 79.

489 '*only an imperfect conception*': WA, IV. 8, p. 320.

489 '*averted my eyes from it*': HABr, ii. 79.

489 *G. A. Camper*: ibid. 76.

489 *classes on perspective*: Bode, *Rom*, p. 86.

489 *J. A. Arens*: Grumach, iii. 185–6.

489 *J. H. Meyer and young Bury*: HABr, ii. 80.

489 *Kayser too arrived*: HA, xi. 435.

489 '*our domestic academy*': ibid. 444.

490 '*entirely among strangers*': HABr, ii. 79.

490 *guide Goethe through his art by playing*: ibid. 70.

490 *in Rome Kayser set about composing*: WA, IV. 8, pp. 351–2 (contra Bode, *Rom*, p. 81; cp. WA, IV. 8, p. 257).

490 *pave the way*: WA, IV. 8, p. 244.

490 *Göschen had expressed himself*: ibid. 279.

490 *intended Cagliostro opera*: ibid. 244–5.

490 *honorarium for which he would need to help pay*: HABr, ii. 81.

490 '*only the title*': WA, IV. 8, p. 342.

490 *gave up the idea of submitting*: ibid.; cp. ibid. 347, 367.

490 '*Whoever leaves Rome*': HABr, ii. 77.

491 '*I am too old*': HA, xi. 434. Trunz's original emendation (insertion of a 'nicht') rightly eliminated in 1988 edn. (cp. p. 517, HABr, ii. 73).

491 '*As an artist*': HABr, ii. 285.

491 *uninterested in Roman administration*: WA, IV. 8, p. 287; Grumach, iii. 192.

491 *now called 'paradise'*: HABr, ii. 61, 88; cp. WA IV. 8, p. 340.

492 *observe the life and circumstances of the Roman people*: HABr, ii. 64, 69.

492 *visited the Borghese gallery*: HA, xi. 526.

492 *offers some reflections*: HABr, ii. 71–2, cp. 76, 78.

492 *'all the importunities'*: ibid. 73–4; cp. HA, xi. 385.

492 *chose never even to call*: WA, IV. 8, pp. 334–5.

492 *'life in Rome so agreeable'*: HABr, ii. 62–3.

493 *'fulfilment of all my desires'*: ibid. 61.

493 *'the promised land'*: ibid. 88.

493 *scheme for a German National Academy*: Herder, *Werke*, xvi. 600–16.

493 *the intellectual class*: ibid. 613 (that the Academy is a substitute for earlier dreams of a national theatre is stated p. 607).

493 *'lack of a livelier literary commerce'*: WA, IV. 8, p. 286.

493 *poor reception of his* Literary Works: ibid. 246, Loram, *Publishers*, p. 21.

494 *1,000 subscribers*: Loram, *Publishers*, p. 153.

494 *'alpha and omega of all art'*: HABr, ii. 64.

494 *'I could venture to live'*: ibid. 79.

494 *'sensible' as never before*: HA, xi. 530, 'vernünftig'.

494 *Lessing, it will be remembered, had noted*: Bode, i. 262, No. 421.

494 *'seed of madness'*: HA, xi. 531.

494 *'subordinating the poetry'*: WA, IV. 8, p. 245.

494 *Seidel objected*: ibid. 354–5.

494 *intended to be sung*: ibid. 336.

495 Egmont *will be performed*: ibid. 244.

495 *Moritz's little treatise*: Boulby, *Moritz*, p. 164.

496 *resulted from their long conversations*: HA, xi. 534.

496 *refreshed by mail from Germany*: ibid. 387–8.

496 *'substantial forces'*: Herder, *Werke*, xvi. 451.

496 *devoured it with passionate enthusiasm*: HA, xi. 393.

496 *'great totality of Nature'*: Moritz, *Bildende Nachahmung*, p. 14.

496 *a 'small divinity in its own sphere'*: Leibniz, *Monadology*, para. 83.

497 *'If Nature herself'*: Moritz, *Bildende Nachahmung*, p. 14.

497 *having accepted the de-Christianized monadism*: cp. HA, xi. 395.

498 *is it Goethe?*: 'Kunst' is not used with this general sense in the essay which was Moritz's starting-point *(Versuch einer Vereinigung aller schönen Künste und Wissenschaften unter dem Begriff des in sich selbst Vollendeten*, pub. *Berlinische Monatsschrift* 5, Berlin, 1785, pp. 225 ff., reprinted by S. Auerbach in Moritz, *Bildende Nachahmung*, pp. 38 ff.). The term, however, was widely used in Weimar in this sense before Moritz's visit (e.g. Knebel to Herder, 7 Nov. 1788, Steiger, ii. 690). Even in 1793 the third edn. of Sulzer's *Allgemeine Theorie* did not know the usage, devoting 25 pages to 'Künste; Schöne Künste', and only two to 'Kunst; Künstlich'.

498   *short article*: see previous note.

498   *products of 'Art'*: Moritz, *Bildende Nachahmung*, p. 18.

499   *'as an artist!'*: HABr, ii. 85.

499   *'just a prosaic German'*: WA, IV. 8, pp. 354–5.

500   *'my* Harmonia plantarum': HABr, ii. 64.

500   'hen kai pan': HA, xi. 395.

500   *single universal divine principle*: ibid. 389.

500   *experiments Seidel was conducting*: WA, IV. 8, p. 320.

500   *'model' which, if taken in conjunction with a 'key'*: HABr, ii. 60.

501   *'formula' which 'explains'*: ibid. 67.

501   *growing out of the flowers*: ibid. 64–5, 67; cp. HA, xiii. 166, 95–6, *CGZ*, Vb, Nos. 67–8.

501   *interest in physiognomy*: HA, xi. 386.

501   *series of gradations*: *CGZ*, VIa, No. 47, III, Nos. 170, 177, 216; cp. ibid. Vb, Nos. 77–8.

501   *shapes of the lips*: ibid. III, Nos. 121–2.

501   *'beautiful sights . . . remain alive'*: HABr, ii. 58.

501   *'indestructible' component*: ibid. 54.

501   *to be carried home*: HA, xi. 323.

501   *obsessive redrawing*: cp. *CGZ*, VIb, p. 34, ad No. 82.

502   *'Four slender little volumes'*: HA, xi. 399.

502   *shadow of the homeward journey*: ibid. 435; cp. HABr, ii. 77.

502   *typical of the whole of human life*: WA, IV. 8, p. 307.

502   *'Periamo noi'*: Grumach, iii. 206; cp. WA, I. 32, p. 451.

502   *Goethe found 'splendid'*: HA, xi. 419.

502   *'quite transported' Moritz*: ibid. 477.

502   *consolations of Leibniz: there is no death in the universe*: Herder, *Werke*, xvi. 565–6; cp. Leibniz, *Monadology*, paras. 69, 73; also Herder, *Werke*, xvi. 541, 540 ('world-spirit').

502   *'permanent destruction of the individual'*: Moritz, *Bildende Nachahmung*, p. 34.

502   *laws of harmonic motion*: Herder, *Werke*, xiv. 225.

502   *culture reaches its 'maximum'*: ibid. 229–30, 247.

502   *arguments Kant had put forward*: Kant, *Werke*, 12 (6), pp. 791–2, 800, 804–6.

502   *'cosmic order'*: Herder, *Werke*, xvi. 540.

502   *Protestant contrast with the Roman 'Babylon'*: HA, xi. 387.

502   Ideas *as the true 'gospel'*: ibid. 415, 417.

502   *vanquish the apologists*: ibid. 416.

503   *'not expecting any Messiah'*: ibid. 415.

503 *'the moment is everything'*: HA, xi. 419. That the comment is not an interpolation, or, if it is, is not an anachronism, is shown by the parallel cited from Moritz's *Reisen*. Cp. also the conclusion of *Das Römische Karneval*, no later than early 1789, and possibly already drafted in 1788, the note 'Ergreifen des Moments' made during Goethe's return journey (WA, I. 32. p. 461), and our analysis of *Egmont*. The word 'vernünftigen' may be a later addition (cp. Bode, i. 358, No. 601).

503 *'a friend of lies'*: HA, xi. 413.

503 *'indeterminate spirituality' of Rembrandt*: WA, IV. 8, p. 304.

503 *"enjoy every beautiful scene"*: Boulby, *Moritz*, p. 183, citing *Reisen eines Deutschen in Italien* (Berlin, 1792), iii. 283 (his translation); cp. Harnack, *Nachgeschichte*, p. 62.

503 *'no sense of any of the ills'*: HABr, ii. 77.

503 *'nothing in common'*: ibid. 69.

503 *'think back to Germany'*: ibid. 65; cp. WA, IV. 8, p. 252; HABr, ii. 82.

503 *His Wilhelmiad*: ibid. 81.

504 *calling in a loan*: WA, IV. 8, p. 247.

504 *'end of all my labours'*: ibid. 302.

504 *'heart readily becomes cold and brazen'*: HABr, ii. 82.

504 *'magnetic mountain'*: WA, IV. 8, p. 371.

504 *'city of the Muses'*: HABr, ii. 89.

504 *'unconditionally happy'*: ibid. 94.

504 'Summa Summarum': WA, IV. 8, p. 348.

504 *'starting a new page'*: ibid.

504 *'Time, alas'*: ibid. 320.

504 *Duke, now on his way back*: Steiger, ii. 650.

504 *refer to haemorrhoids*: WA, IV. 8, p. 336.

504 *length and promptness of Goethe's reply*: HABr, ii. 78–83; cp. WA, IV. 8, pp. 324–36.

505 *Gazzaniga's opera Don Giovanni*: Steiger, ii. 652.

505 *'usually spend my evenings'*: HABr, ii. 76.

505 *agreed with Bertuch*: WA, IV. 8, p. 319.

505 *Schütz would provide the illustrations*: HA, xi. 520.

505 *not just surveying the scene*: WA, IV. 8, pp. 338–9.

505 *pushing his way through on foot*: Steiger, ii. 652.

505 *'far too noisy days'*: WA, IV. 8, p. 340.

505 *for some barrack-room gossip*: ibid. 314–15.

505 *unable to oblige*: ibid. 262.

505 *responded the Duke*: ibid. 346–7.

505 *self-absorption of Goethe's was bad*: HABr, ii. 79.

505 *unrhymed song*: HA, i. 237.

505 *drawings of Hero and Leander*: CGZ, III, Nos. 259, 259Rs, 260.

506 *'cold towards his friends'*: Bode, i. 348, No. 577.

506 *matter-of-fact reports*: many letters are missing, probably destroyed by Goethe in 1829.

506 *before receiving the momentous letter*: cp. the phrase 'die horizontale Lage', from *Tristram Shandy*, HABr, ii. 78.

506 *Faustina Antonini*: Bode, *Rom*, p. 100.

506 *started to make assignations*: Grumach, iii. 206.

506 *under the name*: WA, IV. 8, p. 370.

506 *writing a Roman 'IV' in spilt wine*: HA, i. 168.

506 *he had outlined to Carl August on 29 December*: HABr, ii. 75.

506 *'some agreeable promenades'*: WA, IV. 8, p. 347.

507 *'various measures'*: HA, xi. 528.

507 *laurel, viburnum, and box*: HABr, ii. 89.

507 *old friend well received*: HA, xi. 521–2.

507 *Carl Pieck*: Nohl, *Möller*, p. 117.

507 *new growth of ivy*: HABr, ii. 89.

507 *blossoming gardens*: HA, xi. 544.

507 *setting up of his own gallery*: ibid. 544–7; cp. Steiger, ii. 656.

507 *animals from one of the Egyptian obelisks*: HA, xi. 393.

507 *Olympus to the underworld*: HA, i. 162.

508 *drew his own moonlit tomb*: CGZ, II, Nos. 332–3.

508 *'most beautiful and solemn cemetery'*: Shelley, *Letters*, ii. 60.

508 *consultations with Anna Amalia*: Bode, *Rom*, p. 122.

508 *Duke accepted*: HABr, ii. 501–2. The contents of the letters from the Duke and Herder are reconstructed from indications in Goethe's letters, largely following Bode, *Rom*, p. 122.

508 *taken over by Bertuch*: WA, IV. 8, p. 368.

509 *"I come!"*: HABr, ii. 84–8.

509 *Carl August promulgated*: Steiger, ii. 659.

509 *happy for him to stay*: HABr, ii. 115.

509 *desperately anxious*: Grumach, iii. 214.

509 *'Duke has refused him an extension'*: Bode, i. 349, No. 579.

509 *response to 'a hint'*: HABr, ii. 88.

509 *'evil spirit'*: WA, IV. 9, p. 32.

509 *'driving me out'*: HABr, ii. 103.

509 *Letters went immediately*: WA, IV. 8, p. 422.

510 *'circuit of the prophet'*: HABr, ii. 92.

510 *the doubter,* Nathanael: Bode, *Rom*, pp. 141–3.

510 *completing his correspondence by Tuesday*: HABr, ii. 84. Apart from his regular letter to Frau von Stein and a note to Göschen about a proof-correction, this was the only letter written that week (WA, IV. 8, p. 422).

510 *what he saw*: HA, xi. 530–1.

510 *'liked to be a child'*: HABr, ii. 91.

511 *Raphael's skull*: ibid. 87.

511 *'Fritz the Second'*: WA, IV. 8, p. 350.

511 *watch tearfully*: Grumach, iii. 199.

511 *'departure from Rome'*: HABr, ii. 92.

511 *'wept like a child'*: Grumach, iii. 223.

511 *transfer to a special account*: WA, IV. 8, p. 370; ibid. I, 32, pp. 457, l. 7, 458, ll. 5–7.

511 *full moon illuminated*: the moon was full on Sunday 20 Apr., *Nautical Almanac*, p. 37; cp. HA, xi. 554.

511 *promising himself inwardly*: Bode, *Rom*, p. 148.

511 *present for the Duke*: WA, IV. 8, p. 372.

511 *'pierced my heart'*: Grumach, iii. 199.

511 *gathered round the coach*: ibid. 199–200.

511 *Moritz was at his best*: Biedermann-Herwig 3. 1, p. 113, No. 4681.

511 *'nothing more to say'*: HABr, ii. 93 (cp. WA, I. 32, p. 459, l. 33).

511 *'seven different moods'*: HABr, ii. 92.

512 *Passing through Viterbo*: Steiger, ii. 665.

512 *'more practised eye'*: WA, IV. 9, p. 122.

512 *seen 'almost everything Florence contains'*: ibid. 8, p. 371.

512 *set magnificently fair*: ibid. 370, 372.

512 *remarkable private collection*: Grumach, iii. 207–8.

512 *acquaintance with Giotto*: Steiger, ii. 665.

512 *'a real keystone'*: HABr, ii. 93.

513 *the blue skies, 'the fruit, etc.'*: ibid. 92.

513 *letter to Knebel*: ibid. 93–4.

513 *mood of self-preoccupation*: HA (1988), xi. 704–5.

513 *new material for the play*: e.g. a first draft of the discussion of the lost Golden Age, WA, I. 32, p. 461.

513 *indications from as early as the second day*: CGZ, II, No. 343.

513 *geometrical—even symmetrical—composition*: ibid., Nos. 71, 74, cp. No. 21.

513 *sketches of April and May*: ibid., Nos. 343, 345, 355, 372.

514 *Strong emotions are sometimes suggested*: ibid., Nos. 354, 357, 360.

514 *three minute brush studies*: ibid., Nos. 361–3. See plate 34.

514 *leaving Milan on 28 May*: Steiger, ii. 668.

514 *twenty views of the lake-shore*: CGZ, II, Nos. 373–91.

514 *anticipation of Turner*: e.g. ibid., No. 374.

514 *grasp and fix the last moments*: cp. WA, I. 32, p. 461, ll. 23–4.

The Great Soul: Works 1787–1788

514 *'Finishing my older things'*: HABr, ii. 63.

514 *'priceless moments'*: WA, IV. 8, p. 279.

515 *'great soul'*: l. 1076.

516 *In the first scene*: HA, iv. 373, other instances of the motif pp. 423–4, 426, 430, 433, 438.

516 *recalled by Clärchen*: ibid. 387.

516 *'Now life is coming'*: ibid. 451.

516 *'Do I live only'*: ibid. 399.

517 *manifests itself as destruction*: Moritz, *Bildende Nachahmung*, p. 25.

517 *fires death-dealing arrows*: ibid. 34–5.

517 *almost certainly wrote in Rome*: not only is the theme Roman, but the speech and Egmont's next following speech are readily detachable from their dramatic context, from which they are rhetorically clearly distinct.

517 *'As if lashed'*: HA, iv. 400–1.

517 *Moritz told Caroline Herder*: ibid. (1988) 589–90.

517 *'Youth and fine love'*: ibid. 411.

518 *'I have risen high'*: ibid. 401.

518 *'ending' Egmont*: WA, IV. 8, p. 239, dated by Steiger (ii. 618–19) to 30 June 1787. It could however be the letter to Herder of 23 June (WA, IV. 8, p. 420).

518 *Clärchen's rallying-cry*: HA, iv. 435.

519 *'incredibly carefree nature'*: WA, III. 1, p. 321.

519 *'my situation will be'*: HABr, ii. 45.

519 *'at the highest point'*: Moritz, *Bildende Nachahmung*, p. 25.

519 *'I am ceasing to live'*: HA, iv. 451.

520 *image of what they are living*: ibid. 430 (transfer to the people of the leitmotif of the horse).

520 *'I am dying for the freedom'*: ibid. 453–4.

520 *'Humanity cannot raise'*: Moritz, *Bildende Nachahmung*, p. 31.

520 *'I now live for you'*: HA, iv. 450.

521 *wrote a long critique*: HABr, ii. 90.

521 *recent additions to the last act*: HA, xi. 432, 458–9.

521 *Schiller, in a most acute review*: Schiller, *Werke*, v. 932–42.

521 *'the ethical part'*: WA, IV. 9, p. 37.

521 *'the unpoetical amateur'*: HA, xi. 458.

521 *'ideeller Bezug aufs Wirkliche'*: ibid. 454.

521 *'rather a frigid effect'*: Bode, i. 349, No. 580.

521 *'more as what it could be'*: HABr, ii. 90.

522 *elimination of Olimpia and Bernardo*: WA, IV. 7, p. 168.

522 *'Ohne Thräne'*: ibid. I. 11, p. 329, ll. 885–6.

522 *error in the naming of the dramatis personae*: ibid. IV. 8, p. 363.

523 *reflection on the blank-verse monologue*: Moritz, *Bildende Nachahmung*, pp. 25–6.

524 *hoped to complete the play by Easter 1789*: Steiger, ii. 661.

525 *'it always comes down'*: HA, iv. 399.

525 *In February 1788*: Binder, *Faust*, pp. 112, 10. I largely follow Binder's views. However I take it that 'unbedingt' in l. 1855 is an acknowledgement by Mephistopheles that no agreement in the traditional sense has taken place (similarly ll. 1866–7, if they were written at this stage, could have been in the indicative without metrical difficulty). Binder overlooks that the principal reason for Mephistopheles' defence of 'Vernunft und Wissenschaft' in ll. 1851–2 is to prepare for the 'Schülerszene' that follows, especially l. 2047.

525 *'Und was der ganzen'*: ll. 1770–5.

525 *its 'crowns' as he here calls them*: l. 1804; cp. WA, I. 14, p. 258.

525 *dialogue between Faust and Mephistopheles*: ll. 1776–1867.

525 *intended to follow almost immediately*: Binder, *Faust*, esp. p. 39.

526 *to* be *all that is open*: l. 1774.

526 *to* have *all the satisfaction*: l. 1825.

526 *'straight into the world'*: l. 1829.

526 *bird in the lime*: l. 1862.

526 *his mind was on Junos*: Grumach, iii. 169.

527 *'joys of earth'*: l. 1859.

527 *second scene that Goethe wrote*: Steiger, ii. 656, to Eckermann 10 Apr. 1829.

527 *'How do we set about'*: l. 1834.

527 *through Fuseli and Lavater*: cp. CGZ, I, Nos. 302–3, 305–6, VIb, No. 216, and WA, IV. 8, p. 239.

527 *Carracci's fresco*: see plate 35.

527 *anxious to recover the tone*: HA, xi. 525.

528 *nothing in common with the profound nature-magic*: Butler, *Fortunes*, p. 157.

528 *versification is at times flat*: e.g. ll. 2436–7, 2575–6.

528 *Mephistopheles' identification of himself*: ll. 2481–2515, cp. 1783, 1866.

528 *noticed on its publication*: Gräf 2.2, p. 136.

528 *'Cupid to stir'*: l. 2598.

529 *'Du siehst'*: ll. 2603–4.

529 *something to occupy him in the winter*: HABr, ii. 91.

529 *'and so on a journey'*: WA, IV. 8, p. 369.

529 *scarcely concealed the affinity*: HABr, ii. 90–1; cp. HA (1988), xi. 705.

529 *as with* Egmont: HABr, i. 385.

529 *knowledge of the military intervention*: ibid. ii. 15–16.

529 *'intensified* Werther': HA (1988), v. 504, to Eckermann, 3 May 1827.

530 *'elegiac mood'*: cp. HA, xi. 555 (1988), 705.

530 *'almost completely sacrificed'*: HABr, ii. 94.

530 *'glimpses of the Promised Land'*: ibid. 88.

530 *'Like a coast'*: ibid. 92.

530 *'I took heart'*: HA (1988), xi. 705.

CHAPTER 8

Old and New Faces: June–December 1788

533 *weather deteriorated*: Bode, *Rom*, p. 161.

533 *Via Mala, of which he made an accomplished sketch*: CGZ, II, No. 401.

533 *reached Chur*: Steiger, ii. 668.

533 *letter to await Herder*: HABr, ii. 94–5.

533 *she arrived at the inn*: Grumach, iii. 209–13.

533 *Slim and sunburnt*: ibid. 215.

534 *'concealment of present'*: WA, I. 32, p. 460; cp. Steiger, ii. 671.

534 *arrived in Weimar*: Grumach, iii. 214.

534 *'fighting to get at him'*: ibid. 215.

535 *'Italy and Italy'*: ibid.

535 *'move heaven and earth'*: ibid. 214.

535 *Gore had made his money*: Bode, *Rom*, pp. 189–90.

535 *translate passages from* Werther: Bode, i. 357, No. 600.

535 *'fine enough if'*: WA, IV. 9, p. 10.

535 *'a gaiety he has not had'*: Grumach, iii. 242.

535 *Knebel, who had been in Ilmenau*: Bode, i. 354, No. 592.

535 *'not that unhappy here'*: Grumach, iii. 217.

535 *'I was with Duchess Luise'*: ibid. 216–17.

535 *epoch had ended*: Bode, *Stein*, pp. 309, 312, 300–1.

536 *'there is a premonition'*: Steiger, ii. 675.

536 *'To you I think'*: HABr, ii. 95.

536 *'unfriendly enough reception'*: ibid. 115.

536 *'Pleasure of travellers'*: Grumach, iii. 216.

536 *her dog Lulu*: Bode, *Stein*, pp. 277–8.

536 *'nothing but tedium'*: Steiger, ii. 675–6.

536 *'out of temper'*: Bode, i. 358, No. 602.

536 *'I shall carry on'*: WA, IV. 9, p. 5.

537 *'From Italy, rich in forms'*: HA, xiii. 102.

537 *leave the University of Jena*: Parth, *Christiane*, p. 25.

538 *for 'girls of the middle class'*: Federn, *Christiane*, p. 12.

538 *Goethe first saw her*: cp. 'Der neue Pausias und sein Blumenmädchen' (WA, I. 1, pp. 272–80) and Parth, *Christiane*, p. 17.

538 *Herder, passing through Nuremberg, took pity*: Grumach, iii. 230.

538 *immediately written to August*: WA, IV. 9, p. 385.

538 *looking for a private secretary and tutor*: ibid. 20–2, 38, 45–6.

538 *Goethe sent him money*: ibid. 61–2.

538 *sent him ten dollars*: ibid. 104.

538 *Göschen was asked*: ibid. 107–8, 134–5.

538 *discontented young man*: ibid. 355–6.

538 *further help from Breitkopf*: ibid. 151–2.

538 *Leipzig in October 1789*: Steiger, iii. 44.

539 *Goethe and Christiana celebrated 12 July*: Steiger, ii. 674.

539 *'Morgenklagen'*: HA, i. 239–41 (the poem is possibly a fantasy).

539 *'Der Besuch'*: ibid. 237–9.

539 *pencil-sketch of the scene*: CGZ, IVb, No. 65.

540 *'The weather . . . deadens my spirit*: HABr, ii. 100; WA, IV. 9, p. 19.

540 *Italian heat*: HABraG, i. 104.

540 *'live like a snail'*: WA, IV. 9, p. 38; cp. ibid. 46.

540 *open fire he had laid for him*: ibid. 13.

540 *'in political matters nothing'*: Bode, i. 357, No. 600.

540 *'as good a face'*: HABr, ii. 98.

540 *pure self-indulgence*: ibid. 99.

540 *felt like Epimenides*: ibid. 103.

540 *'he actually is unhappy'*: Grumach, iii. 242.

540 *'pity that he always has his armour'*: ibid. 245.

540 *'thinks he has grown sensual'*: ibid. 225.

540 *'before he went to Italy'*: Steiger, iii. 51–2.

540 *refusal to be reconciled*: Bode, i. 360, No. 606.

541 *to Gross Kochberg*: Grumach, iii. 231.

541 *Frau von Stein had herself rewritten*: HA, i. 545–6. The sequence of the versions is disputed, but the peculiarities of Frau von Stein's are comprehensible only if she was working from Goethe's first version, not his second (e.g. the use of 'veracht' in the last strophe).

542 *the possibility of a new poem*: ibid. 129–30.

543 *the first version*: cited above, p. 262.

544 *Schiller had become emotionally entangled*: Bode, *Stein*, pp. 204–6.

544 *Charlotte von Lengefeld, who was staying*: ibid. 302.

544 *'I have weakened'*: to Körner, 14 Nov. 1788, Schiller, *Briefe*, ii. 145.

545 *emphatically rejected the notion of an after-life*: e.g. Schiller, *Werke*, i. 130–3, 162.

545 *ambition warranted by his abilities*: ibid. 818.

545 *'upshot of all my experiences'*: to Huber, 28 Aug. 1787, Schiller, *Briefe*, i. 394.

545 *'Iphigenia has given me another really good day'*: to C. J. R. Ridel, 7 July 1788, ibid. 85.

545 *Schiller had been looking forward*: Bode, i. 357, No. 599.

545 *'The first sight of him'*: Grumach, iii. 233–4, to Körner, 12 Sept. 1788, Schiller, *Briefe*, ii. 115–17.

546 *Lengefelds attributed*: Grumach, iii. 234–5.

546 *wished to put behind him*: HA, x. 538.

546 *'The Gods of Greece', which he took*: Grumach, iii. 235.

546 *read with approval*: Bode, i. 382, No. 645.

546 *'The public is now'*: Schiller, *Werke*, v. 856.

546 *intention to live by his pen*: to Körner, 7 Jan 1788, Schiller, *Briefe*, ii. 4–5.

546 *discussion of 'The Gods of Greece'*: Grumach, iii. 235–6.

547 *'Finally he said'*: ibid. 235.

547 Künstlers Apotheose: HA, i. 68–77.

547 *echoes of the lessons*: the belief that no one artist can possess the perfection of art (ll. 135–6) is also reflected in the conversation of 8 Sept., Grumach, iii. 235–6.

548 *'fruitful in more senses'*: Grumach, iii. 238.

548 *'it is pretty well decided'*: ibid. 251.

548 *'living as* homme de lettre[s]'*: ibid. 255.

548 *'useful enough to the Duke'*: ibid. 308.

548 *Ilmenau where the task of drainage*: WA, IV. 9, p. 36.

548 *In Jena the old palace*: Steiger, iii. 36.

548 *purchase of land*: WA, IV. 9, p. 39.

548 *rebinding programme*: ibid. 38.

548 *collecting publicly and privately*: ibid. 56.

548 *organized crowd-control*: ibid. 34.

549 *counteract the loss of Eichhorn*: Grumach, iii. 258.

549 *'The best way to learn'*: ibid.

549 *education of the Crown Prince*: cp. HABr, ii. 115; Grumach, iii. 219.

549 *occasional trips away*: ibid. 241, 245.

549 *might one day study there*: ibid. 252.

549 *'I am so happy'*: Steiger, ii. 692.

549 *negotiations to rescue*: HABr, ii. 105, 109.

549 *loan to Merck*: WA, IV. 9, p. 23.

549 *living a 'natural'*: ibid. 37.

550 *borrowing a Propertius*: Steiger, ii. 691.

550 *what he called 'Erotica'*: WA, IV. 9. pp. 46, 57.

550 *'Frau Schardt told me'*: Grumach, iii. 250.

550 *boasting of a visit*: WA, IV. 9, p. 58.

550 *beginning to look around*: cp. also Grumach, iii. p. 213.

550 *'no children to feed'*: WA, IV. 9, p. 15.

550 *'man needs a comfortable home'*: Grumach, iii. 225.

550 *a pronounced jealousy*: Bode, i. 366, 368, Nos. 620, 625.

Rome in Weimar: December 1788–May 1789

550 *'Purgatorial fire'*: WA, IV. 9, p. 59.

550 *'Knebel . . . has many female friends'*: Bode, i. 379, No. 643.

550 *On 3 December*: Boulby, *Moritz*, p. 185.

551 *enjoyment of female company*: HABr, ii. 107.

551 *'interestingly comical'*: Bode, i. 381, No. 645.

551 *introduced into every circle*: Steiger, ii. 694–6.

551 *breath-taking Goethe-worship*: Bode, i. 378–9, No. 642.

551 *'mediocre' productions*: ibid. 381, No. 645.

551 *'dreadful enthusiasm'*: ibid. 383, No. 647.

551 *in Paris the temperatures*: Kington, 'Mapping', p. 32.

551 *Moritz arrived in Weimar, penniless*: Boulby, *Moritz*, p. 185.

551 *already sought to commend Moritz*: Steiger, ii. 673.

551 *'like an orange-tree'*: WA, IV. 9, p. 86.

551 *hoped that Moritz would stay*: Grumach, iii. 256, 263.

551 *Moritz had had in the play*: Steiger, iii. 17.

551 *toying with some ideas*: Grumach, iii. 260.

551 *giving English lessons*: Boulby, *Moritz*, pp. 185–6.

552 *Duke impressed by Moritz's powers*: Grumach, iii. 261.

552 *'deep thinker'*: Bode, i. 371, No. 630.

552 *'my favourite ideas'*: ibid. 381, No. 645.

552 *'sufficient to itself'*: Die Künstler, l. 157 (Schiller, *Werke*, i. 178).

552 *analogous to the one great Artist*: ibid., ll. 329–35, p. 183.

552 *death, 'gently' dealt*: ibid., ll. 312–15, p. 182.

552 *transformed by art*: ibid., ll. 210–19, p. 179.

552 *circulated widely*: Boulby, *Moritz*, p. 18.

552 *'Herderian ways of thinking'*: Bode i. 375, No. 640.

552 *'utterly Goethean'*: ibid. 387, No. 654.

552 *Herder, however, had Duchess Luise*: ibid. 388, No. 654; cp. also Boulby, *Moritz*, p. 189.

552 *'brought back from Italy'*: ibid. 379–81, No. 644.

552 *first reaction to Kant's*: Steiger, iii. 13.

553 *'Naturlehre'*: WA, II. 13, pp. 427–9.

553 *conciliatory reply from Knebel*: the authorship of 'Antwort' (ibid. 429–31) is unclear; cp. Bode, i. 381, No. 644 and DKV, I. 25, p. 871: 'Goethe . . . nahm Knebels Plan einer gedruckten Antwort für sich selbst auf'.

553 *identified the principal weakness*: Grumach, iii. 271–2.

553 *published as a review in the* Teutscher Merkur: WA, I. 47, pp. 84–90.

554 *what the fuss had been about*: Grumach, iii. 286.

554 *'Totalbegriff'*: Bode, i. 385, No. 650.

554 *full of talk about 'art'*: e.g. ibid. 384–5, 391, 394–5, Nos. 650, 659, 667.

554 *'now it is a flower-factory'*: HABraG, i. 114.

554 *'puddles left by a dead sea'*: Bode, i. 390, No. 658.

554 *stylization as of Goethe*: cp. Steiger, ii. 690.

555 *'poet took only as much'*: Bode, i. 384–5, No. 650.

555 *'I have really had a great revelation'*: ibid. 390–1, No. 659.

555 *'I corrected her'*: ibid. 393–4, No. 666.

555 *'Old Harry take'*: ibid. 391–2, No. 661.

556 *'Dearest angel'*: ibid. 394–5, No. 667.

556 *'tailor-made to fit'*: ibid. 390, No. 658.

556 *Moritz, inaccurately, reading the proofs*: WA, IV. 9, p. 117.

556 Das Römische Karneval: HA, xi. 484–515.

556 *edition of 250 copies*: Steiger, iii. 50.

557 *compared to the horses*: HA, xi. 515.

557 *Withdrawn from the 'element'*: Steiger, ii. 677.

557 *'lies . . . intimations'*: HA, xi. 413.

558 *'One becomes a dream'*: HABr, ii. 107.

558 *'Frech und froh'*: HA, i. 241.

558 *' "Then I was given" '*: Grumach, iii. 228.

558 *'turbid efflux'*: Herder, *Werke*, xiv. 292.

558 *'Christianity you have treated'*: HABr, ii. 99.

558 *'Märchen von Christus'*: either generally 'the whole Christian legend', or particularly 'the empty religion about Christ invented by the Church, as distinct from Christ's teaching'. Cp. above, p. 189.

559 *longing for a peasant's 'faith'*: HABraG, i. 110.

559 *Luise, deeply depressed*: Grumach, iii. 29.

559 *wrote in December to F. L. Stolberg*: HABr, ii. 104–5, written the day after Moritz arrived, 'in einer freudigen Stunde'.

559 *'teaching of Lucretius'*: ibid. 109.

559 *'It is my conviction'*: ibid. 113.

560 *philologist, C. G. Heyne*: ibid. 97–8.

560 *'Zur Theorie der bildenden Künste'*: WA, I. 47, pp. 60–6.

560 *anatomy course at Jena*: Bode, i. 369, No. 628.

560 *'physiognomical discoveries'*: HABr, ii. 106–7.

560 *studies of the balding and choleric Knebel*: CGZ, IVb, Nos. 23–9, 44–5.

561 *'Über einfache Nachahmung'*: HA, xii. 30–4.

561 *'incomparable' adjunct*: Bode, i. 385, No. 650.

562 *'Frauenrollen auf dem Römischen Theater'*: WA, I. 47, pp. 269–74.

562 *'one experienced here'*: ibid. 274.

563 *correspondence with J. H. Meyer about a fresco*: see Harnack, *Nachgeschichte*, pp. 81–2; HABr, ii. 102, 109–10; HABraG, i. 100–1.

563 *Circe offering Ulysses*: see plate 35.

563 *'the painter should and can narrate'*: HABr, ii. 102.

563 *patterning of figures on ancient vases*: ibid. 109.

563 *'never a matter of present'*: WA, III. 1, p. 307.

563 *'The Ancients regarded'*: HABr, ii. 110.

563 *attitudes he now attributed*: ibid. 109–10.

564 *'as far as possible'*: HABr, ii. 106.

564 *'Hibernating like a hamster'*: ibid. 112.

564 *heating his rooms*: Bode, *Stein*, p. 443.

565 '*survive the most burdensome part*': HABr, ii. 108.

565 '*conversation with distant friends*': ibid. 96.

565 *impromptu illustrative sketches*: Grumach, iii. 271.

565 Campi Phlegroei: ibid. 279.

565 '*good conversation . . . on some cosy evening*': WA, IV. 8, p. 210.

565 *Gifts to friends*: ibid. 9, pp. 63, 66.

565 *taste for Mediterranean fruits*: Grumach, iii. 242.

565 *Italian seeds and the pine-cone*: ibid. 277.

565 '*flowers have a nice smell*': ibid. 281.

565 '*breath of the South*': HABr, ii. 113.

565 *the most literate of the German artist colony*: WA, IV. 9, p. 30.

565 *from Hirt a proposal*: ibid. 68.

565 *price-list of his work*: ibid. 28–31.

565 *showed off specimens*: Harnack, *Nachgeschichte*, pp. 134–5.

565 *send a substantial order*: WA, IV. 9. 75–6.

565 *borrowing money in Goethe's name*: Harnack *Nachgeschichte*, pp. 111–13.

565 *Tischbein was a more difficult case*: HABr, ii. 111–12.

565 *Herder's trip, unfortunately*: Bode, *Rom*, pp. 215–17.

566 *2,500 dollars*: WA, IV. 9, p. 133.

566 *found his advice helpful*: ibid. 94–5.

566 *need the formal clerical garb*: Bode, i. 364, No. 616.

566 '*suggestions about Rome are worthless*': ibid.

566 '*do not want to live as Goethe*': ibid. 367, No. 623.

566 *well conducted by Filippo Collina*: WA, IV. 9, p. 47.

566 *her first month in Rome*: Harnack, *Nachgeschichte*, p. 102.

566 *and later Hirt*: ibid. 116.

566 *Pope, who presented her with a fine mosaic*: HABraG, i. 113.

566 *all wrote delightedly*: Harnack, *Nachgeschichte*, p. 95, 102, 93.

566 *Herder's temper improved*: ibid. 91.

566 *addressing him as 'Archbishop'*: ibid. 106.

566 *weather proved not to be good*: Harnack, *Nachgeschichte*, p. 120.

566 *Venuti, Hackert, and Hamilton*: ibid. 146.

566 *installed herself in a villa*: ibid. 116.

566 *19 May, when she returned to Naples*: ibid. 169.

566 '*ex-Chamber-Presidential*': WA, IV. 9, p. 106.

567 *ordering paintings*: ibid. 48, 31.

567 *Schütz depicted the occasion*: reproduction in Bode, *Rom*, between pp. 216–17.

567 *buying books, music, coins*: WA, IV. 9, pp. 81–3, 167, 156, 141, 143.

567 *'brought me almost to despair'*: ibid. 105.

567 *felt ill at ease*: Harnack, *Nachgeschichte*, pp. 86–7.

567 *'all the immense labour'*: WA, IV. 9, p. 157.

567 *visited Weimar*: Steiger, iii. 25.

567 *Sophie von la Roche*: WA, IV. 9, p. 133.

567 *Knebel's youngest brother Max*: Grumach, iii. 310.

567 *40–45,000 dollars*: Harnack, *Nachgeschichte*, p. 244.

568 *Tischbein's portrait of Anna Amalia*: see plate 23.

568 *Herder as the Holy Ghost, 'the Comforter'*: HABraG, i. 107.

568 *'Trippel is starting on the bust'*: Bode, i. 388, No. 654.

568 *'much resemblance between the character'*: ibid. 363, No. 610.

568 *expected to join Anna Amalia*: ibid. 360, No. 607.

568 *'would be a very good thing'*: ibid. 370, No. 628.

568 *'flourishes best in Rome'*: ibid.

568 *Herder endorsed*: Harnack, *Nachgeschichte*, p. 119.

568 *Rome in September*: Grumach, iii. 278.

568 Tasso *which was becoming 'like a dream'*: WA, IV. 9, p. 119.

568 *rebuilding, at long last, the burnt-out ducal palace:* ibid. 89.

569 *other members were Voigt, Wedel, and Schmidt*: Steiger, iii. 21–2.

569 *meeting on 25 March*: Grumach, iii. 280–1.

569 *'artifex', Herder called him*: Harnack, *Nachgeschichte*, p. 119.

569 *'None of us, of course'*: WA, IV. 9, p. 89.

569 *sketch plans for Italian bridges*: CGZ, IVb, No. 187; VIa, No. 181.

569 *in 1788 three classical pillars*: Jellicoe, *Gardens*, p. 269; Vulpius, *Goethepark*, p. 22.

569 *retain as much of the shell*: Bode, *Rom*, pp. 283–4.

569 *Arens finally arrived*: Steiger, iii. 31–6.

569 *'enjoy the presence of an artist'*: HABr, ii. 117.

569 *wrote to Lips*: WA, IV. 9, pp. 97–9.

569 *'lesser significance for art'*: Harnack, *Nachgeschichte*, p. 159; cp. WA, IV. 9, p. 115.

570 *listed in the Weimar almanac*: Steiger, iii. 50.

570 *'Now I am covered'*: WA, IV. 9, p. 111.

570 *make another 'acquisition'*: ibid. 115.

570 *hold out to Meyer*: ibid. 110.

570 *confirm the proposal*: ibid. 149–51.

570 *F. W. Facius*: ibid. 139, 156.

570 *'set up a nice little Academy'*: ibid. 167.

570 *'cannot live without artists'*: ibid.

571 *'corpulent little female'*: Steiger, iii. 20.

571 *'Vulpius girl as his Clärchen'*: Bode, i. 392, No. 662.

571 *'careless' Egmont*: ibid. 385, No. 651.

571 *'What you write'*: ibid. 395, No. 668.

571 *study of the heads on ancient coins*: CGZ, IVb, Nos. 14–17.

571 *November 1788 he drew*: ibid. p. 13 ad No. 16.

571 *assimilate her increasingly*: ibid. Nos. 38–43.

571 *studied the form carefully with Knebel*: WA, IV. 9, p. 102, 111.

571 *attempt at hendecasyllables*: ibid. 112.

571 *scarcely more than space-fillers*: ibid. 60–1.

572 *probably Tibullus*: Steiger, ii. 691.

572 *'cannot expunge from my heart'*: HABr, ii. 106.

572 *under the plaster-cast of Raphael's skull*: WA, IV. 9, p. 103.

572 *poems as 'jokes'*: HABr, ii. 119.

572 *'Näher haben wir das!'*: WA, I. 53, p. 4.

573 *until November 1789*: ibid. IV. 9, p. 162.

573 *by introspection and by public obsession*: *Elegien*, 2, ll. 7, 6 (original version, HA (1988), i. 585). Unless otherwise stated, the published text of the *Römische Elegien* is cited from HA, i. 157–173.

573 *rapid start to the liaison*: for this and following details, *Elegien*, 3, 14–15, 18, 8, 9, 6, 19–20.

574 *'Und belehr' ich'*: ibid. 5, ll. 7–10.

574 *Carl August might be surprised*: WA, IV. 9, pp. 115, 120.

574 *eighteen line 'epigram'*: HA, i. 178, No. 17.

574 *in an associated poem*: ibid. No. 16.

575 *invitation to become a member of the Berlin Academy*: Steiger, iii. 19.

575 *'feels more comfortable in Jena'*: Grumach, iii. 247.

575 *reconciling him to Germany*: Steiger, iii. 13.

576 *purchase of his apparatus*: WA, IV. 9, pp. 70, 96.

576 *the poet Bürger*: Grumach, iii. 287–90.

576 *keep Schütz satisfied*: WA, IV. 9, p. 88.

576 *reported to Carl August on Schiller's removal*: ibid. 117.

576 *Bellomo's regular repertoire*: Steiger, iii. 17; Bode, *Rom*, p. 281.

576 *'To be near Goethe'*: Bode, i. 381–2, No. 645.

577 *'one really has too little'*: ibid. 389, No. 656.

577 *'this person, this Goethe'*: Bode, i. 392, No. 663.

577 *'What is universal history'*: Schiller, *Werke*, iv. 749–67.

577 *audience of three to four hundred*: Steiger, iii. 30.

577 *collapse of a dogmatic theological structure*: Schiller, *Werke*, iv. 752.

577 *'middle class' as the sole vehicle*: ibid. 759.

577 *offered Schiller a pension*: Bode, *Stein*, pp. 330–6.

577 *Jena lost a figure*: WA, IV. 9, pp. 155, 357.

578 *advised Herder against considering the chancellorship*: Grumach, iii. 292.

578 *preaching was a magnet*: Bruford, *Culture*, pp. 309–10.

578 *foresaw both the ruin of Jena*: Steiger, iii. 24–5.

578 *only intellectual companion*: Bode, i.357, No. 600.

578 *'tired of the connection'*: ibid. 396–9, No. 672.

578 *'his selfish existence'*: ibid. 383, No. 647.

578 *its library was already so rich*: Grumach, iii. 307–8.

578 *warmly welcomed the offer*: Bode, i. 400–1, No. 676.

578 *'In an academic senate'*: Grumach, iii. 306.

579 *incurred Herder's anger*: Bode, i. 372, No. 633; 386, No. 654; WA, IV. 9, p. 32.

579 *needed a little coaxing*: WA, IV. 9, p. 118.

579 *pay all his outstanding debts*: Bode, i. 402, No. 680.

579 *Carl August and Luise did their bit*: Grumach, iii. 307.

579 *an elaborate scheme*: WA, IV. 9, pp. 131–2, 142.

579 *Dowager Duchess agreed*: Harnack, *Nachgeschichte*, p. 187; cp. Grumach, iii. 307.

579 *back in Weimar 'in good humour'*: WA, IV. 9, p. 139.

579 *appointment was promulgated*: ibid. 356.

579 *'found so many real proofs'*: Bode, i. 399, No. 675.

579 *'Knebel is an unsteady, uncertain reed'*: ibid. 401, No. 676.

'I am a different man': June–December 1789

579 *period of sexual abstinence*: cp. WA, IV. 9, p. 173.

580 *the courtly solution*: e.g. Carl August's later relation with Caroline Jagemann.

580 *as soon as she knew*: Steiger, iii. 26.

580 *taken by 8 May*: WA, IV. 9, p. 111.

580 *evidence of his 'paganism'*: ibid.

581 *first week of May or after 6 June*: cp. Grumach, iii. 291, 295.

581 *'distinguishes the powerful'*: WA, IV. 9, p. 118.

581 *to finish Tasso*: ibid. 111.

581 *Hamilton had at last married*: Harnack, *Nachgeschichte*, p. 115.

581 *Until a suitable house*: Steiger, iii. 34.

581 *not wish to be separated from her aunt*: ibid. 37.

581 'in re incerta': WA, IV. 9, p. 134.

582 *letter went to August Vulpius*: ibid. 135.

582 '*I find myself thinking*': ibid. 138.

582 '*as with an illness*': Bode, i. 396, No. 670.

582 *source of the dissension*: ibid. 388, No. 655.

582 '*very, very unhappy*': ibid. 402, No. 679.

582 *before leaving Weimar she wrote*: Steiger, iii. 27.

583 '*I thank you for your letter*': HABr, ii. 115–16.

584 '*There can have been few*': ibid. 117.

584 '*old former friend*': Bode, i. 404, No. 685.

584 *tormented by 'bad memories'*: ibid. 406, No. 690.

584 '*sank into quiet mourning*': ibid., No. 692.

585 '*living on the income*': ibid. 374, No. 636.

585 '*outwith or beside the law*': HABr, iv. 19; cp. Steiger, iii. 53.

586 '*Freye Liebe sie läßt*': WA, I. 53, p. 13, No. 28.

586 '"*Ach, mein Hals*"': HA, i. 183, No. *41*, possibly the 'nagelneues Erotikon' announced to Knebel on 8 May, WA, IV. 9, p. 112.

586 '*land of the famed mountain-nymphs*': HABr, ii. 119.

587 '*first . . . period of my writing*': ibid. 63; cp. WA, IV. 8, p. 178.

587 '*free man at last*': WA, IV. 9, p. 160.

587 *Tasso was 'on the point*': ibid. 137.

587 *accompanied only by* Lila: ibid. IV. 18, pp. 38, 39.

587 '*publish* Faust *as a fragment*': ibid. 9, p. 139.

587 '*not to publish anything piecemeal*': Hagen–Nahler, i. 55, No. 94.

587 *complete the edition 'for this time*': WA, IV. 9, pp. 159, 160.

587 *had to 'bury himself*': Grumach, iii. 315.

587 *despatch* Faust *to Göschen*: WA, IV. 9, p. 393.

587 *send his publisher the definitive text*: Steiger, iii. 41; cp. Grumach, iii. 305.

587 *entertained a few friends*: ibid.

589 *echo in the peasantry*: Blanning, *Revolution*, p. 52.

589 *bread supplies organized by the elder Voigt*: WA, IV. 9, p. 162.

589 *Archbishop-Elector of Cologne was mobbed*: Blanning *Revolution*, pp. 56–7.

589 *Kant's acquaintances noticed*: Gooch, *Germany*, p. 264.

589 '*greatest action of this century*': Klopstock, 'Die Etats Généraux', l. 13.

589 '*best day since the fall*': Gooch, *Germany*, p. 47.

589 *other poets*: Gooch, Germany, 118–41.

589 *a few, such as Matthias Claudius*: ibid. 54–6.

589 *most emphatic friends*: Gooch, *Germany*, pp. 163, 448–53.

590 Teutscher Merkur *in October 1789*: ibid. 145.

590 *Frau von Stein and Duchess Luise*: ibid. 446–7.

590 *disagreement in November*: Steiger, iii. 47–8.

590 *been in Paris in 1785*: Gooch, *Germany*, p. 381.

590 *apprehensive about the course*: ibid. 214, 383.

590 *Reichardt 'impertinent'*: Grumach, iii. 290.

590 *reading of Saint-Simon*: WA, IV. 9, p. 117.

590 *allegations about the Duke*: Steiger, iii. 19.

590 *new Masonic lodge in Jena*: Grumach, iii. 282–5.

591 *'magnetic' powers*: ibid. 270–1.

591 *a most serious libel*: ibid. 284.

591 *rewrote what is now the second*: WA, I. 1. pp. 412–13.

591 *'modern fable'*: Elegien 2, l. 19.

591 *'storm threatening us from without'*: WA, I. 1. p. 413.

591 *list of tasks for the year 1789–90*: ibid. III. 2, pp. 323–4.

592 *'striving to dissipate'*: ibid. IV. 9, p. 168.

592 *visit of A. G. Werner*: Steiger, iii. 42.

592 *volcanoes were geologically superficial*: WA, III. 1, p. 342.

592 *return to fold*: ibid. IV. 9, p. 153.

592 *essay intended to bridge*: LA, I. 11, pp. 37–8.

592 *general defence of the necessity*: ibid. pp. 35–6 (for the dating cp. DKV, I. 25, p. 875).

593 *writing up his 'System'*: Bode, i. 359, No. 604.

593 *copies of certain works of Linnaeus*: WA, IV. 18, p. 31.

593 *supper-time conversation*: Grumach, iii. 309.

593 *promoting his protégé Batsch's scheme*: ibid. 265.

593 *appointing Batsch*: Steiger, iii. 45.

593 *consultations of a scientific nature*: ibid. 52.

593 *treatise by the botanist C. K. Sprengel*: WA, IV. 9, p. 358 (the treatise eventually appeared under a different title in 1793, LA, II. 9A, p. 537).

593 *draft in a series of paragraphs*: WA, IV. 9, p. 169.

593 *20 December he went to Jena*: Steiger, iii. 50–1.

593 *Göschen would bring it out*: Hagen-Nahler, i. 191.

593 *'not put together'*: HABr, ii. 109.

594 *not using it in conversation*: ibid. 121.

594 *precedents in Linnaeus*: LA, II. 9A, pp. 539–40; for the analogy of plant and insect structure see *CGZ*, Vb, No. 86 of about the same date.

594 *'model' with the notion of a 'key'*: HABr, ii. 60.

594 *'Everything is leaf'*: WA, II. 7, pp. 282–3.

594 *'metamorphosed' into different specific shapes*: HA, xiii. 101, para. 120.

596 *possessed in an edition of 1783*: Nisbet, *Tradition*, pp. 8–11.

596 *the leaf 'is' the root*: Wells, *Development*, pp. 44–5.

596 *'petal-leaf of the calyx'*: HA, xiii. 101, para. 120.

597 what *the shapes are*: cp. ibid., xi. 389.

597 *with Angelica Kauffmann*: Steiger, ii. 624.

597 *borrowed from Büttner*: Steiger, iii. 38.

597 *Jägerhaus to which he had moved*: ibid. 48.

597 *perhaps through Loder*: not Knebel (ibid. 48), who was in Weimar, not Jena, that winter.

597 *displayed only for the first time in May*: WA, IV. 9, p. 116.

597 *large new heating stove*: ibid. 155, 162.

597 *open airy situation*: ibid. 162.

597 Faust *was 'fragmented'*: ibid. 160.

597 *study of Greek*: ibid. 161.

597 *'gently manœuvring'*: Steiger, iii. 47.

598 *a facetious description*: ibid.

598 *'Wonniglich ist's'*: HA, i. 183–4, No. *42*.

598 *Christiana might die*: Grumach, iii. 319.

598 *kingship of Hungary*: AS, ii. 1, pp. 161–5.

598 *named after his uncle*: Bode, *Rom*, p. 300.

599 *'in a very unfortunate'*: Grumach, iii. 322.

599 *'now we can get on'*: WA, IV. 9, p. 160.

599 *France would fall into anarchy*: cp. Gooch, *Germany*, p. 214.

600 *'Ob erfüllt sey'*: WA, I. 53, p. 18.

Summa Summarum: The Edition Completed

600 *'All these recapitulations'*: WA, IV. 8, pp. 347–8.

600 *'never actually get anything done'*: Bode, i. 403, No. 681.

602 *'a Summa Summarum'*: WA, IV. 9, p. 44.

602 *'these youthful scribbles'*: Bode, i. 391, No. 661.

602 *two 'collections' of poems*: reprinted in their original form in DKV, I. 1, pp. 277–368.

602 *also 'polishing' them*: Grumach, iii. 234.

602 *'Einschränkung'*: HA, i. 132.

602 *'Liebebedürfnis'*: ibid. 140–1.

602 *revision of 'On the Lake'*: ibid. 102–3.

603 *'Just as love'*: Moritz, *Bildende Nachahmung*, p. 30.

603 *recurs in some of the as yet unpublished* Elegies: *Elegien* 2, l. 28; cp. WA, I. 53, p. 12, No. 23.

603 *'Willkommen und Abschied'*: HA, i. 128–9.

604 *new versions of 'Rose upon the Heath'*: ibid. 78 (ll. 18, 19).

604 *'Evening Song of the Huntsman'*: ibid. 121–2 (ll. 2, 9–12).

604 *'Bounds of Humanity'*: ibid. 147, (1988) 558–9.

605 *'Ihnen gleiche der Mensch!'*: ibid. 147–8, (1988) 560.

605 *revision created a suggestive ambiguity*: von Loeper (WA, I. 2, p. 314) wrongly thinks the original form ambiguous because he is trying to read into it the sense of the second version; the only possible referent of 'sie' is 'Götter'.

605 *'because of certain relations'*: WA, IV. 18, p. 32.

605 *end on a left-hand page*: ibid. 33.

605 *'Herzog Leopold von Braunschweig'*: ibid. I. 2, p. 123.

605 *'Dem Ackermann'*: ibid.

606 *'With the extreme care'*: ibid. IV. 9, p. 134.

606 *a de luxe edition*: ibid. 18, pp. 37–8.

606 *'is ample reward'*: ibid. 9, p. 147.

606 Torquato Tasso: HA, iii. 73–167 (line references follow this edition).

606 *characters intend to leave*: ll. 42–3, 346–62.

607 *at times allusive and stiff*: ll. 1310–15.

608 *'strebt Die goldne Zeit'*: ll. 974–7.

608 *'das Geheimnis einer edlen'*: ll. 1107–8.

608 *'moderate use of life'*: ll. 1267–8.

609 *'hall of the gods'*: l. 1558.

609 *'I recognize only one lord'*: ll. 2302–3.

610 *'Und eh nun die Verzweiflung'*: ll. 2826–9.

610 *'Ich bin auf ihn'*: ll. 2851–3.

610 *'not everyone serves'*: l. 2939.

611 *'The human being will gain'*: l. 3078.

611 *'So you are going'*: l. 3118.

611 *'Gar wenig ist's'*: ll. 3234–45.

612 *'that feeling which alone'*: ll. 3257–8.

612 *'Er kommt von Sinnen'*: l. 3285.

612 *'bin ich* nichts': ll. 3415–18.

612 *one thing 'remains'*: l. 3426.

612 *'Und wenn der Mensch'*: ll. 3432–3.

612 *'So klammert sich'*: ll. 3452–3.

613 *'I had the life of Tasso'*: to Eckermann, 6 May 1827.

613 *remained in contact since 1777*: Steiger, ii. 676, 692; WA, IV. 9, pp. 48–9.

613 *struggling young Schiller*: HA (1988), v. 512.

613 *sad letters of Merck*: HABraG, i. 105–6, 109–10.

614 *'many days, some spent'*: ll. 574–5.

614 *'Es ist unmöglich'*: ll. 767–70.

614 *'So selten ist es'*: ll. 1906–13.

615 *'ladies here are of quite remarkable sensibility'*: to Körner, 29 Aug. 1787, Schiller, *Briefe*, i. 405.

615 *closeness to Duke Alfonso as co-patron*: ll. 283–6.

615 *'wo ist der Mann?'*: ll. 924–6.

615 *growth in mutual knowledge*: ll. 1863–8, 1121–4.

615 *'I had a friend'*: l. 1821.

615 *'noble women' and of the quality*: ll. 1013–22.

615 *cannot quite appreciate*: l. 1821.

615 *sensually deprived childhood*: ll. 1800–16.

615 *combination of erotic encouragement*: ll. 1109–24; cp. 3265–6.

615 *too violent a stimulus*: ll. 856–8.

615 *'Ich mußt ihn lieben'*: ll. 1889–90.

616 *'healed' the false tendencies*: l. 880.

616 *he entrusts himself*: ll. 963, 1065.

616 *'all that is pleasing'*: l. 994.

616 *Poussin's* Landscape: Blunt, *Poussin*, No. 208, p. 143.

616 *'all that is seemly'*: l. 1006.

616 *illusion fostered by 'the poets'*: l. 998.

616 *central to Tasso's life*: l. 3018.

616 *'seen it with my own eyes'*: ll. 1097–8.

617 *'Es sind nicht Schatten'*: ll. 1103–4.

617 *'the real attracts him'*: ll. 175–6.

617 *'Oh teach me'*: l. 1065.

617 *'moderation and abstinence'*: ll. 1121–2.

617 *double enjoyment of life*: ll. 3092–3.

617 *'an intensified Werther'*: to Eckermann, 3 May 1827.

618 *airy nothings, 'mere' signs*: ll. 2025–9.

618 *meanings, he says, have more effect*: ll. 2279–80.

618 *'daß er hoch'*: ll. 500–2.

618 *unattainable goal*: ll. 2628–9.

618 *'draw together the ultimate bounds'*: ll. 2135–6.

618 *'round his poem'*: l. 275.

618 *'can never finish'*: l. 265.

618 *postpones the moment*: ll. 268–9.

618 *'seemed to have been given'*: l. 1573.

619 *'Zu früh war mir'*: ll. 1574–6.

619 *'halls of the gods'*: ll. 1557–9.

619 *a sudden fall*: ibid.

619 *'So jung hat er'*: ll. 2950–1.

619 *'toil of this life'*: ll. 2948–9.

619 *'Mir ist an diesem'*: l. 443.

620 *'Er ist gekommen!'*: ll. 564–5.

620 *'Laßt mich nur Abschied'*: ll. 3395–8.

620 *'disproportion between talent'*: Bode, i. 393, No. 666.

621 *politically fragmented but aware of its nationhood*: ll. 821–8.

621 *'Das Volk hat jene Stadt'*: ll. 54–7.

621 *'artist' ('Künstler')*: l. 282.

621 *fruitless acknowledgement*: ll. 466–9.

621 *absolutism of the patron*: ll. 2302–4.

621 *to the court he owes*: ll. 428–39.

621 *'An euch nur'*: l. 444.

621 *'Wer nicht die'*: ll. 447–56.

622 *epic for the fame it will bring*: l. 291.

622 *calls 'it in a certain sense my own'*: l. 394.

622 *'Wie reizend ist's'*: ll. 1928–9.

622 *brilliantly illusionistic quality*: ll. 711–41.

622 *art, in Antonio's view*: ll. 665–71.

622 *relaxation after the business*: ll. 2005–10.

622 *'like a flame'*: l. 1841.

622 *world, faded by sickness*: ll. 856–9.

623 *'Denn ihre Neigung'*: ll. 1954–9.

625 *'Nichts gehöret mir'*: ll. 3276–83.

625 *Antonio's firm and compassionate advice*: ll. 3362, 3406, 3420.

626 *in correspondence as 'a tragedy'*: WA, IV. 18, p. 37.

626 *'I see a point'*: HABraG, i. 115.

626 *feigned madness of poetic tradition*: ll. 731–3.

627 *'goods of this life'*: Tasso, ll. 2948–9.

627 *'overleaps the joys'*: Faust, l. 1859.

627 *moments in Antonio's complaints*: Tasso, ll. 2933–4.

628 *Duke's tolerant responses*: ll. 2939–41.

628 *Scene 1 of* Tasso: WA, IV. 9, p. 94.

629 *make up a 'course'*: ll. 2052–4, 2072 (lines from *Faust. Ein Fragment* are identified by their numbering in the HA edition of *Faust I*).

629 *a 'new career'*: ibid.

629 *'Nun, nun! ich lass''*: ll. 3257–8.

630 *one at least of the formulae*: ll. 1692–7.

630 *one of the clerks*: WA, IV. 9, p. 160.

631 *'So tauml' ich'*: ll. 3249–50.

632 *made for its popularity*: Bode, i. 415–16, Nos. 712–14; 417–18, Nos. 716–17, and see also Luden's reminiscences, Gräf, 2.2, pp. 122–60, esp. 123–7.

632 *'easier with the fragmentary manner'*: HABr, ii. 119.

632 *with one exception*: Elegien 13, published 1791, WA, I. 1, p. 411.

633 *referred to in the poem itself*: Elegien 2 (first version), 5, 7, 13, 15, 20.

633 *'Und so gleichen'*: Elegien 4.

633 *'Amor schüret'*: Elegien 5.

633 *telling of 'old stories'*: Elegien 19.

633 *future publication*: Elegien 20.

634 *specifically a 'poet' ('Dichter')*: Elegien 7, 11, 15.

634 *'Einem Dichter zuliebe'*: Elegien 15.

634 *learned ballast*: notably Elegien 3, 13.

634 *quite uneducated woman*: Elegien 2.

634 *much more complex effect*: the interpretation here follows Luke, *Elegies*, pp. 98–9.

635 *'limitation . . . to a sensually graspable'*: F. Klingner, cited HA (1988), i. 584.

635 *'uprushing flame'*: Luke, *Elegies*, p. 51.

636 *love has still to come*: Elegien 3, l. 1.

636 *transfiguration is 'not yet'*: Elegien 1, l. 7.

636 *'that time is over'*: Elegien 4, l. 31.

636 *liaison is already under threat*: Elegien 6, l. 22.

638 *a moment of physical love*: Elegien 5, 9, 11, 12, 17, and the suppressed poems, WA, I. 53, pp. 3–6.

Farewell to Italy: January–June 1790

641 *Christiana could not feed the child*: WA, IV. 9, p. 173.

641 *regarded himself as a godparent*: ibid.

641 *edict of 6 April*: AS, ii. 1, pp. 63–5.

641 *claim to the succession in Lusatia*: ibid. 165–8, 170–1.

641 *diplomatic mission*: ibid. 163–5.

641 *accompany Carl August on the manœuvres*: WA, IV. 9, p. 179.

642 *'Bring your affairs'*: ibid. 174–5.

642 *'revolution for me too'*: HABr, ii. 121.

642 *'full of merriment'*: Grumach, iii. 326.

642 *loyal subjects throughout Germany*: e.g. Blanning, *Revolution*, p. 47.

642 *'For these last few days'*: WA, IV. 9, p. 173.

642 *introducing him in Erfurt and Gotha*: Steiger, iii. 56.

643 *officially inaugurated on 9 April*: Bode, *Rom*, p. 84.

643 *'That in present circumstances'*: WA, IV. 9, p. 173.

643 *Weimar was tiring of them*: Carlson, *Theatre*, pp. 55–6.

643 *permanent professional court theatre*: WA, IV. 9, pp. 177, 179.

643 *'I knew in advance'*: HABr, ii. 120.

643 *'It is not for production'*: Bode, i. 411, No. 702.

643 *Göschen's rejection*: Hagen-Nahler, i. 191, 195–6.

644 *new career as a natural historian*: HABr, ii. 120–1.

644 *'on the* Form of Animals': ibid. 121.

644 *weather was frequently fine*: WA, IV. 9, p. 171.

644 *reported to be 'heavenly'*: ibid. 174.

644 *letter to Einsiedel*: Steiger, iii. 55; cp. WA, IV. 9, p. 178.

644 *phenomena of refraction*: Steiger, i. 265–6.

644 *'I was standing'*: HA, xiv. 259.

645 *possibly J. H. Voigt*: Grumach, iii. 328.

645 *'whatever objections I made'*: HA, xiv. 261.

645 *A recent exegesis*: Schöne, *Farbentheologie*.

646 *comparison of Light suffering*: ibid. 177–8 date to 'Frühjahr 1790', but no reason is given.

646 *'Alles erklärt sich'*: WA, I. 1, p. 325.

646 *'Was mit mir'*: HA, i. 181, No. 30.

647 *see his life as a meaningless failure*: WA, I. 53, p. 12, No. 20.

647 *'no one will come properly to their senses'*: HABr, ii. 99.

647 *more important than all his 'poetical' works*: to Eckermann, 19 Feb. 1829; cp. Schöne, *Farbentheologie*, p. 8.

648 *road to Damascus*: WA, I. 53, p. 8, No. 1.

648 *resumption of marital relations*: ibid. IV. 9, p. 173.

648 membrum virile *'purissimum'*: Steiger, iii. 61.

648 *For the unmarried Prince August*: WA, I. 53, pp. 492, 497.

648 *free exposition in Latin*: ibid. 203–7.

648 *a number of comments*: ibid. 197–202.

648 *This symbol recurs*: WA, I. 53, p. 428; *CGZ*, VIa, Nos. 66–7.

648 *same structure in the reproductive organs of plants*: WA, II. 13, p. 32.

648 *two prominently phallic scenes*: *CGZ*, IVb, Nos. 51–2.

648 *erect garden herm*: ibid. VIa, No. 65.

648 *two Erotica* devoted *to Priapus*: WA, I. 53, pp. 6–7, Nos 3, 4 (there is however no evidence that these two poems were ever considered for inclusion in the cycle of *Elegien*).

648 *'I am studying the ancients'*: HABr, ii. 121.

649 *unlikely to return to Weimar*: Grumach, iii. 334.

649 *'He is yearning'*: ibid. 330.

649 *plenty to occupy his subordinates*: WA, IV. 9, pp. 178–9.

649 *conducted an expert*: Grumach, iii. 327–8; Steiger, iii. 60.

649 *'In order myself as a pagan'*: HABr, ii. 123.

649 *'I have my own way'*: ibid. 122.

649 *'This time I am leaving'*: ibid.

649 *attacked them instead*: Steiger, iii. 63; WA, IV. 9, pp. 188–95.

650 *emissary of the highest authority*: *AS*, ii. 1, pp. 171–2.

650 *without passing the decision back*: WA, IV. 9, p. 185.

650 *pattern of innumerable minor disturbances*: Blanning, *Revolution*, pp. 53–4.

650 *'if I could ever come to hate'*: HABr, ii. 122.

650 *left at a quarter to seven*: WA, III. 2, p. 1.

650 *Wachtel flogged and cashiered*: Bode, *Rom*, p. 289.

650 *drove continuously for two days and two nights*: WA, I. 32, p. 490.

651 *'I shall stay a few days'*: ibid. 491.

651 *'But the veiled crown'*: ibid.

651 *'not yet have an Italian look'*: ibid. IV. 9, p. 197.

652 *'I must by the way confess'*: ibid. 197–8.

653 *'I had a strange enough experience'*: HABr, ii. 97.

653 *'pudenda . . . very stupid'*: WA, IV. 9, p. 9.

653 *'without such an intention'*: ibid. 14.

653 *'forget me not'*: *CGZ*, IVa, No. 1.

654 *'Das ist Italien'*: HA, i. 175, No. 4.

654 *'they are very close'*: HABr, ii. 127–8.

654 *'a little more intolerant'*: WA, IV. 9. p. 198.

654 *'among other laudable things'*: HABr, ii. 124.

654 *'My attitudes are more domestic'*: ibid. 126.

654 *'I long very much'*: WA, IV. 9, p. 201.

654 *Rome was the great lodestone*: ibid. 8, p. 371.

654 *'Glänzen sah ich'*: HA, i. 182, No. *38*.

655 *'For a year one is submerged'*: Grumach, iii. 351.

655 *'sunlight through the whole of your life'*: WA, IV. 9, p. 93.

655 *'Emsig wallet'*: HA, i. 177, No. *11*.

656 *in a roundel form*: CGZ, IVa, Nos. 2–7, VIb, No. 98.

656 *he was 'very worried'*: HABr, ii. 127.

656 *days had to be spent indoors*: WA, III. 2, pp. 17–18.

656 *On Easter Sunday*: ibid. 16–17.

656 *development of Venetian art*: ibid. I. 47, pp. 211–223.

656 *'Nowhere does one see'*: ibid. 427.

656 *concentrating on blue*: LA, II. 3, pp. 18–21.

656 *'ill looking at paintings'*: HABr, ii. 126.

657 *a troupe of acrobats*: Steiger, iii. 78.

657 *scarcely a tree*: HABr, ii. 125.

657 *sea-creatures to observe*: ibid.

657 *boneless agility of the young acrobat*: WA, I. 1, p. 317, No. 37.

657 *a crude anti-Semitic jest*: HABr, ii. 126.

657 *'Characteristic of every vertebra'*: WA, II. 13, p. 237.

657 *banker Zucchi*: HABr, ii. 126.

657 *'I cannot deny'*: ibid. 125.

657 *not added anything since December*: 'das erste *Eroticon*', WA, IV. 9, p. 174; cp. ibid. I, 53, p. 430, 'Emsig fand ich heute'.

657 *now complete*: HABr, ii. 124.

657 *'not even a trace of that vein'*: WA, IV. 9, p. 199.

658 *as rapidly as the piddocks*: ibid. 201.

658 *by 4 May there were 100*: HABr, ii. 125.

658 *'these are fruits'*: ibid. 124.

658 *neither knew nor cared about*: WA, I. 1, p. 325, No. 75.

658 *complaints about the rain*: HA, i. 177, No. *12*.

658 *mud*: WA, I. 1, pp. 312–13, Nos. 23–5.

658 *lack of trees*: ibid. 310, No. 13.

658 *expense of the priests*: ibid. 309–10, Nos. 9, 11.

658 *Papal Nuncio*: HA, i. 176. No. 7.

658 *coffee shops*: WA, I. 53, p. 15, No. 36.

658 *snuff-sellers*: ibid. 1, p. 311, No. 18.

658 *yo-yo*: HA, i. 182, No. 35.

658 *lions at the Arsenal*: WA, I. 1, p. 312, No. 20.

658 *beggars*: ibid. 314, Nos. 30–2.

658 *prostitutes*: ibid. 323–4, Nos. 67–70.

658 *gondolas*: ibid. 309, Nos. 5, 8.

658 *caged singers*: ibid. 53, p. 13, No. 26.

658 *troupe of acrobats Goethe saw*: ibid. 1, pp. 316–19, Nos. 30–45, 47; ibid. 53, pp. 14–15, Nos. 30–5, pp. 350–1, Paral. 13.

658 *'sad fate of France'*: HA, i. 180, No. 22.

658 *'if you would free many'*: ibid. 179, No. 20.

658 *princes and demagogues*: WA, I. 1, p. 320, No. 51.

658 *sometimes equally deceitful*: HA, i. 180, No. 25.

658 *'Mir auch scheinen'*: WA, I. 1, p. 321, No. 57.

658 *old Venetian constitution*: ibid. 311, No. 19.

659 *'Dich betrügt'*: ibid. 53, p. 10, No. 8.

659 *'Jeglichen Schwärmer'*: HA, i. 179, No. 21.

659 *neither is a religion 'for free men'*: WA, I. 53, p. 11, No. 12.

659 *'God, man, and the world'*: ibid. 1, p. 322, No. 65.

659 *'Warum treibt sich'*: HA, i. 176, No. 8.

659 *no deeper secret*: ibid. 182, No. 37.

659 *'Folgen mag ich'*: WA, I. 53, p. 10, No. 10.

660 *disciples stole his body*: ibid. No. 11.

660 *all equally intolerant*: ibid. p. 11, No. 13.

660 *'mingles sense and nonsense'*: ibid. 348, Paral. 6.

660 *'naked pudenda'*: ibid. p. 9, No. 5.

660 *'So griff Lavater'*: ibid. No. 6.

660 *'Welche Hoffnung'*: ibid. p. 17, No. 43.

660 *unacceptable subjects*: ibid. p. 17, No. 45; p. 14, No. 33; p. 8, 14–15, Nos. 1, 34–5; p. 15, Nos. 36–7; pp. 15–16, Nos. 38, 40; p. 16, No. 39; p. 349, Paral. 9.

660 *associates Christianity with the suppression*: ibid. p. 8, Nos. 1, 2.

661 *'Eine Liebe hatt' ich'*: ibid. 1, p. 309, No. 7.

661 *'So go forth'*: conclusion to canto 5 of *Amours de Voyage*, Clough, *Poems*, p. 316.

663 *'congenial to the wanderer'*: HA, i. 174, No. 2.

663 *third epigram*: HA, i. 174–5, No. *3*.

663 *'Alle Neun'*: WA, I. 1, p. 313, No. 27.

664 *'Wißt ihr, wie ich'*: ibid. 320, No. 49.

664 *'sweetest spices of the world'*: HA, i. 184, No. *43*.

664 *'I cannot say how I felt'*: Grumach, iii. 340.

664 *sightseeing began the following day*: the itinerary given here follows Steiger, iii. 81–6.

665 *Ascension Day, 13 May*: Grumach, iii. 522.

665 *Whit Sunday was spent*: the return journey as in Steiger, iii. 86–92.

665 *less than twice the size of Weimar*: Bode, *Rom*, p. 337.

665 *'I yearn longingly'*: HABr, ii. 128.

665 *'future artists' republic'*: Grumach, iii. 343.

666 *out walking with his brother Max*: Bode, *Rom*, pp. 343–4.

666 *'Duchess is well and contented'*: WA, IV. 9, p. 209.

General Index

Recipients of letters are not normally indexed. Cross-references preceded by a Roman numeral, I. to V., are to the subdivisions of the entry 'GOETHE, J. W. VON'. Cross-references without a preceding Roman numeral are to the main sequence of entries. For references to Goethe's works see the separate index. In the *General Index* titles of German works are translated into English. In the *Index of Goethe's Works* they are left in the original.

Index of Goethe's Works

An initial definite or indefinite article is disregarded in determining alphabetical order. Poems are listed by title, where this exists and is original, otherwise by first line. For references to groups of works (e.g. 'Sesenheim poems') and to discussions of general literary issues (e.g. 'political themes') see General Index: GOETHE, J. W. VON: V. WRITINGS

DISCARD